Proceedings of the ASME

INTERNATIONAL MECHANICAL ENGINEERING CONGRESS AND EXPOSITION

– 2014 –

VOLUME 5

EDUCATION AND GLOBALIZATION

presented at

ASME 2014 INTERNATIONAL MECHANICAL ENGINEERING CONGRESS AND EXPOSITION
NOVEMBER 14–20, 2014
MONTREAL, QUEBEC, CANADA

A S M E

Two Park Avenue * New York, N.Y. 10016

ISBN No. 978-0-7918-4950-7

WELCOME FROM THE CHAIRS

Congress offers the unique opportunity to bring together members of academia, industry, and government to discuss a broad range of mechanical engineering topics. The 2014 Congress theme is *Engineering for Global Development* and includes 20 multi-disciplinary tracks over 2400 podium and poster presentations. The breadth of the technical tracks that have been organized by our track chairs and topic organizers is truly amazing and shows the dedication of our volunteers.

Congress will kick off with Sunday's Opening Reception in the Exhibit Hall, giving us a chance to refresh our connections and make new ones. Monday's events begin with the keynote panel presentation on *Engineering for Global Development*, moderated by NPR's **John Hockenberry**. The panel consists of **Paul Polak**, Founder and CEO, Windhorse International and Paul Polak Enterprises, **Shyam Prasad Rajan**, Chief Technology Officer, GE Healthcare – India, **Bryan Willson**, Professor of Mechanical Engineering, Colorado State University, and **Annica Wayman**, U.S. Agency for International Development (USAID).

The Congress-wide breakfast and closing lunch sessions feature exceptional speakers. At Tuesday's breakfast, we will hear **Chris Van Buiten**, Vice President, Sikorsky Innovations, on *Taking on the 'Impossible:' The Sikorsky Innovations Story*. At Wednesday's breakfast session, we are pleased to host **Dennis O'Donoghue**, Vice President, Boeing Test & Evaluation, who will speak on *Testing the Future of Flight — Boeing's ecoDemonstrator Program*. At Thursday's lunch session **Ann Marie Sastry**, co-Founder and CEO; Sakti3, will discuss *The Next Energy Storage Revolution — Cheap, Clean, Solid State Devices*. We are truly honored to have these individuals speak at Congress to give a broader perspective of the challenges and successes that mechanical engineering can provide to the industrial and government arenas.

A new feature of this year's Congress is the Invited Industry Presentations, presented immediately after lunch by distinguished industry engineering leaders. On Tuesday, we will hear **Brian Nolan**, Chief Engineer, Rolls-Royce Energy, on *Extension of the Natural Gas Envelope on Rolls-Royce Industrial Engines and Development of the Industrial Trent DLE*. On Wednesday, we will hear **Yves Rabellino**, Director, Research & Technology, Pratt & Whitney Canada, who will speak on *Trends of the 21st Century in Aero Engines*. On Thursday, our invited industry speaker is **Fassi Kafyeke**, Director, Strategic Technology, Bombardier Aerospace, on *Bombardier Strategic Technology: The Future of Business and Commercial Aviation*. We are honored to have these distinguished individuals give the industry state-of-the-art and perspective on these important topics.

Monday evening's Honors Assembly events spotlight some of the great contributions our society's members have made. Tuesday's featured event is the annual NSF-sponsored Micro/Nano Forum Poster competition, which brings together researchers from across the Mechanical Engineering disciplines to share their findings in this growing area.

Other Congress highlights include Sunday's Student Design Competition, Members and Students Luncheon, as well as an exciting event to help draw and showcase the next generation of engineers to ASME and IMECE: the third annual Undergraduate Expo. This poster session features research performed by undergraduate Mechanical Engineering students from around the world. We encourage everyone to attend and contribute to this wonderful event – you may find the next graduate student or employee amongst the bright minds participating.

Additional events of interest include the ME Department Heads Forum and the Micro and Nano Forum Poster session. The National Science Foundation will continue its active support of Congress by holding the CBET/CMMI Info and New initiatives session and the Session on Proposal Writing. Also, the general poster session will be held on Wednesday afternoon during lunch time with the latest state-of-the-art research posters on a variety of topics encompassing the essence of all track topics. Use the ASME Itinerary Planner to schedule your time for these exciting activities!

There will also be many events hosted by the ASME divisions and committees, including the Heat Transfer Honors and Awards Luncheon and the Applied Mechanics Dinner. Your participation in ASME division and committee meetings is the pathway to helping us continue to improve the Society and future conferences such as IMECE.

An event of this magnitude requires the hard work and dedication of many people, from volunteer organizers to the ASME staff and others too numerous to mention individually here. We want to thank the organizational support of the Congress Steering Committee, our operations team including our local section representatives, and the dedicated ASME staff with whom we have worked closely. The quality and breadth of the conference program is entirely due to the hard work of our track chairs, topic and symposium organizers, session chairs, reviewers and judges. These volunteers have put forth an incalculable amount of effort to ensure that this conference has papers and presentations of innovativeness, substance and quality. Finally, we want to recognize the authors, plenary, and keynote speakers whose contributions and insights we have the pleasure of reading and listening to and whose time here is spent generously discussing and exchanging ideas with us.

While attending the Congress, please take the time to thank the ASME staff and volunteers who make IMECE run. We look forward to meeting many of you in the halls of the convention center and hotel, at the special events, presentations and in technical sessions. We hope you enjoy Congress!

Sincerely,

George Kardomateas
2014 Congress Technical Program Chair

Assimina (Mina) Pelegri
2014 Congress General Chair

Ramakrishna Koganti
2014 Congress Technical Program Vice Chair

Julie Chen
2014 Congress Steering Committee Chair

2014 CONGRESS STEERING COMMITTEE

General Conference Chair
Assimina Pelegri
Rutgers - The State University of New Jersey

Technical Program Chair
George Kardomateas
Georgia Institute of Technology

Technical Program Vice Chair
(2015 Technical Program Chair)
Rama Koganti
JPS Health Network

Steering Committee Chair
Julie Chen
University of Massachusetts Lowell

Steering Committee Vice Chair
Aaron Knobloch
General Electric Global Research

Steering Committee Senate Chair
Pradeep Lall
Auburn University

CIDs and ASME Conference Proceedings

Beginning with the 2013 conference proceedings, ASME will transition to e-first publication of conference proceedings. As a result, instead of traditional page numbers, the online pagination will be replaced by 11-digit citation identifiers (CIDs); For instance, using the CID V###T##A### as an example, the structure of the basic 11-digit citation is defined as follows:

- First four digits (V00#): indicate the **Volume Number**
- Middle three digits (T##): indicate the **Track Number**
- Last four digits (A###) indicate the **Article order** in which the paper is published within the Track.

The CID is followed by the total number of pages in the article. For example: V###T##A###; 3 pages

In the print version, the CID appears on every page of the paper followed sequentially by the page number of each article starting with page one. For example, the pagination for a three page article would use the following sequence:

V###T##A###-1

V###T##A###-2

V###T##A###-3

CONTENTS

EDUCATION AND GLOBALIZATION

APPLIED MECHANICS, DYNAMIC SYSTEMS AND CONTROL ENGINEERING

CURRICULUM INNOVATIONS, PEDAGODY AND LEARNING METHODOLOGIES

DISTANCE/ONLINE ENGINEERING EDUCATION, MODELS AND ENABLING TECHNOLOGIES

EDUCATION RESEARCH INNOVATION AND SUSTAINABLE TRENDS IN ENGINEERING

ENGINEERING ACCREDITATION, DATA COLLECTION, ASSESSMENT AND ABET

FLUID MECHANICS, HEAT TRANSFER, EXPERIMENTS AND ENERGY SYSTEMS

TEACHING LABORATORIES, MACHINE SHOP EXPERIENCE, AND TECHNOLOGY-AIDED LECTURING

This page left blank intentionally.

Proceedings of the ASME 2014 International Mechanical Engineering Congress and Exposition
IMECE2014
November 14-20, 2014, Montreal, Quebec, Canada

IMECE2014-37472

NUMERICAL SIMULATION AS IN INTEGRAL COMPONENT OF DYNAMICS PROBLEM SOLVING

Matthew Stein
Roger Williams University
Bristol, RI, USA

ABSTRACT

The engineering faculty at Roger Williams University are committed to training students to use modern computer-based tools when performing engineering analysis. But achieving this is a tall order, as engineering courses are already jam-packed with essential technical material and any hindrance to delivering this material is unwelcome. Likewise, we routinely pay lip service to the necessity for students to double-check their work, yet we provide students with few tools for systematically accomplishing this. This paper describes an effort by the author to integrate solid modeling into a *Dynamics* course by requiring numerical validation of symbolic solutions to homework problems. The students solve traditional homework problems using free-body diagrams, equations of motion, pencils and calculators; but then must demonstrate that their answers are valid through an independent check. Students construct solid models in SolidWorks© to duplicate the geometric and inertial properties of the problem, and then use the Motion Analysis, a SolidWorks Simulation add-in, to create a motion study duplicating the conditions of the problem. Students may place dynamically updating dimensions to determine distances or may generate graphs, e.g. velocity versus time, to study motion characteristics. As a direct result, students are able to independently validate their symbolic solutions with numerical simulations. This paper will provide a detailed description of the use of SolidWorks in a sophomore level *Dynamics* course offered spring 2012 and spring 2013. This paper will present examples of student work and assess the benefits and challenges associated with this teaching method.

INTRODUCTION

The faculty in our engineering program are committed to achieving ABET objective *k. an ability to use the techniques, skills and modern engineering tools necessary for engineering practice*. However, an isolated, introductory *Computer Applications for Engineering* course in the freshman year does not alone achieve this educational objective. We believe *objective k* is only achieved by four-year vertical integration of computer applications as engineering tools. But this challenging in practice, as engineering courses are already packed with essential technical material and any encumbrance to delivering this material is unwelcome. The classical mechanics course (*Dynamics*) described in this paper is an effort to satisfy the two conflicting goals of building skills while maintaining content. Other authors [i,ii] have described the challenges of achieving this objective.

The literature shows that many instructors use some form of computer-aided simulation of problems [iii] and some further employ problem solving software and interactive computing [iv,v,vi,vii,viii]. *Dynamics* is a required course for all students in the fourth semester of our eight-semester general engineering program. This course is offered to the entire sophomore class in two sections with a population of thirty students each.

In prior reports [ix,x], I described my efforts to completely transform this course into a computer-based learning experience where problems and even class notes were taken via computer, similar to a course described in [xi]. The central vehicle to this effort was MathCAD; a computer aided engineering application that allowed for symbolic manipulation and numeric simulation. The transmogrification of MathCAD in its Version 14 release caused me to abandon its use and return to a traditional chalkboard classroom in spring 2009. Frustrated, but committed to pursuing this objective, I began to employ SolidWorks motion studies in 2009 as a numerical tool to compliment theoretical problem solving. In the subsequent years I developed and enhanced the incorporation of SolidWorks so that it is now an essential and effective component of the course.

CHALLENGES

The primary challenge to this approach is building student competency in SolidWorks. There is scarcely enough time in the course for engineering mechanics content, leaving no class time for SolidWorks training. The engineering program incorporates an *Engineering Graphics and Design* course in the first semester provides sufficient prerequisite skills in SolidWorks for the majority of the population. *Dynamics*, offered in the fourth semester, must overcome gaps in preparation that have opened due to the passing of time from the first to fourth semester and for transfer students without prior experience in SolidWorks.

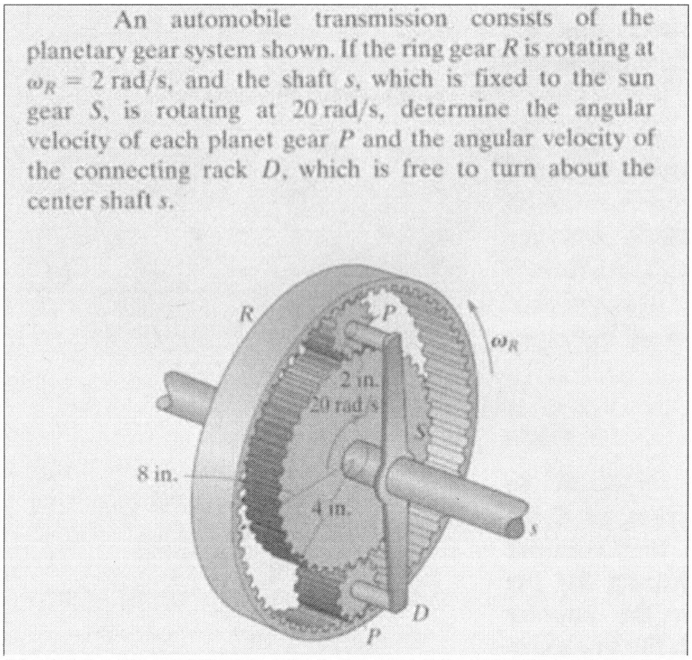

FIGURE 1 PLANETARY GEAR TRAIN HOMEWORK PROBELM

To provide instructional support without dedicating valuable class time, I created a library of narrated video demonstrations. Each video guides the viewer keystroke-by-keystroke through the SolidWorks application to create a SolidWorks simulation of each of the forty-three in-class example problems. The simulation must duplicate the mechanical conditions of the problem and produce a numerical result for an output parameter. This result is compared to theoretical analysis with algebraic manipulation of equations and computation on a hand-held calculator. Each video demonstrates how to extract numerical results from the simulation and compares these to hand-calculated results, finding remarkable agreement. Students follow the videos in the school's computer laboratory listening to the audio on individual headphones. They are often seen repeatedly rewinding and replaying the video while simultaneously constructing a simulation of their own similar but distinct homework problems. To define this term, the problems are similar in that both are, for example, projectile motion problems, but the figures, parameters and unknowns are distinct.

BENEFITS

The Dynamics course is a traditional lecture format in which I perform live demonstrations with mechanical props, derive relevant formulas and work example problems on a chalkboard. Students complete weekly homework sets in groups, with each group assigned five to six problems. As the example in Figure 1 shows, the problems, are relatively challenging; requiring five to seven hours weekly. Each student in the homework group is responsible for completing three "initial solutions" and one "numeric check". The "initial solutions" are traditional paper-and-pencil solutions of the assigned homework problems using the student's hand-held calculator to perform computations. There is little novelty in the initial solutions; student work resembles student work of ten years ago and even resembles the homework I submitted as an undergraduate student quite a few years before that.

The novelty is exhibited in the "numeric checks", where students must simulate homework problems in SolidWorks by constructing solid models duplicating the kinematic and dynamic conditions of the problem. Students must then use this model to check their own work or the answers of their peers. The following paragraphs describe the primary benefits of this activity to student learning.

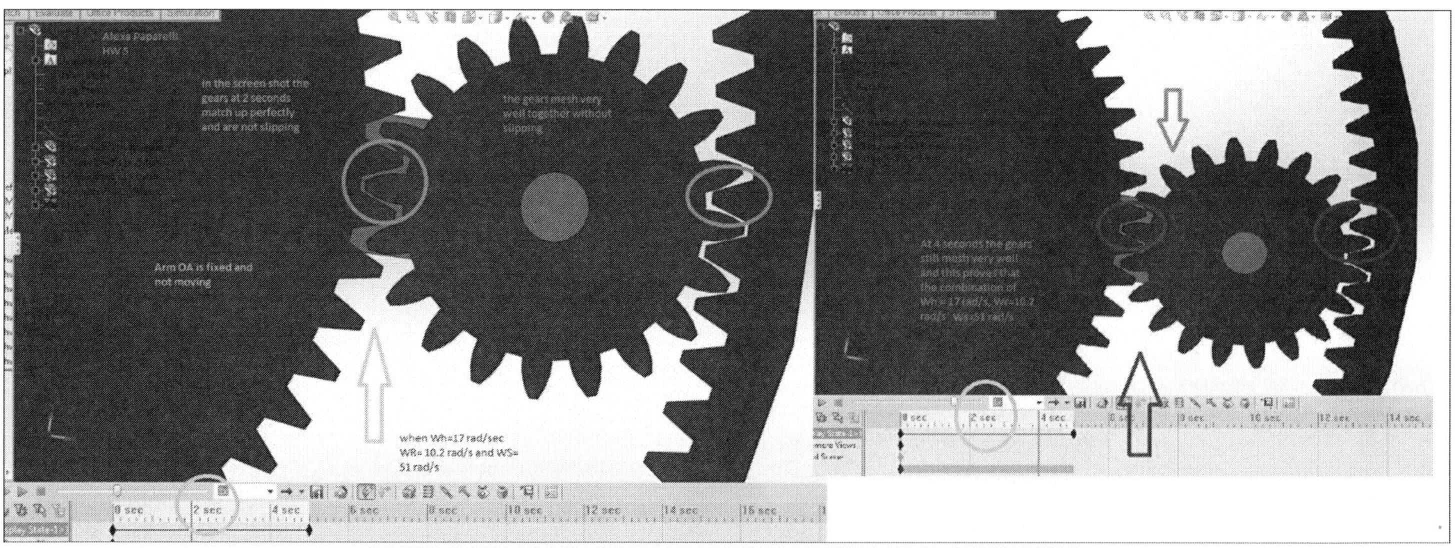

FIGURE 2 PLANETARY GEAR IMPLEMENTED IN SOLIDWORKS

KINEMATIC VISUALIZATION

As a two-dimensional pictorial description is brought to life by the student's own hand; these assemblies are immediately useful for visualization. By dragging the mouse, the student witnesses the motions as gears spin, links move, bodies come into contact and range of motion limits are reached. This is particularly useful in complex kinematic problems as shown in Figure 1. Students often struggle with visualizing the motion of the planetary gear system. The simulation shown in Figure 2 requires several hours to build, and in the process students must select the appropriate kinematic mate to implement the motion. Selection of the wrong mate either causes a simulation error or the gear teeth to pass through each other during simulation. A properly built gear system will be pleasing to watch as gears enmesh and planet gears revolve around the sun gear. I expected that this would in itself be gratifying and educational for the students.

MOTION STUDIES

Once the problem kinematics are established, students move to the *Motion Study*, a rich set of SolidWorks features allowing students to generate motion and analyze the results. Figure 3 shows a student submission of a motion study implementing a multiple pulley problem shown in Figure 4. The inset graphs on Figure 3 are the results of a motion study selected by the student to demonstrate agreement with theoretical results produced by integrating angular acceleration twice with respect to time. In this case, the student has selected two graphs to ascertain "how far has wheel B spun when it stops". A student may not immediately recognize that a body with a positive initial velocity but increasingly negative acceleration will eventually stop, but the motion study shows convincingly that this is the case. The student has selected to show the angular displacement versus time and the angular velocity versus time plots. The student found the time that the velocity crosses zero as the red vertical line on both graphs and reads the angular displacement at that time.

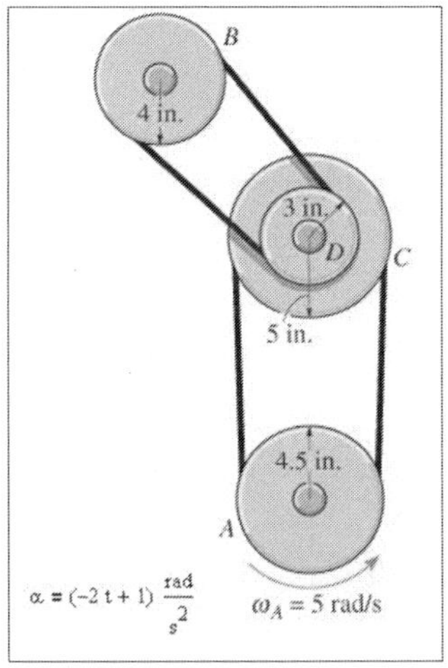

FIGURE 3 PULLEY PROBLEM

BACK OF THE BOOK ANSWERS?

Although we may debate the wisdom of providing "back-of-the-book" answers to homework problems, we would likely agree that back-of-book answers are rarely available in engineering practice. Rather, successful engineers have developed the skill set necessary to employ various techniques to check their own work. In ABET documents we claim a portion of this class satisfies ABET objective *i. a recognition of the need*

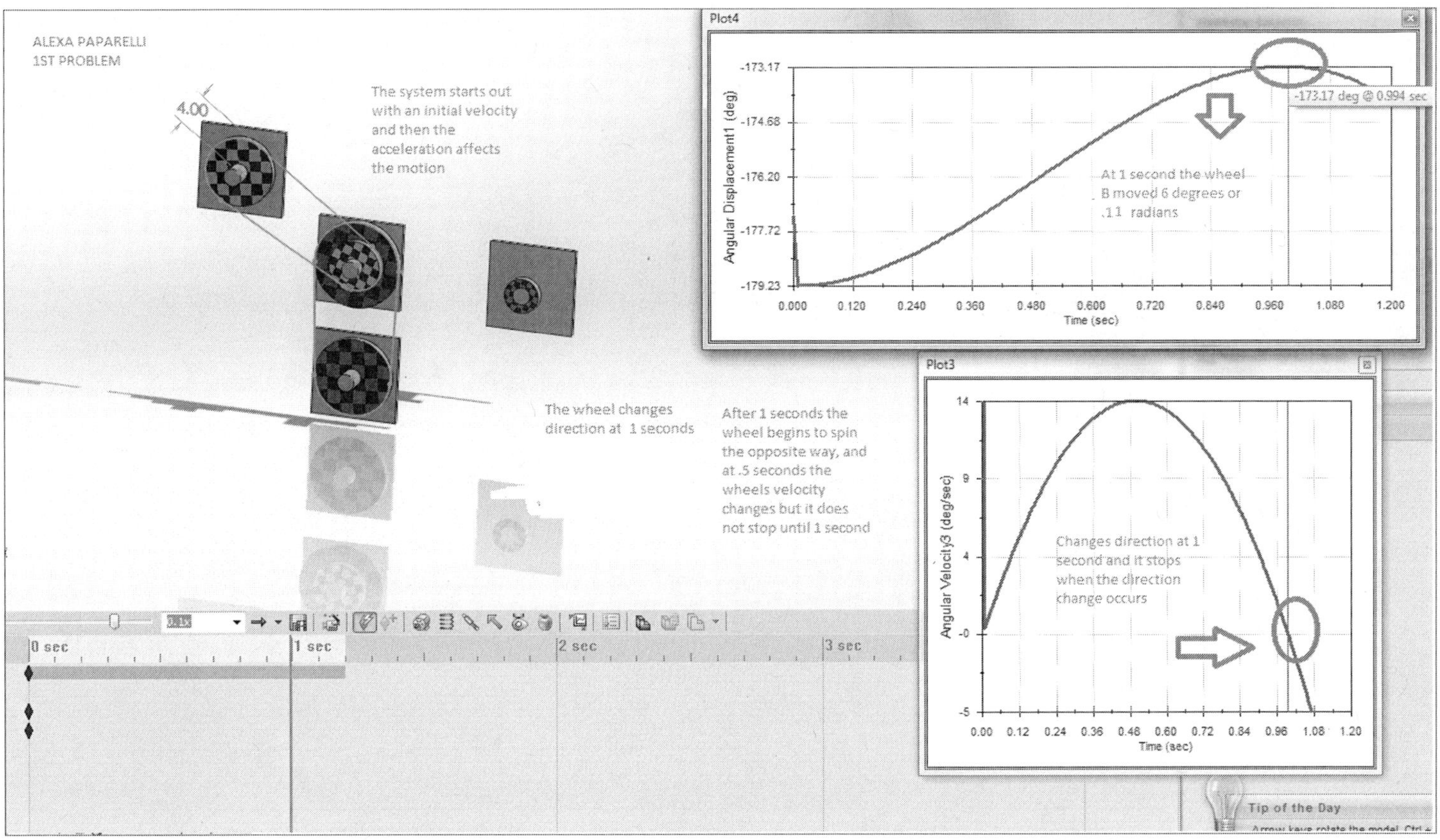

FIGURE 4 MULTI-PULLEY PROBLEM CONSTRUCTED IN SOLDWORKS

for, and an ability to engage in lifelong learning due to this training.

Each group is required to develop a numerical simulation that checks the work of the other students in the group. As shown in Figure 5, a student has simulated the motion from the textbook problem and stopped the simulation at the specified instant in time. Also seen in Figure 5 is a block of text added by the student stating "Tip velocity when bar was horizontal was 172in/s which is 14.25ft/s. Calculated value was 14.32ft/s." Note that Figure 5 is a kinetic simulation of the rod falling under the effect of gravity, whereas the previous figures demonstrated only kinematic simulation.

Training engineers to devise their own means of teaching themselves is critical to an ABET accredited program, yet it is not clear that traditional homework assignments effectively produce this outcome. The numeric simulation is an alternative means of producing an answer, and neither the symbolic solution nor the numerical solution is as convincing in isolation as when the two methods produce the same result. Students recognize the potential of the simulation to produce erroneous results and often initially produce completely different results than they calculate. There is a subtle moment that students experience when they can find nothing else wrong with their simulation and start to believe that the error must be in their calculations. These students have never had the capability to prove themselves right, and I expected they would find this experience exhilarating.

SKILLS

A final benefit is exercising and enhancing our student's skills in SolidWorks. We expect our seniors to use SolidWorks extensively in our capstone senior design, but our students learn SolidWorks formally in their first semester. By the fourth semester *Dynamics* course they have already become a little rusty. If we do not require students to use SolidWorks as an integral part of their intermediate coursework; we should show little surprise when they proclaim that they have forgotten it all as seniors.

RESULTS

The four benefits described above are simply conjectured by the author. A survey was administered to students having just completed the *Dynamics* course in the spring of 2013. The survey was (inexpertly) designed by myself to test the above conjectures by directly inquiry. Figure 6 presents data from an anonymous survey administered to students at the end of the semester. Of sixty-three students registered, fifty-eight

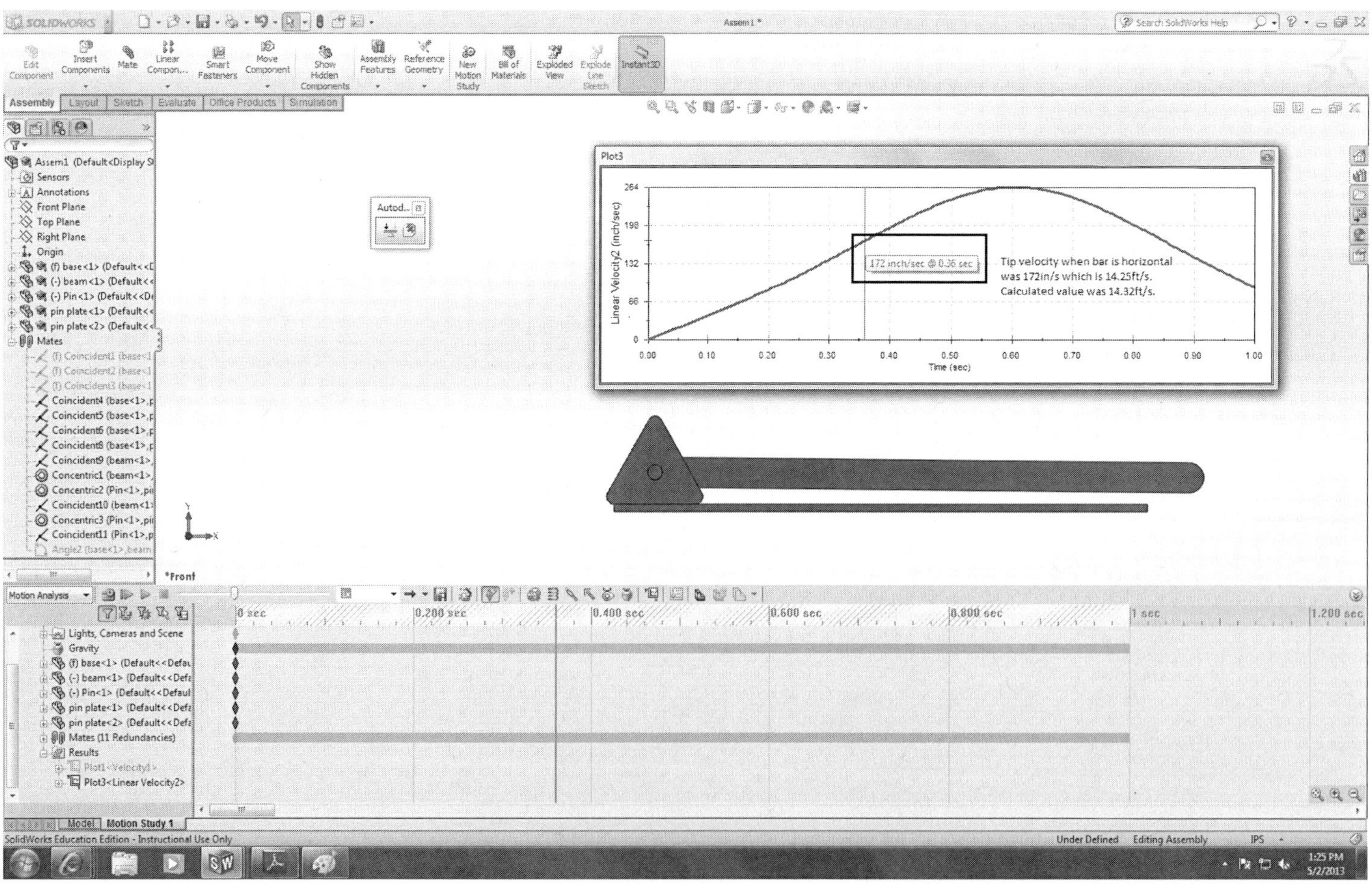

FIGURE 5 SOLIDWORKS MOTION STUDY SHOWING AGREEMENT WITH THEORETICAL SOLUTION.

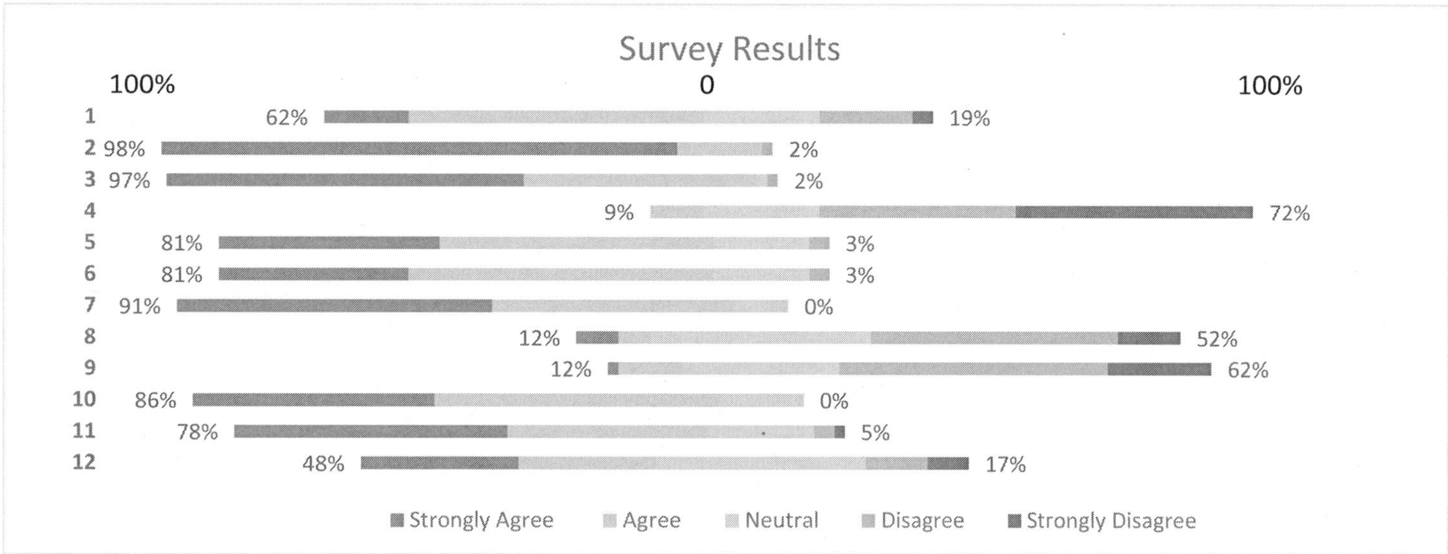

FIGURE 6 LIKERT SCALRE RESPONSES TO TWELVE SURVEY QUESTIONS

voluntarily completed the survey. The survey consisted of twelve Likert scale questions.

Adhering to the format suggested in [xii], Likert scale results are presented as a percentage of respondents centered at "neutral", with agreement with the prompt on the left in green and disagreement with the prompt on the right in red. The total percentage of responses to the left and to the right of "neutral", i.e. the sum of "strongly agree" and "agree" is shown as a label on each side. To limit "Acquiescence bias" one-fourth of prompts were keyed in the negative direction.

The survey only indirectly measures the effectiveness of the video library. As all students utilize the library, the ability to complete the simulations is an indirect measure of the tools assisting them. In retrospect, it may have been valuable to include a direct question about the video library. Students were also asked to report the number of hours they spent on SolidWorks each week, responses were: Mean: 3.70, Mode: 3, Max: 10, Min:1, Standard Deviation: 1.54, N: 58.

This section will consider the questions in order and discuss both the rationale for the question and my interpretation of the responses.

1. *My prior coursework prepared me to use SolidWorks in this course.* For the majority of respondents, prior coursework is a full semester of *Engineering Graphics and Design* three semesters prior and a single project in *Statics* one semester prior. Students indicated agreement by a margin of approximately 3:1, and this is interpreted as general satisfaction with the prerequisite structure but with significant room for improvement.
2. *My use of SolidWorks improved as a result of this course.* This question directly tests one of the conjectured benefits. Overwhelming agreement (98%) indicates to the author that the exercises provide an appropriate challenge to students to expand their proficiency in SolidWorks.
3. *SolidWorks is a good tool for performing numeric simulations.* This question tests the suitability of SolidWorks to this use. Overwhelming agreement (97%) suggests to the author that SolidWorks is an appropriate tool for generating numeric results that are comparable to theoretically derived results.
4. *SolidWorks should be removed from the course or made optional.* This question tests overall validity of this initiative. Students disagree with this statement by an 8:1 margin suggesting that a strong majority believe the course is better with SolidWorks than without.
5. *SolidWorks helped me envision the problems better.* A test of the "kinematic visualization" result, eighty-one percent agreement including thirty-six percent strong agreement suggests that SolidWorks is effective in helping students envision problems.
6. *The act of constructing the model helped me to better understand the mechanics of the problem.* I had imagined the act of construction to be a distinct experience from manipulating the constructed model, but identical results with question 5 suggests that perhaps there is no real distinction in the minds of the students.
7. *I feel confident I will be able to use SolidWorks in future courses because of the use in this class.* No disagreement suggests students have gained sufficient confidence in SolidWorks to use it in courses later in the program.
8. *The time invested in SolidWorks is not proportional to the benefit.* The question is poorly worded with an implied bias towards time spent in SolidWorks *in excess* of benefit. The mean time spent per week reported by the students was 3.7 with a mode of 3 hours per week. The response of only 10% in strong disagreement indicates relatively large ambiguity in the student population. Whereas the previous responses clearly indicate there are perceived benefits, the population is not so well convinced the benefits are worth the significant time investment.

9. *I didn't really trust SolidWorks, I could always fudge a number to get the answer the others got.* This exact sentiment was expressed by some students in prior years and was included here as a quantitative measure of its prevalence. A margin of 5:1 disagreement suggests that this phenomenon is less prevalent than I had feared.
10. *The ability to verify my own solutions was rewarding to me.* Included to test the conjecture that SolidWorks was better than "back of the book" answers. Very large agreement suggests this intended consequence is achieved.
11. *I liked building the SolidWorks models.* Test the conjecture that students enjoyed working in SolidWorks, 15:1 agreement suggests that this activity is less of a burden on students than I originally feared.
12. *At some point in the semester, I started to trust the SolidWorks result more than my hand calculations.* A frivolous prompt placed by the researcher in search of the holy grail of this endeavor. It is not clear what response would be selected by a student whose trusts his or her hand calculations *as much as* SolidWorks, or if this happened at no definable point in the semester, or if it had been the case prior to the semester. Mixed results may indicate a brand-new level of student confidence with SolidWorks but may also indicate that the researcher is an amateur at designing surveys.

CONCLUSIONS

Survey results indicating levels of satisfaction with the activity exceeded the author's expectations and indicate that the project is worth continuing. Despite a significant time investment reported by the participants (an average of 3.7 hours per week), a strong majority believe it should be retained as a requirement in the course (Q4). The ability to check one's own work (Q10) is both a new and important skill that can benefit students significantly when introduced in the sophomore year. All of the conjectural results described in the first section were confirmed by direct survey. The author concludes that the inclusion of SolidWorks not only improves the teaching of Dynamics, but strengthens the entire engineering program by equipping students with the tools for lifelong learning early in their career.

We plan on continuing this use of SolidWorks in the *Dynamics* course and may expand its application to other courses. The results of Question 1 show there is room for improvement in the preparedness of students for this experience. We particularly need to address the special needs of transfer students who may get transfer credit for a graphics course using another application, (e.g. AutoCAD), and therefore will have no preparation for SolidWorks before taking the *Dynamics* course. We will also refine the questionnaire for the spring of 2014 to test the perceived value of the tutorial videos.

REFERENCES

i W. Whiteman, K. Nygren, Achieving the Right Balance: Properly Integrating Mathematica Software Packages into Engineering Education. *Journal of Engineering Education*, July 2000.

ii B. Hodge, W. Steele, A Survey of Computational Paradigms in Undergraduate Mechanical Engineering Education. *Journal of Engineering Education*, October 2002.

iii A. Mazzei Integrating Simulation Software into an Undergraduate Dynamics Course: a Web-based Approach. *Proceedings of the 2003 American Society of Engineering Education Annual Conference and Exhibition.*

iv R. Flori, M. Koen, D. Oglesby, Basic Engineering Software for Teaching ("BEST") Dynamics. *Journal of Engineering Education*, January 1996.

v C. Demetry, J. Groccia, A Comparative Assessment of Student Experiences in Two Instructional Formats of an Introductory Materials Science Course. *Journal of Engineering Education*, July 1997.

vi Phillip J. Cornwell, Dynamics Evolution - Chance or Design. *Proceedings of the 2000 American Society of Engineering Education Annual Conference and Exhibition.*

vii D. M. Fraser, R. Pillay, L. Tjatindi, J. M. Cace, Enhancing the Learning of Fluid Mechanics Using Computer Simulations. *Journal of Engineering Education*, October 2007

viii R Stanley and G. DiGiuseppe, An Efficient Way to Increase the Engineering Student's Fundamental Understanding of Thermodynamics by Utilizing Interactive Web Based Animation Software. *Proceedings of the 2010 American Society of Engineering Education Annual Conference and Exhibition.*

ix M.R. Stein. A Classical Mechanics Course in an Electronic Classroom: Challenges and Rewards. *Proceedings of the 2009 American Society of Engineering Education Northeast Section Annual Meeting.* April 2009. University of Bridgeport, Bridgeport, CT.

x M.R. Stein. A Classical Mechanics Course in an Electronic Classroom: Challenges and Rewards. *Proceedings of the 2007 American Society of Engineering Education Northeast Section Annual Meeting.* April 2007. University of Rhode Island, South Kingston, RI

xi P. Boyle, Electronic Course Packaging for Statics and Dynamics: A review of Effort, Reward and Potential. *Proceedings of the 2007 American Society of Engineering Education Annual Conference and Exhibition.*

xii Naomi B. Robbins, Richard M. Heiberger, Plotting Likert and Other Rating Scales, 2011 Joint Statistical Meetings - American Statistical Association. Miami Florida, July 30th, 2011

Proceedings of the ASME 2014 International Mechanical Engineering Congress and Exposition
IMECE2014
November 14-20, 2014, Montreal, Quebec, Canada

IMECE2014-37726

STRESS ANALYSIS ALONG TREE BRANCHES

Allison Kaminski
Manhattan College, Mechanical Engineering
Riverdale, NY, United States

Simon Mysliwiec
Manhattan College, Mechanical Engineering
Riverdale, NY, United States

Zahra Shahbazi
Manhattan College, Mechanical Engineering
Riverdale, NY, United States

Lance Evans
Manhattan College, Biology Department
Riverdale, NY, United States

ABSTRACT
Efforts have been made to develop various models to calculate the stress due to weight throughout tree branches. Most studies assumed a uniform modulus of elasticity throughout the branch as well as analyzing the branch as a tapered cantilever beam orientated horizontally or at an angle. However, previous studies show that branches located lower on the tree have a greater variance of modulus of elasticity values in the radial direction and that branches located lower on a tree are more curved. Also, different tree species have different morphologies, some with curvy branches. In this work we have developed a model which considers the curved shape and varying modulus of elasticity values in order to determine stress across the tree branches more accurately. To do this the cross sectional area was divided into rings and each ring was assigned a different modulus of elasticity. Next, the area of the rings was transformed according to their modulus of elasticity. We then considered the curved shape of the branch by generating a best fit line for the diameter of the tree branch in terms of distance from the end of the branch. The generated diameter equation was used in the stress calculations to provide more realistic results. Based on equations developed in this work, we have created a Graphical User Interface (GUI) in Matlab, which can be used as a tool to calculate stress within tree branches by biologists without getting involved with the mathematical and mechanical calculations. We also created a Finite Element Model (FEM) in Abaqus and compared results.

INTRODUCTION
A tool used to accurately calculate the stress on tree branches from easily measured dimensions and properties can have multiple applications. This tool will be beneficial to biologists researching the relationships between tree morphology and stress because it is hypothesized that tree branches grow in a specific way which creates a uniform stress throughout the branch [1].

Previous studies calculated the stresses on tree branches by examining them as tapered cantilever beams of either an elliptical or circular cross sections. These stress calculations assumed the tree branch was of a uniform material with a uniform Modulus of Elasticity (MOE) value [1]. The MOE is a measure of the stiffness of a material. Materials with a small MOE bend more easily while materials with a large MOE are stiffer. This study accounts for the varying MOE values within tree branches. Multiple studies have shown the MOE depends on the age of the wood and its location on the tree. As a tree grows, new layers of the tree branch grow on the outermost parts of the branch. This growth pattern makes the outer most part of the branch the youngest while the innermost part of the branch is the oldest [2]. The innermost wood has a small MOE value, and has little mechanical significance to resist bending. The outer wood resists a majority of the bending in a tree branch [3]. Therefore at a given cross section the MOE of the branch varies in the radial direction - as the radius increases the MOE value increases.

This study considers radial variances in MOE when calculating stress. Branches located lower on the tree have a greater variance of MOE in the radial direction than branches located closer to the top of the tree [2]. For this reason it was hypothesized that using the proposed tree branch model will be more necessary for branches located lower on the tree.

As mentioned above, in addition to modeling tree branches as being composed of a uniform material, previous studies modeled tree branches as tapered cantilevered beams [1, 2]. Branches in nature do not resemble straight lines, they have curves. In this study a tool to calculate stresses in a curved tree branch is developed.

To summarize, in this study a tool is developed to analyze the stress in different tree branch models. Two new cases are proposed in this study, the first considering the varying MOE of branches in the radial direction and the second considering the curviness of the branch. Four additional models are used for branches with less complicated geometries and for comparison purposes. These models include a cantilever beam of a fixed circular and elliptical cross section, as well as a tapered cantilever beam of a circular and elliptical cross section. In addition, a Finite Element simulation is developed to verify results.

METHOD

Six different tree branch models were created. Case 1 is a fixed circular cross section, and Case 2 is a fixed elliptical cross section. Case 3 is a tapered circular cross section, and Case 4 is a tapered elliptical cross section. Case 5 is a fixed circular cross section with a non-uniform material where the modulus of elasticity varies in the radial direction. Case 6 is a curved branch of uniform material and a circular cross section.

Equations to calculate stress were derived for each of the six cases. These equations were used to write a code in Matlab that allows measurable dimensions of the branch to be inputted and then a stress analysis at a specified location would be outputted. The Matlab code is able to calculate stress at the top, bottom and sides of any desired cross section along the length of the branch.

Only the stress due to the weight of the branch was considered as no external loads were examined. The weight of the branch was analyzed as a distributed load that acts vertically downward. To perform the analysis the weight was broken into two components, one parallel to the axis of the branch and one perpendicular to the branch (Fig. 1).

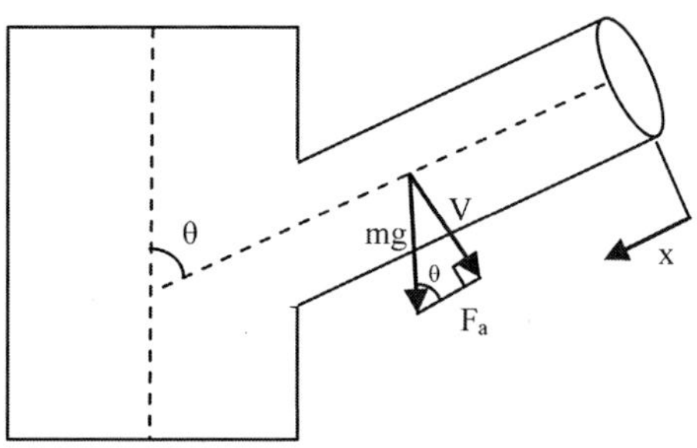

Figure 1. Tree branch model showing the force due to weight acting on the branch along with its components.

The component of the distributed weight that acts perpendicular to the branch and parallel to the cross sectional area is the shearing distributed load, $w_s(x)$. The component of the distributed weight that acts parallel to the branch and perpendicular to the cross sectional area is the axial distributed load, $w_a(x)$. The shearing distributed load contributes to the normal stress due to bending, while the axial distributed load contributes to the normal compressive stress on the branch. These two stresses need to both be considered in order to calculate the total normal stress acting on a given cross section.

General equations were derived to calculate stress that can be used for each of the six cases. All of the equations derived assume a uniform modulus of elasticity (MOE) throughout the branch. To calculate the axial compressive stress acting on a cross section of a tree branch a general equation for the axial distributed load needs to be determined, which is the load per unit length (1).

$$w_a(x) = \frac{mg\cos\theta}{L} = \frac{\rho V(x)g\cos\theta}{x} = \frac{\rho A(x)xg\cos\theta}{x} \quad (1)$$

Where m is the total mass of the branch, L is the total length of the branch, g is the gravitational constant, x is the distance from the tip of the branch where the stress analysis is desired, $V(x)$ is the volume of the branch up to the specified point x, $A(x)$ is the cross sectional area of the branch and θ is the angle of the tree branch with respect to the vertical. For instance, if the branch is perfectly horizontal then θ will be 90 degrees, making the axial distributed load zero. Using Equation (1) the axial force acting at any location from the tip of the branch, $F_a(x)$, can be determined from Equation (2).

$$F_a(x) = -\int w_a(x)dx \quad (2)$$

The normal axial stress due to the axial component of the weight is

$$\sigma_a(x) = \frac{F_a(x)}{A(x)} \quad (3)$$

The normal stress due to bending is caused by the component of weight acting perpendicular to the branch. The shearing distributed load, $w_s(x)$, which acts perpendicular to the length of branch can be calculated using

$$w_s(x) = \frac{mg\sin\theta}{L} = \frac{\rho V(x)g\sin\theta}{x} = \frac{\rho A(x)xg\sin\theta}{x} \quad (4)$$

The variables are the same as for the distributed axial load. The component of weight that acts perpendicular to the length of the branch and parallel to the cross sectional area is the shear force, $V_s(x)$, which can be determined from the shearing distributed load.

$$V_s(x) = -\int w_s(x)dx \quad (5)$$

The bending moment in terms of distance from the tip of the branch (6) can be obtained by integrated the shear force, $V_s(x)$, with respect to distance from the tip of the branch, x.

$$M(x) = \int V(x)dx \quad (6)$$

The normal stress due to bending can be expressed as

$$\sigma_b(x) = \pm\frac{M(x)c(x)}{I(x)} \quad (7)$$

Where $c(x)$ is the radial distance from the center of the cross section to the location in the radial direction where the stress analysis is desired. For instance to obtain the bending stress at the top of the cross section $c(x)$ would be the radial distance from the center of the cross section to the top of the cross section. $I(x)$ is the moment of inertia of the cross section. The normal stress due to bending is compressive (negative) or tensile (positive) depending on the location of the cross section being examined. The top of the branch will be in tension and the bottom will be in compression. The neutral axis is where the stress passes from positive to negative and does not experience the effects of bending. The left and right sides of the cross section lie along the neutral axis and therefore have a bending stress of zero.

The total normal stress acting on a cross section is the sum of the stress due to bending and the axial stress. When adding these two stresses together the sign convention of each stress must be considered.

Equations (1) through (7) are the general equations used to perform the stress analysis for each of the cases. For some of the cases additional analyses are needed. For Case 1 the generic procedure described above can be followed exactly.

Case 2 is a fixed elliptical cross section. The equations described above are used by substituting in the cross sectional area of an ellipse, $A = \pi R_V R_H$, and the moment of inertia for an ellipse, $I = \frac{\pi}{4} R_H R_V^{\ 3}$.

Cases 3 and 4 are tapered; therefore additional calculations must be considered. Case 3 is tapered and has a circular cross sectional area (Fig. 2).

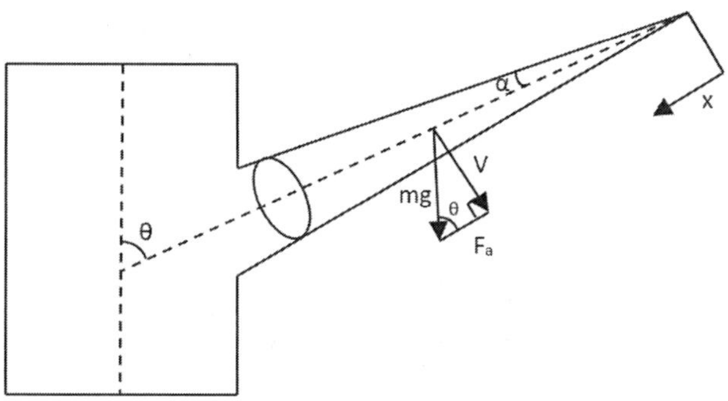

Figure 2. Case 3 tree branch model, tapered with a circular cross section. Where α is the angle of taper.

The angle of taper can be calculated using

$$\tan \alpha = \frac{R(x)}{x} = \frac{R_o}{L} \qquad (8)$$

Where R_o is the radius at the base of the branch. Rearranging for the radius of the branch in terms of distance from the tip of the branch, $R(x)$, gives Equation (9).

$$R(x) = \frac{R_o x}{L} \qquad (9)$$

The cross sectional area in terms of distance from the tip of the branch then becomes

$$A(x) = \pi R(x)^2 = \pi \left(\frac{R_o x}{L} \right)^2 \qquad (10)$$

The moment of inertia also varies with x and must be calculated using $R(x)$. By making these adjustments to the generic equations the stress analysis can be performed for Case 3.

Case 4 models the branch as a tapered elliptical beam. In this case there are two angles of taper, one in the x-y plane and the other in the x-z plane (Fig. 3).

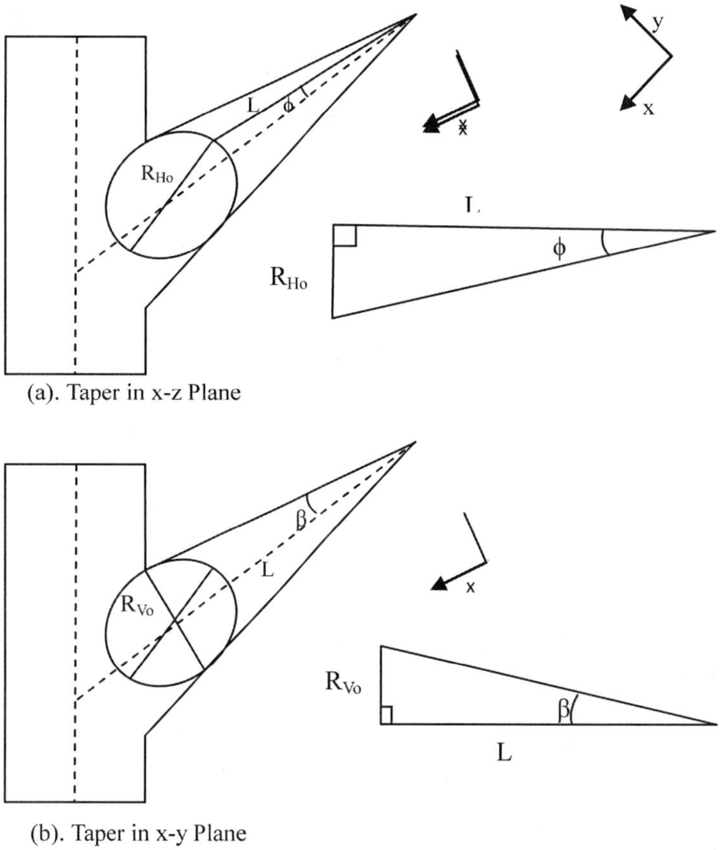

Figure 3. Case 4 taper in the x-y planes and x-z planes. R_{Ho} and R_{Vo} are the radii in the horizontal and vertical directions respectively at the base of the tree branch. ϕ is the angle of taper in the horizontal direction. β is the angle of taper in the vertical direction.

The angle of taper in the x-z plane is determined from the horizontal radius of the ellipse and can be calculated using Equation (11).

$$\tan \phi = \frac{R_H(x)}{x} = \frac{R_{Ho}}{L} \qquad (11)$$

Where ϕ is the angle of taper in the x-z plane, R_{Ho} is the horizontal radius at the base of the branch, and $R_H(x)$ is the horizontal radius with respect to distance from the tip of the branch. Equation (12) provides the horizontal radius at any location, x, along the branch.

$$R_H(x) = \frac{R_{Ho} x}{L} \qquad (12)$$

The angle of taper in the x-y plane can be calculated using

$$\tan \beta = \frac{R_V(x)}{x} = \frac{R_{Vo}}{L} \qquad (13)$$

Where β is the angle of taper in the x-y plane, R_{Vo} is the vertical radius at the base of the branch, and $R_V(x)$ is the vertical radius with respect to distance from the tip of the branch.

$$R_V(x) = \frac{R_{Vo}x}{L} \quad (14)$$

The radiuses of the branch in the horizontal and vertical directions vary along the length of the branch; therefore, the cross sectional area of the branch needs to be expressed in terms of x, the distance from the tip of the branch.

$$A(x) = \pi R_V(x)R_H(x) = \pi\left(\frac{R_{Vo}x}{L}\right)\left(\frac{R_{Ho}x}{L}\right) = \frac{\pi R_{Vo}R_{Ho}}{L^2}x^2 \quad (15)$$

The moment of inertia equation for the branch becomes

$$I = \frac{\pi}{4}R_H(x)R_V(x)^3 \quad (16)$$

Substituting these equations into the generic equations the stress analysis can be performed for Case 4.

Case 5 is a fixed circular cross section with a non-uniform material. The MOE varies in the radial direction (Fig. 4).

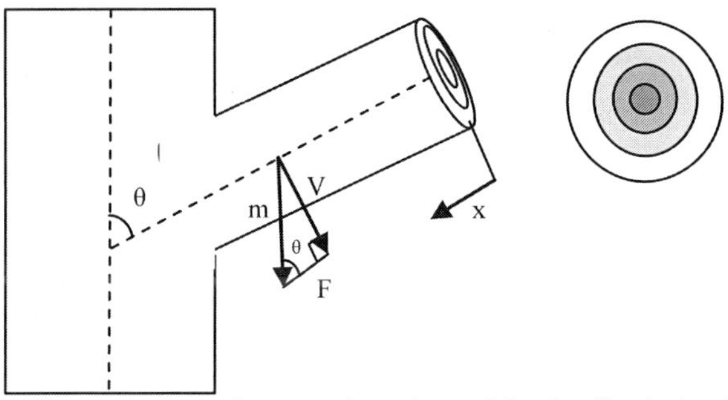

Figure 4. Case 5, a tree branch model of a fixed circular cross section and varying modulus of elasticity in the radial direction.

The cross section of the branch is broken into concentric rings of equal width each having a different MOE value. The derived equations assume the branch is made of a uniform material. In order to use these equations the moment of inertia of each ring must be transformed based on their MOE values. After doing this the branch can be analyzed as being composed of a uniform material and the generic equations for the stress analysis can be used.

The Matlab program for Case 5 was written so that the branch can be divided into any number of rings. Therefore based on the outer diameter of the branch and the desired number of rings inputted, the radius of each ring is calculated. The outer radius of each concentric ring is

$$r_h = h \times \frac{R}{m} \quad (17)$$

Where m represents the total number of rings that the cross section is divided into and h represents the number assigned to each ring. The inner most ring is h=1, while for the outer most ring h=m. R represents the outer radius of the branch. Using the radius of each ring the area of each ring can also be calculated.

The inner most ring has the smallest MOE value and therefore its moment of inertia does not need to be transformed. All the other rings have a MOE value that is greater than the inner most ring. To account for the larger MOE values, the ratio of each ring's MOE value to the MOE value of the inner most ring will be used to transform the moment of inertia of each section. For example, if a ring has a MOE that is two times greater than that of the inner most ring, then the moment of inertia of the outer ring needs to be two times greater as well. This represents how a ring of a larger MOE is able to resist bending more easily and therefore can be examined as having the same MOE as the inner most ring if a larger moment of inertia is used. After transforming the moment of inertia values the branch can be analyzed as a having uniform material. The ratio of the MOE of an outer ring to the MOE of the inner most section is calculated using Equation (18).

$$n_h = \frac{E_h}{E_1} \quad (18)$$

Note for h=1 the ratio becomes 1. The transformed moment of inertia becomes

$$I_h = n_h \frac{\pi}{4}\left(r_h{}^4 - r_{h-1}{}^4\right) \quad (19)$$

The new moment of inertia of the entire cross section is the sum of the transformed moments of inertia for each ring.

$$I_{new} = I_1 + I_2 + \ldots + I_m \quad (20)$$

The stress due to the bending moment on each ring can be calculated using the new moment of inertia and the MOE ratio.

$$\sigma_{b_h}(x) = \pm\frac{n_h M(x) r_h}{I_{new}} \quad (21)$$

The axial stress on each ring was calculated by considering the different MOE values of each ring. First the distributed axial load, $w_a(x)$, and the total axial force, $F_a(x)$, were calculated using the total area of the cross section, $A_{total}(x)$. These variables are independent of the MOE of the branch.

Each ring was of a different area and a different modulus of elasticity value therefore each ring had a different amount of the axial force acting on its section. The sum of the axial forces acting on each ring is equivalent to the total axial force (22).

$$F_1(x) + F_2(x) + \ldots + F_m(x) = F_a(x) \quad (22)$$

Where $F_a(x)$ is the total axial force due to the weight of the branch. Forces $F_1(x)$ through $F_m(x)$ are the compressive forces acting on each ring. The forces on each ring cause the branch to deform along the axis of the branch, making the branch shorter. This deformation can be calculated as

$$\delta_h = \frac{F_h L}{A_h E_h} \quad (23)$$

Where δ_h is the deformation of the ring, F_h is the axial force acting on the ring, A_h is the area of the ring, and E_h is the modulus of elasticity of the ring. Since all of the rings are connected to one another they will have the same deformation.

$$\delta_1 = \delta_2 = \ldots = \delta_m$$

$$\frac{F_1 L}{A_1 E_1} = \frac{F_2 L}{A_2 E_2} = \ldots = \frac{F_m L}{A_m E_m} \quad (24)$$

Using Equation (24) the axial force on each ring can be solved for in terms of the axial force acting on any of the other rings. The example to follow is all done in terms of ring 1, however the same procedure can be followed to determine the axial force for any of the rings. The axial force on any of the rings in terms of the axial force on ring 1 is

$$F_h = \frac{A_h E_h}{A_1 E_1} F_1 \tag{25}$$

The total axial force on the branch in terms of the force on ring 1 becomes

$$F_a = F_1 + \frac{A_2 E_2}{A_1 E_1} F_1 + \ldots + \frac{A_m E_m}{A_1 E_1} F_1 \tag{26}$$

Equation (26) can be rearranged to solve for the axial force acting on ring 1 (27).

$$F_1 = \frac{F_a}{1 + \frac{A_2 E_2}{A_1 E_1} + \ldots + \frac{A_m E_m}{A_1 E_1}} \tag{27}$$

$$F_1 = \frac{F_a}{\frac{1}{A_1 E_1}\left(A_1 E_1 + A_2 E_2 + \ldots + A_m E_m\right)}$$

This procedure was repeated to determine the axial forces acting on each of the rings. Using the axial force acting on each ring the axial stress on each ring can be calculated as

$$\sigma_{ah}(x) = -\frac{F_h(x)}{A_h} \tag{28}$$

The total stress for Case 5 is the sum of the axial and bending stresses.

Case 6 accounts for the curves of a branch in the x-y plane, seen from the side view. The curviness of the branch is determined by measuring the height of the branch above and below the axis of the branch at incremental distances along its length. The axis line is the line that connects the center of the cross section at the base of the branch to the center of the cross section at the tip of the branch (Fig 5.).

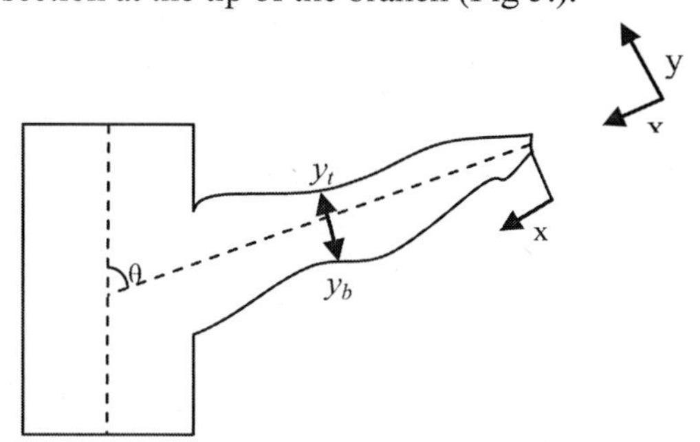

Figure 5. Case 6: curvy tree branch model. Where y_t is the height of the branch on top of the axis, y_b is the height below the axis.

The distance from the axis to the top of the branch is y_t. The distance from the axis to the bottom of the branch is y_b. The diameter of the branch is the sum of y_t and y_b. The branch is assumed to have a circular cross section in this model because the varying heights of the branch were only considered in the x-y plane (side view), not the x-z plane (top view).

To perform a stress analysis for Case 6 calculations were done to determine the stresses at the base of the branch, where $x=L$. At this location the axis of the branch coincides with the center of the cross section of the branch, because the axis was defined earlier as the line that connects the centers of the cross sections at the base of the branch and at the tip of the branch. Performing the stress analysis at the base provides the maximum stresses experienced by the branch.

To account for the curviness of the branch a best fit polynomial equation was determined for y_t and y_b in terms of x. Because the branch is curved, the centroid of the branch may not lie on the branch axis. Therefore the axial component of weight applied at the centroid must be moved to the axis to perform the stress analysis. To account for moving the forces an additional bending moment must be added.

The centroid of the area in the x-y plane was determined by first calculating the centroid of the areas above the axis and below the axis separately. Then the two centroids were combined to obtain the centroid of the entire branch. To do this the area of the top and bottom parts must be calculated separately. This was done by integrating the y_t and y_b equations to find the area under the curve. Using the areas the x-coordinate of the centroid for the top (29) and bottom (30) portions of the branch were calculated using

$$\bar{x}_t = \frac{1}{A_t}\int_A x f_t(x)dx \tag{29}$$

$$\bar{x}_b = \frac{1}{A_b}\int_A x f_b(x)dx \tag{30}$$

Where y_t and y_b are both functions of x, $y_i=f_i(x)$. To determine the y-coordinate of the centroid, two additional best fit polynomial equations were determined for x in terms of y_t and y_b, giving $x_i=f_i(y)$. The y-coordinates of the centroids above (31) and below (32) to the axis line are

$$\bar{y}_t = \frac{1}{A_t}\int_A y f_t(y)dy \tag{31}$$

$$\bar{y}_b = \frac{1}{A_b}\int_A y f_b(y)dy \tag{32}$$

Next the x and y coordinates of centroid, relative to the branch axis, were calculated for the entire branch (33) and (34).

$$\bar{x} = \frac{\sum_i A_i \bar{x}_i}{\sum_i A_i} = \frac{(A_t \times \bar{x}_t) + (A_b \times \bar{x}_b)}{(A_t + A_b)} \tag{33}$$

$$\bar{y} = \frac{\sum_i A_i \bar{y}_i}{\sum_i A_i} = \frac{(A_t \times \bar{y}_t) + (A_b \times \bar{y}_b)}{(A_t + A_b)} \tag{34}$$

To perform the stress analysis, the axial force, F_a, must be moved from the centroid to the center of the branch's cross section. The perpendicular distance that F_a, must be moved is $\overline{y}$. Therefore the additional bending moment that must be added to account for moving the axial force is

$$M_{addional} = F_a \times \overline{y} \tag{35}$$

The bending moment that is due to the shear force (36) is the product of the shear force and the perpendicular distance between the x-coordinate for the centroid and the base of the branch.

$$M_{shear} = V_s \times (L - \overline{x}) \tag{36}$$

Recall that $\overline{x}$ is measured from the tip of the branch, so it must be subtracted by the length to get the distance between the base and the centroid.

The total bending moment is the sum of these two bending moments, when adding them together the signs must be considered. After the total bending moment has been calculated, the stress due to bending can be determined. The axial stress is calculated using Equation (3) after the force has been moved to the center of the cross section. The total normal stress acting on the cross section at the base is the sum of normal stress due to bending and the axial stress. Once again, when adding stresses, the sign convention must be considered.

The equations derived for each of the cases were used to write codes in Matlab that can perform a stress analysis on any branch after measureable dimensions are inputted. The Matlab code provides an exact solution for the stress analysis. To make the code more user friendly a Graphical User Interface (GUI) was generated to allow anyone to easily determine the stresses on a tree branch (Fig. 6).

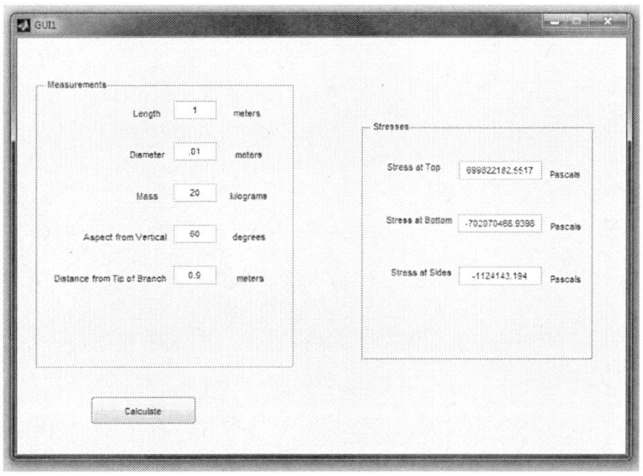

Figure 6. Image of the Graphical User interface for Case 1.

FINITE ELEMENT ANALYSIS

The results were compared to the results obtained from Finite Element Analysis. The Finite Element Analysis was performed using Abaqus CAE. Two and three dimensional tree branch models were created. Material properties were assigned to the models based on literature. Boundary conditions were applied to the base of the branch to keep it fixed. No load was applied to the tree branch other than gravity, which the program applies by using the material's density and shape. Gravity is distributed evenly throughout the model. Once the model is completed, Abaqus performs a stress analysis and provides a contour image displaying stress, strain or displacement throughout the entire branch (figures 7 and 8).

Figure 7. 2D "Simple Beam" branch model

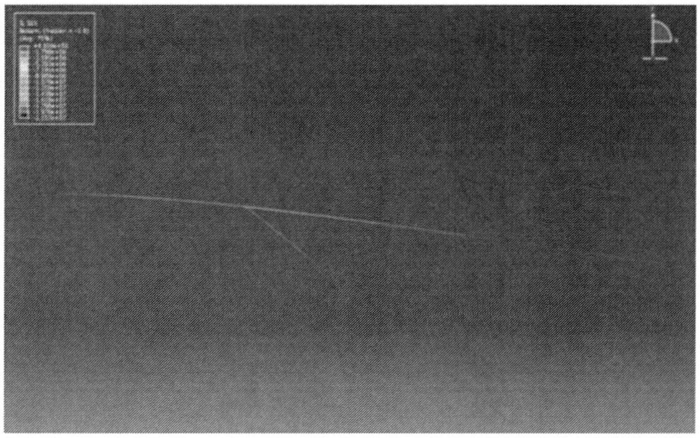

Figure 8. 2D Model containing secondary branches

RESULTS AND DISCUSSION

Results from 2D finite element model and our calculations explained above match exactly (with less than 0.06 % error).

Next, the developed model (6 cases) was used to examine trends that result from changing a single input. For Cases 1 through 5 all of the input variables were kept constant except for one in order to examine how changing that single variable affects the results. The variables considered include branch length, diameter, density and aspect (or angle) with respect to the vertical. For Case 5 changes in MOE were also considered.

For each of the scenarios tested, all of the cases produced the same trends. As the length of the branch increased, so did the mass, because the density remained fixed. However the ratio of length to mass remained fixed because they both grow at the same rate. Therefore as the length of the branch increased so did the maximum stress (Fig. 9).

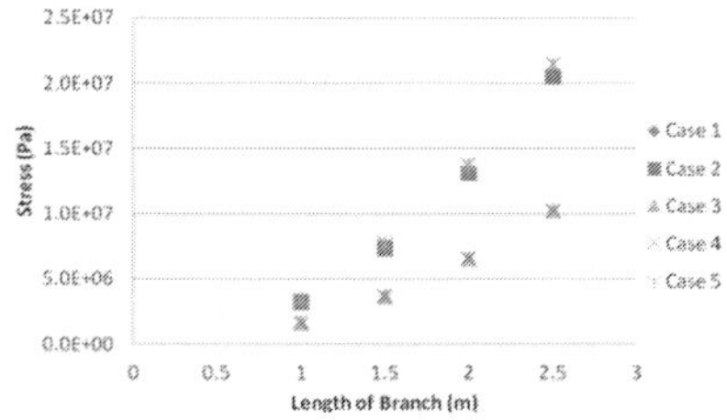

Figure 9. Length of branch versus maximum stress.

As the length increases the maximum stress appears to begin to grow more rapidly, this would be due to the fact that an increased length correlates to an increased bending arm. As for the tree branch models, Cases 1, 2, and 5 all produced nearly identical results. These were the fixed cross sectional area cases that were not tapered. According to the stress analysis the benefits of having an elliptical over a circular cross section appear to be limited. The tapered Cases 4 and 5 had smaller stress values than the non-tapered models. As the length increased the difference between the tapered and non-tapered cases became more apparent. This makes sense because a non-tapered branch is more massive, creating greater stress.

When the diameter of the branch was varied, the mass also varied because the density was kept constant. However, the diameter and the mass did not increase at the same rates. When examining the changes in diameter solely, an increase in diameter appears to produce a decrease in stress (Fig. 10). However, an increase in diameter per unit mass leads to an increase in stress (Fig. 11).

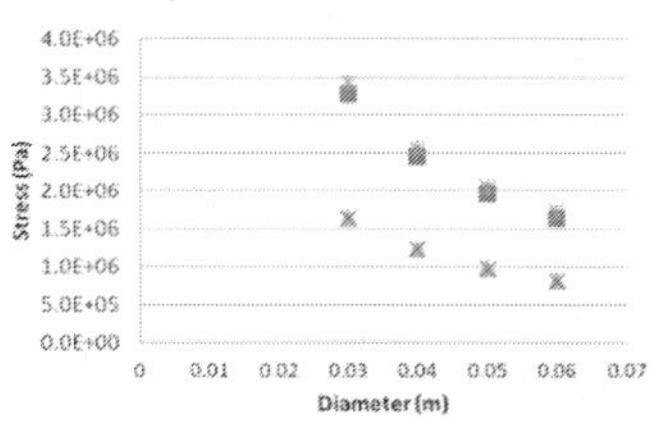

Figure 10. Diameter versus stress.

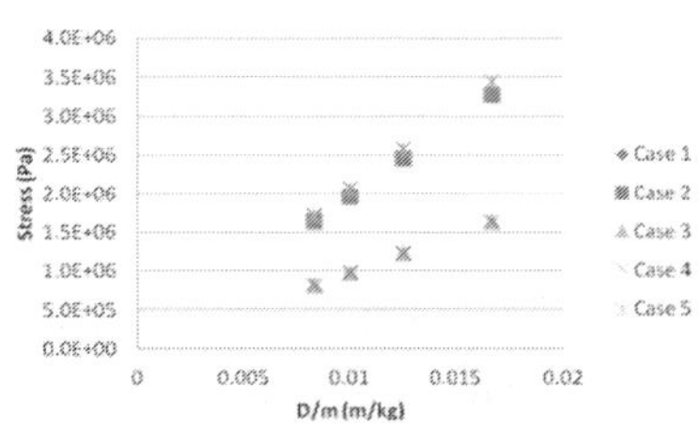

Figure 11. Diameter to mass ratio versus maximum stress.

As the diameter of the branch increases the stress decreases. This shows that when comparing two tree branches from the same tree a thicker branch can resists more stress. As the diameter per unit mass ratio increases the stress increases. This shows that when mass is not considered a branch with a smaller diameter will have less stress. Therefore over a random sampling of tree branches, an increase in diameter results in an increase of stress. The relationships between the different tree branch models remained the same; the tapered models have less stress than the fixed models, and the shape of the cross section had a negligible effect on the stress.

For each of the cases the density and the aspect of the branch with respect to the vertical were also analyzed. The density was varied by keeping the dimensions of the branch constant and varying mass. As the density of the branch increased the stress increased (Fig. 12) and as the aspect of the branch from the vertical increased (became more horizontal) the stress increased (Fig. 13).

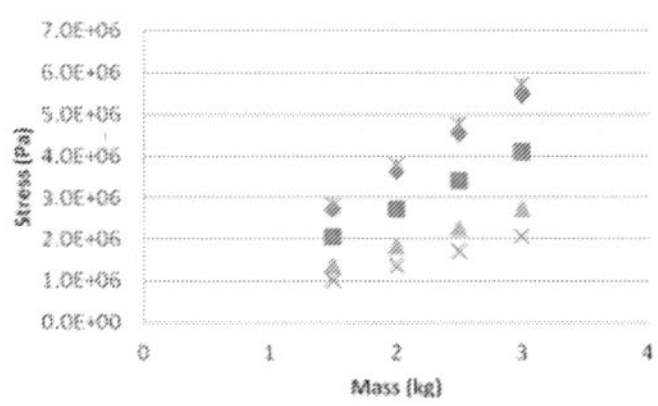

Figure 12. Mass versus maximum stress.

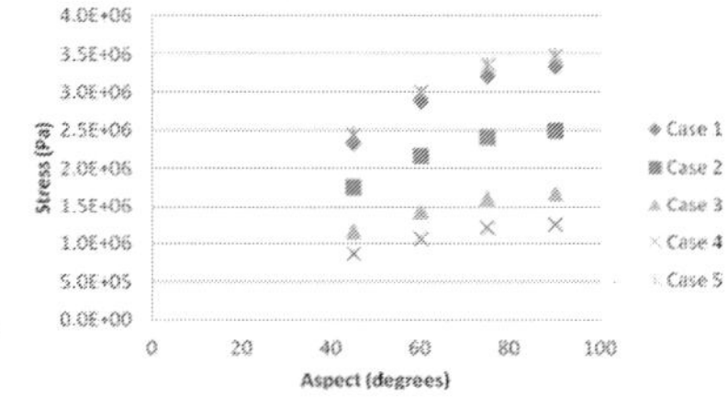

Figure 13. Aspect from vertical versus maximum stress.

Here there is a more noticeable difference between the different cases. Cases 1 and 5, the two cases with fixed circular cross sections produced results that were much greater than any of the other cases. Cases 1 and 5 also produced very similar results. Here Case 2 has significantly lower stress values than Case 1, this shows that when considering branch aspect and mass the elliptical shape of the branch is beneficial for reducing the affects of stress. This is further supported by the significantly lower values produced by Case 4, the tapered elliptical branch, than Case 3, the tapered circular branch. Again the tapered branches produce less stress than the non-tapered branches.

For Case 5 the effects of varying MOE values were analyzed. Case 5 was divided into three rings for this analysis, and MOE values were selected for each ring based on results from previous research. The percent increase in MOE values between the first and second rings was 113%, while the percent increase between the second and third rings was 25%. These percent increases were increased or decreases by 5% until there was either an additional positive or negative 20% from the reference percents (Fig. 14).

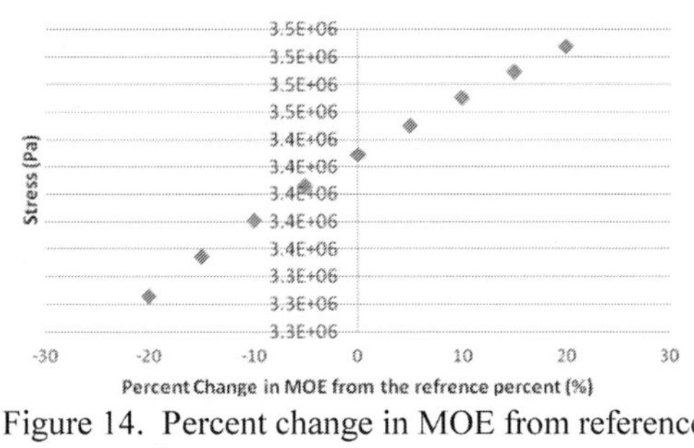

Figure 14. Percent change in MOE from reference percent versus maximum stress.

As the percent change in MOE increased between concentric rings the stress increased. This means that a branch composed of a material that has a more uniform MOE throughout will experience less stress than a branch that has drastic changes of MOE in the radial direction.

Case 6 accounted for the curviness of the branch. Different dimensions were entered into the program to create branches with varying degrees of curves. For consistency all of the branches had the same base diameter and the same tip diameter. First the stress on a branch with no curves was compared to Case 3. They should produce similar results because they both had the same base diameter and get very narrow at the tip of the branch. However, they had noticeably different results (Table 1).

Table 1. Comparison of the stresses in a non-curved branch as calculated by cases 3 and 6. The percent error calculation is based on the assumption that case 3 is more accurate.

Case 3 Stress (Pa)	Case 6 Stress (Pa)	Percent Error
2.73×10^5	3.63×10^5	32.97%

The substantial difference in the stress values could be due to the fact that Case 6 generates a best fit polynomial equation even when the branch may be more accurately represented by a straight line. Based on this Case 6 should only be used when the branch is curved, or when there is an abnormality in the branch which would shift the centroid of the branch away from the branch axis.

Figure 15 shows the shapes of two branches developed by the best fit polynomial equations and their corresponding maximum stress values.

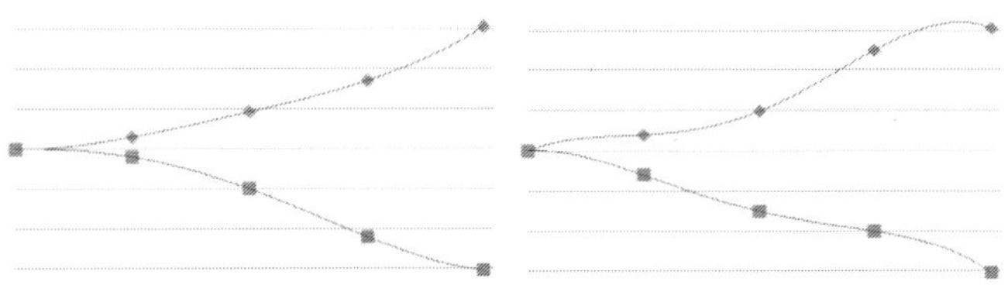

(a). Max Stress= 2.89×10^5 Pa (b). Max Stress= 3.29×10^5 Pa

Figure 15. Branch Shapes generated by polynomial equations and their corresponding stress values.

The tree branch generated in Figure 13a has very slight curves. The stress value calculated for this branch more closely aligns to the stress calculated using Case 3 displayed in Table 2. This could be due to the fact that Fig. 13a has minimal curves so it resembles Case 3 and yet curves are present so a polynomial equation will fit the shape of the branch better than a straight line. Fig. 13b has more prominent curves, and produces a larger stress value. The code was written so that the polynomial fit equation will pass through all of the input points. Therefore whenever a branch has a drastic change in height it should be entered. In addition, input points should be added at locations between the drastic changes to guide the path of the line.

CONCLUSION

In this study we performed exact stress analysis and a finite element simulation on tree branches for 6 different cases. Next, using the developed model, we analyzed the effect of several different variables on the stress within branches.

The cases that had fixed cross sections produced stresses that were greater than those with tapered cross sections. This is due to the fact that non-tapered branches are more massive. When the mass and angle of the branch were varied, having an elliptical cross section proved to be advantageous in resisting stress. Case 5 considered variances of MOE in the radial direction. This produced results very similar to Case 1. Case 5 proved to be most beneficial when there are drastic changes in MOE values between two concentric rings. Case 6 is most beneficial for branches that have curves or abnormalities that shift the center of mass of the branch away from the branch axis.

In the future, with further collaboration with the biology department, the Matlab code will be utilized to examine stress patterns that form in actual trees as they grow. Additional cases will be considered. Secondary branches will be added to the main branches to determine how they affect stress. Also, Case 5 will be modified to account for compression wood and tension wood. Compression wood is able to resist greater compressive stresses, while tension wood is able to resist greater tensile stresses. To resist different types of stresses, the biological center of the wood does not coincide with its geometric center. This causes the tree to develop rings that are off center; therefore, variances in the MOE will be off center [4]. To account for these wood properties additional programs will be written for compression and tensile wood, where instead of having concentric rings of different MOE values, the rings will be off center. In addition, we will expand our FE model to study 3D and more geometrically complicated branches.

REFERENCES

1. Evans, L., Z. Kahn-Jetter, J. Torres, M. Martinez and P. Tarsia. 2008. Mechanical stresses of primary branches: a survey of 40 woody tree and shrub species. Trees 22:283-289.

2. Sone, K., K. Noguchi and I. Terashima. 2006. Mechanical and ecophysiological significance of the form of a young acer rufinerve tree: vertical gradient in branch mechanical properties. Tree Physiology 26:1549-1558.

3. Mencuccini, M., J. Grace and M. Fioravanti. 1997. Biomechanical and hydraulic determinants of tree structures in Scots pine: anatomical characteristics. Tree Physiol 17:105-113.

4. Almeras, T., A. Thibaut, J. Gril. 2005. Effect of circumferential heterogeneity of wood maturation strain, modulus of elasticity and radial growth on the regulation of stem orientation in trees. Trees 19:457-467.

Proceedings of the ASME 2014 International Mechanical Engineering Congress and Exposition
IMECE2014
November 14-20, 2014, Montreal, Quebec, Canada

IMECE2014-38536

DESIGN AND IMPLEMENTATION OF A GENERAL CONTROL SYSTEM PLATFORM

Kyle T. Peerless
Jasmine M. Panosian
Pezhman a. Hassanpour*
Department of Mechanical Engineering
Loyola Marymount University
Los Angeles, California, 90045-2659

ABSTRACT

In this study, the design and implementation of a general control system platform for educational purposes was performed. This project has been designed to facilitate the understanding of control systems in mechanical engineering by creating a foundational system to position-control the rotation of a DC motor, and then employing it as a teaching tool in the undergraduate Control Systems course at Loyola Marymount University (LMU). The objective of this effort was to lay the groundwork for a more "hands-on" control systems education at LMU by designing the general mechanical platform and a pilot on-off controller to illustrate its functionality and feasibility in the classroom. With the foundational stages complete, students in future classes will be able to design and build different controllers for this same device in order to investigate and apply their newly-acquired knowledge of control systems.

INTRODUCTION

A traditional course in Controls emphasizes elaborate concepts such as mathematical models and transfer functions with little discussion on how the theory can be applied to real-world applications. For example, on-off controllers, which are one of the most commonly used controllers, are either overlooked or superficially discussed in most Controls textbooks. On the other hand, many commercial lab setups are designed so that the parameters of the system can be changed through a panel or computer interface. This provides little insight into how transfer functions can be related to an actual dynamic system in operation. A traditional engineering curriculum in the United States will feature integration of physical design experience into the final year of an undergraduate program via capstone design courses. Students often face difficulty with these courses as they are their first so-called "open-ended" physical design problems. This project and general control system platform is classified as a "guided" design experience, which has been shown to be an effective introductory experience to take before, or in this case, in conjunction with a capstone course [1].

The objective of the project presented in this paper is to develop a firsthand understanding of control systems in mechanical engineering by creating a system that engages students on a higher level to apply their knowledge to a real controller design. This intentionality to develop firsthand experience for students is directly aligned with ABET Criterion 3 for Student Outcomes in Engineering Programs [2], namely:

(1) an ability to apply knowledge of mathematics, science, and engineering (outcome a)
(2) an ability to design a system, component, or process to meet desired needs within realistic constraints such as economic, environmental, social, political, ethical, health and safety, manufacturability, and sustainability (outcome c)
(3) an ability to identify, formulate, and solve engineering problems (outcome e)
(4) an ability to use the techniques, skills, and modern engineering tools necessary for engineering practice. (outcome k)

Aside from addressing accreditation concerns, the notion of

*Address all correspondence to this author. Email: hassanpour@asme.org developing an "inverted" class format with more realistic and firsthand experience for students, as opposed to a solely lecture-based course, has been associated with increased academic performance in upper division Control Systems courses [3].

The advantage of this project and other similar self-developed control modules that have been demonstrated [4] is the simplicity of the set-up and low costs of the necessary components. The platform consists of a mechanical system and a controller. The mechanical system includes a DC motor that is connected to a potentiometer with a shaft. The selection consisted of a DC motor as the actuator as opposed to a stepper motor; and a potentiometer was selected instead of an encoder. In this arrangement, sensing the output will force the students to design analog circuits for the controller. In doing this, they will learn how to employ their knowledge of electric circuits, particularly Wheatstone bridges and operational amplifiers, to translate their transfer functions into actual working systems.

APPROACH AND IMPLEMENTATION

Theory of Controller Circuit

The foundation of the electronic circuit used in the control system is a Wheatstone Bridge-design with two resistors and two variably-resistant potentiometers, as illustrated within the context of the whole controller in Figure 1 and more specifically in Figure 2. The theory behind the design is that when the resistance of one of the potentiometers is changed, a voltage difference or "error voltage" is created between the resistors and the potentiometers which is then fed into a comparator. Depending on the sign of the input voltage into the comparator, one of two resulting voltages is fed to the operational amplifier which outputs ± 12 V to turn the DC motor. If the error voltage is positive, the comparator output will be 5 V; and if the voltage is negative, the comparator output will be zero. Using circuit analysis and knowledge of operational amplifiers, correct resistor implementation has been performed in the feedback loop of the amplifier that in turn yields desired output voltages to turn the DC-motor in correct direction. In the other words, the operational amplifier maps the 0 V/5 V input signal into ± 12 V. In terms of additional aesthetics and alterations; a switch is wired in series ahead of the motor allowing the user to further manipulate the operation, direction-specific LEDs have been added to the circuit corresponding to the direction the motor is turning, and an aluminum heat sink is attached to the operational amplifier to help facilitate heat transfer out of that component.

On-Off Controller

The way the system is designed, the operational amplifier is constantly receiving an input voltage whether the potentiometers are turned or the system is at the desired position. This input voltage will then generate an output voltage to turn the motor which will, in turn, cause the potentiometer to rotate generating an error voltage across the Wheatstone Bridge. This recently created error voltage will then pass through the comparator and output a correcting voltage to the operational amplifier until the error voltage is extinguished and the process can start over again. This sequence of events occurs at an extreme pace, which results in the oscillating voltage conditions that characterizes this on-off controller.

Error Voltage

To perform the Wheatstone Bridge analysis, Figure 2, a mathematical relationship needed to be discovered between the displacement of the potentiometers, and the resulting error voltage that was created across the bridge. This relationship would provide valuable insight into both the sign of the error voltage as well as the nature (exponential, linear, etc.) of the relationship between the displacement and the error voltage. After a circuit analysis of the bridge using Kirchoff's equations, it was determined that the two had the following approximate linear relationship [5].

$$V_e = -\frac{V_s \delta}{4R}, \tag{1}$$

where V_s is the source voltage, R is the value of the fixed resistors of the bridge, and δ is the difference between the resistance of the two potentiometers, one for reference and one for the actual position of the motor.

Variable-Resistance Potentiometer Selection

The Wheatstone Bridge analysis that was performed and the resulting relationship above make an assumption that all the resistor values on the bridge are equal. That being said, design considerations needed to be made to recreate the theoretical math scenario that was used. To do this, the potentiometers needed to have a reference value equal to that of the other two non-varying resistors and have a maximum range of displacement with that same resistance value. In our case, the potentiometer needed to have a range from 0 Ω to 200 kΩ, with a reference starting value of 100 kΩ.

Feedback Loop Resistor Ratio Analysis

As explained earlier, the 5 V signal from the comparator must be mapped to -12 V signal at the output of the power amplifier. In addition, the 0 V signal from the comparator leads to 12 V output at the power amplifier. This yields a system of equations which can be solved to determine the feedforward and feedback resistors of the power amplifier. The power amplifier output voltage V_O can be claculated from [5]

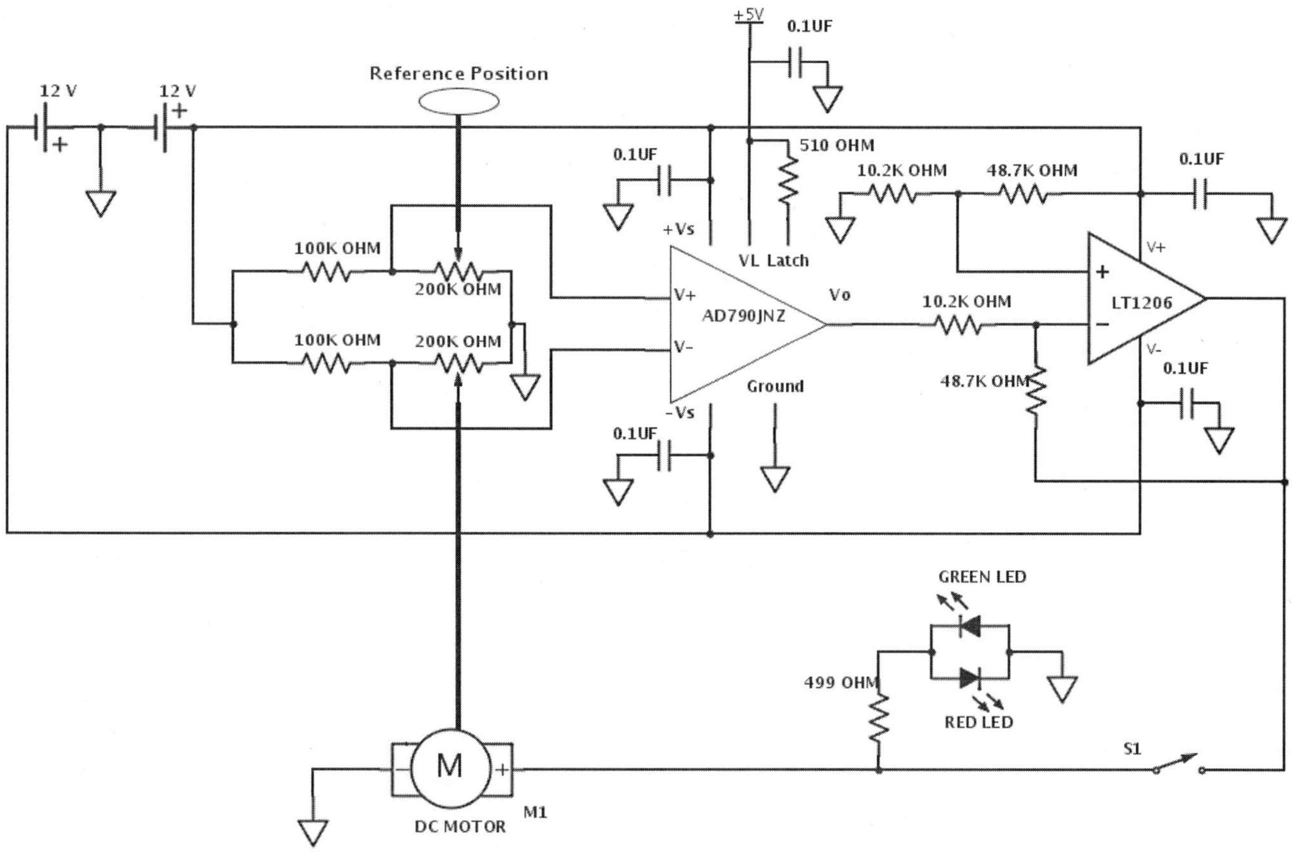

FIGURE 1. Implementation of the position control system for a DC motor

$$V_O = V_2 - G(V_2 - V_1), \quad (2)$$

where V_1 and V_2 are the voltage of the amplifier inputs, and G is the amplifier gain. Applying the input-output requirements for the power amplifier, one can show that $G = 4.8$, and $V_2 = -2.7\,\text{V}$.

From the analysis above, the $48.7\,\text{k}\Omega$ and $10.2\,\text{k}\Omega$ resistors were selected for the power amplifier.

DC Motor

Based on the design of the controller, the DC motor operates only at two states of $12\,\text{V}$ or $-12\,\text{V}$. In this project, nonlinear effects, e.g., DC motor and its built-in gear box delays and backlash, are ignored. Besides, it should be noted that all electronic components are assumed to be ideal. For example, the saturation of the amplifier and the comparator delay are ignored. Due to the introductory nature of the course, and limitations of the components of this platform, the performance of students' controllers will be evaluated more qualitatively rather than in comparison with theoretical transfer functions.

Aluminum Heat Sink and LEDs as Motion-Direction Indicators

An aluminum heat sink is used on the operational amplifier to protect it from overheating and potential failure. Due to the large amount of current flowing through the output of amplifier to the motor, a heat sink alleviates the growing heat that develops in the amplifier. The heat sink is also used to demonstrate a basic application of the 2nd Law of Thermodynamics to students that heat energy must transfer from a higher to lower temperature reservoir when it is being transferred out of the amplifier. As a visual aid, LEDs were added on the circuit to indicate the polarity and direction of the turning motor. These were oriented in such a way so that the LED on the right would turn on if the motor was turned to the right, and vice versa.

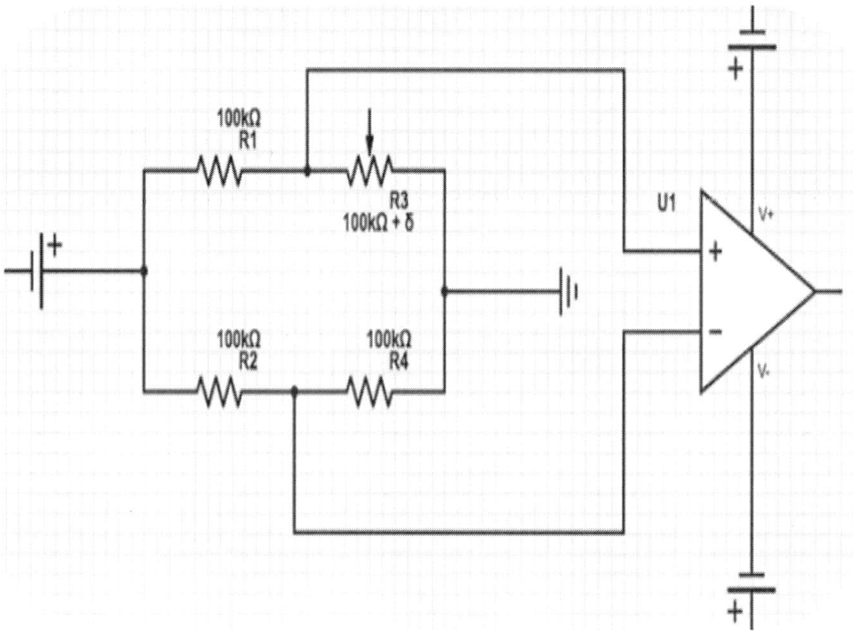

FIGURE 2. Wheatstone Bridge

The Mechanical Platform

The design of the mechanical part of the apparatus needed to effectively hold and display all the necessary components of the control system. Given these design parameters, several concepts were sketched and developed into CAD models using the SolidWorks software. The most effective final design idea was to create a housing with a base and two opposite walls. This layout emerged as the best design because the base could serve as a resting place for the circuit, and then one wall could hold one of the potentiometers, the switch and the motor; with the latter being connected to the potentiometer on the opposite wall via a rod and a pair of coupling joints, see Figure 3. Special flexible coupling joints were used to make the connections on both sides of the robot arm for two key reasons. First, it is nearly impossible to manually drill two perfectly aligned holes on either side of the housing for the potentiometer and motor to be linked by the rod. Therefore, joints that could handle slight differences in these holes were selected so that unnecessary strain could be alleviated from the rod and couplings. Additionally, each of the aforementioned components had different diameters so the joints also had to accommodate this variation while maintaining a sturdy fit. The housing was constructed with an aluminum plate which was then formed and drilled.

CONCLUSION AND FUTURE WORK

With the general platform in place, the project will be expanded in various directions concurrent to the device's use in the classroom starting in Fall 2014. The preliminary results of deployment will be presented at time of the presentation of this paper at the IMECE 2014. The next step is to create and analyze the performance of other types of controllers such as Proportional, Proportional-Integral, and Proportional-Integral-Derivative controllers because the platform created in this research needs to be compatible so that any of the above controllers can be connected and have their performance measured and compared by students. Additionally, given the relevance and past success of incorporating interactive software into the framework of a Controls education [6, 7], the system will be modeled and studied in MATLAB/Simulink to predict certain design behavior and compare the system's outputs to theoretical computer models. Finally, rapid prototyped (RP) parts will be designed and added to the shaft and platform to more clearly illustrate the rotation of the shaft in response to commands given to it by the controller.

This research is significant because in designing a platform that can be used by undergraduate students, they will get to see the application of course fundamentals on a mechanical system. This is important and key to their understanding of the implementations of controls in larger engineering systems. Application of concepts, especially in an interdisciplinary sense, is representative of engineering beyond the classroom in industries and research fields. In integrating various subjects together, this project will help the students to develop a deeper understanding of the topics of mechanical engineering and how they are interrelated.

FIGURE 3. Final implementation of the control system

ACKNOWLEDGMENT

The authors would like to acknowledge the support of the Office of Undergraduate Research at Loyola Marymount University, as well as Mr. Joe Foyos and Mr. John Unalp, Laboratory Facilities Managers in the Seaver College of Science and Engineering and Mr. Ben Horten who contributed to this project.

REFERENCES

[1] Dhaouadi, R., Al-Assaf, Y., and Rehman, H.-u., 2012. "Introducing the guided design experience in control engineering education". *International Journal of Engineering Education,* ***28***(3), pp. 563–571.

[2] ABET, 2012. Criteria for accrediting engineering programs. Accreditation Board for Engineering and Technology.

[3] Mason, G. S., Shuman, T. R., and Cook, K. E., 2013. "Comparing the effectiveness of an inverted classroom to a traditional classroom in an upper-division engineering course". *IEEE Transactions on Education,* ***56***(4), pp. 430–435.

[4] Rubaai, A., 2000. "Training of undergraduate engineering students in sensing and control technologies". *International Journal of Engineering Education,* ***16***(6), pp. 534–543.

[5] Nilsson, J. W., and Riedel, S., 2014. *Electric Circuits*. Prentice Hall.

[6] Dormido, S., Dormido-Canto, S., Dormido, R., Sanchez, J., and Duro, N., 2005. "The role of interactivity in control learning". *International Journal of Engineering Education,* ***21***(6), pp. 1122–1133.

[7] Kozick, R. J., and Crane, C. C., 1996. "An integrated environment for modeling, simulation, digital signal processing, and control". *IEEE Transactions on Education,* ***39***(2), pp. 114–119.

This page left blank intentionally.

Proceedings of the ASME 2014 International Mechanical Engineering Congress and Exposition
IMECE2014
November 14-20, 2014, Montreal, Quebec, Canada

IMECE2014-38676

DESIGN OF A LOW COST APPARATUS FOR TORSION TESTING

Mohammad Mahinfalah, Professor
Department of Mechanical Engineering
Milwaukee School of Engineering
Milwaukee, WI, United States

Joseph Musto, Professor
Department of Mechanical Engineering
Milwaukee School of Engineering
Milwaukee, WI, United States

ABSTRACT

Torsional loading is present in a variety of structures and machines. In designing structures that are subjected to this type of loading it is of paramount importance that the effect of loading on the machines be understood. To reinforce the student's learning in a mechanics of material class a low cost table top apparatus is designed which can be used in a classroom or a laboratory. Many engineering programs attempt to strengthen the students understanding by introducing laboratory experiments. In general equipment used to measure the effect of torsional loading are capital intensive and bulky. As a result, the experiments are run by the instructors or by graduate assistants while students are only observing. This newly designed table-top device by the authors, costs fraction of the equipment that are available on the market. Five devices were made and instrumented to be used by 30 students in ME 207 class at Milwaukee School of Engineering (MSOE) during the spring 2014. The students indicated that this was a very useful tool to reinforce understanding of mechanics of materials concepts taught in the class. Overall, this experiment strengthened the understanding of theoretical relationships discussed in class as related to the torsional loading of circular shafts and tubes.

INTRODUCTION

Transmission of torque from a plane to another plane in the design of machines is a task that the designer must consider. To accomplish this task use of circular shafts is a common practice by engineers. The shaft that is used to transmit the torque experiences stress and strain during its operation. Understanding of these stresses and strains along with the distortion of the shaft is of paramount importance in the safe operation of the machinery that involves this type of loading.

Unlike the previous generation of engineers who mostly came from farming communities, today's students in majority of engineering programs come from mostly urban environment. As a result, they have not experienced repairing machinery and retrofitting design. To prepare the students for design and manufacturing sector, it is necessary that they are exposed to more and more meaningful laboratory (hands on) experience in their education. It is preferred that this experience is gained by direct engagement of student in running of the experiments and analysis of the obtained data.

Unfortunately, the prohibitive cost of testing equipment, lack of available space and increase demand on use of valuable instructional time limits the time that the students spend in acquiring "hands on" skills in the laboratories. As a result, fewer experiments are performed. In most cases, students are observers of these experiments, and later on may get the data to write a report. The newly designed table-top device as shown in Figure 1 costs a fraction of the equipment that are available on the market to perform an experiment in torsion.

This apparatus consists of a base plate which must be clamped to a rigid support such as a table, a torque wrench (0-800 in-lb), an aluminum tube and an angle of twist meter. A rectangular strain gage rosette (CEA-13 120CZ) is installed on the side of the tube and it is arranged such that two of the three gages are at 45° with respect to the axis of the shaft.

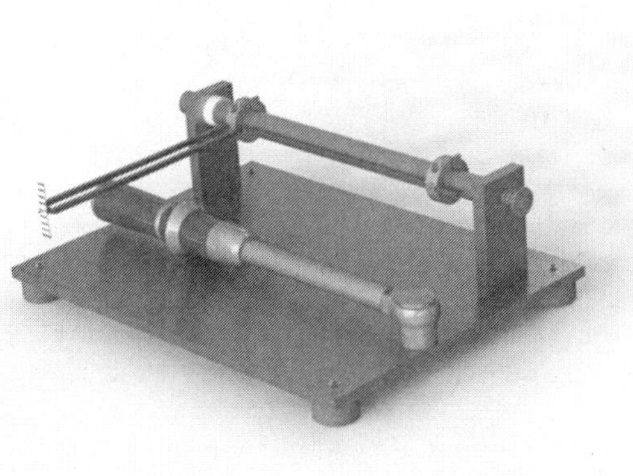

Figure 1: Table-Top Torsion Experiment Device

NOMENCLATURE

θ – angle of twist
ε – measured strain
r – outside radius of the tube
R – distance from the center of the tube to the pointer
H – distance from the outside of the tube to the pointer
S– arc displacement on the outside surface of the tube
S_P – arc displacement at the pointer
G– modulus of rigidity
T– applied torque
τ – shearing stress
J – polar moment of inertia
γ – shearing strain
L – the distance between the two ends of the pointer
d_o – outside diameter of the tube
d_i – inside diameter of the tube

Background

Students who will be exposed to this laboratory experiment have either taken a class in mechanics of materials or are enrolled in the course at the same time. The mechanics of materials class at Milwaukee School of Engineering is a 4-credit class with a weekly two hour laboratory component, and is taken by students from mechanical, industrial and biomedical engineering programs. Typically in all universities the students are exposed to the theory first and later on should be conducting this experiment. The theory and the equations involved can be found in many of the textbooks [1-6] used by the instructors around the world. In this section, the equations that are necessary for determination of stress, strain and the angle of twist are presented.

Once the torque and the dimensions of the tube are measured the shearing stress can be calculated using Equation 1.

$$\tau = \frac{Tr}{J} \qquad (1)$$

To determine the shearing strain two independent variables the angular displacement, S_p, as shown in Figure 2 and the strain readings from the gages are used. The arc length at the pointer tip is used to magnify the displacement (s) at the surface of the shaft. Using this arc length the shearing strain and the angle of twist can be calculated using Equations 2 and 3.

$$\gamma = \frac{S_p r}{RL} \qquad (2)$$

$$\Theta = \frac{S_P}{R} \qquad (3)$$

The shearing strain can be obtained experimentally by measuring the strain at the surface of the shaft at a direction 45° from the axis of the shaft. Two or four gages can be used to eliminate the effect of bending moments that may occur.

Individual readings of these gages can be taken or a half bridge for two gages or a full bridge for four gages can be used. Then, the shearing strain is twice the average reading of each gage as shown in Equation 4.

$$\gamma = 2\varepsilon \qquad (4)$$

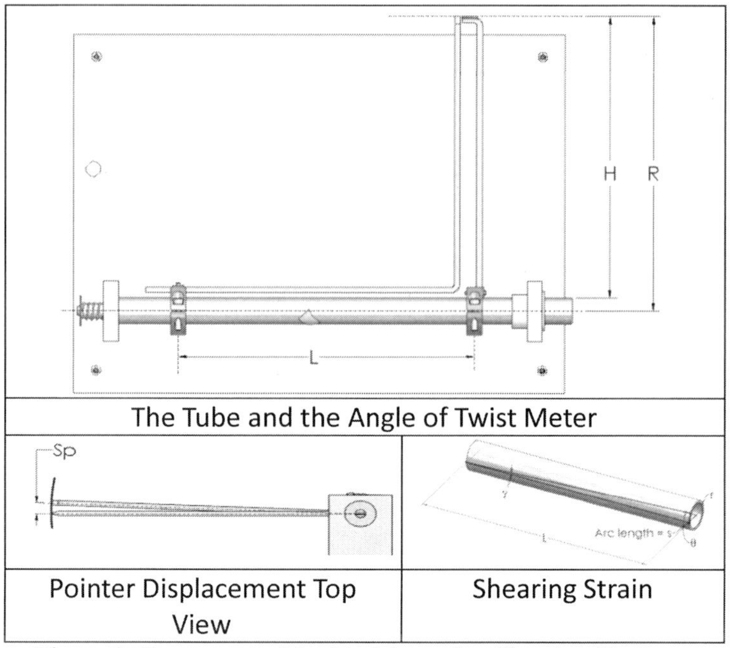

Figure 2. Components Used in Measuring Shearing Strain

Experimental Procedure

Figure 3 shows the setup of the experiment with the torque wrench and Micro Measurement strain indicator. The torque wrench is used to apply and indicate the torsional load. The torsional load can also be applied by using a dead weight to the end of a handle bar or the torque wrench. In both cases the base plate must be clamped to a support (table) and the load be applied.

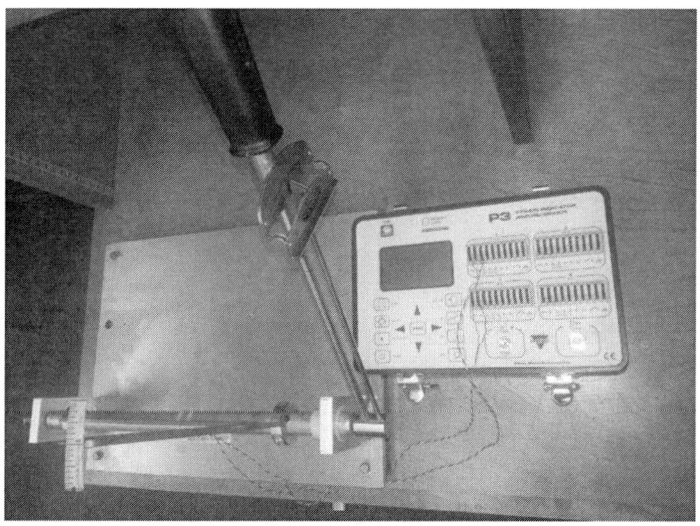

Figure 3. Setup of the Experiment

This apparatus is designed to help to strengthen the understanding of students when it comes to torsion of circular shafts. The device can be used to obtain stress-strain relationship of thin-walled tubes. This can be achieved by using the Torsion Angle Meter or the strain gages independently. Or it can be used as a torque cell if one knows the modulus of rigidity of the tube materials.

Figure 2 depicts the angular displacement at the pointer, and the deformation of the shaft. The following steps are taken to obtain the necessary data for analysis:

1. Clamp the plate to a rigid table or a frame.
2. By pressing the shaft in ward rotate the Angle Meter 90° so it is perpendicular to the plate.
3. Remove the torque wrench and attach it to the end of the tube.
4. Measures d_o, t, R and L.
5. Apply a torque.
6. Measure S_p, and the strains at each torque value.

The toque value can be used to find the shear stress and S_p can be used to find the shearing strain and the angle of the twist for the segment between the two ends of the pointer (L). The reading from the strain gages can also be used to find the shearing strain.

RESULTS AND DISCUSSION

Using the torque wrench a load was applied to the shaft and at a pointer displacement (S_p) increment of 1/16 inch the value of the applied torque and the strain reading of the strain gages were recorded. Table 1 lists the data that was obtained for an aluminum tube with an outside diameter of 7/8 inch and a wall thickness of 0.049 inch. This shaft will yield at an applied torque of 1800 in-lbs. The shaft will not fail during the use by students since it will be impossible to apply this amount of torque to the shaft by hand using the torque wrench used in the lab.

Table1. Sample Data from an Experiment

T in-lb	S_p in	ε_1 micro strain	ε_2 micro strain
75	1/16	-174	156
125	1/8	-328	265
175	3/16	-451	390
225	1/4	-537	478
275	5/16	-737	641
325	3/8	-817	755
375	7/16	-997	934

Using equations 1-4 the data was used to calculate the shear stress and shearing strain. The strains obtained from displacement measurements at the tip of the pointer and shear stresses from the torque measurements are plotted in Figure 4. Adding the magnitudes of the strain gage readings the value of shearing strain is obtained. These values are plotted against the shear stress and the result is shown in Figure 5. From both of the plots using the experimental values a modulus of rigidity of $4x10^6$ psi is observed. This value is exactly the same as the published values for modulus of rigidity of aluminum. To obtain the theoretical lines in Figures 4 and 5 the published value of $4x10^6$ psi is used.

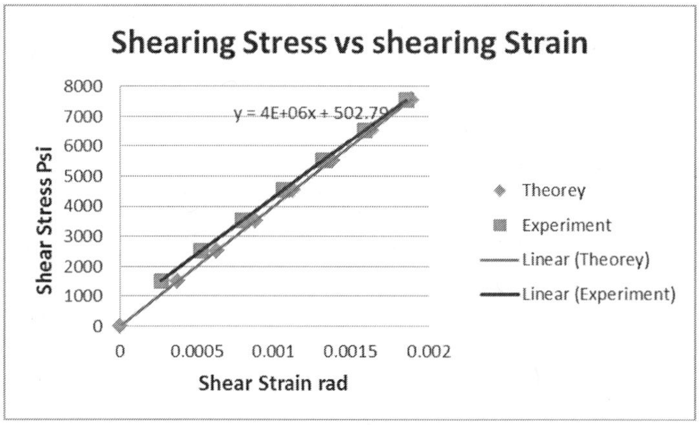

Figure 4. Shear Stress and Stain on the surface of the tube

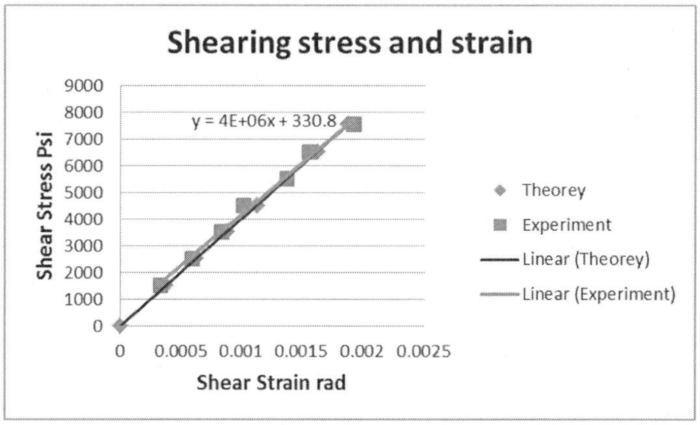

Figure 5. Shear stress and Strain using the strain gages

USE OF THE APPARATUS IN THE LABORATORY

Five devices were made and instrumented to be used by the students in ME 207 class at MSOE during the spring 2014. This lab is a part of ME 207 and it is scheduled for two hours. Overall, there were 30 students in the three sections of this class and each class had a different size. The students were organized in groups of three to five depending on the section that they were in, and there were 8 groups in the three sections.

During the lab each group was asked to pick up a device and a strain indicator box and ready the device for data collection. The students were directed to secure the apparatus and using the torque wrench apply a torque. Once the torque was applied the pointer will indicate a displacement (S_P). Then, the students recorded the applied torque and the two strain gage readings for each 1/16 inch displacement of the pointer. The data collection was continued until a maximum displacement of 0.5 inches or a total of eight data points for each variable.

The maximum measured torque range was 360-455 in-lbs. Using the displacement and strain gage readings students calculated the experimental shearing strains and from the measured torque readings they calculated the shearing stress. Plotting the stress strain diagram the modulus of rigidity was obtained and was compared to the established value of $4x10^6$ psi. Five of the groups had a correct answer and the calculated error was between 0%-7.5%, and three groups had error in calculations and their percent error ranged from 26% to 62%. It is interesting that these groups had wrong calculation of polar moment of inertia as the only or one of the mistakes.

CONCLUSIONS

This paper showed the investigation of a new low-cost apparatus for integrating torsion experiments into an undergraduate mechanics of materials lab. Since the apparatus is low-cost and uses a standard off-the-shelf torque wrench for load application, it is appropriate for undergraduate students to both operate the equipment themselves and perform all measurements. The device is easily replicated, and the experiments can be conducted in just a few minutes in a lab environment or in a classroom.

The device was used in a laboratory by the students and the groups indicated that this was a very useful tool to reinforce understanding of mechanics of materials concepts taught in the class. Using the strain gage data the students were familiarized with the use of strain transformation equations and the concept of Mohr's circle was reinforced. Use of strain gage data helped the students to understand the relationship between the angle of twist and shearing strain. Overall, this experiment strengthened the understanding of theoretical relationships discussed in class as related to the torsional loading of circular shafts and tubes.

ACKNOWLEDGMENTS

Special thanks goes to the technicians of the Mechanical Engineering Department: Richard Philips for mounting gages and Roger Hajny of machine shop who can manufacture what you imagine and draw on a napkin in a short time. The funds were provided by the ME Department.

REFERENCES

1. William F. Riley and L. Zachary, Introduction to Mechanics of Materials, John Wiley & Sons, 1989.
2. Robert L. Mott, Applied Strength of Materials, Prentice Hall, 2007.
3. Russell C. Hibbeler, Mechanics of Materials, Pearson Education, 2013.
4. Ferdinand P. Beer, E. R. Johnston, J. J. DeWolf and D. Mazurek, Mechanics of Materials, McGraw Hill, 2011.
5. James M. Gere and B. J. Goodno, Mechanics of Materials, Course Technology, 2012.
6. Don H. Morris, L. D. Sturges and W. F. Riley, Mechanics of Materials, John Wiley & Sons, 2006.

APPENDIX

Handout to the students
ME 207 Mechanics of Materials
Torsion Test

1. Apparatus
   a. Torsion testing device with the torque wrench
   b. Strain Indicator
   c. Caliper and Tape measure
2. Material tested
   a. An aluminum tube
3. Test procedure
   a. Remove the torque wrench and attach it to the end of the tube.
   b. Measure and record the diameter and the wall thickness of the specimen.
   c. Measure the distance (L) between the pointers.
   d. Measure the distance from the outside of the shaft to the top of the pointer (H).
   e. Attach the lead-wires of the strain gages using the quarter bridge set up shown.
   f. Attach the torque wrench to the end of the shaft.
   g. Apply torsional load and measure the torque and strains every 1/16 inch. Do not exceed ½ inch.
4. Computations
   a. Compute the shear stress for each load.
   b. Compute the shear stress for each load using the measured pointer displacements.
   c. Compute the shear strain corresponding to each load using the strain gage data.
5. Report
   a. Write a report and state the basic assumptions made in your book to obtain the equations discussed in class.
   b. Plot a graph of shear stress and strain using the displacement data and obtain modulus of rigidity.
   c. Plot a graph of shear stress and strain using the strain gage data and obtain modulus of rigidity.
   d. Plot a graph of theoretical stress strain in parts b and c above.
   e. Explain any discrepancy in determining modulus of elasticity in parts b and c.

Please Note Figures 1 and 2 will be provided with this handout, and are omitted for the sake of saving space.

Proceedings of the ASME 2014 International Mechanical Engineering Congress and Exposition
IMECE2014
November 14-20, 2014, Montreal, Quebec, Canada

IMECE2014-38924

DEVELOPMENT OF A LABORATORY EQUIPMENT FOR DYNAMIC SYSTEMS AND PROCESS CONTROL EDUCATION

Rafael E. Vásquez*, Norha L. Posada, Fabio Castrillón, David Giraldo
Grupo de Automática y Diseño A+D
Escuela de Ingenierías
Universidad Pontificia Bolivariana
P.O. Box 56006, Medellín, Colombia
rafael.vasquez@upb.edu.co*, norha.posada@upb.edu.co,
fabio.castrillon@upb.edu.co, david.giraldodu@alfa.upb.edu.co

ABSTRACT

This paper addresses the development of an equipment to teach control engineering fundamentals. The design requirements were determined by users that perform academic, research and industrial training tasks in the area of dynamic systems and process control. Such requirements include: industrial instrumentation; measurement of controlled and manipulated variables, and disturbances; process reconfigurability; different control technologies; several control strategies; appropriate materials for visualization; and compact shape to optimize lab space. The selected process is a tank system that allows one to choose among several dynamic behaviors: first, second, and third order, linear and nonlinear behavior, and dead time; the mathematical model that represents the dynamics of the system is presented. A traditional 3-stage design methodology that includes conceptual, basic and detailed design was followed. The developed equipment allows the user to select from three different technological alternatives to control the system: a PLC, an industrial controller, and a computer. With such flexibility, several control strategies can be implemented: feedback, feedforward, PID, LQG, nonlinear control (gain scheduling, sliding mode, etc.), fuzzy logic, neural networks, dynamic matrix control, etc. The developed system is being used to teach undergrad courses, grad courses, and industrial training. Additionally, the equipment is useful in research projects where grad students and researches can implement and test several advanced control techniques.

*Address all correspondence to this author.

INTRODUCTION

Engineering education has been changing around the world to fulfill, among others, accreditation standards, and society and industry needs [1]. Several successful educational frameworks have been adopted in different universities. For instance, the CDIO initiative puts students inside an scenario in which they work on the conception, design, implementation, and operation of real systems and products [2]. Another example is provided by the ABET (Accreditation Board for Engineering and Technology) model, in which engineering students are expected to address some outcomes in order to fulfill the educational objectives stated by the program [3,4].

There are several curricula that have effectively combined traditional teaching methods with problem or project-based learning, allowing the engineering student to fulfill modern educational requirements [5]. Laboratories play an important role in problem or project-based learning, and to guarantee their effectiveness Feisel and Rosa [6] stated a list of learning objectives that should be considered during the design of lab experiences. Such experiences are important in dynamic systems and process control education, not only because they can help students understand some fundamental concepts, but they can also help in the strengthening of important skills needed by professionals in several disciplines, that can be acquired by using active learning [7]. Such skills include, among others, problem solving, decision making, team skills, written and spoken communication skills, and critical thinking [8].

Control engineering education requires the understanding of fundamental concepts and physical quantities related to dynamic systems and processes [9]. It is possible to find several publications about the development of experimental setups for control engineering education, including systems in which students have physical or remote interaction with the real equipment, and virtual systems [10,11,12,13,14,15,16,17,18,19].

This work addresses the development of a laboratory equipment that can be used not only for undergraduate dynamic systems and process control education, but also for industrial training and research activities. The developed plant is a tank system that offers the flexibility to go from a linear first order system to a nonlinear interacting third order system with variable dead time. Therefore, several control theory works can be tested, allowing the possibility of extending research results from simulations to a real process with industrial instrumentation since both, theory and practice are important in control theory research [20].

DESIGN REQUIREMENTS

To define the laboratory setup, a group of potential users within the School of Engineering at the Universidad Pontificia Bolivariana (UPB), located in Medellín, Colombia, was surveyed. This group, comprised of professors with different backgrounds (mechanical, chemical, electrical and control engineers), defined the following list of design requirements for the new equipment in order to bound the problem and limit the process options:

- Use industrial instrumentation to measure controlled, manipulated and disturbance variables
- Different configurations and system dynamics:
  - First, second and third order
  - Linear and nonlinear behavior
  - Interaction and dead time
- Use different control technologies
- Possibility to implement several control strategies
- Possibility to create remote access
- Use materials that allow visualization
- Compact size to optimize lab space

LAB EQUIPMENT DESIGN AND FABRICATION

The design process was divided into three stages: conceptual, basic, and detailed.

Conceptual Design

There are several options to select a system with the desired design characteristics. However, tank systems are reported in literature as good options due to the reconfiguration possibilities for the process, and their good visualization capabilities [10]. Using the information provided by potential users and a brainstorming process, a draft of the process and instrumentation diagram (P&ID) with basic components was obtained. Figure 1 shows a simplified P&ID of the proposed plant.

The process comprises the upper tank, which is shown as two interacting tanks (Tk1,Tk2) in Figure 1, and the lower tank (Tk3). Level measurements in each tank are indicated with level transmitters (LT). An electrical pump brings water to the system through two different loops; flow transmitters are shown in each loop (FT). The main loop has a control valve to allows flow manipulation, and the secondary loop has a manual valve to introduce disturbances into the tank (Tk3). The upper tank has a removable divider, allowing the user to choose between one big tank, one small tank, and two interacting tanks. Therefore, the system's dynamics (order) can be chosen, depending on the user's selection as shown in Table 1.

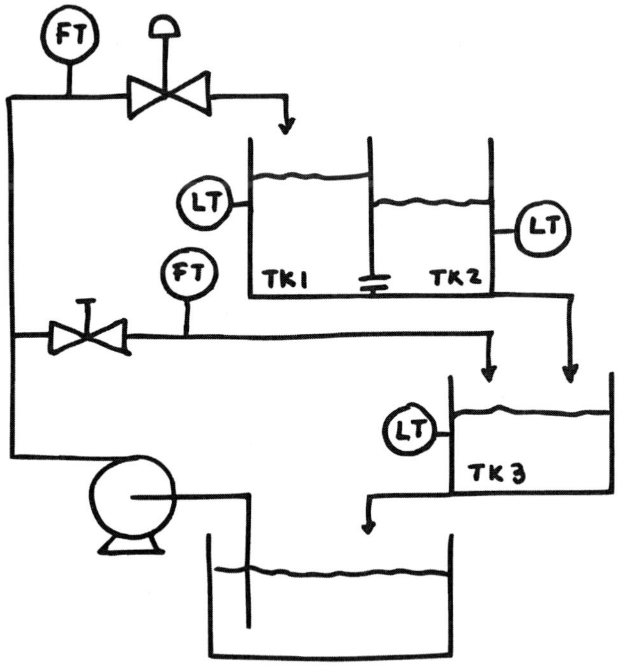

FIGURE 1. Conceptual P&ID of the proposed plant.

TABLE 1. Tank system configurations

Selection	Order	Interaction
1 big tank (Tk1+Tk2)	1	No
1 small tank (Tk2)	1	No
2 series big tanks (Tk1+Tk2), (Tk3)	2	No
2 small interacting tanks (Tk1), (Tk2)	2	Yes
1 small tank (Tk2), 1 big tank (Tk3)	2	No
2 small interacting tanks (Tk1), (Tk2) and 1 big tank (Tk3)	3	Yes

Linear or nonlinear behavior can be introduced with sharp crested open channel weirs used for the output of tanks (Tk2) and (Tk3); different shapes of weirs can be chosen: v-notch, rectangular, linear, etc. The dead time can be created by using a removable pipe for the discharge of (Tk2); such pipe can be selected long enough to create a variable time delay (since it depends on the flow), which represents a challenging control situation if it approaches the dominant time constant of the system.

Basic Design

During this stage possible positions for the system's components were determined. Instrumentation required to perform measurements was selected; measurement and actuation elements are listed in Table 2. Additionally, at this stage different control technologies were evaluated in order fulfill the requirements regarding control strategies and possibility of remote access. Since the system is intended to be used in teaching, research, and training activities, the user can select from three different control technologies when using the equipment:

- Programmable Logic Controller (PLC): Siemens SIMATIC S7–1200. This is a robust element that allows one to manage the process' signals through analog inputs and outputs (A I/O). It can be easily connected to other devices through Ethernet and is intended to be used with a human-machine-interface (HMI). This option is often used in industry, and PID modules are already integrated into the device.
- Industrial controller: Foxboro 762CNA. This type of controller is commonly used in industrial processes. Therefore, it is a valuable tool to train people from industry (technicians, engineers, etc.). This option is good to work in PID tuning, and has the options of lead/lag and dead time compensation, and input prefilter. This controller is connected to the PLC using A I/O.
- Computer control: a laptop with LabVIEW. This control technology allows one to implement almost any control algorithm and state observers. Additionally, it provides the option to save data directly into the computer to be processed off-line. It is connected to the PLC using Ethernet.

Detailed Design

The detailed design involved several calculations and component selection in three main areas: mechanical, electrical, computation. An iterative process was carried out in order to obtain a system that meets the design requirements.

Mechanical components

- Tanks sizing: was performed taking into account the range of the differential pressure transmitters used for level measurement, and the maximum water flow in the system. Transparent plastic was selected to enhance visualization. The reservoir tank's dimension was obtained with the maximum water volume circulating through the system.
- Weirs sizing: linear and nonlinear weirs were dimensioned to be used at the end of the tanks. Figure 2 shows four different weirs selected for the tank system. To include dead time the weir between the upper and lower tanks is replaced by a long pipe which creates a flow delay.
- Structural calculations: a three-level aluminum structure was designed for the system. The pipe routing was developed using 3D CAD software in order to verify interferences and possible manufacturing problems.

TABLE 2. Tank system instrumentation

Element	Range	Brand	Model
Pump	0–6 m^3/h	Pedrollo	CP620D
Control valve (linear)	C_v=7.4 $\frac{gpm}{\sqrt{psi}}$	Spirax-Sarco	LEA31
Pneumatic actuator	20 mm	Spirax-Sarco	PN9120E
Positioner	4–20 mA	Spirax-Sarco	SP5000
Level transmitters	Variable	Siemens	SITRANS P DSIII
Flow transmitters	0–10 m/s	Siemens	SITRANS MAG5100W
Compressor	7 cfm	KTC	431–2525

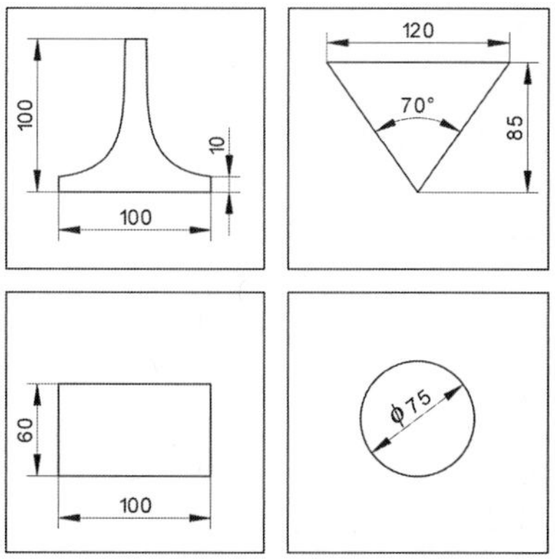

FIGURE 2. Sharp crested weirs: linear, v-notch, rectangular, circular.

FIGURE 3. Dead time device.

Electrical components

- Electrical power source: it was dimensioned considering all the power consumption and voltage/current requirements.
- Pump actuation and speed control: the pump motor has variable speed, and it is manipulated by using a Yaskawa J1000 variable frequency drive.
- Front panel: contains the HMI, start/stop buttons, emergency stop button, the industrial controller, and the variable frequency drive.
- Electrical cabinet: was integrated and located at the upper part of the structure. Two sides were made of transparent plastic material to allow internal visualization of electric components and cabinet's arrangement. Appropriate electrical protections were selected.

Computation components

- PLC application: the PLC is the component that manage signals coming from sensors and signals going to actuators. Additionally, the PLC is one of three technological alternatives to perform closed-loop control in the tank system. Hence, a control application was developed.
- LabVIEW interface: this interface was required to communicate the laptop computer to the PLC. Additionally, several control algorithms can be developed and implemented over this platform.
- Data management: the software interface allows the user to export and save data for further analysis.

Figure 4 shows the virtual prototype of the system, obtained in the detailed design stage. The tank system was manufactured using industrial components that were bought to local providers, and local shops were used for parts fabrication. The assembly was carried out at one of the University's mechanical shops. Figure 5 shows the assembly progress of the developed system.

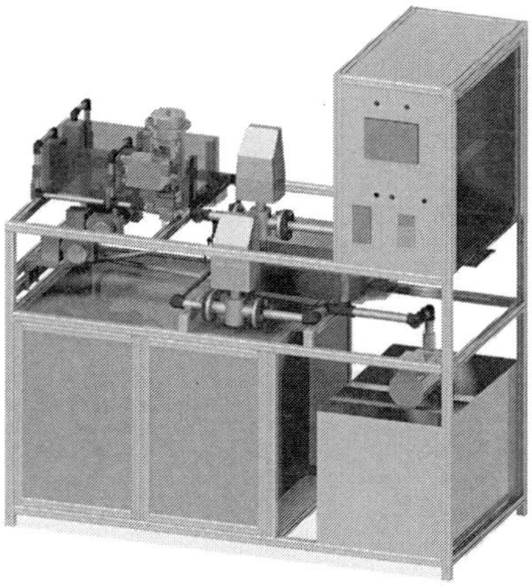

FIGURE 4. Virtual prototype.

FIGURE 5. Tank system assembly in progress.

MATHEMATICAL MODEL

As described in the Conceptual Design section, the equipment offers several possible configuration options for the process' dynamics. Figure 6 describes the closed-loop interaction of the controller, the control valve, and the tanks. As indicated in such diagram, systems described in Table 1, can be divided into four types for modelling purposes: one tank, two series tanks, two interactive tanks, and two interactive and one series tank.

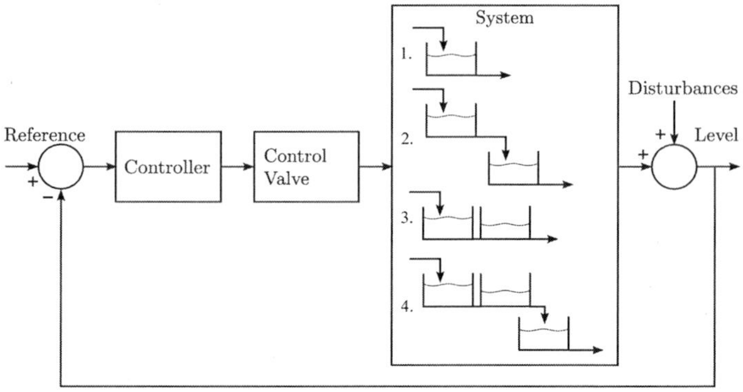

FIGURE 6. Block diagram for the closed-loop system.

Since the controlled variable is the level, and the input of the tank system is the flow provided by the control valve, the mathematical model can be obtained from first principles using a mass balance in each tank. The mass balance can be expressed as

$$\dot{m}_{\mathrm{i}}(t) - \dot{m}_{\mathrm{o}}(t) = \frac{dm(t)}{dt}, \tag{1}$$

where $\dot{m}_{\mathrm{i,o}}(t)$ are the input and output mass flows in a tank, and $\frac{dm(t)}{dt}$ accounts for the time variation of the mass contained in it. Since the fluid is water at environment conditions, the density is considered constant and the mass balance becomes:

$$q_{\mathrm{i}}(t) - q_{\mathrm{o}}(t) = A\frac{dh(t)}{dt}, \tag{2}$$

where $q_{\mathrm{i}}(t)$ and $q_{\mathrm{o}}(t)$ are the input and output volumetric flows in a tank, and $\frac{dh(t)}{dt}$ accounts for the time variation of the level.

Linearity: sharp crested weirs

The output flow depends on the shape of the sharp crested weirs that were designed, hence, $q_{\mathrm{o}}(t)$ can be a linear or a non-linear function of the level $h(t)$. Table 3 contains the output volumetric flow of a tank $q(h)$.

System 1: one tank

When only one tank is considered, the dynamic model of the system is given, using (2), by

$$A_i\dot{h}_i + q(h_i) = q_{\mathrm{i}}(t), \tag{3}$$

where A_i is the corresponding area A_2 or $A_1 + A_2$, see Figure 1. Since two different options can be chosen according to Table 1, one can select two different time constants for the same system configuration. According to Table 3, four different options can be chosen for the output flow, changing the system's linearity as desired.

TABLE 3. Sharp crested weirs equations

Weir type	Linearity	Equation
Linear	Linear	$q(h) = KL\sqrt{a}\left(h + \frac{5a}{8}\right)$
V-notch	Nonlinear	$q(h) = 1.4h^{5/2}$
Rectangular	Nonlinear	$q(h) = 1.838\left(L - \frac{2h_1}{10}\right)h^{3/2}$
Circular	Nonlinear	$q(h) = 1.518D^{0.693}h^{1.807}$

System 2: two series tanks

Using (2) in two series (non-interacting) tanks yields

$$A_i\dot{h} + q_2(h_i) = q_{\mathrm{i}}(t), \tag{4}$$

$$A_3\dot{h}_3 + q_3(h_3) = q_2(h_i), \tag{5}$$

where A_i is the area of the upper tank A_2 or $A_1 + A_2$, A_3 is the area of the lower tank, $q_2(h_i)$ is the output flow in the upper tank, and $q_3(h_3)$ is the output flow in the lower tank. Again, one can select two different time constants for the same system's configuration in the upper tank. According to Table 3, four different options can be chosen for the output flow in each tank, changing the system's linearity as desired.

System 3: two interactive tanks

Using (2) in the two upper interacting tanks yields

$$A_1\dot{h}_1 + k\sqrt{h_1 - h_2} = q_{\mathrm{i}}(t), \tag{6}$$

$$A_2\dot{h}_2 + q_2(h_2) = k\sqrt{h_1 - h_2}, \tag{7}$$

where A_1 and A_2 are the areas of the tanks, $k\sqrt{h_1 - h_2}$ is a nonlinear function that represents the interaction between the tanks, and $q_2(h_2)$ is the output flow in tank 2. According to Table 3, four different options can be chosen for the output flow in the second tank, changing the system's linearity as desired.

System 4: two interactive tanks and one series tank

This system combines the dynamics of the two upper interacting tanks with the lower tank connected in series. Using (2) in the two upper interacting tanks and the lower tank yields

$$A_1\dot{h}_1 + k\sqrt{h_1 - h_2} = q_{\mathrm{i}}(t), \tag{8}$$

$$A_2\dot{h}_2 + q_2(h_2) = k\sqrt{h_1 - h_2}, \tag{9}$$

$$A_3\dot{h}_3 + q_3(h_3) = q_2(h_2), \tag{10}$$

Parameters variation

As it has been described, the developed equipment, which is the scope of this work, provides the opportunity for users to use several configurations. Such changes include variation of parameters, which happens for example in System 1 when a small-area tank is used instead of a big-area tank, reducing the time constant and, therefore, the open loop time response.

MECHANICAL ENGINEERING CONTROL LAB

The control lab course was created to develop abilities that allow the student to conceive, design, implement, and operate automatic control systems for continuous industrial processes. The flexibility offered by the equipment, allows the users to enhance process control and control engineering fundamentals through different activities: components identification, process modeling (time and frequency domains), parameters variation (physical parameters and initial conditions), analyze change in operating points, controller design and implementation (classical, nonlinear, intelligent, etc.), robustness analysis, among others.

Laboratory outcomes

At the end of the control lab, a student should be able to:

- Identify main components, and basic actions performed in an automatic control system, using international standards.
- Describe the characteristics of signals and some communication protocols, and implement a basic DAQ system.
- Describe different measurement elements used in industrial processes, and establish their selection criteria.
- Obtain a linear time invariant continuous mathematical model through static and dynamic plant tests.
- Differentiate between proportional, integral, and derivative modes, and identify different PID algorithms used in industrial controllers.
- Use and compare different tuning methods for PID controllers, using performance and robustness indexes.
- Identify the most relevant aspects for the selection, manufacturing, operation, and maintenance of elements that are included in industrial automatic control systems.

Laboratory sessions

The lab course has been designed for the students to interact as frequently as possible with the developed equipment. The following list contains the weekly laboratory sessions designed for undergrad mechanical engineering students at the UPB:

1. Control systems lab introduction
2. Identification of components ANSI/ISA-5.1-2009 [21]
3. Signals and communication protocols
4. Data acquisition systems
5. Measurement instruments and transmitters
6. Control valves: selection and characterization
7. Control technologies: PLC, industrial, and computer control
8. System identification: static and dynamic plant tests
9. PID control: proportional, integral, and derivative actions
10. PID controller tuning and performance evaluation
11. Computer aided control systems design CACSD
12. Modern control: state feedback and observers
13. Selection of control systems components
14. Safety and reliability in control systems

CONCLUSIONS

This work addressed the development of an equipment to teach dynamic systems and process control fundamentals using industrial components. The plant is used in control engineering courses within the mechanical engineering and chemical engineering programs at the Universidad Pontificia Bolivariana in Medellin, Colombia.

This plant can be used to develop courses that are intended to fulfill accreditation models such as ABET and CDIO. Additionally, it can be used to teach advanced control grad courses. The developed architecture is ready to create remote lab access in future works. The proposed equipment allows one to use it, not only in control engineering education, but also in research activities and industrial training.

Since the plant offers several configurations, from first order linear dynamics to third order interacting nonlinear dynamics with variable dead time, several new control strategies can be developed and tested in a real process. Such strategies include feedforward control, feedback+feedforward control, LQG control, nonlinear controllers: gain scheduling and sliding modes, fuzzy logic, neural networks, genetic algorithms, dynamic matrix control, etc.

Because three different control technologies are used, a lab-scale industrial process such as the developed tank system, constitutes an important tool to compare tuning techniques, control strategies, and to provide technical training for people working in industrial environments, in which automatic control systems are everywhere.

REFERENCES

[1] Kadlowec, J., Bhatia, K., Chandrupatla, T. R., Chen, J. C., Constans, E., Hartman, H., Marchese, A. J., von Lockette, P., and Zhang, H., 2007. "Design integrated in the mechanical engineering curriculum: Assessment of the engineering clinics". *Journal of Mechanical Design,* ***129***(7), pp. 682–691.

[2] Crawley, E. F., Malmqvist, J., Östlund, S., and Brodeur, D. R., 2007. *Rethinking Engineering Education. The CDIO Approach*. Springer.

[3] Felder, R. M., and Brent, R., 2003. "Designing and teaching courses to satisfy the abet engineering criteria". *Journal of Engineering Education,* ***92***(1), pp. 7–25.

[4] Felder, R. M., and Brent, R., 2004. "The abc's of engineering education: ABET, bloom's taxonomy, cooperative learning, and so on". In Proceedings of the 2004 American Society for Engineering Education Annual Conference & Exposition.

[5] Mills, J. E., and Treagust, D. F., 2003. "Engineering education – is problem based or project-based learning the answer?". *The Australasian Journal of Engineering Education,* ***2003***, pp. 2–16.

[6] Feisel, L. D., and Rosa, A. J., 2005. "The role of the laboratory in undergraduate engineering education". *Journal of Engineering Education,* ***94***, pp. 121–130.

[7] Felder, R. M., and Brent, R., 2003. "Learning by doing". *Chemical Engineering Education,* *37*(4), pp. 282–283.

[8] Woods, D. R., Briedis, D., and Perna, A., 2013. "Professional skills needed by our graduates". *Chemical Engineering Education,* ***41***(2), pp. 81–90.

[9] Dormido, S., 2004. "Control learning: present and future". *Annual Reviews in Control,* ***28***(1), pp. 115 – 136.

[10] Horacek, P., 2000. "Laboratory experiments for control theory courses: A survey". *Annual Reviews in Control,* ***24***, pp. 151–162.

[11] Leva, A., 2006. "A simple and flexible experimental laboratory for automatic control courses". *Control Engineering Practice,* ***14***(2), pp. 167 – 176. Special Section on Advances in Control Education Advances in Control Education Symposium.

[12] Keller, J., 2006. "Interactive control system design". *Control Engineering Practice,* ***14***(2), pp. 177 – 184. Special Section on Advances in Control Education Advances in Control Education Symposium.

[13] Aliane, N., Fernández, J., Martínez, A., and Ortiz, J., 2007. "Un laboratorio de ingeniería de control basado en internet". *Información Tecnológica,* ***18***(6), pp. 19–26.

[14] Vásquez, R. E., Correa, J. C., Tamayo, S., and Vásquez, J. M., 2007. "Development of a laboratory equipment for the analysis of mechanisms". In Proceedings of the ASME 2007 International Mechanical Engineering Congress and Exposition, Vol. 7, pp. 179–185.

[15] Márquez, D. A., and Cárdenas, O. O., 2008. "Implementación de un laboratorio virtual para la enseñanza de controladores PID". *Información Tecnológica,* ***19***(3), pp. 75–78.

[16] Betancur, M., Carvajal, J., Congote, L., Díaz, D., and Rincón, E., 2008. "Desarrollo de un equipo didáctico para prácticas de control de tensión en un cabezote textil". In Memorias de XVIII INGENIAR.

[17] Stefanovic, M., Cvijetkovic, V., Matijevic, M., and Simic, V., 2009. "A LabVIEW-based remote laboratory experiments for control engineering education". *Computer Applications in Engineering Education,* ***19***, Febrero, pp. 538–549.

[18] Deschenes, J.-S., Barka, N., Michaud, M., Paradis, D., and Brousseau, J., 2013. "A joint learning activity in process control and distance collaboration between future engineers and technicians". *Chemical Engineering Education,* ***47***(1), pp. 9–14.

[19] He, Z., Shen, Z., and Zhu, S., 2014. "Design and implementation of an internet-based electrical engineering laboratory". *ISA Transactions,* ***In press***(http://dx.doi.org/10.1016/j.isatra.2013.12.035), pp. 1–6.

[20] Rhinehart, R. R., 2013. "ISA transactions 50th year recognition – and some perspectives". *ISA Transactions,* ***52***(1), pp. 2 – 5.

[21] ISA. ANSI/ISA-5.1-2009 Instrumentation Symbols and Identification standard. the international society of automation.

This page left blank intentionally.

Proceedings of the ASME 2014 International Mechanical Engineering Congress and Exposition
IMECE2014
November 14-20, 2014, Montreal, Quebec, Canada

IMECE2014-39665

DESIGN AND ANALYSIS OF AN EXAMPLE LATHE SPINDLE

Rupal Vyasa
Department of Mechanical Engineering
V. G. Engineering College, Chandkheda
Ahmadabad, Gujarat, India

Raghu Echempati
Department of Mechanical Engineering
Kettering University
Flint, MI, U.S.A.

ABSTRACT

This paper discusses the modeling and analysis of an example medium speed medium precision lathe spindle. This and few other similar topics have been assigned as term projects in an introductory senior undergraduate/graduate level finite element analysis course taught at Kettering University. The experiences and the general feedback from the students of the class show satisfactory organization of the course material that includes modeling and analysis of real life examples.

With reference to the specific topic on design of machine tool spindles, it is not a new area, however, it is generally taught at the graduate and research levels. Use of modern computational tools to perform iterative design and analysis calculations of such spindles make the senior undergraduate and/or graduate master students aware of the implications of modeling a real life system using the 1D and 3D finite beam elements and to validate those results by a CAE tool. Final course projects such as this serve as a good learning tool to the graduating engineers. Sample results obtained from various CAE tools such as UG-NX 7.5 are presented in the paper and discussed.

INTRODUCTION

The work presented in this paper is based on the unpublished thesis carried out by the first author [1]. A lot of theoretical and experimental work has been done in the machine tool design area since last several decades which resulted in the development of modern high speed, high precision machine tools [2 to 4]. However, many universities in USA have dropped teaching a course on kinematics and dynamics of machine tools. Due to the nature of the academic calendar at our university, many mechanical engineering students are exposed to either design of or use of machine tools at their work places. Applying the knowledge gained in finite element analysis course to real life applications such as this, helps them appreciate the subject matter better. This paper therefore discusses some technical details about the design and analysis of an example lathe machine spindle and the assessment of the learning outcomes from this and the other similar final projects of this course. Math tools such as MatLab and a variety of CAE tools such as NX I-DEAS, UG NX 7.5, Solid Works and Autodesk have been extensively used by the students of this class. Finite element formulation and solution by direct stiffness method or by energy methods for 1D beam element modeling along with the assessment results are presented and discussed [5 to 7]. Validation of some of the results by simple beam theory and through a Visual Basic Program specially developed for this purpose is presented briefly [1]. It is hoped the outcome of this paper may be beneficial to the academic community and industry partners to understand medium to high-speed machine tool spindle designs.

The main function of the machine tool spindle are to center and hold the work piece (turning) or cutting tool (drilling, grinding, milling and boring) under the effects of weights and cutting forces on one hand and driving forces and torque on the other . Therefore, it fulfills two functions, i.e. it not only locates the work piece or the tool respectively but also drives and guides them with the required accuracy and stiffness in their operational movements (rotation and sometimes axial feed movement). There are several textbooks and research papers in the machine tool design area, however, only a few are cited in this paper with the understanding that this list by no means is complete.

For the purpose of the paper here, an existing lathe machine tool spindle is considered. From earlier machining experiences this unit is considered as critical one from the rigidity and vibration point of view because high surface finish is expected. Machining accuracy and surface finish depend upon the accuracy of rotation of spindle that imparts torque to the wheel. Therefore, the spindle should meet certain requirement in order to produce the high surface finish.

Generally, lathe machine spindle is designed for the requirements of (a) a high degree of accuracy (b) high rigidity and (c) accurate guidance. Under all load and speed conditions the lathe spindle operates smoothly and free from vibrations. Both the end of the spindle is fitted with spindle bearings, which ensure high radial load carrying capacity and rigidity. Therefore, a proper selection of rolling element bearings, their location, pre loading in the case of rolling element bearing and lubrication method will help the spindle to run at higher speeds with high rotational accuracy and stiffness. This paper is concerned with the stress and vibration analysis of a medium speed lathe machine spindle with the aid of CAD/CAE software. As mentioned before, a lot of work has already been carried out by number of authors in the vibration analysis of machine tool spindles and structures [7 to 9].

ANALYSIS

Based on the background and importance of the problem, a computer software program using Visual Basic has been developed for the design of spindle based on strength and in accordance with the existing standards used for machine tools such as BIS and ASTM (including the database of different material) [1]. Typical flow-chart of the developed VB Program is presented in Annexure A. The bearings for the spindle have been selected using the AFBMA and other bearing standards. Several engineering assumptions had to be made while developing the software for the basic design of the spindle.

As is evident from the numerous references, the turning process on a typical lathe involves three stages, which depends on the interference condition of the tool to the work piece, the type of material being cut and the operational conditions of the machine tool. In general, in the first stage, the work piece undergoes elastic deformation due to the load from the cutting tool. At this stage, no chips material particles are taken away from the work piece. In the second stage the work piece is subjected to plastic deformation following the elastic deformation, when the cutting tool is fed against the work piece. In the final stage the cutting tool is further fed deep against the work piece and it produces chips from work piece. To cut precisely with good efficiency, the cutting condition should be of the third stage with as small a cutting force as possible. In addition, high peripheral speed and sharp cutting edges of the cutting tool are two important cutting conditions to improve cutting quality and efficiency.

The mechanics of cutting in lathe operations is well-established and available in the literature [2, 11]. It is also well-known that the dynamics of the flexible cutting tool and the workpiece affects the cutting process sometimes leading to chatter (or self-induced vibrations). Therefore, while cutting, the tool might face a hard spot on the metal surface and start to oscillate, which then results in chatter or instability of the entire cutting operation. The dynamic cutting force (dp) can be expressed as follows:

$$dp = k_1 .\ a.\ ds + (2\pi/\Omega)\ k_2 .\ dr + k_3 .\ d\Omega \qquad (1)$$

Here, ds is the change in the chip thickness, dr is the change in the feed rate, $d\Omega$ is the incremental change in the spindle speed, and k_1, k_2 and k_3 are the corresponding force coefficients. Since the intent of this paper is modeling and analysis of lathe spindle, the cutting mechanics aspects are only included as forces acting on the spindle. Obviously, these forces vary as the operating parameters vary on the machine tool.

The example machine tool used for the work presented in this paper is a Model EP-1330 Mysore Kirloskar Lathe Machine manufactured by Kirloskar Enterprise in India [10]. The main motor power is 2.2 kW with 8 spindle speeds: 54, 90, 140, 224, 315, 550, 775, and 1200 R.P.M. Other specifications of the machine are as follows [1, 10]:

Center height – 175 mm
Swing over Bed – 350 mm
Swing over cross Slide – 200 mm
Swing in Gap – 520 mm
Width of Gap in front of Face Plate – 130 mm
Spindle nose – 4" D1 Cam lock
Morse Taper in Spindle Sleeve – Mt 3/5
Spindle Bore – 41 mm
Longitudinal Feed Range – 0.048 to 0.716 mm/rev.
Transverse Feed Range – 0.0008 to 0.012 inch/rev.

It may be noted that this machine is no longer manufactured as new technology and new materials are available, which is part of the students' exercise to understand where the developments of the modern machine tools are taking place. More detailed specifications of this machine can be found in the reference. Figures B1 to B3 in Annexure B show respectively the photograph view of the headstock drivetrain, the blueprint drawing of the headstock drivetrain assembly and the dimensions of the spindle. The blueprint drawings were provided by the manufacturer but they are not very clear. However, the students were given these figures to carefully study and to extract the critical dimensional data from this information only for carrying out their project work.

The design of lathe spindle is carried out step by step by using a computer program developed specially for this case. However, the developed program can be used for other models and spindles. The program is based on Visual Basic 6 (VB 6) that uses the empirical formulas and data for the beam models used for the spindle. Three different windows are designed in the program: the first one is for the lathe spindle input output specifications data entry. The other windows are designed for material code and data entries and bearing data entries. One example window from VB 6 is shown below in Figure 1.

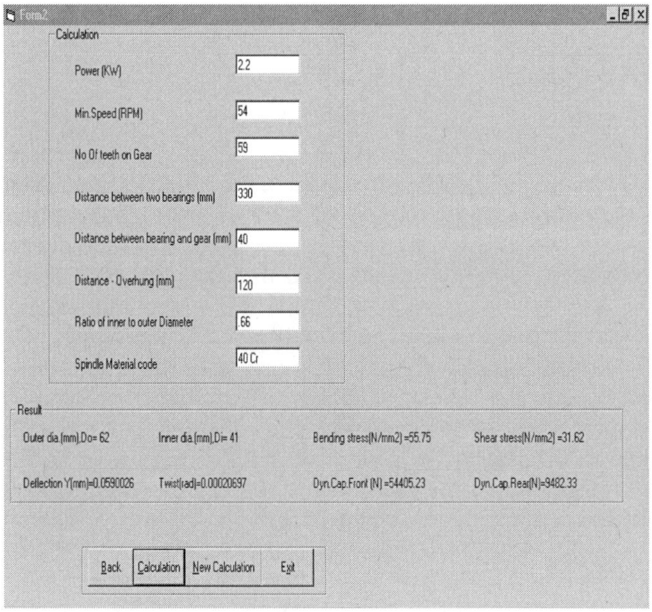

Fig. 1: Input-Output Data Entry Window [1]

Spindles are stepped and hollow to accommodate the work piece. It has a Morse Taper on one side to which the heavy head-stock and chuck assembly are fitted. Taper roller bearings with proper orientations secure the spindle from moving axially while the gear forces from the drive train and the cutting forces act of the spindle. Analyzing the spindle in its present form and as shown in the blueprint drawings of Figures B2 and B3 (of Annexure B), is complex by solid mechanics approach. Understanding the assumptions to simplify the model is one of the students' course learning objectives (CLOs). Carrying out the stress analysis to determine the shaft dimensions by machine design approach is possible to some extent; however, the deflection analysis is complicated. The students of this group carried out the mechanics approach first to verify that the dimensions of the spindle at critical locations are close to what the lead author obtained using the Visual Basic Program. The VB Program is based on the traditional machine design procedure for sizing the members.

Rigidity calculations for the spindles involve the determination of the deflection in bending and in some cases the twist in torsion. In working out the design diagram, the spindle is usually replaced by beam on hinged supports. Such an assumption is valid when there is one ball or roller bearing at each support. In more exact calculations, several ball or roller bearings in a single support are to be regarded as an elastic support, while a spindle running in sleeve bearings is regarded as a beam on an elastic foundation. This last case can also be conditionally reduced to a beam on hinged supports to which a reactive moment M is applied in the support [2]. The value of this moment varies, according to experimental data, from zero (for insignificant loads, as in finishing machines) to 0.3 – 0.35 of the external moment acting in the middle section of the spindle on the support [3]. As lot of literature is available on this aspect of spindle design, it is not discussed in this paper.

Several calculations were made to obtain the design dimensions of the spindle using the developed VB 6 program. Due to space limitations the details of this are not presented in this paper. However, standard textbooks such as those given in references [11] to [13] can be used to understand the details of cutting loads, two-plane bending and torsion analysis of the spindle model, etc., and to obtain the spindle dimensions. However, at the time of the conference, some of the important details of these calculations (also shown in Annexures A & B) will be shown for easy understanding of the work presented in this paper. Few important steps for the design of the spindle are given below following the two-plane bending model shown in Figure 2. SI units have been used; therefore, all loads are in Newton and all linear dimensions are in mm.

The design steps carried out in this work involved:

a. Calculation of cutting forces acting on the Spindle
b. Force analysis of the Spindle
c. Calculation of Spindle diameters by hand calculations
d. Selection of the front and rear bearings
e. Calculation of the total deflection of the Spindle nose by FEA
f. Calculation of torsional twist of the Spindle
g. Calculation procedure for static stiffness of the spindle*
h. Calculation of critical speed of the Spindle (by hand and by FEA)

Some of these items have been explained and the data given to the students. Figure 2 below shows the simplified beam analysis with loads in the vertical (designated by the letter V in front) and horizontal (H) planes. These are used to determine the spindle diameters at critical locations and to determine the deflections at the gear and at the free end in order

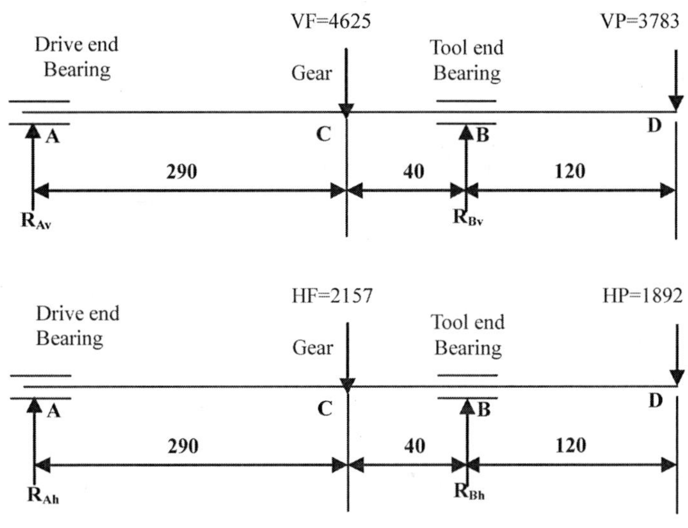

Fig. 2: Forces in Vertical Plane (top) and in Horizontal Plane (bottom)

to calculate the critical speeds of the spindle. As mentioned before, the intention of this paper is to have the students analyze the spindle shown in Figure 2 first by verifying the critical dimensions and to carry out analysis by FEA.

In order for the students to model the spindle using beam elements, the conventional Euler-Bernoulli beam equation is discussed first along with its underlying assumptions and limitations. The out-of-plane displacement w of a beam is governed by this theory as:

$$(d^2/dx^2)\ \{EI(d^2w/dx^2\}=p \qquad (2)$$

Here, p is the distributed loading (load/unit length) acting in the same direction as y (and w), E is the Young's modulus and I is the area moment of inertia of the beam's cross section. In the case of lathe spindle although E does not change, I changes due to various steps in the hollow sections of the spindle. The details of the beam equation and the theory are covered following any mechanics of materials book (such as Beer and Johnston). The Euler beam equation (2) arises from a combination of four distinct subsets of beam theory: the kinematic, constitutive, force resultant, and equilibrium definition equations.

The outcome of each of these segments is summarized below:

Kinematics: $\chi = -\theta = (-\,dw/dx)$ (3)

Constitutive: $\sigma(x,y) = E.\ \varepsilon(x,y)$ (4)

Resultants: $M(x) = \int\int y.\sigma(x,y).\ dy.dz$ (5)

$V(x) = \int\int \sigma_{xy}(x,y).\ dy.dz$ (6)

Equilibrium: $dM/dx = V;\ dV/dx = -\,p$ (7)

When using the finite element analysis, the spindle is assumed to be acted up on by cutting forces which result in reaction forces at the bearing supports. This was shown in figure 2. The students are asked to use the direct stiffness matrix that is based on simple beam theory (for example, presented in Logan's textbook, Reference [6]). The derivation of the stiffness matrix is shown to the students, which is based on several standard steps followed in a typical finite element analysis course. These steps include: choosing an element type (in this case a 1D beam element), assuming a displacement function for the beam deflection (a cubic polynomial), defining the constitutive relationship, which results in equation similar to equation (4) above, and finally, defining the stiffness matrix, k. Following the standard notation used in finite element analysis books, the k-matrix can be derived as [6]:

$$[k]=\frac{EI}{L^3}\begin{bmatrix} 12 & 6L & -12 & 6L \\ 6L & 4L^2 & -6L & 2L^2 \\ -12 & -6L & 12 & -6L \\ 6L & 2L^2 & -6L & 4L^2 \end{bmatrix} \qquad (8)$$

Here, k is the element stiffness matrix, E is Young's modulus and L is the length of each element of the spindle (modeled as 1D beam). Since a conventional spindle has several steps along its length, several beam element equations need to developed and assembled following the standard stiffness method to obtain an overall global stiffness matrix for the entire spindle. Boundary conditions and load conditions are imposed on the global equation of motion following the structural analysis of beams and the equations are solved for unknown quantities. The first quantity solved in FEA problems is the displacement of each node, followed by reaction loads. Stresses are computed using the constitutive equation.

DETERMINATION OF DEFLECTIONS, STRESSES AND CRITICAL SPEEDS OF THE LATHE SPINDLE

Different finite element (FE) models have been developed to perform stress, deflection and dynamic (vibration) analyses. These include the one-dimensional beam elements solved by using MatLab and the full blown 3D CAE models using NX 7.5, NX I-DEAS, ANSYS and SolidWorks for deflection and stress analysis, and Dunkerley and the Rayleigh methods to determine the critical speeds by hand calculations [1, 13]. Students developed the MatLab program to do deflection and vibration analyses using 1D beam elements. While performing these calculations, the weight of the gear (= 40 N) and the weight of the chuck (= 140 N) have been taken in to account in the vertical plane loads. Lot of time and efforts were put in to develope the MatLab and CAE calculations. Table 1 shows consolidation of all the results and discussed in the next section.

Table 1: Deflection, Stress and Critical Speeds of Spindle

Deflection (mm x 10^{-3})	Average Stress (MPa)	Critical Speed (rpm x 10^3)
3.8 (MatLab) 5.6 (NX) 4.0 (ANSYS) 5.3 (SolidWorks)	20.34 (NX) 25.5 (SolidWorks) 21.6 (I-DEAS) 23 (ANSYS)	16.26 (Dunkerley) 20.32 (Rayleigh)

RESULTS AND DISCUSSION

Table 1 shows the deflection, stress and the critical speed results for the spindle carried by different FE models. Even though the same dimensions, loads, and material were used in all these models, the differences in the results between the 1D FE model using MatLab with 7 nodes and 6 elements, and the full blown 3D CAE models containing several thousand of nodes and elements were somewhat different. This is to be expected due to the type and order of power of the elements used. However, the locations of the maximum stresses, displacements and the reaction forces were relatively the same. This indicates that all these models are responding to the applied boundary conditions in generally the same way.

Another possible source for the differences in values can be attributed to the dimensions of the FE model. Since the inside diameter of the spindle has a slight taper, the one-dimensional model, which assumed a constant diameter is different than the three-dimensional CAE model that can more accurately represent the steps and the taper changes in the diameter of the spindle. Although the analysis between the math model and the computer models also differed, the differences were much less. This is expected since the actual model used for both was identical and loading conditions were held constant. The differences between Nastran solver for NX and SolidWorks can be attributed to the differences in the meshing the programs produced and possibly to differences in the material properties stored within these programs.

DISCUSSION OF RESULTS AND CONCLUSIONS

The results presented in Table 1 by classical design method, software and FEA for the stress, deflection and critical speeds show that discrepancy exists in the calculated versus CAE results of stress. The critical speed results show that to avoid resonance, the operational speed of the spindle should be at least 15 - 20% away from the critical speed. Based on Raleigh's method a speed of around 4,000 rpm or 3,200 rpm based on Dunkerley method, are acceptable. Usually, the medium spindle speed such as the case in this paper is well within the calculated values. As mentioned before, the developed VB6 software program calculates the deflections, stresses and the critical speeds of the spindle in a step by step manner.

Sample CAE results from NX 7.5 are shown in Figures 3 and 4. The results shown in the above table are average values obtained from the CAE results obtained in these figures. The figures are self-explanatory. More results will be shown at the time of the conference.

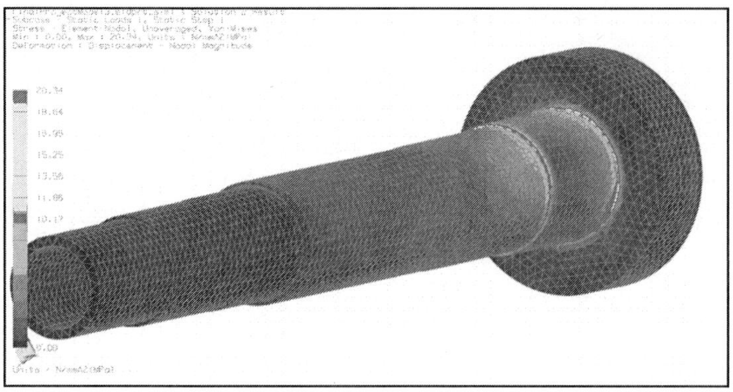

Figure 3: Deflection of the Spindle by NX 7.5

From this study it can be concluded that the stresses and deflections are well within the permissible/ specified limits as given in the various standards, i.e., the stress values are below the permissible values of 137.5 MPa and 80 MPa in bending and torsion respectively, and 0.0660 mm of deflection. Hence, the design is safe from the strength and deflection point of view for the spindle.

The student learning outcomes included study of conventional blue-print drawings to understand the dimensions; modeling aspects of the spindle using mechanics and machine design approach; using simplified 1D and 3D beam models of the real life industrial component; understanding the limitations of modeling, and finally validation of results by CAE and other math tools. Overall, their response was very satisfactory.

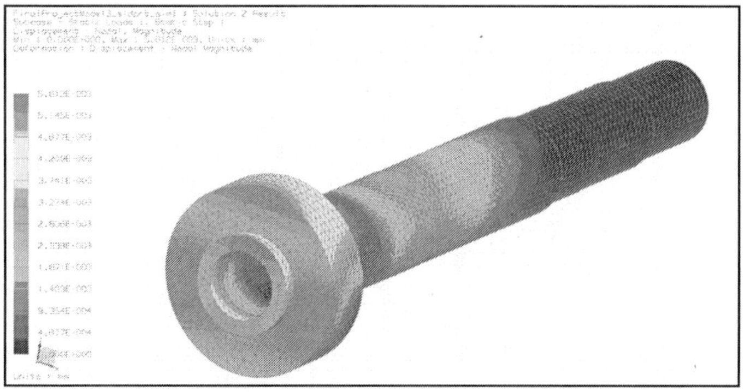

Figure 4: von-Mises stress of the Spindle by NX 7.5

SCOPE FOR FUTURE WORK

From the earlier discussion on spindle design one can say that the overall performance of the machine tool depends upon the performance characteristic of the individual components. The ability to produce a workpiece of required physical features accurately and at minimum cost presents the performance of the machine tool. The spindle is one of the critical components in this consideration. In the present work manual design of the spindle is carried out on mainly rigidity and strength consideration. Moreover, considering the recent advancement in computers and computer languages the software for spindle design is developed. In addition to this the latest CAD/CAE software are used for the spindle design based on strength and rigidity analysis. Design the optimization is considered up to certain extent considering ratio of length to overhang of the spindle.

In future work one can carry out the optimum designing of spindle unit considering space, cost and other aspects such as vibration and chatter. Other CAD/CAE software tools with nonlinear capabilities can be used for the analysis purpose and compare the results with the manual design. Finally, the database of the spindle materials can be updated by adding new materials developed recently.

ACKNOWLEDGMENTS

Part of this work uses Statics Online Concepts Inventory (OLI). This is part of a collaborative grant that the second author received from NSF under their grant number: DUE-0918255. Their support is sincerely acknowledged. The authors wish to acknowledge the experimental and computer laboratory facilities provided by L. D. Engineering College and the V. G. E. College, Ahmadabad, Gujarat (India), and the undergraduate and graduate students at Kettering University, Flint, MI, Mr. Adam Colvin, Mr. Kevin Lucka and Mr. Joel Laber who performed the FE analysis of the spindle as a part of their final project work in the finite element analysis course.

REFERENCES

1. Vyasa Rupal, "Design of Precision Lathe Spindle (ENTERPRISE 1330) Using CAD/CAE Technologies", unpublished Master of Engineering Dissertation, L. D. College of Engineering, Ahmadabad, 2004.

2. S. A. Tobias and F. Koenigsberger, "Advances in Machine Tool Design & Research", Proceedings of the 7th International Machine Tool Design and Research Conference, University of Birmingham, Sept. 1966.

3. G. C. Sen and A. Bhattacharya, "Principles of Machine tools", New Central Book Agency, 1967.

4. N. Acherkan, "Machine tool design", Vol.3, Ch. 5, University Press of Pacific, ISBN-10: 0898750482, 2007.

5. T. R. Chandraputla, "Finite Element Analysis for Engineering and Technology", University Press (India), 2004.

6. D. Logan, "A First Course in Finite Element Method", Cengage Learning, 5th edition, 2011.

7. S. S. Rao, "The Finite Element Method in Engineering", Tata McGraw-Hill Publication, 1989.

8. R. V. Reddy and A. M. Sharan, "The Finite Element Modeled Design of Lathe Spindles: The Static and Dynamic Analyses", Journal of Vibration, Acoustics, Stress, and Reliability in Design, Vol. 109, P. 407, October 1987.

9. T. Sawamoto and K. Konishi., "Development of machine tool spindles with higher rotational speeds", Tribology International, P.189, Aug.1982.

10. Kirloskar Enterprise EP-1330 Mysore Kirloskar Lathe Machine. Information available on the internet at: http://www.machinetools.com/en/models/kirloskar-enterprise-1330/

11. F. Koenigsberger and J. Tlusty, "Machine Tool Structures Volume 1", Pergamon Press, Oxford, 1970.

12. R. G. Budynas and J. K. Nisbett, "Shigley's Mechanical Engineering Design", McGraw-Hill Co., 9th edition, 2012.

13. F. S. Tse, I. E. Morse and Hinkle, "Mechanical Vibrations: Theory and Applications", Allyn and Bacon series in mechanical engineering and applied mechanics, 1978.

ANNEXURE A

FIG. A1: FLOW CHART OF THE SOFTWARE ARCHITECTURE FOR SPINDLE DESIGN AND ANALYSIS [1]

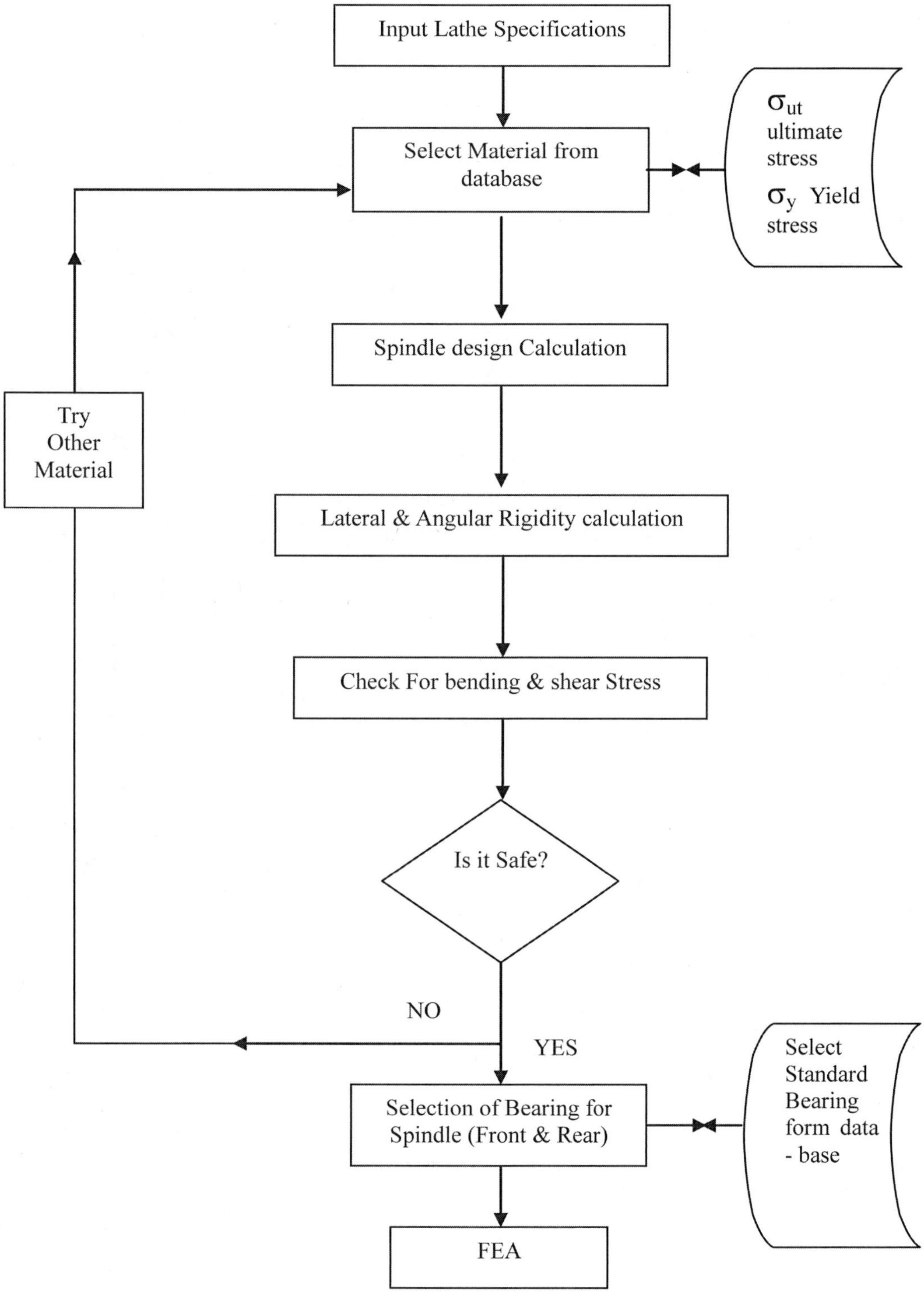

ANNEXURE B

FIG. B1: PHOTOGRAPH OF THE DRIVETRAIN OF THE HEADSTOCK ASSEMBLY [1, 10]

ANNEXURE B

FIG. B2: BLUEPRINT OF ENTERPRISE LATHE DRIVETRAIN ASSEMBLY [1, 10]

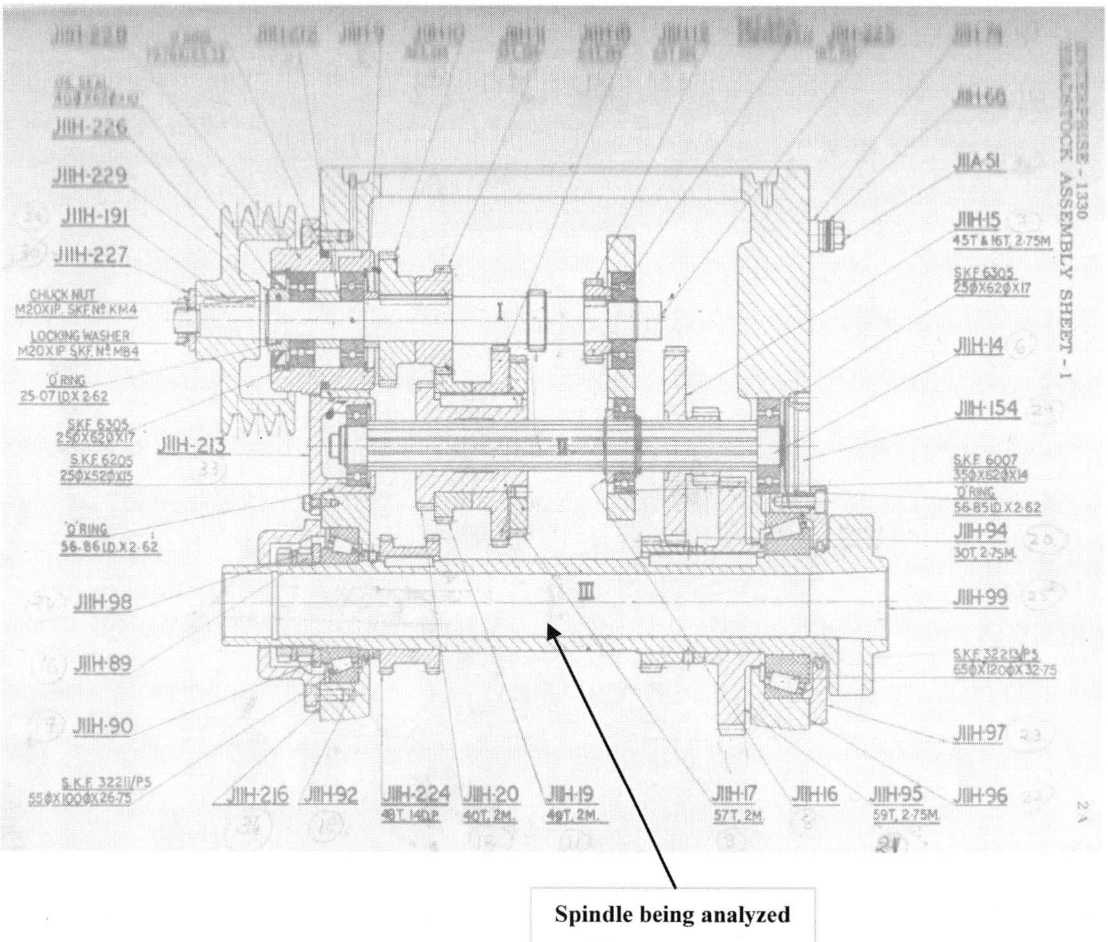

ANNEXURE B

FIG. B2: BLUEPRINT OF THE LATHE SPINDLE WITH APPROXIMATE CRITICAL DIMENSIONS [1, 10]

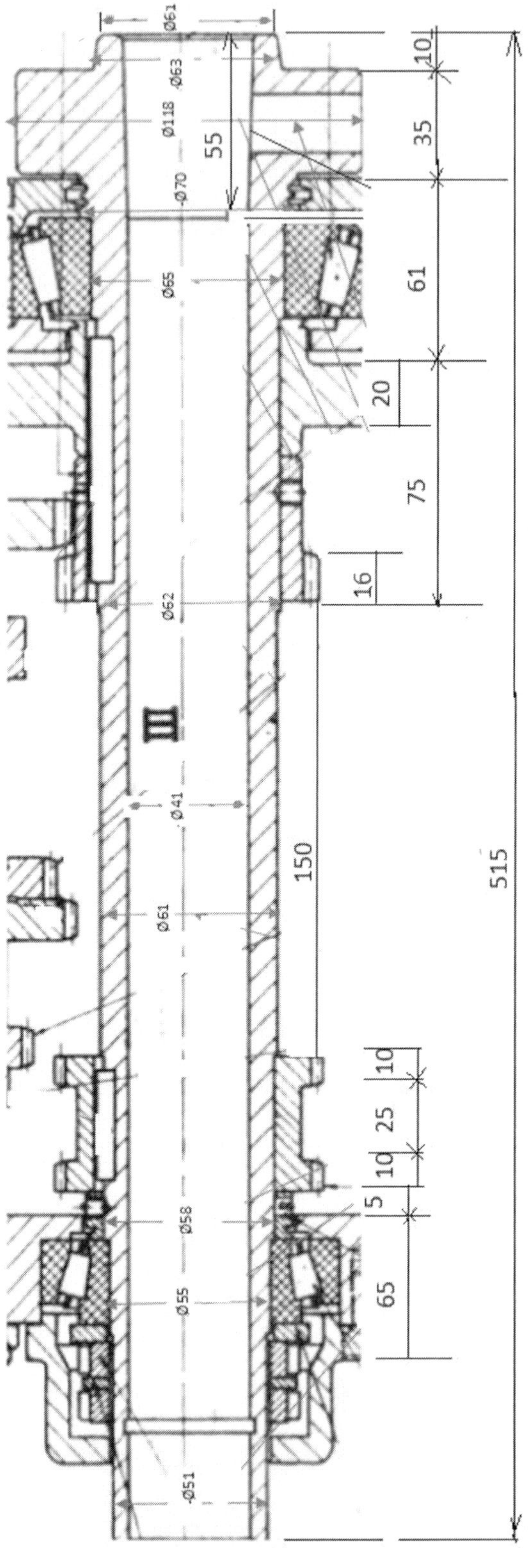

Proceedings of the ASME 2014 International Mechanical Engineering Congress and Exposition
IMECE2014
November 14-20, 2014, Montreal, Quebec, Canada

IMECE2014-39797

A Component-Centric Approach to Structural Analysis of Mechanisms

Christopher Stack
Milwaukee School of Engineering
Milwaukee, WI, United States

Dr. Subha Kumpaty
Milwaukee School of Engineering
Milwaukee, WI, United States
Dr. Mohammad Mahinfalah
Milwaukee School of Engineering
Milwaukee, WI, United States

ABSTRACT

Structural analyses of mechanisms with components that move relative to each other provide a unique problem to the analyst building and running structural models. In these situations, the analyst usually has to either simplify the problem to a point where the results are unusable or maintain multiple models, which will take more effort to maintain and more resources to run the models. If a mechanism is simplified down to just analyzing one component at a time without regard for the other components in the system, the results will not be accurate because the loading effects of the other components will not be accounted for. In cases where all the components are included in the model the loading effects from the other components will be accounted for, however, a separate model will be required for each position. This paper presents a method of breaking down the complex mechanism into a component level model for each part of the assembly, while still accounting for all loading effects of the other components; in the Pivot Method the component under analysis stays stationary and the loading moves around the component to represent the different positions that it can take.

In order to accomplish this task, a simplified model is used to generate loads at each of the joints. Once the pivot loads are known, a spreadsheet can be used to transform the loads to a coordinate system in which the individual component is being modeled. With the pivot loads known and all the loads transformed into the proper coordinate system the structural analysis of the component under investigation can continue.

The intention of this paper is to introduce the Pivot Method and to demonstrate that it provides a good trade off between both the complexity of methods that model the assembly as a system, and those that focus on the component under question alone. To accomplish this, the analysis results of the Pivot Method models will be compared to results obtained from other methods, with the intention of showing that the Pivot Method will provide the same results while requiring less effort to model and less resources to run.

1.0 INTRODUCTION

Many mechanical systems contain mechanisms, which have components that move relative to each other while the mechanism is in use. Oftentimes the relative motion of these components can make structural analysis of these components complicated and difficult. When an assembly consisting of multiple parts is loaded, the load is transmitted from component to component through the joints. When performing structural analysis on the components of a mechanism, the loads transmitted from one component to the next must be accounted for if accurate results are desired. If the components move relative to each other, depending on loading conditions, the orientations of the loads will change resulting in different load conditions and further adding complexity to the structural analysis models.

This paper discusses the development of a new Finite Element Analysis Method for structural analysis of individual components within a mechanism. This new Method is called the *Pivot Method*. The Pivot Method provides a faster solution to the complex problem of the analysis of multiple connected components in a mechanism. To account for the loads transmitted from component to component, the Pivot Method generates loads at each joint and transforms them into a component-level coordinate system. The component-level coordinate system will move with its associated component as it, in turn, moves through the mechanism's movement. This method of calculating and transforming both the input loads and the joint loads into a component-level coordinate system allows the structural analysis of each component to proceed accurately and independently from the other components in the assembly.

The Pivot Method consists of a stick model, a spreadsheet and a three-dimensional structural model. The stick model will be used to generate the joint loads from the overall system-level input loads. The spreadsheet is used to transform the joint loads from the coordinate system of the mechanism to a component-level coordinate system for each component. Once these loads have been transformed, the spreadsheet will display the loads, such that they can be inputted into the three-dimensional structural model. With the loads inputted into the three-dimensional model, the model can be run to provide results for all the load cases and positions in one model.

2.0 BACKGROUND

The Pivot Method is intended to account for all of the loading effects from the other components in the assembly that act on any given component, while still maintaining a relatively simple model. When completed, it is anticipated that this method will have the following benefits over traditional structural analysis methodologies for mechanisms:

To demonstrate the Pivot Method a fictional mechanism will be used. This mechanism is a four bar linkage with applied external loads. This will demonstrate how the Pivot Method accounts for both external forces from actuators and loads applied to the system, as well as internal loads at each pivot point. While this is a simple mechanism with pivot forces that could be solved by hand, it is intended to act as a simple and easy-to-understand example. The Pivot Method can easily be applied to any system that is more complex than the example system. Some mechanisms that are topologically similar to this example system, and are therefore applicable to this method, include suspension systems, lifting mechanisms, material handling equipment and other four bar linkages, among others.

This example system being considered is a two-dimensional system and will be symmetrically loaded. Figure 1 presents the example system being considered in this paper. The example system consists of three structural components: an upper arm, a lower arm, and a middle linkage. The upper arm has a linear actuator connected to it that raises and lowers the system. The lower arm has a hard stop to limit the mechanism's travel and also has a torsion spring to stabilize the system as it moves. The middle link has two external forces applied to it that may vary according to the individual load case under consideration.

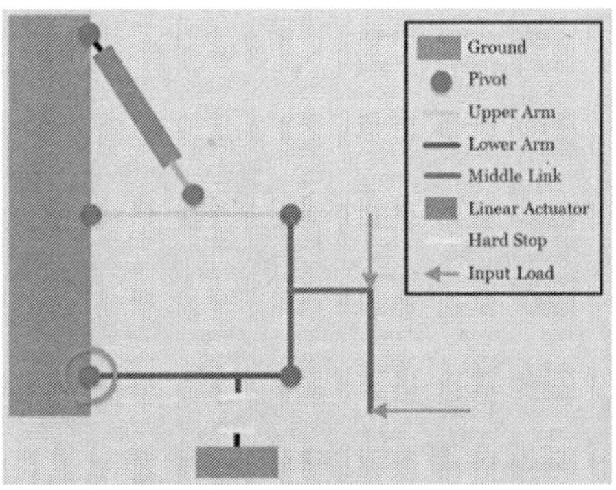

Figure 1: Diagram of the Mechanism Being Used as an Example in This Project

When compared to other modeling methods that take into account the same effects, this method is not intended to provide better results than all the other methods discussed. Instead, it will simply provide the same results in a faster timeline and use less computational resources. To demonstrate the benefits of the Pivot Method, multiple FEA models will be created to represent different modeling methods. The results from these methods will be compared to demonstrate that they all yield similar and valid results. The methods will also be compared to determine the time and resources required to complete each model.

2.1 Introduction to Finite Element Analysis

Finite Element Analysis (FEA) is a method of simultaneously approximating a solution to multiple linked differential equations. FEA is commonly used to calculate stress distributions in structural parts or assemblies. An FEA model consists of a mesh, which is constructed from elements and nodes. Each element is a small geometric representation which, when combined with all the other elements in the model, form the geometric approximation of the problem at hand. The elements are formed by nodes, with one node at each vertex of the element. In a two-dimensional, or shell mesh, the elements can be either triangular or quadrilateral shaped. While in a three-dimensional, or solid, model the elements are either tetrahedron or hexahedron shaped. The choice of which type of model to use and which type of element to use is beyond the scope of this paper and is left to the analyst to decide, given the situation at hand. As Bhavikatti discusses, FEA was originally developed for the aerospace industry, but it has now become a staple in nearly every industry.

> The finite element analysis originated as a method of stress analysis in the design of aircrafts. It started as an extension of matrix method of structural analysis. Today this method is used not only for the analysis in solid mechanics, but even in the analysis of fluid flow, heat transfer, electric and magnetic fields and many others. Civil engineers use this method extensively for the analysis of beams, space frames, plates, shells, folded plates, foundations, rock mechanics problems, and seepage analysis of fluid through porous media. Both static and dynamic problems can be handled by finite element analysis. This method is used extensively for the analysis and design of ships, aircrafts, space crafts, electric motors and heat engines. [1]

The Finite Element Method (FEM) is purely an approximation method by which displacements, stresses, and forces are approximated using a variety of assumptions including, but not limited to, Hooke's Law (a linear stress strain relationship) and the fundamental notion that the deformation in the parts is small. These assumptions, coupled with the discretization of the geometry, mean that the results obtained through this method are purely approximations and will have some error. When using an FEA modeling software package, there are often many methods by which to set up the model; it is up to the analyst to determine the best way to set up the situation at hand. Modern FEA software packages have simplified making of an FEA model. As Prantil, Papadopoulos and Gessler [2] discuss, the structural analyst must make sure to understand the assumptions being made.

> Yet a great hazard of FEM is also that, with the aid of commercial software, it can be too easy to create a model and run an analysis. The ease of operation can foster "computational complacency" [Paulino, 2000] in validating numerical results. It often appears that the myth that "the computer must be right" is alive and well. While, indeed, algorithms in commercial codes are well debugged and are unlikely to contain programming errors, the user is ultimately responsible for making appropriate modeling assumptions and interpretations of the output. [2]

The Finite Element Method works by first solving for the deformed shape of the part. To allow for the calculation of the deformed shape, the CAD geometry is discretized into a meshed model. The meshed model consists of elements, each comprised of a node at the corners. The stiffness of the elements is used to calculate the displacement of each node, and thus the deformed shape of the model can be constructed. From the relative displacements of the nodes on each element, Hooke's Law can be used to calculate the stresses each element has. With the stresses in each element known, the stress distribution for the part as a whole can be determined.

When performing FEA, it is important to understand the boundary conditions that are applied to the model. In some cases, understanding the methods that are used to constrain the model are more important than understanding the loads that are being applied to the model. Narasaiah [3] observes that the components adjacent to the component of interest are equally important to the solution as the component itself.

> Simple supports for the beam imply that the columns do not influence slope of the beam at its free ends. Fixed supports for the beam imply that the slope of the beam at its ends is zero. But, the ends of the horizontal beam are neither simply supported nor fixed. The degree of fixidity or influence of columns on the slope of the beam at its free ends is based on a finite, non-zero stiffness value. Thus, the maximum deflection of the beam depends upon the relative stiffness of the beam and the columns at the two ends of the beam.

In Narasaiah's example of a beam being supported by two columns, it is essential to account for the stiffness effects of the columns on the beam that they support. Without accounting for these effects, one can only bound the solution between the two extremes of fully fixed and simply supported. If the columns were to have an external load applied to them, this effect would be amplified because the analyst would also need to account for the loading transferred through the columns to the beam.

The limitation associated with all of the methods described in this paper is that they are unsuitable for true fatigue analysis. To perform true fatigue analysis, fatigue software such as n'Code would need to be used in conjunction with the FEA solver. The methods presented are, however, suitable to determine how a structure will handle single overload events, as well as pseudo-fatigue load cases, where a few single amplitude loads are assumed to dominate the service life of the mechanism. In this case, these load cases can be run using the methods presented and the stress results can be post-processed to ensure infinite life based on the load cases being fully reversing loads.

> Accurate calculating fatigue life requires considering every significant load in the service life. Loading complexity and resulting stress states mean fatigue analysis is more challenging than simply designing a component to withstand maximum loads. In addition it is too computationally intensive to use finite element method in order to get the fatigue life prediction for the components, which subjected to service loadings. [4]

To perform true fatigue analysis, a structural model can be set up similar to the Pivot Method model. However unit loads would have to be used at all load application points and pivot points to get a stressed state for each unit load independently. With this analysis performed, fatigue analysis software would need to be used along with time history loading data for all the load applications and pivot locations. These time histories would need to be measured in durability testing of a prototype mechanism, and then processed in multibody simulation software to provide the time histories required for fatigue analysis to proceed.

2.2 Review of Literature

A review of available literature shows that much research has been conducted in the modeling and analysis of mechanism performance. Many of the articles reviewed presented new mechanism designs, along with methods to model the projected performance of the proposed system. One of the papers [5] associated with a new design for a six bar suspension system features a discussion of the kinematics of the system, but does not discuss any loading information or structural analysis.

> Analysis of the parallel arm suspension shows that it is capable of providing relative Z motion by change of angle as shown in Figure 4 and can maintain ø = 0 as both joints of the steering axis have equal vertical motion during suspension travel. But the parallel-arm suspension is incapable of satisfying FR 1512: y = 0 during suspension travel. This causes tire-scrub due to suspension travel as illustrated in Figure 4.
>
> In the SLA suspension, we can achieve y=0 (no tire scrub) through assignment of appropriate values to the link lengths, but this doesn't allow ø=0 during suspension travel. This causes camber change and caster change due to suspension travel [5]. A compromise solution for y and ø can be obtained through optimization of the link lengths and joint positions, but we cannot satisfy all three FRs simultaneously using a four bar linkage. Both y and ø can be reduced by increasing the link lengths, but this is limited by the constraints of cost, packaging and unsprung weight of the vehicle. [5]

While the authors do not mention structural analysis in their article on the six bar suspension, the method discussed in this paper is directly applicable to their proposed suspension system.

The Stick Model Method is the most commonly used method to analyze components in a mechanism. This is most likely because the Stick Model is the simplest method to set up. It does not require the initial processing to generate the joint loads, and the loading of the model is easy to understand. Lee & Lee [6] used the Stick Model Method to design an aluminum lower control arm. This method worked well for them, however, it is unclear from their paper if there were multiple positions that were considered or if they only analyzed the suspension system in one position.

> The shock absorber was modeled by a beam element and translational joint, and the damping effect was approximated as the equivalent stiffness with a spring. The mounting bush[ing] was modeled by a spring element. The worst running loads are shown in Table 3 and the loads acted on the contact point of tyre or wheel centre. [6]

In the case of an automobile suspension, if the relative orientation of the components does not change significantly as the system moves through its travel, the Stick Model Method can be used effectively with only one model for all positions. This assumption breaks down if the assembly moves a large amount, such that the angles between the individual components cannot be assumed to be constant throughout the travel of the mechanism, as is demonstrated by the mechanism under consideration in this project.

Once we know the tire contact patch loads we can generate a dynamic model of the whole suspension system. A dynamic model of the system allows us to easily determine the component loads for the system. Individual loads on the components are required to analyze and optimize components individually. Otherwise, all the components must be included in the finite element models. This would make the size of the models and their calculation times impractical. The dynamic modeler used for our analysis was ADAMS, a commercially available software package [7].

> Dynamic models, which use a rigid body approach to modeling a system and determining loads transferred from one component to another, are commonly used in industry. The DANA corporation has used ADAMS (a dynamic modeling software package) to generate loads to be applied to a model for structural analysis and optimization of their upper control arm [7].

In this case, it was recognized that all of the component loads were required to make a valid model. The method employed in this example is similar to the Pivot Method in that it takes up front effort to calculate the joint forces. The benefit the Pivot Method being proposed in this paper has, is that it does not require a third software package to do the dynamic analysis. The Pivot Method is designed to use the same software package to do the initial loads preparation as is used to run the structural analysis model. Using this method, the analysts can do all the work themselves and does not need to learn another software package to perform the analysis.

It is common practice in industry to break a complex system down into simpler models. This is done for a variety of reasons, most commonly to reduce the computational resources required to run the model, but also to allow components to be analyzed without the rest of the mechanism being designed yet – or even as this paper presents – to reduce the number of models needed to account for all the mechanism's configurations.

> Substructure synthesis method is a modeling technique, which permits a complex structure to be represented by a reduced number of degrees of freedom through modal transformation. These techniques are well known and popular in structural dynamic analysis and are very useful when solving large complex structural dynamic problems, especially for structures consisting of several distinct substructures. [8]

Breaking down a model into simpler models is commonly done for multiple types of analysis including structural, vibration, fatigue and even dynamic analysis. Without this practice many mechanisms and situations would be impractical to model.

3.0 METHODS

To demonstrate the advantages of the Pivot Method, the sample system was analyzed using all five methods presented in this paper (Component Method, Over Constrained Component Method, Stick Model Method, Full Model Method, and Pivot Method). The different methods compared in this project were all subjected to the same set of ten loadcases that represent typical loading that would be used to analyze a similar structure for structural suitability. Each of these loadcases was run on all three components using each of the methods presented in this paper. Figure 2 shows where these loads were applied to the mechanism. In the figure, the direction the arrow points is the direction that the load would act in the positive direction.

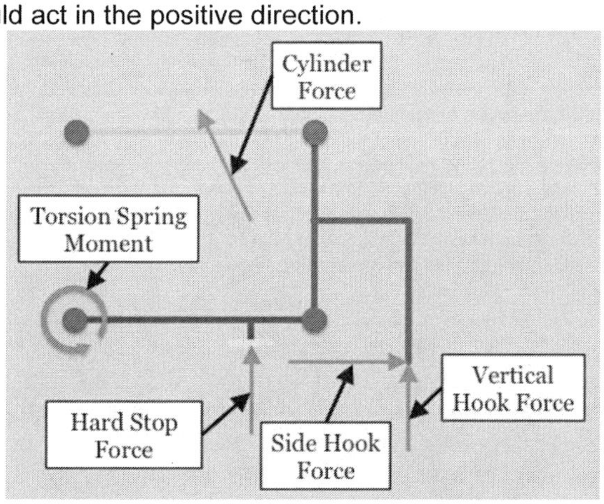

Figure 2: Load Application Diagram

3.1 FEA Modeling

To ensure the results would be comparable from method to method, each of the three components were meshed; then the meshed model was imported into the models for each method. Similarly, the material properties were set up in a template model file and then imported into each model. Using the same mesh and material properties for each method ensures that when comparing the results, the only difference in the results will be the method used to apply the loading. The material properties used in this project are listed in Table 2.

Table 2: Material Properties

Material	MID	Elastic Modulus (psi)	Poisson's Ratio	Density (lbf/in^3)
Aluminum	1	1.00E+07	0.33	0.098
Steel	2	3.00E+07	0.3	0.284

The section properties were also set up using the template model file and imported along with the material properties and the mesh.

3.1.1 Component Method

Each of the components were modeled separately for this method. The components were treated as separate entities without concern for how the components interact with each other. The upper and lower links were treated as simply supported beams with one end treated as fixed in all three axes and allowed to rotate about the pin axis, while the other end was allowed to slide along the length of the part and also allowed to rotate about the pin axis. Figures 3 and 4 show the constraints and loads applied to the upper and lower links respectively.

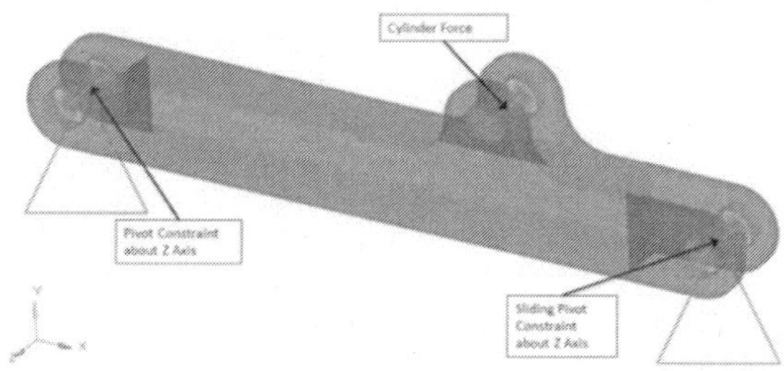

Figure 3: Component Method Upper Link Constraints and Loads

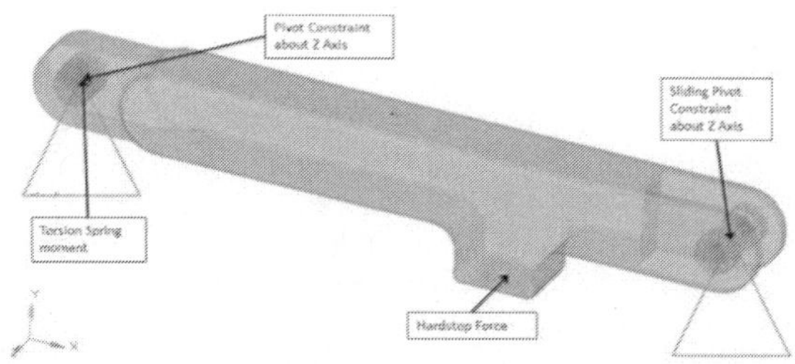

Figure 4: Component Method Lower Link Constraints and Loads

The middle link is treated similar to the other two components, with the upper pivot fixed in all three axes and allowed to rotate about the pin axis, while the other pin is allowed to translate vertically and also allowed to rotate about the pin axis. Figure 5 shows the constraints and loads applied to the middle link.

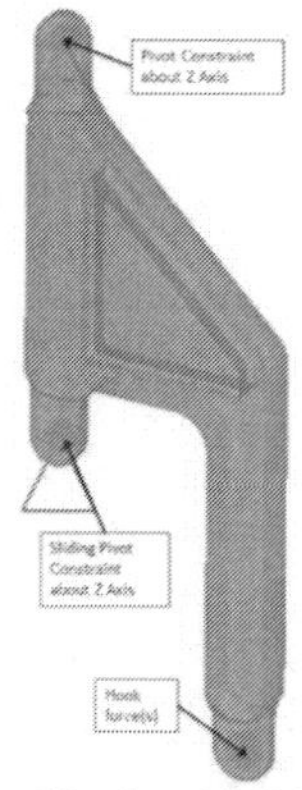

Figure 5: Component Method Middle Link Constraints and Loads

3.1.2 Over Constrained Component Method

Similar to the Component Method, each of the components was modeled separately for this method. The components were treated as separate entities without concern for how the components interacted with each other. The upper and lower links were treated as over constrained beams with both ends fixed in all three axes and allowed to rotate about the pin axis. The middle link was treated similar to the other two components, with both the upper and lower pivots fixed in all three axes and allowed to rotate about the pin axis.

3.1.3 Stick Model Method

The Stick Model Method resulted in the largest number of models. In addition to having a separate model for each component, the Stick Model also needed a separate model for each position. This resulted in a total of nine models in order to account for all the loading conditions on all three components. All nine models were setup in the same way. The upper and lower pivots were constrained in all three translational axes, and allowed to rotate only about the axis of the pin. The cylinder was accounted for by using a beam element that was allowed to pivot about the attachment point on the upper link, and was fixed in all three axes on the other end, however it was allowed to rotate about all axes.

For all the loadcases, the hook loads were applied to the hook end of the middle link. For the middle position, the only input loads were the hook loads. Figure 6 shows how the loads were applied to the middle position models.

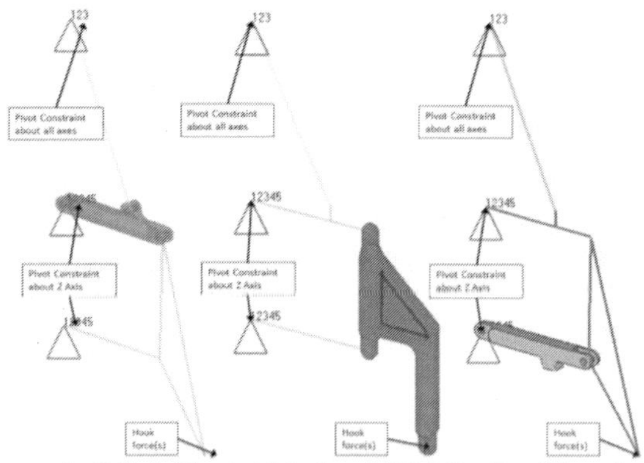

Figure 6: Stick Model Method Middle Position Loads

For all the loadcases, the hook loads were applied to the hook end of the middle link. The torsion spring was accounted for by applying the appropriate torque to the lower link pivot constraint. Figure 7 shows how the loads were applied to the up position models.

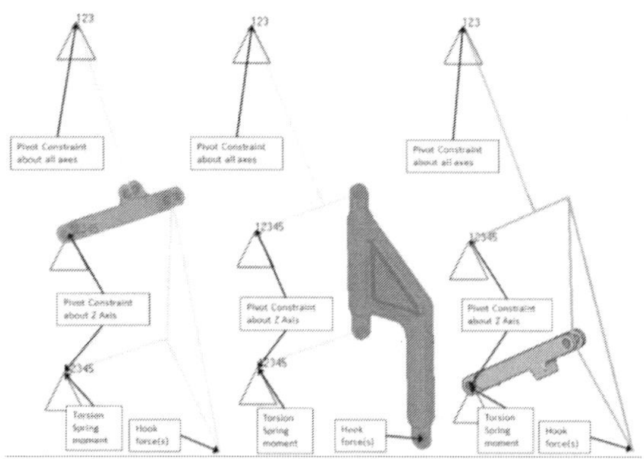

Figure 7: Stick Model Method Up Position Loads

For all the loadcases, the hook loads were applied to the hook end of the middle link. The torsion spring was accounted for by applying the appropriate torque to the lower link pivot constraint. The hardstop was applied to the lower link hardstop location normal to the axis of the part. Figure 8 shows how the loads were applied to the down position models.

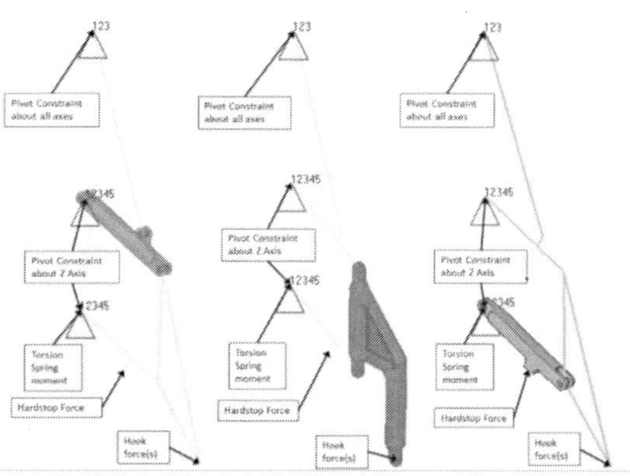

Figure 8: Stick Model Method Down Position Loads

3.1.4 Full Model Method

The Full Model Method required a separate model for each position, however all the components were accounted for in each position model. This means that only three models were required to solve for the stresses in all the components for all the positions. Similar to the Stick Model Method, All three models were setup in the same way. The upper and lower pivots were constrained in all three translational axes, and allowed to rotate only about the axis of the pin. The cylinder was accounted for by using a beam element that was allowed to pivot about the attachment point on the upper link, and was fixed in all three axes on the other end, however it was allowed to rotate about all axes. The loads applied to the full model were set up in the same manner as the stick model.

3.1.5 Pivot Method

The Pivot Method models were set up exactly the same as the Component Method models, with one big exception. In addition to the loading that was applied to the part from the external loads, there was also a force applied to the pivots between the components. The pivot loads were calculated from a model that contained a simplified stick models of the mechanism. The simplified models were set up the same as the Stick Model Methods, with all the components replaced with bar elements. In this pivot model, there was one simplified model for each loadcase in its proper configuration with its appropriate loading. This model ran quickly with minimal resources.

The constraints applied to the Stick Model used to generate the pivot forces were the same as the constraints discussed above used in the Stick Model Method. The upper and lower links were constrained with pivot constraints about the Z axis each on the end that was not connected to the middle link. A bar element representing the lift cylinder was constrained as a ball joint on the end that was not connected to the upper link. The ball joint constraint was chosen to prevent any binding that may have occurred with a pivot constraint.

The loads were applied to the simplified model in the same manner as the loads for the Stick Model and Full Model methods. In this model however, all the loadcases were run at once on individual instances of the simplified model. After running the Pivot Model, the Pivot loads were gathered from the results file, and inputted into the spreadsheet, which converted the loads from the global system level coordinate system to a coordinate system unique for each component.

Each of the components were modeled separately for this method. The components were treated as separate entities while the effects of the other components were accounted for by applying the appropriate load to the pivot. The upper and lower links were treated as simply supported beams with one end treated as fixed in all three axes and allowed to rotate about the pin axis, while the other end was allowed to slide along the length of the part, and also allowed to rotate about the pin axis. The end that was allowed to slide has a force applied to it to account for the effect of the loading from the other components through the middle link.

The middle link was treated similarly to the other two components, with the upper pivot fixed in all three axes and allowed to rotate about the pin axis, while the other pin was allowed to translate vertically and also allowed to rotate about the pin axis. A vertical force was applied to the lower pivot to account for the load that would be transferred between the upper and lower links through the middle link.

4.0 RESULTS AND DISCUSSION

The results presented in the following sections show that the Pivot Method is beneficial from both the standpoint of result accuracy and efficiency. Comparing the output results of all the methods, the Pivot Method displacement agrees mostly with the component method models, while the stresses match most closely with both the stick model and Full Model Methods. These provide the most beneficial results from the standpoint of determining the structural properties of the part. The computational requirements of the Pivot Method show that it takes more computational resources to run the Stick and Full Model Method models than the Component methods, and that the Pivot Method has similar computational requirements as the Component Method models. The modeling effort required for the Pivot Method is slightly higher at the beginning of the project, as the pivot loads must be calculated before any analysis can begin. However, once the loads are known, the Pivot Method is just as simple to set up as the Component Methods. The Stick and Full Model Methods require many more models that must be kept up to date as designs change.

4.1 Analysis Results Comparison

Two different comparison methods were used to compare the analysis output of the different modeling methodologies. The first method was to compare the maximum and minimum values of both displacement and stresses reported for each method with the Pivot Method. The second method of results comparison was to generate a plot of the difference in displacement and the difference in stress for each element, and compare the maximum value reported for each method. The results of these comparison methods are presented in the following sections.

4.1.1 Average Difference in Maximum Displacement

Comparing the displacement results from all of the different modeling methods, the Pivot Method matches much more closely with the Component Method models, than with either the Stick Model or Full Model Methods. As shown in Table 3, the average percent difference between the Pivot Method and the Component Methods for all the components and all the loadcases is 11% compared to 44% for the Stick Model Method, and 46% for the Full Model Method. This is somewhat of an unintended consequence of the Pivot Method. The original intent was to closely match the results of the Stick and Full model methods for all result types. However from a displacement standpoint, matching the Component Method results is more useful for the design of the part. The displacement of the part in the Stick and Full Model methods represents both how the part deforms, and any rigid body displacement caused by the deformation of the other parts. When looking at the displacement results from this standpoint, the rigid body displacement often masks the deformation found in the part. In contrast the results from the Component Method show only the deformation in the part of interest which is oftentimes much more useful information. The maximum displacements do not, however, exactly match the Component Method. This is an indication that while similar to the Component Method, the Pivot Method provides more accurate displacement results than the Component Methods due to its inclusion of the pivot forces.

Table 3: Average Percent Difference in Maximum Displacement

Percent Difference in Maximum Displacement	Upper Link	Middle Link	Lower Link	Average
Component Method	1%	1%	32%	11%
Component Method Overconstrained	0%	1%	30%	11%
Stick Model Method	38%	38%	57%	44%
Full Model Method	38%	36%	65%	46%

This finding is further confirmed by comparing the average difference in maximum displacement values instead of the percent difference. For both the Component methods, the difference in displacement values reported were so low that when rounded to two decimal places, it reported an average difference in displacement of 0.00 inch. The Stick Model and Full Model methods both reported an average difference in maximum displacement of 0.03 inches. These results are presented in Table 4.

Table 4: Average Difference in Maximum Displacement (inches)

Difference in Maximum Displacement (in)	Upper Link	Middle Link	Lower Link	Average
Component Method	0.00	0.00	0.00	0.00
Component Method Overconstrained	0.00	0.00	0.00	0.00
Stick Model Method	0.02	0.03	0.04	0.03
Full Model Method	0.02	0.03	0.04	0.03

4.1.2 Average Difference in Minimum Displacement

Comparing the minimum displacement values reported by each method yields a similar trend as the maximum displacement. However, the percent difference values are much higher. This is because the minimum displacement values are very small; so, for a similar magnitude of difference in the displacement values, the percent difference is high. Table 5 shows that the Component and Overconstrained Component methods have a percent difference of 50% and 49% respectively, while the Stick Model and Full Model methods both result in a percent difference of 96%.

Table 5: Average Percent Difference in Minimum Displacement

Percent Difference in Minimum Displacement	Upper Link	Middle Link	Lower Link	Average
Component Method	33%	48%	70%	50%
Component Method Overconstrained	9%	67%	70%	49%
Stick Model Method	89%	98%	100%	96%
Full Model Method	90%	98%	100%	96%

For all the configurations the percent difference comparison for minimum displacement values showed a higher average percent difference than for the maximum displacement values, however, comparing the difference in minimum displacement values shows again that the Component Methods yield a difference in displacement so small that when rounded to two decimal places it returns 0.00 inches displacement difference. Similarly the difference in minimum displacement values for both the Stick Model and Full Model Method, show 0.02 inches of difference in displacement. This is a slightly smaller difference than for the similar comparison of maximum displacement values, despite the percent differences being much higher. These values are presented in Table 6.

Table 6: Average Difference in Minimum Displacement (in)

Difference in Minimum Displacement (in)	Upper Link	Middle Link	Lower Link	Average
Component Method	0.00	0.00	0.00	0.00
Component Method Overconstrained	0.00	0.00	0.00	0.00
Stick Model Method	0.00	0.05	0.00	0.02
Full Model Method	0.00	0.05	0.00	0.02

4.1.3 Average Difference in Maximum Stress

Comparing the stress results from all the methods, the Pivot Method most closely matches the Stick Model and Full Model Methods. Table 7 reports an average of 0% difference between the stress values reported by the Stick Model and Full Model methods, and those reported by the Pivot Method. In contrast, the Component Method resulted in a difference of 13% and similarly the Overconstrained Component Method resulted in a difference of 14% compared to the Pivot Method.

Table 7: Average Percent Difference in Maximum Stress

Percent Difference in Maximum Stress	Upper Link	Middle Link	Lower Link	Average
Component Method	2%	3%	34%	13%
Component Method Overconstrained	1%	8%	33%	14%
Stick Model Method	0%	0%	1%	0%
Full Model Method	0%	0%	0%	0%

This finding is further confirmed by comparing the average difference in maximum stress values instead of the percent difference. For both the Component Methods, the difference in stress values reported was well over 1000 psi (1901.47 psi for the Component Method, and 1264.13 psi for the Overconstrained Component Method. The Stick Model and Full Model methods reported an average difference in maximum stress of 125.33 psi and 8.73 psi respectively. These results are presented in Table 8.

Table 8: Average Difference in Maximum Stress (psi)

Difference in Maximum Stress Values (psi)	Upper Link	Middle Link	Lower Link	Average
Component Method	848.00	446.00	4410.40	1901.47
Component Method Overconstrained	206.00	828.00	2758.40	1264.13
Stick Model Method	1.00	12.00	363.00	125.33
Full Model Method	1.00	22.00	3.20	8.73

4.1.4 Average Difference in Minimum Stress

The minimum stress values show the same trend as the maximum stress values, however similar to the minimum displacement comparison, the percent difference values are higher. Table 9 reports an average of 14% and 16% difference between the stress values reported by the Stick Model and Full Model Methods respectively, and those reported by the Pivot Method. In contrast, the Component Method resulted in a difference of 50%, and similarly, the Overconstrained Component Method resulted in a difference of 64% compared to the Pivot Method.

Table 9: Average Percent Difference in Minimum Stress

Percent Difference in Minimum Stress	Upper Link	Middle Link	Lower Link	Average
Component Method	34%	29%	86%	50%
Component Method Overconstrained	53%	59%	80%	64%
Stick Model Method	1%	28%	14%	14%
Full Model Method	11%	27%	8%	16%

Similar to the minimum displacement results discussed above, the minimum stress values reported much higher percent difference in stress values, however comparing the minimum stress values shows that for all the methods discussed, the minimum stress values are small. The trend discussed in the maximum stress section above still holds however. The Component Method resulted in a minimum stress for the Component Method of 3.66 psi, and the Overconstrained Component Method showed an even higher difference at 7.02 psi. The Stick Model and Full Model methods only reported a difference of 0.24 psi and 0.20 psi respectively.

Table 10: Average Difference in Minimum Stress (psi)

Difference in Minimum Stress Values (psi)	Upper Link	Middle Link	Lower Link	Average
Component Method	0.91	0.96	9.12	3.66
Component Method Overconstrained	4.17	9.58	7.31	7.02
Stick Model Method	0.05	0.21	0.46	0.24
Full Model Method	0.02	0.50	0.08	0.20

4.1.5 Maximum Difference in Displacement

Comparing the maximum difference in displacement results again confirms that the displacement results of the Pivot Method more closely match with the Component Methods than with the Stick Model Method. In the case of the maximum difference in displacement, the results from the Full Model Method were not able to participate in this comparison method because the node numbering for the components in the Full Model Method were different than for the other methods by virtue of the model having all the components instead of just one. As discussed in the previous section, this is an unintended but beneficial consequence of the Pivot Method.

The Component Methods resulted in a maximum difference in displacement of 0.0015 inches for the Component Method, and 0.0016 inches for the Overconstrained Component Method. The Stick Model resulted in a maximum difference in displacement of .058 inches. Table 11 presents the Maximum Difference in displacement values for each component, along with the average Maximum Difference in Displacement for all three components combined.

Table 11: Maximum Difference in Displacement (in)

Maximum Difference in Displacement (in)	Upper Link	Middle Link	Lower Link	Average
Component Method	0.001877	0.000836	0.001713	0.001476
Overconstrained Component Method	0.001086	0.001844	0.001846	0.001592
Stick Model Method	0.056118	0.057199	0.061285	0.058201
Full Model Method	N/A	N/A	N/A	N/A

4.1.6 Maximum Difference in Stress

The maximum difference in stress comparison tells the same story as the difference in maximum stress comparison. The results of the Pivot Method match much more closely with the Stick Model Method than with either of the Component Methods. As discussed for the maximum difference in displacement results, the Full Model Method was unable to participate in this comparison because of the inherent difference in node numbering.

The Pivot Method resulted in an average maximum difference in stress of only 298 psi for the Stick Model Method. In comparison, the average maximum difference in stress for the Component Method and Overconstrained Component Method was 6121 psi and 6348 psi respectively. Table 12 presents the average maximum difference in stress for each component, as well as the overall average across all components.

Table 12: Maximum Difference in Stress

Maximum Difference in Stress (psi)	Upper Link	Middle Link	Lower Link	Average
Component Method	8580.29	2128.96	7653.60	6120.95
Overconstrained Component Method	4797.97	6001.70	8243.10	6347.59
Stick Model Method	11.11	146.57	734.90	297.53
Full Model Method	N/A	N/A	N/A	N/A

4.2 Computational Requirements Comparison

Comparing the computational requirements shows that the Pivot Method requires approximately the same computational resources to solve as the Component methods. Both the Stick Model and Full Model Methods require more computational resources to solve the models than the other methods. For all the computational resource metrics, the Full Model Method requires the most resources, with the Stick Model Method coming in a close second for all the metrics except the RAM usage. Similarly, for all the metrics, the Component Methods and the Pivot Method use less than half of the resources as the Full Model Method.

4.2.1 Elapsed Time

The elapsed time represents the total amount of time taken to run the model. Table 13 presents the total elapsed time for each method used. To determine the elapsed time for the Component Methods and the Pivot Method, the elapsed time is simply recorded from the output file. To compute the elapsed time for the Stick Model Method, the time from each position model was summed to determine the total elapsed time for all the load cases for each component. Similarly, for the Full Model Method, the elapsed time has been summed for all three of the different position models. For this reason, the total time for all the components is the same as for each component since running one model runs all the components.

Table 13: Elapsed Time Per Method

Elapsed Time	Upper Link	Middle Link	Lower Link	Total
Component Method	0:04:36	0:27:50	0:04:32	0:36:58
Overconstrained Component Method	0:04:31	0:15:14	0:04:38	0:24:23
Stick Model Method	0:12:59	0:45:20	0:12:53	1:11:12
Full Model Method*	1:17:18	1:17:18	1:17:18	1:17:18
Pivot Method	0:04:37	0:15:28	0:04:36	0:24:41

* Solving one model solves all the components. The individual method measurements represent the total for all 3 model runs, therefore the total column is the same as for the component coolumns.

The elapsed time for all the methods shows that the Pivot Method takes about the same time as the Overconstrained Component Method to solve. At one hour and seventeen minutes, the Full Model takes the most time to complete. Following closely behind, the Stick Model Method takes one hour and eleven minutes total to solve all three components. The Component Methods and the Pivot Method all take similar amounts of time. The Component Method takes a little longer than the Overconstrained Component Method because the middle link model of the Component Method has a different constraint method for the hardstop case than for the rest of the loadcases. For this reason, the Component Method has to generate and invert two stiffness matrices instead of only one for the Overconstrained Component Method with the same constraints being applied for all loadcases. Similarly, the Pivot Method uses the same constraints for all the loadcases because the effect of the hardstop is included in the pivot loads that are applied to the lower pivot.

4.2.2 CPU Time

The CPU time represents the total amount of time spent by the computer to solve the model. Table 14 presents the CPU time for each method used. To determine the CPU time for the Component Methods and the Pivot Method, the CPU time is simply recorded from the output file. To compute the CPU time for the Stick Model Method, the time from each position model was summed to determine the total CPU time for all the load cases for each component. Similarly, for the Full Model Method the CPU time has been summed for all three of the different position models. For this reason, the total time for all the components is the same as for each component since running one model runs all the components.

Table 14: CPU Time Per Method

CPU Time	Upper Link	Middle Link	Lower Link	total
Component Method	0:04:31	0:27:37	0:04:28	0:36:36
Overconstrained Component Method	0:04:25	0:14:54	0:04:20	0:23:39
Stick Model Method	0:12:51	0:43:42	0:12:44	1:09:17
Full Model Method*	1:10:08	1:10:08	1:10:08	1:10:08
Pivot Method	0:04:32	0:14:49	0:04:31	0:23:52

* Solving one model solves all the components. The individual method measurements represent the total for all 3 model runs, therefore the total column is the same as for the component coolumns.

The CPU time for all the methods shows that the Pivot Method takes about the same time as the Overconstrained Component Method to solve. At one hour and ten minutes, the Full Model takes the most time to complete. Following closely behind, the Stick Model Method takes one hour and nine minutes total to solve all three components. The Component Methods and the Pivot Method all take similar amounts of time. The Component Method takes a little longer than the Overconstrained Component Method because the middle link model of the Component Method has a different constraint method for the hardstop case than for the rest of the loadcases. For this reason, the Component Method has to generate and invert two stiffness matrices instead of only one for the Overconstrained Component Method with the same constraints being applied for all loadcases. Similarly the Pivot Method uses the same constraints for all the loadcases because the effect of the hardstop is included in the pivot loads that are applied to the lower pivot.

4.2.3 RAM Usage

The RAM usage is an indication of how computationally intensive the calculations being performed are. Table 15 shows the RAM usage for each method and each component. To determine the RAM usage for the Component Methods and the Pivot Method, the RAM usage is simply recorded from the output file. To compute the RAM usage for the Stick Model Method, the RAM usage from each position model was averaged to determine the average RAM usage for all the load cases for each component. Similarly for the Full Model Method the RAM usage has been averaged across all three of the different position models. For this reason, the average RAM usage for all the components is the same as for each component since running one model runs all the components.

Table 15: RAM Usage Per Method

RAM Usage (Mb)	Upper Link	Middle Link	Lower Link	Average
Component Method	489	1219	472	726.67
Overconstrained Component Method	489	1219	727	811.56
Stick Model Method	476	1188	460	708.00
Full Model Method*	2121	2121	2121	2121.00
Pivot Method	489	1219	472	726.67

* Solving one model solves all the components. The individual method measurements represent the average for all 3 model runs, therefore the average column is the same as for the component columns.

Comparing the RAM usage for the different methods, the Full Model Method uses dramatically more RAM on average than the rest of the methods. The average RAM usage for the Full Model Method was over 2000 Mb, while the rest of the methods used between 700 and 800 Mb of RAM on average for running the models. This result is to be expected since the Full Model Method has all three meshed components being solved simultaneously, while all the other methods have only one meshed component at a time. The Pivot Method uses the same amount of RAM to run the model as the component method.

4.2.4 Hard Disk Usage

Comparing the hard disk usage for the different methods presents an indication of the amount of calculations performed. To determine the hard disk usage for the Component Methods and the Pivot Method, the hard disk usage is simply recorded from the output file. To compute the hard disk usage for the Stick Model Method, the hard disk usage from each position model was summed to determine the total hard disk usage for all the load cases for each component. Similarly, for the Full Model Method the hard disk usage has been totaled across all three of the different position models. For this reason, the total hard disk usage for all the components is the same as for each component since running one model runs all the components.

Table 16: Hard Disk Usage per Method

HD Usage (Mb)	Upper Link	Middle Link	Lower Link	Total
Component Method	6115	16726	6043	28884
Overconstrained Component Method	6071	16705	5718	28494
Stick Model Method	17601	49045	17595	84241
Full Model Method*	85691	85691	85691	85691
Pivot Method	6115	16653	6043	28811

* Solving one model solves all the components. The individual method measurements represent the total for all 3 model runs, therefore the total column is the same as for the component coolumns.

Comparing the hard disk usage for the different methods, both the Stick Model and Full Model Methods resulted in the highest hard disk usage of the methods under comparison. Both the Stick and Full Model Methods use more than 80 Gb of hard disk, while the Component Methods and Pivot Method use approximately 28 Gb of hard disk space to perform the calculations. The Stick Model Method has a high total hard disk usage because it uses three different models to solve for each component. Similarly, the Full Model Method solves for all three components at the same time, but still needs three different models to solve for all the positions for each component. For this reason, the hard disk usage shown for each component is the same as the total hard disk usage for all the components. The Pivot Method uses approximately the same hard disk space for solving the solutions as the Component Methods. This data is presented graphically in Figure 17.

4.3 Modeling Effort Comparison

From a modeling effort standpoint, the Component Method models are the simplest. The Pivot Method does not require any more modeling effort while setting up the model of the part to be analyzed, the only additional effort required is to set up the simplified system model and prepare the pivot loads. Comparatively, the Stick Model Method requires the analyst to maintain separate models for each position that the system is to be analyzed in. This does not mean that the analyst has to mesh the same part multiple times when using this method; the analyst should mesh the part once and import the mesh into each position model. This does, however, mean that each model must be set up separately and the imported meshes have to be translated and rotated to match with the rest of the Stick Model. All the different models must be set up the same way to ensure that they all provide equally valid results. Similarly, if there is a change to the design that needs to be analyzed, which happens quite commonly in industry, the analyst must make sure that all the same updates are made to all the separate models in the same manner. If the models are updated differently, this could inadvertently cause one model to provide different results from the others. The modeling effort required for the Full Model Method is similar to the Stick Model Method, except the Full Model Method requires that all the meshed components to be imported into the Full Model and then positioned properly.

The Pivot Method does require extra modeling effort to get the loads setup, however this extra effort is negligible compared to the extra effort required by the Stick and Full Model Methods. Additionally, the initial load calculations for the Pivot Method is only required to be done once, and can be re-used for each subsequent design iteration. The Stick and Full Model Methods will have the same setup required for each of the subsequent design iterations as the first. For these reasons, the Pivot Method requires substantially less modeling effort than either the Stick Model Method or the Full Model Method.

5.0 FUTURE WORK

The method presented in this paper works well for a two-dimensional system operating under quasi-static conditions. To analyze real world systems, it would be helpful to expand the method to handle three-dimensional systems. Instead of using the FEA solver and excel spreadsheet to generate loads, it would be useful to use a multibody dynamics solver to generate loads, especially if dynamic load cases are desired. A multibody solver can also be used to help the Pivot Method solve fatigue cases by utilizing time slice loads for pseudo-fatigue solutions, or by preparing models for full fatigue analysis in a fatigue solver.

6.0 CONCLUSIONS AND RECOMMENDATIONS

As the above sections show, the Pivot Method provides the most useful results of the modeling techniques and does not require multiple models of the same part to analyze all the mechanism's positions. The Pivot Method Models use a substantial amount fewer resources to run than the comparable methods that provide similar stress results and require more modeling effort to build. When performing structural analysis of mechanisms that are topologically similar to the sample system, the Pivot Method described in this paper should be applied to the system under analysis and can be applied directly. For systems that are not topologically similar to the system presented in this paper, the future work section provides some insight into expanding the method to work with three-dimensional systems.

For cases where the Pivot Method is not applicable, this paper is intended to act as motivation to perform a similar method development for the system at hand. As is demonstrated by the Pivot Method and its sample system, putting upfront effort into breaking down complex systems into individual models for each component that account for the effects of the other components is a very worthwhile endeavor. A little up front development work will save both modeling effort and computational resources later, while providing more useful results.

NOMENCLATURE

CAD – Computer Aided Design
CAE – Computer Aided Engineering
CPU – Central Processing Unit
FEA – Finite Element Analysis
Gb – Gigabyte
in – Inch
LBF – Pound Force
Mb – Megabyte
MID – Material Identification Number
PSI – Pounds per Square Inch
RAM – Random Access Memory

REFERENCES

[1] Bhavikatti, S. (2004). *Finite Element Analysis.* Daryaganj, Delhi, India: New Age International.

[2] Prantil, V. C., Papadopoulos, C., & Gessler, P. D. (2013). *Lying By Approximation: The Truth About Finite Element Analysis.* Morgan & Claypool Publishers.

[3] Narasaiah, L. G. (2008). *Finite Element Analysis.* Hyderabad, India: BS Publications

[4] Abdullah, S., Al-Asady, N., Ariffin, A., & Rahman, M. (2008). A Review on Finite Element Analysis Approaches in Durability Assesment of Automotive Components. *Journal of Applied Sciences* , 2192-2201.

[5] Deo, H. V., & Suh, N. P. (2003). *Axiomatic Design of Automobile Suspension and Steering Systems: Proposal for a Novel Six-bar Suspension.* Massachusetts Institute of Technology, Department of Mechanical Engineering. Cambridge: SAE International.

[6] Lee, D.-C., & Lee, J.-L. (2003). Structural Optimization Concept for the Design of an Aluminum Control Arm. *Proceedings of the Institute of Mechanical Engineers* (pp. 647-656). ProQuest.

[7] Demirdogen, C., Ridge, J., Pollock, P., & Anderson, S. (2004, October). Weight Optimized Design of a Front Suspension Component for Commercial Heavy Trucks. *Vehicle Dynamics & Chassis Developments* .

[8] Kim, B. S., Kim, B. S., & Yoo, H. H. (2009). Analysis of Vibration Characteristics of Full Vehicle Model Using Substructure Synthesis Method. *The Sixteenth International Congress on Sound and Vibration* (pp. 1-8). Krakow: ICSV

This page left blank intentionally.

Proceedings of the ASME 2014 International Mechanical Engineering Congress and Exposition
IMECE2014
November 14-20, 2014, Montreal, Quebec, Canada

IMECE2014-36035

INTEGRATING REAL INDUSTRIAL EXPERIENCES INTO THE CURRICULUM THROUGH ROBOTICS APPLICATIONS

Nael Barakat, Ph.D. P.Eng. FASME
Professor and Chair of Mechanical Engineering
School of Engineering
Grand Valley State University
Grand Rapids, MI. 49546 USA
barakatn@gvsu.edu

ABSTRACT

Integrating real industrial applications and bringing actual engineering problems into the curriculum is always a win-win activity. The educational benefits of such activity are countless. In addition to industry, professional engineering organizations can be a great resource to support actual engineering projects and mentor engineering students. Partnering with industry, professional engineering organization, and any similar entity, allows the students a unique opportunity to gain exposure and practice real engineering before graduation.

The challenge with this concept is to find and run the appropriate project or activity that will simultaneously meet the expectations of all parties and fulfill the educational objectives, while conforming to the time limitations of the course. This paper presents and evaluates such experience through two examples of projects utilizing robotics to achieve engineering educational goals through the design and build of functional industry-grade systems. The first project was performed by an interdisciplinary engineering team of students to solve an actual industrial problem. The project was sponsored by an industrial customer and produced a functional prototype that achieved the required goals and was ready to be duplicated at the customer's site. The second project was also performed by a team of engineering students to design and build an industrial scale functional robotic manipulator in partnership with a professional engineering organization. The external organization sponsored the project and provided some professional engineers as mentors. The outcome of the project was a mobile 5-axis robotic manipulator that is used for many engineering demo and recruiting events. These projects are examples of ways in which real engineering exercises can be integrated into the curriculum to enhance the educational process and improve collaboration between academia and other engineering entities.

Keywords

Project based learning, Robotics education, Education and Industry collaboration

INTRODUCTION

Integrating engineering experiences and professionalism into the curriculum is a proven method of enhancing the educational process. The aspects of this integration can take many forms and can be with possible interested organizations. Examples of these activities include project based learning serving industry, professional organizations, the school and university, and the community. Any partnership that provides a focused engineering activity to improve the learning process as well as the relations between academia and external constituents is worth pursuing and investigating.

However, the challenges associated with this concept are plenty. These include the challenge of finding and running the project which is appropriate for the students, the educational objectives, and the resources available. It also includes finding the appropriate location and time in the curriculum as well as finding the interest and expertise among the faculty members to carry out a successful project. These projects are supposed to meet the expectations of all parties involved. Therefore, the subjects of projects are as important as the resources and management aspects of the project. When the right combination of circumstances exists, a project provides countless benefits to engineering students. The main thrust of these projects is that they challenges students to solve actual engineering problems through a design and build process, with all the constraints associated, and utilizing multiple knowledge bases, hard, and soft skills. Problem or project based learning (PBL) where education takes place revolving around an engineering problem under the umbrella of a project have been tested and documented in the literature as best-practice [1, 2, 3].

Robotics based projects are ideal media for comprehensive engineering projects. Robotics are complex systems which require multi-disciplinary knowledge and skills while providing enough flexibility to be utilized in endless applications [4]. It also provides a perfect opportunity for the students to interact in a team and with engineering professionals, which fulfills multiple ABET criteria and allows for practice of soft skills. Therefore, many schools of engineering have integrated robotics projects in its curricula and student activities. However, the immense amount of resources required to carry out robotics projects could sometimes jeopardize the plans to incorporate them in engineering curricula, especially in schools with limited resources [5, 6].

Opportunities to practice real world engineering while having guidance and critique from the instructor and mentor are scarce but very helpful in the engineering educational process. From an instructor point of view, such opportunity helps to identify the strength and weakness points in the curriculum. The benefits go even further when the project requires a multi-disciplinary team of engineering students and involves design and build of a complete engineering system [5].

Another significant benefit is the building of partnerships between academia and industry as well as professional engineering organizations. Engineering projects coming from real industrial needs and sponsored by industrial partners provide ideal activities for students to learn. Partnering with professional engineering organizations to perform engineering projects and allowing the students to mix with professional engineers is also an ideal activity for students to improve their preparedness and to hit-the-ground-running when they graduate.

This paper presents examples of how robotic systems can serve as a vehicle to integrate real engineering applications into the curriculum. These applications are in the form of collaborative projects with two different external entities, industry, and a professional engineering organization. The projects were incorporated in the curriculum at different spots and involved multiple intricate details affecting its directions. These projects are presented here for evaluation of the experience and to demonstrate the process of their integration in the curriculum, the process of executing them, and the benefits gained by the experiences.

EXAMPLE 1: BRAKE-CORE PRODUCTION-AUTOMATION

A team of six students from different engineering disciplines was assigned the brake sand-core production-automation project. This was their senior capstone project over the course of two academic semesters and under the supervision of the author. The customer was a tier-one automotive parts manufacturer.

Project background and scope

Sand cores are primary components in the manufacturing of automotive brake rotors leading to the final product of gray iron brake rotors. The iron rotors are the final product shipped to the customer. The scope of the project included automation of the workcell producing the sand cores only. Sand cores are physically comprised of silica sand and a two-part chemical resin compound. The traditional production process of the sand castings has three stages consisting of: "de-finning", coating, and baking or curing. Three human operators are required to manually execute the process and transport the cores between the different stages. The end result of the core manufacturing process is a finished sand core used in the brake rotors casting process. As a step towards the actual implementation of a fully automated system in the production facility, the manufacturer desired to see a functional prototype of the proposed system designed and built to scale. The goals of the project were to reduce the amount of operators from three to one that only monitors the workcell, and to increase the production rate while maintaining product quality. At that point, the sand cores production was 12,500 units/day/workcell while the project goal was to attempt reaching the 15,000 units/day/workcell mark.

Sand cores are produced by machines in a workcell, as shown by figure 1 and figure 2. The details of the core include a protrusion on one side, known as a "hat," and a symmetrical pattern of spaces or "finns," as shown by figure 3. Detail core production is comprised of four operations. Individual core machines produce two cores approximately every 30 to 45 seconds. A worker manually transports the two cores to the "de-finning" table where the "de-finning" takes place. The worker lowers and then raises the detail core onto the "de-

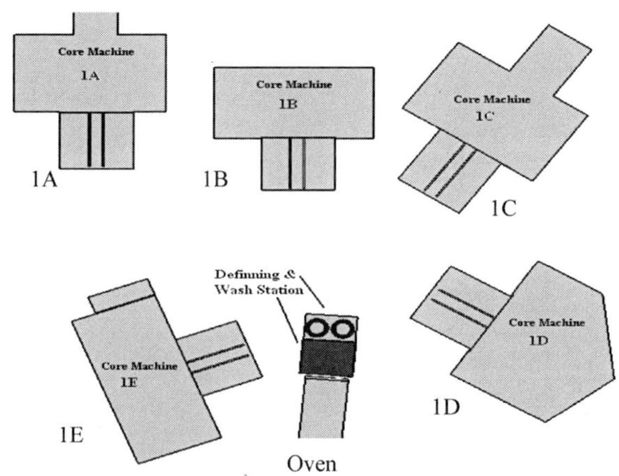

Figure 1 Original arrangement of the workcell producing the sand cores.

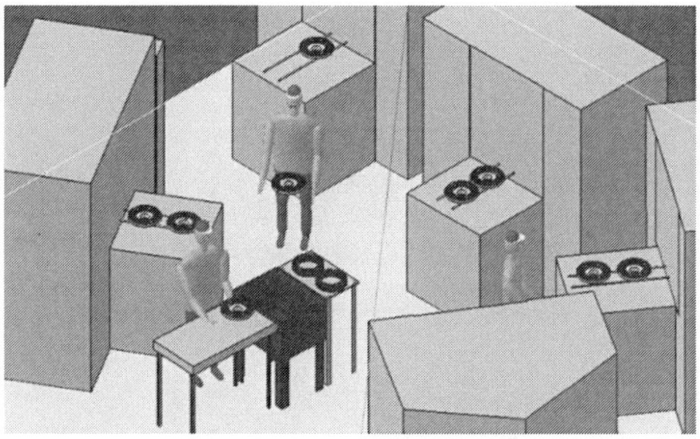

Figure 2 Original workcell condition (before the project) simulated using commercial software.

finning" jig which is a device having the exact opposite image of the actual casting, as shown in figure 4. This causes the excessive material (known as flash) to be removed from in-between the "finns." The worker then performs a visual inspection of the core to ensure that all flash has been removed. The core is then placed hat side down into the core wash tank and is coated with core wash on all outer surfaces where molten metal will later make contact. The wash is a thick refractory paint that is quite toxic to the human body. After this has been completed, the core is placed hat-side down on a conveyor and transported through an oven for curing.

There are a variety of factors and conditions within the workcell that can cause harm to the human body. Reaching out to the core machines themselves to retrieve the cores as well as collision with other workers due to the restricted space imposes one constant chance of harm. The constant noise within the foundry requires that ear protection be worn at all times. Respirators are also required in the workcell at all times. As was previously mentioned, the core wash itself is very hazardous to the human body. Exposure can cause a severe rash while ingestion requires immediate medical attention. In addition, the workcell is always full of dust in the air. It is in the best interest of the customer and the workers that they be removed from the workcell and that the process is automated.

Final system design and functionality

The selected design option for the workcell layout and eventually for the prototype system consisted of two floor mounted FANUC robots. Within this option, the first half of the workcell would be serviced by a FANUC S-500 robot shown by figure 5. The prototype system to be built and tested by the students' team was based on this half. The second half of the workcell included a robot from the same family. These robots would be placed to allow ease of workcell maintenance, minimum modification to the current workcell layout, and the capability to complete the processing tasks. Figure 6 represents the final layout of the automated workcell.

Figure 3 A sample detailed sand core produced by the machines in the workcell.

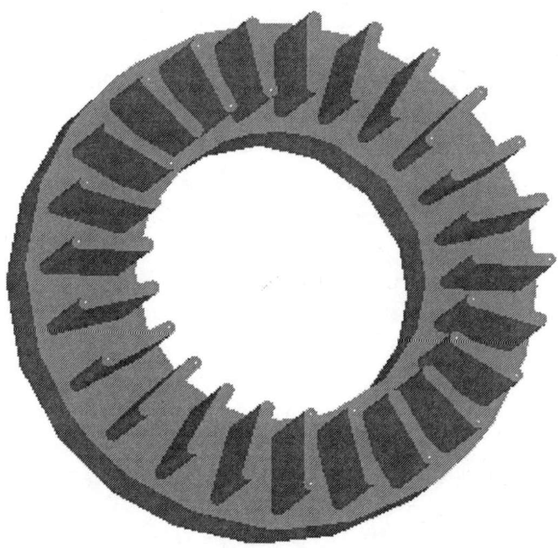

Figure 4 A difinning jig used to clean the spaces between the core ribs.

Simulation and Optimization

Upon finalization and mutual approval of the general design, the workcell and the automated manufacturing processes were simulated using commercial software. This simulation provided a flexible medium to switch between multiple robotic arms and arrive at an optimal workcell design for the lowest cycle time possible, as shown in figure 7. Results of the simulation confirmed that using the proposed design to automate the workcell and based on the first half of the workcell, which includes the less capable robot, a production rate of more than 15,000 units/day/workcell can be

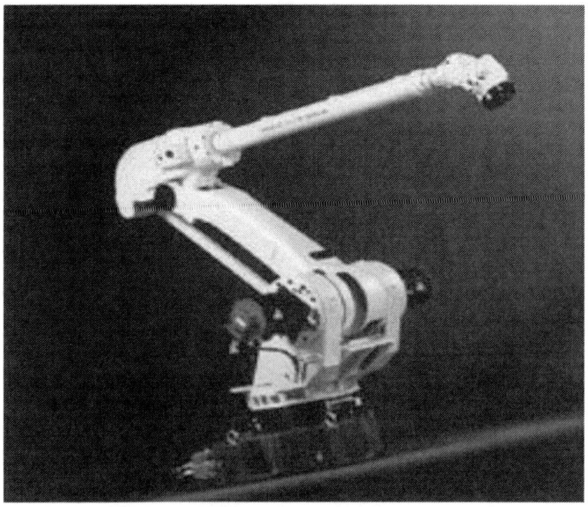

Figure 5 FANUC S-500 robotic arm.

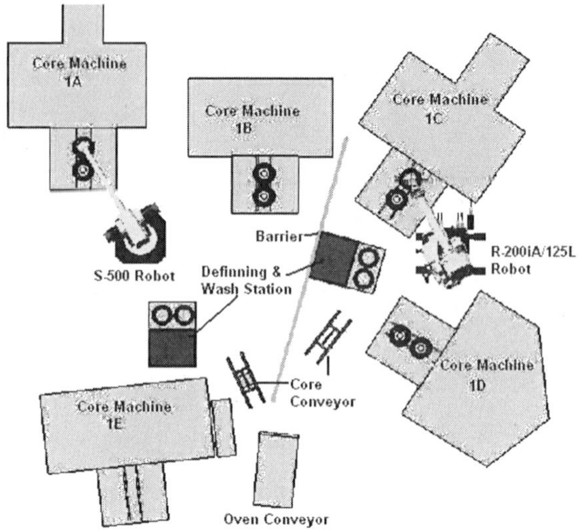

Figure 6 Final design and layout of the workcell.

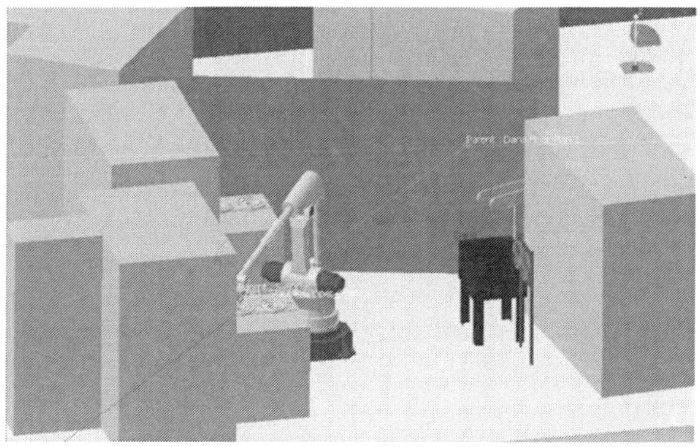

Figure 7 Final design of the workcell simulated using commercial software.

achieved. The cycle time corresponding to this rate would be a maximum of 40.31 seconds.

Experimental Results

As part of the project, several experiments were designed and conducted for many reasons. Example of these reasons was the need to determine the physical properties of the sand cores. The results of these experiments were used to help select the end-of-arm tooling for the robot. For example, in order to determine the minimum force needed to manipulate the core through the workcell, friction testing was conducted between the brake cores and the gripper finger material, yielding a coefficient of static friction of 0.79. In order to determine the maximum force that the gripper could exert upon the cores, strength testing of the cores was conducted. Applying an equal force to the interior of the core's hat was done while performing the strength testing. Force was gradually increased until the core reached the failure point. This testing was performed on a commercial Stress-Strain testing machine. Table 1 summarizes the results of these tests.

Table 1 Properties of the sand cores as determined experimentally.

Property	Value
Average Stress	1013.21 psi
Stress Standard Deviation	50.18 psi
Final Strength	862.66 psi
Static Friction Coefficient	0.792
Dynamic friction Coefficient	0.760

End-of-Arm-Tooling

The end-of-arm-tooling selected was a Techno-Sommer GD1716N three-jaw gripper with sealed, watertight 4-sided guide, as shown by figure 8. This selection was based on the experimental results, cores' hat diameter sizes, weight, and the robot's maximum acceleration. This gripper has a concentric activation, allowing for gripping cores on the inside of the hat and a stroke length which covers the inner diameter range incorporating all required seven sand cores. In addition, the forces exerted by the gripper provided a range that covered and exceeded the required values. The gripper fingers were designed to manipulate the cores by utilizing a step design shown in figure 9. The upper and lower fingertips were designed to handle small and large core diameters respectively. For the prototype system the fingers were machined in-house using 6163-Aluminum for weight reduction considerations.

"De-finning"and Wash Station

The purpose of the "defining" and wash station is to hold the "definning" jigs and core wash for application onto the cores. The station must be sturdy and anchored to the floor, yet removable for performing maintenance on the workcell. The height of the core wash aspect of the station was designed to optimize robotic travel and therefore minimize cycle times as well as match the current wash tank at the foundry. The "definners" are now mounted vertically as opposed to the previous horizontal approach taken by the manufacturer. The "defining" stand was designed to mount all of the "defining" jigs on a common triangular configuration of 3 pins for accurate, repeatable positioning. This positioning must be consistent for the robot to be able to accurately repeat the process. Other benefits of the vertical approach are that all waste material will simply fall to the floor and only require periodic cleanup. The robot will dip the cores in the wash tank, with a pre-determined angle, as they are being rotated with a constant speed to achieve uniform coating and minimum waste, as shown in figure 10.

Building and Testing the Prototype

Upon completion of the design and simulation processes, and the arrival of the robot, a prototype workcell was built. All core machines mockups, workstations, the control panel, wiring, and end-of-arm tooling apparatus, were manufactured at the school by the students. PLCs, sensors, and a switch box for communicating signals to the robot were also built and integrated into the system by the students.

Once the controller receives the signal that a core machine has cores ready for processing; via additional sensors input; the robotic arm will engage and process the cores from the machine that originated the signal. The robot will only process a core from a machine that is sending a signal for cores present. For all cores the cycle of processing operations is the following:

1. Pick core #1 from the core machine
2. Proceed to the "defining" station
3. Lower and raise the core at the "defining" station onto the "defining" jig relating to the core being processed
4. Transfer the core to the wash tank
5. Dip the core into the core wash while rotating it around the center
6. Place the core into the oven conveyor in the inverted position, hat side up

This process is then repeated for core #2 of the same core machine. In a normal cycle of operations, the robot would alternate this process between two core machines, as the cores are produced throughout the day. Figure 11 shows the actual prototype "definer" holding plate and wash tank. While testing the system, some robot motion trajectory had to be modified for smooth and fast processing. Other adjustments to the forms of operation occurring at each part of the process had to be introduced to achieve the desired goals. Finally, testing was carried out for eight continuous hours. This revealed that the average cycle time was 38.0 seconds. This resulted in a production rate of 15,915 units/day/workcell, which passed the targeted mark of 15,000 units/day/workcell.

EXAMPLE 2: 5-AXIS INDUSTRIAL SCALE MANIPULATOR (SPARKY II)

A team of four engineering students taking a robotic-systems-engineering course selected to build a 5-axis industrial scale robotic manipulator as their course project, in collaboration with a professional organization. The course was run by the author and the main ideas of the projects were previously discussed with the Michigan Society of Professional Engineers (MSPE). MSPE was the sponsor of the project in collaboration with the school of engineering (SOE) at Grand Valley State University (GVSU). The name Sparky II was to

Figure 8 Techno-Sommer GD1716N three-jaw gripper selected for end-of-arm-tooling.

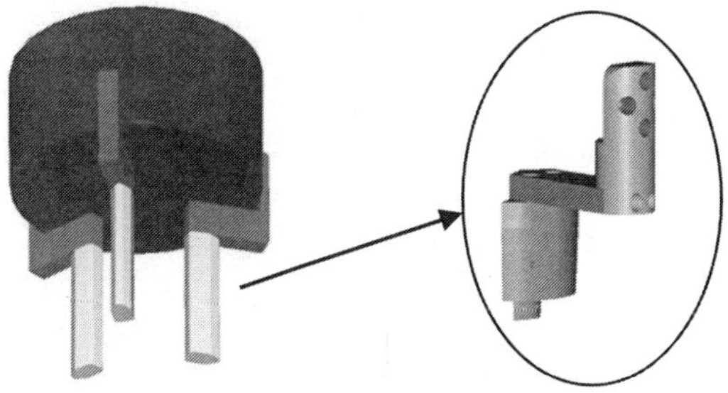

Figure 9 Gripper and finger design to manipulate the cores.

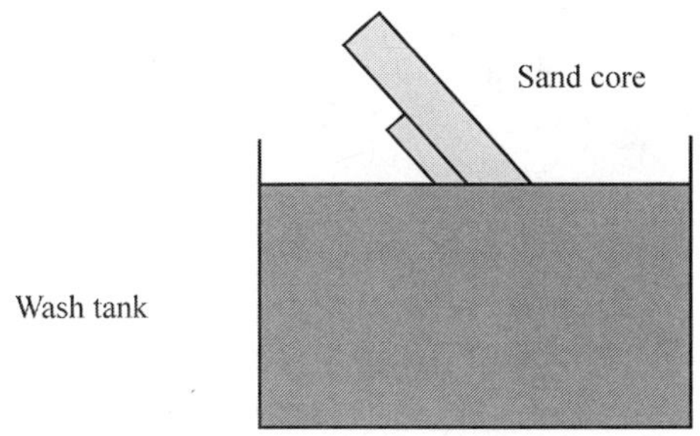

Figure10 Core dipping orientation in the wash tank.

signify a continuation of an earlier Sparky system which was a hydraulic backhoe built by GVSU engineering students as a senior project and was used for demo purposes for a while before Sparky II was built.

Project background and scope

Promoting engineering to prospective students is significantly facilitated by demonstrations using engineering products. Robots are of the most ideal engineering products to aid in this task. This is due to many reasons including the fact that it is a popular topic that has been experienced by almost everyone, it combines multiple engineering disciplines, and it is highly engaging, especially to pre-college students.

MSPE partners with GVSU-SOE every year to provide activities to the public during engineering week. To expand this partnership, MSPE expressed interest in having a portable robot of an industrial size built by engineering students enrolled in the robotics class to be used for promoting engineering and recruiting students. MSPE provided mentoring and a limited funding which was matched in-kind by GVSU-SOE. The final product was designed and built as a semester project for the Robotic Systems Engineering course.

Final system design and functionality

The system requirements were mutually approved with the customer as follows:

1. System will fit through a standard door (32").
2. System will operate in two modes, Demonstration and Manual. A third mode for further development was later included.
3. The only utility available for operation will be a 110 VAC grounded outlet (15A).
4. A primary safety perimeter will be established by restricting the advance of the audience towards the system.
5. System will have the ability to be reprogrammed by a trained technician.

In addition to these constraints, the students had to work with a very small budget ($1000) and a limited time period (one semester). The final design of the system featured a 5-axis counter-weighted robotic manipulator mounted on top of a custom-made steel table that has a foldable side. The table is mounted on four industrial grade castors for ease of mobility as shown in figure 12. The controller of the robot is based on an Allen-Bradley SLC 500 and a BASIC card as shown in figure 13. The first two joints are powered by AC motors controlled by two automatic ½ hp VFDs. The last three joints (towards the end effector) are powered by DC motors.

When the table is folded, it can be moved by one or two persons. The system requires regular 110 VAC to function and has a weight of 575 lbs. Figure 14 shows the system in the folded position while figure 15 shows the system in the unfolded position.

Building and testing

The cover for the controller cabinet was chosen to be transparent and the links of the robotic arm were made out of aluminum 80-20 and left without covers. This design was intended to achieve maximum exposure of the internal components of the system as an engineering product. Most pre-college students become quickly curious to find out about all these components they see. Figure 16 shows Sparky system as it is being operated and tested by engineering students in the lab. For safety purposes, software limitations have been built in the system to avoid collisions. In addition, hardware limitations are also installed in multiple locations of the system

Figure 11 Actual "defining" and wash station of the prototype.

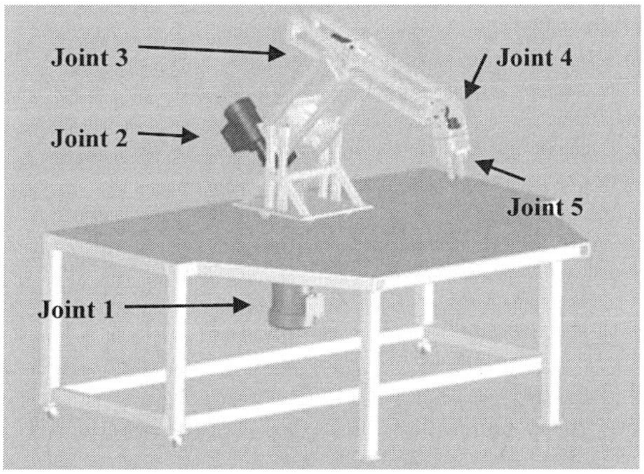

Figure 12 Sparky II final system design.

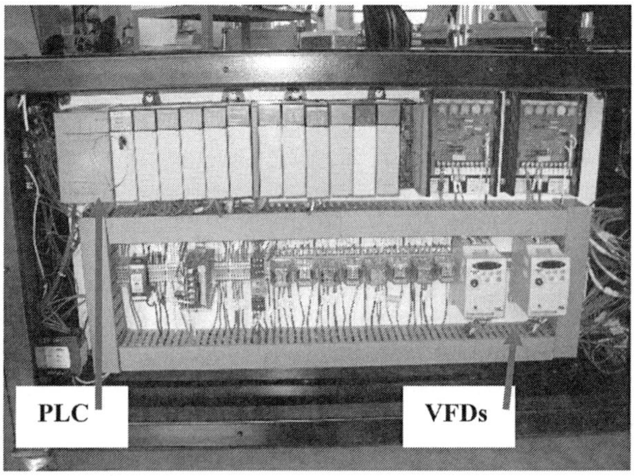

Figure 13 Control cabinet components installed under the table carrying the robot.

to avoid damage to the system or harm to humans surrounding the system. The manual mode, which is usually used during demonstrations by young prospective students, is restricted in speed and reach.

Sparky II is currently still operational for the fifth year and has been used in many recruiting and demonstration purposes. One graduate student is now in the process of refining the mechanical design and investigating an advanced controller for expanded operations by Sparky. This project resulted in a stronger partnership with MSPE. The following year MSPE approached the author for more systems and another project was carried out successfully during the robotic systems engineering course. The outcome of the project was two mobile milling stations capable of engraving on take-away plastic pieces when commanded through the attached touch screens.

OBSERVATIONS, ASSESSMENT, AND LESSONS LEARNED

The previous two examples demonstrate examples of how to integrate industrial engineering projects into the curriculum. One project was integrated as senior project; the other was a course project. Both projects were selected after careful investigation of their suitability from different aspects. These aspects included feasibility, the subject of the projects, and the added value to the curriculum as well as the mutual benefits to all constituents. The fact that they are robotics projects helped satisfy most of the educational benefits sought from such activity. Partnerships were established and others were strengthened through these projects. Success in these projects goes a long way in all directions, especially educating young engineers.

Nevertheless, as much as there are benefits in external projects, there are risks associated with them. Failing in achieving the goals set for these projects, especially with industrial customers, is detrimental to multiple parties and relations. Therefore, thorough inspection of the project goals and outcomes and involvement of many experts to judge the project suitability are vital to the success of the project. It is always better to be conservative and achieve less when accepting these projects than to be aggressive and fail causing a chain of detrimental effects to multiple constituents.

These projects were assessed by each party involved against the set goals. From an engineering education point of view, these were Problem Based Learning tools which involved real external partners as well as multi-disciplinary teams. All of these are standard ABET objectives in an engineering curriculum. Moreover, the students performing the projects were very appreciative of the experience and noted repeatedly in the indirect assessment that this was one of the best opportunities for them to really apply what they have learned in class to an actual product. During direct assessment it was found that all the experience was highly effective in emphasizing multiple program, and ABET objectives through a real experience outside a controlled school environment and with interactions with external customers and mentors. Schools of engineering should always consider partnerships with external entities to augment and improve the curriculum outside the controlled school environment. With the excitement and benefits associated with these opportunities, care must be taken to ensure that the objectives, outcomes, and expectations are clear to everyone involved and that limitations imposed by

Figure 14 Sparky II in the folded position.

Figure 15 Sparky II unfolded and the control panel.

Figure 16 Sparky II system during operation.

resources, time, and the school regulations are taken into consideration.

REFERENCES

1. Savage R. et al. "Integrating Project-based Learning throughout the Undergraduate Engineering Curriculum," Journal of STEM education: Innovation and Research, Jul-Dec 2007, Vol. 8 Issue 3/4, p15-27.
2. Yuqiu Y. "A Project-oriented Approach in Teaching Robotics Application Engineering," Proceedings of the ASEE Annual conference and Exposition, 2009.
3. Egon Moesby, "Curriculum Development for Project-Oriented and Problem-Based Learning (POPBL) with Emphasis on Personal Skills and Abilities," Global J. of Engineering Education, Vol.9, No.2, 2005.
4. Nolfi S. and D. Floreano, "Evolutionary Robotics: The Biology, Intelligence, and Technology of Self organizing machines," MIT press, 2000.
5. Barakat N., "The Ultimate Experience in Learning Robotics: Building Robots in a Robotics Course," Proceedings of the ASME - IMECE, Nov. 2008.
6. Ahlgren D. et al. "Effective Practices in Robotics Education," Proceedings of the ASEE Annual conference and Exposition, 2004.

Proceedings of the ASME 2014 International Mechanical Engineering Congress and Exposition
IMECE2014
November 14-20, 2014, Montreal, Quebec, Canada

IMECE2014-36066

ELEMENTS OF AN UNDERGRADUATE INTRODUCTORY MATERIALS ENGINEERING COURSE: SUCCESSFUL IMPLEMENTATION FOR STUDENTS LEARNING

A. Hossain* and J. Durfee
Department of Engineering and Design
Eastern Washington University, Cheney, Washington – 99004
*Contacting Author, Email: nhossain@ewu.edu

ABSTRACT

The major emphasis of a Materials Engineering course is to provide a broad level of information on different materials used in industries for various engineering applications. In our institution, working on quarter systems, this course becomes very aggressive and challenging in the amount of information that needs to be presented to the students within a 10weeks-time frame. This course involves different laboratory practices, besides regular class lectures, to obtain knowledge on various material properties to reinforce classroom theories. In addition, to increase exposure to different types of materials and material uses, students in the class are required to research an open topic, which is related to this class. Students have to prepare a brief written report and share what they have learned with the class giving an oral presentation. While a course in Engineering Materials is a common offering in many universities, the authors of this paper present the pedagogical approaches undertaken to successfully implement the course objectives to the undergraduate *Engineering* and *Engineering Technology* students. The topics and techniques applied to teach Engineering Materials to enhance student learning outcomes are addressed in this paper.

INTRODUCTION

The "MENG 353-Industrial Materials" is a required course for the Mechanical Engineering (ME), Mechanical Engineering Technology (MET), Manufacturing Engineering, Technology and Construction Management students at Eastern Washington University (EWU). Although, this is a 300-level class; the pre-requisite of this course is basic Chemistry. Therefore, students enrolled in this class vary from freshmen to junior level. As this is a required course for all engineering and technology majors, it is quite likely to get students with different background and strength to conceptually understand the course elements. For example, it is very easy to convey the concepts on different topics for ME and MET students, however they could be difficult to understand for students in other majors. To facilitate the overall learning of all students, this class consists of both lectures and laboratory experiments, and runs over a single quarter at EWU. The students enrolled in this class get exposed to all of the standard material testing procedures including tensile testing, hardness testing, fatigue testing, heat treating, processes of mounting, polishing and etching to view metal samples under a metallographic microscope. In addition, many of the students have not yet had course work in technical writing. So we have added teaching elements of how to write successful lab reports in this course.

There are many books available to teach Materials Science and Engineering for undergraduate and graduate level students. The text by Askeland, Fulay and Wright - *The Science and Engineering of Materials* [1] is one of many examples. This book attempts to provide relationships between the structure, properties, and processing of materials. The authors of this book provided proper balance in breadth and depth for the different subject matters with appropriate level of rigor, and meaningful examples to simulate the intellectual excitement of the reader. The objective of this text book is to describe the foundations and applications of materials science for college-level students as predicated upon the structure-processing-properties paradigm. The book by Dowling [2] – *Mechanical Behavior of Materials* is another example, and may be used as a text for courses on mechanical behavior of materials at the junior or senior undergraduate level. It may also be employed at the first-year graduate level by emphasizing the later chapters. This book covers traditional topics in the area, such as materials testing, yielding and plasticity, stress-based fatigue analysis, and creep. The relatively new methods of fracture mechanics and strain-based fatigue analysis are also considered and treated in some detail. For a practicing engineer, this book provides an understandable reference source on the topics covered. The text by Smith and Hashemi [3] – *Foundation of Materials Science and Engineering* is designed to provide diverse topics in the field with both appropriate breadth and depth. The strength of this book is its balanced presentation of concepts in science of materials and engineering of materials. The science and engineering concepts are integrated through concise textual explanation, relevant and stimulating imagery, detailed sample problems, electronic supplements, and homework problems. This book seems appropriate for an introductory course in materials at the sophomore level and a more advanced (junior or senior level) second course in materials science and engineering.

The authors of this paper use the text by Budinski [4] – *Engineering Materials – Properties and Selection* for the MENG 353- Industrial Materials course and have used it for many years. As mentioned earlier, the MENG 353 is a required course for all majors – including engineering and technology students. Therefore, this text is found the most appropriate, which covers theory and industry-standard selection practices, providing students with the working knowledge to make an informed selection of materials for engineering applications and to correctly specify materials on drawings and purchasing documents. Encompassing all significant material systems–metals, ceramics, plastics, and composites–this text incorporates the most up-to-date information on material usage and availability, addresses the increasingly global nature of the field, and reflects the suggestions of numerous adopters of previous editions.

COURSE STRUCTURE - LECTURES

The major emphasis of the MENG 353 – Industrial Materials class is to provide a *broad* level of information on different materials used in industries for various engineering applications. We are in a quarter system at EWU and the entire course outlines are delivered to students within 10-weeks-time frame. The instructor starts this class presenting "Mechanical Properties" and their roles in selecting engineering materials in Week #1. The pre-requisite of this class is only "General Chemistry". Therefore, many students enrolled in this class have no knowledge on "Strength of Materials". Even students, who have already taken "Strength of Materials", cannot recall the difference between the "stress" and "strength". For example, stress is a state property and depends on force and geometry. On the other hand, strength is a material property and independent of geometry. Subsequently, this chapter helps our students to achieve an understanding of the difference between "stress" and "strength". Students get familiar with typical mechanical testing and the corresponding stresses due to tension, compression, bending, shear and torsion. This chapter also helps to gain a secure knowledge on tensile testing and its products. Students learn how to use mechanical properties in material selection and failure prevention.

The "Mechanical Properties" seems a very vital portion of this course on which many course objectives are built. Students really need solid understanding of Mechanical Properties to conduct many successive labs, addressed later in this paper. Therefore, the instructor continues conversation on this chapter in Week #2 that includes discussion to differentiate the "toughness" and "strength". We discuss how to do hardness testing using different scales, and how hardness is related to tensile strength. We also cover impact testing to measure impact strength, and how it changes with temperature. The chapter concludes explaining the terminologies of endurance limit or fatigue strength, creep and stress rupture. Students receive a thorough understanding of the differences among properties of stiffness, strength and toughness. In the remaining time of Week #2, we discuss "Chemical and Physical Properties in Engineering Materials", and help students to attain a familiarization with the properties that must be reviewed when making material selections.

When students have adequate knowledge on Material Properties, then we move to the chapter discussing "Forming Engineering Materials from the Elements" in Week #3. We present a review of basic chemistry, organic & inorganic materials, nature of the atom, sub-atomic particles, valence electrons, periodic table and others as appropriate. Students achieve an understanding of how the elements are the building blocks for different engineering materials. Students achieve an understanding of how engineering materials, metals, polymers, ceramics and composites are related in origin and structural characteristics. It is important to let students know that some elements, which are pure substances, behave like engineering materials. For example, gold (Au), silver (Ag), titanium (Ti) and platinum (Pt). On the other hand, some elements are combined together to form an alloy to be used as engineering materials. For example, steel is an alloy of iron (Fe), carbon (C) and other hardening elements. Students are also introduced with different types of crystal structures, e.g., simple cubic, body centered cubic (BCC), body centered tetragonal (BCT), face centered cubic (FCC) and hexagonal close-packed (HCP). Students are also introduced to grain size and grain boundaries. Finally, we explain to our students how to strengthen material by different techniques including solid solution hardening, precipitation hardening, quench hardening and mechanical working. The proper knowledge of different crystal structures, grain size, grain boundaries, and material strengthening techniques are important to conduct many lab experiments associated with this class.

In Week #4, students are introduced to "Steel Products" and "Heat Treatment of Steels". During this time, students achieve an understanding of how steels are made. They gain knowledge of the steel products that are commercially available. They learn different steel terminologies used in industries and steel designations by AISI and SAE procedures. Once they have adequate knowledge of steel products, students receive a brief understanding of the metallurgical reactions that occur in heat treatment. Students get familiar with phase diagrams, interstitial & substitutional solid solutions, and different steel morphologies. They learn a spectrum of heat treatment techniques on ferrous metals, reasons of heat treating, and corresponding heat treatment effects. Students gain a thorough understanding of different heat treatment topics including direct hardening (e.g., quenching), diffusion hardening (e.g., carburizing, nitriding, carbonitriding, boronizing), softening (e.g., annealing) and conditioning (e.g., normalizing, stress relieving). Knowledge gained during this week is utilized later to conduct lab experiments associated with this class.

Students are then introduced to "Carbon and Alloy Steels" in Week #5. They recall the knowledge of identification systems used for steel designation. In addition, they learn different letter prefixes and suffixes in steel identification. They achieve an understanding of the differences among various carbon and alloy steels and the effects of alloying elements. Students come across different special alloy steels, e.g., High-Strength-Low-Alloy Steel, H-Steel and Ultra-High-Strength Steel. During this week, students also achieve an understanding of "Stainless Steel", its metallurgy, basic types, designations and differences. The discussion on stainless steel also covers mechanical and physical properties and their fabrication considerations.

After getting familiar with steel products, carbon steel, alloy steels, stainless steel and heat treatment of steels, students are then introduced to "Corrosion" in Week #6. This chapter helps our students to gain an insight to define corrosion, why it happens to metals and non-metals (e.g., plastics and ceramics), effects of corrosion and its various manifestations. Students receive a thorough understanding of various types of corrosion that occurs in nature and gain knowledge of corrosion-measuring techniques. Students are also introduced to different design guidelines and techniques for corrosion control.

The first six weeks of this class mainly deals with many aspects of metals. Students seem a bit eager to learn something on non-metals. Therefore, students are introduced to "Polymeric Materials" in Week #7. Students achieve an understanding of how polymerized organic materials form engineering materials. They learn basic polymerization processes including addition, condensation and cross-linking. They also achieve an understanding of the chemical makeup of polymeric materials and basic techniques that are used to strengthen polymers. Students become able to define Thermoplastic, Thermoset, Atactic, Isotactic, and Syndiotactic polymers. The structural and melting characteristics of crystalline and amorphous thermoplastic polymers are also introduced to students.

The discussion of polymeric materials continues to "Polymeric Families" in Week #8. During this time, students come across the identification of polymer families and relative usage of thermoplastic and thermoset plastics in the United States. They learn many aspects of commodity and engineering thermoplastics and thermosets based on ethenic and non-ethenic monomers. For example, students learn how to get different commodity engineering thermoplastics from ethylene monomers. Similarly, students learn how to get (or form) different engineering thermoplastics and thermosets which are based on non-ethylene monomers. As a whole, students achieve an understanding of polymer chemistry and of how chemical similarities create polymer families. The overall discussion helps our students to get a feeling for the relative differences between polymer families. Week # 8 ends with a discussion of "Polymer Fabrication". Students learn how thermoplastic and thermoset polymers are shaped into desired parts using different fabrication processes. They achieve an understanding of recycling's importance and recycling codes on plastics.

In Week #9, students are introduced to "Composite Materials". They learn many aspects of composites including definition, composite constituents, RVE, anisotropic behavior of unidirectional or bi-directional composites, forming composite laminate, and their advantages. Students learn widely used fiber and matrix materials, and composites fabrication processes used in industries. Finally, students are introduced to "Micromechanics. They learn the mathematical formulations to determine the effective modulus and strength of composites along the fiber direction. Students also solve numerical problems to strengthen their conceptual understanding.

Students are brought back to metal, and introduced to "Aluminum and its Alloy" in Week #10. Aluminum is the 2nd most widely used metal after steel. Therefore, this chapter seems important for engineering students. This chapter helps our students to achieve an understanding of aluminum alloy systems and alloy designation. Students also gain knowledge of heat treatment of aluminum and alloy selection.

COURSE STRUCTURE – LAB EXPERIMENTS

In our institution, working on a quarter system, this course becomes very aggressive and challenging in the amount of information that needs to be presented to the students within 10 weeks-time frame. To facilitate the overall learning, this class requires all students to perform a series of laboratory experiments. Students are required to perform 8 or 9 labs depending on the allocated time. In Lab #1, students do "Tensile Testing of Aluminum" and determine various mechanical properties. In Lab #2, students perform "Hardness Testing" of 4 different steel samples, and find hardness using Rockwell scale. In Week #3, students are involved with "Heat Treatment" of 4 different steel samples used in Lab #2. Two samples are hardened by quenching and the other two samples are kept in the furnace to be annealed. Lab #4 is the continuation of Lab #3, where students determine the "Hardness of Annealed Samples". Students are asked to answer many questions mentioned in the Lab Handouts. In Lab #5, students are required to "Mount Steel Samples with Bakelite". Students learn about Bakelite – a thermoset plastic discussed in formal lecture classes. Lab #6 is continuation of Lab #5, where students "Polish the Steel Samples" using different polishing materials. In Lab #7, students etch the steel samples and finally observe the "Micro-structures using Metallography". In Lab #8, students perform "Impact Testing" of hot-rolled carbon steels at room and subzero temperatures. In Lab #9, students perform "Fatigue Testing" of aluminum rods.

COURSE STRUCTURE - PRESENTATIONS

The Industrial Materials class is designed to give our students some exposure to different engineering materials. To increase students' exposure to different types of materials and material uses, each student in the class is assigned to do some research on an open topic. Then, students share what they have learned with the class by giving an oral presentation. The course instructor usually provides a list of several topics. Each one relates to a specialized use of some industrial material. Students can select a topic that interests him/her and confirm the selection with the instructor over email. The following topics are only *a few* to mention:

- Plastic Piano Parts
- Structural Honeycomb
- Corvette Fiberglass Body
- Natural Fiber Composites
- Aerospace Composites
- Quiet Steel
- SR-71 Titanium
- Shape Memory Alloys
- Liberty Ship Failures
- Structural Failure of Titanic

Students are also allowed to have a topic that is not on the list but directly relates to the nature of this course.

Students are then required to complete a brief report on his/her topic. Students are asked to write the report as if the audience of his/her paper was the class. Depending on which topic students choose, they are either reporting on an interesting new material, or how a material is being used in a different way, or how different treatments, etc. have made additional uses for a material. The report is limited to 2 pages, double-spaced. In addition, a third page with some sort of picture, graph or chart that a student finds while doing his/her research that class members would find interesting. Also, a fourth page that is a list of the sources he/she used to obtain information. Overall, it provides an opportunity to our students how to write a "Technical/Professional Report"

Finally, students are asked to provide an oral presentation, 8-10 minutes long, in class. The course instructor checks the following item during the students' presentation.

- Professional appearance
- Good use of time
- Created interest in audiences
- Good speaking volume
- Provided useful information
- Provided a source of additional useful information
- Relevance to the course

Hence, our students get an opportunity to provide a "Technical/Professional Presentation". It definitely helps our students to be prepared for similar kind of presentation in their professional career upon graduation.

COURSE EVALUATION AND ASSESSMENT

The students enrolled in the MENG 353 class are asked to evaluate the course *instructor* indicating how well he/she taught this class. The evaluation for the instructor is based on four standard questions, and scaled from 1 to 5, where 1 = Poor and 5 = Excellent. The *average* score of these four basic questions, during the last few years, is shown below. All the categories were above the 4.0 threshold.

- The course as a whole was: 4.45
- The course content was: 4.42
- The instructor's effectiveness in teaching the subject matter was: 4.49
- The instructor overall contribution to the course was: 4.53

The students enrolled in the MENG 353 are assessed using several standard formats to check whether they are learning the intended course materials. The following items are used to determine a final letter grade:

- Homework Assignment: 10%
- Quizzes: 20%
- Exams: 40%
- Class Attendance: 10%
- Lab Submittals: 10%
- Written Report: 5%
- Oral Presentation: 5%

This course does not involve solving many mathematical problems. Therefore, homework for this course consists of reading the assigned chapter or topics that helps for quizzes and tests. Students are assigned homework assignments almost every week, which carries 10% of the overall grade. They are also assigned quizzes and tests every other week. For example, students will get Quiz #1 by the end of week #1, and Test # 1 by the end of week #2. Therefore, students get opportunities to correct & better prepare themselves for a Test if they were found wrong in a Quiz. Similarly, Quiz # 2 is assigned by the end of week # 3, and Test # 2 is assigned by the end of week # 4. Similar practices are conducted for the rest of the quarter. Quizzes and tests are given by Chapter topics, and are administered at the end of the week after the Chapter has been covered in class. Students are given 5 quizzes and 5 tests in total. The quizzes and tests carry 20% and 40% of the overall grade, respectively. Students are also required to attend the class lectures, which carries 10% of the overall grade. Students, who miss the class without having a valid excuse, are penalized. For each unexcused class absence, the overall course point will be deducted by 1.25%. Students with more than 10 class absences will receive an F, irrespective of the course performance. On the other hand, students with perfect attendance receive bonus points that help to improve the final grade. Students are also involved with laboratory experiments every week. Therefore, they are required to submit a formal or informal weekly lab report, which carries 10% of the overall grade. The written report and oral presentation, based on Presentation topic, carries the rest of the 10% of overall grade.

Students are also asked to complete a survey indicating how well they feel the course fulfils different course objectives taught in class. The survey is conducted, in absence of course instructor, in a scale of 1 to 4, where 1 = "Poor" and 4 = "Excellent". Students respond this survey considering all aspects of the course including their performances in tests, quizzes, lab experiments and others. The results of students survey in winter 2013 quarter is shown below. All the categories were above the 3.0 threshold. The overall students' response in most cases falls between the "Good" and "Excellent" manner. The course assessment is also related with the coverage of ABET EAC (or ETAC) criteria. The course assessment of the MENG 353 course during winter 2013 quarter is graphically expressed in Figures 1 and 2.

- Objective # 1: Understand the basic chemistry related to industrial materials. 3.3
- Objective # 2: Explain material properties that are important when making a material selection. 3.6
- Objective # 3: List heat treatments for steels and develop a repertoire of steels for use in design. 3.8
- Objective # 4: List various types of corrosion processes and analyze a given design for corrosion issues. 3.8
- Objective # 5: Understanding various concepts of stainless steels 3.5
- Objective # 6: Discussion of a polymer materials, and list different properties of polymer families. 3.7
- Objective # 7: Describe composite materials and their fabrication processes. 3.3
- Objective # 8: Understanding the basic concept of micromechanics related with composite materials. 3.6
- Objective # 9: Demonstrated laboratory procedures for investigating material properties. 3.7
- Objective # 10: Demonstrated communication skills in assigned written and oral reports. 3.7

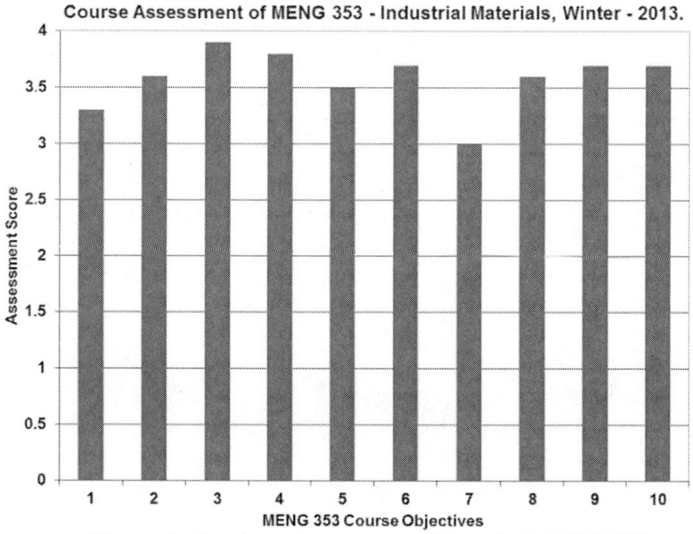

Figure 1: Students Assessment Score in MENG 353, Winter 2013.

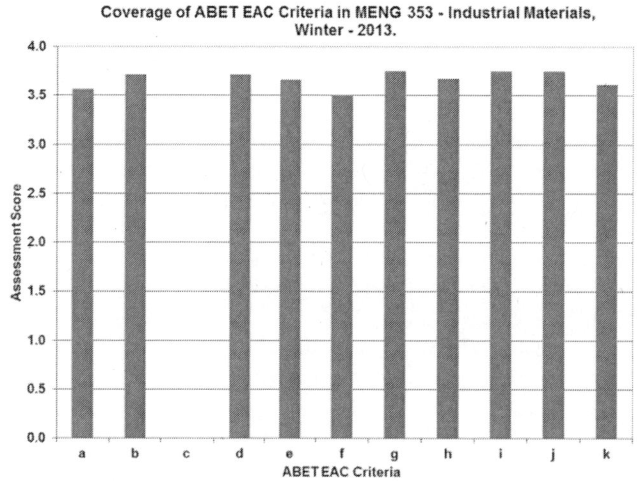

Figure 2: ABET EAC Assessment Score in MENG 353, Winter 2013.

CONCLUSION

The authors have reported different elements of an undergraduate course in Materials Engineering for engineering and technology programs. In summary, this undergraduate course incorporates many useful pedagogical practices, such as homework assignments, quizzes, tests, laboratory experiments and formal presentations that help students to learn the targeted course objectives. The course evaluations, as shown above, confirmed that students learned the intended course outcomes successfully. However, there are a few limitations in this course. For example, this is a required and combined course for **all** majors, including ME, MET, Manufacturing, Technology and Construction Management. Subsequently, it is challenging to meet the expectations of all students. ME students are skilled enough to solve more technical problems, which are mathematically oriented, whereas the MET and others desire to learn something, which is more “hands-on”. Overall, the pedagogical practices outlines in this paper would be good examples to those who are considering teaching Material Engineering course in an undergraduate institution.

REFERENCES

1. Askeland, D. R., Fulay, P. P., and Wright, W. J., “The Science and Engineering of Materials”, Cengage Learning, 6th Edition.
2. Dowling, N. E., “Mechanical Behavior of Materials – Engineering Methods for Deformation, Fracture, and Fatigue”, Pearson, 4th Edition.
3. Smith, W. F. and Hashemi, J., “Foundation of Materials Science and Engineering”, McGraw Hill, 5th Edition.
4. Budinski, K. G. and Budinski, M. K., “Engineering Materials – Properties and Selection, Pearson, 9th Edition.

This page left blank intentionally.

Proceedings of the ASME 2014 International Mechanical Engineering Congress and Exposition
IMECE2014
November 14-20, 2014, Montreal, Quebec, Canada

IMECE2014-36443

IMPLEMENTATION OF MULTI-YEAR PRODUCT INNOVATION PROJECTS

Glenn E. Vallee, Ph.D., P.E.
Western New England University, Department of Mechanical Engineering
1215 Wilbraham Rd., Springfield Massachusetts 01119, USA
(413) 784-1334 gvallee@wne.edu

ABSTRACT

A methodology and case study detailing the implementation of multi-year product innovation projects is presented. A product called the Waterboy, an inexpensive water purification system designed for under developed countries, was developed by three different groups of students over a span of two years. The initial concept was first developed by a six member entrepreneurial team composed of senior level business and engineering students enrolled in a one semester Product Innovation and Development course. This team was responsible assessing the market need, determining product requirements and developing a limited functionality prototype capable of demonstrating the intended product function. A second team consisting of two Mechanical Engineering students continued the project as their one semester Senior Capstone Design project and was charged with the task of developing a fully functional prototype capable of purifying contaminated water. A third student completed the project as a one semester senior level Design Projects course and was charged with the task of modifying the previous design to minimize cost, facilitate ease manufacture and reduced assembly and distribution costs.

In the Fall of 2010, the entrepreneurial team conducted interviews with health professionals and performed research involving a number of world health and philanthropic organizations. They identified the need for an improved water purification device which could purify enough water for a family of four in a reasonable amount of time and at a cost which would make it accessible to people in underdeveloped countries who are at risk of dying from the consumption of contaminated drinking water. They developed a bicycle driven system which used an ultraviolet germicidal bulb to purify water. The team developed a prototype which demonstrated the basic function of the device which was estimated to cost about $80. The project was continued in the Fall of 2011 by the second team of Mechanical Engineering seniors who refined the purification system and function of the device while simplifying the design, resulting in an estimated cost of $49 per unit. The team built and tested a fully functional prototype which confirmed it was capable of reducing water borne bacteria by a factor of 1000. The project was then completed in the Fall of 2012 by a Senior Mechanical Engineering Student who further reduced the cost of the design and improved its portability in order to reduce distribution costs. A partnership with Goodwill Industries was formed to utilize their recycled materials and inexpensive labor force, which reduced the product cost about $24.

INTRODUCTION

The development of new products in the industrial sector typically takes over a year to implement using a staff of full time employees. Many of these companies have significant resources which can be brought to bear in the new product development (NPD) process to help accelerate development. Employees with expertise in several technical disciplines are often involved, including various engineering disciplines, computer and information technology and product testing. Most companies have marketing and sales personnel who can perform in depth consumer research to accurately assess the product need and develop performance requirements which are customer driven. Financial personnel can set pricing and calculate the potential return on investment and many companies have their own fully staffed manufacturing facilities which include QA personnel. These resources are required to be competitive in today's marketplace.

The primary goal of introducing undergraduate students to the NPD process is to give them a working knowledge of the overall tasks undertaken by today's successful companies. Knowledge of engineering, marketing, sales, finance and other business areas is always required. Students who understand and can demonstrate this knowledge are often rated more highly by prospective employers. A second goal is to use the process to actually interface with potential consumers to develop a product which can be manufactured and commercialized, adding much the student's portfolio. The importance of the "voice of the customer" cannot be overstated.

The long term nature of NPD with the aforementioned goals in mind cannot be accomplished in a single semester in academia, nor can it be easily accomplished by a series of individual students who pass the project on to different students from semester to semester. The limited and narrow experience of most students dictates that a more successful approach should utilize teams of students to more quickly advance the development of the product. However, significant obstacles exist when trying to assemble student teams which span multiple semesters. Many engineering and business curricula have little room for multi-year independent project courses and only a small number of electives are typically available, making it difficult to keep a team intact for more than one or two semesters. Students typically do their work, receive a grade and move on. Facilitation of these types of long term, team based projects can in many institutions requires a change in curriculum, particularly in smaller institutions having a limited number of students who are available to pursue these projects. Moreover, once a product is developed to a point where it can be commercialized, many of the participating students have graduated and moved on to their careers. They are not available to take part in the commercialization process.

A number of strategies have been developed in an attempt to address these problems related to the implementation of multi-year student projects. Unfortunately, many are institution specific. For example, Raber and Amato-Henderson [1] describe a methodology for managing large teams of students from various engineering and other STEM disciplines who work for several years in a business-like setting to solve real-world engineering problems supplied by industry. The program is offered as an alternative to the traditional two semester senior capstone design experience offered by most engineering institutions. However, team sizes range from 10 to 70 students due to the limited and varied technical background of team members. Smaller institutions with smaller student enrollment could find these large teams difficult to assemble over several years to typical curriculum constraints which limit the number of electives which can be taken by the students. Bucks, Oaks and Richardson [2] describe a vertical integration approach to multiyear projects which aims to use older engineering students to mentor younger engineering students. Labview tools are used to introduce these younger students to the project though data acquisition and analysis sub-projects. However, significant attention to the business related issues associated with NPD is not always a main requirement for the students.

Vallee [3] discusses a one semester, senior level mechanical engineering course called Product Innovation offered at Western New England University which is designed to provide students with real world exposure to the NPD process. Local companies work with students to develop revolutionary new products and leverage the fact that they typically have no preconceptions in specific product areas. The course culminates with the development of a limited functionality prototype capable of demonstrating product features. Students work with corporate representatives in engineering, marketing, sales and finance, thus gaining a wide range of exposure to those functions which are critical to the NPD process.

This course was expanded in 2009 through a NCIIA grant to include all engineering and business disciplines. This new course, called Product Innovation and Development, is also a one semester course offered only to seniors, but it is open to any business or engineering student. The course was developed to achieve two primary objectives. The first was to give students hands-on insight into the NPD process from concept through commercialization. The second objective was to emphasize the importance of consumer input in all phases of NPD. Students interview potential consumers to help them identify a need, assess the market potential, develop the product and technology to meet the need and fabricate a limited functionality prototype. The course provides a valuable exchange of knowledge between the business and engineering disciplines. Engineering students learn about a number of key business areas such as market analysis, competitive benchmarking, sales forecasting, product absorptions rates, advertising and pricing. Business students learn about the technical constraints involved in the design process, rapid prototyping techniques, manufacturing and quality assurance, as well as ergonomic and safety considerations in the design process. The final prototype is then reviewed by potential consumers to assess its potential for commercial success. Unfortunately, these prototypes are usually not fully functional and cannot be demonstrated to potential investors or companies.

A new approach is therefore needed if these projects are to mature to the point where fully functional prototypes can be consumer tested and be presented to potential investors for commercialization. Due to the relatively small size of the student body at many institutions, a method of continuing the projects started in courses like the Product Innovation and Development course is required to continue the development of the product over several semesters without using large teams of students. The approach must keep students interested in the project while providing meaningful benchmarks and accomplishments for assessment. Students must also receive adequate course credits to make the effort worthwhile, as projects which utilize student volunteers tend to lose momentum as students focus on higher priorities. Grading must reflect the efforts of not only the individual but the team as well so that the effectiveness of the integration of each student's efforts can be assessed. This most accurately reflects how a student will be assessed in industry. Also, an effective method of passing the project to the next student team must be developed to minimize the information lost as the project is passed to the next phase of development. Various student disciplines must be part of the process so students are exposed to areas outside of their comfort zones. Learning the language of another discipline is extremely valuable to the student. It follows that students must gain an appreciation for the entire process and the fact that it takes many different disciplines to develop a successful product. Finally, the issue of IP must be addressed so that all students working on a product and the institution have a stake in the project success.

STRATEGY

The Product Innovation and Development course is used to develop initial concepts. The multidisciplinary teams consisting of four to seven students are assembled from a wide variety of backgrounds including engineering, business, marketing and finance. The initial development of the consumer need and initial product concept can benefit most by teams of students of varying disciplines who can interface with potential consumers and assess the market and market needs. Teams of this size can effectively canvas large numbers of potential consumers while other team members perform benchmarking studies, bench mark testing and an assessment of the current technologies. The product concept can then be converted into a limited functionality prototype which can be reviewed by potential consumers, thus verifying the potential for success in the marketplace.

Once the product concept is well defined and the potential for commercial success is be evaluated, smaller teams of students can be used to focus on the more technical aspects of the product where the focus can be places more on performance and manufacture. The next stage of development therefore focusses on the technical development of the product since the marketing plan has already been defined. The one semester Senior Capstone Design Project platform is used for further development. Here, a team of two engineering students divide the technical problem into two areas, with each student focusing on their own part and the integration of the two areas. Students must develop solutions to problems encountered with the original prototype, design and a fully functional prototype capable of demonstrating all product functions, and thoroughly test the prototype. Refined product cost estimates must also be developed.

The final development is achieved as a single semester Senior Design Project, where a single student refines the concept for ease of manufacture and further cost reduction. Detailed manufacturing and assembly process plans must also be developed in conjunction with a potential manufacturer. This brings the project to the stage where it can be potentially commercialized.

EXECUTION

In order to illustrate the strategy discussed above, a case study involving the development of a human powered water purification system called the Waterboy is discussed as it is developed through a multiyear development process using a total of seven students over the course of 2 years.

A. Initial Concept Development

Product Innovation and Development Course

The Product Development and Innovation course is a 3 credit course that is open to all College of Business and College of Engineering undergraduate seniors. This course can be applied to business degrees as a business elective and to engineering degrees as an engineering elective. Mechanical Engineering students may also apply this course as a design elective. This course discusses new product innovation from both an entrepreneurship and intrapreneurship perspective.

The course has three components: a theoretical component, a practical or clinical component, and an application component. The theoretical component consists of lectures on generating and identifying business opportunities (customer needs), assessing concept ideas from technical, market and financial perspectives; designing and developing new products; testing prototypes from technical and market (meets customer needs) perspectives; and developing a marketing plan including launch, monitoring, and measurement provisions. The practical or clinical component consists of business-engineering student teams called E-Teams. These teams identify new product ideas of their own or select from a faculty-generated list of ideas, some of which are sponsored by local companies. These ideas are chosen on the basis of the importance of the need, the novelty, the challenge, and commercialization potential. Once the E-Teams choose their ideas they develop product concepts, final designs and limited functionality prototypes for their products. In addition, they develop marketing plans to market their new product designs. The application component involves presenting the final designs and plans to an expert panel of business executives, investors, and faculty. This panel of experts evaluates each of the ideas and rates them according to their potential for commercialization. It is the hope of the faculty that some of the ideas will have sufficient merit to warrant the team applying for a provisional patent to protect their intellectual property, and starting a business to commercialize their ideas upon graduation.

The course meets twice weekly for 75 minutes. Two engineering and one business faculty member serve as project managers and serve the same rolls as managers in industry. Each faculty manages two teams. A number of guest speakers from industry are invited to lecture about a number of topics including patent preparation and their own entrepreneurial experiences. A panel of entrepreneurs and venture capitalists are also used to judge a 90 second elevator pitch developed by each team to provide students with meaningful feedback on their product concepts. The students end the course by giving a final presentation which details how the product was conceived from consumer input, the marketing and competitive landscape analyses, the engineering behind the prototype, the technical and consumer testing of the prototype. Representatives from industry are invited to attend these presentations, as are venture capitalists and other members of the local business community.

Thu far, seven provisional patents and three utility patents have been filed on behalf of the students inventions developed in the course. One utility patent has also been issued [4]. In addition, four student inventions developed in the course have been entered in the ASME Innovation Showcase competition and each has earned a place in the top ten most innovative products while competing against some of the largest universities in the country. Additionally, one product line developed in the course involving drink coasters and picture frames manufactured using recycled vinyl records is now in production by Goodwill Industries of Springfield, MA. Currently, students assign their rights to the intellectual property over to Western New England University. If the product is commercialized, the university distributes up to 50% of any royalties collected back to the students.

Students review the course very positively and comment that is one of the few courses having direct real world applications which can immediately help them in their careers. Students also comment on the importance of learning the language of other disciplines that they will be exposed to in the workforce.

Waterboy Concept

Students enrolled in the Product Innovation and development course in the Fall of 2010 were divided into teams consisting of three engineering students and at least one business student. Six such teams were assembled, with two teams being assigned to corporate sponsored projects as intrapreneurial teams, and the remaining four teams as entrepreneurial teams. These entrepreneurial teams may focus on any area they choose, and one such team decided to address some global problem affecting human welfare. They called themselves team "Pure Innovations" and began researching global health issues and quickly determined that about 1.1 billion people in developing countries lack access to clean water, resulting in the deaths of over 1.8 million children annually. The team determined using NationsOnline.org, a web site dedicated to identifying underdeveloped nations in need of basic human nutrition and sanitation, that the regions with the greatest needs included Sub-Saharan Africa, Southern Asia, Eastern Asia, the Middle East and North Africa. The team also discovered that approximately 19 liters of water per day is needed to sustain a family of 4 people in these countries.

The team then conducted an analysis of the competitive landscape. They researched any product capable of purifying water without the need for fossil fuels or electrical power. They found a number of a portable solar stills which slowly purified water using sunlight. These devices are capable of purifying 19 liters of water in 10 days at a cost of about $50 per unit. These devices are entirely dependent on direct sunlight and contain no sediment filtration capability. Another potential competitor involves porous ceramic bowls mixed with activated carbon from charcoal or coconut shells with a filtration range 0.5-50 microns. These device costs $10-12 and can purify 19 liters in 3 to 4 hours. However, they are typically single use devices which must be discarded after use. The team also found portable ultraviolet (UV) purification devices powered by a hand crank. They are capable of purifying 19 liters in 30 minutes but about $95 and cannot filter sediments.

Each team presented their focal area to their project manager for approval. The very large potential market and limited competition in the affordable water purification area resulted in quick approval of this area. The team then discussed the problem of providing clean drinking water to underdeveloped nations with health professionals from Engineering World Health and Engineers Without Boarders, two organizations committed to helping under developed countries. They then formulated the performance criteria which would be used to evaluate potential product concepts. They found that the market demands an inexpensive device (under $30) that operates using no electrical or other fuel source which can be easily transported to the location of the need. The device must also be capable of killing harmful microbes using a proven technology while providing filtration to remove sediment typically found in most water sources. The device must be capable of purifying about 19 liters of water in about 10 minutes if human powered to prevent fatigue and dehydration of the operator in warm and humid climates. Finally, the device must be easily transported and be capable of being assembled and preferably manufactured on site to reduce distribution costs.

The team developed several potential product concepts based on the aforementioned performance criteria. Two devices which used a hand crank to pump water through a series of carbon filters were considered, as was a bicycle powered system which energized a UV germicidal light bulb which could be submerged in a bucket of contaminated water. Two purely solar powered devices were also considered. A screening analysis was then performed by the team to evaluate each product concept based on the performance criteria. The effectiveness of sanitation was weighted most heavily in the screening analysis, followed by cost, ease of manufacture and purification speed.

No single product was found to be superior in all areas when compared to some of the inexpensive solar powered product already on the market. Consequently, the team combined several of their product concepts to arrive at a product concept the team named the "Waterboy", shown below in figure 1.

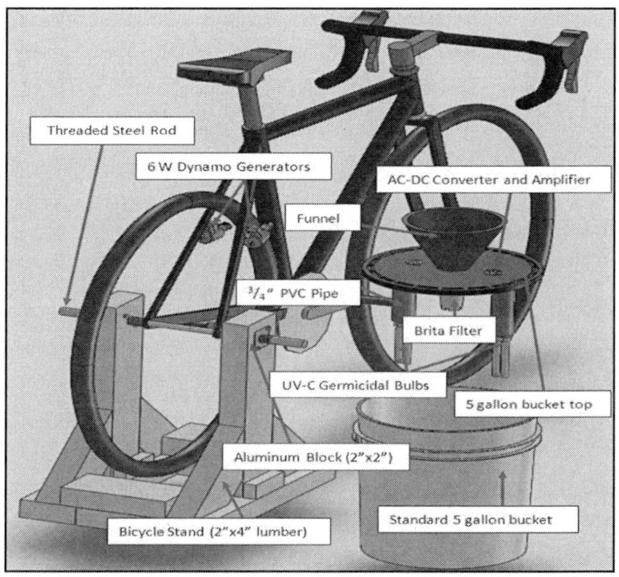

FIGURE 1 – SCHEMATIC OF THE WATERBOY WATER PURIFICATION DEVICE

This device uses a commonly available bicycle to power two 7 watt UV germicidal light bulbs. First, the user must mount the bicycle on the stand. This is accomplished by removing the nuts on the rear axle. Next, the threaded rods are tightened around the axle and compress against the frame of the bicycle. The two dynamo generators must now be attached to the frame of the bicycle using simple brackets tightened using thumb screws . They must be attached so that the dynamo will spin against the sidewall of the tire. Next, the operator must pour water through the funnel where an activated carbon filter removes toxic metals and particles. This step is necessary to ensure the UV rays travel to the microbes and are not blocked by sediments. Once the desired amount of water (not to exceed 19 liters) is poured through the filter, the user may start to pedal the bicycle. The rotation of the rear wheel energizes the two dynamo generators, each powering a UV-C germicidal light bulb using an AC/DC converter and amplifier. The user must pedal at approximately 60 rpm (one pedal rotation per second) to generate enough current to power the bulbs. The user must pedal continuously for 7 to 8 minutes to ensure adequate sanitation. The user must also be knowledgeable of the useful life of the bulb and the filter. The filter must be replaced every 150 liters while the bulbs should be replaced annually. The bulbs must also be cleaned daily to ensure that the UV rays travel through the water.

Prototype Development and Testing

This product was designed and built by the student team, and the finished Waterboy prototype is shown in figure 2. The bicycle stand was fabricated using inexpensive 2 x 3 spruce lumber. The prototype cost using purchased components was approximately \$80. The most expensive components which accounted for over 30% of the total cost were the two germicidal bulbs and the Brita filter. The team estimated that purchasing components in bulk could reduce the product cost to under \$40, although potential unit sales quantities were difficult to estimate due to the unpredictable financial nature of the target governments.

FIGURE 2 – WATERBOY PROTOTYPE

The product was tested by pouring water through the funnel, filling the bucket and pedaling the bicycle to ensure that the germicidal bulbs were energized at a pedaling speed of 60 rpm. While water successfully flowed through the Brita filter and the bulbs did energize, there was not adequate time remaining in the semester to perform any testing to quantify the effectiveness of the sanitation process. This would require the use of water contaminated with a predetermined concentration of a known microbe.

Business Model

An important part of the Product Innovation and Development course is the creation of a business model which can be used to commercialize the product. The team proposed a two pronged approach, the first of which involves contacting the government officials of underdeveloped countries selected from the aforementioned list of nations in most need of purified water for their people. They proposed that in exchange for purchasing a predetermined number of Waterboy systems at a retail price of about \$50.00 per unit, the team would license the rights to the design and work with these governments to set up a manufacturing facility to employ their people to assemble the Waterboy units. The assembly costs in these countries was estimated at about \$2.00 per unit.

The second approach proposed by the team involves developing a relationship with Water Missions International (WMI), a nonprofit, Christian engineering organization serving the water and sanitation needs of people in developing countries and disaster areas. WMI would purchase and distribute the Waterboy units to the nation's people free of charge. WMI would be responsible for on-site delivery as well as on-site education of the need for pure water and demonstration of the proper use and care of the Waterboy, unit. The team would contract the manufacture of the product to on site manufacturers in the host countries and work with them to product the product. Both approaches involve a profit margin for team "Pure Innovations" of about 15% of sales.

All teams enrolled in the Product Innovation and Development course are required to acquire consumer feedback on their prototypes. While it was impossible to recruit potential consumers in under developed countries, the team asked 10 people selected at random to use the Waterboy and assess its perceived performance. Five men and five women were asked to use the device and rate the following categories on a scale of 1 to 10, with 10 being the most favorable response. The results are shown in table 1.

Overall, those surveyed did not feel very comfortable riding the bike on the stand, although they did feel that the stand was stable and safe to use. The majority of those surveyed also felt the device was easy to pedal, even for the required 10 minute sanitation period. However, only 4 of those surveyed felt that such a device would ultimately produce safe drinking water. This highlighted the importance of testing the product for effectiveness and informing potential consumers of the positive results.

No.	M/F	Comfort	Safety	Ease of Pedaling	Ease After 10 Min	Drink Water?
1	M	7	8	8	6	No
2	F	8	8	7	7	Yes
3	M	4	9	7	7	No
4	F	5	7	8	8	No
5	M	5	7	7	6	Yes
6	F	4	3	4	3	No
7	M	6	7	6	6	Yes
8	F	5	6	6	7	No
9	F	4	6	7	7	No
10	M	5	6	8	8	Yes
avg	--	5.3	6.7	6.8	6.5	-

TABLE 1 – RESULTS OF USER SURVEY

The Waterboy was entered in the 2011 ASME Innovation Showcase competition held in Pittsburg, PA, and the product placed in the top 10 in the competition, indicating the promise of the device. Many people attending the competition commented that they were impressed with the prototype which, unlike the other teams in the competition, was developed by a team of undergraduate students without any external funding in only 15 weeks. A provisional patent for the device was filed in March of 2011.

B. Development of a Fully Functional Prototype

While the Waterboy which was developed in the 15 week Product Innovation and Development course showed great promise in addressing the problem of a lack of clean drinking water in underdeveloped countries, the device had a number of deficiencies. Most importantly, the device was never tested to quantify the reduction in microbe population. The concept of submerging two germicidal bulbs in a bucket of water without any agitation meant that water in the bucket located farthest from the bulbs experienced the least amount of UV radiation and might not be adequately sanitized in 10 minutes of pedaling. The method of sediment filtration was crude and utilized gravity feed flow through a Brita filter which increased product cost. The wooden bike stand was bulky and cumbersome to ship to underdeveloped countries and was not adjustable for different sized bikes. Also, the product at this point was estimated to cost over $40 per unit even if manufactured in large quantities.

Senior Capstone Design Project Platform

Since the market analysis and business plan had now been developed, the next step toward commercialization involved the development of a fully functional prototype which could be tested for effectiveness. It was therefore decided that this project would be offered as a one semester, 3 credit integrated Senior Capstone Design project. In this course, two mechanical engineering students were charged with continuing the technical development of the Waterboy with the goal of improving the effectiveness of the filtration and UV sanitation, quantifying the effectiveness of the device and reducing the product cost by at least 25%.

Each student was responsible for their own specific deliverables, as well as the integration of these deliverables into the final product. One student focused on redesigning the bike stand for easier assembly, adjustability and improved rigidity, as well as lower cost. The stand would also serve as a mount for all components in the sanitation system. The second student focused on the UV sanitation and filtration systems, as well as quantifying sanitation effectiveness. Each student kept their own individual project log books but all presentations and the final report were developed by both. The course culminated in a final oral presentation given by both students. However, each was graded in part on their own work and areas of responsibility. The author served as project manager and met with the students together for about one hour per week.

Waterboy II Development

The team was charged with the task of developing a fully functional prototype which could be manufactured for about $30. The system must more precisely control the exposure of the contaminated water to the UV radiation than the previous prototype and the actual reduction in contaminating microbes must be precisely determined. The device must accommodate a wide range of bicycle sizes and must be more easily manufactured and distributed.

1. Redesign of the UV Sanitation system

In order to more precisely control the exposure of contaminated water to the UV radiation, the student developed the system shown in figure 3.

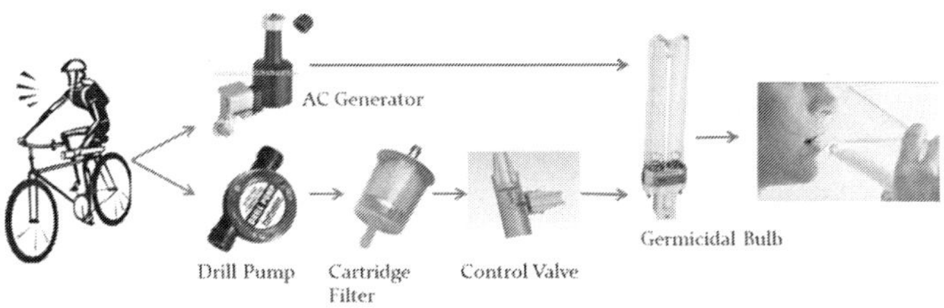

FIGURE 3 – SCHEMATIC OF THE IMPROVED UV SANITATION SYSTEM

The bicycle is used to drive both AC generator used to generate electricity, and a drill pump used to pump water through the system. An inexpensive cartridge filter is used to remove sediment and a control valve was designed to prevent contaminated water from passing through the system if the rear wheel of the bicycle is not spinning at a velocity great enough to adequately energize the UV germicidal bulb for proper sanitation. A housing was designed to enclose the bulb and minimize the clearance between the bulb and water.

Bulb Housing Design

A polymeric housing shown in figure 4 was designed to minimize the maximum distance allowed between the UV bulb and the contaminated water flowing around it. Equation (1) [5].was used to determine the ultraviolet radiation intensity emitted from a source at any given radial distance. For purposes of this project, a theoretical analysis of a 7 watt germicidal bulb was performed using equation (1).

$$I = \frac{P}{4\pi r^2} * e^{-Ar} \quad (1)$$

Where: I = UV intensity at distance r
r = Radial distance from bulb
A = Absorbance of the water based on water quality
P= Power emitted from bulb (watts)

Plots of intensity vs. distance showed that a significant decrease in intensity was seen at a radial distance above 10 mm, so the allowable clearance was set at 9mm. The purification time was then calculated using equation (2) as:

$$Purification\ Time = \frac{Required\ UV\ Dosage}{Intensity\ at\ Purification\ Distance} \quad (2)$$

where the required dosage of UV radiation required to deactivate 99.9% of the most common microbes found in contaminating drinking water is about 100 Ws/mm^2 as found in [6].

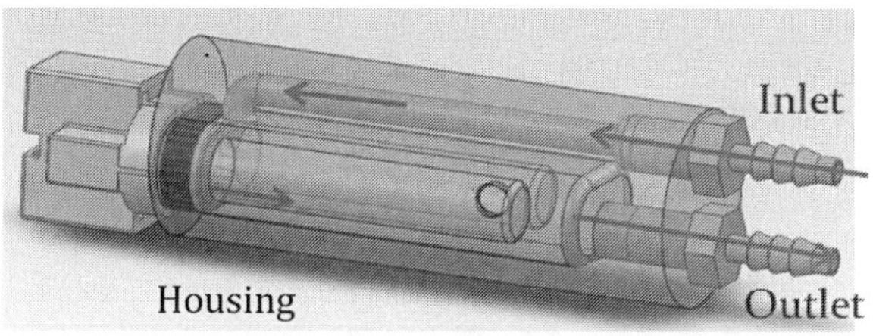

FIGURE 4 – UV BULB HOUSING

The allowable flow rate around the bulb was calculated to ensure the minimum required purification time found using equation (2) for the 7 Watt bulb which produces 100 W/mm^2 at a distance of 9 mm. The result was a flow rate of 2.5 liters per minute, which equates to a purification time of about 8 minutes for 19 liters of water. This was well within the desired requirements and the capability of the drill pump.

Power Supply

A schematic of the method of electrical power generation is shown in figure 5. An inexpensive 6W AC generator was connected to a transformer to boost the voltage to the UV bulb.

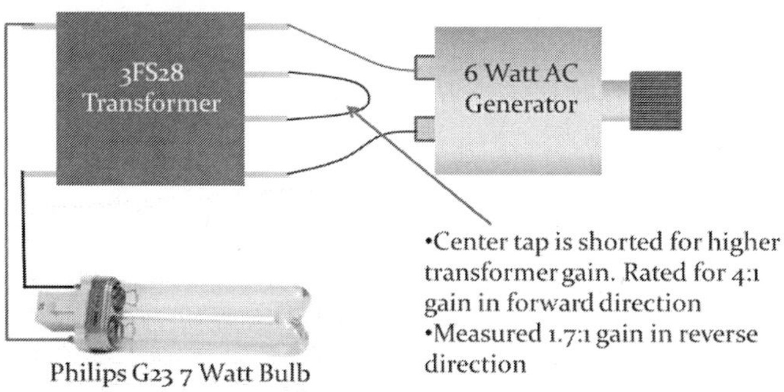

FIGURE 5 – SCHEMATIC OF THE METHOD OF ELECTRICAL POWER GENERATION

Control Valve

A critical component of the improved sanitation system was the flow control valve shown in figure 6. The water flow must be cut off when the bicycle is not pedaled fast enough to illuminate the bulb at the intensity required for sterilization. Two elastomeric umbrella valves having different stiffnesses were used to control the flow through the system. If the flow through the main valve does not develop enough pressure to open it, the water passes through a bleeder valve and back into the source bucket. The umbrella valves were selected such that the flow was bled away from the sanitation system if the drill pump speed fell below 1200 rpm, or a pedaling speed of 60 revolutions per second.

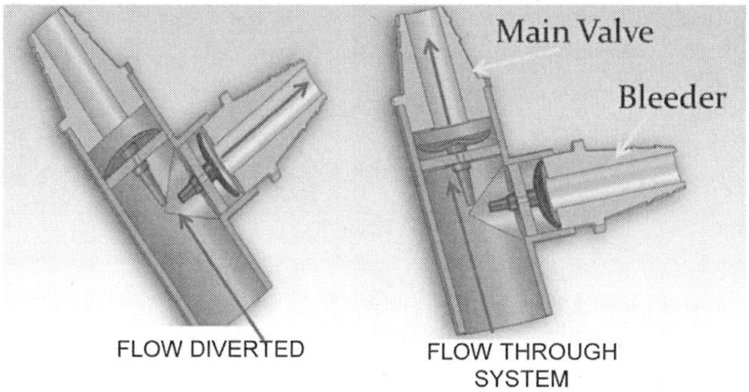

FIGURE 6 – SECTION VIEW OF THE FLOW CONTROL VALVE

2. Redesign of the Stand

The goals of the stand redesign were to reduce manufacturing cost while providing a stable mount for all components of the UV sanitation system. The stand must accommodate a wide variety of bicycle sizes and assembly must be quick and user friendly. The design is shown in figure 7.

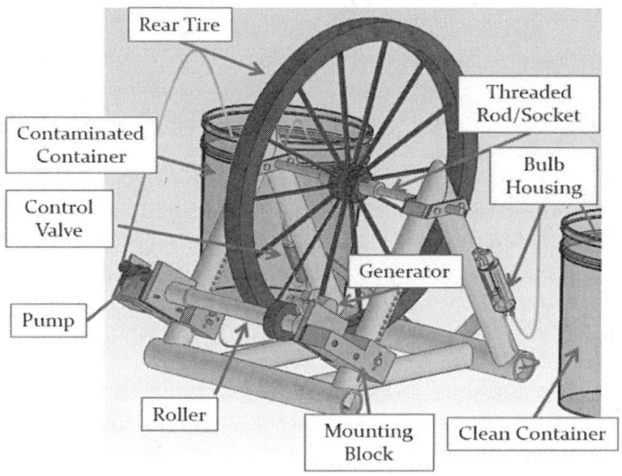

FIGURE 7 – SCHEMATIC OF THE WATERBOY II

An inexpensive stand made from commonly available 1 ½" PVC pipe was designed to support the bicycle. The structure was analyzed using finite element analysis to confirm that it would be stiff enough to support the bicycle and rider during operation. A bicycle mount was then designed so the bicycle could be quickly fastened to the stand without the use of tools. These components were made of steel for rigidity. A number of evenly spaced holes were drilled into the side members of the stand at the rear so the mounting block containing a roller assembly made from PVC pipe could be adjusted to contact various wheel diameters.

3. *Integration and Testing*

The generator and drill pump were mounted to the roller discussed above which was then mounted to the PVC stand using the wooden mounting block. The roller assembly could be moved up and down using the evenly spaced holes in the stand, allowing tire sizes ranging from 380 to 810 mm in diameter to be accommodated. Both the control valve and bulb housing were mounted to the front of the stand as shown in figure 8.

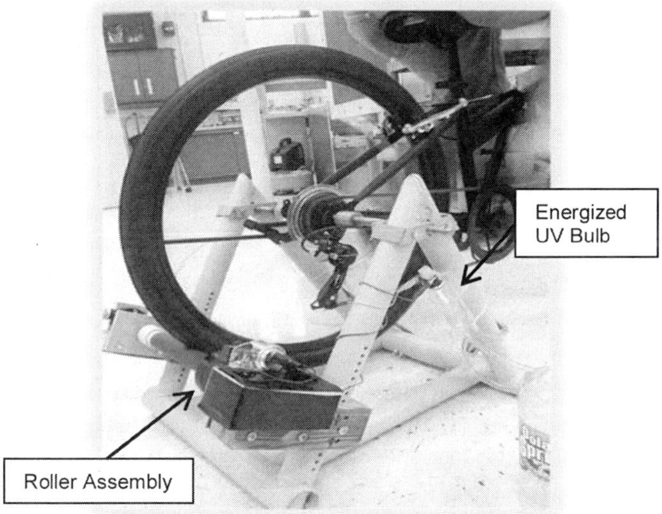

FIGURE 8 – THE WATERBOY II PROTOTYPE

The Waterboy II is operated for testing as follows. The rider mounts the bicycle on the stand by tightening the sockets against the nuts on the wheel hub. This is accomplished by simply tightening the threaded rods. The roller assembly is then raised until it contacts the rear tire and the mounting blocks are pinned in place using the appropriate predrilled mounting holes in the PVC stand. The inlet tube is then placed in the contaminated water container and the outlet tube placed in clean water container. The rider then pedals bike at approx. 8 mph for 8 minutes. The rotating tire spins the roller which in turn spins dynamo to energize the UV bulb and water pump to initiate flow. Figure 8 shows the actual prototype in operation during testing.

A sample of water and silt was heavily contaminated with a known concentration of e-coli bacteria. The sample was passed through the Waterboy II device and analyzed by the Biology department at Western New England University. Using a most probable number analysis, it was determined that the treated sample showed a 1000 fold reduction in microbe counts, making the concentration below the safety thresholds for drinking water.

4. *Improvements in New Design*

The Waterboy II prototype cost $47 to manufacture. However, the team determined that a number of organizations actually pay to have PVC pipe removed from their premises. The team decided to leverage this problem and use recycled PVC in the design. This resulted in an estimated cost of materials of about $35, a reduction of nearly 40%, bringing the cost within the target requirements. A utility patent was filed in 2012 for the Waterboy II device and is expected to be issued in 2014.

C. Refinement for Manufacture

While the Waterboy II represents a significant improvement over the original design, it was still not well suited for manufacture or shipment overseas. The PVC bicycle stand was still rather large and tended to deflect as the rider operated the bicycle. It was also uncertain if the device could in fact be manufactured for the estimated cost of $35 due to the complexity of the UV bulb enclosure and flow control valve. It was therefore deemed necessary to continue to improve the design before it could be commercialized.

The course called Design Projects was therefore used as a platform to continue improvements. This 3 credit course, typically conducted in the fall semester to prevent overlap with the Senior Capstone Design projects, allows a student to undertake a design project which is executed much like the Senior Capstone Design project discussed earlier. The project was proposed to the Senior ME class, and a single student agreed to work on the project in the Fall of 2012.

The goal of this final project was to refine the design to both lower cost and correct the deficiencies of the Waterboy II design. A key factor in reducing cost was a partnership developed between Western New England University and Goodwill Industries of Springfield, MA. Goodwill Industries is a non-profit organization founded in 1902 which provides job training and employment to the disabled and people who are otherwise unemployable. They are also a major collector of recycled goods which are donated by the general public. It was determine that Goodwill could provide a work force capable of manufacturing the device at a greatly reduced cost and could also provide a wide variety of recycled materials which could be used in its manufacture. Production of the Waterboy III would also result in the creation of jobs for the employees of Goodwill.

1. Redesigned Bicycle Stand

Goodwill collects a very large number of aluminum walkers which are provided to people recovering from surgery and the elderly who need assistance walking. These walkers are discarded when newer ones are provided, and many are donated to the Goodwill facility. Few are sold back to the public by the Goodwill retail stores, resulting in a stockpile of walkers which Goodwill typically pays to discard. These walkers are made of lightweight aluminum and resist corrosion. They are durable and foldable, allowing for much less expensive shipment to under developed countries and are preassembled. It was therefore desired to modify the stand design to incorporate this resource.

A bracket fabricated out of corrosion resistant aluminum was designed to mount the roller assembly containing the water pump and generator to the walker, as shown in figure 9. The design may be modified to incorporate scrap aluminum received by Goodwill. A bracket is fastened to both sides of the walker frame using two bolts. The rear wheel of the bicycle fits easily into a receiving groove in the inboard side of the bracket and may be secured using a plate fastened to the bracket using thumb screws.

FIGURE 9 – WHEEL MOUNT BRACKET USED FOR ATTACHMENT TO WALKER

2. *Redesigned Control Valve and Bulb Housing*

The design of the bulb housing and control valve were also modified to allow the use of inexpensive PVC fittings to further reduce cost. The redesigned bulb housing and control valve are shown in figure 10.

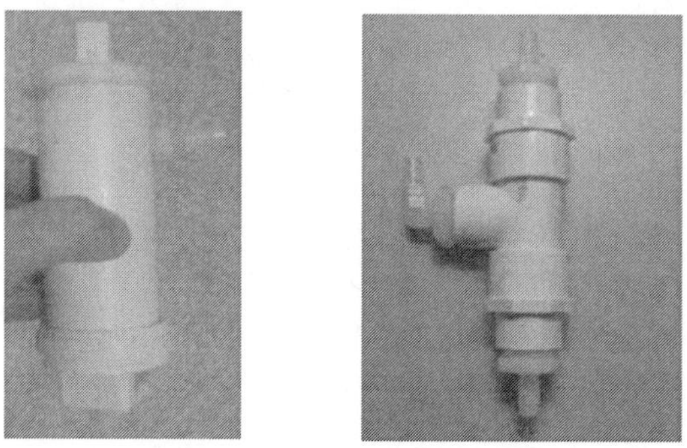

BULB HOUSING FLOW CONTROL VALVE

FIGURE 10 – MODIFIED BULB HOUSING AND FLOW CONTROL VALVE

An aluminum mount was also designed to fasten the roller assembly to the opposite side of the walker as shown in figure 11. This mount can be slid and fastened at various heights in order to accommodate wheels of varying diameters. A commercially available and inexpensive grip tape was wrapped around the roller to improve contact between the tire and roller.

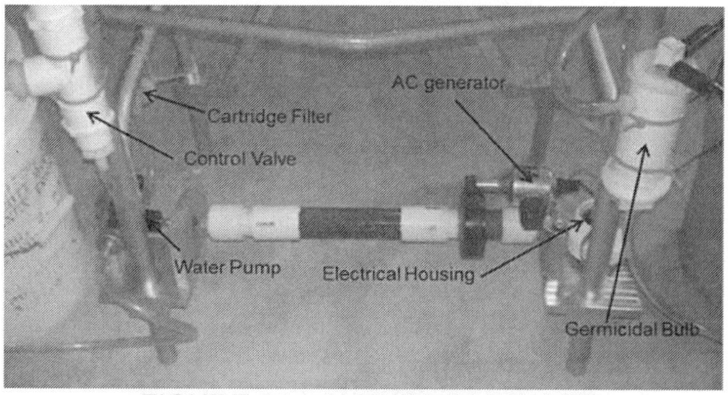

FIGURE 11 – ALUNIMUM ROLLER ASSEMBLY MOUNT

The aforementioned modifications to the design resulted in an estimated cost of $24, below the target price of $25.

3. *Testing*

The Waterboy III was tested to determine performance. The device was far more stable than the previous design due to the rigid aluminum walker frame. Contaminant testing revealed the same reduction in microorganisms as the previous design.

4. *Manufacturing Process Plan*

The final part of the project involved the development of a detailed process plan for the manufacture of the device. Such a plan was developed for each component. Figure 12 shows the final assembly plan for manufacture of the device.

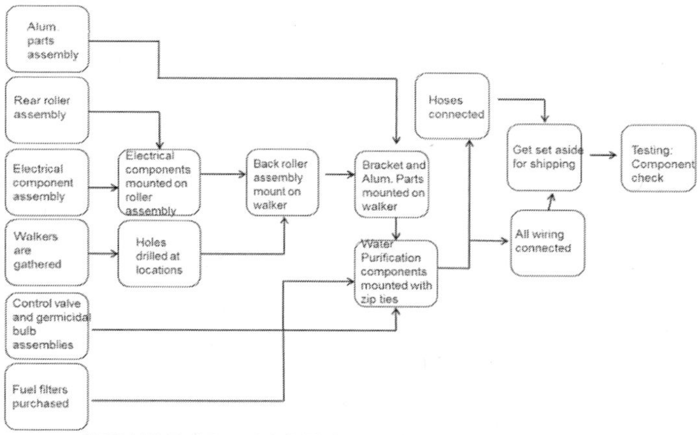

FIGURE 12 – ASSEMBLY PLAN FOR THE WATERBOY III

NEXT STEPS

Western New England University is currently in the process of seeking investors willing to support the manufacture and distribution of multiple Waterboy III units for field testing. Successful field testing is required to demonstrate that users from these under developed countries can successfully use and maintain the device. A low cost kit which can be used on site to verify that water processed using the Waterboy III is in fact safe to drink is also being pursued. Once field testing is complete, the process of developing a relationship with the governments of underdeveloped countries and organizations like Water Missions International will begin.

CONCLUSIONS

The methodology presented for the development of the Waterboy water purification device demonstrated how a product can be developed using several groups of students over a span of several years. The Product Innovation and Development course platform which teamed business and engineering students was used to successfully develop a product concept and underlying technology through a market and competitive landscape analysis. The Capstone Senior Design Project platform was then used to refine the device, improve its performance and develop a fully functional prototype which verified the effective of the device. The Senior Design Project platform was then used to refine the product for ease of manufacture, lower cost and developed a detailed process plan for fabrication and assembly. This methodology is particularly well suited to a smaller institution where curricular restrictions and a relatively small pool of students limit the number students who are available to work on the project.

The methodology discussed in this work was received very positively by both students and faculty. Students in the Product Innovation and Development course commented that the knowledge of the product development process acquired in the course would be very important to their careers and that this course was more valuable than others because of its direct applicability to the real industrial world. They also learned valuable lessons as they worked through the difficulties experienced when working with students from other disciplines. They found that working on a product that could eventually reach the commercialization stage provided an additional motivation and sense of pride that they had not experienced in other courses. The seniors who continued the project commented that they felt a sense of ownership and responsibility for providing the market with an effective and reliable new product.

Faculty from the Colleges of Business, Engineering and Law who were familiar with the project were impressed with the progress made by the students over time. They recognized that this was the first product which was developed by different teams of students over several semesters that was actually ready for commercialization. Faculty also pointed out that the students appeared to be excited about the product and were anxious to demonstrate its potential for saving lives in underdeveloped nations.

ACKNOWLEDGEMENTS

The author would like to acknowledge the students from Western New England University who worked on developing the Waterboy: Micah Bowen, Joseph Conti, Amanda Fortus, Andrew Labrie, Mike Grinaski, Mike McCormack and Brian Hamel. The author would also like to acknowledge Western New England University for their financial support throughout the course of this project.

NOMENCLATURE

I - ultraviolet intensity at distance r
r - radial distance from the UV bulb
A - absorbance of the water based on water quality
P - power emitted from bulb (watts)

REFERENCES

[1] Raber, M., Amato-Henderson, S. L., 2012, "Assessing the impact of faculty teaching, advising and mentoring in an alternative multi-year, interdisciplinary capstone design program", 119th ASEE Annual Conference and Exposition; American Society for Engineering Education.

[2] Bucks, G., Oakes, W., Richardson, J., 2009, "Facilitating vertically integrated design teams", 2009 ASEE Annual Conference and Exposition, American Society for Engineering Education.

[3] Vallee, G. E., 2007, "A Novel course in product innovation", 2007 ASME International Congress and Exposition, American Society of Mechanical Engineers.

[4] Vallee, et. al., April 8, 2014, Portable Drill Bit Sharpener, United States Patent No. US 8,690,643 B2.

[5] Snicer, Gregory A.. 2000, "UV inactivation of viruses in natural waters", AWWA Research Foundation and American Water Works Association, Denver, Colorado.

[6] Johnson, K.M., Anil Kumar, P Ponmurugan, and Mythili Gananamangai, 2010, "Ultraviolet Radiation and its germicidal effect in drinking water Purification", Journal of Phytology **2.5**, pp. 12-19.

Proceedings of the ASME 2014 International Mechanical Engineering Congress and Exposition
IMECE2014
November 14-20, 2014, Montreal, Quebec, Canada

IMECE2014-36471

A RULE BASED METHOD TO CONSTRUCT THE MOHR'S CIRCLE FOR PLANE STRESS

Swami Karunamoorthy
Department of Mechanical
Engineering & Materials Science
Washington University in St.Louis
Saint Louis, Missouri

ABSTRACT

Mohr's Circle is a graphical representation of the state of plane stress at a point. It is an important concept to learn for the understanding of plane stress transformation. However, it remains a challenge for both engineering as well as engineering technology students to understand the principles of Mohr's Circle. In this paper, an emphasis is given to identify the source of pedagogical difficulty and a simple rule based method is presented as a new approach for effective teaching as well as learning. It is a viable method to understand the Mohr's Circle and its application to plane stress transformation.

INTRODUCTION

The topic on Plane Stress or Stress Transformation is introduced in a course on Mechanics of Solids or Mechanics of Materials or Strength of Materials at junior level for the undergraduate students in engineering or technology programs. The sign convention for the normal stress is well understood that tensile stress is positive and compressive stress is negative. The shear stress on an element is described as positive if the right diagonal stretches as shown in Figure 1(a).

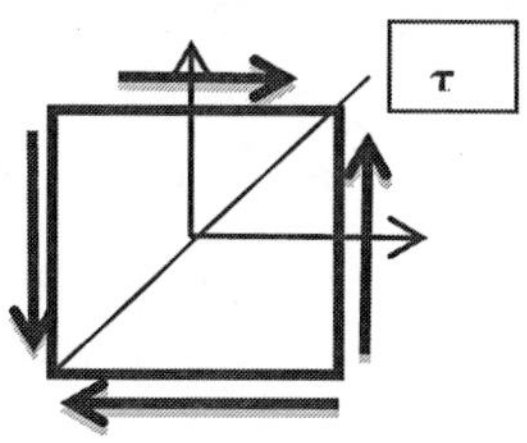

Figure 1(a): Positive shear stress on an element

Similarly, if the left diagonal stretches as shown in Figure 1(b), the shear stress on an element is considered as negative.

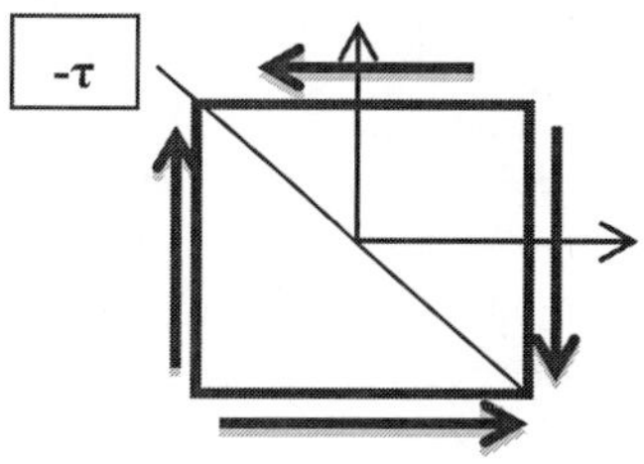

Figure 1(b): Negative shear stress on an element

However, additional sign conventions are required to draw the Mohr's Circle [1] and they are not unique. Different text books adopt different sign conventions for the shear on each plane (or face) of the element. Several text books are used as comparison for this purpose.

NOMENCLATURE

A_1 – Top plane on the element
B_1 – Adjacent plane (90° counterclockwise to A_1)
A_2 – Bottom plane on the element
B_2 – Adjacent plane (90° counterclockwise to A_2)
X-Y Reference axes
X_1-Y_1 New position of reference axes after rotation
σ – Normal stress
σ_1, σ_2 – Principal stresses
τ – Shear stress
Θ - Arbitrary angle of rotation
Θp – Principal angle

ANOMALY AMONG TEXT BOOKS

In Reference [1], the author states, "To employ Mohr's circle, we must set forth additional sign conventions. A shear stress will be taken as positive on a face (A_1) of an element in regard to Mohr's circle when it yields a clockwise moment about the center point at the element. A shear stress yielding a counterclockwise moment (B_1) about the point will then be taken as negative." It is shown in Figure 2.

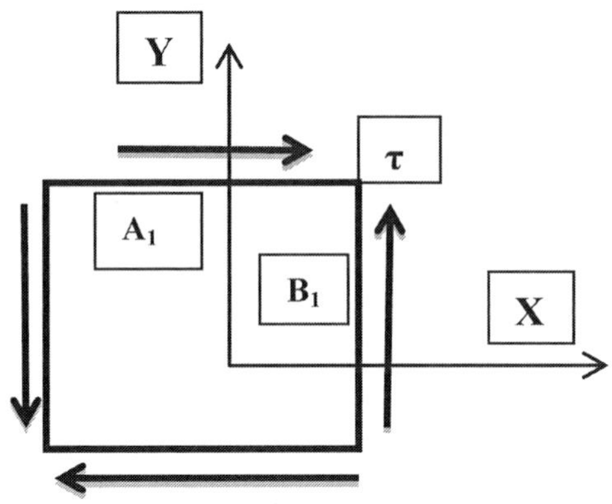

Figure 2: Sign convention for shear on Planes A_1 and B_1 [1]

This method creates a conflict with conventional method for moment based on right hand thumb rule. Based on this rule, counterclockwise moment is positive and clockwise moment is negative. For a student it results in a dilemma of Clockwise or Counterclockwise? Which one results in positive moment? However, this method for Mohr's circle [1], [2], [3] leads to correct solution while it is a pedagogical challenge for an instructor to teach something that is opposite of right hand thumb rule. Also it is a challenge in learning for a student.

In Reference [4], the author states, "An alternative sign convention for shear stresses is sometimes used when constructing Mohr's circle. Instead of thinking of the vertical axis as having negative shear stresses plotted upward and positive shear stresses plotted downward (which is a bit awkward), we can think of the vertical axis as having clockwise shear stresses plotted upward and counterclockwise shear stresses plotted downward." It is given in Figures 3(a) and 3(b).

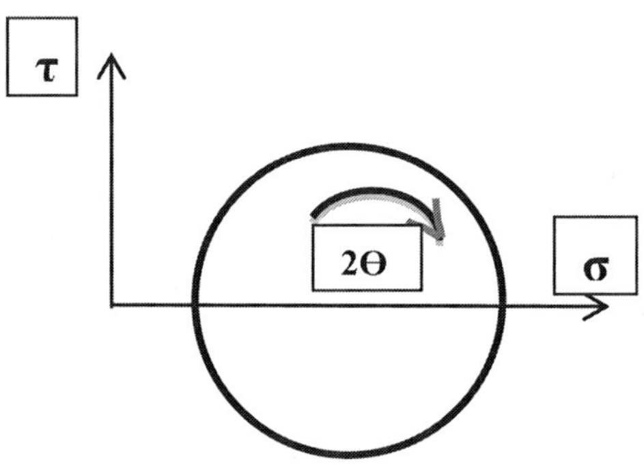

Figure 3(a): Shear axis positive upward [4]

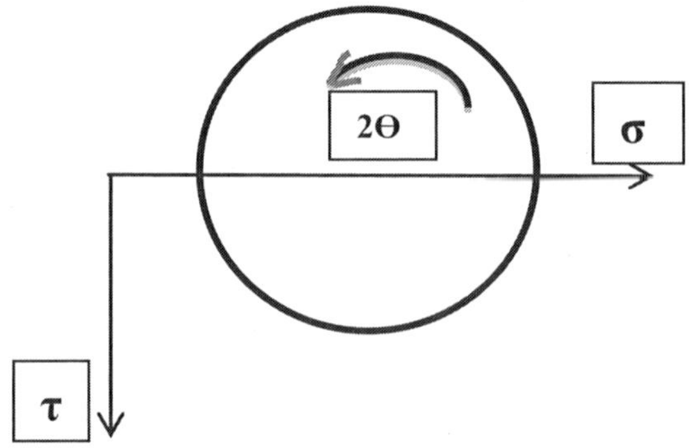

Figure 3(b): Shear axis positive downward [4]

If the shear axis is positive upward, clockwise is considered as positive while if the shear axis is positive downward, counterclockwise is considered as positive. Again it is conceptually correct; however it is a challenge in learning to a student.

In Reference [5], the shear stress is assumed positive in the positive y & y'-directions as shown in Figures 4(a) and 4(b).

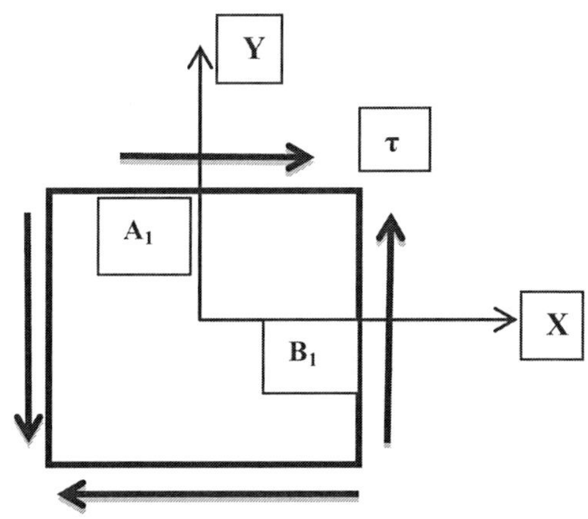

Figure 4(a): Initial position of reference axes

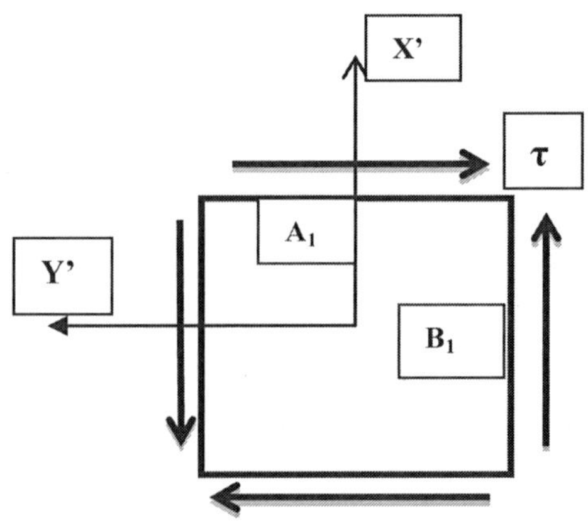

Figure 4 (b): Reference axes rotated 90° counterclockwise

Based on this assumption, the shear stress on Plane-B_1 is positive since it is acting along positive Y-direction as shown in Figure 4(a); while the shear stress on Plane-A_1 is negative since it is acting along negative Y'-direction as shown in Figure 4(b). The X-Y axes are rotated by 90 degrees counterclockwise to X'-Y' axes to justify the sign of shear stress. The author states, "In other words, a rotation Θ of the X' axis on the element will correspond to a rotation 2Θ on the circle in the same (counterclockwise) direction." This is true for considering τ-axis positive downwards in the construction of Mohr's circle. The author adds, "If instead the τ-axis were constructed positive upwards, then the angle 2Θ on the circle would be measured in the opposite direction to the orientation Θ of the X' axis." Although this approach leads to correct transformation of stresses, it is a challenge in learning and comprehension of Mohr's circle.

In Reference [6], the author does not distinguish positive or negative shear stress; however he states, "If the shear stress acting on a face of the stress element tends to rotate the stress element in a clockwise direction, then the shear stress is plotted above the σ-axis. If the shear stress tends to rotate the stress element in a counterclockwise direction, then the shear stress is plotted below the σ-axis." This method eliminates the need for whether τ-axis should be positive upward or downward. Absolute values of shear stress with subscripts of 'cw' or 'ccw' (τ_{cw}, τ_{ccw}) is plotted above or below the σ-axis. Similar treatment on sign convention is encountered in Ref [7], [8], [9], [10], [11], [12], and [13]. All these approaches lead to challenges in learning the concept especially to an undergraduate student in both engineering and technology programs.

RULE BASED METHOD

Let us assume a shear is applied on top plane A_1 in the positive x-direction as shown in Fig. 5(a). For force equilibrium, an equal and opposite shear is applied on bottom plane. For moment equilibrium a shear couple is applied on right plane (B_1) and left plane.

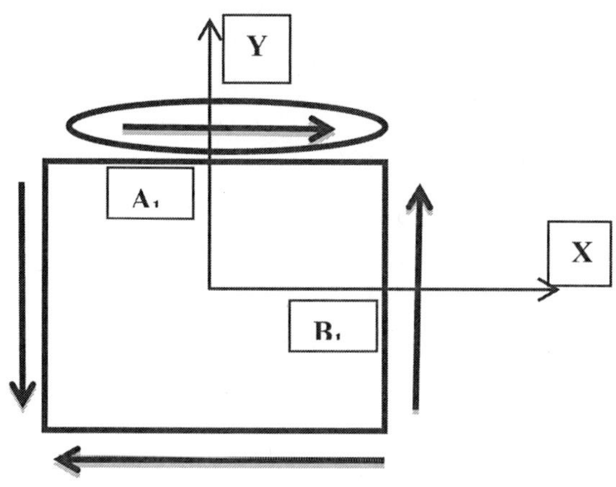

Figure 5(a): Positive shear on Plane-A_1

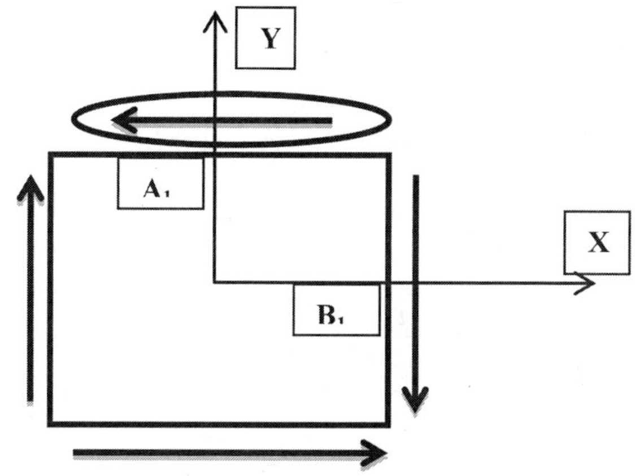

Figure 5(b): Negative shear on Plane-A_1

Since the applied shear on Plane A_1 is in the positive x-direction, it is intuitive to assume this shear is positive. The shear on adjacent plane B_1 (90° counterclockwise to A_1) is applied to counter the moment action of shear on plane A_1. So, the sign of shear on plane B_1 is negative of the sign of shear on plane A_1. Therefore, if the shear on plane A_1 is positive as shown in Fig. 5(a), the shear on plane B_1 is negative. Similarly if the shear on Plane A_1 is negative as shown in Fig. 5(b), the shear on Plane B_1 is positive. Although there are four planes (Top, Bottom, Right and Left) with given shear, the sign can be determined by observing the direction of shear on Top Plane alone. Therefore a set of rules can be developed for the sign convention of shear as given below:

Rule #1: If a shear on Plane-A_1 is acting in the positive X-direction, it is a positive shear stress
Rule #2: If a shear on Plane-A_1 is acting in the negative X-direction, it is a negative shear stress
Rule #3: The shear on Plane-B_1 will be opposite in sign to the shear on Plane-A_1

These rules can be easily applied to construct a Mohr's circle with shear axis positive upwards. If the shear axis is positive downward, the rules are applied on planes A_2 & B_2 (Fig. 6).

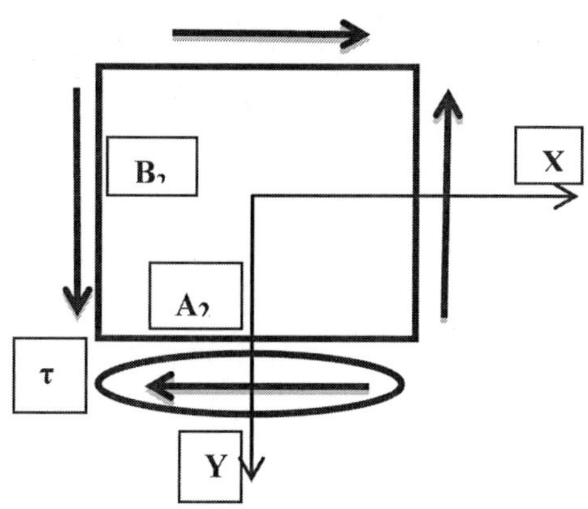

Figure 6: Shear sign convention for τ-axis positive downward

The rules on these planes would appear as:

Rule #1: If a shear on Plane-A_2 is acting in the positive X-direction, it is a positive shear stress
Rule #2: If a shear on Plane-A_2 is acting in the negative X-direction, it is a negative shear stress
Rule #3: The shear on Plane-B_2 will be opposite in sign to the shear on Plane-A_2

With reference to Fig. 6, the shear on Plane A_2 is negative since it is acting in the negative direction of X-axis, and hence the shear on plane B_2 is positive. To construct a Mohr's circle, a diameter is required and to draw the diameter the coordinates of two end points of the straight line is required. If the shear axis is positive upwards, apply the rules on Planes A_1 and B_1 to find the stress coordinates for Mohr's circle diameter. If the shear axis is positive downwards, apply the rules on Planes A_2 and B_2.

MOHR'S CIRCLE

Mohr's circle is a graphical representation of the given state of plane stress on an element. If the element is rotated through an angle Θ, the resulting new stresses on the element can be easily obtained from the Mohr's circle. This stress transformation is shown on the rotated element in Figure-7. If the element is rotated through an angle equivalent to the principal angle Θp, the shear stress on the element vanishes while the normal stresses would be equivalent to maximum and minimum value of stresses known as principal stresses. This is shown in Figure-8 for an element rotated through an angle equivalent to principal angle.

In the construction of a Mohr's circle we are plotting the normal and shear stress values on two adjacent planes of a stress element. The graph has normal stress along the horizontal axis and shear stress along the vertical axis. In general horizontal axis is positive to the right and vertical axis is positive upwards. If we have the stress values on two adjacent planes (A_1-B_1) as coordinates, they can be represented on the graph as a diameter of the circle. With this diameter on the graph, a circle can be easily constructed. The rules 1, 2, & 3 help to determine these coordinates easily without any ambiguity. If the shear axis is chosen as positive downward, the rules can be easily applied for stresses on Planes A_2-B_2, to determine the appropriate coordinates.

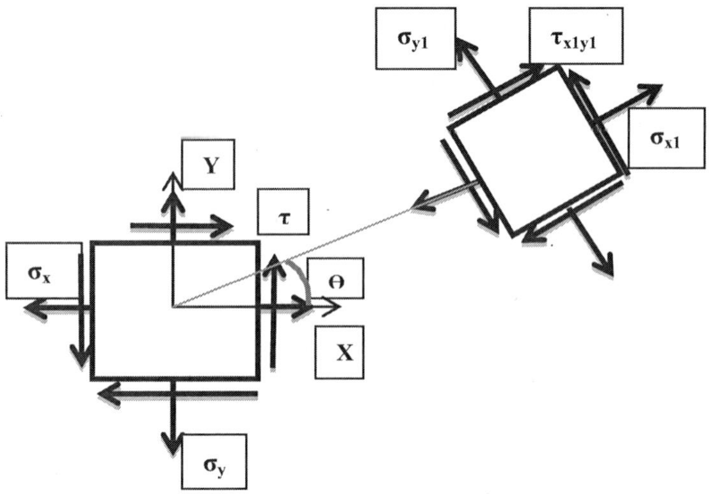

Figure 7: Stresses on an element rotated through an angle, Θ

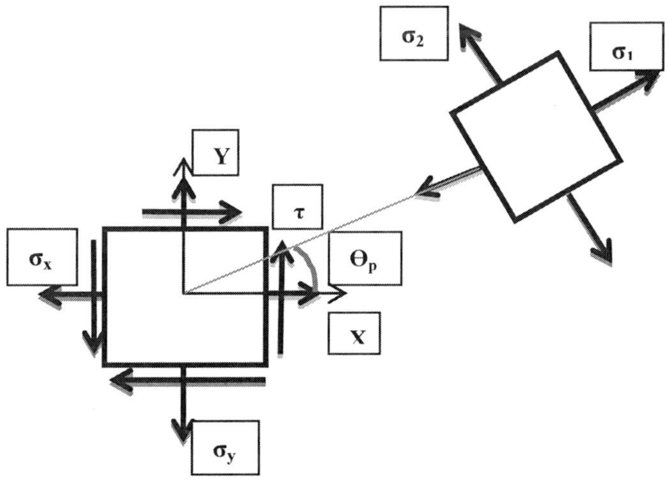

Figure 8: Stresses on an element rotated through an angle, Θ_p

For construction of Mohr's circle, the state of stress on an element is given in Figure 9. The stress coordinates on each plane is given below:
Plane A_1: (σ_y , τ_{xy})
Plane B_1: (σ_x , - τ_{xy}).
Plane A_2: (σ_y , - τ_{xy})
Plane B_2: (σ_x , τ_{xy}).

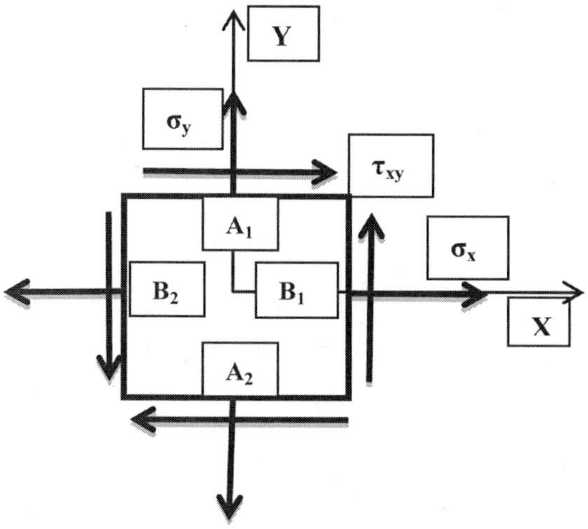

Figure 9: A general element with plane stress

EXAMPLE

To illustrate the application of rule based method, an example problem [14] is given below in Figure 10.

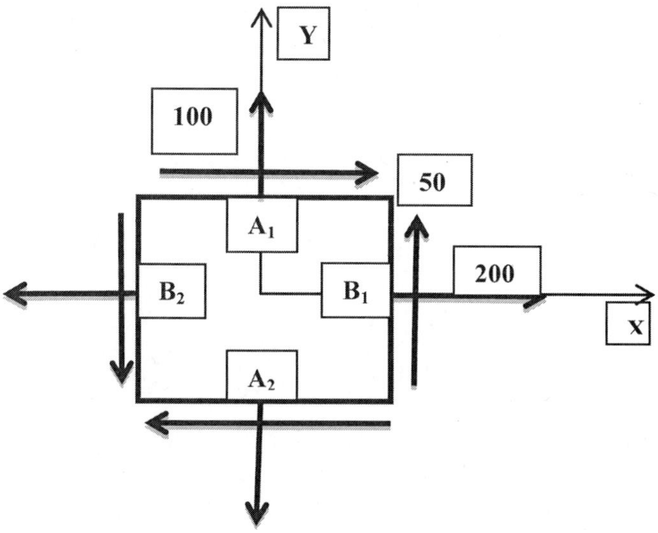

Figure 10: A typical element with given plane stress

$\sigma_x = 200$ psi, $\sigma_y = 100$ psi, $\tau_{xy} = 50$ psi

The normal stresses are tensile stresses and hence they are positive. The shear stress on the element is positive because the shear on Plane A_1 is acting in the positive X-direction (Rule #1). The Mohr's circle is constructed with the shear axis positive upwards as well as positive downwards.

To construct a Mohr's circle with the shear axis positive upwards, the stress information on Planes A_1 and B_1 are:

A_1 (100, 50); B_1 (200, -50)

To construct a Mohr's circle with the shear axis positive downwards, the stress information on Planes A2 and B2 are:

A_2 (100, -50); B_2 (200, 50)

The Mohr's circle is presented in Figure 11.

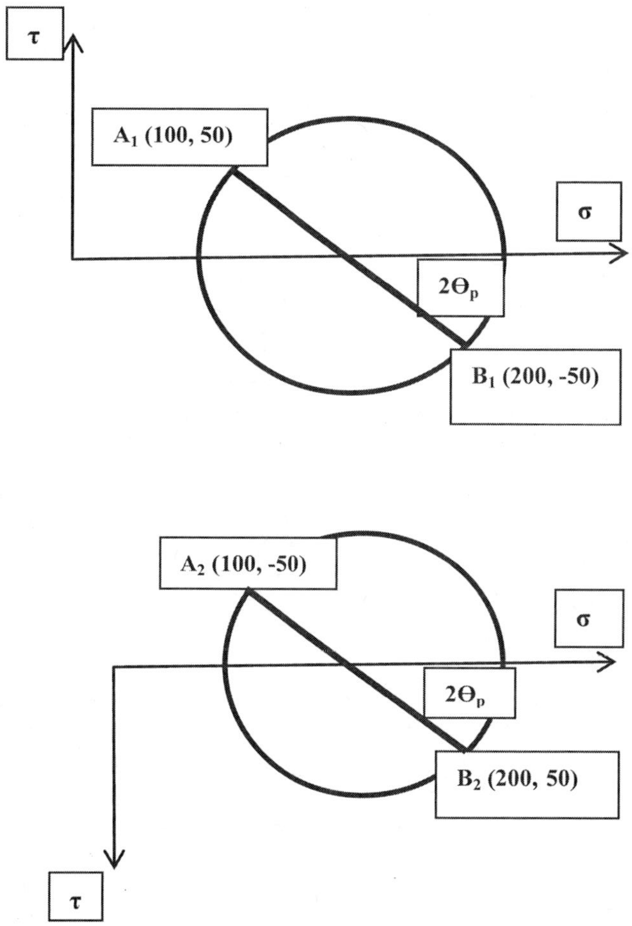

Figure 11: Mohr's circle for the example problem

IMPACT ON EDUCATION

The method from traditional textbooks was used for several years in teaching Mohr's circle and about fifty percent of students understood the concept. In the last few years, the rule based method was used in the classroom resulting in remarkable improvement in understanding the concept. About ninety five percent or above students did well in the test on Mohr's circle. For the benefit of students, this concept was posted on "You

Tube" under "Swami Classroom" [14]. Some of the comments received are summarized below:

(i) "Explained better than all my books!"
(ii) "This was the only topic that was annoying! Thanks for the video, got relieved for tomorrow's exam in a minute"
(iii) "Straight forward teaching of the concept…"
(iv) "Only video that explains in a very clear way"
(v) "Simple presentation, very useful"

The students in classroom appreciated the rule-based method in learning the concept of Mohr's circle and also they used the geometric tools in constructing the Mohr's circle to the scale. It helped them to measure the principal stresses, principal angle, and maximum shear stress from the graph.

CONCLUSION

The construction of Mohr's circle has been presented by several different methods by different authors in the literature. Although all the methods lead to correct stress transformation, it is a pedagogical challenge to present it in such a way that students can learn effectively in both engineering and technology programs. The primary challenge is to define the shear on each plane of the stress element without any ambiguity. A simple rule based method is presented in this article to account for the shear and determine the coordinates for constructing the Mohr's circle. This method eliminates the need to know whether the shear rotate the element in clockwise or counterclockwise direction. The example problem illustrates the application of the rules and demonstrates the process for easy construction of Mohr's circle. The student performance showed significant improvement based on learning by "Rule-Based" method compared to the traditional methods from the textbooks. The rule based method is a viable approach for effective learning of stress transformation and understand the principles of Mohr's circle.

REFERENCES

[1] Shames,I.H..Introduction to Solid Mechanics, Prentice Hall

[2] Lagace,P.A., "Mohr's Circle", Unified handout on Materials and Structures (#M-6), MIT

[3] Ugural, A.C., Mechanics of materials, McGraw Hill

[4] Gere, J.M., Goodno, B.J., Mechanics of Materials, Cengage Learning

[5] Hibbeler, R.C., Mechanics of Materials, Pearson Prentice Hall

[6] Philpot, T.A., Mechanics of Materials, John Wiley & Sons, Inc

[7] Beer, F.P., Johnston, E.R., DeWolf, J.T., Mechanics of Materials, McGraw Hill

[8] Davis R.O., Selvadurai, A.P.S., Elasticity and geo-mechanics, Cambridge university press

[9] Holtz, R.D., Kovacs, W.D., An introduction to geotechnical engineering, Prentice Hall

[10] Parry, R.H.G, Mohr circles, stress paths and geo-technics, Taylor & Francis

[11] Timoshenko, S.P., Goodier, J.N., Theory of Elasticity, McGraw Hill

[12] Timoshenko, S.P., History of Strength of Materials, Dover Publications

[13] Jaeger, J.C., Cook, N.G.W., Zimmerman, R.W., Fundamentals of rock mechanics, Wiley-Blackwell

[14] Karunamoorthy, S., "Mohr's Circle", Swami Classroom – You Tube Video

Proceedings of the ASME 2014 International Mechanical Engineering Congress and Exposition
IMECE2014
November 14-20, 2014, Montreal, Quebec, Canada

IMECE2014-37004

PEDAGOGICAL EXPERIENCE IN ENGINEERING EDUCATION: STUDYING CASES OF PATENT DISPUTE AS A METHOD FOR BOTH INNOVATION AND LEGAL EDUCATION

Rong-Jer Lai
Department of Mechanical Engineering
National Kaohsiung University of Applied Sciences
Kaohsiung, Taiwan

ABSTRACT

The issue of intellectual property is inevitable for an engineering design course today. Traditionally, it would be discussed separately from the technical or the legal perspective. But they are two sides of the same coin. In the course of *comparative analysis of patent dispute*, the case method was used to treat the issue involved from both perspectives at the same time. Patent disputes are mainly concerned with the issue of patentability or infringement, and normally both involved. For analysis of a dispute, it is normally required to compare the elements of the invented object with either the prior arts or the alleged infringement, depending on the cases. For such legal analysis one can instead take design methods such as TRIZ (an acronym in Russian standing for "theory of inventive problem solving") and SCAMPER (an acronym standing for "substitute, combine, adapt, magnify or minify, put to other uses, eliminate or elaborate, and rearrangement or reverse") to check the design concepts or the inventive principles behind the objects, so that the students can learn the innovation methods and the legal procedure at the same time. In this paper pedagogical experience with concrete examples was demonstrated.

INTRODUCTION

Intellectual property refers to creations of the mind, and is traditionally divided into two categories: industrial property and copyright. The former includes patents for inventions, trademarks, industrial designs and geographical indications, and the latter covers literary works, films, music, artistic works and architectural design [1]. The issue of intellectual property is inevitable for an engineering design course, particularly at the present time when the number of patent applications continues the significant upward growth, and big cases of patent litigation often make headlines, for example, patent wars among mobile phone makers [2-4].

Traditionally, the issue of intellectual property would be discussed separately from the technical or the legal perspective. For example, the topic of "Patents and Intellectual Property" appears in the course of product design to study the prior arts for the concept generation, and to file a patent application for protection of inventions [5]. On such occasions legal aspect of patent dispute is scarcely considered. On the contrary, the principles of patent law are usually taught by using the cases of patent dispute as raw materials in the classrooms of law school [6], where the holdings are derived and applied to other possible situations, for instance, application of the "teaching-suggestion-motivation" (TSM) test from the decision of *KSR Int'l Co. v. Teleflex, Inc.* to the issue of obviousness [7].

But they are actually two sides of the same coin, and for the education of engineering design the legal issue of intellectual property should not "just leave it to the attorneys or the lawyers". Patent disputes are mainly concerned with the issue of patentability or infringement, and normally both involved. For legal analysis of a dispute, it is normally required to follow the conventional procedures to compare the elements of the invented object with either the prior arts or the alleged infringement, depending on the cases. Instead one can take design methods such as TRIZ and SCAMPER to check the design concepts or the inventive principles behind the objects, so that the students can learn the innovation methods and the legal procedure at the same time. In this paper pedagogical experience in such a course was demonstrated.

BACKGROUND OF COURSE DEVELOPMENT

Since the issue of intellectual property began to be negotiated in the multilateral trading system in the late 1980s, reform of domestic patent system has frequently appeared on the agenda of legislature. An insight into the balance of payments of Taiwan shows that the deficit on the transactions of royalties and license fees has increased to a great extent since 1980s (Fig. 1). To foster the industrial product design, the government has initiated various programs. The design-related fields attract more and more young people, and innovation becomes the slogan of different kinds of campus activities.

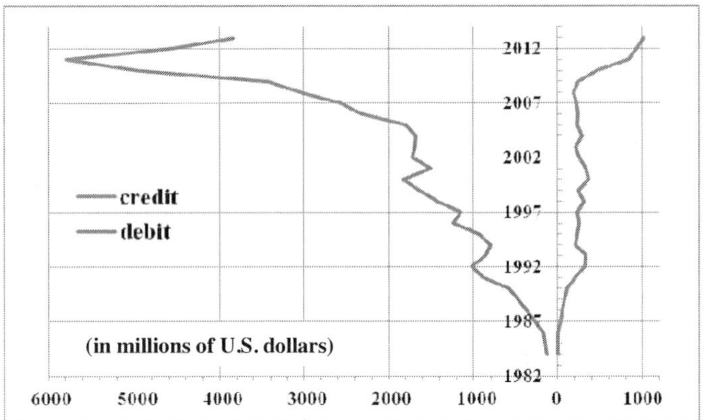

Data Source: Central Bank of Taiwan

Fig. 1: Transactions of royalties and license fees on the balance of payments of Taiwan (1983~2013)

With such a background several patent-related courses have been offered for the undergraduate and graduate students majored in mechanical engineering in the past years (Table 1). These selective courses have been developed gradually and modified with the time. At the beginning, it was a course with the title "*systematic design methodology*" in 1997, which focused on the systematic design process, mainly following the approach of Pahl and Beitz [8] and included the methods applied at the different stages of design process. The topic of intellectual property, particularly patent was discussed in a condensed manner at the stage of conceptual design. The course was offered for the senior students, including the students from the college of continuing and extension education. The latter highly appraised the topic of patent analysis out of their daily experiences.

Table 1: Subjects of the patent-related courses

undergraduate courses	year	graduate courses
systematic design methodology	1997	
patent analysis	1999	
creative design methods		
	2006	patent strategy
		comparative analysis of patent dispute
	2014	patent design-around

Two years later the course was divided into two independent subjects: *patent analysis* and *creative design methods* for further details on the individual topics. For the course *patent analysis*, the contents include history of patent system, rules of patent law, patent documents, patent search, claim formats and patent maps. Some litigation cases and the domestic judiciary as well as mechanisms for the resolution of disputes are also introduced briefly. On the other hand, innovation and various available design methods, for example, brainstorm, morphological chart, Quality Function Deployment (QFD), etc. are discussed and demonstrated in the course *creative design methods*.

In the meantime, the course *intellectual property rights* has been offered for the general education purpose, which covers general concepts of patent, trademarks, copyright, trade secrets and so on. Any student who has interest in such a topic can select the course, no matter what her/his major is.

For the graduate students, the patent-related courses began in 2006. The course *patent strategy* focuses on the management aspect, where Taiwan Intellectual Property Management System (TIPS) as well as the processes based on ISO 9001:2000, i.e. Plan-Do-Check-Action (PDCA) model (Fig. 2) are introduced [9], and patent strategies for different purposes are discussed.

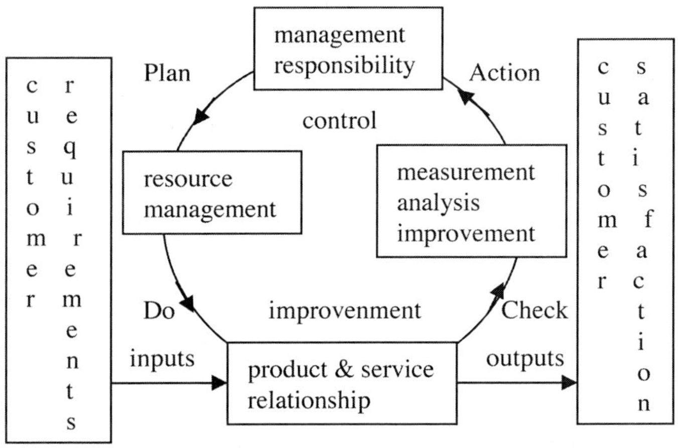

Fig. 2: Model of process-based IP management system

In the course *comparative analysis of patent dispute* the tangled issues from technical and legal perspectives are discussed at the same time, what will be explained later.

In 2014 the course *patent design-around* was opened. Historical cases that constitute the legal bases for design-around are explained first, for example, *Markman v. Westview Instruments, Inc.* [10] for claim construction, and *Warner-Jenkinson Company, Inc. v. Hilton Davis Chemical Co.*, [11] for doctrine of equivalents. Then, examples of design-around are demonstrated, and finally the students are required to complete a case study for practice. Basically, the contents could be considered as an extension of the previous course in 2006 with emphasis on the avoidance of patent infringement.

THE COURSE OF *COMPARATIVE ANALYSIS OF PATENT DISPUTE*

In spite of some criticism of the case method [12], studying cases of patent dispute was taken as the instructional method in this course, since concrete cases were more attractive and easier to understand than abstract legal principles for the students.

The course has been offered for the graduate students, including those of on-job master's degree program. Since 2006 there have been 173 students who attended the course (Table 2). 81 of them belong to the regular students, and 92 are on-job. Most students who attended the course *comparative analysis of patent dispute* had engineering backgrounds, some even have had experience of applying a patent before, but all lacked formal legal education in the field of intellectual property rights.

Table 2: Student population of the course

	'06	'07	'08	'09	'10	'11	'12	'13	sum
R.	12	14	11	5	X	7	27	5	81
OJ.	X	19	X	35	X	13	X	25	92
sum	12	33	11	40	X	20	27	30	173

(R.: regular program; OJ.: on-job program)

To compensate the situation, the learning materials of the course include legal analysis and design methods [13], as shown in the course organization (Fig.3).

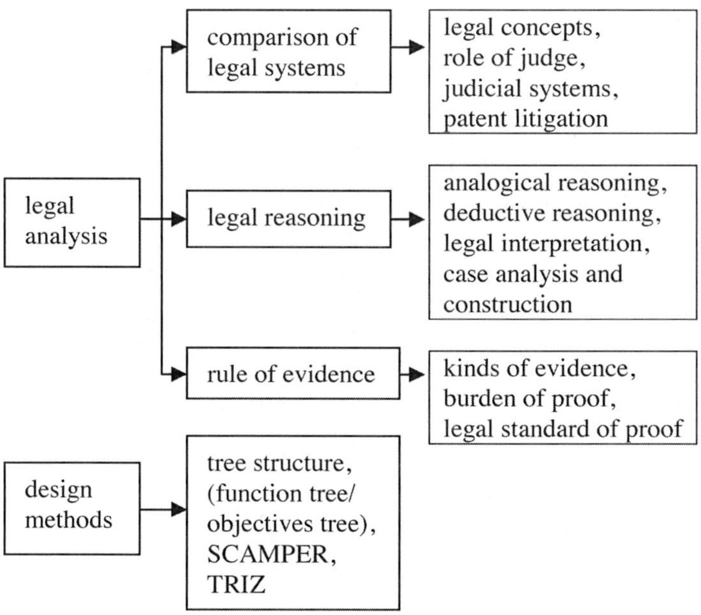

Fig. 3: Course organization

By the part of legal analysis, the students learn the legal knowledge required, especially the topics of legal reasoning and rule of evidence, including the process of analogical reasoning and the syllogism by deductive reasoning [14-15], rules of statutory construction, burden of proof as well as legal standard of proof.

In addition, the legal concepts and judicial processes for patent infringement and patentability are compared, for example, novelty vs. all-element rule, non-obviousness vs. doctrine of equivalents and so on [16]. Also the distinctive features of civil law and common law systems, as well as the legal processes of different judiciaries are demonstrated.

Before discussing a concrete case of patent dispute, the claims must be interpreted properly. For the students to understand the meaning of claims, various formats of patent claims are elucidated in the class [17]. Depending on the complexity of case, more analysis of invention and claim drafting might be carried out, if necessary [18-19].

For comparison of the patent at issue with prior arts or alleged infringement, each claim in dispute is analyzed and reformulated in a tree form, so that the elements and the limitations in question would be identified. For example, as shown in Fig. 4 is the tree structure of claim 1 of a utility model patent (TW M276939) with a title of "A mounting base for video camera". From the viewpoint of design method, the tree structure showing the hierarchical relationships and interconnections of the claim elements displays the same characteristics of a function tree or objectives tree [20].

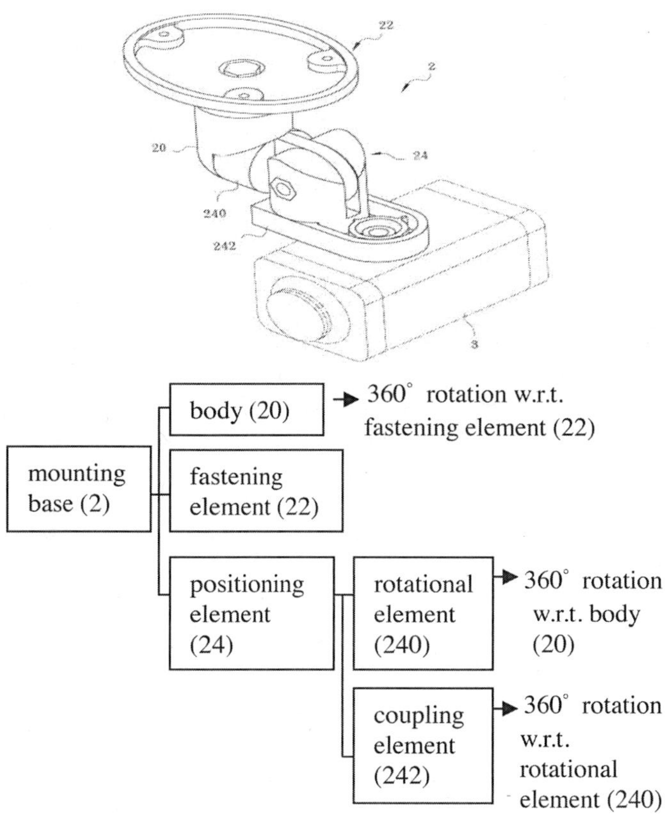

Fig. 4: Tree structure of claim elements

The students who attend the course must complete the assignments twice, i.e. midterm report about dispute of patentability and final report about patent infringement according to the flow chart for case analysis (Fig. 5).

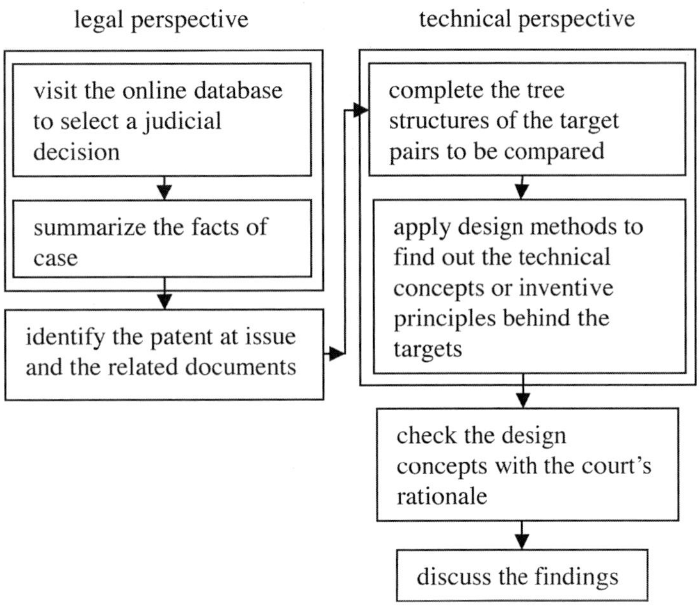

Fig. 5: Flow chart for case analysis

Each time they should visit the online database to select a judicial decision of patent dispute. Then the facts of the case, including procedural and evidentiary facts, rules of law, issue(s) presented to the court to resolve, summary of the various arguments raised, the court's holding on each issue, and the court's rationale that supports its holding, as well as the court's judgment must be summarized and prepared in order. So far are the works involved in the case analysis from the legal perspective.

After identifying the patent at issue and the prior arts or the alleged infringement, depending on the assignment, the students must search out the related documents, and complete the tree structures of the target pairs, i.e. the elements of claim at issue and the corresponding objects to be compared. Here the design methods such as SCAMPER and TRIZ [21] would be applied to find out the technical concepts or inventive principles behind the invention, prior arts and the alleged infringement, respectively. These works involved in the case analysis would be attributed to the technical perspective.

When using TRIZ, the problem to be solved and the parameters involved, which are described in the patent specification, must be analyzed, and the corresponding contradictions of parameters would be identified and explicitly remarked.

Three models of applying TRIZ have been introduced for the above-mentioned purpose:

Model 1: Start with the patent at issue and identify the problem to be solved that was described in the background of patent specification. Next, use the contradiction matrix of TRIZ to find out the possible innovation directions or inventive principles, say X and Y, and then examine the means used in the patent at issue, say Z, to check if Z belongs to the predicted X or Y.

Model 2: If the problems to be solved in prior arts are different from those in the patent at issue, carry out the corresponding individual analysis like Model 1. Then compare the predicted inventive principles resulted from both sides.

Model 3: If the problems to be solved in prior arts coincide with those of the patent at issue, carry out the analysis like Model 1, and check if both technical concepts belong to the predicted.

The ideas behind the above-mentioned models would be linked to the "problem-and-solution approach" taken by EPO for assessing the inventive step. In the problem-and-solution approach, there are three main stages: (i) determining the "closest prior art", (ii) establishing the "objective technical problem" to be solved, and (iii) considering whether or not the claimed invention, starting from the closest prior art and the objective technical problem, would have been obvious to the skilled person [22].

Finally, follow the process of legal reasoning to check the court's holding on each issue as well as the court's rationale that supports its holding with the design concepts that were found based on the design methods. If there were any inconsistency to be found, i.e. different opinions or doubt about the court's holding, encourage the students to speak out their reasons and discuss them.

EXAMPLE OF PATENT DISPUTE CASE

In an example of patent infringement case, the patent at issue (TW 462363) was an improvement of non-slip grip pad, and the alleged infringement having similar form functioned similarly, as shown in Fig. 6.

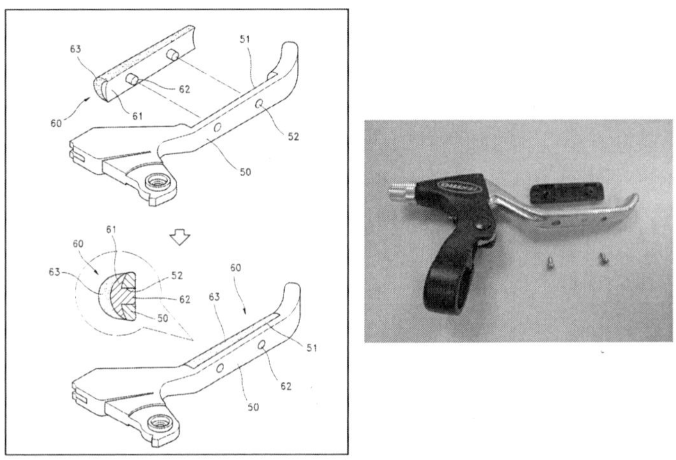

Fig. 6: Patent at issue (left) and the alleged infringement (right)

As shown in Fig. 7, the procedure of patent infringement verification begins with the interpretation of claim at issue. The technical features of the claim and the corresponding ones of the alleged infringement would be analyzed next. Then the all-element rule would be applied to see if all elements of the claim could be read on the alleged infringement. If invalid, the doctrine of equivalents should be called upon to examine the in-coincidence. Now, if the doctrine of equivalents were invalid, the alleged object would be proved as non-infringement. On the contrary, if it were valid, prosecution history would be reviewed to check if any inconsistency exists, or any prior arts have been practiced. If it were the situation, then it turns into non-infringement. Otherwise, it falls into the scope of equivalents of claim, i.e. infringement.

On the other hand, when all elements of a claim are read on the alleged infringement, the reverse doctrine of equivalents would be taken to examine if the claim has been extended beyond the fair scope of the invention. If it were true, the alleged object would be declared as non-infringement. Otherwise, it would fall into the scope of the claim, i.e. infringement.

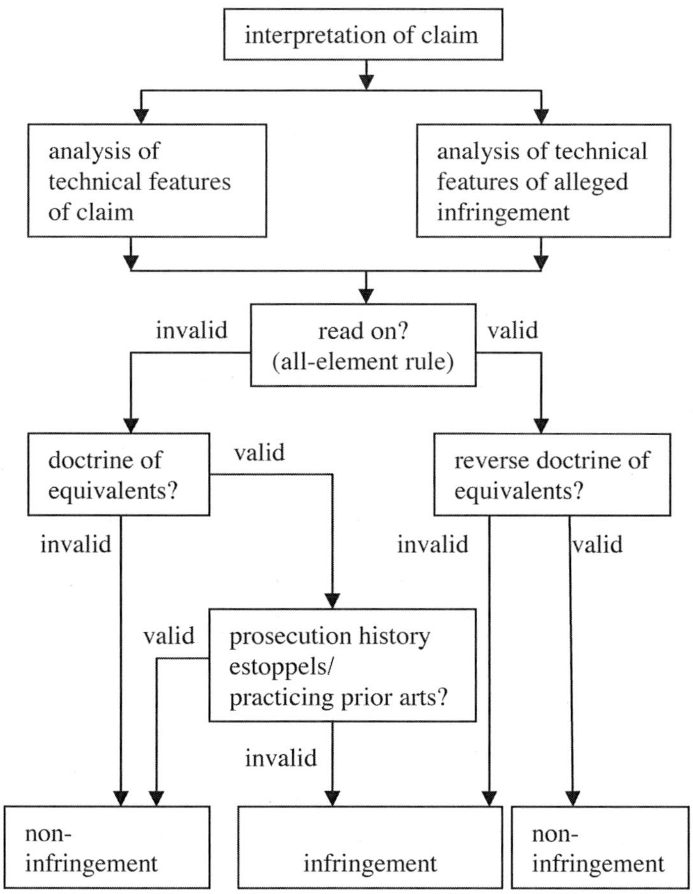

Fig. 7: Procedure of patent infringement verification

The technical feature of the patent was the non-slip pad (60), which included a hard thin bottom layer (61), and a soft non-slip top layer (63). On the contrary, the alleged infringement was a solid body made of soft non-slip material. From the viewpoint of design method SCAMPER, the principle *substitute* and the principle *combine* were applied. That is, the combination of top and bottom layers was substituted by a solid body, and two kinds of materials were replaced by a single one. It seems to be definite that the solid one be declared as an infringement, but the decision of court was "non-infringement", which was based on the claim interpretation.

CONCLUSION

For the engineering design education, the legal issue of intellectual property should not "just leave it to the attorneys or the lawyers". According to the course organization presented here, both the technical and the legal perspectives could be discussed at the same time by studying the cases of patent dispute. In this paper pedagogical experience in the course of *comparative analysis of patent dispute* was demonstrated to express the development of such a course and the learning materials. The experiences in the past years show that the course helps the students to understand the complete facts involved in the real world of patent dispute and the principles that exist in different fields, particularly for the students of regular program. It is also true that the students of on-job master's degree program appraised the course more than those of regular program.

Finally, just as Mr. Giles S. Rich said in 1998, "Don't accept everything you read in this book just because it is in print or written by some court. Judges, like other authors, are not infallible. Controversial and disputed material is included to further one of the objectives of the authors — to make you THINK." [23], the experience shows the same challenge for the students as well as for the teacher in such a course. One of the author's aims is to invite more discussion, and now, please.

REFERENCES

[1] WIPO, 2004, "WIPO Intellectual Property Handbook: Policy, Law and Use", 2nd edition, World Intellectual Property Organization (WIPO), Geneva.

[2] WIPO, 2014," US and China Drive International Patent Filing Growth in Record-Setting Year ", Geneva, March 13, 2014, PR/2014/755

[3] Barry, C., Arad R., Ansell, L., and Clark, E., 2013, "2013 Patent Litigation Study", PricewaterhouseCoopers LLP, Delaware, http://www.pwc.com/us/en/forensic-services/publications/2013-patent-litigation-study.jhtml

[4] The Guardian, 2014, "Smartphone patent wars", http://www.theguardian.com/technology/smartphone-patent-wars

[5] Ulrich, K.T., and Eppinger, S.D., 2012, "Product Design and Development", 5th edition, chapter 16: Patents and Intellectual Property, McGraw-Hill, New York.
[6] Chisum, D.S., Nard, C.A.,Schwartz, H.F., Newman, P., and Kieff,F.S.,2004, "Principles of Patent Law", 3rd edition, Foundation Press, New York.
[7] KSR Int'l Co. v. Teleflex, Inc., 550 U.S. 398 (2007)
[8] Pahl, G. and Beitz, W., 1995, "Engineering Design: A Systematic Approach", 2nd edition, Springer.
[9] Science and Technology Law Institute, 2007, "Taiwan Intellectual Property Management System (TIPS)", Institute for Information Industry, https://www.tips.org.tw/
[10] *Markman v. Westview Instruments, Inc.*, 517 U.S. 370 (1996)
[11] *Warner-Jenkinson Company, Inc. v. Hilton Davis Chemical Co.*, 520 U.S. 17 (1997)
[12] Moskovitz, M., 1992, "Beyond the Case Method: It's Time to Teach with Problems", Journal of Legal Education, **42,** pp. 241-270
[13] Lai,.R.J., 2007,"A Comparative Analysis of Patent Dispute", 1st edition, Sharing Publishing, Taipei.
[14] Burton, S.J., 1995, "An Introduction to Law and Legal Reasoning", 2nd edition, Little, Brown and Company, Boston.
[15] Wellford, R.S., 2002, "Legal Reasoning, Writing, and Persuasive Argument", 2nd edition, Matthew Bender & Company.
[16] Oake, R.G., 1997, "The Calculus of Comparison: Obviousness and Equivalency Principles in Patent Law", www.oake.com/page2/assets/Thesis%20DC.pdf
[17] WIPO, 2010, "WIPO Patent Drafting Manual", www.wipo.int/freepublications/en/patents/867/wipo_pub_867.pdf
[18] Slusky, R.D., 2007, "Invention Analysis and Claiming: a patent lawyer's guide", ABA Publishing, Chicago, Illinois.
[19] Faber, R.C., 2008, "Faber on Mechanics of Patent Claim Drafting", 6th edition, Practising Law Institute, New York City.
[20] Cross, N., 2008, "Engineering Design Methods: Strategies for Product Design", 4th Edition, Wiley.
[21] Bukhman, I., 2011, " TRIZ Technology for Innovation", Cubic Creativity Company.
[22] EPO, 2013, "Guidelines for Examination in the European Patent Office", Part G - Patentability, Chapter VII – Inventive step, 5 Problem-and-solution approach, http://www.epo.org/law-practice/legal-texts/html/guidelines/e/g_vii_5.htm
[23] Rich, G.S., 1998, *Foreword* in "Principles of Patent Law", by Chisum, D.S. et al., 3rd edition, Foundation Press, New York.

Proceedings of the ASME 2014 International Mechanical Engineering Congress and Exposition
IMECE2014
November 14-20, 2014, Montreal, Quebec, Canada

IMECE2014-38192

FOSTERING SUSTAINABLE DEVELOPMENT THINKING THROUGH LEAN ENGINEERING EDUCATION

Anabela C. Alves
University of Minho
Guimarães, Portugal
anabela@dps.uminho.pt

Franz-Josef Kahlen
University of Cape Town
Cape Town, South Africa
fj.kahlen@uct.ac.za

Shannon Flumerfelt
Oakland University
Rochester, Michigan, USA
flumerfe@oakland.edu

Anna Bella Siriban-Manalang
De La Salle University
Manila, The Philippines
beng_manalang@yahoo.com

ABSTRACT

An implementation plan for the UN Decade of Education for Sustainable Development (ESD) by the implementing agency UNESCO, launched in 2005 to build partnerships with various partners mainly aimed to reflect the great diversity of interests, concerns, and challenges for ESD. Embedded in this UN program is the challenge to academic engineering programs to develop a technical workforce which is capable of designing and continually developing sustainable processes and products. ESD therefore requires engineers to be able to learn and perform in an interdisciplinary work environment as critical thinkers and problem solvers, to be value-driven and to practice participatory and transparent decision making. Taking these challenges back into curriculum development, the authors of this paper propose that their Lean Engineering Education (LEE) initiative be examined and scrutinized as a contribution to ESD. Thus, this paper reviews ESD roots and partnerships and, then, present, succinctly, LEE as a curricular innovation for engineering courses that promotes the behavior intended for future engineers, to foster a ESD and the Sustainable Development thinking.

Keywords: Engineering education; Education for Sustainable Development; Lean Engineering

INTRODUCTION

Sustainability is a word on the agenda of the United Nations Decade of Education for Sustainable Development (UNESD) and education is the engine to integrate this concept in the young people´s mindsets. The Resolution 57/254 of the UN General Assembly on the Decade of Education for Sustainable Development (DESD) 2005-2014 is ending. The main goal was to involve all stakeholders in committing to a sustainable life for all generations. To achieve this, the UNESCO was the agency responsible for outlining an international implementation scheme [1] in order to implement DESD. In this decade, it was expected that ESD would become the focus of a high quality education, with characteristics such as [1]:

- Interdisciplinary and holistic thinking: learning for sustainable development embedded in the whole curriculum, not as a separate subject;
- Values-driven approaches: sharing the values and principles underpinning sustainable development;
- Critical thinking and problem solving: leading to confidence in addressing the dilemmas and challenges of sustainable development;

- Multi-method approach: word, art, drama, debate, experience, … different pedagogies for modelling processes;
- Participatory decision-making: learners participate in decisions on how they are to learn;
- Applicability: learning experiences are integrated in day to day personal and professional life;
- Local relevance: addressing local as well as global issues, and using the language(s) which learners most commonly use.

Such characteristics demand the integration of values inherent in sustainable development into all aspects of learning and to encourage changes in behavior that lead to a more sustainable and just society for all. This, in turn, has led to an understanding of the need to rethink and to revise education from nursery school through university to include a clear focus on the development of knowledge, skills, perspectives and values related to sustainability. Particularly, the training of future engineers is a critical concern since they impact society and nature directly. Five years after the DESD launching, UNESCO [2] felt the need to highlight the important role of engineers in society. This report discussed four main needs regarding engineering education:

1) develop public and policy awareness and understanding of engineering, affirming the role of engineering as the driver of innovation, social and economic development;
2) develop information on engineering, highlighting the urgent need for better statistics and indicators on engineering (such as how many and what types of engineers a country has and needs);
3) transform engineering education, curricula and teaching methods to emphasize relevance and a problem-solving approach to engineering; and
4) more effectively innovate and apply engineering and technology to global issues and challenges such as poverty reduction, sustainable development and climate change – and urgently develop greener engineering and lower carbon technology.

These ideas resemble some of the characteristics referred by ESD above. So, perhaps, at least, in engineering education, ESD continued without the needed characteristics already referred. Meanwhile, the reports about engineering education curriculum reform keep coming [3],[4]. For example, The Royal Academy of Engineering [4] pointed out the gap between what universities teach and what industry needs. In this, is also evident that the "how" universities teach, many times without the appropriate link between theory and practice, increases enormously this gap. The consequence of this is that a desynchronized education results in poor performance and inadaptability of young engineers in the marketplace and workplace. To overcome this gap and develop the right competencies in engineering students and future engineers, the authors of this paper proposed Lean Engineering Education as a curricular innovation to develop systems, sustainability and ethics competency mastery coupled with content mastery in a double helix DNA [5]. Originating from Toyota's Education Model, which is promoted in their companies and global subsidiaries, Lean Engineering Education (LEE) seeks to provide a pathway to develop engineering competencies and to tap into engineers' potential creativity and innovation capabilities. Developed out of a sentiment that the core tenets of Lean Engineering should be practiced not just in commercial production or service environments, but in engineering education as well, the authors transposed these Lean Engineering tenets into engineering education and explain in this paper this transposition.

SUSTAINABLE DEVELOPMENT AND DECADE OF EDUCATION TO PROMOTE SD

Sustainable Development (SD) was highlighted by the World Commission on Environment and Development (WCED) in 1987: "Humanity has the ability to make development sustainable to ensure that it meets the needs of the present without compromising the ability of future generations to meet their own needs." [6]. To achieve this, interests must align with interests of society, environment and economy, also as referred to as the "triple-bottom line" or 3P (People, Planet, Profit). Since then, some conferences have been devoted to sustainable development, including:

- 1992: UNCED - UN Conference on Environment & Development, Rio de Janeiro, Brazil, also known as Rio Conference or Earth Summit. Major outcomes: a climate-change agreement that led to the Kyoto Protocol, Agenda 21, a Convention on Biological Diversity, new international institutions, e.g. Commission on Sustainable Development and United Nations Framework Convention on Climate Change (UNFCCC) [7].
- 2002: UNCSD - World Summit on Sustainable Development (WSSD), Johannesburg, South Africa, also known as Rio+10 or Earth Summit 2002. Major outcomes: Johannesburg Declaration and almost 300 international partnership initiatives to help achieve the Millennium Development Goals [8].
- 2012: UNCSD - United Nations Conference on Sustainable Development, Rio de Janeiro, Brazil, also known as Rio+20. Major outcomes are contained within a 49-page document called "The Future We Want" where, mainly, the 192 heads of State and Government and high-level representatives renewed their commitment to sustainable development and insure the promotion of an economically, socially and environmentally sustainable future for the planet and for present and future generations [9].

A common concern behind these conferences is education. This concern started in 1972 in the UN Conference on the Human Environment (UNCHE) in Stockholm [10], and since then, some conferences have addressed education. Chapter 36 of Agenda 21 of UNCED [7] in 1992 conference is dedicated to *"Promoting Education, Public Awareness and Training"*. The main aim was to promote formal and non-formal education so that people have the capacity to assess and address their sustainable development concerns. Education is indispensable in changing people's attitudes and to achieve environmental and ethical awareness, values and attitudes, skills and behavior consistent with sustainable development and for effective public participation in decision-making, making informed choices. Environment and development education should be integrated in all disciplines, and should employ formal and non-formal methods and effective means of communication.

Parallel to these conferences, some reports and debates were prepared in thinking how to educate for a sustainable future, e.g.:

- *Educating for a Sustainable Future - A Trans disciplinary Vision for Concerted Action*, an action-oriented paper for preparation of the conference on the same year: *International Conference on Environment and Society: Education and Public Awareness for Sustainability* by UNESCO in 1997 [11];
- *Preparing for a Sustainable Future: Higher Education and Sustainable Human Development* by The United Nations University (UNU) in the World Conference on Higher Education: *Higher education in the Twenty-first Century: Vision and action* in 1998 [12].

These conferences were followed by the United Nations Decade of Education for Sustainable development (DESD) 2005-2014: Draft International Implementation scheme launched in 2005 [1]. This scheme idea was to work as an instrument of mobilization, diffusion and information with the purpose of promoting and guiding initiatives to achieve DESD objectives. The seven interlinked strategies for DESD are: 1) advocacy and vision building; 2) consultation and ownership; 3) partnership and networks; 4) capacity building and training; 5) research and innovation; 6) information and communication technologies; 7) monitoring and evaluation. These strategies would help to achieve a high quality education characterized by interdisciplinary and holistic thinking; values-driven approach; critical thinking and problem solving; different pedagogies in the learning process; participatory decision-making; learning experiences integrated in day to day personal and professional life and local relevance. During the decade, three reports were to be prepared by UNESCO to monitor and evaluate progress of DESD actions and initiatives and two of them were already published [13],[14]. Some other reports have been done in order to identify indicators for DESD implementation [15]. This monitoring and evaluation are important to verify the change in DESD priority areas and ensure the relevance and effectiveness of ESD policies, programs and activities. Additionally, conferences have been organized to share experiences, discuss DESD new developments and restate the DESD objectives such as UNESCO World Conference on Education for Sustainable Development 2009 [16], whose main document was the Bonn declaration [17]. In this declaration was reaffirmed the need of an education of quality that provides the values, knowledge, skills and competencies for sustainable living and participation in society and decent work. To foster the ESD, the UNESCO Education sector launched some Learning and Training Tools to put ESD in action, see Table 1.

Table 1. UNESCO Learning and Training tools

Year	Number	Title
2006 [18]	Learning & Training Tools nº 1	Education for Sustainable Development Toolkit
2010 [19]	Learning & Training Tools nº 2	Education for Sustainable Development Lens: A Policy and Practice Review Tool
2012 [20]	Learning & Training Tools nº 3	Exploring Sustainable Development: A Multiple-Perspective Approach
2012 [21]	Learning & Training Tools nº 4	Education for Sustainable Development source book

These constitute fundamental tools to implement ESD in all levels and types of education (formal and non-formal). Other initiatives include the Education for All (EFA) and Millennium Development Goals (MDG). Additional resources include climate change starters book for educators and practitioners [22] and climate change and lifestyles guide books [23], among others that can be found at the UNESCO site: http://www.unesco.org/new/en/education/themes/leading-the-international-agenda/education-for-sustainable-development/.

DESD IN ENGINEERING EDUCATION

ESD assumes particular importance in engineers training, since engineers have a crucial contribution to wake in society and in developing the world [2]. This training must include societal, environmental and economic aspects in contributing efficiently to sustainable development. This means that engineers must be socially responsible using the knowledge they acquire. More knowledge means more power and this power could be used, consciously or unconsciously, for good or for bad purposes (e.g., wind energy or atomic bomb) as history shows us. Therefore, having engineers who are aware of and guided by social responsibility is a demand of society.

According to this, Engineering Education programs must be reoriented to sustainability. To achieve this, it is important to break barriers and compartmentalized knowledge between fields of expertise, promoting inter-disciplinarity and the sharing of values between them [11]. These changes are not occurring nearly as rapidly as would be desired, in spite of efforts by organizations such UNESCO [2], and accreditation programs from ABET [24] that include relevant criteria for student outcomes:

...

(c) an ability to design a system, component, or process to meet desired needs within realistic constraints such as economic, environmental, social, political, ethical, health and safety, manufacturability, and sustainability

...

(h) the broad education necessary to understand the impact of engineering solutions in a global, economic, environmental, and societal context

...

(j) a knowledge of contemporary issues

Also, EUR-ACE [25] states that transferable skills for European Engineering students must include: demonstrate awareness of the health, safety and legal issues and responsibilities of engineering practice, the impact of engineering solutions in a societal and environmental context, and commit to professional ethics, responsibilities and norms of engineering practice. Educational systems from other regions and countries have been aware for this need [26], e.g., South Africa has made an effort to integrate questions of resource efficiency into engineering curricula through course work [27] and Philippine engineering schools are in the process of transforming their systems into outcomes based in ethics, sustainability and environmental engineering. Incorporating sustainability issues in the curriculum has been a concern and universities are including these in Engineering programs. The way they are integrating these, particularly at USA institutions, was a theme for a project that consisted in a benchmarking [28].

Independent of a new program or restructure of a traditional one [29], this process implies changes in the educational paradigm. These changes must reflect a new kind of education and focus on acquiring and developing skills by active involvement of students in the learning process. According to the last UNESCO report [14] "Teaching and research are placing a new emphasis on real-world challenges to sustainability in the communities that surround campuses." Learning methodologies suitable for such development are based on active learning, cooperative and participatory activities that use student-centered approaches like Problem and Project-Based Learning (PBL) [30]–[32], service-learning projects through NGO like Engineers without Borders (EWB) or projects with support by governmental agencies like U. S. Environmental Protection Agency (EPA) [33],[34].

The change process could be more disruptive and appeal to a different learning taxonomy. It was the change wanted by Zhang et al. [35] who described a project enabling engineering faculty to more easily incorporate sustainability approaches into curricula through a proposed transformational learning practice and peer- to-peer networks. These authors incorporated and leveraged Fink's taxonomy [36] of significant learning in their textbook and the course design. They explain Fink's levels in the following manner: Foundational knowledge; Application; Integration; Human Dimensions; Caring and Learning how to Learn where the levels are interactive in a spiral and significant learning results require all of these different kinds of learning, faculty as coach and mentor, student-centred and peer-to-peer networks. To achieve this, the learning methodologies referred to above are vital for the transformation learning.

In spite of these efforts and the numerous papers which have been published about integrating sustainability issues in EE curricula, Colombo et al. [37] verified, in a review of 170 papers published in some well-known EE conferences proceedings and EE journals, that DESD was not clearly mentioned. This seems strange, and, as a result, the authors wonder if DESD has been discussed enough and/or disseminated.

DESD IN INDUSTRY

In a discussion of DESD, industry represents the economy. After all, industry has been and will be the provider of all goods and services which allow a society to enjoy a comfortable and healthy life. But improvement in economic conditions has a high price, not realized until recently. The steep price – including a diminution in resources – must be addressed [6]-[10], [12]-[16].

Particularly important for ESD in industry has been the intervention of UNEP. UNEP is the UN agency responsible for environmental activities and it was funded after the 1972 UNCHE conference [10]. It has contributed resources and reports to help industry produce in a "cleaner" and eco-efficient way. Some of these resources are Cleaner Production, in 1996 [38], Life Cycle Management: a Business Guide to Sustainability [39], Assessing Biofuels [40], Life Cycle Management [41] and decoupling natural resource use and environmental impacts from economic growth [42]. The concept of Cleaner Production was introduced by UNEP Industry and Environment in 1989. Cleaner Production is the continuous application of an integrated preventive environmental strategy applied to processes, products, and services to increase eco-efficiency and to reduce risks for humans and the environment. It applies to: 1) production processes: conserving raw materials and energy, eliminating toxic raw materials and reducing the quantity and toxicity of all emissions and wastes; 2) products: reducing negative impacts along the life cycle of a product, from raw materials extraction to its ultimate disposal, and 3) services: incorporating environmental concerns into designing and delivering services. Cleaner production requires changing attitudes, responsible environmental management, creating conducive national policy environments, and evaluating technology options [43]. Cleaner production tools include Environmental Impact Assessment (EIA), Life Cycle Assessment (LCA) among others [38].

Meanwhile, in 1995 companies concerned with the industrial impact in the environment decided to join their efforts and formed the World Business Council for Sustainable Development (WBCSD). From 34 countries and more than 20 major industrial sectors, 120 international companies shaped a global network of national and regional business councils and partner organizations united by a shared commitment to the environment and to the principles of economic growth and sustainable development. Allied with UNEP, they published a document, "Eco-efficiency and Cleaner Production: charting the course to sustainability" [43], to support the United Nations Commission on Sustainable Development (UNCSD) annual meeting held in April-May 1996 in New York. This joint effort resulted from their understanding that they had been working in the same concepts like cleaner production and eco-efficiency.

The concept of Eco-efficiency concept was defined by the Business Council for Sustainable Development (BCSD) in 1992 as "The delivery of competitively priced goods and services that satisfy human needs and bring quality of life, while progressively reducing ecological impact and resource intensity throughout the life cycle, to a level at least in line with the Earth's estimated carrying capacity." [43]. Eco-efficiency translates into the simple idea of "creating more with less" by: (i) reducing materials intensity; (ii) minimizing energy intensity in both products and services; (iii) reducing the quantity and the dispersion of toxic substances and decreasing the level of toxicity of such substances; (iv) promoting recycling and the use of renewable energy; (v) extending the durability of products, and; (vi) increasing service intensity.

WBCSD objectives were also to present and disseminate voluntary actions undertaken by industry to implement these concepts 'from the laboratory into the market' that would bring economic value to corporations and shareholders. To achieve this, they had been producing reports [44],[45] and eco-efficiency learning modules [46] that are available on their website: http://www.wbcsd.org/home.aspx. Fortunately, many companies are practicing the "Walk the talk" [47], and are transforming "Green to Gold" [48], and it is necessary that more following the example.

Recently, WBCSD launched the *"The Vision 2050 – The New Agenda for Business"* [49] report. This report lays out a pathway leading to a global population of some 9 billion people living well, within the resource limits of the planet by 2050. This intends to address the dual areas of sustainability and the main current challenge facing humanity: the high human development with minimal ecological impact.

In 2007, UNEP launched the imminent International Resource Panel (IRP), a group of top global scientists and experts "to provide independent, coherent and authoritative scientific assessment on the sustainable use of natural resources and the environmental impacts of resource use over the full life cycle." The IRP aims to "contribute to a better understanding of how to decouple human development and economic growth from environmental degradation and intends to contribute knowledge which are policy relevant and support policy framing, policy and program planning, and enable evaluation and monitoring of environmental policy effectiveness."

Meanwhile, some concepts and news ways to remake things such as Industrial Ecology, Green Production, Sustainable Engineering, Design for the Environment, Industrial Metabolism, Natural capitalism, Cradle to Grave or Cradle to Cradle have emerged in various spheres appealing to a change of attitudes and behaviors and to positively contribute to sustainable development.

DESD RELEVANCE TO PARTNERSHIP IN ENGINEERING EDUCATION AND INDUSTRY

The discussion above has set out the needs to integrate sustainability in Engineering Education curricula, in the industry engine and in business' mind-set to foster sustainable development. It is now possible to the relevance of some concerns about Sustainable Development in EE and industry. Additionally, there has been a concern to promote some links to prepare people for the labour market. One such link is discussed in UNESCO reports about DESD in 2009 [13] and in 2012 [14] that reported the need for Technical and Vocational Education and Training (TVET) as education for the world of work. These provide learning and life-skills programmes for young people and adults. According to these reports, TVET is essential for the expansion of skills and development of competencies necessary in rapidly changing labour markets. TVET is seen as one of the solutions to poverty reduction and as a means of supporting socio-economic development. Besides technical knowledge and the capacity for productive team work, they also recognize the need for people to have preparation that goes beyond basic literacy and numeracy to include both vocational and social skills, together with values that help build harmonious societies. Increasingly, TVET is considered a life-long learning programme that takes place not only in schools but also in the workplace. WBCSD, among other organizations, promotes this training. UNESCO Report from 2012 [14] noticed that business and industry are pushing TVET engagement as a need to update workers' capacities in order to respond better to the challenges of environmental crises, resource scarcity, policies requiring reductions of carbon-prints and changing consumer demands plus new green technologies.

However, much more is needed because DESD is not reaching its objective and the decade is almost over. Industry continues producing a lot of environmental waste that it is defined by US-EPA [50]: "Environmental waste is an unnecessary or excess use of resources or a substance released to the air, water, or land that could harm human health or the environment. Environmental wastes can occur when companies use resources to provide products or services to customers, and/or when customers use and dispose of products". Unfortunately, images like the one in Fig. 1, stubbornly, continue to appear in our neighbourhoods and/or in the media. "Coloured" rivers (resulted from discharges from factories) or garbage dumped on the roads (e.g., construction civil wastes) are commonly observed.

Fig. 1: Urban wastes from a city, deposited in an open land-fill

Moreover, individual behaviours, habits and activities (e.g. trash thrown out the window of the car, recycled products that aren´t conveniently separated by ignorance or by laziness in putting the recycled in the convenient container, knowledge absence about value of the discharged products such as cans, electronic, old clothes among others, a exacerbated consumption of material things) continue to reveal that many people – particularly the youngest –need education in order to understand the concepts and purposes associated with SD (see Fig. 2). These are unacceptable behaviours that are against society and nature, so DESD is not reaching all because the intended skills for SD are missing [51].

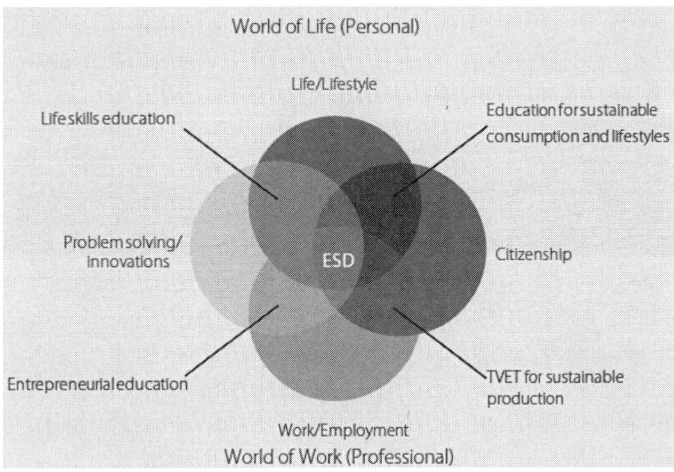

Fig. 2: Skills for Sustainable Development [51]

This puts in educators (relatives, faculty, employers,…) "shoulders" a bigger responsibility: to well-educate youngest, after all they are the "lungs" that permits society refreshment (Fig. 3).

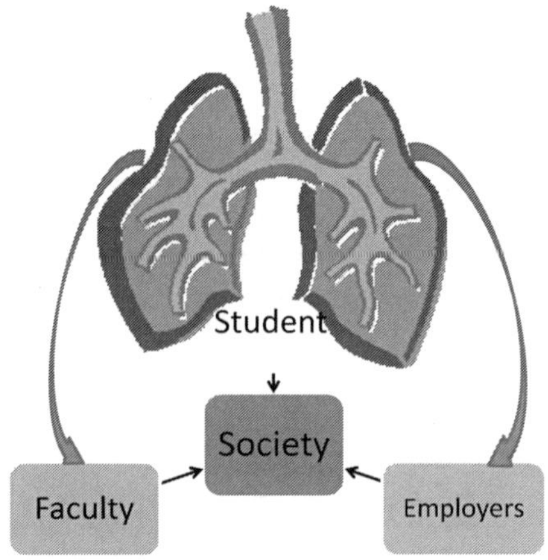

Fig. 3: Students as "lungs" of society

In higher education, where society´s future leaders and decision-makers are educated, faculty must develop programs aligned with sustainability and prepare students to change the working world towards sustainability. This presupposes faculty who have knowledge of business and industry and can established partnerships that provide a smooth transition for students. This is possible if faculty and employers work together sharing the same vision and united by the same concern. In fact, "Nowadays a world without interactions between Higher Education Institutions (HEIs) and business is inconceivable.", as reported to the European Commission by Science-to-Business Marketing Research Centre [52].

This transition is provided by some major corporations, such as Toyota or Bosch who arrange for new graduates to take part in training that educates them in the company´s management principles and tools. In most major corporations, such management principles are inspired by the Lean Production model management system, [53] that had its roots in Toyota Production System (TPS) [54],[55], and Lean Thinking principles [56] adopted to their own mind-sets and cultures. So, for the authors of this paper it is clear that Lean is well accepted internationally among companies. For this reason, the paper's authors believe that Lean provides the ideal platform to educate engineers for the workplace, providing competencies to them that industry needs. So, industry has an important role to this educational process, which the authors call Lean Engineering Education.

LEAN ENGINEERING EDUCATION

Lean Engineering Education derives from the Toyota Education Model that was taught in all factories of Toyota in Japan and abroad. Under this model, Lean is taught using Socratic methods versus didactic methods, allowing for learner-centred methods and problem-based learning. The Toyota Education Model therefore, is regarded as the mechanism which helped to build the fundamental mind-set [57],[58] for continuous improvement, also recognized by others companies that rapidly adopted such principles and training methods. Even so, faculty, traditionally tied to classical programs and naturally resistant to changes have had difficulties in accepting a different model, especially one that comes from industry. Nonetheless, the authors advocate that this is a suitable model for educating engineering graduates.

Definition

Lean Engineering Education (LEE) is [5] as:

> "A systematic, student-centered and value-enhanced approach to educational service delivery that enables students to holistically meet, lead and shape industrial, individual and societal needs by integrating comprehension, appreciation and application of tools and concepts of engineering fundamentals and professional practice through principles based on **respect for people and the environment** and continuous improvement." .

With LEE, the authors believe that students are enabled to develop problem solving skills, to think systemically and ethically, as they consider the sustainability of their designs or engineering interventions [59]. Beyond the benefits for the students, LEE also offer benefits for the academy that include the improvement of course design/delivery based on problem/project-based learning and of the overall quality of the learning experience based on student-centeredness [59],[60] and competence based in rubrics assessment [5].

It is posited that Lean Engineering Education will provide student outcomes that include three essential competencies: 1) ethics, 2) systems thinking and 3) sustainability. The mastery of these competencies must be interrelated with the content mastery in a way that resembles a double helix DNA (Fig. 4).

Fig. 4: Content and Competency Mastery combined in a double helix DNA

Ethics and system thinking competency developed in LEE engineering student graduates were discussed by Flumerfelt et al. [61], [62] and [63]. The concern here is to demonstrate how LEE students graduates foster sustainable development and how this could be the "seed" for a company endeavour into the road of sustainability.

Sustainability competency pulled by LEE

The authors begin to assuming that Lean Engineering is the natural evolution of the application of the principles of Lean Thinking to Engineering. Consequently, the authors believe that all engineers must have an education based on Lean Thinking (LT) principles. Lean Thinking [56] principles are: 1) to create value for the customer, 2) to identify the value stream, 3) to create flow, 4) to produce only what is pulled by the customer, and 5) to pursue perfection by continuous identification and elimination of waste. Waste is defined as "all activities that no contribute with added value to the products, by the point of view of the client". According to Ohno [55], there are seven types of waste: transport, inventory, motion, waiting, overproduction, over-processing and defects. These have a direct impact on environmental performance as their presence in the production system implies more use of energy and resources.

Sustainable modes of production and consumption - Overproduction is considered the worst waste as it reflects an increase of the others, i.e., more production means more demand of materials, energy, water, people, warehouses for inventory, and more entropy (motions, transports) in the system with consequences in the products 'quality (more defects), machine maintenance and repairs, more work stress, and more emissions of pollutants for air, land and water (environmental wastes). Additionally, overproduction increases the financial risks and the accidental risks when such products are not wanted by customers and are maintained in the warehouses for a "just-in-case" opportunity, exposed to fires and natural catastrophes. Naturally, efforts are made to sell these unwanted products to customers, promoting in this way consumption.

The systematic application of LT principles guides companies to a best-practice of sustainable modes of production (lean production). Lean Production is the perfect step to Cleaner Production (CP), whose definition was upgraded by WBCSD for "the international term for reducing environmental impacts from processes, products and services by using better management strategies, methods and tools. In doing so, the consumption of natural resources, pollution and waste are reduced, bringing both financial and environmental benefits" [46].

Nonetheless, a lean production mode is not enough if society continues demanding *"this world and the other"* for satisfying its whims, i.e., an unbridled consumption. So, it is important that provider and consumer be aligned, with the first (provider) supplying exactly what the second (consumer) wants, where and when he/she wants without wasting his/her time by solving his/her problem permanently. This implies that consumers must also change their consumption pattern, and, instead of a mass consumption behavior, they must adopt lean consumption [64] and be aligned with a lean production. Educating engineers with this thought will provide them with an adequate understanding that will also contribute to a reduction in the depletion of raw materials and avoid this depletion. This is aligned with one of the four goals of Chapter 36 of Agenda 21 of UNCED conference in 1992 [7] "Train the workforce: Continuing technical and vocational education of directors and workers, particularly those in trade and industry, will be enriched to enable them to adopt sustainable modes of production and consumption. This includes a social component (e.g. equity, human rights)."

Lean-Green synergy - Additionally, Lean Production provides the synergic link necessary to join the Green and the Social, discussed in International workshop, "Green and Social: Managing Synergies and Trade-Offs" [65] and avoid the depletion of scarce resources. In this workshop, von Weizsäcker [65] discussed the need to economic well-being from resource consumption, also referred in UNEP report [42]. In response to questions on the role of growth, he observed that growth will occur and could be destructive without policy interventions to decouple growth from resource consumption. The Lean-Green relationship has been widely discussed and advocated by many authors and publications, reviewed in Moreira et al. [66].

As Kidwell [67] points out, "Lean strategies coincidentally benefit the environment, without the need for special "environmental" toolkits or a separate focus on environmental considerations". So, educating engineers in LEE promotes an awareness of environment and this does not necessarily need to be addressed as a content mastery. Future engineers will develop a commitment to the symbiotic relation of Lean production and the environment through the development of eco-efficient systems [66]. It is worthwhile to emphasize the resemblance of key-ideas of Lean Production, "doing more with less", and eco-efficiency, "creating more with less".

At this point, it is also important to fit a quotation from the business side. For example: Pasquale Pistorio, president and CEO, ST Microelectronics, observes in [46], "Our experience shows that sustainability does not require sacrifice or deprivation. Rather it means exploiting the latest technologies, reducing waste, reusing resources, adopting lean production and better logistics practices and making sound investments in efficient productive capacity. In fact, eco-efficiency and sustainability are not free; for a market leader they are actually more than free because they provide a real competitive advantage.".

Socially responsible strategy - Beyond the Green, the "Social" side of sustainability competency is not overlooked by LEE, as this promotes in engineering students a waste-free way of life, adopting a different attitude in their own individual practices that will continue for their work life. At the same time, they learn to question established practices, i.e., adopting a system thinker attitude [68], eliminating in this way the eight waste that it is the untapped human potential. Lean Production is totally akin to a socially responsible strategy and it call for a dramatic reduction of all kinds of wastes as part of a whole-system thinking [69].

To conclude this discussion, the competency of sustainability is underlined in the understanding of waste, the eight identified above and the various aspects of environmental waste. Sustainability competency is a major societal concern. Such a concern underscores LEE as an excellent platform to achieve sustainable development that must be in the vision and mission of all companies (Fig. 5).

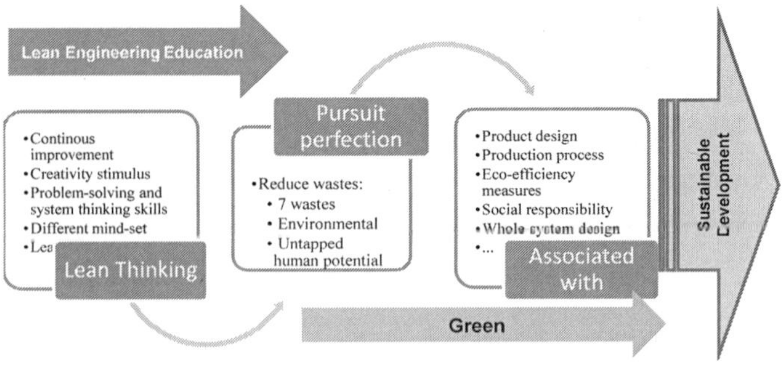

Fig. 5: LEE guides to a Sustainable Development

Sustainable development and climate change are the first of the 15 Global Challenges of the Millennium Project (MDG) [70] facing humanity and accomplishing sustainability will improve enormously life for all on the planet. Climates change is a reality that no one can escape [71] and due to this, the way of working must change, requiring the learning of adaptation of all, individuals and companies. Companies must be flexible in order to respond quickly to their customers. By applying Lean tools, companies produce only what is needed, at the right quantity and on the right time (JIT production). Other Lean tools are necessary to implement JIT production, the most important tool being the engagement and motivation of people. With engaged and thinking people the companies will have the agility for adapting to changes that are occurring [72].

CONCLUSION

This paper presents LEE as a suitable platform to educate engineers in sustainability issues. Sustainability, more than content, should be a competency in engineers' education that provides them with the necessary skills to foster Sustainable Development. Adopting LP as a strategic methodology for companies and LEE for engineering curricula, industry and faculty will be aligned in the same purposes, benefiting from this model. Industry benefits from Lean Engineering projects twice: by reducing cost and increasing productivity/profits, by reducing/eliminating waste and having people more prepared and motivated to assume a system-level thinking, ethical, sustainability and entrepreneurship attitude (continuously unsatisfied with the *status-quo*). The Academy benefits from improvement of course design/delivery and from the overall quality of the learning experience. Because of its global recognition, LEE could provide a common platform for Engineering Education in all countries and regions and, at the same time, be accepted internationally among many companies. LEE is a venue that translates well culturally, corporately and educationally.

ACKNOWLEDGMENTS

Prof. Alves is grateful to the Portuguese Foundation for Science and Technology, under Strategic Projects PEst-OE/EME/UI0252/2011.

NOMENCLATURE

ABET	Accreditation Board for Engineering and Technology
DESD	Decade of Education for Sustainable Development
DNA	Deoxyribonucleic acid
EC	European Commission
EE	Engineering Education
EFA	Education for All
EWB	Engineers Without Borders
ESD	Education for Sustainable Development
EU	European Union
EUR-ACE	European Accredited Engineering programs
LEE	Lean Engineering Education
MDG	Millennium Development Goals
NAE	National Academy of Engineering
NGO	Non-Governmental Organization
OEM	Original Equipment Manufacturer
PBL	Project-Based Learning
UK	United Kingdom
UN	United Nations
UNESCO	United Nations Educational, Scientific and Cultural Organization
UNEP	United Nations of Environment Program
USA	United States of America
US-EPA	US Environmental Protection Agency

REFERENCES

[1] UNESCO, *United Nations Decade of Education for Sustainable development 2005-2014- Draft International Implementation scheme*. UNESCO, 2005.

[2] UNESCO, *Engineering: Issues, Challenges and Opportunities for Development*. UNESCO, 2010.

[3] ASME Board on Education, "Vision 2030: Creating the Future of Mechanical Engineering Education," 2012.

[4] R. Graham, "Achieving excellence in engineering education: the ingredients of successful change," The Royal Academy of Engineering, 2012.

[5] S. Flumerfelt, F.-J. Kahlen, A. C. Alves, and A.-B. Siriban-Manalang, *Lean Engineering Education: DNA for change*. ASME Press, 2014.

[6] United Nations, "Our Common Future: Report of the World Comission on Environment and Development," 1987.

[7] United Nations, "United Nations Conference on Environment & Development: Agenda 21," in *United Nations Conference on Environment & Development:*, 1992, no. June.

[8] United Nations, *Report of the World Summit on Sustainable Development*, no. September. Johannesburg, South Africa, 2002.

[9] United Nations, "Report of the United Nations Conference on Sustainable Development Report of the United Nations Conference on," 2012.

[10] United Nations, "Declaration of the United Nations Conference on the Human Environment," 1972.

[11] UNESCO, *Educating for a Sustainable Future A Transdisciplinary Vision for Concerted Action*. 1997.

[12] The United Nations University (UNU), "Preparing for a Sustainable Future: Higher Education and Sustainable Human Development," in *World Conference on Higher Education: Higher education in the Twenty-first Century: Vision and action*, 1998, no. October, pp. 5–9.

[13] UNESCO, "Review of contexts and structures for Education for Sustainable Development 2009," 2009.

[14] UNESCO, "Shaping the Education of Tomorrow 2012 Report on the UN Decade of Education for Sustainable Development , Abridged DESD Monitoring and Evaluation," UNESCO, 2012.

[15] D. Tilbury and S. Janousek, *Development of a National Approach to Monitoring, Assessment and Reporting on the Decade of Education for Sustainable Development Stage 1 : Identification of National Indicators Summarising Documented Experiences on the Development of ESD Indicators*. Sydney: Australian Research Institute of Education for Sustainability and Australian Government Department of the Environment and Water Resources, 2006.

[16] UNESCO, "UNESCO World Conference on Education for Sustainable Development," in *Proceedings of UNESCO World Conference on Education for Sustainable Development*, 2009, no. April.

[17] UNESCO, "Bonn Declaration," in *UNESCO World Conference on Education for Sustainable Development*, 2009, no. April.

[18] UNESCO, "Education for Sustainable Development Toolkit," 2006.

[19] UNESCO, "Education for Sustainable Development Development Lens: A Policy and Practice Review Tool," 2010.

[20] UNESCO, "Exploring Sustainable Development: A Multiple-Perspective Approach," 2012.

[21] UNESCO, *Education for Sustainable Development in Action*. 2012, p. Learning & Training tools n° 4.

[22] UNEP/UNESCO, *Climate Change: Starters's Guidebook: An issues guide for Education planners and practitioners*. 2011.

[23] UNESCO/UNEP, *Youth change: climate change and lifestyles guidebook*. 2011.

[24] ABET, "Criteria for accrediting engineering programs: Effective for Reviews During the 2012-2013 Accreditation Cycle.," Baltimore: U.S.A., 2011.

[25] EUR-ACE, "Framework Standards. EUR-ACE Framework Standards," 2008. [Online]. Available: http://www.enaee.eu/app/download/4213072563/A1_EUR-ACE_Framework-Standards_2008-11-05.pdf?t=1286469542.

[26] A. C. Alves, S. Flumerfelt, F.-J. Kahlen, and A. B. S. Manalang, "Comparing Engineering Education Systems among USA, EU, Philippines and South Africa.," in *Proceedings of the ASME 2013 International Mechanical Engineering Congress & Exposition (IMECE2013), November 15-21, 2013, S. Diego, California, USA, 2013.*, 2013.

[27] Engineering Council of South Africa, "Background to Accreditation of Engineering Education Programmes," 2008.

[28] D. Allen, B. Allenby, M. Bridges, J. Crittenden, C. Davidson, S. Matthews, C. Murphy, D. Pijawka, and C. Nolt-helm, "Benchmarking Sustainability Engineering Education: Final Report: EPA Grant X3-83235101-0," 2008.

[29] C. F. Murphy, D. Allen, B. Allenby, J. Crittenden, C. I. Davidson, C. Hendrickson, and H. S. Matthews, "Sustainability in engineering education and research at U.S. universities.," *Environ. Sci. Technol.*, vol. 43, no. 15, pp. 5558–64, Aug. 2009.

[30] M. Prince, "Does active learning work? A review of the research," *J. Eng. Educ.*, vol. 93, no. 3, pp. 223–231, 2004.

[31] E. Graaff and A. Kolmos, *Management Change: implementation of Problem-Based and Project-Based Learning in Engineering*. Sense Publishers, 2007.

[32] P. C. Powell and W. Weenk, *Project-Led Engineering Education*. Utrecht: Lemma., 2003.
[33] B. P. J. Bosscher, J. S. Russell, and W. B. Stouffer, "The Sustainable Classroom : Teaching Sustainability to Tomorrow ' s Engineers 1," 2005.
[34] D. Pines and B. Gallant, "Providing Engineering Students a Global Perspective through a Project for Developing Communities Lessons Learned at the University of Hartford," in *Proceedings of the 2006 American Society for Engineering Education Annual Conference & Exposition*, 2006, no. May, pp. 2006–1188.
[35] Q. Zhang, J. Zimmerman, J. Mihelcic, and L. Vanasupa, "Civil and Environmental Engineering Education (CEEE) Transformational change: tools and strategies for Sustainability integration and assessment in Engineering Education," in *Proceedings of the 2008 American Society for Engineering Education Annual Conference & Exposition*, 2008, p. AC 2008–1670.
[36] L. D. Fink, *Creating Significant Learning Experiences: An Integrated Approach to Designing College Courses*, 1st ed. New York:: Jossey-Bass, 2003.
[37] C. R. Colombo, A. C. Alves, N. van Hattum-Janssen, and F. Moreira, "A study on impacts of the UN Decade of Education for Sustainable Development on Industrial Engineering Education," in *Proceedings of ICIEOM 2014 (Submitted)*, 2014.
[38] UNEP, *Cleaner Production: a training resource package*. 1996.
[39] UNEP, *Life Cycle Management: A Business Guide to Sustainability*. 2007.
[40] UNEP, *Towards sustainable production and use of resources: assessing biofuels*. 2009.
[41] UNEP, *Life Cycle Management: How business uses it to drecrease footprint, create opportunities and make value chains more sustainable*. 2009, pp. 181–202.
[42] UNEP, *Decoupling Natural Resource Use and Environmental Impacts from Economic Growth*. 2011.
[43] WBCSD and UNEP, *Eco-Efficiency and Cleaner Production: Charting the Course to Sustainability*. World Business Council for Sustainable Development, 1996.
[44] WBCSD, "Environmental performance and shareholder value," 1997.
[45] WBCSD, *Thinking globally, acting locally*. 2010.
[46] WBCSD, *Eco-efficiency: Learning module*. 2005.
[47] C. O. Holliday, S. Schmidheiny, and P. Watts, *Walking the talk: the business case for sustainable Development*. Greenlaf Publishing, 2002.
[48] D. C. Esty and A. Winston, *Green to Gold: How Smart Companies Use Environmental Strategy to Innovate, Create Value, and Build Competitive Advantage*. Wiley, 2009.
[49] WBCSD, "Vision 2050: the new agenda for business," 2010.
[50] US-EPA, "The Lean and Environment Toolkit," 2007.
[51] UNESCO, "Promoting Skills for Sustainable Development," 2012.
[52] T. Davey, T. Baaken, V. G. Muros, and A. Meerman, "The State of European University-Business Cooperation: Final Report - Study on the cooperation between Higher Education Institutions and public and private organizations in Europe," 2011.
[53] J. P. Womack, D. T. Jones, and D. Roos, *The Machine That Changed the World: The Story of Lean Production- Toyota's Secret Weapon in the Global Car Wars That Is Now Revolutionizing World Industry*, First edit. New York: Free Press, 1990, p. 342.
[54] Y. Monden, *Toyota Production System: practical approach to production management*, First edit. Industrial Engineering and Management Press, Institute of Industrial Engineers, 1983.
[55] T. Ohno, *The Toyota Production System: beyond large-scale production*. Productivity Press, 1988.
[56] J. Womack and D. Jones, *Lean Thinking: Banish waste and create wealth in your corporation*, 1st editio. New York, USA: Simon & Schuster, 1996.
[57] H. Takeuchi, E. Osono, and N. Shimizu, "The contradictions that drives Toyota´s success," *Harv. Bus. Rev.*, no. June, pp. 98–104, 2008.
[58] Y. Yamamoto and M. Bellgran, "Fundamental mindset that drives improvements towards lean production," *Assem. Autom.*, vol. 30, no. 2, pp. 124–130, 2010.
[59] F. J. Kahlen, S. Flumerfelt, A. B. Sinban-Manalang, and A. Alves, "Benefits of lean teaching," in *ASME 2011 International Mechanical Engineering Congress and Exposition, IMECE 2011*, 2011, vol. 5, pp. 351–358.
[60] Pawley Lean Institute, "Mission [website]. Retrieved May 17, 2011 from:," 2011. [Online]. Available: http://www.oakland.edu/?id=2461&sid=12.
[61] S. Flumerfelt, A. C. Alves, F.-J. Kahlen, and A. B. Siriban Manalang, "Why Require Ethics in Engineering?," in *Volume 5: Education and Globalization; General Topics*, 2012, vol. 5.
[62] S. Flumerfelt, A. C. Alves, and F.-J. Kahlen, "What Lean teaches us about ethics in engineering," *Proc. ASME 2013 Int. Mech. Eng. Congr. Expo. (IMECE2013), Novemb. 15-21, 2013, S. Diego, California, USA, 2013.*, 2013.
[63] F.-J. Kahlen, S. Flumerfelt, A. C. Alves, and A. B. S. Manalang, "The möbius strip of Lean Engineering and Systems Engineering," in *Proceedings of the ASME 2013 - IMECE2013*.
[64] J. P. Womack and D. T. Jones, *Lean Solutions: How Companies and Customers Can Create Value and Wealth Together*. London: Simon & Schuster UK Ltd, 2005.
[65] I. I. for S. D. (IISD), "A summary report of the International Workshop 'Green and Social: Managing synergies and trade-offs,'" *Green and Social Bulletin*, pp. 1–9, 2014.
[66] F. Moreira, A. C. Alves, and R. M. Sousa, "Towards Eco-efficient Lean Production Systems," in *IFIP Advances in Information and Communication Technology, Volume 322, Balanced Automation Systems for Future Manufacturing Networks*, 2010, pp. 100–108.
[67] M. Kidwell, "Lean Manufacturing and the Environment," *Target*, vol. 22, no. 6, pp. 13–18, 2006.
[68] A. C. Alves, J. Dinis-Carvalho, and R. M. Sousa, "Lean production as promoter of thinkers to achieve companies' agility," *Learn. Organ.*, vol. 19, no. 3, pp. 219–237, 2012.
[69] A. B. Lovins, L. H. Lovins, and P. Hawken, "A Road Map for Natural Capitalism," H. B. R. on G. B. Strategy, Ed. Harvard Business School Press, 2007.
[70] United Nations, "The Millennium Development Goals Report 2011," 2011.
[71] I. I. for S. D. (IISD), "Summary of the 10th session of Working Group II of the Intergovernmmental Panel on Climate Change (IPCC) and the Thirty-eight session of the IPCC," *Earth Negotiations Bulletin*, vol. 12, no. 596, pp. 1–20, 2014.
[72] L. C. Maia, A. C. Alves, and C. P. Leão, "Sustainable work environment with lean production in textile and clothing industry," *Int. J. Ind. Eng. Manag.*, vol. 4, no. 3, pp. 183–190, 2013.

Proceedings of the ASME 2014 International Mechanical Engineering Congress and Exposition
IMECE2014
November 14-20, 2014, Montreal, Quebec, Canada

IMECE2014-40249

SHORTEN THE MATH GAP FOR PRE-ENGINEERING STUDENTS WITH INTENSIVE SUMMER BRIDGE PROGRAM

Yong Zhou
School of Engineering and Computational Science
University of Texas at Brownsville
Brownsville, TX, USA

Nazmul Islam
School of Engineering and Computational Science
University of Texas at Brownsville
Brownsville, TX, USA

Cheng-Chang (Sam) Pan
College of Education
University of Texas at Brownsville
Brownsville, TX, USA

Sanjay Kumar
School of Engineering and Computational Science
University of Texas at Brownsville
Brownsville, TX, USA

ABSTRACT

Engineering Summer Bridge (ESB) program at the University of Texas at Brownsville (UTB) is designed to prepare the engineering freshmen intellectually for an early introduction to the engineering culture and mathematics and science expectation. The program curriculum and content were specifically designed to prepare underrepresented Hispanic students for their success in the coming science and engineering study at UT-Brownsville. More than 92% of the targeted students are underrepresented Hispanic, and English is the second language for 86% of them. Most of these targeted students are academically below the top 10% in their high school graduating classes due to the pre-selection of TOP10 Texas House Bill. The ESB program at UTB cultivates a diverse community of engineering and pre-engineering students and intensively enhances their mathematics preparation in Pre-Calculus and College Algebra. Statistics data from 2012 and 2013 ESB program indicates that more than 81% of the participants in both years did not take Pre-Calculus in high school. Another finding is that 71% of the ESB participants with at least an attendance rate of 50% earned a grade higher than a B in their Calculus I class later on, while only 43% from the group with an attendance rate lower than 50% earned a grade higher than a B in the Calculus I class. Students seem more successful in their Calculus I study if they attend the classes more frequently. It is also found the early contact with engineering faculty through Summer Bridge Programs, together with an early and longer engineering orientation seminar during the program, are successful ways to assist in the retention of engineering freshman [1-2].

INTRODUCTION

Mathematics and science are the two of the primary tools of engineering. The earlier the student learns to master mathematics and science at an appropriate level, the better he/she can manage the tools and apply them to problem solving. Lack of preparation in mathematics is one of the main factors contributing to the student dropout rate in engineering at UTB. Internal data at UTB have shown that more than 60% of entering freshman engineering students are not Calculus ready. Knowing that the UTB engineering program assumes incoming engineering students are Calculus ready, the fact that 70% of entering freshmen engineering students are under-prepared for the program due to deficiency in mathematics is deemed an issue. Studies [3] have indicated that the performance in the first mathematics class for those students retained to their sophomore year is significantly different to those who are not. In fact this is a common concern in many engineering programs and special mathematics sessions are usually included in most of the successful summer bridge programs [4-5] to address the deficiencies in entering freshmen preparedness for the engineering curriculum. As the result of the deficiency in mathematics skills, students tend to perform poorly in mathematics-related engineering classes and take longer time to graduate, which leads to low six-year graduation rates (as measured by the Integrated Postsecondary Education Data System, IPEDS) from engineering. Calculus courses are fundamental to the engineering curriculum as they provide a critical foundation for future engineering courses. Therefore, success in Calculus is important to succeed in engineering. Prior studies support the fact that freshman students entering engineering programs lack the ability to progress through the Calculus sequence [6-7]. Furthermore, research indicates that those students who failed had entered engineering programs without the ability to solve multi-step problems or higher-order thinking skills necessary to solve engineering problems. Those students, unfortunately, have yet to develop a questioning or critical-thinking attitude necessary for the fundamental understanding of the meaning behind the mathematics concepts presented. These essential skills are required at all levels of the curriculum to enhance success in engineering. The one-week mathematics bridge program at Purdue allowed students to discover strengths and weaknesses and improve preparation in mathematics. This program was one of the most effective pre-college retention programs offered by the schools of engineering. Some other bridge programs are 10 weeks long and the participants received course credits for participation [4]. Many of the programs offered were free of charge and included scholarships. The University of Maryland ran engineering retention program [5] and tried to align high school preparation with expected standards of engineering freshman majors. This program focused on improving academic success and included life-skills and social and motivational components. The students who attended this program benefited largely in terms of adapting and succeeding in the first semester.

The Rio Grande Valley of South Texas is one of the fastest growing economic regions in the United States. There is an increasing need for graduates well trained in science, technology, engineering, and mathematics (STEM) fields. For years, UTB STEM graduates have contributed to the employee diversity that the state s high-technology companies need to stay nationally- and globally-competitive and innovative. However, underrepresented minority engineering discipline graduates, are still notably few; it is also a nationally-recognized fact [8-10]. The US Hispanic population is predicted to triple and steadily grow up to 30% of the total population in 2050 [11]. However, statistics indicates that only 7.2% of engineering bachelor s degrees were earned by Hispanic graduates in 2008, and many of these students are the first generation in their families to attend college and typically come from a lower socioeconomic background.

UTB is an exception to national statistics because it has the potential to support more Hispanic students than is possible at many other Hispanic-serving institutions (HIS). This is because 86% of entering freshmen declare STEM majors, and most of them are in Engineering or Medical Sciences. This has been attributed to the students receiving an overly abundant amount of advice from school counselors, peers, and family members about becoming an engineer or pursuing any STEM-related career. Most of the advice is given to students from people that are unaware of the amount of mathematics that is necessary for engineering studies. In fact, more than 60% of engineering freshmen enter the university with engineering enthusiasm but unfortunately without having taken Pre-Calculus at high school.

Since the UTB Engineering Physics program earns ABET accreditation in 2010, the engineering enrollment has kept increasing by 15% annually. However, most of the engineering students at UTB are academically under the top 10% in their high school classes, due to the pre-selection of the top 10% students by the Texas flagship universities (Texas House Bill 588). Statistics also indicates that less than 30% among those who claim engineering majors are still attending engineering classes beyond the sophomore year. In an analysis of those students who have declared engineering as their majors, the results indicate that they either drop out or change majors before they start taking upper level engineering courses. While the UTB engineering curriculum is designed based on the Pre-Calculus entry level, for years the UTB faculty covering various courses has recognized the mathematics obstacles to most of the students that performed poorly in their early college study. Many of the mathematics topics and skills that should have been covered and mastered in high school are found missing. For the engineering curriculum, these topics include trigonometric functions, logarithms, exponentials, and complex numbers and operations, etc. This lack of mathematics preparation may have both lengthened the time required for their college study or led to their early exit even before they are officially admitted to an engineering program, thus will apparently affect the retention rate of the first year.

The intention of the work introduced in this paper is to develop effective and efficient approaches to the preparation of incoming freshmen during the period of matriculation for their college and engineering readiness in mathematics and science. The term of matriculation is defined by Glenn and Landis [12] as activities and period between the time the students are admitted and their first semester of enrollment. In this context, the UTB Engineering Summer Bridge (ESB) program is designed to bridge and prepare high school seniors before their first semester in engineering study.

Curriculum and content of the ESB program at UTB is specifically designed to mathematically prepare the underrepresented Hispanic students for their success in the coming science and engineering study, to early alert on the mathematic challenges in their engineering study, and to provide an early introduction of the university s expectations and culture. The program will focus on university readiness of the incoming students. Besides, the ESB program also cultivates a diverse learning community of engineering students and prepares them to engage actively with faculty and fellow students. First-year university students that require developmental education in English reading, writing, or mathematics will become university-ready in those subject areas through the designed ESB summer program. Specific goals of this summer bridge programs include:

- Reduce the mathematic gap between high school and university engineering curriculum in the areas of College Algebra, Pre-Calculus and Introduction to Engineering.
- Introduce engineering expectations and provide early emphasis for incoming freshmen in engineering curriculum.
- Reduce the need for developmental education for students that almost university-ready by passing the COMPASS or Pre-Calculus Test administered by the UTB Mathematics Department;

IMPLEMENTATION OF PROGRAM

The UTB summer bridge program is offered for four weeks from July to August during the summer since 2012. The programs are offered free of charge under the support of Minority Science and Engineering Improvement Program (MSEIP) from US Department of Education. The bridge program actively engages the students with the pre-calculus topics and the modern engineering design concepts and tools. The ESB program at UTB was primarily created to ease the transition into the first mathematic course-Calculus I, and the first introductory engineering class-Introduction to Engineering (ENGR 1101). The bridge program was offered directly by the UTB engineering faculty members that have first-hand experience on the mathematic weakness in their engineering major course teaching. Because of that, the program content is uniquely designed to interface with the integrated engineering curriculum. The program consists of a series of workshops intended for high school graduates who are interested in pursuing a degree in engineering from the University of Texas at Brownsville. It also serves as a way to further motivate students to seek a career in engineering while helping them prepare for their engineering studies. In order for a student to be eligible to participate, the student must be enrolled as a UTB student and must be intent on seeking a degree in engineering. The program was held on a daily basis Monday through Friday from 9AM to 2PM. The meetings consist of two types of the activities:

Lectures

The students were exposed to various basic and essential math topics in engineering, including functions, polynomial equations, logarithms, trigonometry, complex numbers and operations. Mathematics was covered on a daily basis and in details to allow students to be easily engaged and participate in the lecture. Engineering topics were also included in the daily lesson plan. The students became acquainted with engineering ethics, the various engineering degrees and their respective statistics. Educational videos were occasionally shown in class to illustrate the history and importance of engineering in the accident Aztecs and Greek civilization. The students were also introduced to thc basic concepts of forces and electricity through visual. PowerPoint presentations were used to demonstrate accurate representation of engineering signals and how they relate to sine and cosine, and how to obtain the respective frequency and period of the signal. Lectures also cover mathematics expectations of the engineering program for incoming freshmen in the areas of College Algebra, Pre-Calculus, and Introduction to Engineering.

Assignments and Testing Tools

Students were handed math homework assignments to practice the lesson covered for the day and were asked to demonstrate their knowledge through several quizzes. Other assignments involved report writing on a given subject and present their work orally to the class. This allowed the student to conduct oral and written communication skills.

Laboratories and Field Trips

The students were given a tour of all the engineering laboratories and had the opportunity to learn firsthand experience by working in the electronics lab. There students learned how to construct the basic circuit and use the digital multimeter to measure AC voltage and current across a resistor and display the sinusoidal waveform on the oscilloscope screen. MATLAB was introduced and practiced daily. This software is an essential tool for engineers when they need to solve complex problems. The basic principles of flight were introduced in class and a wind tunnel in the Fluid Mechanics lab demonstrated how an airplane wing experiences less drag and more lift. Other activity includes adopting real-life Introductory engineering projects practice and use mathematics in real life engineering projects: Fly a plane in the wind tunnel, test and fire a rocket engine on a model rocket. Field trips to local industry and power plant were taken to show students what engineers do in their respective areas. These real world engineering practices are designed to emphasize the significance of mathematic knowledge in engineering.

The program data collected from ESB summer 2012 and 2013 is elaborated as follows:

No. of the students registered for the ESB program:

- There were 11 students that signed up for the ESB 2012 program.
- There were nine students that signed up for the ESB 2013 program.

No. of the participants declared a major in Engineering during or after ESB:

- 91% of the students declared a major in engineering during or after ESB 2012 program.
- 89% of the participants declared a major in engineering during or after ESB 2013.

No. of the participants in both cohorts (years) that did not take Pre-Calculus in high school:

- 81.8% of the participants in the 2012 program reported that they did not take Pre-Calculus in high school.

— None of the participants in the 2013 program took Pre-Calculus in high school.

No. of the participants took Calculus I during or after ESB (summer) 2012 and 2013:

— 73% of the participants took Calculus I during or after ESB (summer) 2012.

— At least 67% of the participants took Calculus I during or after ESB (summer) 2013.

No. of the participants who declared a major in engineering waited for at least a year before they took Calculus I:

— 55% of the participants (as an engineering major) in the ESB 2012 program waited for at least three semesters (including summer) before they took Calculus I.

— Over 11% of the participants in the ESB 2013 program waited for at least three semesters (including summer) before they took Calculus I.

No. of the participants passed the Pre-Calculus Test/Exam:

— 36% of the 2012 participants had passed the Pre-Calculus Test.

— 33% of the 2013 participants had passed the Pre-Calculus Test.

No. of the female participating students in 2012 and 2013, respectively:

— In terms of sex (gender), the male to female ratio in 2012 was 9 to 2. The male to female ratio in 2013 was 8 to 1.

The program attendance rate by year and by sex:

— 46% of the participants in 2012 reported an attendance rate over 49%.

— 67% of the participants in 2013 reported an attendance rate over 49%.

— 55% of the participants in 2012 with no lower than a 50% attendance rate were male. Female students attendance rate was below 50%.

— 83% of the participants in 2013 with no lower than a 50% attendance rate were male. The female student s attendance rate was above 49%.

Cross-tabulation with Attendance rate and Calculus I Grade:

— 75% of those with at least an attendance rate of 50% in ESB 2012 earned a grade higher than a B in the Calculus I class. Only 40% of those with less than a 50% attendance rate in ESB 2012 earned a grade higher than a B in Calculus I.

— 67% of those with at least an attendance rate of 50% in ESB 2013 earned a grade higher than a B in the Calculus I class. And, 50% of those with less than a 50% attendance rate in ESB 2013 earned a grade higher than a B in the same class.

— With the two programs combined, 71% of the ESB participants with at least an attendance rate of 50% earned a grade higher than a B in the Calculus I class. In the group with an attendance rate lower than 50%, 43% earned a grade higher than a B in the Calculus class.

The descriptive data presented above are intended to give preliminary findings. When the quantitative data from the two years are collapsed, students seem more successful in Calculus I if they attend the classes more frequently. It is worth noting that given the small number of the participating students and the non-mandatory nature of summer programs, the findings are limited for further interpretation. To portray a fuller picture of how effective the programs really are may require a longitudinal investigation.

CONCLUSION

First-generation college-goers experience a variety of challenges as they enter and move through higher education. The Engineering Summer Bridge provides students with specific types of resources and support to ensure that they move into and through engineering study smoothly and to shorten the time for their engineering study. Current data from 2012 and 2013 ESB program have suggested an 28% more of the students who receive a grade higher than a B in the Calculus I class come from the group with at least a 50% attendance rate than those with an attendance rate lower than 50%. Assessment of student strengths in mathematics and other engineering-related areas will be further carried out through administering first- and second-year retention rate and the ESB program in next years to further confirm the student success as well as the program s efficacy.

ACKNOWLEDGMENTS

This work is partly supported by the Department of Education of the United States through the MSEIP program. The authors would like to thank Itzitzy Godinez for gathering the data necessary for the writing of this paper and for her willingness to continually respond to our requests for additional program information.

REFERENCES

[1] Reyes, M., Anderson-Rowland, M., & McCartney, M., Freshman introductory engineering seminar course: Coupled with bridge program equals academic success and retention , The ASEE/IEEE Frontiers in Education Conference, Tempe, Arizona, 1998

[2] Klingbeil, N. W., Mercer, R. E., Rattan, K. S., Raymer, M. L., & Reynolds, D. B., Rethinking engineering mathematics education: a model for increased retention, motivation and success in engineering . The American Society for Engineering Education Annual Conference & Exposition, Salt Lake City, Utah, 2004

[3] Anderson-Rowland, Mary R., Retention: Are Students Good Predictors? Proceedings of Frontier in Education Conference, Pittsburgh, Session 3553, Pennsylvania, Nov. 1997

[4] Reichert, Monty, Taking another look at educating African engineers. Journal of Engineering Education 86, No. 3: 241 253, 1997

[5] Nagchaudhuri, A, and G. Singh, Summer engineering bridge program at the University of Maryland Eastern Shore: Objectives and enrichment activities. American Society for Engineering Education Annual Conference and Exposition, 2001

[6] Sathianathan, D, STavener, K Voss, Armentrout, P Yaeger, and RMarra, Using applied engineering problems in calculus classes to promote learning in context and teamwork. Frontiers in Education: 12-14, 1999

[7] Fowler, D A, D. A. Maxwell, and J. E. Froyd, Learning strategy growth not what expected after two years through engineering curriculum. ASEE Conference Proceedings. Nashville, Tennessee, 2003

[8] National Science Foundation, Women, Minorities, and Persons with Disabilities in Science and Engineering , 2008

[9] Michael F. Summers and Freeman A. Hrabowski III, Preparing Minority Scientists and Engineers , Science 311: 1870-71, 2006

[10] Raymond B. Landis, Retention by Design: Achieving Excellence in Minority Engineering Education , White Plains, New York: National Action Council for Minorities in Engineering, 2005

[11] US Census Bureau, National Population Projections , 2008

[12] Glenn, R., and Landis, R., Matriculation and Summer Bridge Programs, NACME Inc., New York, 1985.

This page left blank intentionally.

Proceedings of the ASME 2014 International Mechanical Engineering Congress and Exposition
IMECE2014
November 14-20, 2014, Montreal, Quebec, Canada

IMECE2014-37064

ALGORITHM MODIFICATION APPROACH TO IMPROVE THE KINECT'S PERFORMANCE IN POINT CLOUD PROCESSING

Mingshao Zhang
Stevens Institute of Technology
Hoboken, New Jersey, USA

Zhou Zhang
Stevens Institute of Technology
Hoboken, New Jersey, USA

Sven K. Esche
Stevens Institute of Technology
Hoboken, New Jersey, USA

Constantin Chassapis
Stevens Institute of Technology
Hoboken, New Jersey, USA

1. ABSTRACT

Since its introduction in 2010, Microsoft's Kinect input device for game consoles and computers has shown its great potential in a large number of applications, including game development, research and education. Many of these implementations are still in the prototype stages and exhibit a somewhat limited performance. These limitations are mainly caused by the quality of the point clouds generated by the Kinect, which include limited range, high dependency on surface properties, shadowing, low depth accuracy, etc. One of the Kinect's most significant limitations is the low accuracy and high error associated with its point cloud. The severity of these defects varies with the points' locations in the Kinect's camera coordinate system. The available traditional algorithms for processing point clouds are based on the same assumption that input point clouds are perfect and have the same characteristics throughout the entire point cloud.

In the first part of this paper, the Kinect's point cloud properties (including resolution, depth accuracy, noise level and error) and their dependency on the point pixel location will be systematically studied. Second, the Kinect's calibration, both by hardware and software approaches, will be explored and methods for improving the quality of its output point clouds will be identified. Then, modified algorithms adapted to the Kinect's unique properties will be introduced. This method allows to better judge the output point cloud properties in a quantifiable manner and then to modify traditional computer vision algorithms by adjusting their assumptions regarding the input cloud properties to the actual parameters of the Kinect. Finally, the modified algorithms will be tested in a prototype application, which shows that the Kinect does have the potential for successful usage in educational applications if the according algorithms are design properly.

Keywords: Microsoft Kinect, Kinect limitation, Algorithms modification, Calibration.

2. INTRODUCTION

2.1. OVERVIEW OF KINECT RELATED DEVELOPMENTS

Since its release in November 2010, the Kinect has not only become one of the most popular game consoles and has been widely accepted by consumers, but it also captured the attention of researchers and developers. A large number of Kinect-based applications have emerged since then and the Kinect has been proved to have potential in various areas. The most common applications can be categorized into two groups based on the methods used to process the Kinect's raw data.

Applications of the first kind are known as human skeleton tracking, which are popular for two reasons. First, the Kinect was originally designed as a game interface that uses gestures of the human body as command inputs to control games. Second, Microsoft provided a Software Development Kit for the Kinect, in which the information on the major joints of the human skeleton is available to users, thus easing the difficulty of the development work. These applications include human detection [1], physical rehabilitation [2], sign language recognition [3], etc.

The applications in the other group mostly deal with the point cloud generated by the Kinect directly. Based on the hardware components, the Kinect is essentially a structured-light 3-D scanner. This kind of applications is mostly motivated by the intention to use the Kinect as a substitute for commercial 3-D scanners, due to their expensive price and unavailability to consumers. Among these applications, there are still some that focus on human body related developments, such as hand gesture tracking [4] and face scans [5]. The other ones are similar to traditional 3-D scanner applications, such as indoor environments modeling [6,7], 3-D surface reconstruction [8,9,10], alternative educational sensor [11,12], etc.

Developing applications based on human skeleton recognition and tracking is still a very popular topic and many related developments are reported every year. However, the number of reports about Kinect-based research that is directly related to point cloud processing has been decreasing lately, and furthermore a lot of such applications stop at the prototype stage. One of the main reasons that cause the difference between these two kinds of development is that the Kinect has a lower resolution and accuracy as well as a higher error rate than traditional 3-D scanners. When calculating skeleton notes from the point cloud (e.g. in full-body tracking), coordinate information of 20 skeleton notes are extracted from a much larger number of points (e.g. 20,000 points). As one result, some influence of random error in certain local regions of the point cloud gets eliminated. Also, skeleton-related projects usually have lower requirements for the accuracy of the results, and the quality of the Kinect output point clouds can still satisfy these requirements.

2.2. KINECT'S POTENTIAL IN DISTANCE ENGINEERING EDCATION

Upon the development of suitable image processing algorithms, the Kinect has the potential for becoming a versatile and adjustable 3-D range sensor with numerous options for educational and industrial usage. Taking educational laboratories as an example, the Kinect will be able to meet some fundamental requirements, such as a short time for learning the use of the system, ease of operation, the ability to add resources as the students' knowledge grows and robustness to prevent certain kinds of error [13]. Furthermore, the ability to track the position, velocity, acceleration and even deformation of multiple moving objects in three-dimensional space would make the Kinect superior to most of the existing data acquisition systems in educational laboratories.

2.3. OBSTICLES FOR KINECT'S FURTHER IMPLEMENTATIONS

As a consequence of the Kinect being a structured light scanner, it inherits their functional and practical operating limitations. Among these limitations, several have a direct effect on common depth imaging applications, such as range, surface properties and shadowing [14].

The minimum range of structured light sensors is determined by the range at which the reflected light pattern exceeds the camera's field of view or is not sufficiently bright to saturate the image sensor. The maximum range depends on the sensor's ability to resolve light reflected from distant objects. For the Kinect, the angular range is 57°×43° (horizontal × vertical) and the recommended distance range is 0.6-6.0 m. The surface properties of the resulting range images partially depend on the properties of the detected object surfaces. Highly reflective surfaces (e.g. mirrors or shiny metal) produce specular highlights while highly absorptive surfaces (e.g. fluffy or dark materials) prevent the return of light patterns to the scanner. Both cases result in "holes" in the acquired images. Shadowing is caused by the pattern light projector being positioned slightly to the side of the camera. All regions that are "blocked" by objects in the scene from the projector's point of view but are still "visible" to the depth camera lose their depth value and become "holes" in the resulting images.

As with any structured light system, the performance in acquiring depth data can degrade in difficult lighting conditions. Bright fluorescent lighting can increase the noise in the data, while low illumination can cause failure in the depth data acquisition. The second generation of the Kinect was introduced with the release of the new game console – the Xbox One. It uses timing and a dual ported sensor array simultaneously to obtain depth data and neutralize problematic lighting conditions [15]. It is believed that the new Kinect can have improved performance under difficult lighting conditions.

The limitations outlined above should all be taken into consideration when designing Kinect-related applications. However, they are either unquantifiable or can only be improved by redesigning the Kinect's hardware. Among all the critical limitations that the Kinect exhibits, there are two kinds that are worth mentioning, namely the lens distortion of the Kinect's IR and RGB cameras [16] and random measurement error in the sensor's depth direction [17]. One of the consequences of these two error sources is the fact that points throughout the Kinect's resulting point cloud have different characteristics regarding resolution, accuracy, etc. While most of the existing 3-D computer vision algorithms are constructed based on the assumption that input point clouds are either perfect or all the points have the same properties across the entire point cloud, it is to be expected that if this conflict between the assumptions of traditional computer vision algorithms and the Kinect's unique characteristics can be resolved, the quality of Kinect applications that are based directly on point cloud processing can be greatly improved.

In Section 3 below, the two error sources mentioned above will be studied, quantified and their influence on the quality of point cloud processing discussed. In Section 4, existing methods for calibrating the Kinect and enhancing the point cloud quality will be introduced. In Section 5, a systematic method for modifying traditional computer vision algorithms according to the Kinect's unique characteristics will be introduced and some improved results will be presented. These modified algorithms aiming to enhance the Kinect's performance have the potential to facilitate its application in educational and industrial areas.

3. STUDY OF THE KINECT'S TECHNICAL SPECS

As mentioned above, the first kind of quantifiable error source are lens distortions of the Kinect's IR and RGB cameras, especially "barrel distortion", which causes images to be either spherised or inflated. Barrel distortion is a side effect of the round glass elements within the lens, which "bend" the light beams near the edges of the lens more than the ones near the center of the lens. To assess the effect of lens distortion, a well-known model developed by Brown [18] is often used. This is also the model that has been used in a lot of Kinect-related research [16,17,19].

The assessment result reported by Khoshelham et al. [17] indicated that the overall root mean square (RMS) error in image space is 0.14 pixel for the IR images and 0.09 pixel for the RGB images. The combined effect of radial and decentered distortions for both the IR and RGB cameras can be as little as none when the point is close to the 2-D image center and as large as 9 pixels when it is located at the image corner. Similar results were also published by other researchers. It is worth noting that for both RGB and IR cameras, the lens distortion effects are not statistically significant. From the theoretical camera model, it is known that the image resolution in the x and y directions (57×43 degree view angle and default mode of 640×480 pixels, although the camera hardware is capable of producing 1280×1024 pixels) yields:

$$
\begin{aligned}
q_x(x,z) &= \frac{57\pi/180}{640}\sqrt{x^2+z^2} = 0.001554\sqrt{x^2+z^2} \\
q_y(y,z) &= \frac{43\pi/180}{480}\sqrt{y^2+z^2} = 0.001564\sqrt{x^2+z^2}
\end{aligned}
\tag{1}
$$

Here, the x-y coordinate center is aligned with the camera image center. Based on this equation and Brown's model, the effect of lens distortion from both the RGB and IR cameras can be quantified.

The other error source is the random error of the depth measurements, which is assumed to be a random variable with a normal distribution and standard deviation [17]:

$$\sigma_z(z) = q_z(z)\times\sigma_d \tag{2}$$

Here, σ_d is the standard deviation of the measured normalized disparity and $q_z(z)$ is the depth resolution (i.e. the minimum depth difference that can be measured). The depth resolution is a quadratic function of z and the interpolated result tested by Smisek et al. [16] shows that it follows the rule:

$$q_z(z) = 2.73z^2 + 0.74z - 0.58 \tag{3}$$

Here, z is expressed in meters while the result $q_z(z)$ is expressed in millimeters. The minimal range is 0.5 m. Since the depth is involved in the calculation of the planimetric coordinates (x and y coordinates in this case), by propagating the depth measurement error and ignoring the x, y coordinates' error caused by lens distortion, the random error of the x and y coordinates can be obtained as:

$$
\begin{aligned}
\sigma_x(x,z) &= \frac{x}{f}q_z(z)\sigma_d \\
\sigma_y(y,z) &= \frac{y}{f}q_z(z)\sigma_d
\end{aligned}
\tag{4}
$$

Here, f represents the focal length of the infrared camera. The error of the depth measurements is proportional to the square distance from the sensor to the object, and it increases quadratically from a few millimeters when the distance is close to about 4 cm when the distance is 5 m. The error caused by the camera lens distortion is comparatively smaller than that caused by the depth measurement, especially for points that are located near the 2-D image center (i.e. have small x, y values).

4. CALIBRATION AND IMAGE ENHANCEMENT APPROACHES TO IMPROVE THE KINECT'S OUTPUT QUALITY

The intrinsic parameters (e.g. focal length, center of projection, distortion parameters, etc.) of both the RGB and IR cameras of the Kinect and their relative poses are pre-calibrated by the manufacturer and stored in the firmware, based on the average performance of the Kinect sensors. However, these camera parameters usually vary between devices, and furthermore, the factory calibration is not accurate enough for many applications. This has resulted in many developments of calibration methods specifically for the Kinect. Canessa et al. [20] proposed a method for obtaining 3-D measurements for a multi-Kinect setup through the correction of the depth measurements. The traditional checkerboard calibration method failed to work well with the Kinect because its corner feature cannot be reliably detected by the depth camera. Zhang et al. [21] improved this method by using a maximum likelihood algorithm to calibrate the color and depth sensors jointly. Herrera et al. [22] proposed using the Kinect in conjunction with an external high-resolution color camera to correct the depth distortion of the Kinect's depth sensor. This approach was then further improved by Raposo et al. [23] for fewer input image frames and faster run times. Kim et al. [24] also proposed a calibration method for correcting the triangulation error by using the Kinect together with an external color camera. Yamazoe et al. [25] proposed to improve the results of the traditional methods by including the calibration of the Kinect's projector-related parameters, instead of just calibrating the camera-related parameters.

The output range images of the Kinect are noisy, which can be the result of inadequate calibration, sensor noise or round-off error during normalization. The point cloud also exhibits many missing values due to the properties of the detected surfaces and shadowing [14]. Researchers have begun to explore the sensor characteristics of the Kinect [16,17,26], and various methods have been proposed for denoising and estimating the missing depth values (also known as hole filling approach) at the same time. Camplani et al. [27] proposed to use an adaptive joint-bilateral filter with a depth-color fusion strategy to reduce the depth noise and interpolated the missing depth values. Shen et al. [28] proposed a method that uses a depth layer labeling method and a probabilistic model to capture various types of uncertainties in the depth measuring process, which can then be used to denoise and fill in the missing values of the depth maps. Simone et al. [29] also proposed a joint denoising and interpolation approach for the Kinect sensor.

The Kinect produces depth images with comparatively low resolution (640×480 pixels), which is inadequate for many applications. While many depth map up-sampling approaches can be used to resolve this issue, most of them require an external color camera for producing a registered high-resolution texture image as reference. The algorithms used in these approaches include joint bilateral up-sampling [30], guided image filtering [31], non-local means filtering in a MRF-based up-sampling scheme [32], cost-volume filtering [33,34] the NL-means formulation for denoising, up-sampling [35], etc.

Regarding the concerns of long computation times for 3-D image processing, Wasza et al. [36] developed a GPU based pipeline for processing dense 3-D range image streams, which was designed to enable real-time pre-processing of high-resolution range data.

5. MODIFICATION OF TRADITIONAL ALGORITHMS USING THE KINECT'S PROPERTIES

5.1. OVERVIEW OF APPROACHES

For a lot of applications that deal with point clouds directly (e.g. 3-D environment reconstruction, object recognition and motion tracking), the Kinect shows its inadequacy in accuracy, error rates and noise level. These limitations can be overcome by either calibration or depth map enhancement (i.e. denoising, up-sampling and hole filling as mentioned above). However, most of these methods require additional hardware setup modifications (e.g. an additional external color camera), which adds to the operational complexity for developers, in both the hardware and software aspects. Modifications of traditional algorithms with the consideration of the Kinect's unique technical aspects have captured our attention, which hopefully can achieve acceptable results in those applications as mentioned using only the Kinect's low quality raw depth and color images, without the help of any hardware modifications.

As shown in Section 3, compared to the RGB and IR camera distortions, the random error associated with the depth measurements has a larger impact on the resulting range image quality. Therefore, modifications of traditional computer vision algorithms regarding the influence of the Kinect's depth measurement error have been chosen as the main focus in the work presented here. As examples of this algorithm development, three kinds of common computer vision algorithms were chosen, including noise removal, segmentation and keypoint detection.

5.2. NOISE REMOVAL ALGORITHMS

Points that have error in the depth measurement become sparse outliers (i.e. noise), which have a large impact on the result of many point cloud processing algorithms and cannot be ignored. One main reason for this is that sparse outliers complicate the estimation of the local point cloud characteristics such as surface normal or curvature changes, thus leading to erroneous values. One commonly used approach for estimating the correct depth value of those sparse outliers is based on using their local information (e.g. interpolation methods). However, these methods (e.g. polynomial interpolation) are very costly in terms of computation time.

Some of these irregularities can be resolved by performing a statistical analysis on each point's neighborhood, and trimming those points that do not meet certain criteria (also known as statistical outlier removal filtering [37]). For each point, the mean distance $\bar{x}$ from it and its K (number preset by the user) nearest neighbors is computed, and all points whose mean distances are outside of an interval (defined by the global distances' mean value μ and standard deviation σ) can be considered as outliers and trimmed according to:

$$\bar{x} > \mu + n\sigma \tag{5}$$

Here, n is a standard deviation multiplier that is selected by the user. This statistical outlier removal filter is a very practical approach for trimming sparse outliers in point clouds. As an example, a point cloud of a laboratory was captured by the Kinect and filtered using this method, with all the parameters set according to our past experience in point cloud processing.

The result can be seen in parts (c) and (d) of Figure 1. In part (d) of Figure 1, especially in the upper left corner of the image, holes appear where there is supposed to be a flat wall. This means that many points originally belonging to the wall were trimmed as outliers, which obviously is not desirable. This error is introduced by the fact that traditional statistical filters assume that all points across the whole point cloud have the same characteristics (i.e. resolution, error rate and noise level). These assumptions do not hold in Kinect applications for two reasons. First, as shown in Equations (1) and (3), it is seen that the resolution of the Kinect's output point cloud is coordinate dependent, meaning that the average distance between two nearest points varies when the points are located in different parts of the point cloud. When calculating for each point the average distance to its K nearest neighbors in the first step of statistical outlier removal methods, it is assumed that the resulting distance distribution is Gaussian with a mean value and a standard deviation. Second, the algorithms used for noise removal throughout the whole point cloud only function well when the noise characteristics are coordinate independent (i.e. normal distribution with same mean and standard deviation or random distribution with same maximum). However, as indicated in Equations (2) and (4), the error level varies for different coordinates. In fact, it increases quadratically with increasing distance. The effect of this property can be observed clearly in part (a) of Figure 1. The increasing measurement error in the depth direction distorted what should be a flat wall into numerous local point clusters, which then fall into several layers of depth.

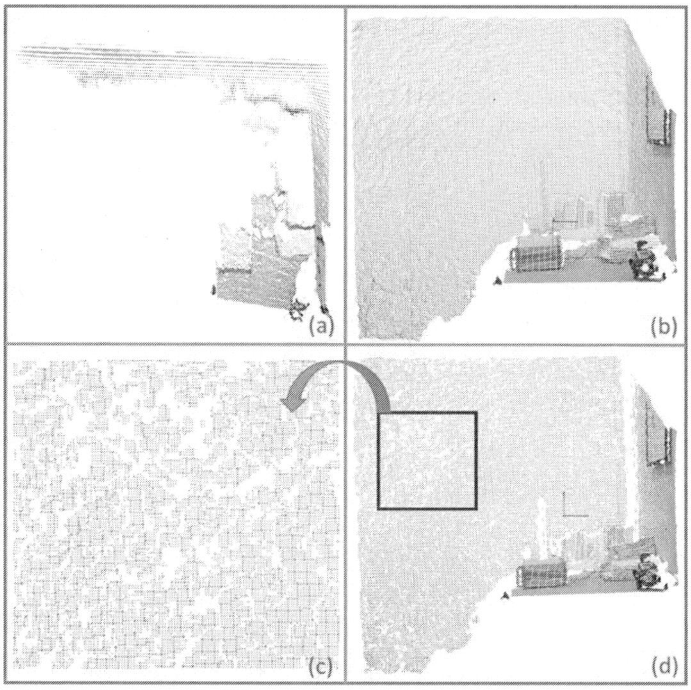

FIGURE 1: (A) AND (B) TWO VIEWS OF UNFILTERED BACKGROUND POINT CLOUD, (C) AND (D) POINT CLOUD FILTERED USING STATISTICAL OUTLIER REMOVAL METHOD

A Kinect study [17] shows that the depth direction z has a larger impact on the points' properties than the other two directions (x and y). Inspired by Shen et al. [28] and other Kinect-related work, it can be assumed that points in the same depth layer (i.e. with similar z values) have similar resolutions and noise levels. The traditional statistical outlier algorithm can be adopted by changing only the outlier classification criterion as follows:

$$\bar{x} > \begin{cases} \mu + n\sigma & z \le 1 \\ \mu \dfrac{q_z(z)}{q_z(z=1)} + n\sigma \dfrac{\sigma_z(z)}{\sigma_z(z=1)} & z > 1 \end{cases} \tag{6}$$

The result of statistical outlier removal after considering the quadratic decrease in the point cloud resolution and increasing error magnitude can be seen in Figure 2. Some of the boundary points of local clusters were inevitably filtered. However, the main part of the wall was preserved.

The statistical outlier removal filter presented here requires two iterations of the entire point cloud and calculation of the K nearest neighbors for each point. Its computational complexity hampers its broad usage in many applications that require high computational speed. In these cases, a radius outlier removal filter can be used, which works by iterating through the entire point cloud only once. In such a method, the number of neighbors N within a certain radius R of each point is retrieved, and the point is classified as outlier and trimmed if it has fewer neighbors than a certain preset threshold K.

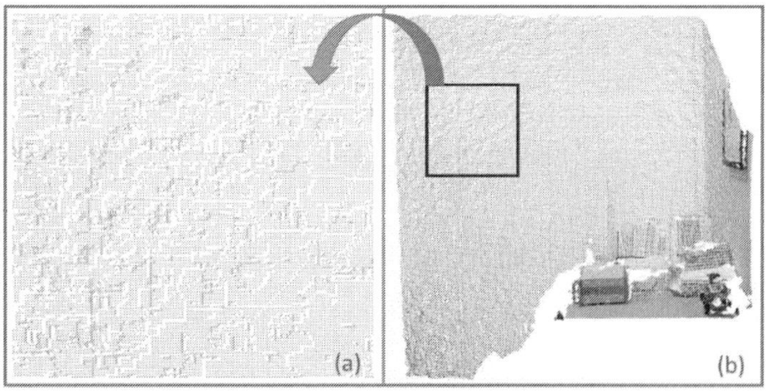

FIGURE 2: (A) AND (B) POINT CLOUD FILTERED USING MODIFIED STATISTICAL OUTLIER REMOVAL METHOD

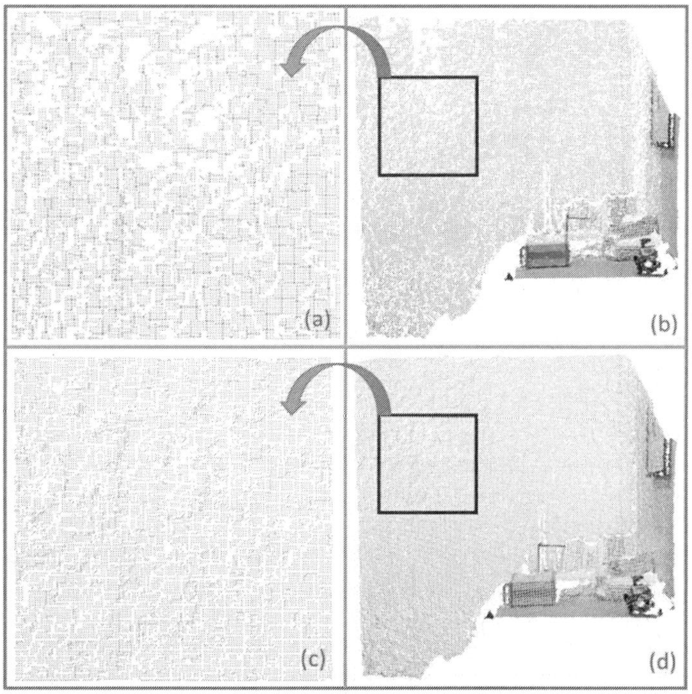

FIGURE 3: (A) AND (B) BACKGROUND POINT CLOUD FILTERED USING TRADITIONAL RADIUS OUTLIER REMOVAL; (C) AND (D) SAME POINT CLOUD FILTERED USING MODIFIED RADIUS OUTLIER REMOVAL METHOD

The result of this filter being applied to the same indoor laboratory environment used above is shown as parts (a) and (b) in Figure 3. While this filter works well on points close to the Kinect camera, holes are introduced erroneously for points far away from the camera. This phenomenon is explicable by the decreasing resolution and increasing error magnitude for increasing z. According to the Kinect's unique properties, the adopted radius outlier filter can be modified by changing the search radius as follows:

$$R(x,y,z) = m\left\{\begin{array}{l}\left[q_x(x)+\sigma_x(x,z)\right]^2 \\ +\left[q_y(y)+\sigma_y(y,z)\right]^2 \\ +\left[q_z(z)+\sigma_z(z)\right]^2\end{array}\right\}^{\frac{1}{2}} \quad (7)$$

Here, m is an adjustment factor. The result of the modified filter is depicted in parts (c) and (d) of Figure 3, in which the noise is filtered out while the integrity of the scene is preserved at the same time.

5.3. SCENE SEGMENTATION AND CLUSTERING

One of the most important processes in automatic point cloud processing is segmentation, which can be interpreted as labeling each measurement in a point cloud such that the points belonging to the same surface or region are given the same label. Different segmentation methods are usually distinguished by the criterion used to measure the similarity between the points of a given set. The evaluated similarities are then used for making the grouping decision. The first common kind are edge-based segmentation approaches, which detect borders of different regions in the point cloud and then group the points inside of each boundary as final segments [38,39]. The second kind are surface-based segmentation methods, which merge together the points that are spatially close and have similar surface properties (e.g. parameters of surface interpolation) [40,41]. The third kind are scanline-based segmentation algorithms, which are suitable for laser scanner data and consider each row of the range image as a 3-D line and group the adjacent lines with similar properties to form surface segmentations [42,43,44].

Traditional segmentation methods that are based on curvature or its other higher-order derivatives often lead to over segmentation, which subsequently necessitates significant manual editing to finalize the segmentation. Furthermore, the Kinect produces noisy point clouds, which can induce undesired errors by fitting higher-order surfaces. In these cases, segmentation using a smoothness constraint based only on the local surface normal and point connectivity becomes suitable [45]. This approach was then improved by Rusu et al. [46,47] by combining the influence of sudden changes in color, curvature and normal orientation, which renders the algorithm more robust and suitable for Kinect applications.

Among those widely used segmentation algorithms, Euclidean cluster segmentation is popular because it is theoretically straightforward and easily to implement. While the implementation can vary in different scenarios, the criteria are simple. If two points have a Euclidean distance of less than a preset threshold r, then they should belong to the same cluster. The result of implementing the original Euclidean clustering on a Kinect point cloud is illustrated in part (a) of Figure 4, in which every color represents a different cluster. While this point cloud should be classified as a single cluster due to the continuity in the scene, the wall far away from the camera got divided into numerous clusters due to the decreasing resolution and increasing error magnitude.

The original Euclidean clustering segmentation method can be improved by changing the criterion of Euclidean distance between any two points as follows:

$$d < \begin{cases} r & z \le 1 \\ r\dfrac{R_z(z)}{R_z(z=1)} & z > 1 \end{cases} \quad (8)$$

$$R_z(z) = q_z(z) + \sigma_z(z)$$

The definitions of $\sigma_z(z)$ and $q_z(z)$ can be found in Equations (2) and (3), respectively. The result of the modified Euclidean clustering is shown in part (b) of Figure 4. The continuous scene is grouped as a single cluster, even for the points far away from the Kinect camera.

FIGURE 4: (A) ORIGINAL EUCLIDEAN CLUSTERING; (B) MODIFIED EUCLIDEAN CLUSERING

5.4. 3-D KEYPOINT DETECTOR

Recently, research on 3-D keypoint detectors has started. This is motivated by the desire to accelerate the diffusion of low-cost 3-D sensors and by the growing importance of applications such as 3-D shape retrieval and object recognition. One of the main research foci is the detection of repeatable and distinctive 3-D surface keypoints, which are those points of the shape that are prominent according to a particular definition of interestingness and saliency. According to Tombari et al. [48], the detection methods for 3-D keypoints can be divided into two categories, namely fixed-scale and adaptive-scale detectors. The first kind of detectors find distinctive keypoints at a specific constant scale, which prunes the input data by thresholding a certain quality measurement and then extracts keypoints based upon a saliency measure (which can be either point-wise or region-wise) computed at each point. This kind of algorithms include Local Surface Patches [49], Intrinsic Shape Signature [50], KeyPoint Quality [51], Heat Kernel Signature [52], etc. The main difference of the second kind of detector compared with the first kind is that they adopt a 2-D automatic scale selection method [53] into 3-D space, followed by extracting keypoints and pruning. These detection methods include Laplace-Beltrami Scale-space [54], MeshDoG [55], Salient Points [56], etc.

In this section, the 3-D range image feature detection method called Normal Aligned Radial Feature (NARF) [57] was used as an example. Compared to other keypoint detectors, this method is very robust against noise and changes in resolution. It explicitly uses object boundary information and extracts NARF features in areas where the surface is stable (to ensure a robust estimation of the surface normal) but exhibits significant change (represented by principal curvature direction) in the vicinity. The algorithm first detects borders in the range image and then determines a dominant direction of surface change for every point on the border. For each point, an interest value that evaluates the neighborhood's dominant direction changes and surface changes is calculated. The Non-Maximum Suppression (NMS) method is then used (after smoothing these values) to find the final keypoints. The results of employing the original NARF keypoint detector on a laboratory scene (see part (a) of Figure 5) is depicted in parts (c) and (d) of Figure 5. Due to the increasing random error, one flat far-away wall was distorted into numerous small local clusters, and thus, many undesired NARF keypoints were extracted along the local borders of these clusters.

The NARF keypoint extraction algorithm was modified here by changing the method of computing the 3-D Euclidean distance of any two points from their x, y and z coordinates. The distance calculation method remains as defined before, except when it is used as criterion for detecting borders in the range image. At this stage, the coordinates of certain points (originally (x, y, z)) are replaced by a range (x', y', z') defined as follows:

$$\begin{aligned} x' &\in [x-\sigma_x, x+\sigma_x] \\ y' &\in [y-\sigma_y, y+\sigma_y] \\ z' &\in [z-\sigma_z, z+\sigma_z] \end{aligned} \tag{9}$$

The distance between two points is defined as the minimum of all possible Euclidean distances when these points are located in their pre-defined range. The result of the modified NARF keypoint detector is shown in part (b) of Figure 5. The number of keypoints decreased dramatically due to the decreasing number of detected borders.

6. POTENTIAL IMPACT ON EDUCATIONAL APPLICATIONS

Recently, the potential of using the Kinect in educational applications has begun to be studied at various educational institutions. Among those educational applications of the Kinect, some are using it as a user interface that identifies the users' gestures from the Kinect's skeleton tracking results and uses them as commands to control the educational system [58]. In this case, 20 skeleton nodes are extracted from a comparatively larger number of points, which provided results with acceptable accuracy.

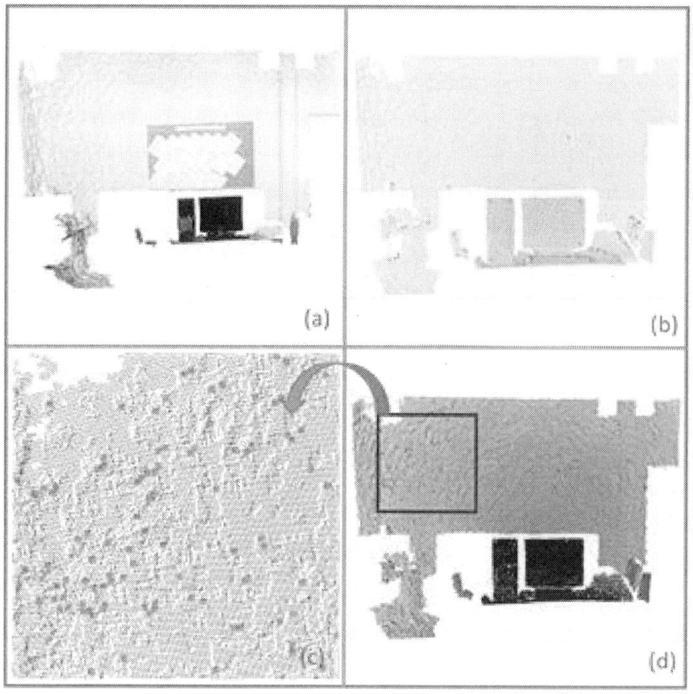

FIGURE 5: (A) ORIGINAL SCENE; (B) RESULT OF MODIFIED NARF KEYPOINT EXTRATION; (C) AND (D) RESULT OF ORIGINAL NARF KEYPOINT EXTRATION

In other educational applications, the point clouds generated by the Kinect are analyzed directly and useful information is extracted from them [59]. If the problems associated with the Kinect's hardware limitations and the preset assumptions of traditional computer vision algorithms discussed in this paper are remedied, the potential certainly exists for the Kinect to be employed for a wide range of purposes other than as an alternative user interface. For instance, the Kinect can be used as a universal range data acquisition system for engineering experiments [11,12] or as a real-time reconstruction tool for game-based virtual environments [7]. Thus, the algorithm modification approaches discussed above can be considered as measures aimed at facilitating advanced educational technologies and applications.

7. CONCLUSIONS AND FUTURE RESEARCH

In this paper, several technical specifications of the Kinect and their influence on potential applications were discussed. The technical obstacles caused by the Kinect's unique properties and assumptions of traditional algorithms for processing data point clouds that could hamper the Kinect's wide application in various areas were studied. Traditional approaches such as calibration and depth map enhancement were briefly reviewed and several traditional algorithms (including denoising, segmentation and keypoint extraction) were modified by taking the Kinect's coordinate dependent resolution and error level into consideration. Finally, the performance of these modified algorithms was tested using several examples and found to be effective in achieving acceptable results in point cloud processing using only the Kinect's raw output data. Compared to traditional methods, these approaches are more computational efficient and more robust in the presence of the Kinect's unique errors.

The next stages of research should be focused on three parts. First, the influence of the lens distortion (of both the Kinect's IR and RGB cameras) should be taken into consideration when modifying the algorithms further. Second, more 3-D point cloud processing algorithms should be included in this systematic modification effort. Third, suitable prototype applications should be chosen to prove the validity and impact of this work.

8. ACKNOWLEGEMENT

The authors wish to thank Mr. Yizhe Chang, Dr. El-Sayed Aziz and Dr. Serdar Tumkor for many stimulating discussions on the topic.

9. REFERENCES

[1] Xia, L., Chen, C.-C. and Aggarwal, J., 2011, "Human Detection Using Depth Information by Kinect," *2011 IEEE Computer Society Conference on Computer Vision and Pattern Recognition Workshops*, Colorado Springs, Colorado, USA, pp. 15-22.

[2] Chang, Y., Chen, S. and Huang, J., 2011, "A Kinect-Based System for Physical Rehabilitation: A Pilot Study for Young Adults with Motor Disabilities," *Research in Developmental Disabilities*, **32**(6), pp. 2566-2570.

[3] Zafrulla, Z., Brashear, H., Starner, T., Hamilton, H. and Presti, P., 2011, "American Sign Language Recognition with the Kinect," *Proceedings of the 13th International Conference on Multimodal Interfaces*, Alicante, Spain, pp. 279-286.

[4] Oikonomidis, I., Kyriazis, N. and Argyros, A. A., 2011, "Efficient Model-Based 3D Tracking of Hand Articulations Using Kinect," *Proceedings of the 22nd British Machine Vision Conference*, Dundee, Scotland, UK, pp. 1-11.

[5] Zollhöfer, M., Martinek, M., Greiner, G., Stamminger, M. and Süßmuth, J., 2011, "Automatic Reconstruction of Personalized Avatars from 3D Face Scans," *Computer Animation and Virtual Worlds*, **22**(2-3), pp. 195-202.

[6] Henry, P., Krainin, M., Herbst, E., Ren, X. and Fox, D., 2012, "RGB-D Mapping: Using Kinect-Style Depth Cameras for Dense 3D Modeling of Indoor Environments," *The International Journal of Robotics Research*, **31**(5), pp. 647-663.

[7] Zhang, Z., Zhang, M., Chang, Y., Aziz, E.-S., Esche, S. K. and Chassapis, C., 2013, "Real-time 3D model reconstruction using Kinect for a game-based virtual laboratory," *Proceedings of the ASME International Mechanical Engineering Conference & Exposition IMECE'13*, San Diego, California.

[8] Izadi, S., Kim, D., Hilliges, O., Molyneaux, D., Newcombe, R., Kohli, P., Shotton, J., Hodges, S., Freeman, D. and Davison, A., 2011, "Kinectfusion: Real-Time 3D Reconstruction and Interaction Using a Moving Depth Camera," *Proceedings of the 24th Annual ACM Symposium on User Interface Software and Technology*, Santa Barbara, California, USA, pp. 559-568.

[9] Lim, Y.-W., Lee, H.-Z., Yang, N.-E. and Park, R.-H., 2012, "3-D Reconstruction Using the Kinect Sensor and its Application to a Visualization System," *International Conference on Systems, Man and Cybernetics*, Seoul, Korea, pp. 3361-3366.

[10] Ozbay, E. and Cinar, A., 2013, "3D Reconstruction Technique with Kinect and Point Cloud Computing," *Global Journal on Technology, Vol. 3,* pp. 1748-1754.

[11] Zhang, M., Zhang, Z., Aziz, E.-S., Esche, S. K. and Chassapis, C., 2013, "Kinect-based universal range sensor for laboratory experiments," *Proceedings of the ASME International Mechanical Engineering Conference & Exposition IMECE'13*, San Diego, California, USA.

[12] Zhang, M., Zhang, Z., Esche, S. K. & Chassapis, C., 2013, "Universal range data acquisition for educational laboratories using Microsoft Kinect," *Proceedings of the 2013 ASEE Annual Conference & Exposition*, Atlanta, Georgia, USA.

[13] Yáñez, J., Quintana, D., Quintáns, C., Fariña, J. and Rodríguez-Andina, J., 2005, "FPGA-Based System for the Education in Data Acquisition and Signal Generation", 31st Annual Conference of IEEE, November 6-10, 2005.

[14] M. T. Draelos, 2012, "The Kinect up Close: Modifications for Short-Range Depth Imaging," Master's Thesis, North Carolina State University, Raleigh, North Carolina, USA.

[15] http://www.hotchips.org/, accessed on June 19, 2014.

[16] Smisek, J., Jancosek, M. and Pajdla, T., 2013, "3D with Kinect," *Consumer Depth Cameras for Computer Vision*, Springer, London, pp. 3-25.

[17] Khoshelham, K. and Elberink, S. O., 2012, "Accuracy and Resolution of Kinect Depth Data for Indoor Mapping Applications," *Sensors*, **12**(2), pp. 1437-1454.

[18] Brown, D., C., 1971, "Close-range Camera Calibration," *Photogrammetric Engineering*, **37**(8), pp. 855-866.

[19] Menna, F., Remondino, F., Battisti, R. and Nocerino, E., 2011, "Geometric Investigation of A Gaming Active Device," *Proceedings of SPIE Optical Metrology (8085). Videometrics, Range Imaging, and Applications*, Munich, Germany.

[20] Canessa, A., Chessa, M., Gibaldi, A., Sabatini, S. P. and Solari, F., 2014, "Calibrated Depth and Color Cameras for Accurate 3D Interaction in A Stereoscopic Augmented Reality Environment," *Journal of Visual Communication and Image Representation*, **25**(1), pp. 227-237.

[21] Zhang, C. and Zhang, Z., 2011, "Calibration between Depth and Color Sensors for Commodity Depth Cameras," *2011 IEEE International Conference on Multimedia and Expo (ICME),* Barcelona, Spain, pp. 1-6.

[22] Herrera, C., Kannala, J. and Heikkilä, J., 2012, "Joint Depth and Color Camera Calibration with Distortion Correction," *IEEE Transactions on Pattern Analysis and Machine Intelligence*, **34**(10), pp. 2058-2064.

[23] Raposo, C., Barreto, J. P. and Nunes, U., 2013, "Fast and Accurate Calibration of a Kinect Sensor," *2013 International Conference on 3D Vision*, Seattle, Washington, USA, pp. 342-349.

[24] Kim, J. H., Choi, J. S. and Koo, B. K., 2013, "Simultaneous Color Camera and Depth Sensor Calibration with Correction of Triangulation Errors," *Advances in Visual Computing*, Springer Berlin Heidelberg, pp. 301-311.

[25] Yamazoe, H., Habe, H., Mitsugami, I. and Yagi, Y., 2012, "Easy Depth Sensor Calibration," *2012 21st International Conference on Pattern Recognition (ICPR)*, Tsukuba, Japan, pp. 465-468.

[26] Nguyen, C. V., Izadi, S. and Lovell, D., 2012, "Modeling Kinect Sensor Noise for Improved 3D Reconstruction and Tracking," *2012 Second International Conference on 3D Imaging, Modeling, Processing, Visualization and Transmission (3DIMPVT)*, Zurich, Switzerland, pp. 524-530.

[27] Camplani, M., Mantecón, T. and Salgado, L., 2013, "Depth-Color Fusion Strategy for 3-D Scene Modeling With Kinect," *IEEE Transactions on Cybernetics*, **43**(6), pp. 1560-1571.

[28] Shen, J. and Cheung, S. C. S., 2013, "Layer Depth Denoising and Completion for Structured-Light RGB-D Cameras," *2013 IEEE Conference on Computer Vision and Pattern Recognition (CVPR)*, Portland, Oregon, USA, pp. 1187-1194.

[29] Milani, S. and Calvagno, G., 2012, "Joint Denoising and Interpolation of Depth Maps for MS Kinect Sensors," *2012 IEEE International Conference on Acoustics, Speech and Signal Processing (ICASSP)*, Kyoto, Japan, pp. 797-800.

[30] Kopf J., Cohen M. F., Lischinski D. and Uyttendaele M., 2007, "Joint Bilateral Upsampling," *ACM Transactions on Graphics*, **26**(3) Article 96.

[31] He, K., Sun, J. and Tang, X., 2010, "Guided Image Filtering," *Computer Vision–ECCV*, Springer Berlin Heidelberg, pp. 1-14.

[32] Park, J., Kim, H., Tai, Y. W., Brown, M. S. and Kweon, I., 2011, "High Quality Depth Map Upsampling for 3D-TOF Cameras," *2011 IEEE International Conference on Computer Vision (ICCV)*, Barcelona, Spain, pp. 1623-1630.

[33] Petschnigg, G., Szeliski, R., Agrawala, M., Cohen, M., Hoppe, H. and Toyama, K., 2004, "Digital Photography with Flash and No-Flash Image Pairs," *ACM Transactions on Graphics*, **23**(3), pp. 664-672.

[34] Cho, J. H., Ikehata, S., Yoo, H., Gelautz, M. and Aizawa, K., 2013, "Depth Map Up-Sampling Using Cost-Volume Filtering," *2013 11th IEEE IVMSP Workshop*, Seoul, Korea, pp. 1-4.

[35] Huhle, B., Schairer, T., Jenke, P. and Straßer, W., 2010, "Fusion of Range and Color Images for Denoising and Resolution Enhancement with A Non-Local Filter," *Computer Vision and Image Understanding*, **114**(12), pp. 1336-1345.

[36] Wasza, J., Bauer, S. and Hornegger, J., 2011, "Real-Time Preprocessing for Dense 3-D Range Imaging on the GPU: Defect Interpolation, Bilateral Temporal Averaging and Guided Filtering," *2011 IEEE International Conference on Computer Vision Workshops*, Barcelona, Spain, pp. 1221-1227.

[37] Rusu, R. B., Marton, Z. C., Blodow, N., Dolha, M. and Beetz, M., 2008, "Towards 3D Point Cloud Based Object Maps for Household Environments," *Robotics and Autonomous Systems*, **56**(11), pp. 927-941.

[38] Bhanu, B., Lee, S., Ho, C. C. and Henderson, T., 1986, "Range Data Processing: Representation of Surfaces By Edges," *Proceedings of the 8th International Conference on Pattern Recognition*, pp. 236-238.

[39] Sappa, A. D. and Devy, M., 2001, "Fast Range Image Segmentation by an Edge Detection Strategy," *Proceedings of 3rd International Conference on 3-D Digital Imaging and Modeling*, Quebec City, Quebec, Canada, pp. 292-299.

[40] Parvin, B. and Medioni, G., 1986. "Segmentation of Range Images into Planar Surfaces by Split and Merge," *Proceedings of International Conference on Computer Vision and Pattern Recognition*, pp. 415-417.

[41] Xiang, R. and Wang, R., 2004, "Range Image Segmentation Based on Split-Merge Clustering," *Proceedings of the 17th International Conference on Pattern Recognition*, **3**, Cambridge, United Kingdom, pp. 614-617.

[42] Jiang, X. Y., Bunke, H. and Meier, U., 1996, "Fast Range Image Segmentation Using High-Level Segmentation Primitives," *Proceedings of the 3rd IEEE Workshop on Applications of Computer Vision*, Washington, DC, USA, pp. 83-88.

[43] Khalifa, I., Moussa, M. and Kamel, M., 2003, "Range Image Segmentation Using Local Approximation of Scan Lines with Application to CAD Model Acquisition," *Machine Vision Applications*, **13**(5-6), pp. 263-274.

[44] Sithole, G. and Vosselman, G., 2003, "Automatic Structure Detection in a Point-Cloud of an Urban Landscape," *Proceedings of 2nd Joint Workshop on Remote Sensing and Data Fusion over Urban Areas*, Berlin, Germany.

[45] Rabbani, T., van den Heuvel, F. and Vosselmann, G., 2006, "Segmentation of Point Clouds Using Smoothness Constraint," *International Archives of Photogrammetry, Remote Sensing and Spatial Information Sciences*, **36**(5), pp. 248-253.

[46] Rusu, R. B., Marton, Z. C., Blodow, N., Dolha, M. and Beetz, M., 2008, "Towards 3D Point Cloud Based Object Maps for Household Environments," *Robotics and Autonomous Systems*, **56**(11), pp. 927-941.

[47] Rusu, R. B., Blodow, N., Marton, Z. C. and Beetz, M., 2009, "Close-Range Scene Segmentation and Reconstruction of 3D Point Cloud Maps for Mobile Manipulation in Domestic Environments," *International Conference on Intelligent Robots and Systems*, Chicago, USA, pp. 1-6.

[48] Tombari, F., Salti, S. and Di Stefano, L., 2013, "Performance Evaluation of 3D Keypoint Detectors," *International Journal of Computer Vision*, **102**(1-3), pp. 198-220.

[49] Chen, H. and Bhanu, B., 2007, "3D Free-Form Object Recognition in Range Images Using Local Surface Patches," *Pattern Recognition Letters*, **28**(10), pp. 1252-1262.

[50] Zhong, Y., 2009, "Intrinsic Shape Signatures: a Shape Descriptor for 3D Object Recognition," *IEEE 12th International Conference on Computer Vision Workshops*, Kyoto, Japan, pp. 689-696.

[51] Mian, A. S., Bennamoun, M. and Owens, R. A., 2010, "On the Repeatability and Quality of Keypoints for Local Feature-Based 3D Object Retrieval from Cluttered Scenes," *International Journal of Computer Vision*, **89**(2-3), pp. 348-361.

[52] Sun, J., Ovsjanikov, M. and Guibas, L., 2009, "A Concise and Provably Informative Multi-Scale Signature Based on Heat Diffusion," *Computer Graphics Forum*, **28**(5), pp. 1383-1392

[53] Lindeberg, T., 1998, "Feature Detection with Automatic Scale Selection," *International Journal of Computer Vision*, **30**(2), pp. 79-116.

[54] Unnikrishnan, R. and Hebert, M., 2008, "Multi-Scale Interest Regions from Unorganized Point Clouds," *IEEE Computer Society Conference on Computer Vision and Pattern Recognition Workshops*, Anchorage, Alaska, USA, pp. 1-8.

[55] Zaharescu, A., Boyer, E., Varanasi, K. and Horaud, R., 2009, "Surface Feature Detection and Description with Applications to Mesh Matching," *IEEE Conference on Computer Vision and Pattern Recognition*, Miami, Florida, USA, pp. 373-380.

[56] Bronstein, A. M., Bronstein, M. M., Bustos, B., Castellani, U., Crisani, M., Falcidieno, B., Guibas, L. J., Kokkinos, I., Murino, V., Ovsjanikov, M., Patané, G., Sipiran, I., Spagnuolo, M. and Sun, J., 2010, "SHREC 2010: Robust Feature Detection and Description Benchmark," *EUROGRAPHICS Workshop on 3D Object Retrieval*. Norrkoping, Sweden.

[57] Steder, B., Rusu, R. B., Konolige, K. and Burgard, W., 2011, "Point Feature Extraction on 3D Range Scans Taking into Account Object Boundaries", *IEEE International Conference on Robotics and Automation*, Shanghai, China, pp. 2601-2608.

[58] Hsu, J., 2011, "The potential of Kinect in education," *International Journal of Information and Education Technology*, **1**(5), pp. 365-370.

[59] Tölgyessy, M. and Hubinský, P., 2011, "The Kinect sensor in robotics education," *Proceedings of 2nd International Conference on Robotics in Education*, Vienna, Austria.

Proceedings of the ASME 2014 International Mechanical Engineering Congress and Exposition
IMECE2014
November 14-20, 2014, Montreal, Quebec, Canada

IMECE2014-37149

AN EFFICIENT METHOD FOR CREATING VIRTUAL SPACES FOR VIRTUAL REALITY

Zhou Zhang
Stevens Institute of Technology
Hoboken, New Jersey, USA

Mingshao Zhang
Stevens Institute of Technology
Hoboken, New Jersey, USA

Yizhe Chang
Stevens Institute of Technology
Hoboken, New Jersey, USA

Sven K. Esche
Stevens Institute of Technology
Hoboken, New Jersey, USA

Constantin Chassapis
Stevens Institute of Technology
Hoboken, New Jersey, USA

1. ABSTRACT

A virtual space (VS) is an indispensable component of a virtual environment (VE) in virtual reality (VR). Usually, it is created using general tools and skills that are independent of the users' specific applications and intents. Creating a VS by surveying the real world with traditional measuring tools or creating virtual features with CAD software involves many steps and thus is time consuming and complicated. This renders the construction of VEs difficult, impairs their flexibility and hampers their widespread usage.

In this paper, an efficient method for creating VSs with a handheld camera is introduced. In this approach, the camera is used as a measuring tool that scans the real scene and obtains the corresponding surface information. This information is then used to generate a virtual 3D model through a series of data processing procedures. Firstly, the camera's pose is traced in order to locate the points of the scene's surface, whereby these surface points form a point cloud. Then, this point cloud is meshed and the mesh elements are textured automatically one by one. Unfortunately, the virtual 3D model resulting from this procedure represents an impenetrable solid and thus collision detection would prevent the avatars from entering into this VS. Therefore, an approach for eliminating this restriction is proposed here. Finally, a game-based virtual laboratory (GBVL) for an undergraduate mechanical engineering class was developed to demonstrate the feasibility of the proposed methodology. The model format used in Garry's Mod (GMod) is also found in other VEs, and therefore the method proposed here can be straightforwardly generalized to other VE implementations.

Keywords: VR, GBVL, GMod, Kinect

2. INTRODUCTION

VR is defined as an environment that models the real world by a computer-simulated virtual representation. It is now becoming increasingly popular and widely used in education, research, training and the military and has the potential to alleviate the existing constraints on various natural and social resources. The creation of virtual systems instead of physical ones can make the creation of multiple copies of physical devices unnecessary, thus reducing the consumption of scarce natural resources. In addition, virtual systems are inherently safer and less failure prone than physical ones and can also be shared remotely by multiple users. There are several important forms of VR, including desktop VR, immersive VR, distributed VR and augmented VR. VR is often used in training systems. Such systems can spare most of the human and material resources, make them accessible remotely for the participants and simulate dangerous situations and hazardous environments safely (e.g. firefighter training, military training, disaster relief training, etc.).

The development of VR mainly involves creating a VS, building virtual models of physical devices, customizing user avatars for this VS and devising plots for the activities carried out by the avatars in the VS. While the plot design is based on knowledge corresponding to the specific application and implements the users' intent, the remaining tasks are used to generate the VE. Some of the challenges associated with the development of VR include the creation of VSs in real time, the generation of the models of physical devices in real time, the seamless integration of different platforms and systems as well as the facilitation of efficient interactions between humans and VEs including 3D sound synthesis, stereoscopic 3D display technologies and perception feedback [1]. The visualization of the real world involves the creation of VSs and the generation of models of physical devices. A VS represents a map in which all of the interactions between models can be implemented. The process of creating a VS is often performed manually using CAD software. It is complicated and tedious and thus time-consuming and requiring patience. With the development of new electronic components, emerging electro-optical technologies, powerful algorithms and data processing tools, improving the efficiency in implementing a VR is becoming possible. An innovative method for creating virtual models of physical devices based on a real-time 3D reconstruction technique was introduced previously [2]. The models generated by this method represent impenetrable solids. Thus, collision detection would prevent other VR components from entering into such models. Therefore, these models are not suitable for VSs. In this paper, an efficient method based on 3D reconstruction is proposed to create VSs.

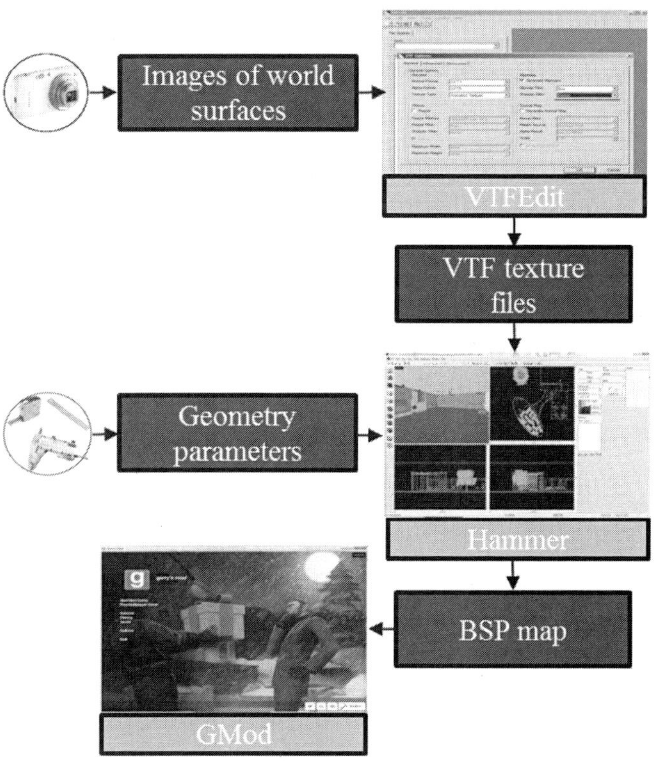

FIGURE 1: TRADITIONAL MAP CREATION PIPELINE OF GBVLS IN GMOD

In this 3D reconstruction method, surface data are obtained by surveying the real world with scanners. Then, the corresponding virtual surfaces are generated by deploying a series of image processing and computer vision algorithms. The virtual world created by 3D reconstruction can be rendered in specific virtual platforms, or it can directly be exported into a 3D model file that is recognizable by other popular 3D software. The acquisition and the processing of the shape information are the core steps in reconstructing the real world [3]. It has been a few years since 3D reconstruction techniques were first introduced, but their potential for deployment in various fields of application (e.g. archaeological research [4], medical applications [5] and recovery of 3D shapes of deformable surfaces [6]) makes them the subject of ongoing research. With the development of structured light cameras (laser scanners, infrared cameras, etc.), 3D reconstruction techniques have been improving quickly, and many innovative applications of 3D reconstruction have been reported recently, for instance 3D voxel reconstruction of human motions [7], 3D reconstruction of faces of people [8] and reconstruction of urban scenes from videos [9]. There are two main approaches for obtaining the shape information of real objects. The first approach is to retrieve the surface information from 2D still images. For instance, '3DSOM Pro' and 'insight3d' can be used to produce 3D models through a series of alignment pictures [10,11]. The other approach is to register the surface information of real objects by dynamic scanning devices. For example, 'ReconstructMeQt' can be employed to build 3D surfaces with the Kinect and to export 3D surface files to other software [12]. The 'KinectFusion' is one of the tools for developing 3D reconstruction applications. It can be used to generate 3D models in real time from the shape information of objects that was acquired by the 'Microsoft Kinect for Windows' [13,14].

In order to validate the method introduced here, a VS of a laboratory was created and imported into a GBVL implementation as a map. GBVLs are based on multi-player computer game engines [15,16,17]. They represent one of the implementations of VR and are used at various levels of education, spanning from elementary schools to universities. The game engine used in the GBVL presented here is endowed with various basic ready-to-use functions such as graphics rendering, sound generation, networking, physics modeling, game logics, artificial intelligence and user interactions [18,19]. These characteristics of the game engine render the GBVL collaborative, distributed and immersive. One of the important advantages of GBVLs is that users can create their own avatars, models, maps and other virtual features. This advantage makes GBVLs customizable and extensible [1]. The GBVL presented here was built based on GMod, which is a modification of the 'Source' game engine and has been developed into a user-friendly multi-player computer game platform that permits users to create their own applications, program these applications with G_LUA and plot their own stories [20,21]. The common method used to build maps in GMod can be generalized into the following steps: (i) obtain the geometry parameters of the real world by surveying it with traditional measuring tools such as a measuring tape, (ii) plot maps of the real world with built-in tools such as 'Hammer' or third-party software such as '3ds Max', (iii) take pictures of the real world and use 'VTFEdit' to generate VTF texture files [22,23], (iv) apply textures to the surfaces of the maps and (v) compile the 'Valve' map format (VMF) file to generate ready-to-use maps of GMod. VMF is a proprietary format that stores raw map data used by the 'Valve Hammer Editor' to save production-stage maps and premade fabrications (prefabs). Prefabs are separately saved objects, usually consisting of multiple parts. They allow users to easily store and reuse pieces from their own levels [24]. They contain information on all pieces of maps and entities in a key value format, stored with a '.vmf' extension [25]. Figure 1 illustrates the common mapping process of GBVLs in GMod.

3. INOVTATIVE METHOD FOR CREATING VIRTUAL SPACES

3.1. Feasibility Analysis

Currently, the efficient creation of VSs is hampered by the methods for acquiring the geometric parameters of the real world and by the selection of the mapping tools. Therefore, if there are some sensors that can replace the traditional measuring devices, the laborious work of surveying the real world can be completed very quickly. Moreover, if there is some middleware that can process the acquired raw data and generate the ready-to-use map automatically, the whole process of creating VSs can be simplified significantly.

3.2. Comparison of New vs. Traditional Data Acquisition Tools

Various kinds of non-contact measurement devices such as 3D depth scanners and 2D scanners may be considered as alternative tools for surveying the real world. 3D depth scanners can be classified into two main types. One type is infrared depth scanners, which are commonly used in less demanding circumstances [26]. The other type is laser depth scanners, which are often used in high accuracy applications [27]. 2D scanners are often simply common cameras, which are suitable for applications whose requirements for both accuracy and real-time are not too strict. The inconvenience of employing common cameras is that only 2D information of the world can be obtained. Therefore, much more complicated algorithms must be used for generating 3D virtual geometries, thus prolonging the time necessary for creating the VSs dramatically. Therefore, 3D depth scanners are more appropriate for improving the efficiency of the creation of VSs compared with common cameras. On the other hand, laser depth scanners are usually more expensive than infrared depth scanners, and their operation also requires specialists. Infrared scanners such as the Microsoft Kinect are much more affordable, accessible and flexible when being compared with laser depth scanners. Furthermore, Microsoft has provided a 'software development kit' (SDK) for the Kinect that facilitates its application. For instance, the Kinect was recently applied as a data acquisition device for range data [28].

Computer games have their own specific characteristics. From the users' perspective, the most important aspect is the design of the game plots but not the resolutions of the avatars, maps and models. Therefore, the components with a slightly distorted appearance compared with the real world can still meet the expectations of the players. Hence, the Kinect can satisfy the requirements of GBVLs in spite of its low resolution. Thus, when taking into account the requirements of the game players and the cost of the device, the Kinect becomes a true option for replacing the traditional measuring tools.

The Microsoft Kinect for Windows sensor was designed mainly for the Windows operating system platform. It is composed of two sensors (cameras), an accelerometer, a structured light projector and a microphone array. The two sensors are a 'color (RGB)' sensor and an 'infrared (IR) depth sensor'. The color sensor captures the color information of the object with a resolution of up to 1280×960 in pixels and the IR sensor acquires the depth information of the object with a resolution of up to 640×480 in pixels [29,30]. The tolerance of the depth values is no less than 1 mm, and therefore, the tolerance of the virtual world created with the Kinect is more than 1 mm. The Kinect for Windows SDK enables users to extend the function of the Kinect, and it includes device drivers, development tools, development APIs, device interfaces and code samples. Frequent updates of this SDK are adding improvements for more control and deeper access to the sensor data.

3.3. Practicability of Kinect for Creation of Virtual Spaces

There are two possible ways for scanning an object with the Kinect. One is to move the object relative to the Kinect, and the other is to move the Kinect relative to the object [2]. Since the world cannot be moved, the second method is the only choice for surveying the real world. A hand-held Kinect was used here for scanning the real world. Since the Kinect is moving in this process, its pose must be tracked. The algorithms used to track the pose of the Kinect are more complicated than those used when the Kinect is kept still. This method for surveying the world with a hand-held Kinect is fairly flexible and failsafe. In order to obtain the pose of the Kinect, a fast stereo matching algorithm was employed. A 3D point cloud of the real world was obtained once the coordinates of the Kinect had been determined.

The next problem to be solved is to create the VS from this point cloud. In order to create a penetrable map, the point cloud is divided into several sub-point clouds. Each sub-point cloud represents one of the 6 orientations of the world. Then, every sub-point cloud is meshed and textured, and each element of the meshed sub-point cloud represents one side in the VMF file. The side is a kind of class used in the VMF structure that describes its orientation (by its normal), the texture attached to it and the properties of material, light and so on. If the number of mesh elements and the resolution of the textures are limited, the processing time for the mapping can be kept at a reasonable level. After completing the above steps, the VS is composed of many simple 3D solids (cubes, cylinders, etc.), which are referred to as brushes inside the 'Hammer editor' [31]. Both 'VTFEdit' (the conversion tool for textures) and the VMF compilation tool can be run from the command line. Thus, using the API provided by the Windows shell, the above two executable files can be executed automatically [32]. Therefore, it is feasible to make a middleware for creating VSs automatically.

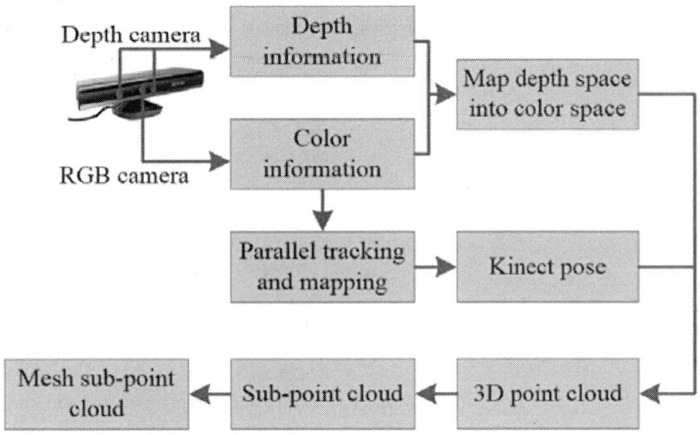

FIGURE 2: FLOW CHART OF DATA ACQUISITION AND PROCESSING

3.4. Acquisition and Processing of Real World Data

The main steps of the data acquisition and processing are depicted in Figure 2. As illustrated in this flow chart, a set of sequential color frames is captured to determine the pose of the Kinect using the 'parallel tracking and mapping' (PTAM) method [33]. The PTAM method splits the tracking and mapping process into two threads that run in parallel.

Being a kind of camera, the Kinect has extrinsic and intrinsic parameters. The extrinsic parameters represent the camera's location and orientation in the world coordinates. The intrinsic parameters represent the relationships between the camera coordinates and the pixel coordinates. Both of these parameters compose the project matrix P (see Eq. 1). Points with homogenous coordinates $(x,y,z,1)$ in world coordinates are mapped into 2D image coordinates with homogenous coordinates (u,v,w). The intrinsic parameters are represented by 3×3 matrix $K_{3\times3}$ and the extrinsic parameters are represented by 3×4 matrix $[R_{3\times3}\ \ t_{3\times1}]_{3\times4}$ [34].

$$\begin{cases} P = K_{3\times3}[R_{3\times3} \quad t_{3\times1}]_{3\times4} \\ \begin{bmatrix} u \\ v \\ w \end{bmatrix} = P\begin{bmatrix} x \\ y \\ z \\ 1 \end{bmatrix} \end{cases} \tag{1}$$

If matrix M represents the motion of the camera, then the transformation from world coordinates $(x,y,z,1)$ to camera-centered coordinates $(x',y',z',1)$ can be performed using a 4×4 camera pose matrix C_p (see Eq. 2).

$$\begin{bmatrix} x' \\ y' \\ z' \\ 1 \end{bmatrix} = C_p\begin{bmatrix} x \\ y \\ z \\ 1 \end{bmatrix} \tag{2}$$

The projection from camera-centered coordinates to image coordinates is then expressed by Eq. 3.

$$\begin{bmatrix} u \\ v \\ w \end{bmatrix} = PC_p\begin{bmatrix} x \\ y \\ z \\ 1 \end{bmatrix} \tag{3}$$

The pose matrix C_p is changed correspondingly when the Kinect is moved. Then, the change of the Kinect's pose can be described by Eq. 4. The motions of the Kinect include 3 rotations about the 3 axes and 3 translations along the 3 axes. Thus, they can be expressed by a 6-dimensional vector v_m using the exponential map of Eq. 4 [35]. Once v_m has been determined, the pose of the Kinect can be obtained. The procedures for tracking the pose of the Kinect are described in detail elsewhere [33].

$$C_p' = MC_p = e^{v_m}C_p \tag{4}$$

3.5. 3D Point Cloud Generation, Subdivision and Triangular Meshing

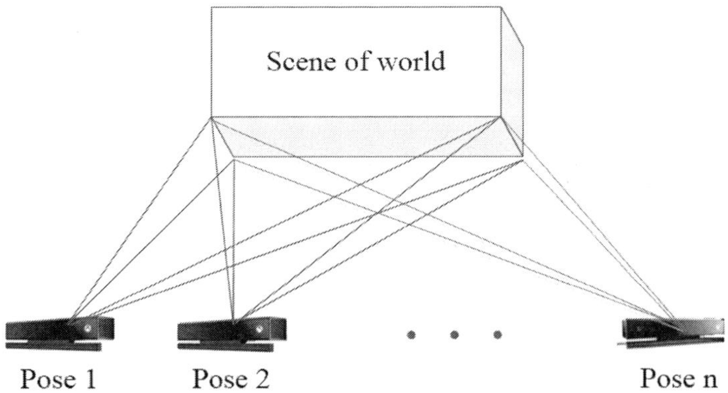

FIGURE 3 OBTAINING COORDINATES OF POINTS BY TRACKING KINECT POSES

When surveying the real world, the first pose of the Kinect is taken as the reference for the entire world space. The coordinates of the points of the world surfaces can be identified relative to the reference pose of the Kinect. Therefore, the point cloud representing the world surfaces can be obtained after the pose of the Kinect has been obtained. Figure 3 depicts n poses of the Kinect. Pose 1 is taken as the reference. The relative positions in the camera coordinate system of the other poses can be obtained by Eq. 4. Then, the point coordinates in the world coordinate space can be obtained by left-multiplying the inverse of the camera pose matrix C_p^{-1} onto both sides of Eq. 2.

The point cloud of the world surfaces is divided into 6 sub-point clouds according to the 6 directions of the world (top, bottom, left, right, front and back). Then, every sub-point cloud is meshed with the 3D Delaunay triangulation algorithm [36]. Finally, the sub-point clouds generate 6 independent parts of the map, and these parts are not combined into one single solid, which ensures that the VS created is a penetrable solid.

3.6. Generation of Textures and Maps

After triangulation, each side of the VS is composed of a set of triangles. The textures are then defined as the color pixel points, which are located both in the triangular domain and on the boundary of the domain. Next, all these triangular texture patches are combined into 6 images, which correspond to the 6 directions of the VS. After converting these 6 images into 6 VTF texture files using 'VTFCmd.EXE', all the information about the textures and vertices is stored in a VMF file. Finally, this VMF file is compiled into BSP map format using three commands. First, 'VBSP.EXE' renders the geometry, then 'VVIS.EXE' renders the texture of the geometry, and finally, 'VRAD.EXE' renders the light [22]. The entire procedure is illustrated in Figure 4.

4. VALIDATION BY CREATION OF GBVL MAP

4.1. Application in Undergraduate Engineering Education

The GBVL described here is currently used by students in an undergraduate fluid mechanics course. The experimental devices from a traditional hands-on flow-rig laboratory were virtualized into corresponding models for usage in the GBVL. This experiment is used to familiarize the students with some basic principles of air flow and specifically with the characteristics of air distribution systems. The components of the experimental setup are depicted in Figure 5. The Pitot tubes located at different cross sections of the flow pipe are used to measure the pressure distribution, from which the velocity distributions are then determined. Both the students and the instructor are represented as avatars. They interact with the virtual experimental setup through keyboard and mouse [1,37].

4.2. Specific Implementation of VS Creation in GBVL

In order to demonstrate that the method proposed here is feasible, a map of the GBVL described above was created using the procedure described in this paper (see Figure 5).

A straightforward method for reconstructing the VS is to detect all of the surfaces in the real world, mesh all of these surfaces into triangles and texture all of the triangles. However, this approach is too costly in terms of time and computer resources, especially considering the many unnecessary details of the world. In the map illustrated in Figure 5, every side of the laboratory was simplified into 2 large triangles, and the image of every side of the real laboratory was directly attached to the corresponding side. This processing method for the sides of the VS saves a lot of time and resources. Therefore, in order to decrease the number of vertices used for meshing and texturing, the keypoints of the laboratory were extracted from the sub-point cloud (e.g. corner coordinates of one wall or one big cabinet of the laboratory). These keypoints can represent the 6 sides of the laboratory from 6 directions coarsely. Although this map is not perfect, it is sufficient for the purpose of the implementation of experiments.

When creating the map of the laboratory with traditional methods, the first step would be to measure the width, length and height dimensions of the room, and then, the dimensions of the objects in the laboratory would also need to be calibrated. This process would be laborious. The second step would be to take pictures of the world from 6 different orientations. Note that due to the limited camera viewing frustum, it may be necessary to take several pictures of each side and to subsequently stitch these pictures together with image processing software such as 'Photoshop'. The third step would be to create the map with a CAD software such as '3ds Max', which may take several hours, even when using very powerful computers. The forth step would be to convert the pictures of the world to VTF format with a software such as 'VTFEdit'. The fifth step would be to texture the map with CAD software and to compile the raw map file into ready-to-use map format for VR. Therefore, the total time consumed by this multi-step procedure would be over half of a day with current computers.

In contrast, the only work necessary when using the method proposed here is to scan the world with a hand-held Kinect, and then the map of the real world is created automatically. This procedure may take only about 30 to 60 minutes with the today's computers of average configuration. In addition to speeding up the process, the proposed method is also less complicated and requires fewer tools. Therefore, anyone even without any measuring and modeling experience can easily and quickly create their own maps for VR.

At present, the method proposed here has only been applied for creating a simple VS since the algorithm employed to track the pose of the Kinect is limited to applications in small spaces, which is owed to the fact that the data processing capability and the storage capacity of the devices (i.e. CPU, GPU and memory) are limited. In addition, according to Moore's law, the limitation is that once transistors can be created as small as atomic particles, then there will be no more room for growth in the CPU market as far as speeds are concerned. This means that the limitation may be overcome only by employing multiple devices, for example multiple GPUs [38]. The quality in resolution and accuracy of the VS generated is fairly low because there is a limitation in the number of pixels in the created VS, which also results from the limit in the resources of the devices. At present, this can only be improved by increasing the number of devices.

5. CONCLUSIONS

In this paper, a method for creating virtual spaces automatically with a hand-held Microsoft Kinect was proposed.

Firstly, the conventional mapping method applied in GMod was summarized. Secondly, the steps of the proposed mapping approach for data acquisition and processing were discussed in detail, including tracking the pose of a hand-held Kinect, subdividing and meshing the point cloud and generating the final map. Finally, the map of a virtual laboratory was created, which was used in a GBVL and was then taken as the VS for a fluid mechanics experiment.

The method presented here does not have the ability to recognize objects since it only acquires the surface information of the real world. These limitations can be overcome by applying advanced 3D computer vision algorithms that can improve the resolution and accuracy and by pattern recognition algorithms, which can render the mapping process more sophisticated and enable it to identify the scanned objects. Future work will be focused on improving the precision of the modeling, the quality of the models and the ability to recognize specific objects. Although there remain some problems to be solved, the proposed method shows promise for improving the efficiency of VR creation.

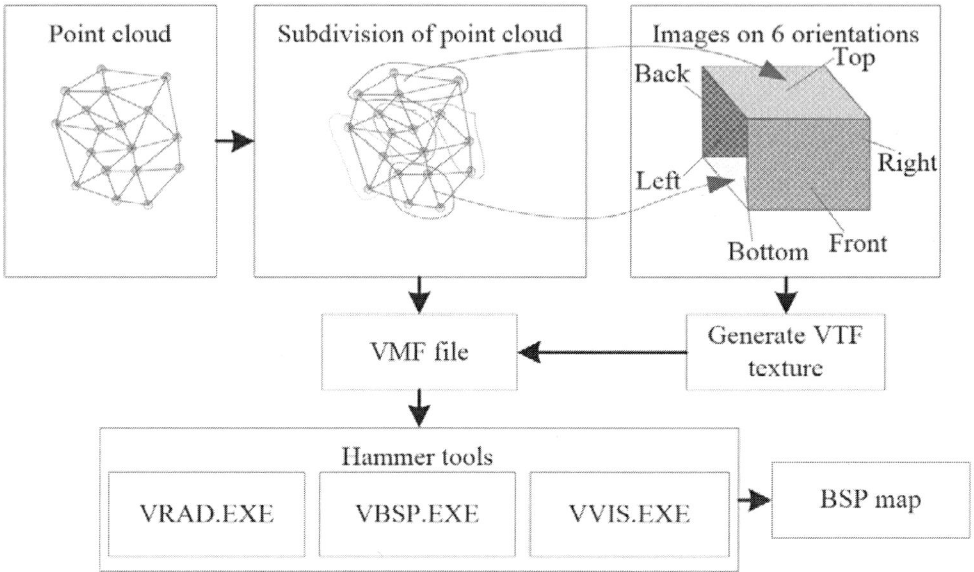

FIGURE 4: MESHES, TEXTURES AND MAP GENERATION

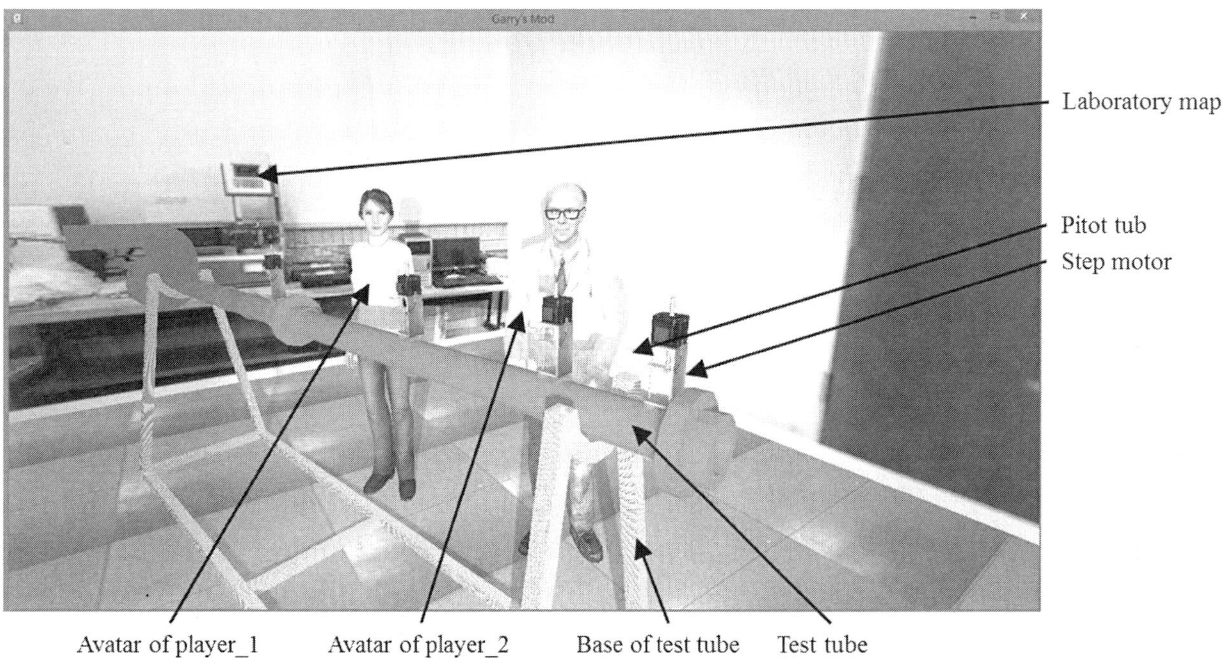

FIGURE 5: LABORATORY MAP WITH FLUID EXPERIMENT SETUP

6. REFERENCES

[1] Zhang, Z., Zhang, M., Tumkor, S., Chang, Y., Esche, S. K. & Chassapis, C., 2013, "Integration of physical devices into game-based virtual reality", International Journal of Online Engineering, Vol. 9, No. 5, pp. 25-38.

[2] Zhang, Z., Zhang, M., Aziz, E.-S., Esche, S. K. & Chassapis, C., 2013, "Real-time 3D model reconstruction and interaction using Kinect for a game-based virtual laboratory", Proceeding of 2013 ASME International Mechanical Engineering Congress & Exposition, San Diego, CA, USA, November 15-21, 2013.

[3] Mouragnon, E., Lhuillier, M., Dhome, M., Dekeyser F. & Sayd, P., 2006, "Real time localization and 3D reconstruction", Proceedings of the 2006 IEEE Conference on Computer Vision and Pattern Recognition, New York, NY, USA, June 17-22, 2006, Vol. 1, pp. 363-370.

[4] Fiz, I. & Orengo, H. A., 2007, "The application of 3D reconstruction techniques in the analysis of ancient Tarraco's urban topography", Proceeding of the 35th International Conference on Computer Applications and Quantitative Methods in Archaeology, Berlin, Germany, April 2-6, 2007.

[5] Hibbard, L. S., Grothe, R. A., Arnicar-Sulze, T. L., Dovey-Hartman, B. J. & Page, R. B., 1993, "Computed three-dimensional reconstruction of median eminence capillary modules", Journal of Microscopy, Vol. 171, pp. 39-56.

[6] Varol, A., Shaji, A., Salzmann, M. & Fua, P., 2012, "Monocular 3D reconstruction of locally textured surfaces", IEEE Transactions of Pattern Analysis and Machine Intelligence, Vol. 34, No. 6, pp. 1118-1130.

[7] Cheung, G. K., Kanade, T., Bouguet, J. Y. & Holler, M., 2000, "A real time system for robust 3D voxel reconstruction of human motions", Proceedings of the 2000 IEEE Conference on Computer Vision and Pattern Recognition, June 13-15, 2000, Hilton Head, SC, USA.

[8] Choi, J., Medioni, G., Lin, Y., Silva, L., Regina, O., Pamplona, M. & Faltemier, T. C., 2010, "3D face reconstruction using a single or multiple views", Proceeding of International Conference on Pattern Recognition, Istanbul, Turkey, August 23-26, 2010.

[9] Pollefeys, M. et al., 2008, "Detailed real-time urban 3D reconstruction from video", International Journal of Computer Vision, Vol. 78, No. 2, pp. 143-167.

[10] http://www.3dsom.com/, accessed in April 2014.

[11] http://insight3d.sourceforge.net, accessed in April 2014.

[12] http://reconstructme.net/projects/reconstructmeqt, accessed in April 2014.

[13] http://msdn.microsoft.com/en-us/library/dn188670.aspx, accessed in April 2014.

[14] Izadi, S., Kim, D., Hilliges, O., Molyneaux, D., Newcombe, R. A., Kohli, P., Shotton, J., Hodges, S., Freeman, D., Davison, A. J. & Fitzgibbon, A. W., 2011, "KinectFusion: real-time 3D reconstruction and interaction using a moving depth camera", Proceedings of the 24th Annual ACM Symposium on User Interface Software and Technology, Santa Barbara, CA, USA, October 16-19, 2011, pp. 559-568.

[15] Aziz, E.-S., Chang, Y., Esche, S. K., & Chassapis, C., 2013, "A multi-user virtual laboratory environment for gear train design", Computer Applications in Engineering Education, DOI: 10.1002/cae.21573.

[16] Barham, W., Preston, J. & Werner, J., 2012, " Using a virtual gaming environment in strength of materials laboratory", Proceedings of Computing in Civil Engineering, June 17-20, 2012, Clearwater Beach, FL, USA, pp. 105-112.

[17] http://www.virtualgamelab.com, accessed in April 2014.

[18] Baba, S. A., Hussain, H. & Embi, Z. C., 2007, "An overview of parameters of game engine", IEEE Multidisciplinary Engineering Education Magazine, Vol. 2, No. 3, pp. 10-12.

[19] Thorn, A., 2010, "Game Engine Design and Implementation", Chapter 1, 1st Ed., Jones & Bartlett Publishers.

[20] http://garrysmod.com/, accessed in April 2014.

[21] http://source.valvesoftware.com/, accessed in April 2014.

[22] https://developer.valvesoftware.com/wiki/Valve_Hammer_Editor, accessed in April 2014.

[23] https://developer.valvesoftware.com/wiki/Studiomdl_Data, accessed in April 2014.

[24] https://developer.valvesoftware.com/wiki/Prefab, accessed in April 2014.

[25] https://developer.valvesoftware.com/wiki/Valve_Map_Format, accessed in April 2014.

[26] Khoshelham, K. & Elberink, S. O., 2012, "Accuracy and resolution of Kinect depth data for indoor mapping applications", Sensors, Vol. 12, No. 2, pp. 1437-1454.

[27] Hoffman, A., Goetz, M., Vieth, M., Galle, P. R., Neurath, M. F. & Kiesslich, R., 2006, " Confocal laser endomicroscopy: technical status and current indications", Endoscopy, Vol. 38, No. 12, pp. 1275-1283.

[28] Zhang, M., Zhang, Z., Aziz, E.-S., Esche, S. K. & Chassapis, C., 2013, "Kinect-based universal range sensor for laboratory experiments", Proceeding of 2013 ASME International Mechanical Engineering Congress & Exposition, San Diego, CA, USA, November 15-21, 2013.

[29] http://msdn.microsoft.com/en-us/library/jj131033.aspx, accessed in April 2014.

[30] Karayev, S., Jia, Y., Barron, J., Fritz, M., Saenko, K. & Darrell, T., 2011, "A category-level 3-D object dataset: putting the Kinect to work", Proceedings of the IEEE International Conference on Computer Vision Workshops, November 6-13, 2011, Barcelona, Spain, pp. 1167-1174.

[31] https://developer.valvesoftware.com/wiki/Basic_Construction, accessed in April 2014.

[32] http://msdn.microsoft.com/enus/library/windows/desktop/bb762154(v=vs.85).aspx, accessed in April 2014.

[33] Klein, G. & Murray, D., 2007, "Parallel tracking and mapping for small AR workspaces", Proceedings of 6th IEEE and ACM International Symposium on Mixed and Augmented Reality, Nara, Japan, November 13-16, 2007, pp. 225-234.

[34] Szeliski, R., 2010, "Computer Vision: Algorithms and Applications", Springer, 2010.

[35] Hall, B., 2003, "Lie Groups, Lie Algebras, and Representations: An Elementary Introduction", Springer, 2003.

[36] Devillers, O., 2012, "Delaunay triangulations, theory vs. practice", Proceedings of the 28th European Workshop on Computational Geometry, Assisi, Italy, March 19-21, 2012.

[37] Tumkor, S., Zhang, Z., Zhang, M., Chang, Y., Esche, S. K. & Chassapis, C., 2012, "Integration of a real-time remote experiment into a multi-player game laboratory environment", Proceedings of the 2012 ASME International Mechanical Engineering Conference & Exposition, Houston, TX, USA, November 9-15, 2012.

[38] http://www.mooreslaw.org/, accessed in June, 2014.

Proceedings of the ASME 2014 International Mechanical Engineering Congress and Exposition
IMECE2014
November 14-20, 2014, Montreal, Quebec, Canada

IMECE2014-37915

A WEB-BASED ONLINE COLLABORATION TOOL FOR FORMULATING SENIOR DESIGN PROJECTS

Sainath Varikuti **Jitesh H Panchal** **John M Starkey**
School of Mechanical Engineering
Purdue University
West Lafayette, Indiana 47907

ABSTRACT

A well formed senior design project is known to have significant benefits in terms of project outcome, student motivation, team cohesiveness, engagement, and student learning. Defining a good problem statement, forming a team of compatible and appropriately skilled students, and selecting an appropriate faculty mentor are critical aspects of project formation. Therefore, students in Mechanical Engineering at Purdue University are encouraged to suggest project ideas, form teams, and have them approved by the course coordinator before the semester starts. While there is significant literature on senior design projects, most of the existing work is focused on activities after the problem is defined and the teams are formed. There is a lack of mechanisms and tools to guide the project formation phase in senior design projects, which makes it challenging for students and faculty to collaboratively develop and refine project ideas and to establish appropriate teams. To address this challenge, we have implemented an online collaboration tool to share, discuss and obtain feedback on project ideas, and to facilitate collaboration among students and faculty prior to the start of the semester. Through an online survey and questionnaire to students, we are exploring the impact of the collaboration tool on the senior design project formation process. In this paper we present the design of the tool and the results from our ongoing study in the senior design class at Purdue Mechanical Engineering.

Keywords: Online Collaboration, Senior Design, Mechanical Engineering

1 INTRODUCTION

Engineering design projects provide real-life design process experience to students [1]. Research shows that project formation process is one of the most challenging tasks associated with engineering design capstone courses [2]. The foremost task is framing good project definitions. Significant research has been conducted to understand the characteristics of a senior design project. According to Dutson et al. [3], a good quality project statement should:

- be challenging,
- simulate a real life problem,
- involve engineering design work,
- emphasize engineering theory and application, and
- meet specified standards and safety criteria.

Formulating a problem for a project is a challenge for both faculty and students. One of the main reasons is that an individual does not posses complete knowledge. As explained by Chandrasegaran et al. [4], it requires both formal and tacit knowledge to formulate a good problem statement. While the former type of knowledge is present in documents, repositories etc., the latter is only gained over a period of time with learning and experience. Thus, it is necessary to promote knowledge sharing activity among faculty and students.

Since design projects are team oriented projects, team formation is important. Different team formation processes are adopted by various faculty members to form engineering design teams. The most practiced schemes include [5]:

- letting the students self-select the team members,

- random assignment of team members by the instructors,
- team member selection based on prior performance, and
- group selection based on a heterogeneous mixture.

Regardless of the adopted scheme, individual team members' motivation and enthusiasm about the project is one of the critical factors in the success of a project [6]. Instructors often conduct questionnaires about student interests, academic strengths and experiences in an attempt to form effective design teams [3]. However, there are generally groups of students whose individual interests do not align with the assigned project and hence, makes the process challenging.

Finally, selecting an appropriate faculty adviser is another important task in the project formation process [3]. Studies show that, for various reasons, faculty members are interested or motivated in mentoring only specific design projects [3]. Therefore, sometimes there are many projects and student teams but very few faculty members who are interested in advising them. Thus, there exists a need for a collaborative system where faculty and students can exchange ideas, form teams, select mentors and execute potential projects.

Over the past few years, several researchers have reported tools and systems to formulate projects and teams for successful project execution [6, 7]. For instance, Wyard-Scott et al. [8] have implemented an online tool to foster communication among instructors and student team members. Similarly, Kanai [9] created a web-based bulletin board system to help faculty, students and industrial sponsors to stay connected. Lo et al. [10] have proposed an online collaboration system for senior design project grading and document review. However, these systems are focused on the activities after the problem is defined and the teams are formed. There is a lack of mechanisms and tools to guide the project formation phase, which makes it challenging for students and faculty to collaboratively develop and refine project ideas and to establish appropriate teams. The question that guided this study is: *Does an online collaboration tool improve the project formation process in engineering design courses?*

In this paper we present the design of an online collaboration tool and report the results of our ongoing exploration of the impact of this tool on senior design project formation process. The rest of the paper is organized as follows: the following section discusses the motivation followed by the design and implementation of the tool. We then present the evaluation of the tool. Finally, the results of an ongoing study and future work are discussed.

2 MOTIVATION

A senior design or capstone project is an essential part of many Mechanical Engineering programs. In the School of Mechanical Engineering at Purdue University, the Engineering Design course is offered to provide practical guidance to students in integrating various engineering sciences into practical engineering design projects. This allows students to apply and refine their design skills, teamwork skills and communication skills. Also, this course plays a vital role in assessing the ABET (Accreditation Board for Engineering and Technology) student learning outcomes [11].

In our engineering design course, a team of senior engineering students work on a design problem. Students, faculty and industry sponsors are the major sources of project definitions. The course coordinator has to ensure that the projects simulate a real-life project/problem and student team is motivated to solve the design problem. As a result, the students are advised to suggest project ideas, form teams and have them approved by the course coordinator before the course starts. The key activities that take place during this phase are:

1. problem definition and scoping,
2. team formation, and
3. faculty adviser selection.

During this phase, the course coordinator iterates with the student proposing the project to develop the problem statement until it meets the educational objectives of senior design. Then the students look for potential team mates, and the course coordinator looks for a faculty member to act as an adviser for the project in the coming semester. The entire process can be sub-categorized into three phases including the idea initialization phase, idea development phase, and idea finalization phase. Each phase has corresponding activities for carrying out the development to achieve a high quality problem definition and motivated student team. The key stakeholders in the process are the students, faculty, the instructor in charge and the industry sponsors. Figure 1 presents a bird's eye view of activities and interactions among stakeholders during each phase of the process.

The lack of mechanisms and tools makes it difficult for the individual stakeholders to interact with each other. Current interaction methods such as face-to-face meetings and e-mail interactions are not effective for communication and successfully guiding a large number of students. Inadequate interactions among stakeholders at this stage result in problem definitions with poorly defined scope and incomplete capturing of requirements, thereby, significantly affecting the quality of problem definition. Other issues include:

1. Students often find it difficult to look for team mates who are enthusiastic and motivated about the specific project idea.
2. Students are limited by the number of faculty they can approach for guidance.
3. Good projects may be missed because the course coordinator is the single point of contact for the problem statement approval process.
4. At times, a faculty member can come up with a good senior design project which might take more than a semester

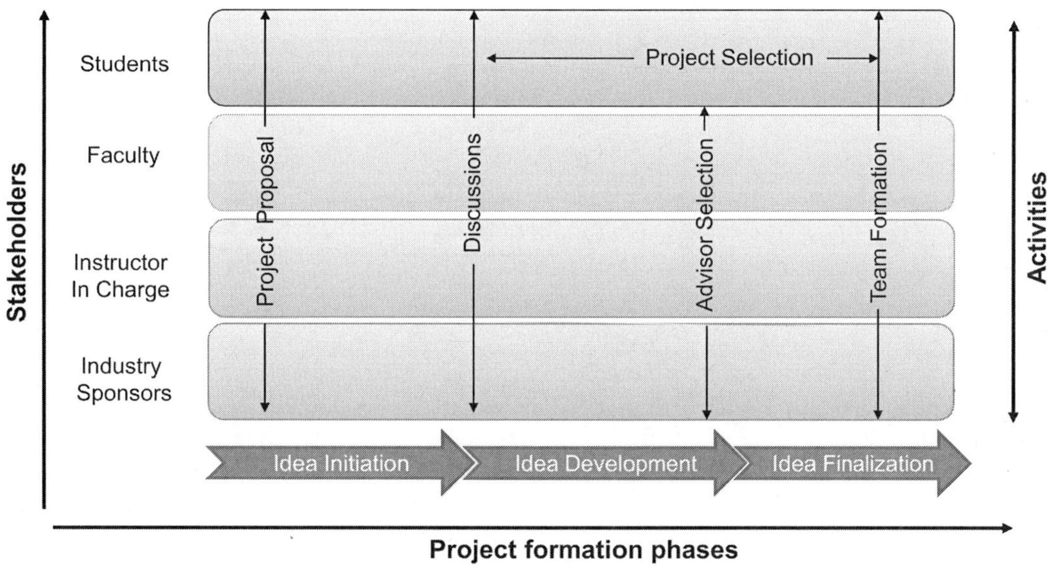

FIGURE 1: DIFFERENT PHASES, ACTIVITIES AND INTERACTIONS AMONG STAKEHOLDERS DURING SENIOR DESIGN PROJECT FORMATION

to complete. So, the faculty member might want to communicate this to the students and want to work with the team of interested students a little earlier.

Therefore, we hypothesize that implementing an online collaboration platform that enables participation and interaction among stakeholders will improve the project formation process. We also observe that creating a public collaboration platform will facilitate collaboration links that might otherwise not be made. The platform described here is intended for identifying high quality problem definitions and motivated student teams through increasing interactions among student and faculty.

3 ONLINE PLATFORM : REQUIREMENT ANALYSIS

In this section we present the critical requirements of the online platform. The requirements of the online platform are categorized into a) functional requirements and b) software requirements.

The functional requirements of the platform are as follows:

1. All stakeholders including students, faculty, and industry sponsors should be able to propose new project ideas.
2. The platform should allow all stakeholders to discuss and comment on the shared project ideas.
3. Students and faculty members should be able to register for interested projects.
4. The platform should allow students to communicate with each other and propose their own team.
5. Students should be able to archive their projects ideas in order to motivate students enrolled in future semesters.
6. The platform should allow the lead instructor to administer the project formation activity.

The software requirements of the system that are essential for choosing a system are:

1. The system should be able to provide access to all stakeholders.
2. The system should have password restricted accessibility and provide data security.
3. Currently, there are four different stakeholders in the system. The system should facilitate multiple user roles and permission levels.

In fact, knowing who the stakeholders should be and forming collaborative connections is a challenge. The use case diagram in Figure 2 illustrates the design details. As the diagram shows, the critical aspects of the system are collaboration, communication and information sharing among different actors.

4 IMPLEMENTATION

Existing learning management systems available at Purdue University, such as Blackboard/WebCT [12], do not satisfy several of the unique needs of the system. For instance, Blackboard limits accessibility to only students and faculty of a particular course and single section. As a result, students and faculty who belong to the same course but in a different section cannot collaborate through Blackboard. Additionally, industry

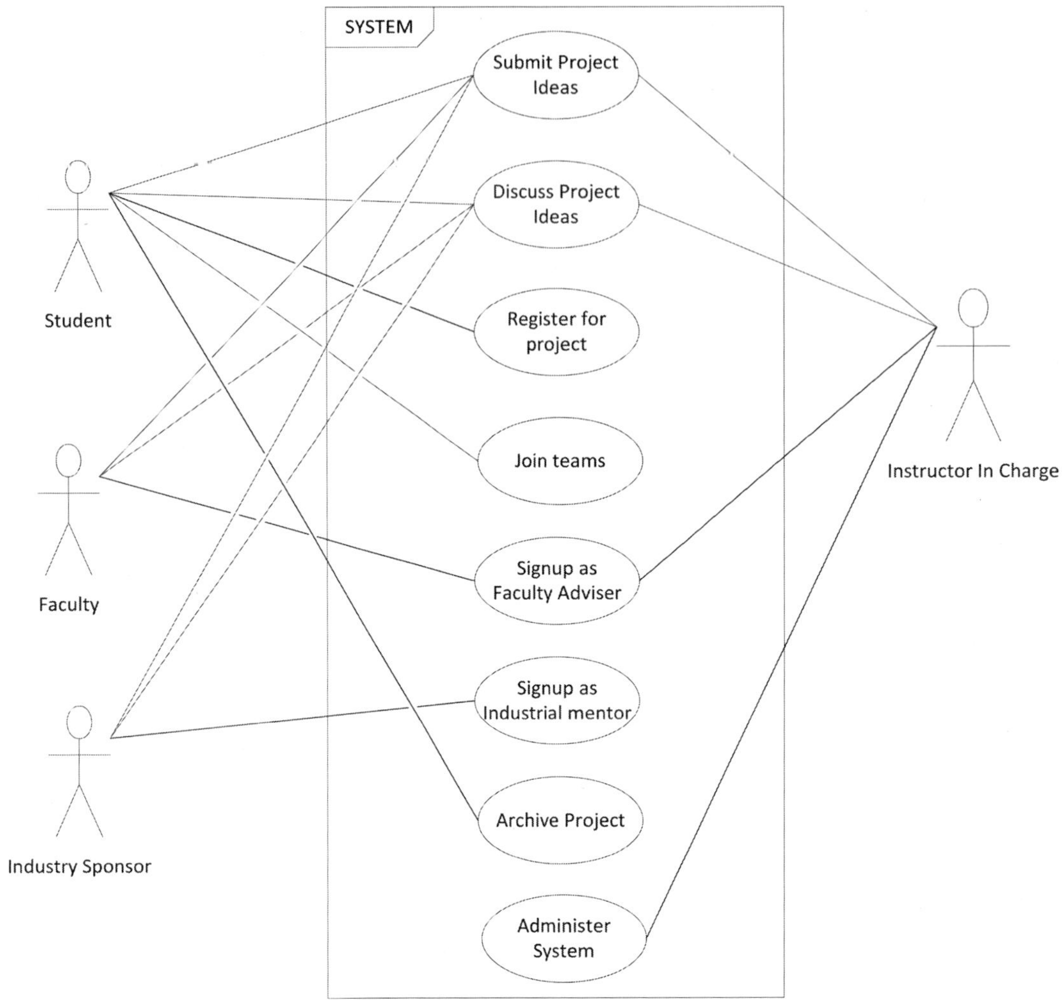

FIGURE 2: USE-CASE DIAGRAM OF THE PLATFORM

sponsors cannot have access to Blackboard. Also, other software requirements such as multiple user roles and permissions are not supported by Blackboard. Other available online platforms such as GlobalHub [13] which is developed using the HUBZero platform, is limited by the number of user roles available. Therefore, the online platform is implemented using Drupal, an open source web-based content management system [14]. Table 1 gives a comparison of existing online platforms and Drupal with the needs of the system.

The online platform is built on the LAMP [15] environment. The Drupal CMS is installed on top of apache web server. The complete platform architecture is illustrated in Figure 3(a). Drupal is built upon open source community contributed add-ons called modules. Each module is built to render a specific functionality to the web interface. In Drupal, each item of content is defined as a node and each node is defined as a specific content type. The platform is implemented by using modules and creating custom content types that meet the functionality of the system. The functional diagram mapping necessary modules [16, 17] and content types to the requirements is illustrated in Figure 3(b). The detailed design and current implementation of the online platform is available at [18].

5 EVALUATION

Several researchers suggest that interaction is an essential element to student learning [19, 20]. Hillman et al. [21] consider interactions among students, and between instructor and students

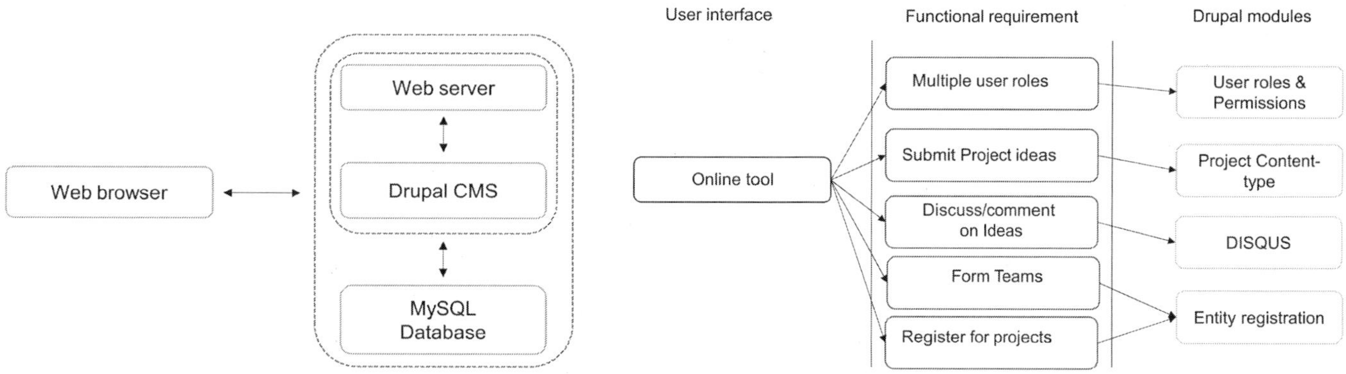

(a) PLATFORM ARCHITECTURE

(b) REQUIREMENT MAPPING

FIGURE 3: SYSTEM IMPLEMENTATION

TABLE 1: COMPARISON OF EXISTING ONLINE PLATFORMS

Requirements	Blackboard	Global-Hub	Drupal
Multiple user roles	Not Supported	Not Supported	Supported
All stakeholders accessibility	Not Supported	Supported	Supported
Discussion & chat	Supported	Supported	Supported
Teams & Groups	Not Supported	Supported	Supported

as "educational transaction". Research suggests that the amount of interactions is a performance measure of online platforms in academic setting [22]. To evaluate the effectiveness of the online platform we measured the amount of interactions among students and faculty both before and after the implementation of the online platform.

The study was conducted during Spring 2014 in which two groups of mechanical engineering senior design students were surveyed. The survey was administered in two phases. The first phase was carried out prior to the implementation of the online platform. During the first phase, a group of 194 Mechanical Engineering design students enrolled in the Engineering Design course were surveyed using the questionnaire presented in Table 2. This group of students relied on face-to-face meetings and email communication for formulating the engineering design projects i.e., forming teams and formulating problem statements.

During the second phase, the online platform was implemented and the students intending to take the course in Summer/Fall 2014 were encouraged to submit the project ideas and form teams through the online platform. This group consisted of 48 students. Students were informed about the online platform and were given a period of one month to frame the problem statements and form teams. After one month the second group of students was surveyed using the same questionnaire presented in Table 2. In comparison with the first group, the number of students were much lower in the second group because lesser number of students take senior design in the Summer/Fall semester compared to the Spring semester of an academic year in the school of Mechanical Engineering at Purdue University.

5.1 QUESTIONNAIRE

A questionnaire measuring the amount of interactions among students and faculty both before and after the implementation of the platform is conducted. The questionnaire addresses student-student interactions, student-faculty interactions and problem definition quality. Table 2 enumerates the metrics used and corresponding survey questions. A five point Likert-type scale, ranging from 1 as "strongly disagree" to 5 as "strongly agree" is used for the items 1 to 6 in Table, 1 as "no refinement" to 5 as "extensive refinement" is used for item 7 in and 1 as "very poor scope" to 5 as "very well defined scope" is used for item 8.

1. *Student - Student interactions*: This scale is developed to measure the extent to which students shared and discussed project ideas with other students and identified team members for the project. Questions 1-3 in Table 2 measure the student-student interaction.
2. *Student - Faculty interactions*: This scale is developed to measure the extent to which students shared, interacted and obtained feedback from faculty members on project ideas and identified interested faculty members for the project. This is measured using a 3 item scale. Questions 4-6 in Table 2 measure the student-faculty interactions.
3. *Problem statement quality*: The amount of refinement and

TABLE 2: SURVEY VARIABLES AND MEASURES

Metrics	Survey Questions
Student – Student interaction	1. I was able to share my project ideas with other students. 2. I was able to receive feedback from other students on my project idea. 3. I was able to identify team members who were interested in my project idea.
Student – Faculty interaction	4. I was able to share my project ideas with the faculty members 5. I was able to receive feedback from the faculty members on my project idea. 6. I was able to identify faculty members who were interested in my project idea.
Problem Statement quality	7. Rate the amount of refinement you made to your problem statement. 8. How focused is the scope of the problem statement.

problem statement scope is measured using this scale. This is measured using a 2 item scale. Questions 7 and 8 in Table 2 measure the problem statement quality.

6 RESULTS

6.1 UTILITY OF THE ONLINE TOOL

First, students' adaptability to the new system was examined. Figure 4 shows the frequency of use of the online platform by the second group of students who intend to take senior design course in Summer/Fall 2014. Although, some students never accessed the online tool, most of the students accessed the tool at least once a week. An active participation from students was observed in sharing project ideas and participating in discussion with the faculty. After implementation of the online tool, in a period of one month a total of 16 project ideas were proposed. 4 among 16 ideas were put forward by the faculty and 12 were proposed by the students. Additional web statistics observed for a period of one month i.e. from 1st March to 31st March 2014 is provided in Table 3. Faculty and students collectively participated in refining the problem statements. Students who accessed the online tool reported that the use of online tool was simple and useful for collaboration.

6.2 STUDENT-STUDENT INTERACTION

Table 4 presents the response of the student - student interaction. A comparison of the average student-student interactions both before and after implementing the online tool is performed. Figure 5 draws a comparison of interactions among students. An increase of 23% (i.e., from 28% to 51%) is observed in the amount of student-student interaction. After tool implementation, more than half of the students (56-65%) agreed that they were able to share their project ideas with their fellow students, as compared to (25-33%) before the implementation. A

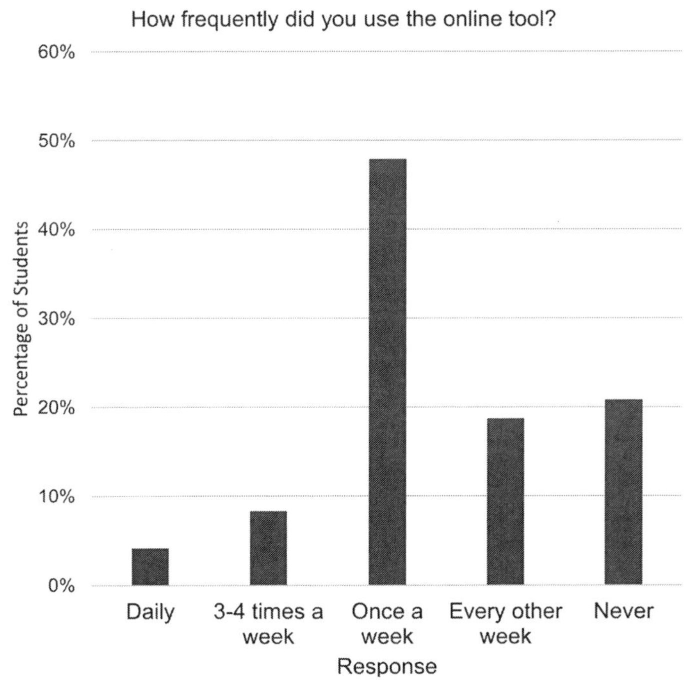

FIGURE 4: FREQUENCY OF USING THE ONLINE TOOL

TABLE 3: ONLINE PLATFORM STATISTICS

Component	Students	Faculty	Total
Registered users	48	4	52
Proposed Project ideas	12	4	16
Comments Posted	13	19	32

TABLE 4: SURVEY RESPONSE

Options	1	2	3	4	5	1	2	3	4	5
Student-Student interaction	**Before tool implementation (N=194)**					**After tool implementation (N=48)**				
I was able to share my project ideas with other students (1= Strongly disagree, 5 = Strongly agree)	24%	14%	29%	22%	11%	4%	4%	27%	59%	6%
I was able to receive feedback from other students on my project ideas (1=Strongly disagree, 5 = Strongly agree)	32%	19%	26%	13%	10%	4%	12%	38%	42%	4%
I was able to identify team members who were interested in my project idea (1 = Strongly disagree, 5 = Strongly agree)	28%	16%	27%	16%	13%	4%	23%	31%	36%	6%
Student-Faculty interaction	**Before tool implementation (N=194)**					**After tool implementation (N=48)**				
I was able to share my project ideas with the faculty members (1 = Strongly disagree, 5 = Strongly agree)	28%	20%	27%	15%	10%	6%	9%	29%	52%	4%
I was able to receive feedback from the faculty members on my project idea (1 = Strongly disagree, 5 = Strongly agree)	30%	23%	27%	12%	8%	6%	6%	40%	38%	10%
I was able to identify faculty members who are interested in my project idea (1 = Strongly disagree, 5 = Strongly agree)	37%	20%	27%	8%	8%	6%	13%	44%	35%	2%

two sample statistical t-test is performed to determine whether or not the difference between mean amount of student-student interaction in both the groups of students is statistically significant, by testing $\{H_0 : \mu_1 = \mu_2 \text{ vs. } H_1 : \mu_1 \neq \mu_2\}$ with the level of significance $\alpha = 0.05$. Here, μ_1 is the mean amount of student-student interaction of the group that used the online platform and μ_2 is the mean amount of student-student interaction that did not use the online platform. Table 5 provides descriptive statistics of the amount of student-student interaction for both the groups. The p-value of the t-test is observed to be 0.001 and is less than the significance level (0.05). Therefore we accept the alternate hypothesis and claim that there is a statistically significant difference in the mean amount of student-student interaction.

TABLE 5: DESCRIPTIVE STATISTICS OF STUDENT-STUDENT INTERACTION.

Variable	Group	Group Size	Mean	Std. dev.
student-student interaction	1	48	3.34	0.77
	2	194	2.67	1.23

6.3 STUDENT-FACULTY INTERACTION

The responses of student - faculty interaction both before and after the implementation of online tool are shown in Table 4. As shown in Figure 6, a comparison of average student-faculty interactions before and after the online tool suggests an increase of 27% (i.e., from 21% to 48%) in the amount of student-faculty interaction. Also, some students reported that the system made it easy to discuss and collaborate with faculty, eliminating the need for scheduling meetings with them. A two sample statistical t-test is again performed to determine whether or not the difference between mean amount of student-faculty interaction in both the groups of students is statistically significant, by testing $\{H_0 : \mu_1 = \mu_2 \text{ vs. } H_1 : \mu_1 \neq \mu_2\}$ with the level of significance $\alpha = 0.05$. Here, μ_1 is the mean amount of student-faculty interaction of the group that used the online platform and μ_2 is the mean amount of student-faculty interaction that did not use the online platform. Table 6 provides descriptive statistics of the amount of student-faculty interaction for both the groups. The p-value of the t-test is observed to be 0.001 and is less than the significance level (0.05). Therefore we accept the alternate hypothesis and claim that there is a statistically significant difference in the mean amount of student-faculty interaction.

TABLE 6: DESCRIPTIVE STATISTICS OF STUDENT-FACULTY INTERACTION.

Variable	Group	Group Size	Mean	Std. dev.
student-faculty interaction	1	48	3.31	0.88
	2	194	2.45	1.21

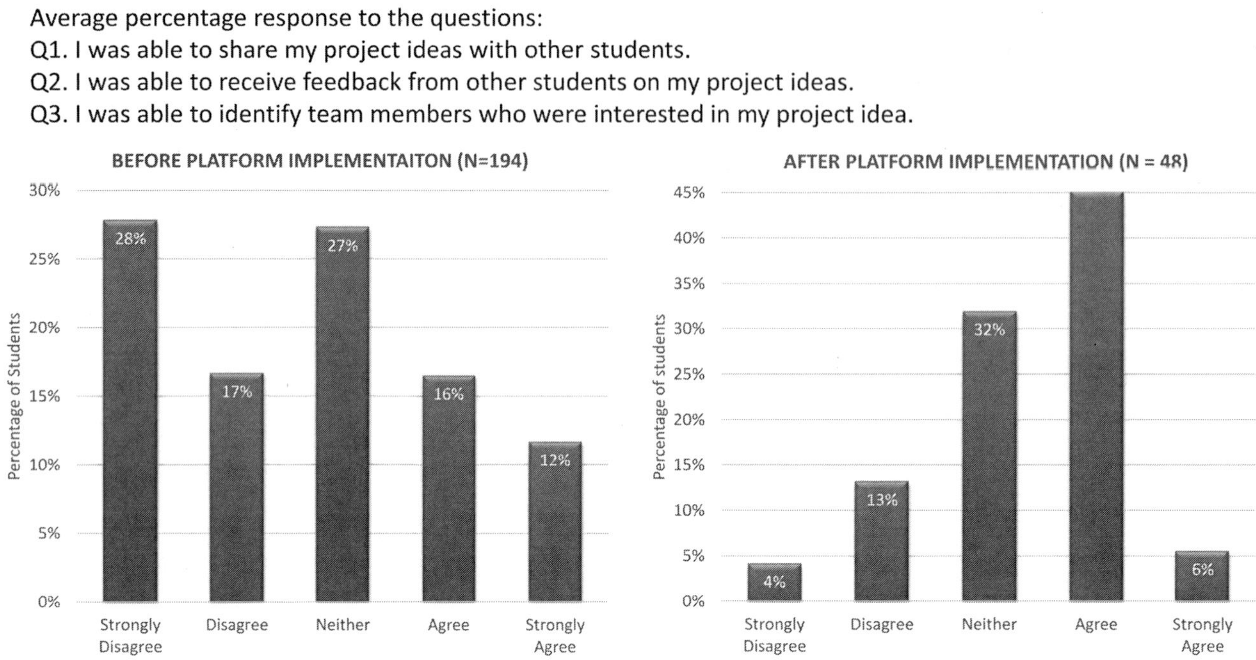

FIGURE 5: AVERAGE STUDENT-STUDENT INTERACTION.

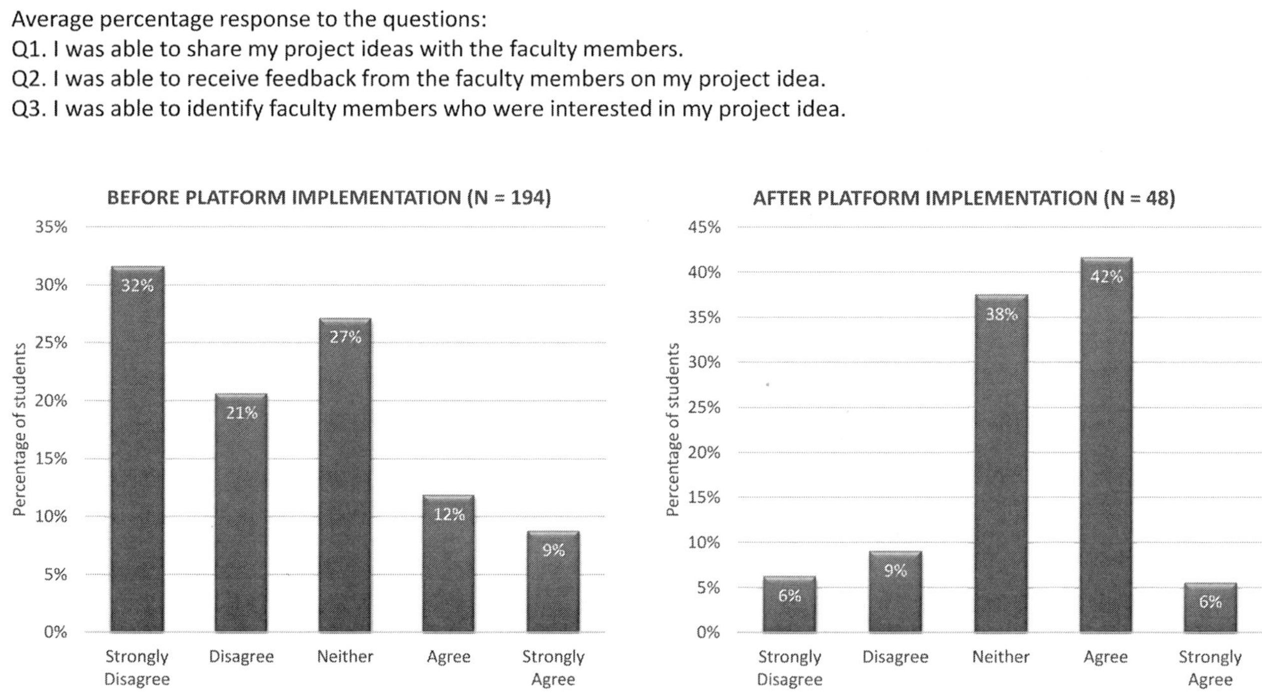

FIGURE 6: AVERAGE STUDENT-FACULTY INTERACTION

7 DISCUSSION

After implementation of the online tool, a total of 16 project ideas were proposed. 4 among the 16 ideas were put forward by the faculty and 12 were proposed by the students. Faculty and students collectively participated in refining the problem statements. Initial findings show a statistically significant increase in the amount of student-student interaction and student-faculty interaction.

Next observation is that the frequency of using the online tool was lower than expected. This can be observed in Figure 4, which show that the majority of the students only used the online tool once per week. We expected that the students would atleast use the online tool 3-4 times per week. This may be attributed to the fact that it requires more time for any social platform to become popular, establish a community of users and promote frequent interactions between them [23]. The response to the statement "I was able to receive feedback from other students" shows that less than half of the students (42-46%) agreed to the statement. This may be because the same discussion space was provided to both students and faculty to provide feedback. This can be inferred from the students suggestion for a separate discussion space for students and faculty in the future versions of the tool. Also, it was anticipated that the students would use the discussion section to express their interest on a specific project, and this would aid the team building exercise. Since less discussion activity by students was observed, this explains the response to the statement "I was able to identify team members on my project idea", as only 42% of students agreed to the statement.

Finally, students' perception in using the online tool was also collected. We observed that 81% responded with affirmative that the online tool was a useful tool for collaboration. Although most of the students agreed that the online tool was useful for collaboration, only 26% agreed to the statement "I believe, I can collaborate better through the online tool than through face-face meetings". This may be because of the low frequency in using the online platform or it could be because of delays in receiving a reply for a post or comment as mentioned in the written feedback by one of the students. This lack of spontaneity in response may hinder building a sustainable online community. Therefore to address this, some type of synchronous communication tool must be integrated in the future version of the online tool.

When asked to state a better alternative for forming engineering design project and teams, a large portion of the students stated that the online tool was probably one of the best alternatives. Other responses for better alternatives included meeting in person and sharing ideas. While there are some technical deficiencies with the current implementation, students generally found the tool to be useful.

8 CONCLUSION AND PATH FORWARD

The impact of social tools in an virtual organization is emphasized in [24]. The author states that the social tools allow us to create networks of connections and aid in finding people with right expertise at the right time. By utilizing these tools, there exist plethora of opportunities for us to explore our current environment and create new ways for experiencing it. One such opportunity is improving the engineering design project formation process. An online collaborative platform should promote a sense of community, social presence, and frequent interaction [23]. The research question we tried to answer in this study is: *Does an online collaboration tool improve the project formation process in engineering design courses?*.

In the present study we observed that the online tool promotes collaboration by connecting appropriate stakeholders and building a network of these stakeholders, which might not be made through face-to-face meetings and e-mail communications. Initial findings show a significant increase in the amount of interactions among students and faculty. Students found the tool simple to use. While the tool appears to be very useful for collaboration, more time is required for it to become widely used. There appears to be a cultural change for the stakeholders to fully covert from email communication to using online collaboration tool. The suggestions for improving the tool generally included features that were not necessarily design specific, indicating that the underlying concept of this online platform provides a strong basis and can be extended for future online tools.

Figure 7 illustrates the difference in the process of formulating senior design projects before and after implementing the online tool. As illustrated in the Figure 7, the online tool provided an opportunity for connecting these stakeholders at one platform which would have otherwise not been made. Although we observed an increase in interaction quantitatively, a qualitative analysis observing the responses and project statements should be performed in order to understand the quality of these interactions. Establishing successful online communities typically span an extended time frame [25]. As a result it is difficult to capture the effectiveness of these collaborative platforms by just observing for a period of one semester. We believe that more insights can be obtained if further analysis is carried over a period of time observing the online community evolving. Through further studies, the existing platform may be improved, and it could ultimately enhance the way the engineering design is carried out.

ACKNOWLEDGMENT

The authors gratefully acknowledge the financial support from the National Science Foundation through the CAREER grant 1265622.

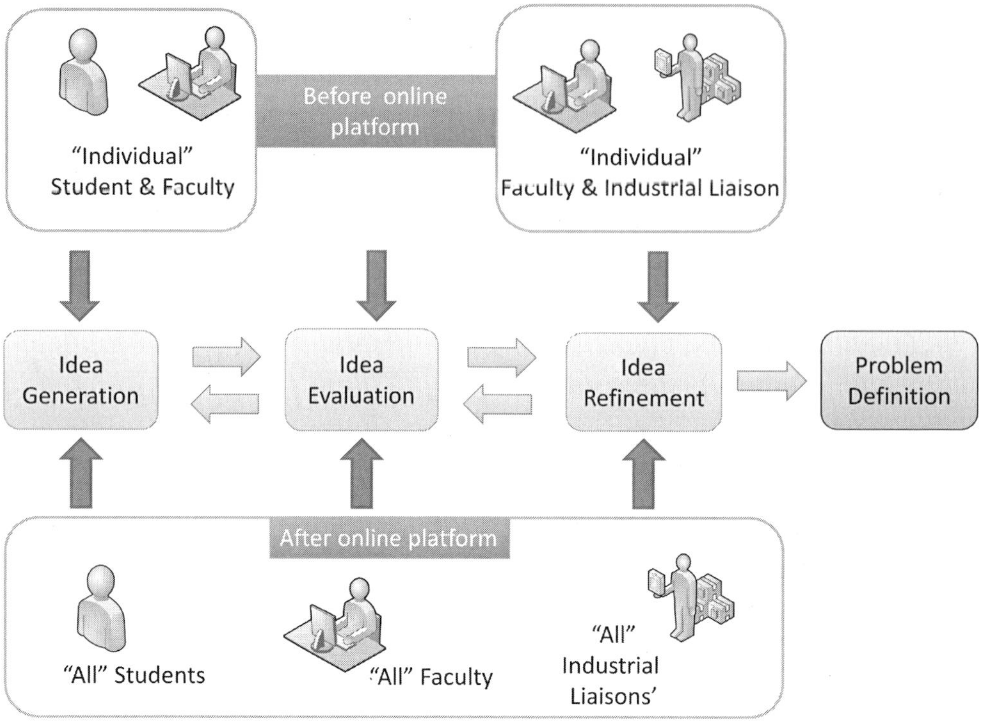

FIGURE 7: COMPARISON OF PROJECT FORMATION PROCESS BEFORE AND AFTER IMPLEMENTATION OF ONLINE TOOL.

REFERENCES

[1] Cano, J. L., Lidon, I., Rebollar, R., Roman, P., and Saenz, M. J., 2006. "Student groups solving real-life projects. a case study of experiential learning". *International Journal of Engineering Education,* ***22***(6), p. 1252.

[2] Lasher, W. C., and Ford, R. M., 2008. "Workshop-effective practices for team formation and faculty involvement in capstone design courses". In Frontiers in Education Conference, 2008. FIE 2008. 38th Annual, IEEE, pp. W2B–1.

[3] Dutson, A. J., Todd, R. H., Magleby, S. P., and Sorensen, C. D., 1997. "A review of literature on teaching engineering design through project-oriented capstone courses". *Journal of Engineering Education,* ***86***(1), pp. 17–28.

[4] Chandrasegaran, S. K., Ramani, K., Sriram, R. D., Horváth, I., Bernard, A., Harik, R. F., and Gao, W., 2013. "The evolution, challenges, and future of knowledge representation in product design systems". *Computer-Aided Design,* ***45***(2), pp. 204–228.

[5] Lumsdaine, E., Loukus, J., Dreyer, J., Chenoweth, S., and Lumsdaine, M., 2009. "Forming and managing project teams in large capstone design courses". In American Society for Engineering Education, American Society for Engineering Education.

[6] Paliwal, M., and Sepahpour, B., 2012. "A revised approach for better implementation of capstone senior design projects". In American Society for Engineering Education, American Society for Engineering Education.

[7] Novak, M. ; Dolšak, B., 2011. "How to define projects to teach design?". In the 13th International Conference on Engineering and Product Design Education, Proceedings of E and PDE, pp. 239–244.

[8] Wyard-Scott, L. "An online tool for capstone design course communication". In Proceedings of the Capstone Design Conference 2010.

[9] Kanai, J., 2008. "Web-based collaboration tool in engineering design courses". *Online]. ASEE Zone,* ***1***, p. 2008.

[10] Lo, D. C.-T., and Karam, O., 2013. "Enhance capstone projects with a new online collaboration system". In Proceedings of the 2013 IEEE 13th International Conference on Advanced Learning Technologies, IEEE Computer Society, pp. 217–218.

[11] Davis, D., Beyerlein, S., Harrison, O., Thompson, P., Trevisan, M., and Mount, B., 2006. "A conceptual model for capstone engineering design performance and assessment". In Proceedings of Annual Conference of American Society for Engineering Education.

[12] `http://www.blackboard.com/`.

[13] `https://globalhub.org/`.

[14] https://drupal.org/.
[15] http://en.wikipedia.org/wiki/LAMP_(software_bundle).
[16] https://drupal.org/project/disqus.
[17] https://drupal.org/project/registration.
[18] https://engineering.purdue.edu/PEARL/.
[19] Bruning, K., 2005. "The role of critical thinking in the online learning environment". *International Journal of Instructional Technology and Distance Learning,* **2**(5), pp. 21–31.
[20] Burnett, K., Bonnici, L., Miksa, S., and Kim, J., 2007. "Frequency, intensity and topicality in online learning: An exploration of the interaction dimensions that contribute to student satisfaction in online learning". *Journal of Education for Library and Information Science*, pp. 21–35.
[21] Hillman, D., Willis, D., and Gunawardena, C., 1994. "Learner-interface interaction in distance education: An extension of contemporary models and strategies for practitioners". *American Journal of Distance Education,* **8**(2), pp. 30–42.
[22] Sher, A., 2009. "Assessing the relationship of student-instructor and student-student interaction to student learning and satisfaction in web-based online learning environment.". *Journal of Interactive Online Learning,* **8**(2).
[23] DeSchryver, M., Mishra, P., Koehleer, M., and Francis, A., 2009. "Moodle vs. facebook: does using facebook for discussions in an online course enhance perceived social presence and student interaction?". In Society for Information Technology & Teacher Education International Conference, Vol. 2009, pp. 329–336.
[24] Bertoni, M., and Chirumalla, K., 2011. "Leveraging web 2.0 in new product development: Lessons learned from a cross-company study.". *J. UCS,* **17**(4), pp. 548–564.
[25] Lave, J., and Wenger, E., 1991. "Situated learning: Legitimate peripheral participation".

This page left blank intentionally.

Proceedings of the ASME 2014 International Mechanical Engineering Congress and Exposition
IMECE2014
November 14-20, 2014, Montreal, Quebec, Canada

IMECE2014-38451

EMBEDDING SPECIALIZED ONLINE LEARNING MODULES IN COURSES

Kumar V. Singh
Mechanical & Manufacturing Engineering Department,
Miami University
Oxford, OH, USA

Fazeel Khan
Mechanical & Manufacturing Engineering Department,
Miami University
Oxford, OH, USA

ABSTRACT

In recent years, online learning modules and interactive tools have been developed for integrating synergistic computational and experimental activities into several courses in the mechanical and manufacturing engineering curriculum. These modules may complement existing labs, introduce experiments to lecture-based courses, or help reinforce the understanding of concepts within a course through case studies, computational modeling and its validation through experimental data analysis. Within a given undergraduate course in engineering, the traditional approach is to cover the fundamental concepts and their applications in problem solving and design. However, often some advanced concepts pertinent to the course material cannot be included in great details due to the restrictions on total number of credit/teaching hours. The exposure to such advanced material is considered very important for establishing a broader appreciation of the relevance of the course material and potentially instilling an interest in successive classes and, most importantly, research projects.

The project described in this paper has focused in the development of learning modules which can be used in various courses to introduce advanced topics, analyses and applications related to the fundamental course content. In this paper the development, implementation and assessment of two modules which feature enhanced content is presented. One module has been designed to be used as a tutorial on rotor dynamics and has been embedded in an introductory course on mechanical vibrations. The second is a module being developed for introducing the static and dynamic characterization of viscoelastic polymers. This module will be associated with an advanced mechanics of materials course, as well as the course on vibrations. These two separate concepts/topics are not formally a part of the undergraduate level courses but the underlying concepts can be easily related to the associated course learning outcomes. Such modules can, furthermore, be modified and used with various other courses as assignments and/or as a pre-requisite to a given case study. These modules can also be used for training and support of undergraduate researchers.

INTRODUCTION

The pedagogical practices adopted by educators are evolving to societal needs and technological changes and the engineering courses are now being taught using different teaching styles and various forms of instruction, such as active learning, project-based learning, and experiential learning etc. [1-2]. In recent years, several online instructional materials, online labs and interactive simulation and computational resources in engineering science have been developed so support these different instruction styles. For example, simulation tools and interactive modules for teaching various concepts such as dynamics signal processing, chemical engineering etc. [3-8], virtual labs [9-11], online conceptual modules for specific topics and course [12-13]. A collection of online educational resources for teaching in the Science, Technology, Engineering and Mathematics (STEM) area could be found in the National Science Digital Library [14]. Several assessment tools and instruments are also developed to evaluate the effectiveness of these modules and to assess the student understanding of a given concept using such online educational resources [15-17]. Some favorable attributes, such as the ability to complete online modules at one's own pace, availability of modules to part time students and for distant education, and understanding of student learning through the continuous monitoring of the student progress [18-19]. Some challenges of online modules and courses and criticism regarding its inability to facilitate instant feedback to students for their queries are also noted [20-21].

At Miami University, online learning modules and interactive tools have been developed in recent years for integrating synergistic computational and experimental activities into several courses in the mechanical and manufacturing engineering curriculum [22]. A brief outline of the five studio areas developed for this initiative, its architecture and details of the studio styled modules with the associated grouping of courses is highlighted in [23]. The motivation, format and pilot assessment results for the Materials studio were presented in [24]. In order to compliment these studios, additional online and interactive learning modules on polymeric biomaterials were developed and their details can be found in [25]. The details of the assessment methodology which are designed to enable continuous improvement of these modules with some assessment data were presented in [26]. The details of integrating vibration experiments and data analysis through online modules are highlighted in [27].

The usefulness of the learning modules extends beyond their use as supporting material in courses. The creation of more specialized learning modules not only introduces students to more advanced applications but more importantly serves to increase the preparation for collaborating with faculty on research based or capstone design projects. The modules could be used by the students to independently learn the subject specific knowledge or practice in the use of various computational and data analysis tools. In this paper, the development, implementation and assessment of two such specialized learning modules is presented. One module is developed towards parameter estimation and material characterization using dynamic testing. Another tutorial module on rotor dynamics is developed. These modules are embedded as individual and group assignments in an introductory course on mechanical vibrations. In the next section, some features of these learning modules are detailed. It is followed by the brief summary of the assessment process and survey instruments. Both of these modules are implemented this semester (Spring 2014) and the assessment plan and sample assessment data is presented in the last section of the paper.

LEARNING MODULES FOR STRUCTURE AND MATERIAL PARAMETER ESTIMATION FROM DYNAMIC TESTING

The learning objective for this module is to introduce students various techniques for the estimation of unknown system parameters (mass, stiffness and damping) from forced vibration testing and dynamic characterization of polymeric material parameters which may be utilized for vibration suppression applications. In achieving this objective, it is anticipated that this module will provide students:

(i) exposure to forced vibration experiments conducted for estimating the performance measures of a given system and facilitate the process for extracting and analyzing experimental data in both time and frequency domain,

(ii) knowledge of identifying key characteristics of step response, such as maximum overshoot, settling time, peak time, and rise time etc. of the system's response,

(iii) an understanding of how forced vibrational response (in time domain) and associated transfer functions (in frequency domain) obtained during forced vibration experiments can become the basis for unknown model parameter estimation,

(iv) several comparative tools (numerical/analytical) which can be used for computing and validating the response of vibrating systems subjected to excitation force, and,

(v) exposure to the state-of-the-art dynamic material characterization techniques for polymeric materials and understanding of their potential engineering applications in the areas of vibration suppression.

In order to achieve such a wide range of outcomes, the activities are divided into two parts. First the students as a group learn about the fundamentals of forced vibration testing through an experimental setup of a single degree of freedom force vibration testing using an ECP rectilinear plant [28]. This is done by providing a schematic illustration, an embedded test video and details of a given forced vibration test and its relation to the associated mathematical model, as shown in Figure 1.

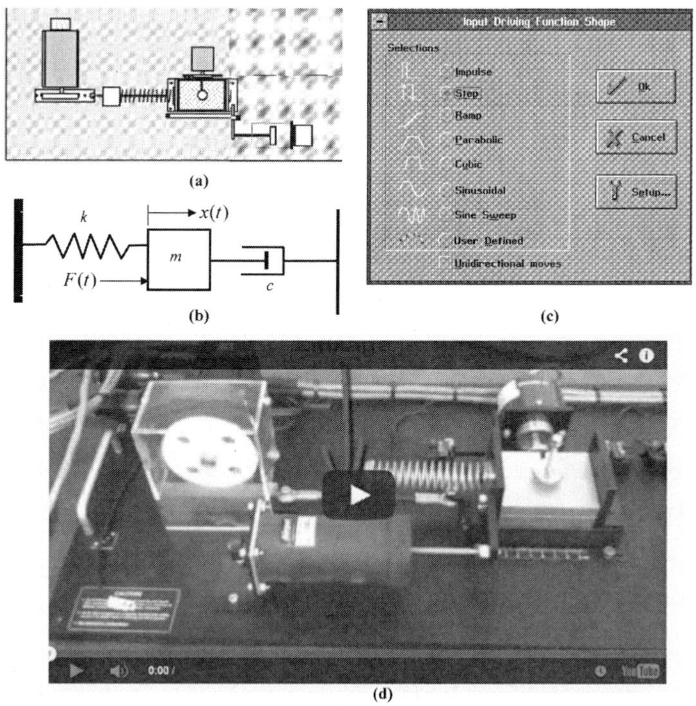

Figure 1: (a) Experimental configuration for Single Degree of Freedom (SDoF) system (associated with Figure 1) to be identified, (b) Schematic of an equivalent system, (c) Representative input forcing function available for excitation, (d) Embedded video explaining the experiment

The students are then introduced to the time and frequency domain characteristics of a forced vibrating system. To this end the fundamental terminologies associated with the resulting frequency response functions are presented, as shown in Figure 2. This schematic relates the mathematical model to the measurable frequency response functions and their parameters. In frequency domain they can then estimate the modal parameters such as natural frequencies and damping. The students are provided time domain data and by utilizing the concepts of discrete Fourier transform in class, they can obtain the associated frequency response functions needed for estimation.

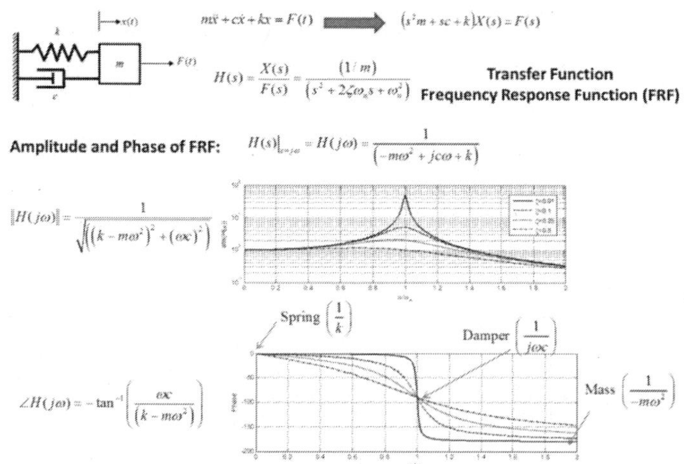

Figure 2: Frequency Response Function of a single degree of freedom system

Students validate these findings by estimating the same parameters in time domain. For example, they can estimate several dynamic characteristics of the system and may completely identify the unknown system parameters from the step-response, as shown in the Figure3.

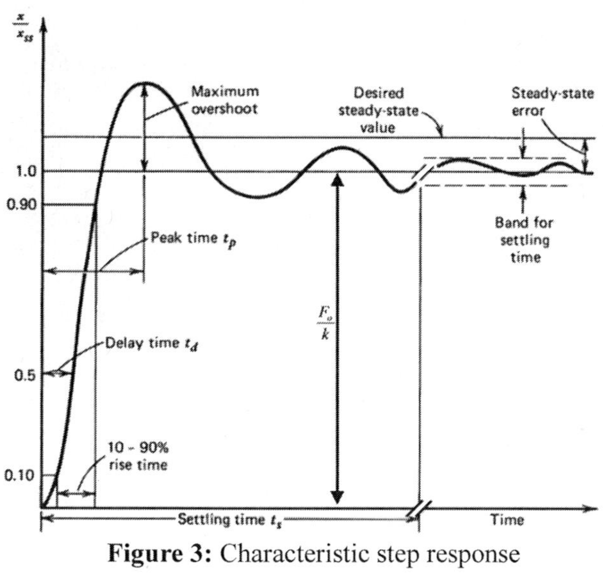

Figure 3: Characteristic step response

Students then compare the estimated parameters from time and frequency domain analysis. Furthermore they compute the response of the system (analytically or numerically) by using the estimated parameters and they validate the accuracy of their estimation by comparing them with available experimental response. Students are asked to compare two different estimation approaches (time and frequency domain) and to critique their findings by summarizing the errors between the estimated response and experimental response. They also perform a simulation using identified parameters and observe how these parameters influence the various dynamical characteristics of the system. They use these simulations in designing isolators (spring and damping) for a desired behavior (for a prescribed overshoot and settling time). These activities are aligned with the learning outcomes (i)-(iv) associated with the module.

Another favorable attribute of this module, is that it can help students relate how these principles can be used and applied in real life engineering applications. To this end they are asked to replace the purely elastic and viscous isolators by more commonly used polymeric materials. In order to select and/or design with polymers they need to understand how the material properties (complex modulus and loss factors) are characterized through dynamic testing and how the stiffness and damping of such materials may be estimated. In order to have this practical understanding they use another studio which is developed for static and dynamic characterization of polymeric materials. This studio consists of three modules:

A. An interactive module on the fundamental of viscoelastic polymeric materials [29],
B. An interactive module for constitutive viscoelastic-visco-plastic modeling of polymers and biomaterials, and,
C. A module on dynamic characterization of polymeric materials.

Figure 4: Integrated module for dynamic characterization

The assignment directs students to a new website, as shown in Figure 4, for the module on the dynamic characterization of polymeric materials.

In this module students are introduced to the differences between the dynamic characteristics of elastic, viscous and viscoelastic materials. For the viscoelastic materials the basic formulation and method for estimation of the complex modulus and associated damping estimation from the stress and strain curves are provide. Some video links highlighting the fundamentals of dynamical mechanical testing (DMA) and the capabilities of DMA instruments and testing protocols are provide to students for their reference.

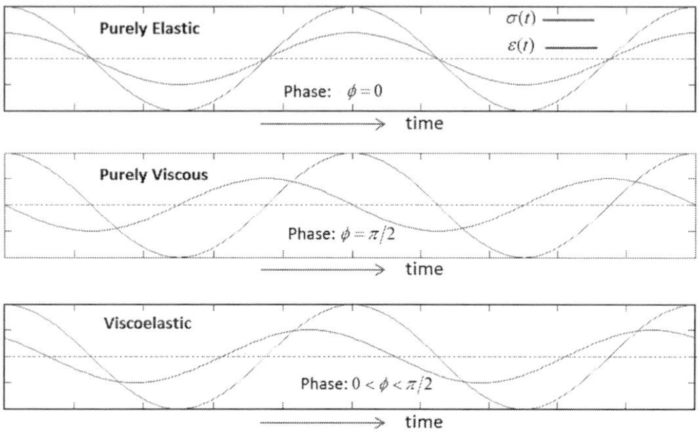

Figure 5: Dynamic characteristics of materials (Stress-Strain as a function of time)

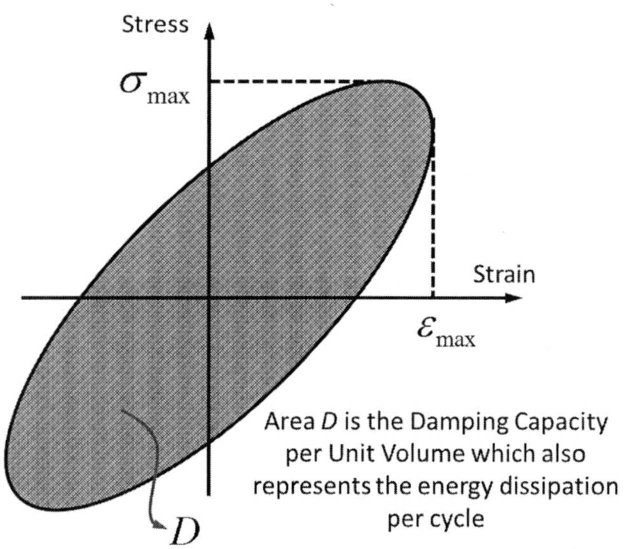

Figure 6: Hysteresis loop associated with the cyclic loading

In the assignment, the students are provided with cyclic test data associated with two set of polymers: ultra-high molecular weight polyethylene (UHMWPE) and silicone. In order to highlight the dependency of material parameters on frequency and temperature two different scenarios are presented. Questions asked in the assignment are derived from the learning outcomes of the module. For example, the students are asked to perform data analysis and summarize their observations on the dynamic properties of viscoelastic materials. These observations include understanding of:

a) the relationship of complex stiffness of viscoelastic materials from the estimated complex modulus and damping,
b) the cycle dependent effect of material properties (stiffening and softening of materials),
c) the effect of temperature on the complex stiffness of the materials, and,
d) the potential applications where usage of such materials is advantageous compared to pure linear elastic/viscous isolators.

TUTORIAL MODULE IN ROTORDYNAMICS

A tutorial module on rotor dynamics is being developed for use at the undergraduate level Mechanical Vibrations course. This module is also intended to provide basic training to undergraduate and graduate researchers and students involved in senior design projects. In this tutorial with the help of an experimental setup using "GE/Bentley Rotor Kit-RK4" fundamental definitions, terminologies and calculations are introduced. The learning outcomes of this module are focused on facilitating:

(i) Understanding of the basic components of and terminologies used for rotordynamic analysis such as critical speeds, Campbell diagram, rotating unbalance, order analysis, whirl etc.
(ii) Simplified mathematical models used for rotordynamic analysis and basic calculations associated with the performance of simplified shaft-disk system
(iii) Exposure to the experiments in the field of rotordynamic dynamic analysis
(iv) Relevant applications of rotordynamic and available tools for advanced rotordynamic analysis

In this module with a brief description of the importance of rotordynamic analysis, the main components of a rotor dynamic system such as shaft, disk and bearings are introduced. A basic understanding of the transverse vibration of shaft disk systems due to rotating unbalance, approximation to an equivalent single degree of freedom system, as shown in Figure 7, and governing differential equations of motions is provided.

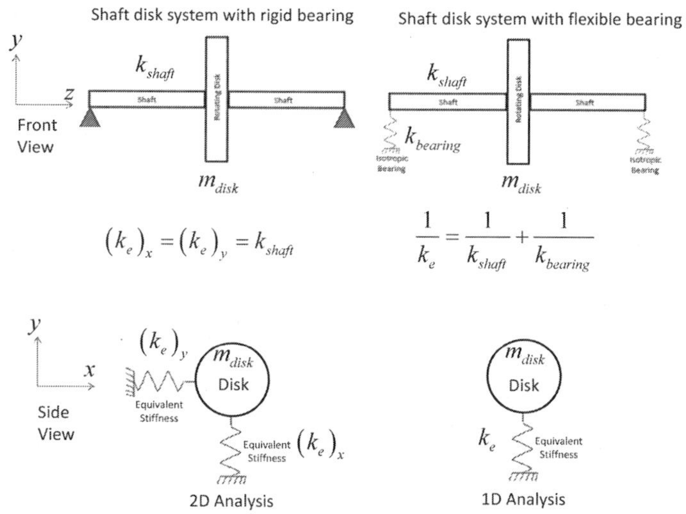

Figure 7: Simplified modeling approximation for a rotor

Unique terminologies relevant in the rotordynamic analysis are introduced with the help of simplified models. For example, the concept of *critical speeds* of shaft-disk systems supported by bearings is introduced. The phenomenon of whirling is explained. Graphical representation of the characteristics of rotordynamic systems in the form of Campbell Diagram is highlighted, as shown in Figure 8.

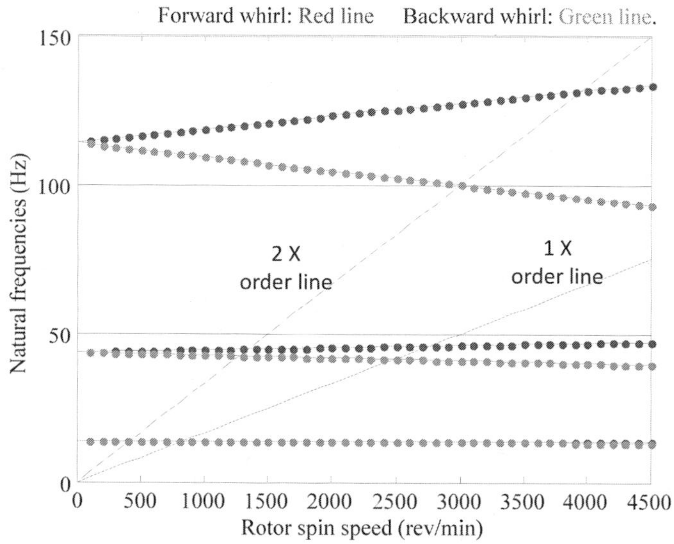

Figure 8: Campbell diagram

With the help of the system schematic and an embedded video, an experimental setup for an academic RK4 test rig, as shown in Figure 9, is introduced to the students. The components such as sensors, driving motors, shafts and bearing locations yield a dynamical system whose properties can be modeled as shown in Figure 7. Data extraction procedures are described to the students and the emerging data is presented in industry-standard plots such as cascade and waterfall plots, as shown in Figure 10. The first critical speed in Figure 8 is now clearly observed in Figure 10 during experiment. Students are then asked to predict the performance characteristics based on a supplied set of machine parameters. Multiple choice type quizzes will be used to access the attainment of the learning outcomes by the students.

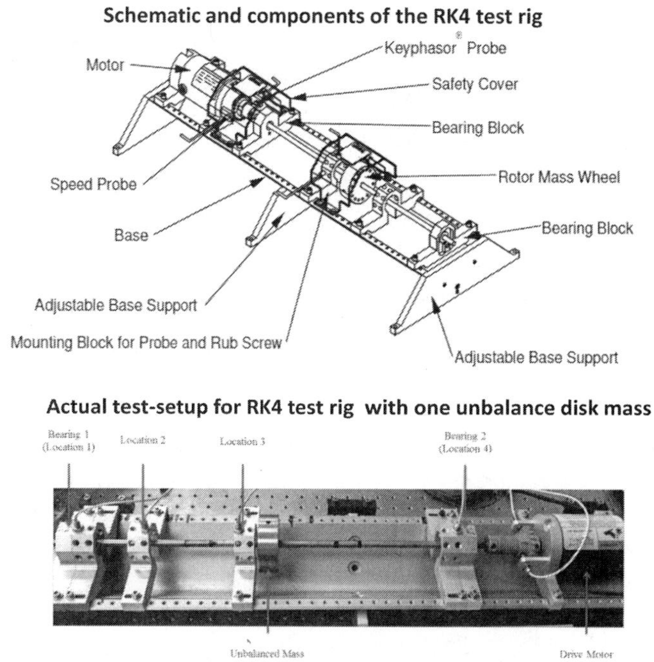

Figure 9: Experimental test rig

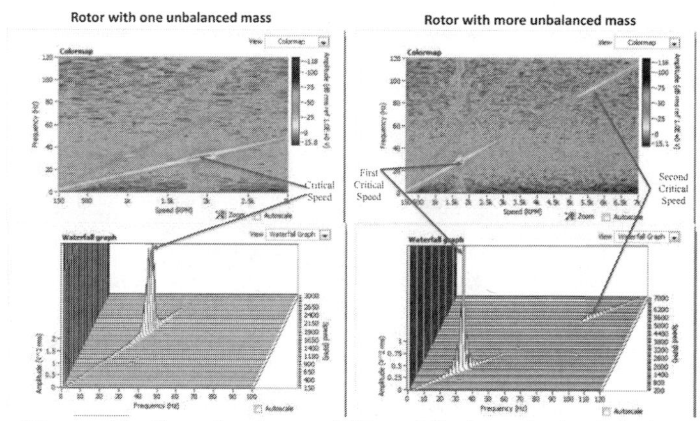

Figure 10: Cascade plots and Waterfall plots obtained during experiments

ASSESSMENT PLANS AND RESULTS

The assessment instruments for these modules have been developed with the help of the Ohio Evaluation and Assessment Center [30] at Miami University. For example an assessment survey was given to students along with the assignment associated with the "learning modules for structure and material parameter estimation from dynamic testing". The survey form and the results from spring 2014 semester are presented in Appendix A. The main categories in which survey data is collected pertain to: (i) attainments of the learning outcomes, (ii) quality of multi-media content and (iii) overall effectiveness of the assignments. To expand briefly on these categories, questions pertaining to the learning outcomes specifically attempted to quantify the extent to which the learning outcomes associated with a given module have been achieved. Understanding student perception of the multimedia content is particularly important since these tools are used as part of the active and passive learning exercises.

In spring 2014, a total of 44 students participated in the survey. They were assigned to complete the module as a group; however the survey was to be submitted by the entire student in the class. The survey questions were divided in three categories. First set of questions (Appendix A. I) targeted to assess the quality, presentation and clarity of the learning module, attainment of learning outcomes and overall perception of the students (level of clarity, amount of detail, amount of guidance, amount of work, and applicability) regarding this module. The results are summarized in Figure 11 in the Appendix. It is found that student found agreement that the learning objectives and outcomes were reasonably clear (3.59/5.0), supporting materials in the modules were available (3.86/5.0) and suitable guidance was provided to them (3.11/5.0) towards the assignment. However, they felt that even in the group setting the amount of work was high (2.50/5.0). They also felt that such modules may be useful beyond the coursework (3.45/5.0). Second set of questions were targeted to assess the overall effectiveness of the modules and its usefulness perceived by the students. The survey questions (Appendix A. II) and the results are presented in Figure 12 in the Appendix. The score in this category ranges from 3.5-4.0 out of 5.0. It is encouraging to see that student generally agree that such a module help improve their understanding of various concepts associated with vibration testing and material characterization. The last set of survey questions (Appendix A. III) were intended to assess the usefulness of the active and passive components of the modules such as images, videos, text documents, experimental data files, details of experiments etc. Students found these components to be moderately useful to very useful (average response ranging from 2.30-2.56 out of 3.0), as shown in Figure 13.

This data highlights areas of improvement such as modification of this learning module for two part assignment (one for vibration testing and other for dynamic material testing) and providing more guidance to students. Additional multiple choice questions as well as more experimental data linked with real life engineering case studies needs to developed to facilitated independent learning beyond this coursework (research or capstone design projects). We hope that data obtained from subsequent semesters will enable continuous improvements and guide the creation of improved material.

CONCLUSIONS

The structure and content of two specialized learning modules introduced in the mechanical vibration course has been presented in this paper. The studio consists of discrete learning modules and case studies. The learning modules integrate validation studies of computation models with experimental data. The objective is to engage students in the process of dissecting complex systems, selecting from among several computational approaches, and identifying means of verifying their results. The modules are customizable in that the contained expansive experimental data sets can be used in conjunction with various problem scenarios a faculty member might choose to create. The modules, in the area of rotor dynamics and polymer materials, were also born of the need to support a steady number of student research projects in which the task of background preparation could be expedited with targeted module-assisted student learning. Details of two learning modules, for estimating the system and material parameters from dynamic data, and a tutorial module in the applied area are presented. Plans for assessment are summarized. The assessment of these modules will be completed by the end of this semester and the initial data of student perception on the effectiveness, usefulness and quality of these modules will be presented during the conference. It is anticipated that with continuous improvement such a resource may be useful for improving the course material, introducing more case studies and research problems in the course and potentially developing an online course in vibration for greater dissemination.

ACKNOWLEDGMENTS

This research is supported in part by the National Science Foundation (NSF) "TUES: Development and Integration of Synergistic Computational and Experimental Activities within the Mechanical and Manufacturing Engineering Curriculum", Grant #1044698, MathWorks® Academic Support for Curriculum Development and Miami University's Tech-Free Grant 2012-13.

REFERENCES

[1] Prince, M.J. and Felder, R.M., "Inductive Teaching and Learning Methods: Definitions, Comparisons, and Research Bases," Journal of Engineering Education, 95 (2), pp. 123-138, 2006.

[2] Felder, R. M., and Silverman, L. K., "Learning and Teaching Styles", Engineering Education," Engineering Education, 78(7), 674-681, 1988.

[3] Llado, J. and Sanchez, B., "A Computer-based Tool to Foster Engineering Student's Interest in Dynamics," Comput. Appl. Eng. Educ, vol. 19 (3), pp. 475-486, January 2011.

[4] Dolecek, G., "MATLAB-Based Program for Teaching Autocorrelation Function and Noise Concepts," IEEE T. Educ., vol. 99, pp. 1-8, 2011.

[5] Ali, E. and Idriss, A. "An Overview of Simulation Module, PCLAB, for Undergraduate Chemical Engineers in Process Control," Comput. Appl. Eng. Educ., vol. 18 (2), pp. 306-318, June 2010.

[6] Koe, C. et al. "Investigating the Effect of 3D Simulation-Based Learning on the Motivation and Performance of Engineering Students," J. Eng. Educ., vol. 99 (4), pp. 237-251, 2010.

[7] Aliane, N. "A Matlab/Simulink-Based Interactive Module for Servo Systems Learning," IEEE T. Educ., 53 (2), pp. 265-271, 2010.

[8] Goeser, P., Johnson, W., Hamza-Lup, F. and Schaefer, D., "View – A Virtual Interactive Web-based Learning Environment for Engineering," Adv. Eng. Educ., pp. 1-24, 2011.

[9] Balamuralithara, B., and Woods, P.C., "Virtual laboratories in engineering education: The simulation lab and remote lab", Computer Applications in Engineering Education, 17(1), pp. 108–118, 2009.

[10] Shyr, W-J. "Internet-Based Laboratory Platform for Distance Learning in Engineering Education," Int. J. Eng. Educ., vol. 25 (4), pp. 693-700, 2009.

[11] Virtual Laboratories, Johns Hopkins University, http://www.jhu.edu/virtlab/virtual-laboratory/, (last accessed: December 2013).

[12] Maiorana, C., Sgarbossa, L., and Salomoni, V., "New Methodologies in Teaching e-Structural Mechanics Using WWW", Computer Applications in Engineering Education, 16 (3), pp. 189-210, 2008.

[13] Aziz, E-S. "Teaching and Learning Enhancement in Undergraduate Machine Dynamics," Comput. Appl. Eng. Educ., vol. 17 (1), pp. 61-79, 2009.

[14] National Science Digital Library Serving science, technology, engineering, and mathematics education, http://nsdl.org/ (last accessed: April 2014).

[15] Taraban, R., Anderson, E., DeFinis, A., Brown, A., Weigold, A. and Sharma, M.P., "First Steps in Understanding Engineering Students' Growth of Conceptual and Procedural Knowledge in an Interactive Learning Context," J. Eng. Educ., 96 (1), pp. 57-68, 2007.

[16] Warren, S., Nidhi, T. and Bennett, A. "Lessons Learned from the Application of Online Homework Generation Modules in a Signals and Systems Course," Proc. 38th ASEE/IEEE Frontiers in Education Conference, Saratoga Springs, NY, pp. TB4-17 – TB4-22, October 2008.

[17] Yang, D. et al. "Using Computer-based Online Learning Modules to Promote Conceptual Change: Helping Students Understand Difficult Concepts in Thermal and Transport Science," Int. J. Eng. Educ., vol. 28 (3), pp. 686-700, 2012.

[18] Hoic-Bozic, N., Mornar, V., and Boticki, I., "A Blended Learning Approach to Course Design and Implementation", IEEE Transactions on Education, 52 (1), pp. 19-30, 2009.

[19] Steif, P. and Dollar, A., "Study of Usage Patterns and Learning Gains in a Web-based Interactive Static Course". Journal of Engineering Education, 98 (4), pp. 321-333, 2009.

[20] Nistor, N., and Neubauer, K., "From participation to dropout: Quantitative participation patterns in online university courses", Computers & Education, 55(2), pp. 663-672, 2010.

[21] Sara, M., and Fredericka, B., "Modern challenges in converting a traditional engineering class into a fully online course", Proceedings of the ASME 2012 International Mechanical Engineering Congress & Exposition, IMECE2012, Paper Number IMECE2012-88750, November 9-15, 2012, Houston, Texas, USA.

[22] http://comex.csi.muohio.edu/ (last accessed: April 2014).

[23] Khan, F., Birchfield, N. and Singh, K. V., "Revitalizing the engineering curriculum through studio based Instruction", Proceedings of the ASME International Mechanical Engineering Congress and Exposition (IMECE-12), Paper Number: IMECE2012-88628, pp. 1-8, November 9-15, Houston, TX, USA, 2012.

[24] Rogers, N., Singh, K. V., and Khan, F. J. "Development of Educational Techniques for Computational-Experimental Analysis," Frontiers in Education Conference, FIE/ IEEE 2013, pp. 553-557, October 23-26, 2013, Oklahoma City, OK, USA.

[25] Singh, K. V., and Khan, F., "Interactive Module for Biomaterial Modeling and Characterization", Proceedings of the 121st ASEE Annual Conference, June 15-18, Indianapolis, IN, USA, 2014. (To appear)

[26] Khan, F. and Singh, K. V., "Embedded Learning Modules for the Mechanical Engineering Curriculum", Proceedings of the 121st ASEE Annual Conference, June 15-18, Indianapolis, IN, USA, 2014. (To appear)

[27] Singh, K. V., Birchfield, N. and Khan, F., "Integrating experimental data analysis through online modules", Proceedings of the ASME International Mechanical Engineering Congress and Exposition (IMECE-13), Paper Number: IMECE2013-65413, November 15-21, San Diego, CA, USA, 2013.

[28] http://www.ecpsystems.com/dynamics_recplant.htm (last accessed: April 2014).

[29] http://polymodmw.csi.muohio.edu/ (last accessed: April 2014).

[30] Ohio Evaluation and Assessment Center-Miami University, www.ohioeval.muohio.edu/ (last accessed: April 2014).

APPENDIX A

REPRESENTATIVE SURVEY DATA FORM FOR LEARNING MODULE ON STRUCTURE AND MATERIAL PARAMETER ESTIMATION FROM DYNAMIC TESTING

I. Please indicate your level of agreement on the following aspects of the learning module you have completed.

	Strongly Disagree	Disagree	Neutral	Agree	Strongly Agree
The objectives and outcomes of the assignment were clear	○	○	○	○	○
Supporting materials (Matlab tips, data analysis tips etc.) pertaining to the work required was provided	○	○	○	○	○
Suitable amount of guidance was provided by materials	○	○	○	○	○
The amount of work required was reasonable	○	○	○	○	○
These online modules may be used for your research, senior design resources and review of specific concepts	○	○	○	○	○

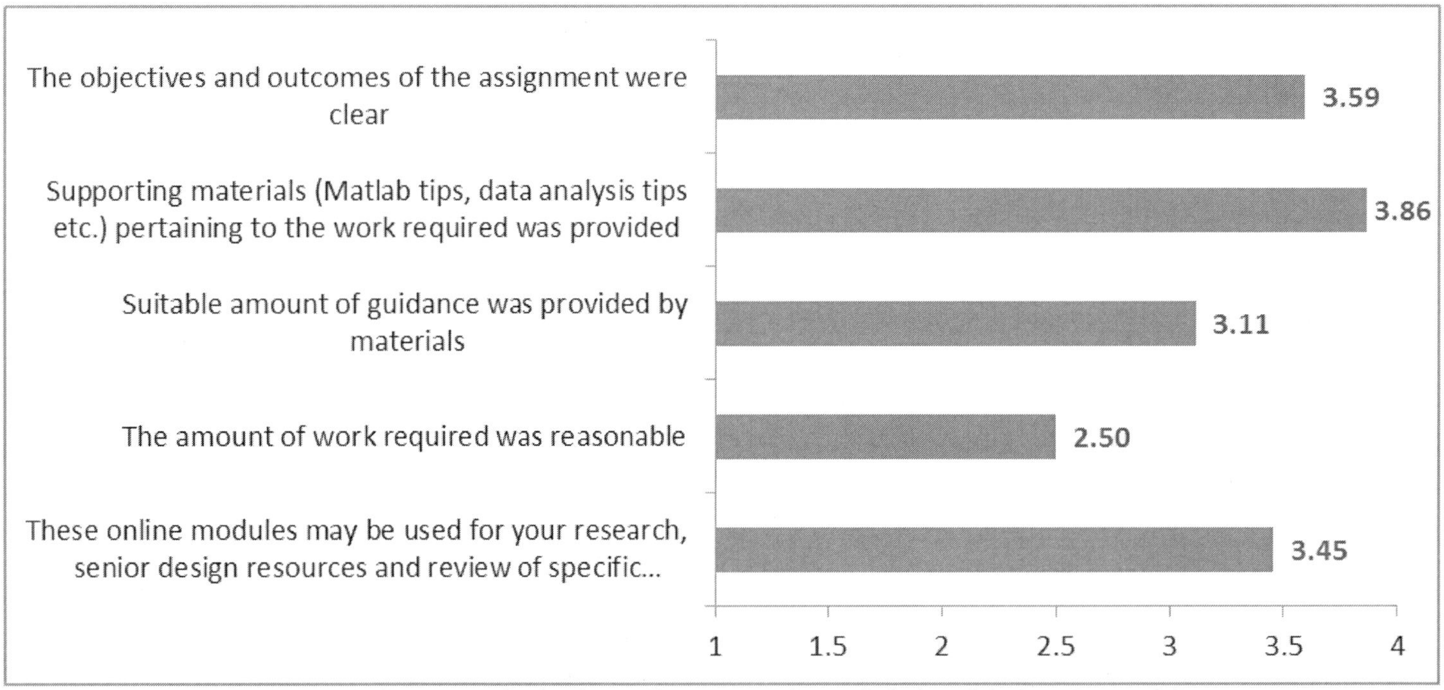

Figure 11: Survey results for the quality, presentation and clarity of the learning module (Scale: 1-Strongly Disagree and 5-Strongly Agree)

II. Please indicate your level of agreement with the following statements regarding the activity you have completed.

	Strongly Disagree	Disagree	Neutral	Agree	Strongly Agree
This module has improved my understanding of the usage of dynamic testing in the area of vibration.	○	○	○	○	○
This module has improved my understanding of the usage of dynamic testing in the area of material testing.	○	○	○	○	○
This module has improved my understanding of the dynamic data analysis in both time domain and frequency domain.	○	○	○	○	○
This module has provided the opportunity to learn and apply new techniques in estimating structural parameters (mass, stiffness, damping) from experimental data.	○	○	○	○	○
This module has provided the opportunity to learn and apply new techniques in estimating material parameters (modulus, complex stiffness, loss factor) from experimental data.	○	○	○	○	○
This module has introduced me to application of vibration testing in multidisciplinary field of dynamic characterization of polymers (bioengineering).	○	○	○	○	○

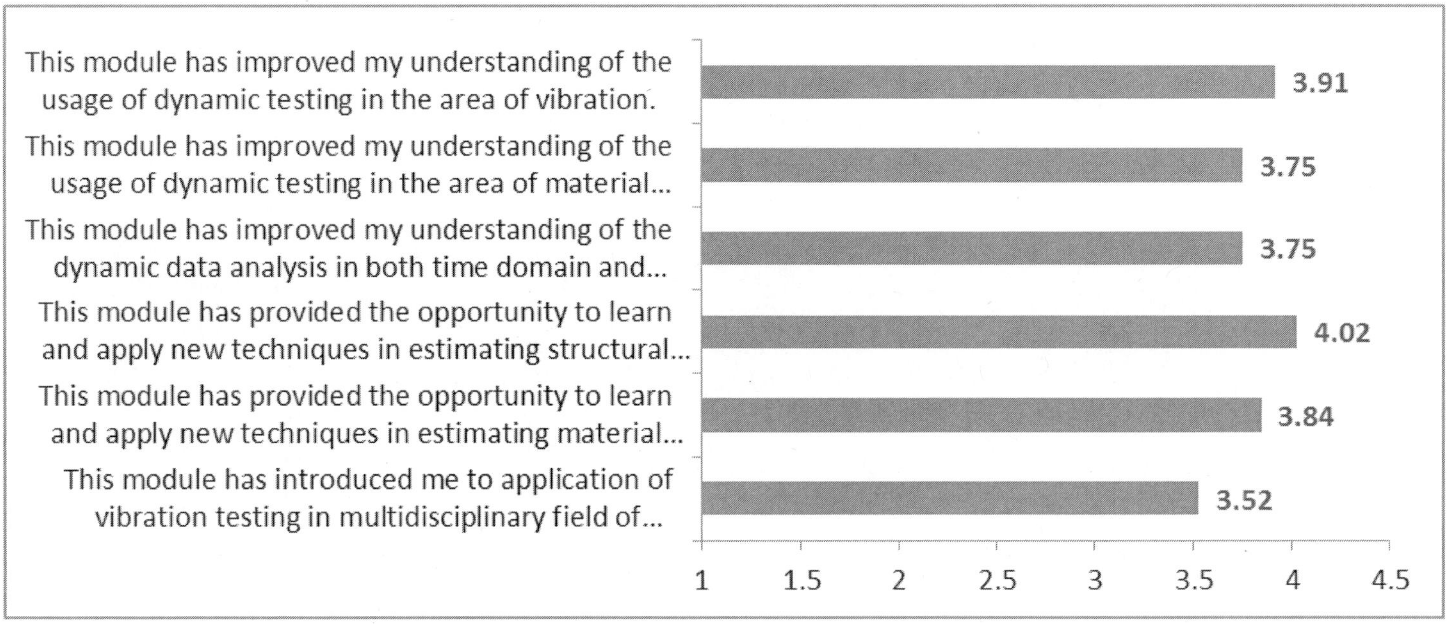

Figure 12: Survey results for the effectiveness of the learning module (Scale: 1-Strongly Disagree and 5-Strongly Agree)

III. Please rate the usefulness of different components which have been included in the learning module. Please select Not Applicable if the component was not included in the module activities you completed.

	Not at all Useful	**Moderately Useful**	**Very Useful**	**Not Applicable**
Text document providing background information	○	○	○	○
Text document describing the problem, activity, or case	○	○	○	○
Module of dynamic characterization of polymeric materials	○	○	○	○
Still images and its relevance to discussion	○	○	○	○
Raw Experimental Data Files	○	○	○	○
Description of Experimental Data	○	○	○	○

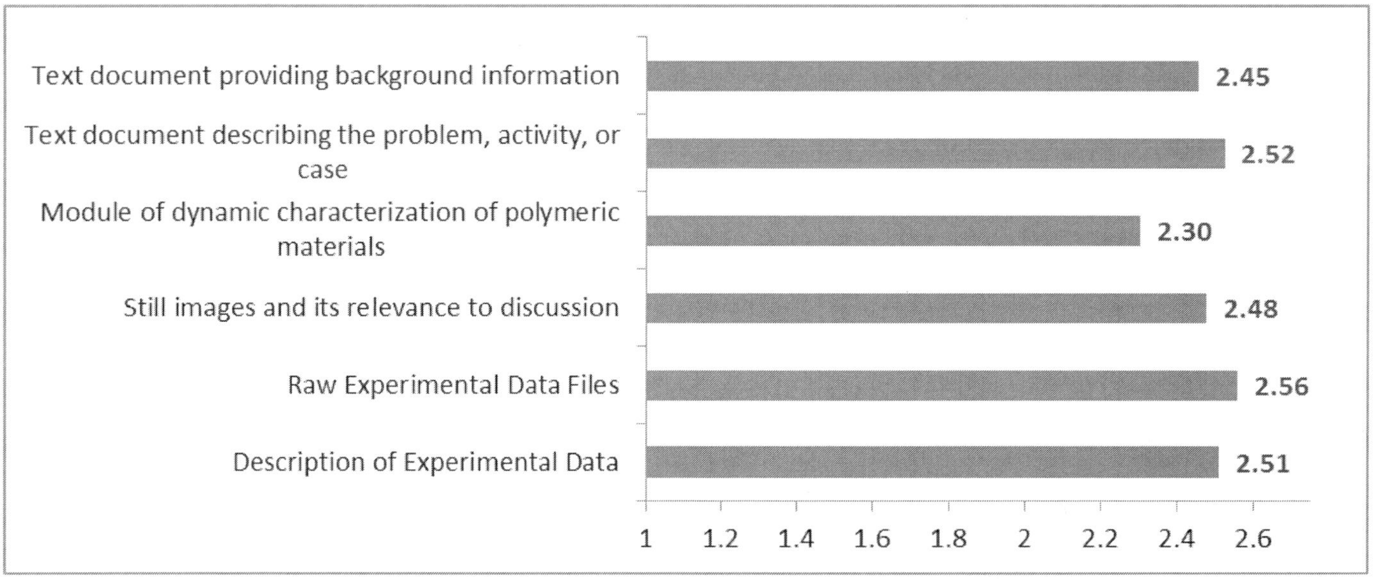

Figure 13: Survey results for the usefulness of the active/passive components of the learning module (Scale: 1-Not at all useful and 3-Very useful)

Proceedings of the ASME 2014 International Mechanical Engineering Congress and Exposition
IMECE2014
November 14-20, 2014, Montreal, Quebec, Canada

IMECE2014-38606

A Platform for Mechanical Assembly Education Using the Microsoft Kinect

Yizhe Chang
Stevens Institute of Technology
Hoboken, NJ, USA

El-Sayed Aziz
Stevens Institute of Technology
Hoboken, NJ, USA

Zhou Zhang
Stevens Institute of Technology
Hoboken, NJ, USA

Mingshao Zhang
Stevens Institute of Technology
Hoboken, NJ, USA

Sven Esche
Stevens Institute of Technology
Hoboken, NJ, USA

Constantin Chassapis
Stevens Institute of Technology
Hoboken, NJ, USA

ABSTRACT

Mechanical assembly activities involve multiple factors including humans, mechanical parts, tools and assembly environments. In order to simulate assembly processes by computers for educational purposes, all these factors should be considered. Virtual reality (VR) technology, which aims to integrate natural human motion into real-world scenarios, provides an ideal simulation medium. Novel VR devices such as 3D glasses, motion-tracking gloves, haptic sensors, etc. are able to fulfill fundamental assembly simulation needs. However, most of these implementations focus on assembly simulations for computer-aided design, which are geared toward professionals rather than students, thus leading to complicated assembly procedures not suitable for students. Furthermore, the costs of these novel VR devices and specifically designed VR platforms represent an untenable financial burden for most educational institutions.

In this paper, a virtual platform for mechanical assembly education based on the Microsoft Kinect sensor and Garry's Mod (GMod) is presented. With the help of the Kinect's body tracking function and voice recognition technology in conjunction with the graphics and physics simulation capabilities of GMod, a low-cost VR platform that enables educators to author their own assembly simulations was implemented. This platform utilizes the Kinect as the sole input device. Students can use voice commands to navigate their avatars inside of a GMod powered virtual laboratory as well as use their body's motions to integrate pre-defined mechanical parts into assemblies. Under this platform, assembly procedures involving the picking, placing and attaching of parts can be performed collaboratively by multiple users. In addition, the platform allows collaborative learning without the need for the learners to be co-located. A pilot study for this platform showed that, with the instructor's assistance, mechanical engineering undergraduate students are able to complete basic assembly operations.

1. INTRODUCTION

1.1. BACKGROUND

As a form of distance education, online education has become more and more popular. In the United States, an independent survey found that in 2003 around 1.6 million post-secondary students (representing 9.6% of the total school enrollment) took online courses. In 2008 over 3.9 million of post-secondary students participated in online courses, which made up 22% of the total post-secondary school enrollment [1]. Meanwhile, statistics from the United States Department of Education show that the number of distance education participants in 2008 was 4.2 million [2]. These figures show that in 2008 online education became a mainstream form of distance education. A recent survey found that the number of online education students in 2013 increased to 6.9 million, which was around 32% of the post-secondary enrollment [3]. Commercially, several organizations have proved the profitability of online education. For instance, Massive Open Online Course (MOOC) provider Coursera hit revenues of $1 million in 2013 [4].

Although many course subjects are covered by current online education providers, the pedagogical methods of these MOOCs are limited. While most of the courses are offered in the form of lectures and homework; a few of them may also include simple simulated laboratory work. For engineering education, this is far from enough, as engineering education demands large amounts of fieldwork, which requires not only complicated devices but also collaboration between students.

1.2. BUILDING A MECHANICAL ASSEMBLY EDUCATION PLATFORM USING A GAME ENGINE

Mechanical assembly education is one of the subjects that demands extensive practice. In order to offer this type of experience online, a complete assembly procedure simulation platform must be constructed. However, such a platform shows the complexity in the integration of multiple features for various needs for both educators and students. These needs are listed in Figure 1 On one hand, in order to provide a user-friendly pedagogical assembly design software for educators, the platform must be adaptable to modern computer-aided design software. On the other hand, in order to enhance the learning effectiveness for students, an immersive and interactive 3D environment with additional support of mechanical assembly is essential. In addition, as an education system, the cost must be affordable for both students and educators.

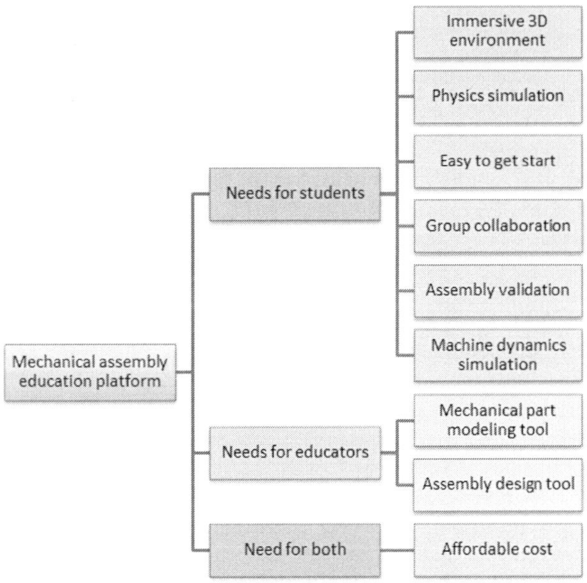

FIGURE 1: FUNCTIONALITIES NEEDED FOR MECHANICAL ASSEMBLY EDUCATION PLATFORM

While there are various types of commercially available computer-aided design software suitable for fulfilling the needs of educators, the needs for students are hard to be satisfied by simply amalgamating existing commercial products, because these needs are diversified. Luckily, desktop virtual reality (desktop VR) technology may partially provide a solution by serving as the foundation of the platform. Unlike expensive immersive VR equipment, the desktop VR can be run on personal computers with standard I/O devices such as desktop monitor, keyboard, mouse, etc. 3D game engines, especially first-person-shooter (FPS) game engines, are a type of desktop VR [5]. With the widespread availability of low-cost computer graphics hardware, screen-based 3D game engines provide realistic, flexible, interactive, and easy-to-learn tools for people to navigate through 3D environments [6]. A basic 3D game engine includes graphics, physics, audio and artificial-intelligence engines as well as a human-computer interaction module. In most cases, small-scale multiplayer capabilities are also included in a game engine [7]. Since most game engines provide software development kits (SDKs) for programmers to make modifications, there are reports of integrating assembly validation components and machine dynamics simulation components into game engines to meet the needs of students [8].

1.3. NATURAL HUMAN-COMPUTER INTERACTION

For many years, the primary input devices for computers have been mouse and keyboard. This combination is practical in daily computer usage such as document editing, webpage surfing, etc. However, for 3D environment navigation, these devices fall short due to their unrealistic mapping of computer operation and real-world activities. For instance, most first-person-shooter game engines set the keys “A”, “S”, “D”, and “W” to control translational movements in 3D environments and use the mouse to control rotational movements. Some other in-game commands (e.g. crouching down and walking slowly) are given through tapping or holding other keys. Experienced game players can perform these operations well. However, for beginners these operations are hard to complete. Since the level of gaming experience of students varies dramatically, the question arises whether it is possible to provide them with a versatile input method to promote their overall simulated assembly learning performance.

New natural human-computer interaction devices are making this possible. Driven by the entertainment industry, motion sensors such as the Nintendo Wii Remote, the PlayStation Move controller, the Microsoft Kinect Sensor, etc. are becoming more and more popular for providing natural human-computer interaction at low cost. Based on different physical principles, these devices each have respective advantages. The Wii and the PlayStaion Move controller are hand-held devices that can only track the motion of the player’s hand with high accuracy.

Although simple in structure, the Kinect has the advantage of being capable of dynamically capturing the movement of the whole human body, thus making it the most suitable choice for assembly education among these new devices. This is not only because immersive simulations in desktop VR need the kinematic information of the whole body (a human avatar without feet would look really strange), but also because assembly operations usually require the cooperation of one’s multiple body parts.

The Kinect comprises three sensors: a high definition video camera for image capture, an infrared sensor (IR) for depth sensing, and a microphone array for receiving voice commands. Among these sensors, the video camera and the infrared sensor can work together to provide 3D point clouds. These point clouds are then further processed by the Kinect motion capture software to extract the positions of all the parts of a human-body (arm, leg, head, torso, etc.). There are reports showing that the Kinect can be used effectively in simulating assembly activities [9] [10]. These reports addressed the natural assembly operation that can only be realized by the Kinect. Some of these reports also included assembly validation and machine dynamics simulation modules designed specifically for assembly simulations. However, these systems cannot be applied directly to an educational platform because of their lack of desktop VR, real-world physics simulation and group collaboration.

The work presented here aims to overcome the difficulties in constructing a mechanical assembly education platform by using a game engine and the Kinect. After a short review of the literature on the usage of the Kinect in education and the usage of the VR for assembly simulations, major challenges such as human-avatar synchronization, voice commanding, part picking and assembly simulation with the Kinect are discussed.

1.4. USING GARRY'S MOD AS ENGINE OF ASSEMBLY PLATFORM

GMod was selected for implementing the assembly platform described here. As a physics sand box of the "Source" game engine, GMod allows users to create novel assemblies, called "contraptions" in the game community, and real-world simulations. In addition, this sand box allows game developers to use the script language "lua" to program customized scenarios. There are reports of utilizing this sand box to build an assembly procedure simulation system, which was proven to be feasible for simulations and viable as an alternative for training and education [11]. The details of the implementation of the assembly system were described in reference [12].

2. PRIOR WORK

2.1. KINECT APPLICATIONS BEYOND ENTERTAINMENT

Using the Kinect in education has become a hot topic since its introduction in 2011, when it was only sold together with the Xbox for entertainment purposes. However, online communities soon cracked the Kinect code in order to utilize the functionalities that are not officially released by Microsoft. In 2012, Microsoft released the Kinect for Windows with an open-source Kinect software development kit (Kinect SDK), thus allowing researchers and teachers to develop their own applications with less technical difficulty.

The Kinect can potentially serve as a lecture tool. During lectures, it would be easier for instructors to use their gestures to control computers rather than to click a mouse or tap a keyboard again and again. For instance, an instructor can use a combination of gestures to control a slide show [13] [14]. For example, when an instructor wants to demonstrate the kinematics of a robot in a robotics class with the help of the Kinect, he/she can guide the robot by waving his/her arms, a method that is faster than operating the robot via a traditional interface and thus saves lecture time [15].

The Kinect can also be applied to strengthen student learning by facilitating kinesthetic interactions. For instance, research showed that Kinect-assisted reading helps youths remembering more words [16]. When being combined with a virtual environment, the Kinect can also promote the social competences and executive functions of students with Asperger syndrome [17]. For autistic children, playing Kinect-based games may benefit their skill learning [18].

Furthermore, the Kinect can also serve as an intelligent student supervision tool. Marquardt et al. used its body tracking function to monitor the gestures of ballet students during practice [19]. Schofield et al. tried to combine the Kinect with augmented reality (AR) technology to help young performing artists to improve their stage performance [20].

2.2. VIRTUAL ENVIRONMENT FOR ASSEMBLY PROCEDURE SIMULATION

As mentioned in Section 1, the assembly education platform presented here was constructed on desktop VR. Therefore, in this section, prior applications of VR technology for assembly procedure simulation, assembly planning, assembly design, and assembly design review are briefly summarized.

The early projects generally focused on constructing virtual environments for assembly procedure simulation purposes. In these projects, basic virtual environments that included an assembly facility, mechanical parts and assembly tools were authored. The assembly procedures were also regulated for computer simulation purposes. Some of them considered physics simulations, especially collision detection, while some others chose to neglect them. Jayaram et al. authored the Virtual Assembly Design Environment (VADE) that covered the functionalities of virtual assembly facility, assembly tools, hand picking of objects, collision detection, etc. The VADE used keyboard and mouse as input devices and employed a normal monitor as output device [21] [22].

New VR technology provides new input and output devices. Haptic devices are popular as hand movements can be accurately tracked and their operation is easy to learn. 3D glasses or head-mounted displays (HDM) enhance the immersive experience. Vance et al. used a handle-like haptic device as the input and 3D glasses to enhance the output 3D effect. Imbedding a physics module, their research featured the force feedback to the hand via the haptic device. Their models were stored by voxel representation, which facilitates the collision detection [23] [24]. Brough et al. authored an assembly training system that uses a hand-held console and an HDM. Besides the OpenGL-based graphics, their environment included a virtual agent as the mentor who can automatically instruct trainees [25].

In all the above projects, the virtual assembly operator was represented by a pair of hands. Such a simplification critically undermines the immersive experience of the operator, especially when the assembly simulation includes group activities. In order to improve this, humanized avatars were brought in. Qiu et al. modeled a fully animated human avatar, which can be controlled by data gloves [26]. Zhao et al. also brought human avatars into a virtual environment for assembly simulation. They attached five position sensors to an operator to sense the full-body motion. A cyber glove was used to precisely sense the motion of the operator's hand. Their human avatars were implemented through a commercially available program for human simulation and ergonomic analysis called Jack® [27].

3. SYNCRONIZATION OF HUMAN BODY MOTION AND AVATAR

3.1. HUMAN RAGDOLL IN GAME ENGINE

Garry's Mod was designed as a sandbox for the "Source" game engine, which was employed to developed many first person shooter games. It provides delicate human avatars, which can either serve as game player characters or non-player characters (NPC).

The models of these avatars can be animated and are called "ragdolls" by the developer community. A ragdoll model has two types of meshes. The first type is the geometry mesh for rendering, and the other type is the physics mesh for physics simulation (see Figure 2). The term "ragdoll" indicates that, like dolls, these models have virtual bones that cannot bend and joints that allow connected bones to rotate relative to each other within certain limits. An avatar's gestures are actually simulated by controlling the movement of the corresponding joints. For example, when simulating head nodding, the game engine simply makes the joint between the "neck bone" and the "spine bone" repeatedly flex and extend.

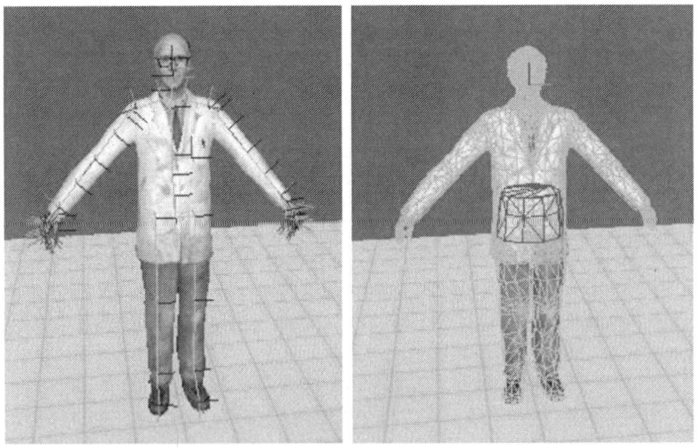

FIGURE 2: (LEFT) AVATAR OF GMOD WITH BONES LABELLED, COLORS OF BONE REPRESENT HIERARCHY OF BONES;
(RIGHT) GEOMETRY MODEL MESH IN BLUE AND PHYSICS MODEL MESH FOR LOWER TORSO IN RED

However, synchronizing a human body and an avatar by controlling the "bones" in GMod is unrealistic, because a human body ragdoll typically has around 70 bones. In addition, these bones are not directly accessible for game developers.

What can be accessed are the physics models of a ragdoll. A physics model mesh usually has the shape of a part of the geometry model mesh and contains several bones. They share the same physics properties, such as collision bounding box, mass, moment of inertia, friction, etc. Therefore, most physics animations are based on the physics model rather than the geometry model. By dividing a ragdoll, different parts of it can behave differently. For example, when an avatar is moving forward and its right hand hits a desk but its torso does not, the avatar can still move forward on the previous route with the hand changing its course, instead of the whole avatar stopping. Sometimes, physics models may be simpler than the corresponding geometry models because the computational cost of a complex physics model is significantly higher than that of a simple model.

3.2. SKELETAL TRACKING BY KINECT SDK AND SKLENTON IN GARRY'S MOD

One of the most exciting features of the Kinect is its body tracking. Based on the depth data, the machine learning algorithm that supports the Kinect is capable of extracting the positions of the human body's major joints at very high rates and with promising accuracy [28]. The Kinect SDK, released by Microsoft, enables third party software developers to employ the Kinect for tracking the human body's motion using the C# and C++ languages.

Currently, the Kinect SDK supports the tracking of 20 body joints: 4 for each arm, 4 for each leg, 2 for the torso, 1 for the head and 1 for the neck. The position of each joint in 3D skeleton coordinates (x, y, z), which is a Cartesian coordinate system with its origin at the Kinect, can be derived in real time.

In GMod, the physics models can also form such a skeleton. There are 15 physics models for a human avatar in the GMod, most of which are at the same position as their corresponding Kinect SDK joints. The major differences between these two skeletons are that the GMod skeleton does not have ankle joints, wrist joints and the neck joint. Furthermore, the positions for the chest joint and the pivot joint are defined at different places. Figure 3 shows these two skeletons. For the Kinect skeleton, the joints that are not presented in the GMod skeleton are not shown.

3.3. ANIMATING HUMAN AVATAR FROM KINECT'S SKELETAL FRAME

Knowing the joint positions provided by the Kinect, animating an avatar so as to let it display the motion of a human seems straightforward. After the Kinect acquires the 3D joint positions of the human, they are transmitted to GMod through their shared memory [29]. The corresponding joints can then be set to these positions in GMod to synchronize the human and the avatar. However, this method does not work properly. One problem is that the animated avatar does not move smoothly, and another problem is the error caused by the differences between the Kinect's skeleton and GMod's avatar.

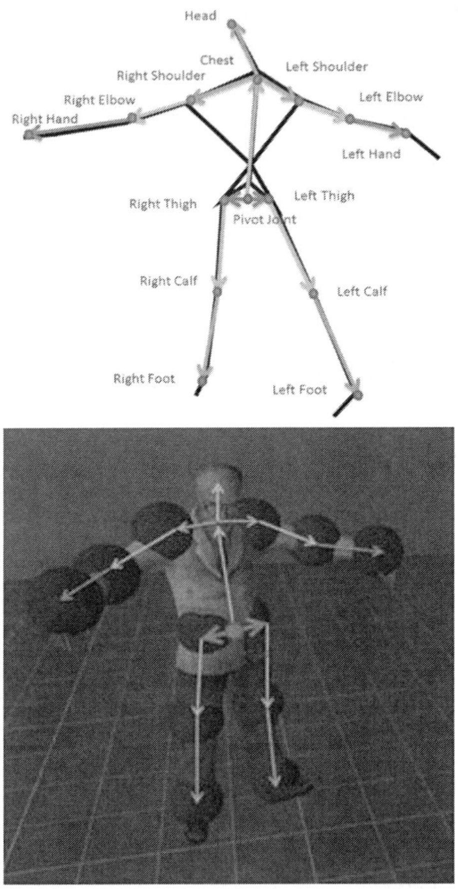

FIGURE 3: (TOP) SKELETON AND BONES IN GAME-ENGINE, RED NODES ARE ORIGIN POINTS OF RESPECTIVE PHYSICS MODEL;
(BOTTOM) CORRESPONDING SKELETON DERIVED FROM KINECT SKELETON FRAME.

When the position of each physics model is assigned directly based on Kinect-acquired position information when animating the human avatar, the animation appears to be discontinuous. This is caused by the difference in the frame rate of the Kinect and that of the game engine. The internal rendering frame rate of the game engine is more than 200 fps; while that of the Kinect is only around 20 fps. What makes this situation worse is the fact that the communication between the Kinect SDK and the game engine also takes time. Thus, the skeleton frame rate that GMod can receive is further reduced. Therefore, to animate a human avatar, a better method is to specify the velocity of the selected physics model, as the game engine can automatically process the velocity and render the models at a higher rendering frame rate. The velocity for a physics model is:

$$v = \frac{\boldsymbol{P_t} - \boldsymbol{P_c}}{\Delta t} \tag{1}$$

In this equation, $\boldsymbol{P_t}$ is the target position that the Kinect SDK acquired from the actual person, $\boldsymbol{P_c}$ is the current position of the physics model and Δt is the time between two Kinect frames. Typically, Δt is 0.08 *s*.

Another challenge lies in the differences between the Kinect skeleton and the avatar. In the Kinect skeletal frame, the joint positions vary depending on the place where the human stands and so does the length of each bone. Also, the lengths of the bones extracted by the Kinect are disproportionate, i.e. the closer to the Kinect the bone is located, the longer it becomes.

The solution to this problem is converting the Kinect joint positions to bone orientations and normalizing the length to unity as follows:

$$\boldsymbol{N} = \frac{\boldsymbol{P_{j1}} - \boldsymbol{P_{j2}}}{\|\boldsymbol{P_{j1}} - \boldsymbol{P_{j2}}\|} \tag{2}$$

In this equation, the $\boldsymbol{P_j}$s are the positions of the two joints of a bone given by the Kinect SDK and the 3D vector $\boldsymbol{N}$ is the orientation of the bone. The position of a joint in the game engine is:

$$\widetilde{\boldsymbol{P}}_{j2} = \widetilde{\boldsymbol{P_{j1}}} + \boldsymbol{L_b N} \tag{3}$$

In this equation, $\tilde{P}_{j2}$ is the position of the joint to be derived, P_{j1} is the position of the known joint in the game engine and L_b is the length of the bone connected by these two joints. The lengths of these bones are pre-set by the designer who modeled the avatar.

The fundamental joint of the whole body is the pivot joint. The positions of all joints are represented relative to this joint as the origin. However, although the Kinect SDK provides the position of this joint, its position is derived by finding the center point of the left and right thighs. This is because in the game engine, the position of the pivot point is closer to the center of the thigh joints.

3.4. TRANSLATION AND ROTATION OF AVATAR

In the assembly platform presented here, the avatar is capable of navigating around the virtual environment rather than only standing at a certain point. Therefore, the skeleton as a whole must be able to move in the plane and make turns. The best way to achieve this is to differentiate the world coordinate system and the avatar coordinate system.

The world coordinate system is fixed to the virtual environment. It is defined by the virtual environment map author and cannot be changed. The avatar coordinate system is attached to the avatar, with the *x*-axis pointing to the right, the *y*-axis pointing up and the *z*-axis pointing front. The origin point is on the pivot point. Since the turning always occurs about the vertical *y*-axis, a joint's position in the world coordinate system is obtained from its avatar coordinates as follows:

$$\begin{bmatrix} x_1 \\ y_1 \\ z_1 \end{bmatrix} = \begin{bmatrix} \boldsymbol{cos\theta} & \boldsymbol{0} & \boldsymbol{sin\theta} \\ \boldsymbol{0} & \boldsymbol{1} & \boldsymbol{0} \\ \boldsymbol{-sin\theta} & \boldsymbol{0} & \boldsymbol{cos\theta} \end{bmatrix} \begin{bmatrix} X_1 \\ Y_1 \\ Z_1 \end{bmatrix} + \begin{bmatrix} x_p \\ y_p \\ z_p \end{bmatrix} \tag{4}$$

In this equation, a joint's position in the avatar coordinate is described by the (upper case) coordinates (X_1, Y_1, Z_1) where (x_p, y_p, z_p) are the pivot point's (small case) world coordinates and θ is the direction that the avatar looks at in world coordinate system.

3.5. DATA FLOW OF HUMAN-AVATAR MOTION SYNCRONIZATION

The last question for the synchronization is how to enable multiple students to work collaboratively in the assembly environment. Since the platform that this paper presents is for multiple user online education, from network framework perspective, there are one server and several client computers. For a virtual assembly activity in which multiple students and instructors are participating, the position data are collected at the client computers, processed by both client and server computer, and finally sent back to the client side for rendering. The data flow diagram for the process described in this section is shown in Figure 4.

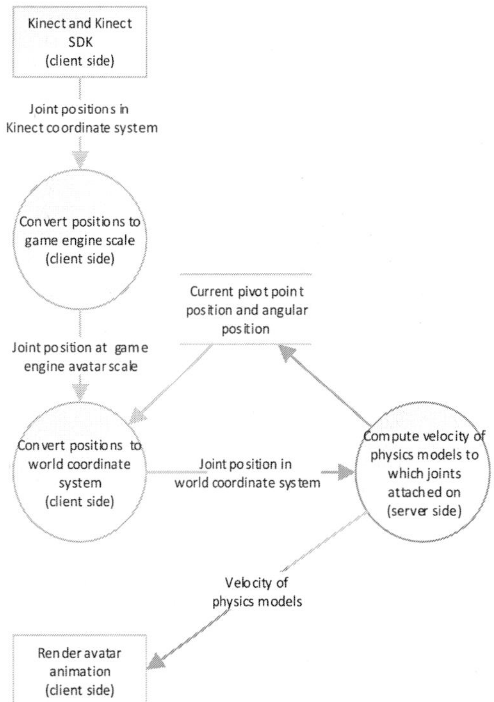

FIGURE 4: DATA FLOW DIAGRAM FOR MOTION SYNCHRONIZATION. FUNCTIONS LISTED ON LEFT (YELLOW) COLUMNS ARE EXECUTED ON CLIENT SIDE; FUNCTION LISTED ON RIGHT (BLUE) COLUMN IS EXECUTED ON SERVER SIDE

4. VOICE COMMANDING

4.1. COMMANDS AND SPEECHES

As mentioned in the previous section, in the assembly platform, the students should be able to navigate through the 3D virtual environment and complete tasks such as picking a part, answering a question, etc. Therefore, they need some way to command the avatar to move. The game engine supports the keyboard and mouse as the main input devices. Using these devices in combination with the Kinect is impractical since the students are not able to carry a keyboard or a mouse when they make the assembly by using their body motion.

Input using simple gestures or hand signals is a way to replace these traditional input devices [30] [31]. For instance, by traffic law, lifting the left arm toward left signals turning left. This inputting method can be utilized in the virtual environment, too. While for an application with only a few commands, this approach is acceptable, for an assembly platform that involves more than 10 commands, it is unlikely that the students can manage these gestures easily. Thus, the question is why we must give commands silently by to learn assembly?

Using voice commands can greatly reduce the students' difficulties with getting started with the assembly platform. The Kinect contains an array of 4 microphones for collecting voice data. These data can then be processed and recognized by the Kinect SDK and by the Microsoft Speech API. The speech recognition can be accomplished at the same time as the body tracking.

A command is an order that students or instructors can give. In the mechanical assembly platform presented here, there are currently 49 voice commands in 4 categories: 9 for navigation, 2 for assembly, 2 for GUI operation and 36 for pop-up menu operation. The number of commands will be expanded when more functionality is integrated into the platform. Examples of these commands are listed in Table 1.

TABLE 1: EXAMPLES OF VOICE COMMANDS

Category	Command	Speech	Command effect
Navigation	Forward	Forward Forwards	Avatar moves forward
Navigation	Accelerate	Accelerate Faster	Avatar moves faster
Assembly	Drop	Drop Dropping	Drop a part from the hand
GUI	Spawn	Spawn Create	Pop up list of parts
Choice selection	"A"	Alpha Apple	Select "A" option

As can be seen in Table 1, a voice command consists of a command effect and several speeches. A speech is a word or phrase that the computer program can recognize from pronunciation. That is to say, the students or instructors speak these words or phrases to give a command. In order to make the commanding convenient, a command usually includes several speeches whose semantics are close to the meaning of the command in natural language. For instance, the words "accelerate" and "faster" both mean move faster. Thus, these two words are chosen for the speech of "accelerate".

4.2. VOICE COMMAND PROCESSING

Like motion tracking, the voice command processing involves both the server and client computers. A voice signal is detected and recognized on the client side. The client then reports the command to the server. The server responds to the command accordingly and sends its response back to the client for rendering. Since there are 4 categories of voice commands, the flowchart of processing these commands is more complicated than that for processing the body tracking information. A simplified flowchart is shown in Figure 5. When a voice command is received, the assembly platform first checks whether there is currently a pop-up window on the student's screen. If so, the server only accepts pop-up menu operation commands.

When a navigation command is received, the avatar moves on the floor or makes turns accordingly. When the avatar collides with an object that it cannot pass through, it stops. If the student still gives this navigation command, the avatar rejects it. For example, when the avatar moves forward and hits an obstacle in the virtual environment, an additional "forward" command is not executed until the student orders "backward" to move the avatar away from the wall (see Figure 6)

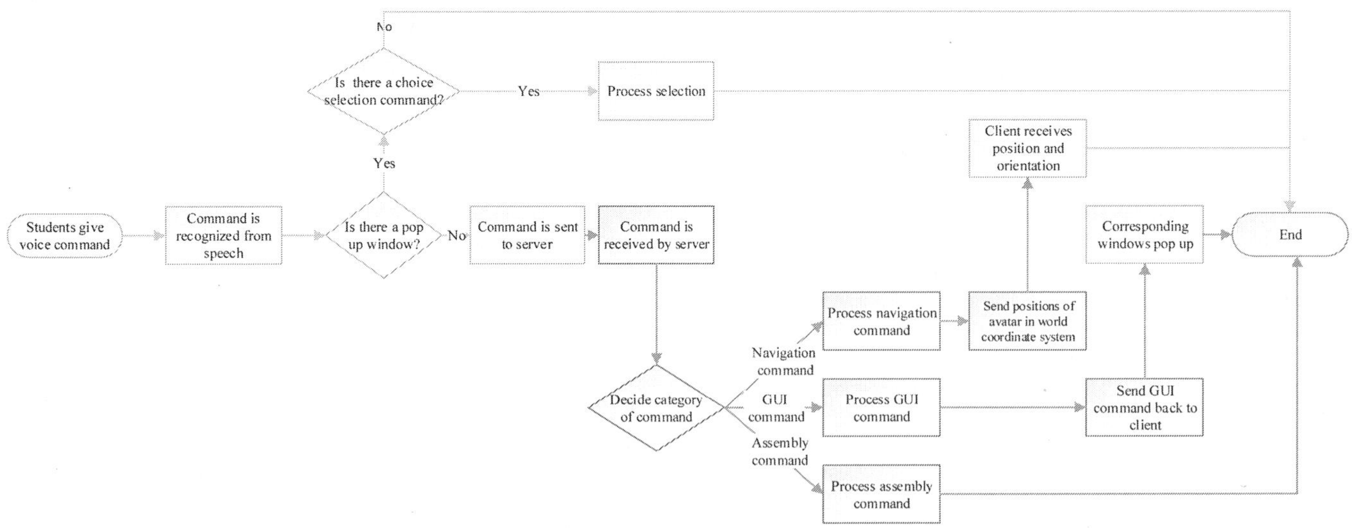

FIGURE 5: FLOW CHART FOR VOICE COMMAND PROCESSING.

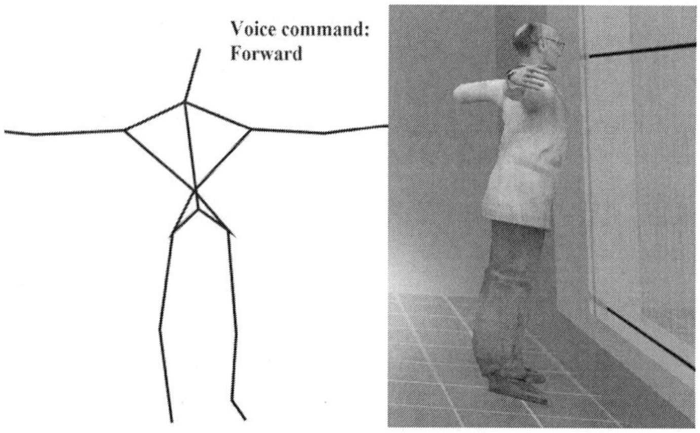

FIGURE 6 (LEFT) KINECT SKELETON IS CAPTURED WITH "FORWARD" VOICE COMMAND RECOGNIZED;
(RIGHT) IN GAME, AVATAR HITS WALL AND FURTHER "FORWARD" VOICE COMMAND IS NOT EXECUTED

5. PART PICKING AND ASSEMBLY

In the current platform, a part can be picked and manipulated by one hand or by both hands. In this way, the picking appears to be realistic. Generally, for larger parts, two-hand picking is used while for smaller parts, one hand suffices.

The picking positions are virtual handles for avatars to hold during the assembly. Of course, these "handles" do not necessary have the shape of a handle. These picking positions are pre-defined by the assembly educator in the part's local coordinates. Figure 7 shows a gear with two "handles" on its shaft for two-hand picking. A local coordinate system is attached to the right end of the shaft.

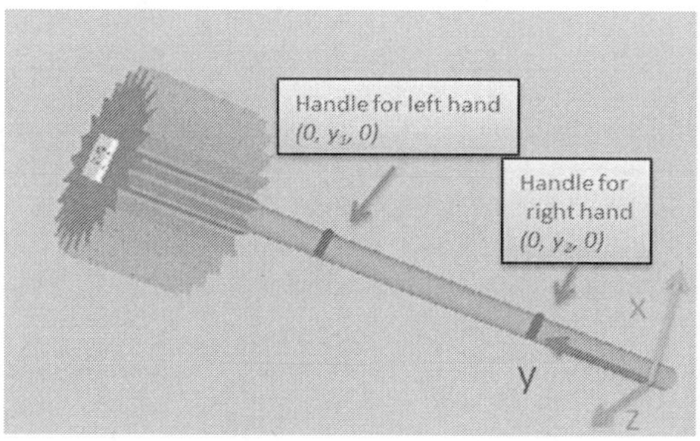

FIGURE 7: PART DEFINED WITH TWO "HANDLES" FOR TWO-HAND PICKING

The picking procedure was simplified in the platform described here. When the physical model of an avatar's hand hits a part, a picking-validation process is triggered. This validation process is show in Figure 9.

If the validation is passed, a kinematic constraint is imposed between the physics model and the part at the place of the "handle". For one-hand picking, the constraint imposed is a "welding" constraint, which removes all six relative degrees of freedom between the hand and the part. For two-hand picking, the constraint imposed is of the type "ball socket", which removes only the three relative translational degrees of freedom. In this way, the avatar can rotate the part with flexibility by grasping the part with both hands. Figure 8 shows an avatar using both hands to pick up a gear.

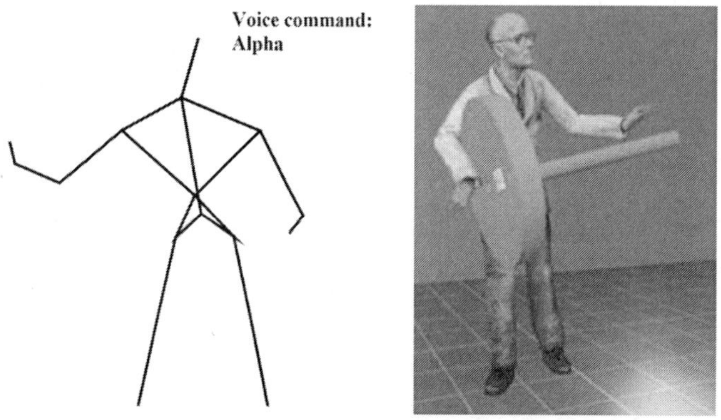

FIGURE 8 (LEFT) SKELETON CAPTURED BY KINECT;
(RIGHT) AVATAR PICK UP A PART IN GMOD BY BOTH HANDS

6. ASSEMBLY SIMULATION

The final step of the procedure simulation is creating an assembly. The assembly is designed by the educator in advance. In the assembly education platform, an algorithm for versatile assembly representation was developed. The representation uses a topological model to describe assemblies, allowing the virtual environment to simulate the different assembly procedures in a unified fashion, thus reducing the educators' time required for designing a pedagogically effective assembly. This representation as well as its implementation is described in detail in reference [32].

In order to simplify the assembly education platform, an assembly validation process is triggered whenever a part held by a student approaches or touches another part. Based on this validation process, the system decides whether these two parts are available to be mated. If these two parts have mating features that allow them to be connected, an auto-guidance process is triggered so that pre-defined constraints are imposed. At the same time, the avatar releases its hand(s), thus ceasing to grasp the part.

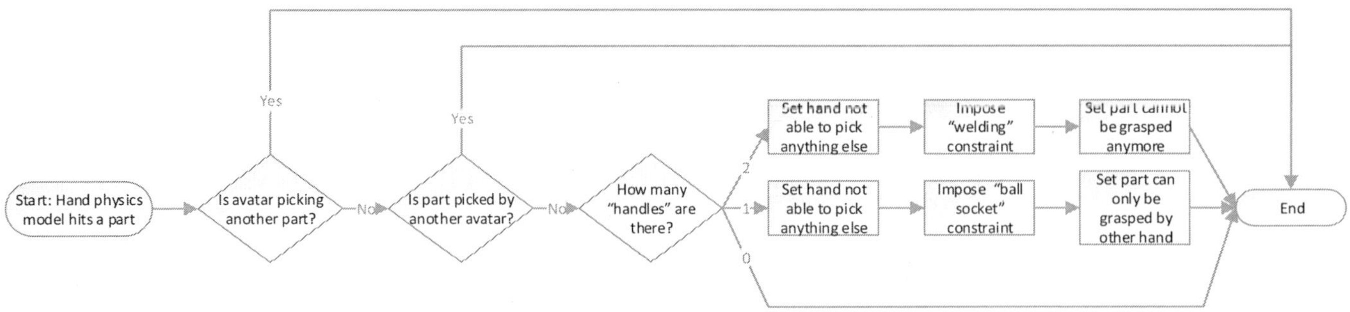

FIGURE 9: FLOWCHART FOR VALIDATION PROCESS OF PICKING.

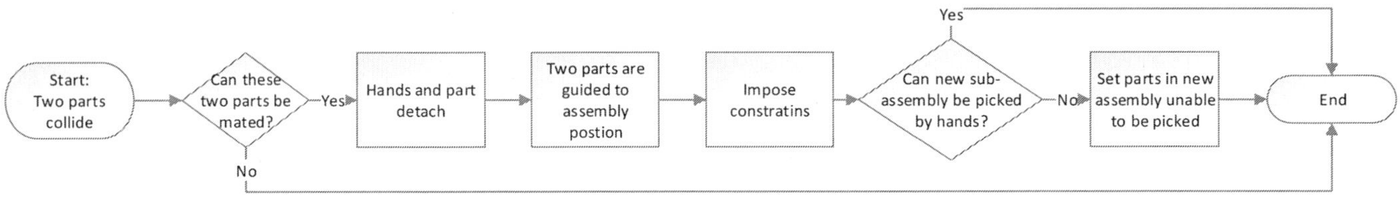

FIGURE 10: FLOWCHART FOR ASSEMBLY PROCEDURE

After the assembly is completed, the assembly platform determines whether the new sub-assembly is eligible to be picked by this avatar. For instance, when a part of the sub-assembly is fixed to the ground, the new sub-assembly cannot be picked again. In some other cases, the sub-assembly can be picked again and manipulated. The flowchart of the assembly procedure is shown in Figure 10.

7. CONCLUSIONS

In this paper, a platform for mechanical assembly education was described. This platform consists of two main features: (i) a game engine as the foundation of desktop VR and (ii) the Kinect as the only input device. A prototype of this platform has been implemented through integration of C# based Kinect SDK and physics sand box Garry's Mod (GMod).

Under this platform, students who learn about mechanical assemblies can use their bodies' motion to control fully animated avatars and complete assembly tasks. They can also use voice command to interact with computers. Compared to simulations controlled by keyboard and mouse, simulations with the Kinect interface are more realistic. In addition, compared to assembly simulations based on immersive VR technology, the platform described here is more affordable.

The technical details of the platform were described, the major challenges and solutions were presented and flowcharts for these solutions were provided. These challenges include human-avatar synchronization, voice commanding, part picking and part assembly. Through the integration of these components, the platform exhibits robustness in creating various assemblies and ease in designing pedagogically effective assembly processes.

Previous studies on assembly platforms based on game engines with keyboard and mouse showed that mechanical engineering undergraduate students overall had positive attitudes toward this form of education. In the next step, studies about the usability and learning effectiveness will be conducted to identify the advantages and the disadvantages of the platform presented here.

REFERENCES

[1] E. Allen and J. Seaman, "Staying the Course: Online Education in the United States, 2008," The Sloan Consortium, 2008.

[2] S. Aud, W. Hussar, G. Kena, K. Bianco and L. Frohlich, "The Condition of Education 2011," National Center for Education Statistics, 2011.

[3] E. Allen and J. Seaman, "Changing Course: Ten Years of Tracking," The Sloan Consortium, 2013.

[4] K. M. Heussner, "Coursera hits $1M in revenue through verified certificates," Gigaom.com, 2013.

[5] J. Jacobson and M. Lewis, "Game engine virtual reality with CaveUT," *Computer,* vol. 38, no. 4, pp. 79-82, 2005.

[6] L. Ausburn and F. Ausburn, "Desktop virtual reality: A powerful new technology for teaching and research in industrial teacher education," *Journal of Industrial Teacher Education,* vol. 41, no. 4, 2004.

[7] J. Gregory, "Engine differences across genres," in *Game Engine Architecture*, Boca Raton, Florida, CRC Press, 2009, pp. 13-15.

[8] Y. Chang, E.-S. Aziz, S. Esche and C. Chassapis, "A framework for developing collaborative training environments for assembling," *Computers in Education Journal,* vol. 23, no. 4, pp. 44-59, 2013.

[9] R. Radkowski and C. Stritzke, "Interactive hand gesture-based assembly for augmented reality applications," in *ACHI 2012, The Fifth International Conference on Advances in Computer-Human Interactions*, 2012.

[10] H. Haggag, M. Hossny, S. Nahavandi and D. Creighton, "Real time ergonomic assessment for assembly operations using Kinect," in *2013 UKSim 15th International Conference on Computer Modelling and Simulation (UKSim)*, 2013.

[11] Y. Chang, E.-S. Aziz, S. Esche and C. Chassapis, "A game-based laboratory for gear design," *Computers in Education Journal,* vol. 22, no. 1, pp. 21-31, 2012.

[12] Y. Chang, E.-S. Aziz, S. K. Esche and C. Chassapis, "A multi-user virtual laboratory environment for gear train design.," *Computer Applications in Engineering Education Journal,* p. In press, 2013.

[13] L. Andrew , V. Ng and C.-H. Chan, "Gesture-Based interaction for seamless coordination of presentation aides in lecture streaming," in *8th International Conference on Information and Communication Technology in Teaching and Learning*, Hong Kong, 2013.

[14] Y.-p. Guan and P. Jing, "Human-computer interaction using pointing gesture based on an adaptive virtual touch screen," *International Journal of Signal Processing, Image Processing & Pattern Recognition,* vol. 6, no. 4, pp. 81-91, 2013.

[15] L. Cheng, Q. Sun, H. Su, Y. Cong and S. Zhao, "Design and implementation of human-robot interactive demonstration system based on Kinect," in *2012 24th Chinese Control and Decision Conference (CCDC),*, 2012.

[16] B. D. Homer, C. K. Kinzer, J. L. Plass, S. M. Letourneau, D. Hoffman, M. Bromley, E. O. Hayward, S. Turkay and Y. Kornak, "Moved to learn: the effects of interactivity in a Kinect-based literacy game for beginning readers," *Computers & Education,* vol. 74, pp. 0360-1315, 2014.

[17] G. Lorenzo, J. Pomares and A. Lledó, "Inclusion of immersive virtual learning environments and visual control systems to support the learning of students with Asperger syn," *Computers & Education,* vol. 62, pp. 88-101, 2013.

[18] L. Bartoli, C. Corradi, F. Garzotto and M. Valoriani, "Exploring motion-based touchless games for autistic children's learning," in *Proceedings of the 12th International Conference on Interaction Design and Children*, 2013.

[19] Z. Marquardt, J. Beira, N. Em, I. Paiva and S. Kox, "Super Mirror: a Kinect interface for ballet dancers," in *Proceedings of the 2012 ACM Annual Conference on Human Factors in Computing Systems*, Austin, TX, 2012.

[20] T. Schofield, J. Vines, T. Higham, E. Carter, M. Atken and A. Golding, "Trigger shift: participatory design of an augmented theatrical performance with young people," in *Proceedings of the 9th ACM Conference on Creativity & Cognition*, 2013.

[21] S. Jayaram, H. I. Connacher and K. W. Lyons, "Virtual assembly using virtual reality techniques," *Computer-aided Design,* vol. 29, no. 8, pp. 575-584, 1997.

[22] S. Jayaram, U. Jayaram, Y. Wang, H. Tirumali, K. Lyons and P. Hart, "VADE: a virtual assembly design environment," *IEEE Computer Graphics and Applications,* vol. 19, no. 6, pp. 44-50, 1999.

[23] S. Abhishek, H.-J. Su and J. M. Vance, "SHARP: a system for haptic assembly and realistic prototyping," in *ASME 2006 International Design Engineering Technical Conferences and Computers and Information in Engineering Conference*, 2006.

[24] B. M. Howard and J. M. Vance, "Desktop haptic virtual assembly using physically based modelling," *Virtual Reality,* vol. 11, pp. 207-215, 2007.

[25] J. E. Brough, M. Schwartz, S. K. Gupta, D. K. Anand, R. Kavetsky and R. Pettersen, "Towards the development of a virtual environment-based training system for mechanical assembly operations," *Virtual Reality,* vol. 11, pp. 189-206, 2007.

[26] S. Qiu, X. Fan, D. Wu, Q. He and D. Zhou, "Virtual human modeling for interactive assembly and disassembly operation in virtual reality environment," *The International Journal of Advanced Manufacturing Technology,* vol. 69, no. 9-12, pp. 2355-2372, 2013.

[27] W. Zhao and V. Madhavan, "Virtual assembly operations with grasp and verbal interaction," in *Proceedings of the 2006 ACM international conference on virtual reality continuum and its applications*, Hong Kong, 2006.

[28] J. Shotton, T. Sharp, A. Kipman, A. Fitzgibbon, M. Finocchio, A. Blake, M. Cook and R. Moore, "Real-time human pose recognition in parts from single depth images," *Communications of the ACM,* vol. 56, no. 1, pp. 116-124, 2013.

[29] Z. Zhang, M. Zhang, S. Tumkor, Y. Chang, S. K. Esche and C. Chassapis, "Integration of physical devices into game-based virtually reality," *International Journal of Online Engineering,* vol. 9, pp. 25-38, 2013.

[30] R. Francese, I. Passero and G. Tortora, "Wiimote and Kinect: gestural user interfaces add a natural third dimension to HCI," in *Proceedings of the International Working Conference on Advanced Visual Interfaces*, Naples, Italy, 2012.

[31] M. N. K. Boulos, B. J. Blanchard, C. Walker, J. Montero, A. Tripathy and R. Gutierrez-Osuna, "Web GIS in practice X: a Microsoft Kinect natural user interface for Google Earth navigation," *International Journal of Health Geographics,* vol. 10, no. 1, pp. 1-14, 2011.

[32] E.-S. Aziz, Y. Chang, S. Esche and C. Chassapis, "Capturing assembly constraints of experimental setups in a virtual laboratory environment," in *Proceedings of the ASME International Mechanical Engineering Conference & Exposition*, Houston, 2012.

This page left blank intentionally.

Proceedings of the ASME 2014 International Mechanical Engineering Congress and Exposition
IMECE2014
November 14-20, 2014, Montreal, Quebec, Canada

IMECE2014-38771

PAIR: THE REMOTE INDUSTRIAL AUTOMATION TRAINER

Mário Silva
R&D Centro Algoritmi, Department of Industrial Electronics, School of Engineering
University of Minho
Guimarães, Portugal

Filipe Pereira
R&D Centro Algoritmi, Department of Industrial Electronics, School of Engineering
University of Minho
Guimarães, Portugal

Filomena Soares
R&D Centro Algoritmi, Department of Production and Systems, School of Engineering
University of Minho
Guimarães, Portugal

Celina P. Leão
R&D Centro Algoritmi, Department of Production and Systems, School of Engineering
University of Minho
Guimarães, Portugal

José Machado
R&D CT2M, Department of Mechanical Engineering, School of Engineering
University of Minho
Guimarães, Portugal

Vitor Carvalho
R&D Centro Algoritmi, Department of Industrial Electronics, Engineering School, University of Minho,
Guimarães, Portugal
IPCA-EST, Barcelos, Portugal

ABSTRACT

This paper presents the project PAIR – Remote Industrial Automation Trainer. Its goal is to provide an environment where learners can improve their skills in Industrial Automation and practice their theoretical knowledge. In order to achieve this, PAIR will allow the users to explore automation concepts in some real scenes, as the control of an *intelligent* house or an industrial process. PAIR will also allow simulated virtual experiences associated to the real world scenarios.

Keywords: Automation Education, Industrial Automation, Virtual and Remote Labs.

INTRODUCTION

The current technological development allowed the reduction of unskilled labor posts, evolving for absorbing highly skilled and specialized labor positions.

The key issue in industry is Automation: it can be a *passport* to employment in any company and/or any part of the world. It is up to the citizens, education and training institutions to understand this necessity, as a mean of seeking solutions to the problem of employment under vocational and academic training. Given that vocational training should be geared to the job market, it is important to implement a student-centered education, preparing them to the challenges that they will face. To this goal, it is essential to explore the use of new technologies in course syllabus. This criterion is focused on the individual learning leading the learner to seek answers by their own initiative, rather than waiting to absorb information provided by the teachers.

In an era marked by rationalization of resources, the availability of suitable hardware for training in Industrial Automation may represent a critical point of this whole procedure. It is in this context that the PAIR, Portuguese acronym for Platform for Industrial Automation Trainer, assumes its largest objective.

Laboratory and practical components, especially in professional and higher education in technological areas such as electronics and automation, is becoming increasingly important, and a requirement for the training of good professionals [1].

The use of remote, physical or virtual platforms as a complement to existing theoretical learning is becoming a reality even on the subject of industrial automation.

A virtual laboratory may be described as software with the ability to simulate real scenarios based on mathematical and physical analysis. For example, several software allow users to implement electrical circuits just by dragging figures which represent real components and read data as current or resistance, without wiring any physical component [2]. On the other hand, a remote laboratory is structured and built over physical parts and components, and hardware and software for the controller device. Usually, the software provides connection between the Users and the hardware, allowing Users to send their commands and receive the corresponding data [2].

These web-based learning environments help students and teachers engagement in the knowledge acquisition process. These web environments may be a very important complement to the classroom lectures and exercises [1].

In courses where physical part of the system implies high cost, it is difficult to install and maintain several working positions in a laboratory and it may be even harder to create a home laboratory.

To overcome this problem the virtual and remote laboratories concepts are used. If real labs are accessible online, learners may use them anytime and anywhere, which is a suitable tool to help autonomous learning [1].

Logically, under educational perspective, replacing or complementing traditional classes with these online tools may represent new challenges. The success of these labs is strongly related to a good understanding, not only from the technical issues, but also from the pedagogical aspects, so the virtual/remote lab's utilization may be appealing, efficient and challenging [1].

The paper is organized in four main sections as follows: section 1 states the work's motivation. Section 2 refers to some virtual and remote laboratories. Section 3 presents PAIR platform. Finally, conclusions and future perspectives of the work are detailed.

VIRTUAL AND REMOTE LABS

There are several examples of remote platforms to aid learning in various domains. The Massachusetts Institute of Technology (MIT) offers the iLabs, a project started in the late twentieth century. This platform allows students (remotely and at any time) to develop experiences in the areas such as electronics, control, physics or telecommunications. Currently, this project does not end with the offering developed at MIT, adding projects developed at other universities such as the University of Queensland in Australia, or the Makerere University in Uganda [3]. In India, for example, the Virtual Labs network adds a number of universities providing remote, virtual and physical laboratories as learning support in the areas related to science (physics, chemistry and computing) and engineering (electronics, mechanical or civil, for example) [4]. In Spain, the University Network of Interactive Labs platform aims to take full advantage of the features of the different entities participating in the project and make them available to students, even if they are in different geographical locations. The student simply connects to the system in the period allocated to him/her and applies his/her knowledge on Programmable Logic Controllers (PLC) programming [5]. An integral part of this project is the Remote Laboratory of Automation at Universidad de León. With this tool, users can develop the following experiments, among others: control the level of water in four tanks, process control with four variables, robotic cells control, PLCs control, control of variable speed motor and energy management [6]. Also in Spain, the Universidad de Deusto has the WebLab - Deusto. This provides testing in areas like physics, digital electronics and robotics [7]. From this platform it is possible to access OpenLabs Electronics Laboratory, based at Blekinge Institute of Technology in Sweden, a remote platform dedicated to the support of learning electronics [8].

In Portugal we can point out the example of Remote Industrial Automation Laboratory at the University of Lisbon. Users connect to this remote laboratory from a Virtual Private Network (VPN) and can program the robots available and view the results through an IP (Internet Protocol) web camera [9, 10]. The University of Porto, Portugal, also has a remote online laboratory where it is possible to undertake various tests such as mechanical characterization of materials or monitoring of a weather station [11]. With this project, it is intended to provide vocational training to academic institutions and enterprises as a complete tool for training in Industrial Automation.

PAIR – THE INDUSTRIAL AUTOMATION TRAINER ACCESSIBLE TO ALL

The PAIR platform aims to be an open-access tool available to assist learning in industrial automation, available for students from technical schools and universities, and also for industrial technicians. After performing an initial training, learners need to apply and test their knowledge in specific cases.

To this end, PAIR provides a series of exercises, ranging from whom is still prone to start learning this field to higher levels for more experienced users. To this end, trainees can access PAIR web platform and examine the different challenges proposed. Selecting one, the user can study it, prepare a solution and submit it to PAIR. Then, he/she may observe and analyze the result of the proposed solution in real time.

PAIR – What is offered?

In order to make this a more complete and comprehensive platform, PAIR provides virtual exercises and real world-problems. In both situations, problems with different levels of complexity can be worked out, even though, often, the baseline may be the same.

Thus, PAIR will provide physical experiments as the control of a smart home (e.g., temperature control, lighting control or automatic doors), control of an industrial process for coating an object (subject to certain conditions, the object can move forward and be worked or is retained until it meet these optimal conditions), control barriers, control of a lift or control a light. In the case of industrial process control, in addition to the physical aspect (which is being implemented), it is also provided a virtual component that will be detailed later.

As it can be understood from Figure 1, users can remotely connect to the platform PAIR, control it and see the results, once PAIR allows users to observe the action performed in the different case studies from a webcam. This platform is also responsible for managing access to their experiences, avoiding conflicts between different users and ensuring that each can have full access to this tool. In order to achieve that, each user must login to PAIR's webpage. Each login is only available on a scheduled period of time. So, two different users/institutions cannot use this platform at the same time. However, if, for example, the login is made by a school class, all experiments will be available to the class, although each case study can only be performed one at a time (this means that student A can perform the industrial process and student B can perform the intelligent house at the same time, but they cannot perform simultaneously the industrial process or the intelligent house).

Physically, this platform offers several OMRON® PLCs, being concentrated in a cabinet or locker, designed for this purpose, as can be seen in Figure 2. Sensors and actuators are naturally distributed across different scenarios. The platform allows communication over several protocols such as RS232, RS485, DeviceNet, Profibus, Ethernet and AS-Interface [12-22].

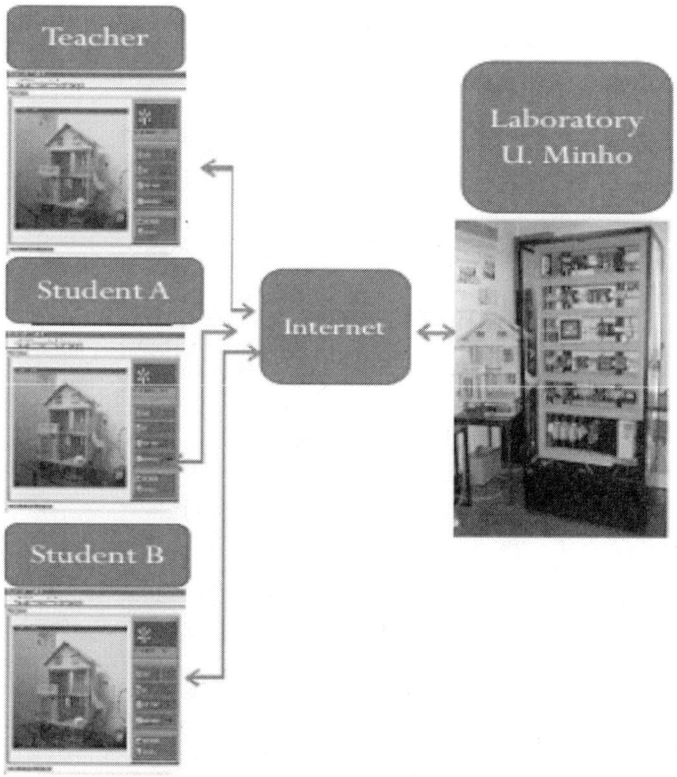

FIGURE 1. PAIR CONCEPT

PAIR – Virtual environment

As already mentioned, one of the scenarios offered by PAIR refers to an industrial process control. In this case study, it is intended that a particular object proceed along a working line, under certain conditions, with the goal of bathing with a specific chemical reagent.

FIGURE 2. ANL – AUTOMATION NETWORK LOCKER

As it can be observed in Figure 3, the system can be divided into four main sections: the entrance (where the identification of the material is made), the pre-heating chamber (allows a gradual heating up to temperatures near the component of the ideal temperatures process), the main chamber (where the component is plated with a chemical reagent) and a standby chamber (where the component may cool gradually to room temperature, avoiding thermal shocks).

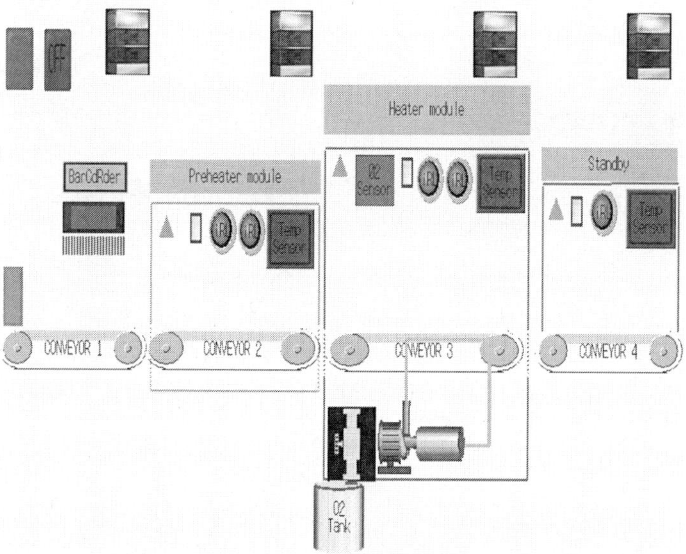

FIGURE 3. INDUSTRIAL CONTROLLING PROCESS: PAIR'S VIRTUAL CASE STUDY

In order to allow that the chemical reagent acts correctly on the component it should be checked that optimum conditions of temperature and oxygen concentration (maximum, minimum parameters and acceptable tolerances for each case can be defined) are present in the main chamber.

When the system (Figure 3) is switched on by pressing the ON button, the chambers of pre-heating, main heating and standby cameras come into play, triggering infrared lamps and occupancy sensors in each of the components and the oxygen sensor and actuator in the main chamber. Guaranteed these ideal conditions, the Conveyor1 and barcode reader (both the system input) comes into operation. When the barcode reader performs a reading code, this information is sent to the main chamber, entering the operating system that circulates the reagent. Depending on the code read, different reagents may be used. Next, the component enters the preheating chamber, where it should remain until an optimum temperature is obtained and only then proceed to the main chamber. In the main chamber, the component is plated with the chemical reagent and then advances to the camera standby, which will make a gradual cooling (Figure 4). Figure 5 illustrates the flow chart concept of the process described.

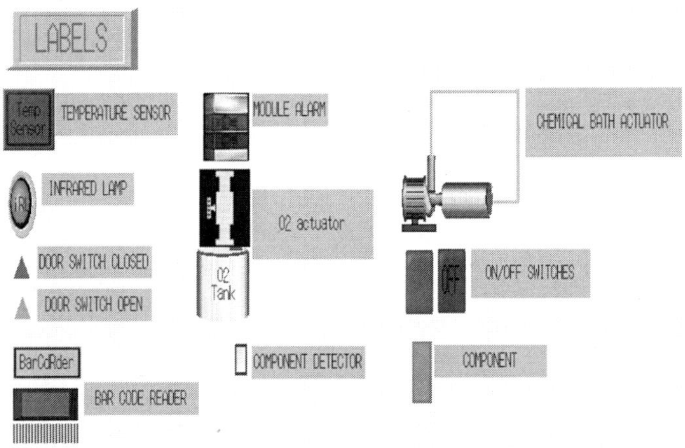

FIGURE 4. LABELS FROM INDUSTRIAL CONTROLLING PROCESS: PAIR'S VIRTUAL CASE STUDY

Logically, starting from this basic idea, it is possible to make the case study more or less complex, according to the target audience. Challenges such as limited pieces per sector, the speed with which the conveyors must rotate, the acceptable parameters for temperature and oxygen or the action to perform in case of incorrect reading of the bar code reader can be introduced. The conditions to access the system, with the opening of the doors or lock access to the system can also be programmed. Information about errors elapsed through the activation of bulbs that can indicate if the process conditions are within or outside of the parameters can also be managed.

The virtual system will be available online, hosted on a server. From a web interface, users can access the proposed problems, observe the features of the case study and perform the assessment. On the webpage it is indicated which sensors and actuators can be used and which records of the PLC are associated to them. When programming a solution to the case-study, users only have to assign to the system components the same ID that appears on the webpage (for example, the ON button is associated to the record 00:01 and the OFF button is associated with the record 12:02) and upload the file with the resolution developed (which can be built in ladder programming language [23]). Afterwards, it will be shown the result in a graphical animation, depending on the state of each sensor and actuator. Furthermore, the system also allows a dynamic behavior since it is possible to move the piece along the working line, as it will fulfill the different process steps.

Since the manufacturer OMRON® designs the available PLCs, the virtual model was developed in CX - One tool, in particular the CX – Programmer, CX – Designer and CX - Simulator application (among other applications), property of the same company [24-27].

The CX-One allows users to create, configure and program devices manufactured by OMRON, not only as PLCs and HMIs, but also sensors and actuators. Since the software already models these devices, the integration complexity and configuration is an automated system becomes highly simplified [24-27].

In this software, we can find also the CX -Programmer application. With this application, the development of a program for OMRON PLCs is simplified. When launching the software, the user starts choosing which PLC he/she wants to use, from the typical components of Ladder language (associating each contact to the respective records) and build the desired program. However, if for some reason it is needed to change the PLC, the program easily is converted to the new device in use [25].

After the coding, it is necessary to verify that it works in accordance to the requirements of each problem. Thus, the CX - Simulator application arises. Whose purpose is to simulate a virtual way of the program performed [26].

However, in order to make this simulation process more attractive and easy to understand, it is important to use the CX - Designer application [27]. One of the strengths of this software lies in the possibility of establishing a virtual man-machine interface. According to the solutions (and variables) defined in the CX - Programmer and CX – Simulator, it is possible to design a graphical animation of the case study. It is worth notice that the variables defined in CX - Designer devices are associated to the same records used in the CX - Programmer. In practice, this tool shows the progress of the program depending on the state of the components of the process [24-27].

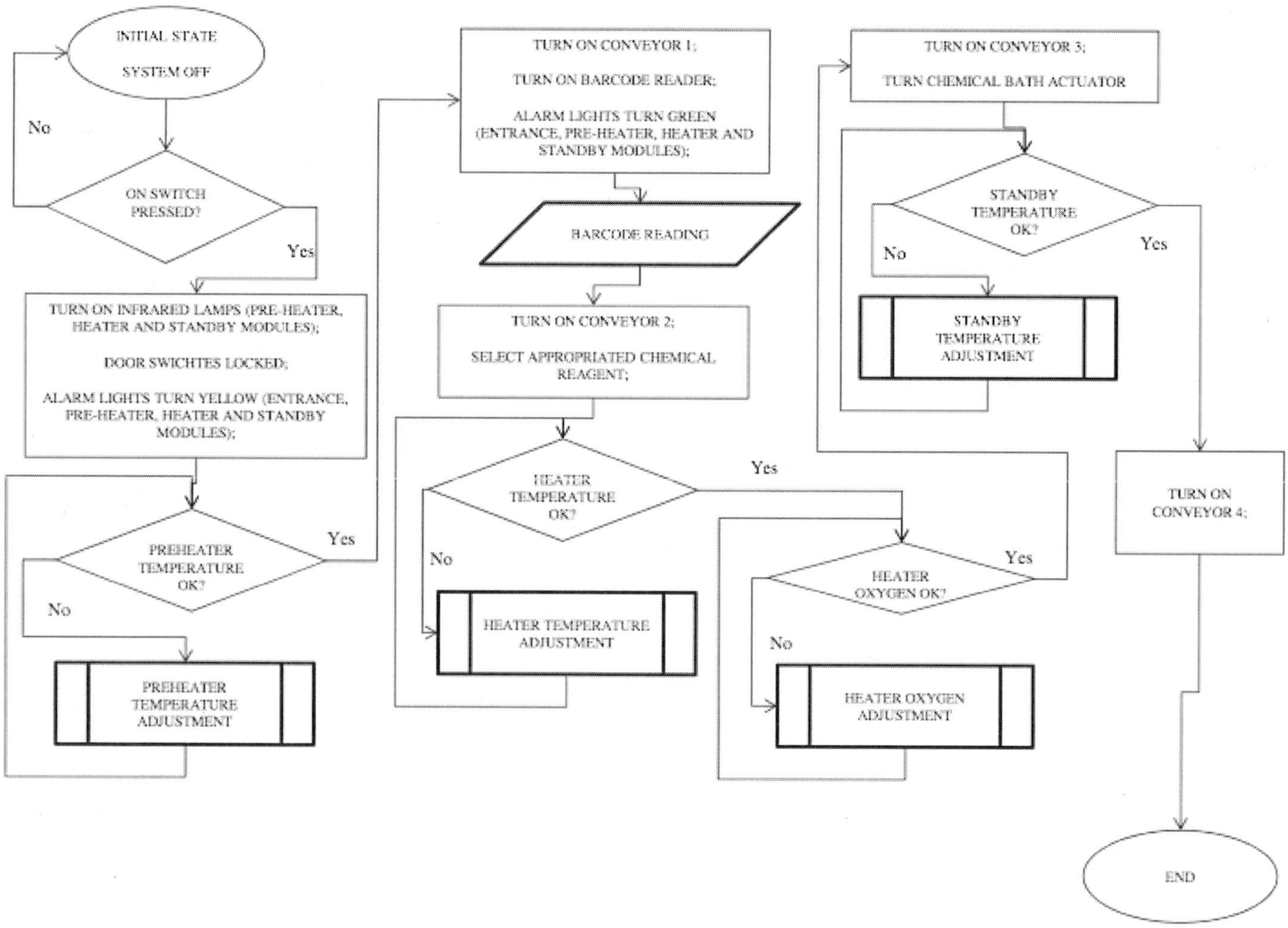

FIGURE 5. PROCESS FLOW CHART CONCEPT

CONCLUSION AND FUTURE WORK

In an era marked by technological change and the need for highly skilled and trained labor, it is urgent to make the process of learning and training approaches as close as possible to the needs of the labor market.

However, the requirements for achieving these objectives may arise certain constraints due creating physical laboratories that have generally high costs associated.

PAIR platform is intended to be an effective answer to these questions. On one hand, it aims to be an appealing training tool, able to motivate learners in industrial automation (and not only) for this area. On the other hand, it aims to overcome the difficulty that may represent replication of training in industrial automation laboratories.

Currently, the platform already has the hardware selection, concerning controllers, such as PLCs; also the selection parts and components includes the adequate sensors and actuators to use in the various case studies to be implemented. A *closet* that allows accommodation of the various PLCs and a model of an intelligent house with an integrated control system (temperature control, lighting, alarm, access to the house) are already available.

The next physical case study to be implemented will be a line of industrial process, whose goal is to bath a piece with a chemical reagent. Directly linked to this scenario, a virtual platform emerges. This allows PAIR users to develop several exercises under an academic way, developing knowledge in creating Grafcets [28] and Ladder programming, even before worrying about all the variables that a physical system can entail.

When PAIR's real and virtual environments are concluded, this platform will need to be tested. Then, several learners, from technical schools, universities and professionals will be invited to use this platform. Their perception will be registered and analyzed so PAIR's impact can be better understood and the platform focus will evolve.

ACKNOWLEDGMENTS

The authors of this paper are grateful to national funds by FCT – Fundação para a Ciência e Tecnologia in the scope of the project: PEst-OE/EEI/UI0319/2014.

REFERENCES

[1] Soares, F., Leão, C.P., Carvalho, V., Vasconcelos, R.M., Costa, S. (2014). Automation and Control Remote Laboratory: A Pedagogical Tool. International Journal of Electrical Engineering Education, 51(1), 54-67.

[2] Nedic, Z., Machotka, J., Nafalski, A. (2003) Remote Laboratories versus Virtual and Real Laboratories. 33rd ASEE/IEEE Frontiers in Education Conference. Boulder, Colorado, USA: IEEE 2003. T3E-1.

[3] MIT iLabs - https://wikis.mit.edu/confluence/display/ILAB2/iLabs (acceded on January, 25th, 2014)

[4] Virtual Labs - http://www.vlab.co.in (acceded on January, 25th, 2014)

[5] University Network of Interactive Labs - http://unilabs.dia.uned.es (acceded on January, 25th, 2014)

[6] Laboratorio Remoto de Automática - http://lra.unileon.es/es (acceded on January, 25th, 2014)

[7] WebLab-Deusto - http://www.weblab.deusto.es/website/ (acceded on January, 25th, 2014)

[8] OpenLabs Electronics Laboratory - http://openlabs.bth.se/electronics/index.php/en?page=StartPage# (acceded on January, 25th, 2014)

[9] Laboratório Remoto de Automação Industrial - http://web.ist.utl.pt/ist11390/aind.htm (acceded on January, 25th, 2014)

[10] Remote Lab FCT - http://www.remotelabfct.pt.vu (acceded on January, 25th, 2014)

[11] Laboratórios Remotos @ UP - http://remotelabsup.fe.up.pt (acceded on January, 25th, 2014)

[12] http://www.profibus.com/technology/profibus/ (accessed on February, 3rd, 2014)

[13] http://www.odva.org/default.aspx?tabid=66 (accessed on February, 3rd, 2014)

[14] http://www.as-interface.net/knowledge-base (accessed on February, 3rd, 2014)

[15] Kurose, J. F., Ross, K. W.: Computer networking: a top-down approach. 5th ed.: Boston: Pearson-Addison Wesley (2010).

[16] Tanenbaum, A. S.: Computer networks. 4th ed. Upper Saddle River: Pearson Education International. (2003).

[17] Tang, A., Scoggins, S.: Open networking with OSI: Englewood Cliffs: Prentice Hall. (1992).

[18] Omron: Serial Communications Boards and Serial Communications Units, downloads.industrial.omron.eu/.../W336-E1-10+CS-CJ-Series+OperManual.pdf (accessed on February, 6th, 2014).

[19] Leão, C.P., Soares, F., Machado, J., Seabra, E., Rodrigues, H. (2011) Design and development of a Industrial Network Laboratory. International Journal of Emerging Technologies in Learning (iJET), 6, 21-26.

[20] Costa, J., Carvalho, N., Leão, C.P., Soares, F.O., Machado, J. (2010) Industrial Network Platform for Monitoring and Control of Automated Manufacturing Systems. In: Proceedings of IFAC 2010 - IFAC Workshop on Intelligent Manufacturing Systems, Lisboa, Portugal, pp. 168-173.

[21] Costa, J., Carvalho, N., Soares, F.O., Leão, C.P., Machado, J. (2010) Remote Access to e-Labs. In: Proceedings of CONTROLO 2010 - 9th Portuguese Conference on Automatic Control, Coimbra, Portugal, pp. 10-11.

[22] Costa, J., Carvalho, N., Soares, F.O., Leão, C.P., Machado, J. (2009) The FINS protocol for complex industrial applications: a case study. In: Proceedings of ICINCO 2009 - International Conference on Informatics in Control, Milan, Italy, 5 pp.

[23] IEC Standard 61 131-3: Programmable Controllers - Part 3, 1993.

[24] CX-ONE - http://industrial.omron.pt/pt/products/catalogue/automation_systems/software/configuration/cx-one/default.html (accessed on April, 12th, 2014)

[25] CX-Programmer http://industrial.omron.pt/pt/products/catalogue/automation_systems/software/configuration/cx-one/cx-programmer.html (accessed on April, 12th, 2014)

[26] CX-Simulator http://industrial.omron.pt/pt/products/catalogue/automation_systems/software/configuration/cx-one/cx-simulator.html (accessed on April, 12th, 2014)

[27] CX-Designer http://industrial.omron.pt/pt/products/catalogue/automation_systems/software/programming/cx-one/cx-designer.html (accessed on April, 12th, 2014)

[28] IEC 60848 Ed. 2 Specification language GRAFCET for sequential function charts, 2000.

Proceedings of the ASME 2014 International Mechanical Engineering Congress and Exposition
IMECE2014
November 14-20, 2014, Montreal, Quebec, Canada

IMECE2014-38362

SUSTAINABLE AND RENEWABLE ENERGY UNDERGRADUATE RESEARCH

Radian Belu
Drexel University
Philadelphia, Pennsylvania, USA

Richard Chiou
Drexel University
Philadelphia, Pennsylvania, USA

Tzu-Liang (Bill) Tseng
University of Texas El Paso
El Paso, Texas, USA

ABSTRACT

Energy is a continuous driving force for the social and technological prospective developments and a vital and essential ingredient for all human transactions. The world is facing an energy "crisis", due to limited fossil fuel resources, growing energy demand and population. All these facts led to and increased interests in renewable energy sources and green manufacturing. Equipping engineering students with the skills and knowledge required to be successful global engineers in the 21st century is one of the primary objectives of academic educators. Enabling students to practice self-directed learning, find design solutions that are sustainable, and helping them recognize that they are part of a global community are just of few of our educational goals. Project-based learning provides the contextual environment making learning exciting and relevant, providing opportunities to explore technical problems from systems-level perspectives, with an appreciation for the inter-connectedness of science principles. The quest for knowledge is the driving force behind education no matter what field is being studied. This means a lot of reading from textbooks, completion of assignments, exams, lectures but quite little of this work involves original research. Active research experience is one of the most effective ways to attract and retain talented undergraduates in science and engineering. At our institutions, we are regularly modifying curriculum content to embrace sustainability and green energy concepts in learning outcomes. However this crosses over between a numbers of multi-disciplinary, multidimensional study areas that include philosophy and ethics. Consequently a major challenge for us is to encourage engineering students whose primary focus is purely technical to include sustainability and renewable energy topics in their designs. To join into this effort of equipping the future engineers and technologists with renewable energy background, we developed a set of project-based courses related to these topics and include them also in our senior project design course sequence. The main objectives of these curricula changes are to provide students with theoretical and practical knowledge reinforced by hands-on experience. These projects are also good examples of multi-disciplinary cooperation of different engineering disciplines as well as providing valuable hands-on and research experience. This paper presents the changes in the course structure, sample of projects, student survey of the course, as well as plans and expectations for future success. We are also discussing here the project team structure, plan and management, component selection, system simulation, and experimental result.

INTRODUCTION

The manufacturing industry uses vast quantities of natural resources (energy, water, materials and land), and produces products and services. Sustainability is a societal challenge; that also requires the contribution of engineers and technologists [1, 2]. The sustainability application of in engineering is about sustainable product and process implementation for the benefit of wider community. An important aspect of the transition to future energy sources involves public understanding and acceptance of new emerging energy technologies as safe and reliable sources for electricity generation, transportation, and energy storage. People at all levels must be educated about the need for new energy technologies, the uses for them, and their role in the energy solutions of the future. In addition, it will be imperative to create a highly educated workforce who can contribute to overcoming future energy challenges [3-10]. One method of supporting workforce development in future energy solutions is to incorporate new and emerging energy technology directly into required undergraduate coursework. Engineering educators must develop new curriculum solutions in advanced energy technologies to fill the gaps in existing coursework and prepare the next generation of students to support global sustainability.

Equipping engineering students with the skills and knowledge required to be successful global engineers in the 21st century is one of the primary objectives of university educators. Enabling students to practice self-directed learning, to find solutions to design problems that are sustainable, and helping them to recognize that they are part of a global community are just a few of our educational goals. Consequently, traditional engineering curriculums were entrenched in the industrial needs of the 19th century; today the new 21st century engineering education model needs to incorporate new principles and ideals including sustainability in the engineering curriculums [6-16. However the increasing knowledge needed to practice as a professional engineers and the accelerating rate of change within the discipline suggest that traditional learning models may not address the requirements of learners. One of the major objectives of an engineering or technology program is to prepare graduates to be technically competent and productive in industry once employed. The skills acquired in school can lead them to a very successful and rewarding career. The quest for knowledge is the driving force behind undergraduate education no matter what field is being studied. This means a lot of reading from text books, completion of homework assignments, exams, lectures but so little of this work involves original research. An active research experience is one of the most effective ways to attract talented undergraduates and retain them in careers in science and engineering. Active participation in real-life research has always been considered a high motivation for the undergraduates. The way the engineering technology program can prepare graduates with the necessary design skills is through the research and design projects. They provide opportunities for students to explore technical problems from a system-level perspective and to develop an appreciation for the inter-connectedness of science and engineering principles. Our senior design project course sequence consists of three courses taken in the final year by all engineering and engineering technology students in our university. They are intended to bridge the gap between theory and real world practice.

Additionally, the senior project design can serve as an excellent culminating point in the program of study where it will focus on projects that have practical value to either consumers or to industry. The topic of alternative energy is not only relevant to a multitude of issues today it is also an effective vehicle for developing instruction that applies across a variety of disciplines and academic standards. Since many of the issues associated with alternative energy are open-ended, alternative energy also lends itself to project-based and problem-based instruction. Our engineering technology program has been recently updated to include strong energy efficiency, green design and renewable energy components [16-18]. These projects are a good example of the multidisciplinary cooperation of different engineering disciplines as well as providing valuable hands-on experience to the students. In addition to providing useful lessons in teamwork and project management, the projects will provide a working demonstration of the wind and solar energy conversion systems. Global environmental concerns, the ever-increasing need for electrical power generation, and steady progress in power deregulation have created increased interest in environmentally conscious distributed power generation. Of particular interest are alternative energy distributed generation (AEDG) systems such as wind, solar/photovoltaic (PV), and fuel cell power generation devices with zero (or near zero) emission of greenhouse and hazardous gases. These generation devices can be used in stand-alone configuration or be connected to the power network for grid reinforcement. Given the rapid progress in AEDG development and utilization, there will be a great need for trained professionals with adequate knowledge in this area to be able to plan, design and operate AEDG systems, and perform analytic evaluation of their impact on power systems to which they are connected. On the other hand, electric power systems, electric transmission and distribution systems are undergoing rapid changes due to deregulation, the penetration of dispersed and distributed energy resources, renewable energy generation and power electronics technologies, and the adoption of efficient computation, communications and control mechanisms. Due to this monumental growth in the use of renewable energy for electricity generation and in the interest of keeping students abreast of the current scientific and technological developments and trends, we believed that it was important and timely to develop upper-level undergraduate courses on renewable energy systems and to include renewable energy projects in power electronics and senior project design courses in our program [16-18]. There is a well-documented demand and need in offering program study, courses and training in the areas of renewable energy. Future engineers must be taught to be creative, flexible and imaginative. There should be meticulous attention given to team work and on the challenges of sustainable development, including cultural, economic, environment and social imperatives. Future engineering curriculum should be built around developing and increasing skills and technical knowledge. The topic of alternative energy is not only relevant to a multitude of issues today, it is also an effective vehicle for developing instruction that applies across a variety of content disciplines and academic standards. Since many of the issues associated with alternative energy are open-ended, alternative energy also lends itself to a project-based and problem-based instruction.

Engineering and engineering technology programs must offer a relevant and validated curriculum that prepares students for post-graduation success. Courses that cover traditional subject matter in mathematics, the sciences, engineering economics and other related topics provide the foundation of knowledge upon which specific skill sets are added. However, it is critical for engineering/technology to transition from theoretical work in the classroom towards experiential learning with applications of technology and design. The main objective of senior design courses in engineering and engineering technology curricula is to bridge the gap between academic theory and real world practice. Accordingly, the proposed senior projects should include elements of both credible analysis and experimental proofing as discussed in ABET criteria [19, 20]. The senior design project can serve as an excellent culminating experience in the program of study when it focuses on research and design projects that have practical value to consumers or to industry. For our program, the senior design course is a year-long educational journey (three quarters) that takes an idea generated by a student or an industrial sponsor and culminates in a product or project [18, 21-24]. This course is an excellent capstone experience, which requires both teamwork and individual skills in solving a modern industrial problem. Senior design seminars in fall and spring quarters bring the students, faculty, and industrial partners together to see the student's results and to give them the additional experience of public presentation of their work.

The purpose of this paper is to describe the renewable energy projects involved in our capstone senior design project. The lessons learned are presented and the ways to improve project management are discussed. Our senior design project course is a 3-term core course usually taken by the students during their terminal year in our program. This paper describes the issues related to the inclusion of the renewable energy projects in senior design courses, undergraduate research and sample of such projects. These projects were offered since 2011-2012 academic year. The topics for the projects are suggested by the faculty involved in this project or are proposed by the students themselves.

IMPORTANCE OF RESEARCH PROJECTS IN THE UNDERGRADUATE EDUCATION

The importance of project work in the curriculum of our undergraduate engineering programs is well understood. Students and faculty need to understand the objectives of doing project work, to have mechanisms for selection of attractive topics, focus and scope of work, estimation of efforts required and facilities needed, etc. Students want and need to know the best practices also [15-27]. What is the focus of engineering educations is an important question that needs to be asked and introspected upon frequently by all of us. An engineer is a problem solver - designing large systems like power, gas, water and communications networks or large industries, building large systems like power plants, dams, server banks, etc., operating large engineering systems like communications services, wide area networks, power systems, etc.; or managing systems and services. Therefore, an engineer should be a designer, thinker, innovator and systems integrator. Hence, the educational system should inculcate into its students various aspects such as engineering principles, standards and practices, design methodologies, modeling and optimization capabilities, systems analysis and integration techniques and new technologies and research areas. An engineer can become a thinker, a creative person or an innovator only if he is allowed to independently put together all aspects of learning to solve a practical problem and look at alternatives. Lectures, tutorials and laboratories allow students to acquire engineering and scientific knowledge. However, only project work gives them the opportunity to become a problem solver or an innovator. Hence, project work is a very important constituent of our engineering curriculum. It is necessary to understand the role and importance of project work in the professional world. Project work provides several opportunities to students to learn several aspects of importance to an engineer that cannot be taught in a class room or laboratory.

For an engineer in industry, a project is a sequence of tasks required to reach an objective. Typically, the objective is to design a device or process that has value to a customer (user). The project begins by defining a performance problem associated with an application and ends with a design solution. The problem drives the learning required to complete the project. Managing the project requires the engineer to demonstrate effective teamwork, clear communication and the ability to balance the social, economic and environmental impacts of the project. Project-based learning (PBL) is based on the practice of solving problems. The concept of problem-based learning was first developed in the medical field in the mid-1950's. It has since been adopted in a variety of educational disciplines [16-18]. Traditionally, the educational process involves students first learning the fundamentals and then utilizing "total recall" to apply these facts to solve a problem; learning objectives are set by the instructor and principles are presented to the students through lectures. Assignments are given to reinforce the application of the concepts, but often students merely "learn" what is necessary to pass the test or "repeat-back" information to satisfy the instructor. In contrast, this approach employs a problem as the driving force for learning the fundamental principles that are required to find a solution. Moreover, this approach provides a context that makes learning the fundamentals more relevant and, hence, results in better retention by students. For clarity, we view problem-based learning as pertaining to the development of knowledge based on the fundamental principles of science and mathematics and project-based learning to include mastering the engineering skills required to implement a design solution. Active participation in real-life research has always been considered a high motivation for the undergraduates. Undergraduate research projects enrich a student's education, being also a way to prepare her/him for graduate school or the workforce. Through undergraduate research, students actively participate in the solution of engineering problems, learn how engineering theory is applied to solve problems, work side by side with faculty and graduate students, and improve their communication skills. Faculty members get a chance to influence to a higher degree the careers of students, encouraging them to excel in their jobs or to continue graduate studies. Undergraduate research is also a way for faculty to integrate research and teaching, a model proposed to balance the many duties professors have. Faculty involved in this project expanded their research to include suitable topics for the students with minimal background in this area of engineering. To enhance the learning and the capability of succeeding in this research the students were given specific weekly plan with outcomes and evaluation incorporated.

ABET defines Engineering Design as: "The process of devising a system, component, or process to meet the desired needs [19, 20]. It is a decision making process, in which the basic sciences, mathematics, and engineering sciences are applied to convert resources optimally to meet stated objective. Among the fundamental elements of the design processes are: the establishment of the objectives and criteria, synthesis, analysis, construction, testing and evaluation". In our senior design classes we have placed this definition at the core of our courses. First we focus on objectives and criteria and ask the student to write a short proposal stating these objectives, principles, and the decisive factors to reach the stated goals. These projects involved elements of structural design, wind and solar energy resource assessment, as well as the electrical, electronics and computer engineering system design. The second step is conceptualization and laying down the foundation for how to achieve the stated objectives. At this stage the students are encouraged to draw a block diagram showing different components of the system they want to design [21-27]. A set of questions are posed to students to further understand the task at hand. These are typical questions:

1. What are the inputs to the system and, what are their characteristics and magnitudes?
2. Do the inputs require conditioning?
3. What is the medium through which inputs are interfaced to the system under consideration?
4. Do the inputs dictate to the system to be designed how to behave, or just activates the system?
5. What is the voltage, current and power requirements for the load?
6. Is it a single output or multi-output system?
7. Are there feedback loops in the system?
8. Do the loads require separate power supplies?

Once the students compile the answers to these questions, they are directed to perform system analysis, design, component purchase and fabrication, building and testing of the prototype, as well as the overall design improvements. From the very beginning, this course sequence was organized following the ABET guideline for capstone and/or senior project design courses. The senior design class is organized in a very structured form.

Teams: All students have to work in teams of three or four. We consider this to be the optimum team size. A team of two may result in distress in cases where one of the students was not able to do his or her share of the work, while for teams larger than four may have difficulties to choose projects which were challenging enough for such a big group of students and still could be finished within three-quarter time frame.

Self and Peer Review: A very simple self and peer review system has been introduced. The students must evaluate their own and their team members' performance on a scale of 5. The main challenges we faced were that we never had anything similar to this and were inexperienced in how to adequately give feedback to the students.

Industrial Advisors: Some of the department's advisory board members are also serving as industry advisors for the senior design class. They are reviewing reports, listen to presentations and give feedback on those and are also serving as judges for the Senior Design presentations.

Reports: All teams must hand in a proposal, two midterm design reviews, and final report. Various faculty and industry advisors review all these reports and the students are provided feedback on their projects as well as on their report writing.

Presentations: All teams must present their proposals and first quarter design review, and a final presentation in Spring quarters. The audience for these presentations is the class, faculty members and some of the industry advisors. They are also required to prepare a professional poster to display with their projects, during Drexel research day, usually in mid-April.. Some of the faculty and industry members also serve as judges. The teams are judged on the projects' technical content. The individual students are also judged on their presentation style. These ratings have a two-fold purpose: they will be used as a part of the students' final grades and for a ranking of projects/teams. The winning team receives a small prize and team members' names will be engraved on plaques to be displayed in the department.

UNDERGRADUATE REAEARCH PROJECTS WITH INCRESEAD RENEWABLE ENERGY CONTENT

For the last two years, our focus shifted towards incorporating renewable energy topics in our senior project design courses. The primary goal of the projects is to provide students with a deeper understanding of the need for renewable energy sources, the advantages and disadvantages of a particular energy type, and the global and economic impact of the chosen energy type. The renewable energy-focused projects, discussed in detail and with respect to ABET program criteria [19, 20], have been delivered for five semesters and continue on an ongoing basis. An overview of the projects is provided again here for reference. These projects are a good example of multi-disciplinary cooperation of different engineering disciplines as well as providing valuable hands-on experience to the students [14-18, 21-27]. In addition to providing useful lessons in teamwork and project management, the project will provide a working demonstration of a wind and solar energy system. At the project assignment, students briefly investigate multiple renewable energy sources to determine their focus area. For the duration of the project, students complete three major parts for their chosen energy source: (1) research on the history and implementation, (2) development of an experimental application to an energy conversion process, and (3) an economic analysis. First, students investigate the background of their chosen energy type, producing a summary of the history, real and specific examples of implemented systems, insight into political or legislative issues, discussion of the environmental impact, and advantages and disadvantages of the chosen energy type. The second project component requires students to design and build an apparatus that converts the chosen energy source to useful energy output, typically electrical or mechanical work. An experiment must also be created and performed to determine the conversion efficiency of the system built. Finally, students complete an economic analysis of the chosen energy type, either based on their actual experimental apparatus. The economic analysis must quantify all initial and recurring costs over an assumed product lifespan, and must present a resulting estimate of cost per kilowatt-hour or similar final economic quantity.

Two examples of undergraduate research projects are presented in the following section of this paper. During the first month of the fall quarter section of the course, each team is given partial specifications for the project. Five deliverables are distributed throughout the semester. In the proposal, students choose a renewable energy source to study and brainstorm possibilities for an experimental apparatus using the chosen energy source. A few weeks later, students identify parts and instrumentation needed to implement their experiment and submit a purchase request (each team receives small funds per project group, from the authors' grant, but they are allow to an unlimited own spending). Each team demonstrates the finished project to the entire class and then a written report summarizing the project is handed. This process synthesizes all of the basic materials in the core courses and can also be used as part of the requirements of the senior project requirements for each student. Examples of the renewable energy senior design projects included in this course are:

1. Analysis and Design of Single-Phase PV Inverters
2. Power Conditioning Units for PV Water Pumping
3. PV Maximum-Power-Point-Tracking Controller
4. Design and Implementation of a Savonius Micro-Wind Turbine
5. Control and Power Electronics of a Small Wind Power for Battery Charging.
6. GPS-based Dual Axis Sun Tracking Systems

The design also includes the prototype testing, test results and analysis, optimization and design alternatives.

PROJECT SAMPLES

Two samples of such undergraduate research projects are presented in the following subsections of the paper.

Project 1: System Design of a PV Power System Project

The multi-disciplinary team was tasked with designing, assembling, and testing a 500 W photovoltaic setup that they could simulate, assemble, and test in order to validate their simulation results. This gave the students an opportunity to reinforce the classroom lessons on solar energy and to apply it in a real-life situation using the same equipment as in large scale solar installations that are becoming more prevalent across the world. Students worked with suppliers and manufacturers to acquire equipment and testing supplies, troubleshoot devices, and to better understand how solar power is harnessed in large scale operations. The students came up with testing scenarios to test different system aspects, and designed system that could test different variables including panel orientation, battery charging algorithm, and charge controller configuration. Students were given approximately three weeks of classroom background on PV systems before being asked to purchase components. With that knowledge in hand, they were assigned components and asked to work together to design a system that will meet the required objectives. Main advantages of photovoltaic power are: (1) short lead time to design, (2) highly modular, (3) static structure, no moving parts, hence, no noise, (4) high power capability per unit of weight, (5) longer life with little maintenance due to no moving parts, (6) highly mobile and portable because of light weight. In this project a 12V, 10W industrial grade solar panel is used as the main solar panel17-20.The output current is rated at 0.60 A. Depending on the light conditions, the solar panel outputs 2.5 Ah to 4 Ah a day. Due to its small dimensions and light weight, it can be easily installed. To improve the system performance, a controller is added between the PV panel and the battery as shown in Figure 1.

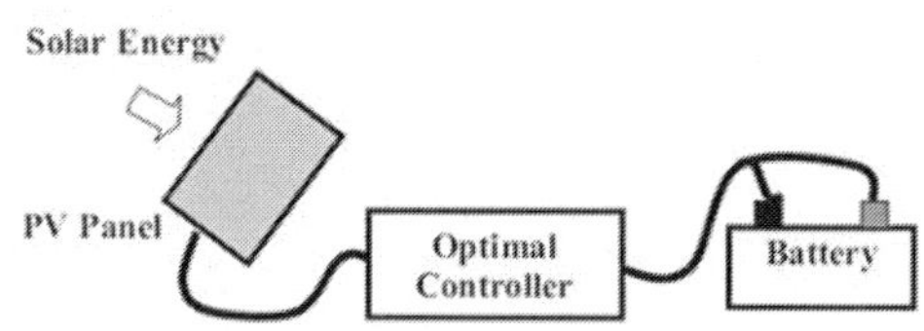

Figure 1: Schematic Diagram of a Solar Powered Battery

The controller is usually a DC/DC converter. It should be a buck/boost to take into account conditions where the PV panel voltage is higher or lower than the battery voltage. In such a system an optimal control algorithm can be implemented to convert the maximum available solar power into electricity and charge the battery. The battery charger is an important device that is used to manage the electrical energy from solar arrays.

To manage the electrical energy from the photovoltaic power system, the power switching circuit is applied in the battery charger of the photovoltaic power system due to the fact that it doesn't consume more energy to operate.

In power switching circuit design, higher switching frequencies allow reduction of the magnetic component size with switching converter. However, increased switching frequencies can cause higher switching losses and greater electromagnetic interference. The switching losses include the current and voltage overlap loss during internal switching and capacitance loss during turn on. DC-DC boost converter is a good switching converter and it was selected for this project because of its simplifications in power circuit and control circuit implementations. The switching losses of an active switch device of the boost converter, which is typically a MOSFET, consist of turn-off and turn-on losses. These losses lead to a significant decrease in converter efficiency. In order to reduce the losses caused by hard switching problems, active auxiliary circuit implementations were required. These switching loss reduction methods were performed by managing the operation of main switch and auxiliary switches. So, the additional control circuitry of the active switches was necessary. In this project, a circuit implementation that required only one active switch, four passive switches, one resonant inductor, and two resonant capacitors for operating the switching losses reduction was proposed in the switching battery charger of PV power system shown in Figure 2.

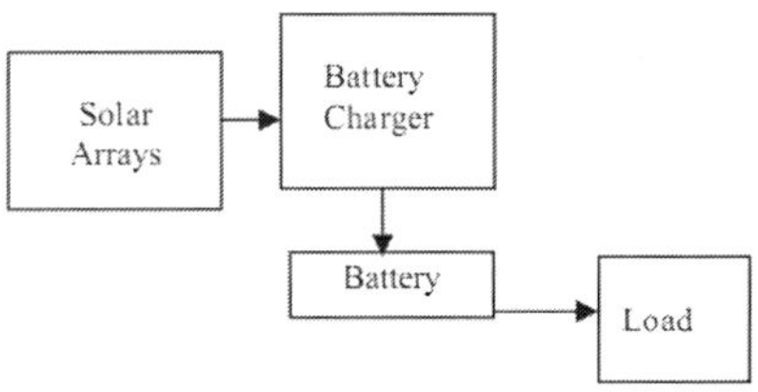

Figure 2: Implementation of photovoltaic power system

A lead-acid battery, commonly used in solar power systems is used. The battery is composed of several single cells connected in series. Each cell produces approximately 2.1 V, thus a 6-volt battery having only three single cells, will, when fully charged, produce an output voltage of 6.3 V. A 12-volt battery has six single cells and produces an output voltage of 12.6 V when fully charged. The lead-acid battery has various versions. Deep cycle rechargeable batteries are designed to provide a constant voltage over a long period of time. They are suitable for repeated full charge and discharge cycles. A deep-cycle rechargeable battery is used for energy storage in this project. The DC output voltage is 12 V and the maximum charging current is 1.25 A. The internal resistance of the battery is 30 mΩ. The operating temperature is 4 ° to 140 °F.

The major objective of this project is to design a solar powered battery charger with an optimal controller. The goals of the proposed system are: 1) to convert the solar power into electricity as much as possible under varying weather condition; 2) to charge the battery as fast as possible in accordance to the battery life cycle conditions. The application of the proposed system can be light electrical vehicles such as golf carts, scooters, airport utility vehicles, as well as other renewable power stations using batteries as energy storage. The initial requirements for this project are as follows: 100 W solar panel, design structure strong enough to support the about 15 lbs, design enclosure to protect electrical equipment, controller and power electronics subsystems, electrical and mechanical design to rotate the solar panel, algorithm to implement tracking and control concepts, and portable system design.

A charge controller is needed because the solar panels alone cannot effectively or safely charge the batteries in the system. A component that regulates the proper charging of the batteries is critical to ensure safe battery charging that maximizes the life of the batteries and also controls the power flow. The charge controller is a component that goes inline between the solar panels and the battery bank. This unit takes the power generated from the panels and manipulates it to safely and efficiently charge the batteries. The charge method of most charge controllers is either a pulse-width-modulation, or maximum-power-point tracking (MPPT) algorithm. When selecting a charge controller for a specific application, these are the two primary specifications to focus on. The controller has selectable operating voltages of 12 V, 24 V, and 48 V. The unit's rated current is 45 A, which far exceeds the requirements of the system, while also providing clearance for additions to the system in the future. This unit was selected over others because it incorporated PC connected monitoring and data logging capabilities, having also a stand-alone logging feature.

Project 2: Modeling and Simulation of a High Performance Wind-Electric Battery Charging System

The major goal of this project is to design, model and analysis of a wind energy based battery charger. The system development, analysis and the system performance are part of the project, while the improvements that can be achieved by altering the system configuration to better match the load to the wind turbine are taken into consideration during to initial project stages. By understanding the basic characteristics of the components many of the performance limitations of the system can be remedied and the optimization of the system can be explored. The approach considered is to insert a load controller between the generator and the load. Different types of controllers were investigated, during this project including a power converter and a capacitor-compensated system.

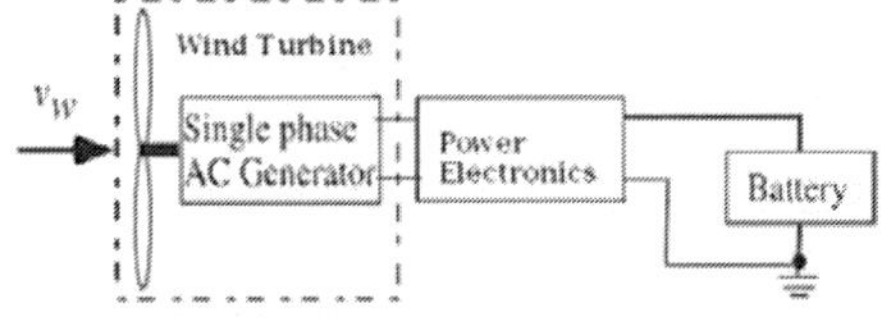

Figure 3: Physical Diagram of the Wind-Turbine-Battery System

The physical diagram of the system under investigation can be shown in Figure 3. The alternator is connected to a rectifier, the output of which is connected to a battery bank. In the conventional system, there is no active control used to adjust the energy produced by the wind turbine; therefore, the power flow to battery is dictated solely by the wind speed and the passive interaction of the various system components. The benefits of this project include but are not limited to:

- Cost effective power generation in remote areas;
- Upgradeable parts ensure the possibility of ongoing student involvement;
- Clean, renewable energy source;
- Can operate in stand-alone application or can be attached to any power grid with minimal modifications.

The features of this project will include: a) Power generated in AC mode, 60 Hz output; b) Modular design for easy upgrades; c) Places to install both a commercial and hand-made generator; d) Hinged 30 foot or shorter tower for easy installation; and e) Low power loss from generation to transmission. Project block diagram is shown in Figure 3, major component included.

IMPACT OF RENEWABLE ENERGY PROJECT

The new evaluation tool for the renewable energy projects was designed with three specific goals in mind. The first goal was to evaluate whether students felt the projects had an impact on their understanding of renewable energy, a subjective evaluation. The second purpose, addressed with the factual-based questions in both the pre-project and post-project surveys, was to attempt to determine objectively whether the projects increased student knowledge of energy topics. And finally, it was desired to evaluate whether students were retaining specific results of their projects and the completed experiments (17-19, 23-27). The results indicate that some success was achieved in each of these three goals. Student perceptions of the renewable energy projects clearly indicate the positive impact, as shown in the table 1.

The integration and extension of the undergraduate research projects in our program, using the new teaching and learning approach was first time offered in the 2011-2012 academic year respectively at the main campus of our university and at one of the partner college. At the end of each quarter, all students have been requested to answer (with a five point scale: 1-very poor, 2-poor, 3-satisactory, 4-good and 5-very good) an anonymous questionnaire as shown in Table 1. According to the results, the new project-based approach received a 3.9/4.1/3.8 on 5.0 point ratings, for all academic years when the course was offered, compared with an average rating of 3.4 for the all courses and years at our program. The results from the students' feedback have been extremely positive with the regard to the renewable energy-related projects and the experiments provided during the laboratory sessions. The majority of students felt that such projects enhanced their understanding of the theoretical materials and made the course more interesting.

Table 1 Questionnaire for the research project evaluation

Q1	Research project are challenging and interesting?
Q2	Have you learn more than expected?
Q3	Is the team project useful to you?
Q4	What was the level of "hands-on" experience?
Q5	Please, provide an overall evaluation of the project learning experience

Essential parts of a successful project are balanced student expertize, team-work and good project management. As part of this process the class instructors must make sure that the students with the correct background are assigned to each project. Ideally, knowledge of individual students' leadership and communication abilities would be provided by each departmental advisor. Then the project teams (3-4 students) are assigned with specific project components based on tasks that need to be accomplished. To facilitate managing these tasks, the design includes discussions, literature search on issues of project management, problem solving, and system level thinking. Project advisors and faculty involved in the project will provide information and active learning sessions to enhance students' capabilities to understand, work, and lead the groups. To make such large groups functional and productive, the teams' advisers should be closely involved with but not dominating the project management. Weekly reporting from teams and frequent peer evaluations are vital, but we have to make it easy both for teams to submit and for mentors to review. This also requires active advisor involvement, meeting at least 1 hour per week or more for large teams and providing and making use of feedback from industry evaluations. The faculty mentors who are involved need to be familiar with multiple engineering disciplines. In addition a group of college wide faculty from different disciplines should be available as team advisors and as consultants available to any team. The faculty mentors who are involved need to be familiar with multiple engineering disciplines. In addition a group of college wide faculty from different disciplines should be available as team advisors and as consultants available to any team.

CONCLUSIONS

There is a proved need at all levels in industry, from large utilities to small consulting firms, for prepared professionals that can deal with energy issues at the utility side as well as at the customer side, in the fields of sustainability, green manufacturing, energy conservation, and especially in the field of renewable energy. It is important to improve the education of engineers in energy-related areas in order to ensure qualified professionals are educated to face today's energy challenges. This is accomplished by integrating undergraduate research to enhance undergraduate education. In response to these needs for curricula in renewable energy, sustainability and green manufacturing topics, research undergraduate projects in these areas have been proposed, included and delivered during Fall

2012 semester and quarter in our programs. The projects involve detailed investigation of renewable energy sources, the design, build, and test an experimental renewable energy conversion system, including not only technical and engineering aspects, but also economic analysis. Assessments of the impact of the renewable energy projects was positive and showed increased student understanding of renewable energy sources, sustainability and green design but fell short of formally evaluating the contribution of the projects to the energy education needs of the future. The overall success and value of the renewable energy projects has been demonstrated further using the expanded evaluation tool. The results presented will be used to further develop the renewable energy projects and the next version of the evaluation tool. The design experience develops the students' lifelong learning skills, self-evaluations, self-discovery, and peer instruction in the design's creation, critique, and justification. Students learn to understand the manufacturer data sheets, application notes, and technical manuals and component specifications. The experience of teamwork, prototype design and test, which would be difficult to complete individually, gives the students a sense of satisfaction and accomplishment that is often lacking in many engineering courses, not including projects. Furthermore, the design experience motivates student learning and develops skills required in industry. The students were able to make satisfactory estimations and calculations of these projects. Their results reflect that they have understood well all the basic ingredients of the modeling techniques and design of the renewable energy systems. Our experience with the incorporation of renewable energy topics in the senior project design courses demonstrated that the abstract knowledge acquired by the students during their first three years of studies was put into practice. They also gained extensive knowledge of in various fields of engineering, electrical, environmental, mechanical, structural, etc.. They learned to identify a problem, conduct research on a particular project, and compare their findings with other similar projects. In addition, the student results identify possible areas for further emphasis in undergraduate engineering and technology curricula. These projects are easily implemented at other institutions, to help advance the education in sustainability and renewable energy.

NOMENCLATURE

Sustainability, renewable energy, student research project, team work, project management, green design.

ACKNOWLEDGMENTS

This work has been supported by the US Department of Education under the joint DHSIP Program with the University of Texas at El Paso and Drexel University, PR/Award No.: P031S120131. The authors wish to express sincere gratitude for their financial support.

REFERENCES

1. Their Future is Green: The Clean-Energy Economy Promises: An Engineering Jobs Bounty – Training Graduates With Right Skills, American Society for Engineering Education, PRISM, pp. 38-41, 4/2010.
2. "Shaping the future: New Expectations for Undergraduate Education in Science, Mathematics, Engineering, and Technology": A Report on the Review of Undergraduate Education from the Committee for the Review to the National Science Foundation Directorate for Education and Human Resources (1996); (http://www.ehr.nsf.gov/EHR/DUE/documents/review/96139/start.htm).
3. On the Road to Energy Security – Implementing A Comprehensive Energy Strategy: A Status Report. (2006). U.S. Department of Energy, Retrieved September 24, 2006, from http://www.energy.gov/about/EPAct.htm.
4. Annual Energy Outlook 2013 with Projections to 2050 (Early Release) – Overview. Energy Information Administration, U.S. Department of Energy, from http://www.eia.gov/forecasts/aeo/
5. Armarego J., and S. Clarke, "Problem-based Design Studios for Undergraduate SE Education," in 18th Conference on Software Engineering Education & Training (CSEET'05), 2005, pp. 249-254.
6. Rosentrater, K. A. and Al-Kalaani, Y., "Renewable energy alternatives – a growing opportunity for engineering and technology education," The Technology Interface, 6(1), 2006, pp. 10-18.
7. Vanasupa L., and F. G. Splitt, "Curricula for a Sustainable Future: A proposal for integrating environmental concepts into our curricula," Materials Research Society, Warrendale, PA 2009.
8. Jennings, P., New directions in renewable energy education. International Journal of Renewable Energy, 34, 435-439, 2009.
9. Allen, D. T., and Shonnard, D. R. (2003). Green engineering, Prentice Hall, Upper Saddle River, NJ, 397416, 2003.
10. Cooper, H. L. Undergraduate renewable energy projects to support energy solutions of the future. Proceedings of the 2006 ASME International Mechanical Engineering Conference and Exposition, Chicago, IL, 2006.
11. Felder, R. M., and R. Brent, (2004a), "The Intellectual Development of Science and Engineering Students. Part 1: Models and Challenges," Journal of Engineering Education, 93(4), pp. 269-277.
12. Felder, R. M., and Brent, R., "The Intellectual Development of Science and Engineering Students. Part 2: Teaching to Promote Growth," Journal of Engineering Education, 93(4), pp. 269-277.
13. McNamee, G., "Careers in Renewable Energy," PixyJack Press LLC, 2008.
14. Walker, S, "Sustainable by Design: Explorations in Theory and Practice," London: Earthscan Publications Ltd, 2006.

15. Clift, R., "Sustainable development and its implications for chemical engineering," Chem Eng Sci, 61: 4179-4187, 2006.

16. Belu R.G. and I. Husanu - An Undergraduate Course on Renewable Energy Conversion Systems for Engineering Technology Students, 2011 ASEEE Conference & Exposition, June 26 - 29, Vancouver, BC, Canada (CD Proceedings).

17. Belu R.G. – Embedding Renewable Energy into the Engineering Technology Curricula, ASEE Mid-Atlantic Sec., Fall Meeting, Philadelphia, USA, Oct. 28-29, 2011 (CD Proceedings).

18. Belu R.G. and I. Husanu - Embedding Renewable Energy and Sustainability into the Engineering Technology Curricula, 2012 ASEEE Conference & Exposition, June 10 – 13, 2012, San Antonio, Texas (CD Proceedings).

19. Criteria for accrediting engineering programs: Effective for evaluations during the 2010–2011 accreditation cycle, ABET, Baltimore. Accreditation Board for Engineering Technology (ABET), Inc. (2010).

20. ABET, Criteria for Accrediting Engineering Programs, 2013,http://www.abet.org/accreditation-criteria-policiesdocuments/,

21. Paulik, M. J., Krishnan, M., "A competition-motivated capstone design course: The result of a fifteen-year evolution", IEEE Transactions on Education, vol. 44(1), 2001, pp 67 - 75.

22. Belu R.G., and A.C. Belu, A Decision Support Software Application for Design of Hybrid Solar-Wind Power Systems- as Teaching-Aid, 2007Annual ASEE Conference, and Exposition, Honolulu, Hawaii (CD Proceedings).

23. Belu R.G., Design and Development of Simulation System for Renewable Energy Laboratory, 2010 ASEEE Conference & Exposition, June 20 - 23, Louisville, Kentucky (CD Proceedings).

24. Belu, R., and D. Koracin, E-learning Platform for Renewable Energy Sources, 2010 ASEEE Conference & Exposition, June 20 - 23, Louisville, Kentucky (CD Proceedings)

25. Belu, R.G., A Project-based Power Electronics Course with an Increased Content of Renewable Energy Applications, June 14-17, 2009 Annual ASEE Conference and Exposition, Austin, Texas (CD Proceedings).

26. Belu, R.G., R. Chiou, and B. Tseng - Life Cycle Assessment – A Useful Tool in Green Manufacturing and Energy Education, 6 th International Conference on Manufacturing Science and Education - MSE 2013 (Smart Sustainable Development), June, 12-15, 2013, Sibiu, Romania (CD Proceedings).

27. Chiou, R., R.G. Belu and B. Tseng – Infusion of Green Energy Manufacturing into Engineering and Technology Curricula, ASME 2013 International Mechanical Engineering Congress & Exposition, November 15-21, 2013, San Diego, California (CD Proceedings).

This page left blank intentionally.

Proceedings of the ASME 2014 International Mechanical Engineering Congress and Exposition
IMECE2014
November 14-20, 2014, Montreal, Quebec, Canada

IMECE2014-38501

ADVANCING SUSTAINABLE ENGINEERING PRACTICE THROUGH EDUCATION AND UNDERGRADUATE RESEARCH PROJECTS

Radian Belu
Drexel University
Philadelphia, Pennsylvania, USA

Richard Chiou
Drexel University
Philadelphia, Pennsylvania, USA

Tzu-Liang (Bill) Tseng
The University of Texas at El Paso
El Paso, USA

Lucian Cioca
Lucian Blaga University
Sibiu, Romania

ABSTRACT

Major challenges such as energy, food, water, environment, health and so many more have never been more prominent than they are today. Engineers and educators, as problem solvers should be addressing these issues and challenges in sustainable ways. They have an enormous opportunity to help create a more sustainable world. Technology problems interconnecting sustainability challenges such as climate change, loss of biodiversity, environmental pollution, economic and social instability are becoming increasingly major concerns for mankind. However, the engineers and scientists have failed on large extend to fully address the sustainability issues. It was also found that engineering graduates do not possess necessary skills to tackle sustainability related problems. Engineering practice and education are changing as social expectations and conditions for engineering practice change too. Students have the responsibility and opportunity to continue improving our life while reducing or even reversing the negative impacts that our industrial society is having on the environment. Current engineering curricula are not equipping them to properly deal with these challenges due to little integration of sustainable and green design strategies and practice. Transforming higher education curricula for sustainable development is a tough challenge, dealing with the complexness of sustainability concepts and integration into engineering education. Teaching students the sustainability principles and equipping them with necessary tools help them to make better choices on materials and energy use, or design. These concepts and methods are still relatively new to engineering curriculum and are not an established practice for most of such programs. Meanwhile, today's students have a strong desire to improve the world through their work, and sustainability connects with these interest and motivations. However, students' hunger for knowledge often outstrips what is available in their courses and the experiences of their professors. Furthermore, to make sustainable design compelling to a wider base of engineering students, we need to craft sustainable design in terms of mainstream design problems that are important, cutting-edge, and achievable. Then we need to help them how to effectively deal with environmental and societal needs and constraints as part of their core design process. The paper highlights the process required for embedding sustainability and green design into our programs, curriculum design, implementation and impediments to surmount for sustainability and green design in engineering education. This was done through a project-based approach, developing three new courses and appropriate changes in a number of existing courses. The skill requirements were studied and finally the list of subjects, topics, teaching and learning methods are identified and discussed in this paper.

INTRODUCTION

Humanity is at an unprecedented crossroads, due to increasing energy demands, depletion of fossil fuel reserves, environmental pollution or population increase [1-4]. Despite all the work we have done on environmental protection, all living systems are declining at an increasing rate, we are severely disrupting the stability of the climate, and there are huge social, economic and public health challenges worldwide. On the other hand, there is a growing awareness among manufacturers of the need to consider the economic, industrial, societal, and environmental performances and the recognition that environmental challenges represent both business opportunities, as well as societal responsibility. Sustainable development is a contemporary issue for everyone to embrace, especially engineers, engineering technologist, architects, designers, manufacturers, etc. Sustainable development became a common practice in many developed or developing countries. However, the concept has not been fully mainstreamed into engineering education. Several engineering societies within the U.S. or abroad have made declarative statements about their commitments to sustainable development [4-8]. Corporations, governmental agencies, and non-governmental organizations have shifted focus away from waste treatment and remediation towards energy conservation and efficiency, waste minimization, and pollution prevention initiatives. Given the increasing importance of sustainability, it is incumbent upon academia to educate future engineers on sustainability topics, i.e. incorporate sustainability into engineering curricula.

Declining environmental, economic, and social conditions worldwide require responses from the engineering and higher education communities, since many of the solutions to these problems are related to the discipline, particularly engineering design for sustainability. We have come to understand that the greatest and most immediate *sustainability problems* that humans face are related to our relationship with the natural world and the populations occupying the planet. Many universities are leading the way by incorporating sustainability into their university mission, and their curricula. In addition, many engineering departments have been exploring opportunities to incorporate sustainability and green design into their programs in novel and multifaceted ways. Recently, there has been an intense, global focus on the rising costs of non-renewable energy sources and the negative impacts that our societies have on the global environment. As fuel costs rise, newer technologies are driving down the costs of many of the more expensive renewable energy sources. These two factors are combining to make the renewable energy economically feasible for many, and the sustainable development a must. Furthermore, with the current economic and military situations, there is a renewed understanding that increased use of some non-renewable fuels is increasing national trade deficits and supporting undesired entities. Finally, scientific evidence continues to accumulate that shows climate changes that may affect our life standard. To many, these reasons signal that significant changes to the status quo are imminent. Embedding sustainability education into engineering requires addressing engineers' contribution towards net positive outcome by focusing on minimizing negative impacts on the environment and society. On these regards there are urgent needs to address the issues of sustainability and green design in engineering processes, projects and products, and one of the important ways of doing this is by incorporating the topics into higher education. These are main goals and objectives of our project.

The paper provides an overview of embedding sustainability and green into our curricula and programs. It begins with a general discussion on sustainability and education for sustainable development, then examine at the major driving factors that have motivated us to prioritize sustainability education. We then look at our project implementation as well as results from other institutions in order to elucidate the best approaches for embedding sustainability into our curricula. The last section concludes the paper.

LITERATURE REVIEW, ISSUES AND CHALLENGES

There is increasing motivation to incorporate concepts of sustainability into the undergraduate engineering curricula. The Accreditation Board for Engineering Technology, Inc. (ABET) [9,10], requires that graduates be able "to design a system, component, or process to meet desired needs within realistic constraints such as economic, environmental, social, political, ethical, health and safety, manufacturability, and sustainability" [5-7,11-21]. There are several initiatives to promote the sustainability of engineered systems and energy efficiency, while still improving the quality of life, looms more immediate. Not only accreditation boards and professional organizations but many corporations are concerned with sustainable development to reduce costs and liabilities and to create products that help improve the quality of life here and abroad [11-24]. Unfortunately, not too many institutes are offering courses or training related to the energy efficiency, sustainability, green design and renewable energy fields. However, many schools with undergraduate engineering programs are working to include environmental sustainability and sustainable design into their curricula [11-15]. The approaches for doing so can be subject- and/or problem-oriented, can be based on case studies, or could be part of a capstone experience. Sustainability can also be an opportunity to satisfy the general education component of ABET criteria.

Part of our project an extensive literature research on sustainability, green manufacturing, energy efficiency and renewable energy inclusions into the engineering and technology curricula, was conducted by the authors and the graduate students [5-7, 11-28]. The main findings were that courses concentrate primarily on smaller systems, particularly those limited to the firm (design for environment) or product (life cycle analysis/assessment), while less are addressing larger systems that examine relationships between different industrial sectors (industrial ecology) or between industrial and non-industrial sectors (cultural and social dimensions). The authors also found that sustainability engineering materials were taught to classes of predominantly upper division undergraduate and graduate students, and while discrete sustainable engineering courses seemed to be the most common approach, material is also incorporated throughout programs. The curriculum developed at our universities will provide the students with instruction about low-impact material and process selection (renewable, environmental impacts, embodied energy), energy efficiency (transportation, product processing and use), quality and durability, design for reuse/recycle, and social relevance/value, consumer value, and responsible production and procurement (fair labor). While the authors are aware of the inter-relationship of these categories, we have isolated each one to make them easier to teach, explain, and assess.

Sustainable engineering is, in many ways, more related to green engineering as it focuses on design that requires fewer natural resources, produces less (or no) waste, and reduces, reuses, or recycles products. Several approaches have been proposed clear educational objectives for incorporating sustainable development concepts into the undergraduate curriculum. The ABET accreditation criteria for engineering and technology programs also calls on programs to directly address issues of sustainability and environment [9, 10]. However, in an already crowded engineering curriculum, it is difficult to envision where and how these concepts can fit. In this paper, we refine the recommendations of others into yearly goals and objectives for the undergraduate curriculum [12-24]. Given the increasing importance of sustainability, it is incumbent upon academia to educate future engineers and other decision makers on such topics, i.e., incorporate sustainability into engineering and technology curricula. The sustainable development, green manufacturing and renewable option incorporates energy efficiency and conservation, energy sustainability, renewable energy, and pollution management will be related to industrial processes, processes of commercial, and light commercial and residential facilities. Students learn a combination of technical and management theories designed to prepare them for careers in the green economy sectors. Program offers practice based theory in the classroom, laboratory experimental design, and hands-on experience. Ideally, graduating engineers would be able to create a process flow and a life-cycle inventory for competing product designs with the aid of software tools. From the inventory, they are able to assess the relative merits of the competing designs based on ecological and resource indicators. These advanced skills draw on the students' knowledge of their discipline. But they also require a set of analytical approaches that encourage holistic, system thinking. In systems thinking, one pays attention to the inputs and outputs of the system and their effects on the surroundings.

These approaches can be integrated into the existing engineering curricula. Promoting sustainability through curriculum design involves articulating what students will take away from a learning experience and connecting this outcome to some kind of assessment. Curriculum developers are usually using systematic, consistent frameworks to organize learning outcomes, activities, and assessments. The curriculum products include syllabus, course outlines, learning modules, and assessments. These documents should be presented to learners upfront prior to any sustained learning experience whether it is a credit course, a workshop, or a seminar. Additionally, after a program and its associated courses are designed, curriculum maps show where program outcomes or other broad core abilities appear across courses. These deliverables are the vehicles by which sustainability is evidenced and communicated to learners and the public. In order to incorporate sustainability and green design principles and methods into engineering and technology curricula, we must understand the barriers that need to be overcome, which include: i) accreditation focus on quantity as opposed to process improvement, ii) conventional thinking of some faculty, and iii) company expectations and recruiting trends. While progress has been made in surmounting or reducing some of these issues, improvement is still needed.

The green manufacturing, renewable energy technology and sustainability skills and knowledge are required to meet the following objectives, energy efficiency and conservation, sustainable facilities, planning and design, sustainable green construction, sustainable manufacturing, and renewable energy sources such as biofuels, solar and wind energy technologies. These are selected because they represent the green technologies with the highest potential impact in our areas. Development of a workforce skilled in these areas is essential toward sustaining a green economy. Several interdisciplinary, sustainability-related courses have been proposed, underway to be developed or developed and offered to our undergraduate students, including renewable energy courses. Life cycle engineering and green design and manufacturing seminars are also proposed to be part of the endeavor. Although these courses have created a quite high level of awareness among our students, they are not required of all students, and no pervasive sustainability theme has been established in our curricula. The authors believe that more information and real-world learning experience is required in each curriculum to incorporate a sustainability theme. However, care should be taken and new innovative strategies should be explored.

Our literature review suggested that there is an inherent inextricability between sustainability as taught in a classroom and as practiced in the field makes curricula embedding both challenging for the institution and particularly relevant to the lives of students after graduation. We found that there is a growing interest in sustainability-related topics in universities, which may indicate a shift in the demands of students and employers. Many employers are looking to hire graduates that are socio-environmentally responsible. For the promotion, recruitment, and training of staff in terms of social and environmental responsibility, larger businesses were most demanding. According to the U.S. Bureau of Labor Statistics, employers need for graduates educated in concentrations related to sustainability is expected to increase at least 20 percent between 2008 and 2018 [3,5]. It is not unreasonable to assume that broad trends will remain for at least a decade. While many universities have signed declarations pledging support of sustainability initiatives, most have fallen short in regards to implementation. Most failed attempts to embed sustainability cite a lack of leadership, incentives, knowledge, or resources when trying to implement new programs or add them to existing curricula. We also believe that support from university for the sustainability embedding is necessary to maintain a cohesive plan and implement it successfully; however, other universities have also yielded positive results from "bottom-up" approaches [5-7, 11-22]. Professional development of academic staff is a major factor identified as contributing to a high success rate among schools that attempt sustainability implementation. Resistance from academic staff was most common when it became necessary to remove existing materials from curricula in order to make room for sustainability education.

Engineering practice and education are changing as social expectations and conditions changed. Skepticisms of public about professional engineers, national review of engineering education, requirement to communicate effectively, effective functioning in multi discipline and multi-cultural environments, involving ethical, social, professional, business, economic, cultural, global responsibilities that are changing engineering education processes. Threatened by the declining public opinion and respect for engineering profession, because of the effects of unsustainable solutions offered to the societies over the centuries, professional engineering societies (ASEE, IEEE, and others) across the globe have forced their members to follow green engineering principles in their professional practice. Pressure towards globalization is being reflected into new course accreditation criteria and higher education structures. Engineering educators have to live with the challenges that emphasize problem solving towards sustainable development. There is a consensus among professional organizations that sustainability is important to engineers and that they have to take it into account in their professional practice, but there has been no agreement on how this has to occur. Professional associations encourage all engineers to become knowledgeable of sustainability principles and to be continuously trained about current sustainable technologies applicable to their work. It is not quite easy to include sustainable modules into existing engineering, technology or management courses. The easy part is to identify a discipline and integrate concepts and applications of sustainability into existing courses. Many instructors are just not aware of what sustainability is about and/or how it relates to their discipline(s). The difficult part is to capture the attention of the faculty member and to make them aware of sustainability concepts, methods and practices in their research areas of and why it is important for solving problems today not in the future.

Engineering programs that include sustainability, green manufacturing and design are radically different from traditional programs in content, teaching and methodology, underpinning focus on graduates attributes than content knowledge. They aim to develop generic skills, attitudes and attributes of professional engineers. Sustainability underpins all areas of the course and addressed through core courses and practically through engineering practice-school components. Sustainability is central engineering concern. It is seen as a useful theme in addressing the recognized need for much greater attention to inter and trans-disciplinary, socio-technical interfaces and ethics, as a positive response to growing community concerns/distrust of engineering. Incoming students have to be oriented to new paradigm through introductory courses on engineering for sustainability, through which faculty's vision is defined. Subjects were proposed as the part of the core design and preliminary planning began on their objectives and their linkages with field of practice. However, achieving sustainable development involves inclusion of sustainability and equity in design criteria. Engineers should involve in problem framing. Engineers should appreciate the social, economic, political and environmental effects of technology they develop. Some of the sustainability issues fall in political spheres, while the traditional engineers are not adequately prepared and are commonly uncomfortable. Changes in engineering practice should address this issue. Future engineers should offer acceptable solutions by recognizing sustainability. They should work with professionals of other disciplines. Preparing such an engineer is the challenge of engineering and technology education. This may also be the case with any other discipline. Engineering and technology education has to prepare engineers who accept sustainability as a basic design requirement in the development of product and processes, and as policy criteria in future industrial developments. Engineers not only satisfy obligations to the clients, strive for excellence but also ensure that real needs for present end users are met and recognize impacts on environment and future generations. Sustainable development can be used to define wider design space. Wider design space leads to realization of holistic solutions. Engineers are mainly concerned with designing sustainable products and services. Irrespective of the faculty students are in, everyone working towards sustainability or trying to solve problems of sustainability should understand several principles underlying the concept of sustainability.

CURRICULUM CHANGES

Nowadays, society needs a change of paradigm towards a sustainable society. It is an urgent change that requires a commitment for all the involved stakeholders, public, industry, academia and government. To achieve this change, universities are playing an important role, having a huge responsibility. Engineering and technology graduates must contribute with effective solutions to the present energy, manufacturing and environmental problems, by developing new production ways to cover present and future needs for humankind, means that allow to reach and to maintain a fair and sustainable society. We believe that teaching green design, energy efficiency, sustainability, and renewable energy to engineering and technology students in integrated courses requires new approaches to course planning, lesson development, pedagogy, student expectations, and assessment. Equipping engineering students with the skills and knowledge required to be successful global engineers in the 21st century is one of the primary objectives of undergraduate educators. Enabling students to practice self-directed learning, to find solutions to design problems that are sustainable and to recognize that they are part of a global community are just of few of our educational goals. However, transforming higher education curricula for sustainable development is a tough challenge to the curriculum developer and course designer. They will have to deal with the complexness of the sustainability concept and integrate that into engineering education. Self-directed learning can define an individual's ability to practice life-long learning. It places the responsibility on the individual to initiate and direct the learning process and can enable an individual to adapt to change. Engineers have an enormous opportunity to help create a more sustainable society. By teaching engineering students sustainability principles, and equipping them with the tools to incorporate sustainability and green design principles and methods into their work, they can make better choices on materials use, energy use, and the lifecycle of their designs [16-33]. However, sustainability is still relatively new to engineering curriculum and is not an established practice for most of the engineering disciplines. Meanwhile, today's students have a strong desire to improve the world through their work, and sustainability connects with these interest and motivations. These students' hunger for knowledge often outstrips what is available in their courses and the experiences of their professors. Therefore, both students and professors are in search of easily accessible, actionable, low-cost sources of information on sustainability. Furthermore, to make sustainable design compelling to a wider base of engineering students, we need to craft sustainable design in terms of mainstream design problems that are important, cutting-edge, and achievable. Then we need to show them how to deal with environmental and societal needs and constraints as part of their design processes.

A two-level approach was taken in our curriculum changes to incorporate sustainability, green design and renewable energy subjects. A first level such topics, subjects and problems were introduced by the faculty involved in this project in their courses, where it is appropriate [22-33]. Projects for the senior design project capstone course sequence were proposed and directed by the project investigators. Limited financial support was also provided, via grant funds for these senior design projects. At second level several courses on green manufacturing, industrial energy systems, and renewable energy were proposed, developed or underway to be developed. However, the efforts to integrate sustainability, green manufacturing, and renewable energy into our curricula have met with some resistance, such as: 1) opposition to adding new courses or integrating sustainability, green design, and renewable energy contexts into already tight curricula, 2) the faculty inter-disciplinary expertise necessary to teach new sustainability, green design or renewable energy, 3) the resistance to revising existing senior design project courses, and 4) new laboratory infrastructure.

In our project developed or we intend to develop one problem solving (project-oriented) green energy course comprising different modules, and other two on renewable energy and industrial energy systems. In these courses students from University of Texas at El Paso (UTEP) and Drexel University (DU) work with engineers from local industry to form teams to work on problem-focused projects. The regional engineers may attend various module classes to earn the certificates for each designated topic. Moreover, access of virtual facilities and software will allow engineering and engineering technology students to explore the principles, characteristics and operation of various energy sources, storage devices, and energy conversion systems. The new learning materials will allow undergraduate students at DU and UTEP, to "see and feel" the emerging green energy technology through interactive laboratory experiments and learning modules. The new courses will address the needs of industry and the course mission in sustainable design technology and environmental systems: 1) To make clear the grand challenges driving the need for sustainable design, and to provide tools and examples useful for tomorrow's engineers to meet them; 2) Describe grand challenges technology poses for water, energy, and climate change along with the incentives and inhibitors to implementing sustainable design to meet these challenges; 3) Teach modeling approaches for estimating impacts of technology design choices on water pollution, air pollution, and global warming potential. Introduce life cycle assessment and predictive life cycle assessment as concepts along with software tools that facilitate their practice; and 4) Learn about sustainable design case studies that focus on the spectrum of product life cycle stages, as well as integrated end-use design, transportation, manufacturing, and consumer products that demonstrate sustainable design principles. A design project will have a positive impact on sustainability education or practice.

During this academic year for the next one we are in the process to develop or to offer two new courses, one in the area of renewable energy technology, while the second one is focusing on industrial energy systems, energy efficiency, and sustainability. The former one developed by one of the author was offered as a special topics course in 2009-2010 and 2010-2011 academic years [27-33]. The course mainly focuses on the wind energy, solar energy, photovoltaics and to a lesser extent on the other renewable energy sources and related technologies. Since this course deals mainly with the analysis and the components of the wind and solar energy conversion systems, as well the analysis of integration and interconnection to the power system grid, the desired prerequisites include a course in energy conversion, and co-requisites a course in power electronics and power system analysis. Students are expected to be well around in general renewable energy issues and technologies. They are expected to be particularly skillful in analyzing and solving wind and solar power systems and related problems. The aims and goals of the second course are to train students to use process integration methods and tools necessary for identifying and designing efficient industrial energy systems. Besides understanding technical and economic issues, students are achieving understanding of the impact of industrial process energy usage on the environment, and the role that industrial energy systems can play with respect to meeting pollutant emissions reduction targets. The course addresses use of methods to identify the cost-optimal mix of different process energy technologies to satisfy a given process energy demand. One important aspect is how future energy policy instruments are influencing these optimal solutions. Technical systems encountered in the course include heat exchangers, boilers, heat pumps, combined heat and power, and thermal separation units.

Sustainable engineering is, in many ways, more related to green engineering as it focuses on design that requires fewer natural resources, produces less (or no) waste, and reduces, reuses, or recycles waste products. Several approaches have been proposed clear educational objectives for incorporating sustainable development concepts into the undergraduate curriculum. The ABET accreditation criteria for engineering and technology programs in the United States calls on programs to directly address issues of sustainability and environment. However, in an already crowded engineering curriculum, it is difficult to envision where and how these concepts can fit. In this paper, we refine the recommendations of others into yearly goals and objectives for the undergraduate curriculum. Given the increasing importance of sustainability, it is incumbent upon academia to educate future engineers and other decision makers on sustainability topics, i.e., incorporate sustainability into engineering and technology curricula. The sustainable development, green manufacturing and renewable option incorporates energy efficiency and conservation, energy sustainability, renewable energy, and carbon management of systems will be related to industrial processes, processes of commercial, and light commercial and residential facilities. Students learn a combination of technical and management theories designed to prepare them for careers in the green economy associated with energy efficiency and conservation, renewable energy, and green manufacturing. This offers practice based theoretical knowledge in the classroom, laboratory experimental design and hands-on experience. Ideally, graduating engineers would be able to create a process flow and a life-cycle inventory for competing product or process designs with the aid of computational tools. From the inventory, they would be able to assess the relative merits of the competing designs based on ecological and resource indicators. These advanced skills draw on the students' knowledge of their discipline. But they also require a set of analytical approaches that encourage holistic, systems thinking. In systems thinking, one pays attention to the inputs and outputs of the system and their effects on the surroundings. These approaches can be integrated into the framework of existing engineering curricula.

Promoting sustainability through curriculum design involves articulating what students will take from a learning experience and connecting this outcome to some kind of assessment. Curriculum developers are using systematic and consistent frameworks to organize learning outcomes, activities, and assessment tasks. The curriculum products include syllabi and learning modules with associated assessment tasks. These documents should be presented to learners upfront prior to any sustained learning experience whether it is a course, workshop, or seminar. They are the vehicles by which sustainability is communicated. In order to incorporate sustainability and green design principles into our curricula, we must understand the barriers that need to be overcome. These barriers include: i) accreditation focus on "bean counting" as opposed to process improvement, ii) conventional thinking of some faculty members, and iii) company expectations and recruiting trends. While progress has been made in surmounting or reducing some of these barriers, much improvement is still needed.

In the authors vision, the green manufacturing, renewable energy technology and sustainability skills and knowledge are required to meet the following objectives, energy efficiency and management, sustainable facilities, planning and design, sustainable green construction, sustainable and green manufacturing, and renewable energy sources such as biomass, biofuels, solar power, and wind energy technologies. These components are selected because they represent the green technologies with the highest potential impact in our areas. Development of a workforce skilled in these areas is essential toward sustaining a green economy in these regions. Several interdisciplinary, sustainability-related courses have been proposed, underway to be developed or developed and offered to our undergraduate students, including renewable energy courses. A life cycle engineering and environmentally responsible design & manufacturing seminar is also proposed to be part of the endeavor. Although these courses have created a quite high level of awareness among our students, they are not required of all students, and no pervasive sustainability theme has been established in our curricula. The authors believe that more information and real-world learning experience will be required in each curriculum to incorporate a sustainability theme. However, care should be taken and new innovative strategies should be explored.

STUDENT PROJECTS

Engineering senior or capstone design courses fill a critically important role in the engineering and technology curriculum, forming a bridge between school and the workplace and have been extensively researched. These courses bring to the forefront many of the ABET outcomes such as lifelong learning, design, teamwork, and contemporary issues. On the other hand, even the sustainability is included in ABET's description of considerations for design but seldom incorporated into student projects [9, 10]. For the past years, our focus has been shifted towards incorporating more renewable energy, sustainability and green manufacturing topics in our senior design course sequence, as well as in adding research projects on these topics in several upper division courses, such as power electronics, energy conversion, power systems, etc. In the first senior design project term, we assign to our students the project topics related to renewable energy, power systems, or green energy manufacturing. These projects provide multi-disciplinary collaboration and valuable hands-on experience to the students. In addition to useful lessons on teamwork and project management, the projects provide working demonstration of green energy manufacturing systems. In the following will focus our discussion to the senior project courses, a three quarter sequence (24,28, 30-33). During the first fall quarter of the senior design course sequence, each team is given with guidelines for the senior design projects. Each team demonstrates the senior design proposal to the entire class and then a written proposal summarizing the proposed activities is handed-in as one of requirements of the senior project design course. Examples of the green energy manufacturing projects for the senior design course are:

1. High Efficiency Photovoltaic Charger Systems;
2. Indoor Solar Harvesting Energy for Sensor Network;
3. Solar Cell Surface Imaging for Quality Measurement;
4. Hybrid Energy Integrator for Green Factory
5. GPS-based Dual Axis Sun Tracking

To enhance the hands-on experience, the senior design course Sequence has been restructured as a project based course. Students are required to analyze, design, simulate, and build a completely functional system by the end-of-term project. The goals of all the projects are to explore and enhance student understanding of the fundamental concept of design-for-environment and hands-on learning of green manufacturing.

CONCLUSIONS AND FUTURE WORK

This paper provides an overview of steps to initiate and implement green engineering, renewable energy, sustainable and green design concepts and topics into our engineering and engineering technology programs. More than one approach is used or will be used to incorporate these concepts into our curricula. This consists of course adaptation to include the new concepts into the old curricula, developing new courses and laboratory experiments, proposing design projects with sustainability, green energy and design topics, not only into the senior design project course sequence but also into the project-based upper level courses. Assessments are underway or planned to be conducted all over the project duration. Lessons learned will be available for interested parties, instructors or researchers. While the present efforts are still in its early stages of our project, we believe that there is much to expand upon, within both engineering and STEM programs, considering the critical importance of understanding the complex reciprocal influences each sustainability and energy efficiency contexts have on the others aspects on engineering practice and education.

ACKNOWLEDGMENTS

This work has been supported by the US Department of Education under the joint DHSIP Program with the University of Texas at El Paso and Drexel University, PR/Award No.: P031S120131. The authors wish to express sincere gratitude for their financial support.

REFERENCES

1. Their Future is Green: The Clean-Energy Economy Promises: An Engineering Jobs Bounty – Training Graduates With Right Skills, *American Society for Engineering Education PRISM*, pp. 38-41, 4/2010.

2. Annual Energy Outlook 2013 with Projections to 2050 (Early Release) – Overview. Energy Information Administration, *U.S. Department of Energy*, from http://www.eia.gov/forecasts/aeo/

3. Rosentrater, K. A. and Al-Kalaani, Y., "Renewable energy alternatives – a growing opportunity for engineering and technology education," *The Technology Interface*, 6(1), 2006.

4. Allen, D. T., and Shonnard, D. R. (2003). Green engineering, *Prentice Hall*, Upper Saddle River, NJ, 397416, 2003.

5. Fokkema, J., L. Jansen, and K. Mulder, "Sustainability: necessity for a prosperous society," International Journal of Sustainability in Higher Education, vol. 6, pp. 219-228, 2005.

6. Fenner, R. A., C. M. Ainger, H. J. Cruickshank, and P. M. Guthrie, "Embedding sustainable development at Cambridge University Engineering Department," International Journal of Sustainability in Higher Education, vol. 6, pp. 229-241, 2005.

7. Lundholm, C., "Learning about environmental issues Postgraduate and undergraduate students' interpretations of environmental contents in education," International Journal of Sustainability in Higher Education, vol. 6, pp. 242-253, 2005.

8. *U. S. Energy Information Administration, Annual Energy Review*, 2010, http://www.eia.gov/energyexplained/index.cfm?page=renewable_home

9. Criteria for accrediting engineering programs: Effective for evaluations during the 2010–2011 accreditation cycle, ABET, Baltimore. *Accreditation Board for Engineering Technology* (ABET), Inc. (2010).

10. ABET, Criteria for Accrediting Engineering Programs, 2013, http://www.abet.org/accreditation-criteria-policiesdocuments/,

11. Felder, R. M., and R. Brent, (2004a), "The Intellectual Development of Science and Engineering Students. Part 1: Models and Challenges," *Journal of Engineering Education,* 93(4), pp. 269-277.

12. Felder, R. M., and Brent, R., "The Intellectual Development of Science and Engineering Students. Part 2: Teaching to Promote Growth," *Journal of Engineering Education,* 93(4), pp. 269-277.

13. McNamee, G., Careers in Renewable Energy, *PixyJack Press LLC*, 2008.

14. Walker, S, Sustainable by Design: Explorations in Theory and Practice, *London: Earthscan Publications* Ltd, 2006.

15. Clift, R., Sustainable development and its implications for chemical engineering, *Chem Eng Sci,* 61: 4179-4187, 2006.

16. Jennings, P., New directions in renewable energy education, Renewable Energy, Vol. 34, 2009, pp. 435-439.

17. Mulder, 2006, Engineering Curricula in Sustainable Development, An evaluation of changes at Delft University of Technology, European Journal of Engineering Education-Vol. 31, No 2, pp 133-144.
18. Carew, A.L. and CA Mitchell, 2002, Characterizing undergraduate engineering student's understanding of sustainability, European Journal of Engineering Education, 2002, 27(4), 349-361.
19. Holmberd J., et al, 2008, Embedding Sustainability in Higher Education through interaction with Lecturers : Case studies from three mEuropean Technical Universities, European Journal of Engineering Education, Vol.33(3), pp. 272-282.
20. Nehrir, M.H., A course on alternative energy wind/PV/fuel cell power generation, 2006 IEEE Power Engine ering Society Meeting, Montreal, Canada, 2006.
21. Santoso, S., and W.M. Grady, Developing an Upper-Level Undergraduate Course on Renewable Energy and Power Systems, Proceedings, 2005 IEEE PES General Meeting, San Francisco, CA, June 12-16
22. Belu R.G., Design and Development of Simulation System for Renewable Energy Laboratory, 2010 ASEEE Conference & Exposition, June 20 - 23, Louisville, Kentucky (CD Proceedings).
23. Belu R.G., and D. Koracin, E-learning Platform for Renewable Energy Sources, 2010 ASEEE Conference & Exposition, June 20 - 23, Louisville, Kentucky (CD Proceedings)
24. Belu, R.G., A Project-based Power Electronics Course with an Increased Content of Renewable Energy Applications, June 14-17, 2009 Annual ASEE Conference and Exposition, Austin, Texas (CD Proceedings).
25. Belu R.G., and A.C. Belu, A Decision Support Software Application for Design of Hybrid Solar-Wind Power Systems-as Teaching-Aid, 2007Annual ASEE Conference, and Exposition, Honolulu, Hawaii (CD Proceedings).
26. Belu R.G. – *Design and Development of Simulation System for Renewable Energy Laboratory,* 2010 ASEEE Conference & Exposition, June 20 - 23, Louisville, Kentucky (CD).
27. Belu R.G and D. Koracin – *E-learning Platform for Renewable Energy Sources*, 2010 ASEEE Conference & Exposition, June 20 - 23, Louisville, Kentucky (CD Proceedings)
28. Belu R.G. - *Renewable Energy Based Capstone Senior Design Projects for an Undergraduate Engineering Technology Curriculum,* 2011 ASEEE Conference & Exposition, June 26 - 29, Vancouver, BC, Canada (CD Proceedings).
29. Belu R.G. and I. Husanu - *An Undergraduate Course on Renewable Energy Conversion Systems for Engineering Technology Students,* 2011 ASEEE Conference & Exposition, June 26 - 29, Vancouver, BC, Canada (CD Proceedings).
30. Belu R.G. – *Embedding Renewable Energy into the Engineering Technology Curricula*, ASEE Mid-Atlantic Sec., Fall Meeting, Philadelphia, USA, Oct. 28-29, 2011 (CD Proceedings).
31. Belu R.G. and I. Husanu - *Embedding Renewable Energy and Sustainability into the Engineering Technology Curricula,* 2012 ASEEE Conference & Exposition, June 10 – 13, 2012, San Antonio, Texas (CD Proceedings).
32. Belu, R.G., R. Chiou, and B. Tseng - *Life Cycle Assessment – A Useful Tool in Green Manufacturing and Energy Education*, 6 th International Conference on Manufacturing Science and Education - MSE 2013 (Smart Sustainable Development), June, 12-15, 2013, Sibiu, Romania (CD Proceedings).
33. Chiou, R., R.G. Belu and B. Tseng – *Infusion of Green Energy Manufacturing into Engineering and Technology Curricula*, ASME 2013 International Mechanical Engineering Congress & Exposition, November 15-21, 2013, San Diego, California (CD Proceedings).

Proceedings of the ASME 2014 International Mechanical Engineering Congress and Exposition
IMECE2014
November 14-20, 2014, Montreal, Quebec, Canada

IMECE2014-39698

IMPLEMENTATION OF SIMILARITY FLOODING ALGORITHM TO SOLVE ENGINEERING PROBLEMS USING DIAGNOSTIC SKILLS TRAINING TECHNIQUE

A. Mehran Shahhosseini *
Associate Professor
Department of Applied Engineering
and Technology Management
Indiana State University
Terre Haute, IN, USA
Email: ashahhosseini@indstate.edu

Haisong Ye
Ph.D. Student
Department of Curriculum, Instruction
and Media Technology
Indiana State University
Terre Haute, IN, USA
Email: Haisong.Ye@indstate.edu

George Maughan
Professor
Director, Ph.D. in Technology Management
Indiana State University
Terre Haute, IN, USA
Email: George.Maughan@indstate.edu

Tad Foster
Professor
Department of Human Resource Development
and Performance Technologies
Indiana State University
Terre Haute, IN, USA
Email: Tad.Foster@indstate.edu

ABSTRACT

This article focuses on the development and field testing of instructional design concepts and educational software to teach diagnostic skills necessary to identify and solve problems in complex technical systems. Using concept mapping software along with expert-system programs, the overall software package enables technical workers and students in technology and engineering fields to benefit from personalized, iterative interactions that permit them to design visual maps of a diagnostic strategy and to allow direct and automatic comparison of their visual map to an expert's map.

The computer-based modules are developed in the Lectora authoring system and incorporate the VUE concept mapping software. The self-paced, interactive modules include an introduction stage, a visual mapping tutorial stage and a technical system and problem stage. Two technical systems will provide the context for a technical problem, the systems are: 1.) electrical power grid, and 2.) heat exchanger used in a waste plastic pelletizer machine. The educational software can be tailored to include other technical systems and technical problems.

This article demonstrates implementation of Similarity Flooding Algorithm (SFA) to solve engineering problems. SFA is used to match nodes and links between learner's and expert's concept maps. To compare two process maps, SFA needs to consider both the relations (links) between nodes and the content of the nodes.

In this article, we describe four improvements to the original SFA code to improve the comparison between two different concept (or process) maps: 1) Similarity of two strings is calculated by comparing the two strings character by character, which means that two strings get lower similarity if they are not the same exactly. During the comparison, WordNet® thesaurus is used to accurately evaluate the content of nodes; 2) Base similarity includes absolute similarity for each paired nodes, according to their links and content, but the overall similarity of the maps is calculated based on the relative similarity of each pair; 3) In a process map, the importance of each node could be different and important nodes have more weighting in overall similarity; 4) We consider two threshold values in the comparison algorithm. One is similarity threshold for the similarity based on the connections (links) of the nodes; the other is synonym threshold for the semantic similarity of

* Address all correspondence to this author.

content. If the similarity is lower than the thresholds, it could be considered as discarded. At the end, we analyze the collected data and show the effectiveness of the proposed technique to solve educational and training problems.

INTRODUCTION

It has long been understood that the best diagnosticians of problems in technical systems have both content expertise and diagnostic skills. In this paper researchers want to develop diagnostic skills within learners who have limited content expertise. The experimental process used in this paper will expose students to visual mapping and then allow them to develop visual maps to address simulated technical problems. After the students map out their proposed process, their map will be compared with an expert's visual map. It is hoped that the mapping process and exposure to expert reasoning will result in improved diagnostic thinking in general. Therefore, the focus of this paper will be how well a computer program can compare a learner's visual map with an expert's visual map and give the learner positive feedback using a process known as similarity flooding algorithm.

- **Understanding complex technical systems**

A technical system is a collection of human-made elements (hard, firm or soft). These artifacts function together, following a specific architecture (a typical hierarchy structure from large to small is: system, sub-system, component) within a specified boundary, to accomplish a goal(s) [1]. A system as such is a construct defined by individuals. For example, the boundaries of the simple technical system that makes up a refrigerator could be specified by the metal cabinet that encloses the components. The electrical power from a wall outlet, although primary to the function of the unit, can be excluded. If the entire refrigerator is considered to be a system, then the control components would be a sub-system as would the compressor elements, and so on. Because the main goal of the refrigerator as a technical system is to subtract heat from items placed inside, any ancillary components, such as a digital clock on the door which does not contribute to the main goal of the technical system, are not considered to be part of the refrigerator system. Boundaries could be identified around the technical components of a natural gas transmission system, an electrical power grid, a rail transportation system, or an information technology system thereby creating complex technical systems. There are no natural technical systems. Boundaries must be identified to envelop a system. The behavior of a technical system (simple or complex) is based on a machine logic whereas all components perform within their design parameters and all interactions among components perform within their design parameters.

A technological (socio-technical) system includes knowledge as well as man-made artifacts and involves humans as a critical component along with the natural environment, social, economic and other institutional elements which are inter-related and operates in a complex manner toward the achievement of goals, some of which are unspecified. System boundaries/environment is loosely defined. Some believe that characteristics of contemporary technological systems include autonomy and universalism, and that the existence of such systems is irreversible when compared to the specified goals of a technical system [1,2,3,4,5].

This project focuses on strategies that advance the diagnostic skills of technical workers who possess limited content expertise who are performing an analysis on problems in complex technical systems. A problem in a complex technical system exists when the performance of the system, subsystem, assembly or component is outside of the acceptable (specified) range of performance of that system, subsystem, assembly or component in the pursuit of the primary goal(s) of the system, and results in failure or substandard performance. Such problems are "hard" problems which can be identified within the context of the efficiency of means toward a defined end [6].

- **Diagnosing technical problems**

Engineers and technicians have long used decision-trees or trouble-shooting guides to assist the diagnostic process while identifying problems in technical systems. Some contemporary technical systems have built-in aids to monitor and ensure performance or to identify problems in the system, its components and/or the interaction among the components. Typically these embedded control tools sense performance and make adjustments. Some tools identify issues such as low oil pressure, increased bearing temperature, or slow data transfer speeds. However there are times when the condition of unspecified problems, cascading problems, or root-causes can remain undetected. Embedded performance control/problem identification elements may also experience difficulty in recognizing performance trends to anticipate problems. A common belief is that the best diagnostic process for complex technical systems involves using embedded performance reporting tools as well as human diagnosticians who possess both content expertise and diagnostic skills.

Technical system problems are analytical problems that require a single solution [7]. Creative problems, designing a bridge for example, usually has multiple solutions. It is the ability of technical workers to create a somewhat agile diagnostic strategy that has good potential to identify problems in technical systems so that these problems can be solved correctly and with efficiency. Jumping directly to problem-solving can lead to inefficient trial and error or large-scale parts/component replacement.

- **Instructional design and visual mapping**

This research focused on the development of interactive computer-based training modules which required the learner to create a diagnostic plan to identify a problem in a complex technical system. The learner created and submitted a process map (concept map) to illustrate his/her diagnostic plan using VUE concept mapping software (an open source project based at Tufts University: http://vue.tufts.edu) within an instructional shell created in Lectora. The concept mapping technique was selected as the main way to communicate diagnostic strategies because of its flexibility in creation and use, and that it forced learners to think in agile ways not typically practiced in the performance of technical work. These factors also provide a means for learners to consider alternative causes of problems they know little about but have observed small indications. In this case, these observations cues can lead to a diagnostic conclusion [8].

Cognitive development theory associated with higher-order thinking skills and problem-solving was integrated into a hierarchical thinking strategy used to assist learners to understand abstract and process concepts. Feedback theory was applied to the instructional design so that learners had the opportunity to receive feedback from experts, rubrics and through iterative feedback messaging.

This project creates situations within a case that require learners to discover and organize knowledge about the performance of a technical system to create a diagnostic strategy to identify problems. This approach is primarily based on Bruner's [9,10] concept of discovery learning and his connection of cognition to a theory of instruction which leads to concept attainment. The three simultaneous processes employed were, (1) the acquisition of new information about the technical system and the diagnostic practice, (2) applying this new knowledge to the case, and (3) checking their results (learner created diagnostic visual map) with a map from an expert and other feedback cues. The concept attained is actually the process concept, diagnosing.

Another connection to the theory of cognition and instruction employed in this project is Butcher and Sumner's work [11] on self-directed learning and sense making. In this case, learners use a self-paced computer-based training program to acquire a content overview of technical systems and diagnostic strategies. These learners then engage in deep-thinking to process and apply this new knowledge to create a diagnostic strategy.

Learners were provided an overview of the technical system in question including performance data about selected components, subsystems or interconnections. Integrated within this data are symptoms (cues) of a technical problem.

- **Intelligent feedback to learners and concept mapping comparison**

In the past decade, researchers were interested in using concept maps as a strategy for assessment [12] in variety of subjects, such as psychology courses [13] computer programming [14], and problem-based medical curriculum [15]. Results from these researchers show that concept maps are a valid means of assessment. However, inefficient manual evaluation of concept maps is an obstacle for the application of concept mapping in computer-based instructional situations.

Matching of the concept maps is a time-consuming and complex task and different techniques are used to automate this task. For example, for the matching one person could use SQL data types to determine what would be the similarity, or another person could use XML data matching which works based on the hierarchical relationships between data elements.

Because concept maps include nodes and links, both the relationships and the content of the nodes are important. Kuo-En, Yao-Ting, Rey-Bin, and Shui-Cheng [16] proposed a weighted concept, whose propositions are given a weight value from 0 to 1. The higher weighted value a proposition is assigned, the more important the proposition is. By comparing a student's map with a teacher's (expert's) map, each node in both maps gets a closeness index. Then based on the closeness index and weighted value of each node, a score of similarity index is calculated for each node. Using the similarity index, learner's comprehension of the node can be ranked into learned, partially learned, or misconception [16]. The limitation in this method is that the nodes in the concept maps are predefined. Students use predefined concepts and links to construct their maps, so the content of the nodes and links are not considered as a factor in comparison. The maps in this research are open – ended maps, so the algorithms of closeness index and similarity index are not applicable. However, the weighting mechanism is helpful to identify the importance of nodes in expert's map.

Melnik & Rahm [17] presented a method called Similarity Flooding Algorithm (SFA). SFA takes two graphs as input, and matches the nodes in both maps. The similarity relies on the intuition that nodes from two maps are similar when their neighbor nodes are similar. The output is a list of corresponding nodes with a similarity value. After a filter selection, the most optimized pairs are considered as the best matched nodes. SFA works on many types of graphs, such as data schemas, catalogs, xml ontologies, and concept maps. Particularly, SFA support open-ended nodes, which are required for this research. This algorithm works for directed labeled graphs, like a process map, which has arrows on links between nodes to indicate the order of the steps of a diagnosis strategy. The authors evaluated the accuracy of SFA and conclude that overall labor saving is above 50%, and actual savings might be higher [17].

Simpson and Dao [18] present an approach of string comparison with the meaning of the words – semantic similarity. The approach uses WordNet®-based Semantic Similarity Measurement (WSSM) as the database for synonyms. WordNet® is a database of English, an open source project based at Princeton University [19]. The authors use five steps to compute a somatic similarity for two sentences. The steps are (a) separating sentence into a list of tokens, (b) disambiguating part-of-speech, (c) stemming words, (d) finding the most appropriate sense, and (e) computing the similarity. Although Simpson and Dao [18] note there might be many limitations in the method and need to be improved, their method works fine for this research, because of the target learners in this project are trained to use terms in their process maps.

Based on the weighting mechanism, SFA, and semantic similarity of two strings, a combination of these three approaches is adopted in this research.

COMPARISON PROCESS

A process map contains nodes and links between nodes. To compare two process maps, the comparison needs to consider both the relations (links) between nodes and the content of the nodes. The researchers used a SFA (Open source java code developed by Melnik et al. [17] converted the java code to C#, pronounced as "see sharp") to match the nodes based on their relationships. The SFA (Figure 1) represents two input maps semantically in code first, then creates an initial map for the product of each node in both maps, calculates their similarities based on the links, and finally generates a list of best paired nodes according to the similarity of each pair. [17].

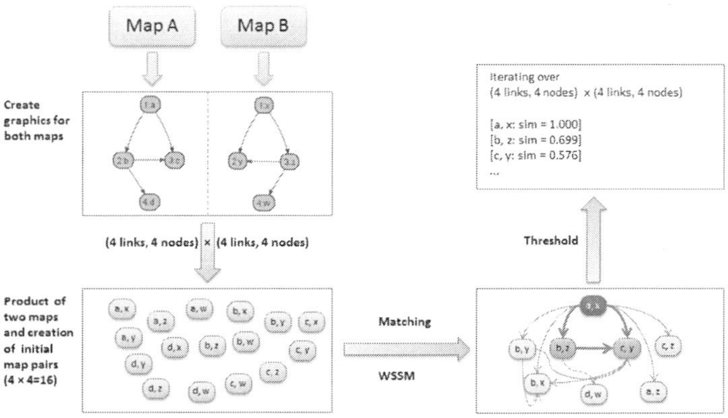

Figure 1: The process of SFA.

SFA Modifications

Four following modifications were made to the original SFA in order to utilize it for our application.

1. Dictionary

In the original code of SFA, the similarity of two strings is calculated by comparing the two strings character by character, which means that two strings will get lower similarity if they are not the same exactly. One improvement is to use synonyms in comparing strings. WordNet® provides a database of English words. While comparing two strings, the code can treat synonyms as similar based on WordNet®. The similarity of two words presents the degree of synonyms. We use a synonym threshold to filter pairs with lower similarity. If the similarity is lower than the target threshold, the two words will be considered as not similar. For example, when we compared "Cut the engine" and "Turn of the motor", we got a similarity of 0.29 in the original SFA method and a similarity of 0.8 in the WordNet® based method (synonyms threshold = 0.75). This method is more effective in comparing short phrases than in comparing long sentences. Therefore, we should encourage learners use single word or short phrase in each node when they develop their visual map.

2. Relative Similarity

In the case of the original SFA, the nodes near center appear higher similarity. Even the same map cannot get all notes paired at full (1.0) similarity (Figure 2).

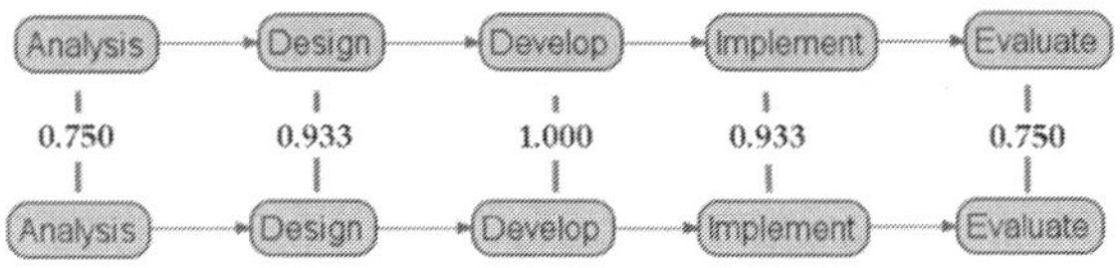

Figure 2: The result of the original SFA for the same map

Our solution is to generate a relative similarity based on the reference result from the existing expert maps. The process for comparing process maps is shown in Figure 3. The first step in the prototype program is to generate a reference similarity that is calculated based on the result of comparing an expert's map with the expert's map itself. The reference similarity includes absolute similarity (s_{aee}) for each paired nodes, according to their links and content. The second step is to compare a student's map with the expert's map. The results of comparison include absolute similarity (s_{ase}) for each node, which is considered as a matched pair with one node in the expert's map. The third step is to calculate the relative similarity (s_{rse}) for each paired nodes by the percentage of s_{ase} and s_{aee}. And, the overall similarity of the map is calculated based on the relative similarity of each pair.

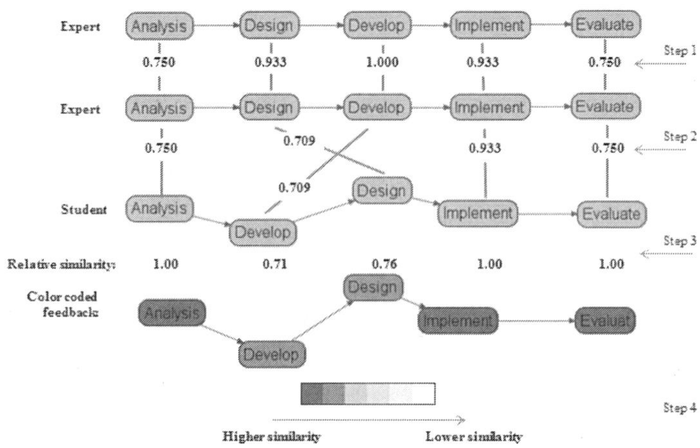

Figure 3: The process of comparison for process maps

The fourth step is to generate feedback. The feedback includes two parts. One part is a summary that shows the overall similarity, number of nodes in expert's map, number of nodes in learner's map, percentage of matched nodes, and the similarity range of matched nodes, the other part is a color-coded similarity, darker color stands for higher similarity and lighter color stands for lower similarity. A new map is generated based on the learner's map. Sometimes, learner's map does not have a high overall similarity because the learner's map has fewer nodes than the expert's, but the feedback still could be positive because the similarities of matched pairs are strong.

Figure 4 is an example of a summary feedback. The overall similarity is 26.21%, because the learner's map contains only 6 nodes and the expert's map has 14 nodes. However, the similarity range of matched nodes is from 40.07% to 98.58%. This means the nodes in the learner's map have high quality as to the nodes in expert's map. When learners get feedback, they can review and analyze which part of their maps might need to be adjusted.

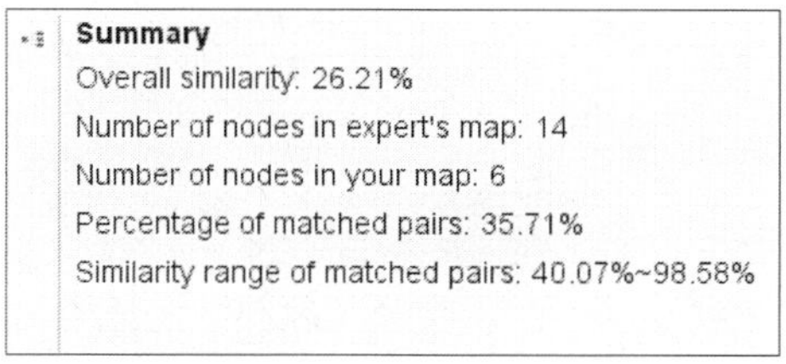

Figure 4: The feedback summary.

3. Weighting

In a process map, the importance of each node could be different. Important nodes can have more weighting in overall similarity.

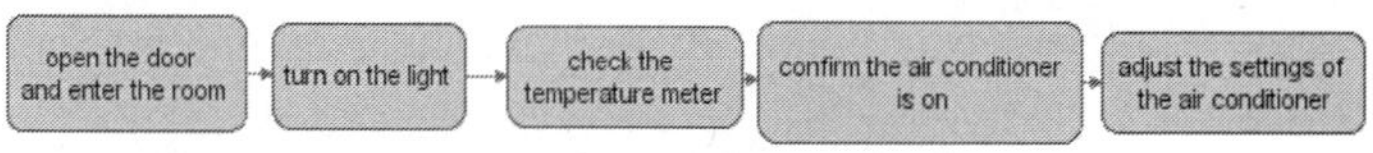

Figure 5: A sample process map

Take this as an example. As shown in Figure 5, if we think the first step is more important than the rest of the nodes, we can increase the weight of the first node. We used the coefficient value in SFA to control the weight of each node. The default value is 1; the highest value we tested is 2. By increasing the weights of the first and second nodes, the similarities of the first and third nodes increased (Figure 6).

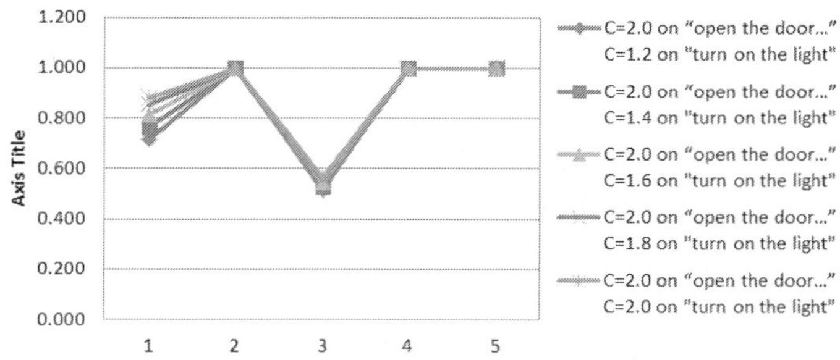

Figure 6: the similarity of expert's maps

Higher reference score needs higher similarity score to match. The coefficient value could be used for weighting, but more tests will be needed.

4. Threshold

There are two threshold values in the comparison algorithm. One is *similarity threshold* for the similarity based on the connections (links) of the nodes; the other is *synonym threshold* for the semantic similarity of content. If the similarity is lower than the threshold, it could be considered to be discarded. For example, if the similarity threshold is 0.25 and the similarity of a pair of nodes is 0.2, then the two nodes are considered as not similar at all. In the formative test, the similarity threshold was 0.0, and the synonym threshold was 0.75.

We had three students' maps in the formative test. The similarity values that came from the program, from high to low, is map1, map2, and map3 (Figure 7, original score). However, our experts thought the order should be map2, map3, and map1. We tried to adjust the threshold values for the same maps. The similarity threshold was changed to 0.25, and the synonym threshold was changed to 0.90. Based on the content of nodes, this pair of values worked better than the values in the formative evaluation, because more paired nodes with lower similarity were excluded. Figure 7 is the summary of the overall similarity values changed before and after adjusting threshold values. The result of user #1, which was not valid before adjustment, becomes reasonable; the overall similarity of user #2 and user #3 are just going down slightly and will not affect original result. To test the validity of the threshold values, more comparison data are needed.

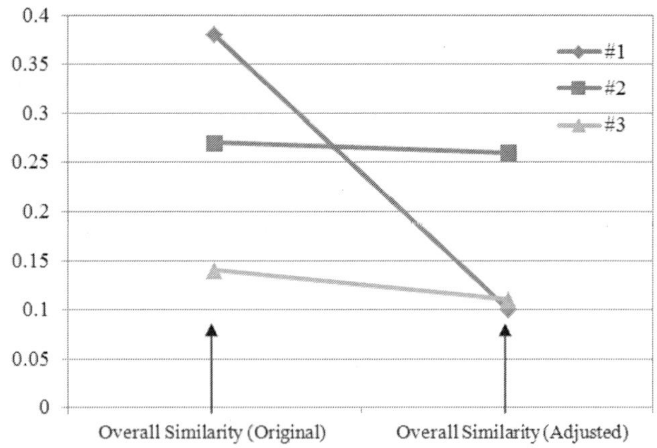

Figure 7: The overall similarity values before and after adjusted threshold values

RESULT AND DISCUSSION

The diagnostic skills training software is undergoing field testing at multiple universities. Data collection is still in process to increase the validity and reliability of the data. Preliminary indications show discriminating differences among the treatment group that does not get expert feedback and those who do. In addition, there is an overall positive result for the computer-based diagnostic skills training software and the ability of this computer-based training package to function as an interactive, stand-alone training package was found to be high.

Data Analysis - Case Discussion

Case 1

The first case is a technical problem in an electrical power grid. The problem was a technical problem in the power grid, but it was not an obvious solution for it. Students were asked to develop a visual map to explain how to diagnose the problem with the data presented by the training program. Figure 8 is the expert's map that was used in the program.

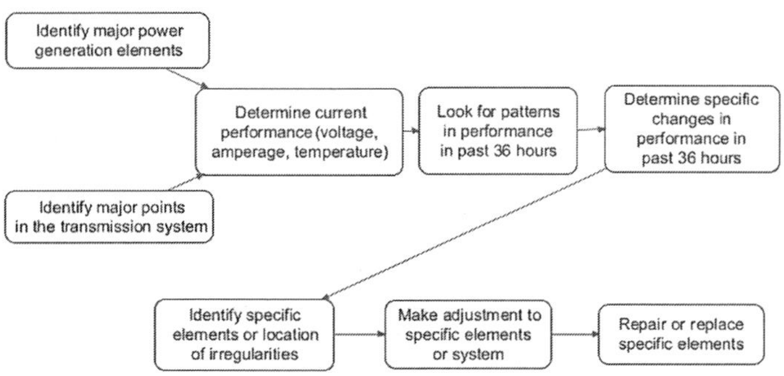

Figure 8. The expert's map for the electrical power grid problem in the training program

In this case, the first map comparison had an 18% overall similarity and 38% of nodes had considerable similarities to the expert's map (range from 29% to 70%). Figures 9 and 10 show the summary and the result. In Figure 10, there are three nodes with green colors. One is darker and the other two are lighter. Student should know which parts of her/his map were similar to the expert's map. This case was in the control group, which means that the student can only see the comparison feedback and make the second map.

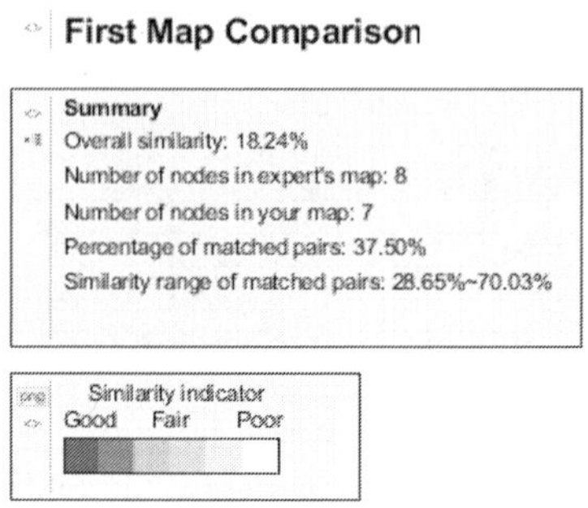

Figure 9. Summary of the first comparison

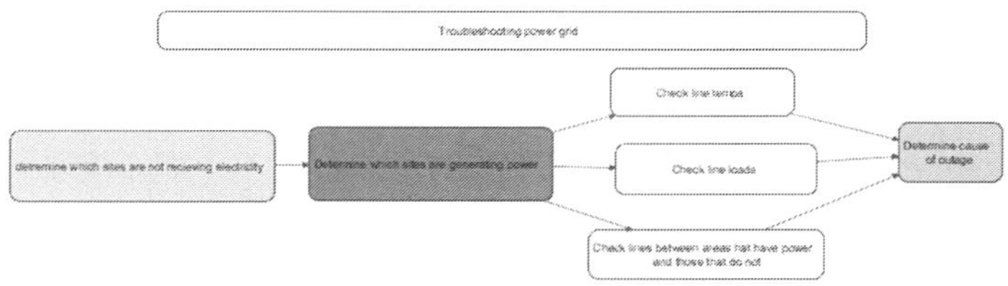

Figure 10. Result of the first comparison (large size figure is in Appendix)

Figures 11 and 12 show the summary and the result of the second map. The second map did not have a very high similarity, but the similarity increased by 14% (from 18% to 32%) and the matched nodes increased by 25% (from 38% to 63%). Although the highest similarity dropped down, the overall similarity was better. In the second map, the student changed a branch structure to a linear structure (Figure 12), which made the map with higher similarity to the expert's map that mainly had a linear structure.

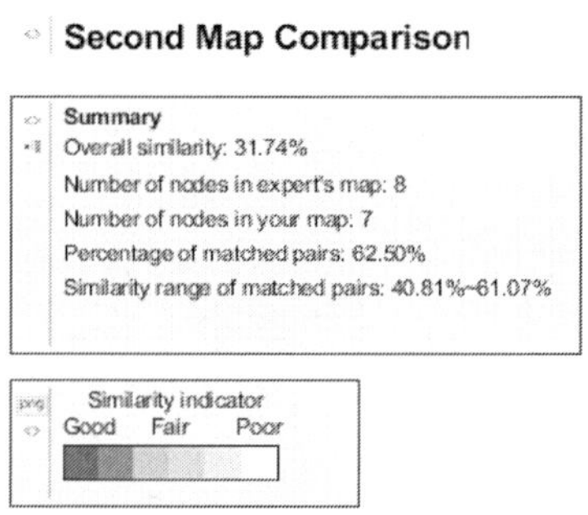

Figure 11. Summary of the second comparison

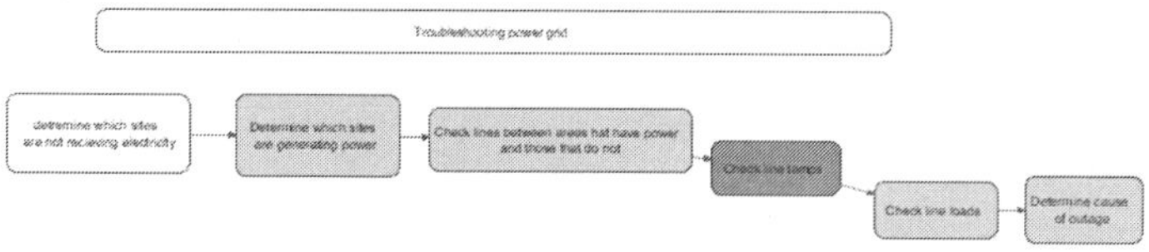

Figure 12. Result of the second comparison (large size figure is in Appendix)

Case 2

The second case is a technical problem of a heat exchanger in a waste plastic pelletizer. The problem was that the water bath system is malfunctioning and there were multiple reasons that could had caused the problem. The task is to develop a process map to diagnose the root cause(s) of the problem. Figure 13 is the expert's map used in the training program.

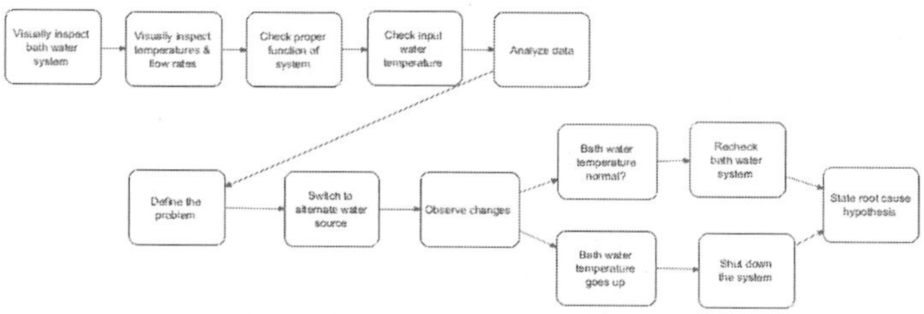

Figure 13. The expert's map for the heat exchanger problem in the training program (large size figure is in Appendix)

In this case, the first map comparison had a 36% overall similarity and 46% of nodes had considerable similarities to the expert's map (range from 51% to 88%). Figures 14 and 15 show the summary and the result. This map only has 6 nodes while the expert's map has 13 nodes. This case was in the treatment group 1, which means that the student can see the comparison feedback and access the expert's map, and then make the second map.

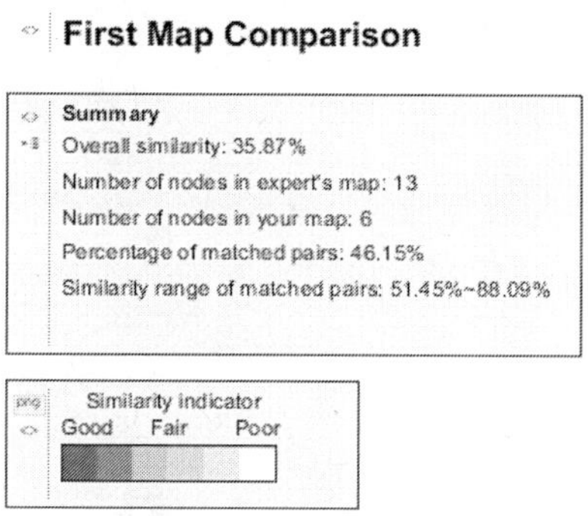

Figure 14. Summary of the first comparison

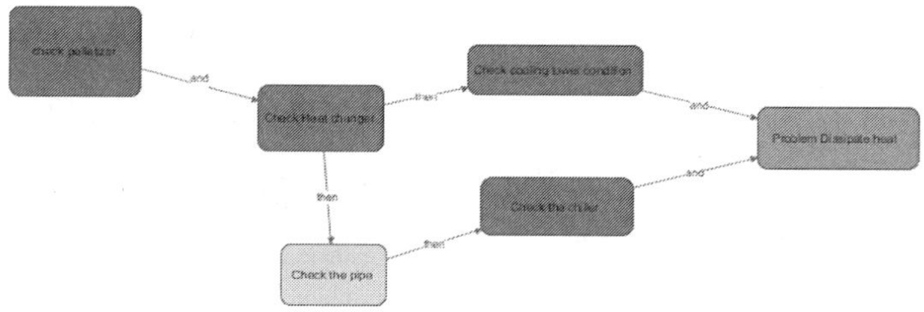

Figure 15. Result of the first comparison (large size figure is in Appendix)

The second map contained three additional nodes that were similar to the nodes in the expert's map. The comparison summary (Figure 16) shows that the overall similarity increased by 25% (from 36% to 61%) and the matched nodes increased by 23% (from 46% to 69%). The highest similarity of an individual node was 100%, which was "Check input water temperature" (Figure 17).

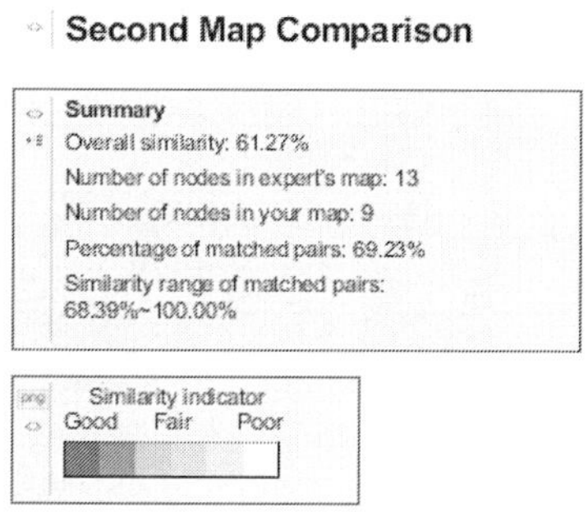

Figure 16. Summary of the second comparison

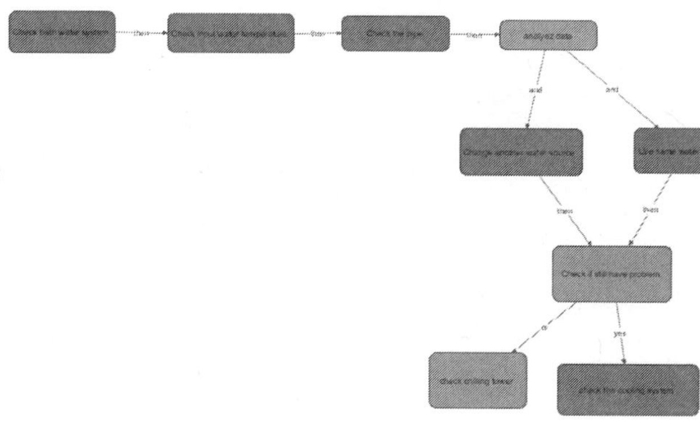

Figure 17. Result of the second comparison (large size figure is in Appendix)

CONCLUSIONS

Visual map is widely used in instructional activities, but the evaluation of visual map is time consuming and an instant feedback is not possible through manual evaluation. This paper presents a possible solution by integrating SFA and WSSM for automatic map comparison and providing instant feedback for students. It provides insight into one possible application of visual maps. The task of comparing two process maps is a challenge. It is difficult to gain a high accuracy score or degree on how much two maps are similar automatically by a computer program. Limitations in machine learning and human language understanding might be problems. One word could have different meanings in different situations and one line between nodes could represent contrast ideas. Because of the number of subjects in the formative evaluation, besides the algorithms, we found that numbers of nodes, correct connections, relative similarity, character similarity, weighting effect, and threshold values are factors that affect the results.

Although fairly high similarity between the two maps was achieved, additional research needs to be done to assess the full potential of the techniques used in this project. Overall, this project establishes a stepping-off position for those offering training in technical system analysis and trouble-shooting.

ACKNOWLEDGEMENT
The authors would like to offer their gratitude to College of Technology – Indiana State University and National Science Foundation. Partial support for this work was provided by the National Science Foundation's Transforming Undergraduate Education in Science, Technology, Engineering and Mathematics (TUES) program under Award No. 1140677.

REFERENCES

1. Moses, J. (2004) Foundatioal Issues in Engineering Systems: A framing Paper. Engineering Systems Monograph, MIT/esd.
2. Ellul, J. (1964) *The Technological Society.* New York: Vintage Books,
3. Churchman, C.W. (1971) *The Design of Inquiring Systems, Basic Concepts of Systems and Organizations*, Basic Books, New York.
4. Winner, L. (1977) *Autonomous Technology: Technics-out-of-Control as a Theme in Political Thought*, M.I.T. Press.
5. Hughes,T.P, and Wiebe E. Bijker and Trevor J. Pinch, eds. (1987) *The Social Construction of Technological Systems: New Directions in the Sociology and History of Technology*. Cambridge, MA: M.I.T. Press.
6. Checkland, P. (1981, rev. 1999) *Systems Thinking, Systems Practice*, Wiley.
7. Oaks, D. (2010). Common problems with basic problem solving. *Quality*, September, 36-40.
8. Waldmann, M. (2000). Competition among causes but not effects in predictive and diagnostic learning. *Journal of Experimental Psychology,* 26(1), 53-76.
9. Bruner, J. S. (1961). The act of discovery. *Harvard Educational Review*, 31, 21-32.
10. Bruner, J. (1960). *The Process of education.* Cambridge, Mass.: Harvard University Press.
11. Butcher, K.R. & Sumner, T. (2011) Self-directed learning and the sensemaking paradox. *Human-Computer Interaction*, 26, 123-159.
12. Williams, M. (2004). Concept mapping -- a strategy for assessment. *Nursing Standard, 19*(9), 33-38.
13. Jacobs-Lawson, J. M., & Hershey, D. A. (2002). Concept Maps As an Assessment Tool in Psychology Courses. *Teaching of Psychology, 29*(1), 25-29.
14. Keppens, J., & Hay, D. (2008). Concept map assessment for teaching computer programming. *Computer Science Education, 18*(1), 31-42. doi: 10.1080/08993400701864880.
15. Kassab, S. E., & Hussain, S. (2010). Concept mapping assessment in a problem-based medical curriculum. *Medical Teacher, 32*(11), 926-931. doi: 10.3109/0142159X.2010.497824
16. Kuo-En, C., Yao-Ting, S., Rey-Bin, C., & Shui-Cheng, L. (2005). A New Assessment for Computer-based Concept Mapping. *Journal of Educational Technology & Society, 8*(3), 138-148.
17. Melnik, S., Garcia-Molina, H., & Rahm, E. (2002). *Similarity flooding: a versatile graph matching algorithm and its application to schema matching.* Article presented at the Data Engineering, 2002. Proceedings. 18th International Conference.
18. Simpson, T., & Dao, T. (2010, February 8). *WordNet-based semantic similarity measurement*. Retrieved from Code Project: http://www.codeproject.com/Articles/11835/WordNet-based-semantic-similarity-measurement
19. http://wordnet.princeton.edu/wordnet/

APPENDIX

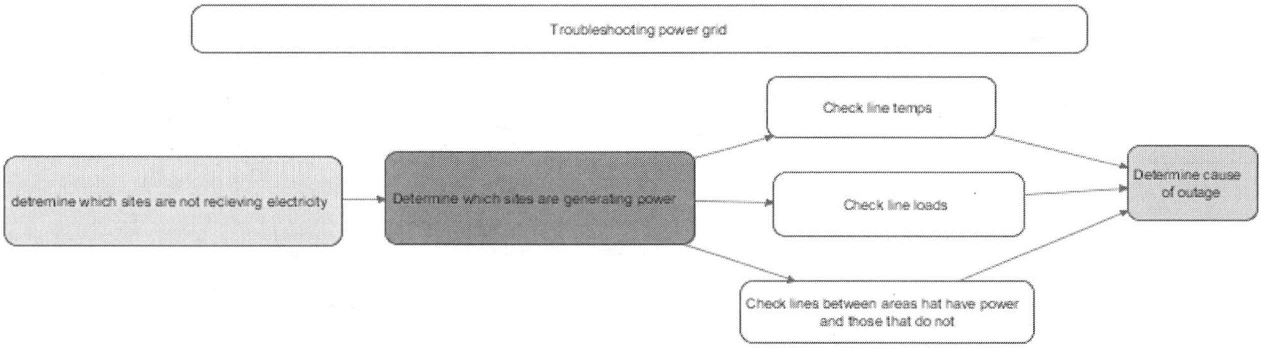

Figure 10. Result of the first comparison

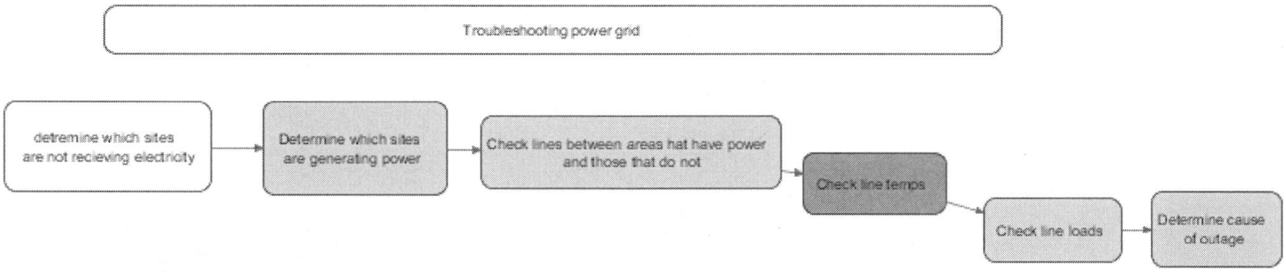

Figure 12. Result of the second comparison

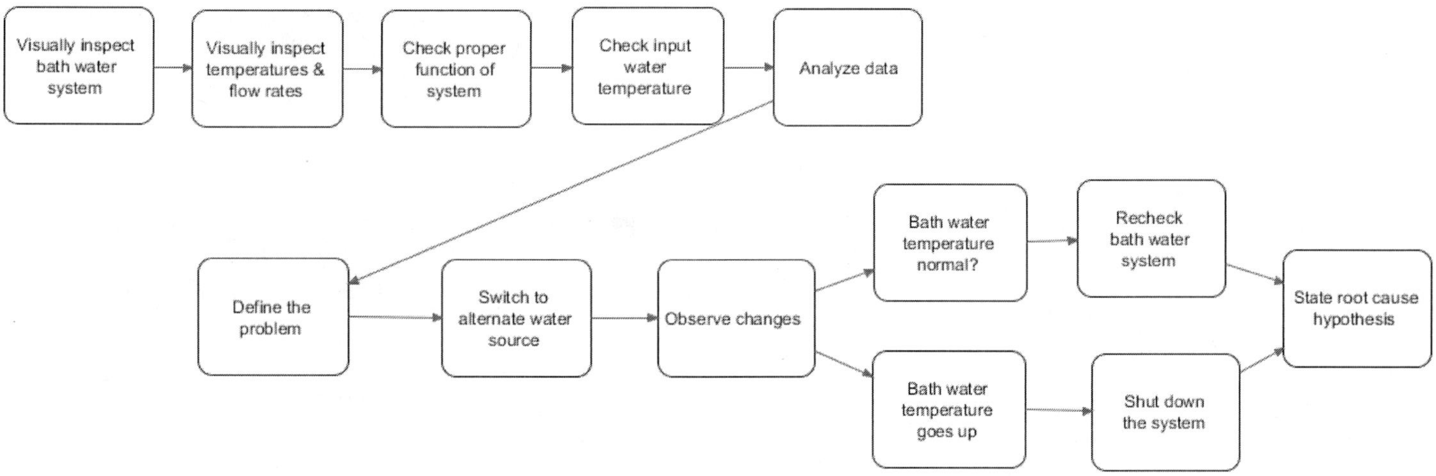

Figure 13. The expert's map for the heat exchanger problem in the training program

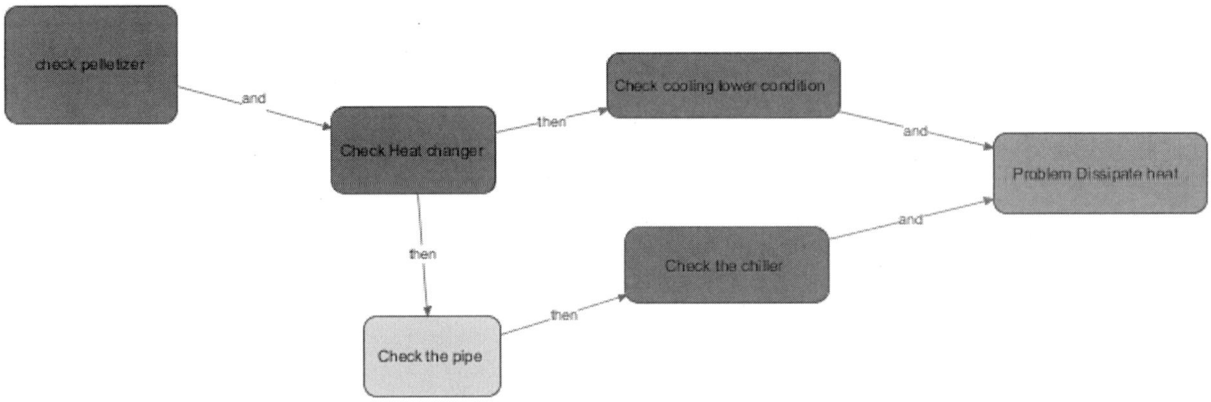

Figure 15. Result of the first comparison

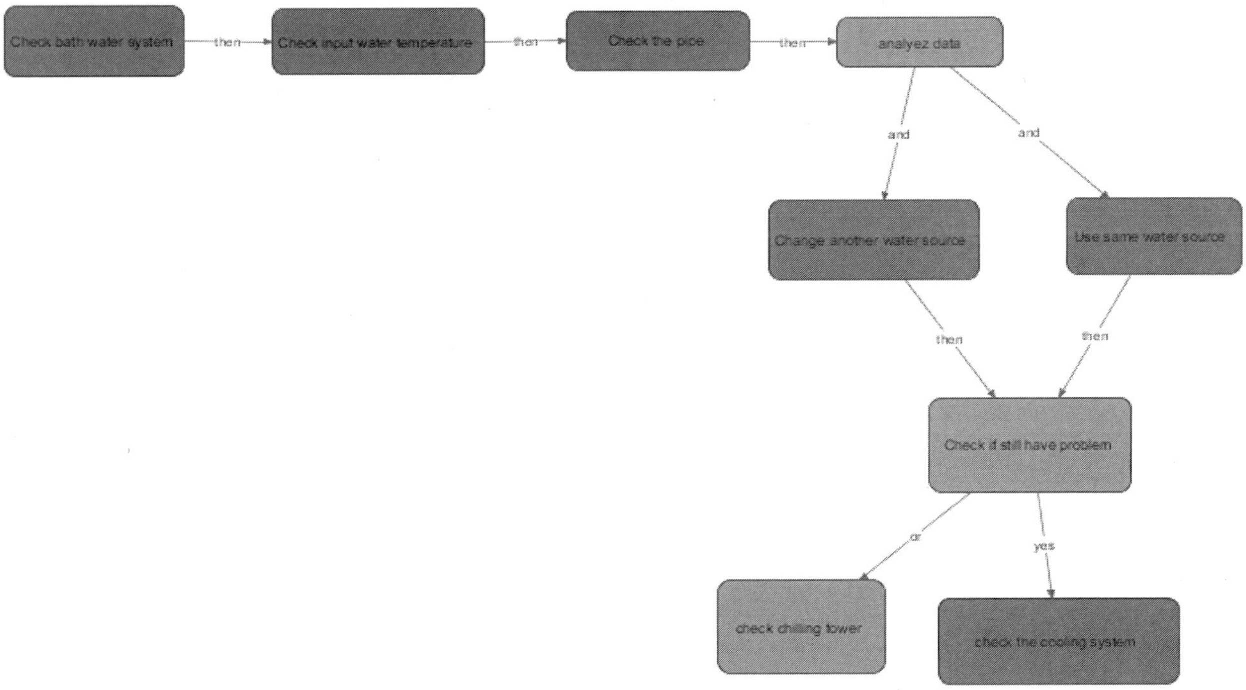

Figure 17. Result of the second comparison

Proceedings of the ASME 2014 International Mechanical Engineering Congress and Exposition
IMECE2014
November 14-20, 2014, Montreal, Quebec, Canada

IMECE2014-39922

DESIGN OF A HUMAN POWERED FLOUR MILL FOR EDUCATIONAL AND COMMUNITY EVENTS

Halsey Ostergaard, John P. Parmigiani
Oregon State University
Corvallis, OR, USA

ABSTRACT

Health education organizations use fairs and events to engage and educate the public about health and nutrition choices. Many organizations have interactive exhibits to attract and hold the attention of attendees. One successful example is a bicycle powered blender. Attendees take turns pedaling to produce fruit and vegetable smoothies that are given out as samples. Whole grain foods are also being promoted as healthy choices. This work describes the design, implementation, and testing of two prototype human powered flour mills for use at health education and community events. Because the mills will be used at public events, it must be safe and usable for a broad range of participant ages, sizes, and physical abilities. The mills should also be easy to transport, setup, and clean by volunteers unfamiliar with the mills. To facilitate teaching conversations, each prototype includes user power input feedback. One mill is based on a reclined bicycle format, constructed with a custom steel frame and standard bicycle components. Power feedback is accomplished with a ball proportionally levitated in a column of air. The second mill is based on a modified commercial rowing machine, and adapts the original power feedback system. The prototypes were mechanically tested, and evaluated by focus groups. Three commercial flour mills were evaluated on milling efficiency and suitability to this project. Recommendations are made for future versions.

INTRODUCTION

Many organizations, both public and private, are trying to address the continuing problems of obesity and ill health. Educating the public about making healthy food choices and increasing physical activity levels is critical to this effort. This outreach happens both in the classroom, and at public events such as health fairs. Organizations commonly employ interactive tools and free samples to gain and hold the attention of participants while engaging them in a meaningful conversation about their nutrition and health goals. Outreach groups have successfully employed devices like the Fender Blender, a bicycle powered blender (Rock the Bike, Berkeley, CA, USA). This device is mechanically simple, easy to setup and clean, and adaptable to a wide range of users. Encouraging participants to choose whole grain foods when possible is an important part of nutrition outreach. In order to engage the public specifically about whole grain nutrition and exercise, it was decided to build a human powered flour mill.

There are many nominally human powered flour mills currently for sale. They are typically hand crank operated table top models based on flat or conical metallic or stone burrs, which require high torque and operate at low speeds. One notable exception is optional bicycle power kit for the GrainMaker No. 99 mill (GrainMaker, Stevensville, MT, USA), but it is designed for adult sized bicycles, and places the mill close to the ground. Several organizations and individual craftspeople have constructed bicycle powered mills based on commercial hand crank mills or custom solutions. Most of these designs are too big to easily transport or do not meet minimum safety standards.

This project aims to design a human powered flour mill that is tailored to educational use, safe, easy to transport, and attractive. Two undergraduate capstone design teams were tasked with designing and implementing prototype mills with help from a graduate advisor. The development of a set of targets for the project is discussed. The project is broken into four subsystems for which several options are discussed. The two prototype designs were constructed using different options and are described below. Three flour mills are evaluated. Testing results and user comments are discussed, and suggestions for further refinement are made. Educational outcomes for both the advisor and capstone teams are summarized.

METHODS

Conversations with educators who will conduct events with the mill were crucial to setting design parameters. Before bringing in the capstone design teams, the project advisor met with the sponsoring health educators to verify the feasibility of the project, and determine the necessary features, timeline, and budget. The capstone teams were then brought in to develop their own set of requirements, which are described below.

The mill is intended for use by the general public at health education events attended by adults and children as young as 10, posing both ergonomic and safety requirements. In order to prepare food samples from the participants' milled flour, the flour product needs to be food safe, consistently fine enough for the desired recipe, and produced in enough quantity for each user to consume some. A production goal of 8g/minute of fine hard wheat flour was established for a young or minimally fit user at moderate exertion. Using the mill as an educational conversation piece imposed additional requirements. The milling process should be visible to both the user and surrounding participants. A power output indicator is included to facilitate conversations about food energy content and exercise energy consumption, as well as to encourage competition between users. To attract attention at events, the mill must also be visually appealing.

Four major subsystems were identified by the teams, and options for each are discussed in following sections:

1. Human Power Input and Transmission
2. Flour Milling
3. Work Input Feedback
4. Frame and Conveyance

HUMAN POWER INPUT AND TRANSMISSION

The style of human power input determines much of the overall layout of the mill. The maximum power output of girls of approximately the minimum target age is 97w [1]. For the purposes of engaging users, hard exercise is not desired, so the design power was set at 25w for minimum flour production. Several methods of human power input were explored: upright and reclined bicycles, hand crank, and rowing. The hand crank format was rejected based on preliminary experiments that indicated the force required to operate a direct drive hand crank mill with moderate crank size is too great for young users to sustain, even with the low minimum power requirement.

Upright and reclined bicycle formats utilize leg muscle groups, which are typically stronger and more adapted to continued use. The reclined bicycle format was chosen over the upright format based on the range of users expected. The upright format requires both handlebar reach and seat height adjustment to accommodate all target users, while the reclined format requires only seat distance adjustment, since the hand grips are attached to the seat.

Finally, a rowing type format was considered. Most rowing machines and competitive boats use a sliding seat design which allows users' legs, back and arms to provide power in a single pull. A one way clutch and spring winds the cord back around the drive axle for the next pull. While less familiar than a hand crank or bicycle format, the rowing motion is simple and potentially attention grabbing.

The two prototypes constructed utilized the reclined bicycle and rowing format, and can be seen in Figures 1 and 2 respectively. The bicycle format prototype uses a standard fixed ratio chain drive. The rowing prototype was based on a commercial rowing machine, and reused the one way bearing and spring mechanism from the original drive system. The resistance fan was removed, and a V-belt pulley was added to drive the mill.

FLOUR MILLING

Several methods exist for milling flour. Historically, grain was milled between two disc shaped stones, called burrs. One stone is rotated against the other. Grain is fed into the center, and is transported by radial or angled grooves towards the outside. The grooves and/or the surface texture of the burrs shear the grain into smaller pieces. Modern industrial milling is typically done by crushing grain with high speed rollers. Fine, consistent flour textures require multiple crushing operations with progressively smaller clearance, which makes roller mills unsuited for this project.

The grinding efficiency and mechanical suitability of three commercially available mills was evaluated for this project: the Wonder Junior Deluxe (The WonderMill Company, Pocatello,

Figure 1: The reclined bicycle prototype.

Figure 2: The rowing machine prototype.

ID, USA), Country Living Grain Mill (Country Living Products, Stanwood, WA, USA), and GrainMaker No. 99. Grain type, desired flour consistency, burr design, and burr size all impact the power required for milling. During preliminary testing, it was determined that the optional stainless steel burrs on the WonderMill model did not produce acceptably fine flour, so the mill was fit with the standard aluminum oxide grit burrs. Several grains were initially evaluated for required milling torque. Hard red winter wheat required the most torque, so it was used for further testing. The mills were initially adjusted to approximate commercial all-purpose whole wheat flour. Flour fineness was evaluated by passing the flour through a certified #30 sieve and calculating the percentage that passed by weight. Maximum torques required over one revolution were measured for each mill, and used to calculate a minimum flour production efficiency for each. The mills were also evaluated as part of a system: the bicycle prototype uses the GrainMaker model, while the rower uses the Country Living Grain Mill.

WORK INPUT FEEDBACK

To engage the user, both prototypes and the final design incorporate a power feedback system. Several bicycle fitness products are available that directly measure input power, based on strain gauge instrumented crank arms, axles, or wheel hubs. These products are expensive and have complicated displays and interfaces unsuited to casual use by many individuals. Exercise equipment typically includes work input feedback based on a much simpler velocity sensor, depending on pre-programmed resistance measurements to convert to power or energy. To engage users in a brief conversation, neither precise nor accurate values required, so it was decided velocity based power measurements would be acceptable.

The bicycle prototype uses a unique solution. A small fan is driven via a step up pulley from an auxiliary gear. The fan blows air into a clear tube with a series of small holes drilled along the length. Air volume increases with fan speed, lifting a table tennis ball contained in the tube. At a given pedaling speed, the ball will reach an equilibrium height dependent on the number of exposed holes in the tube.

The rowing prototype utilized the calorie counter that came with the machine. Drive ratios were chosen so that the torque applied to the mill approximated that of the original resistance unit.

FRAME AND CONVEYANCE

The mill will be frequently transported by volunteers to events, so it is important that the mill design be compact enough to fit in a variety of vehicles, light enough to easily load, and easy to roll from parking to event. The frame also needs to be robust enough for heavy users, and stable for particularly active users.

The cycle prototype frame is made from steel and split in two pieces, joined by pin at the front diagonal braces. The seat slides on an aluminum track and has a spring loaded pin stop to hold adjustments. Pedal mounts and bearings are standard bicycle parts. Casters are attached to the front leg and contact the ground when the rear end is lifted. The mill base slides into machined slots and is retained by pins.

The rowing machine that the prototype is based on was designed with a folding joint in the center of the beam, and integrated wheels below the mechanism housing. These features were retained along with the ergonomic geometry of the original. The rower design is inherently adjustment-free. The mill mount is fabricated from aluminum and includes a belt tensioning assembly.

RESULTS

The results of the milling efficiency measurements are presented in Table 1. The three mills produced similar quality flour as evaluated by the percentage passing through the sieve. Several issues were identified during testing. The Country Living model did not hold its grind setting through several experiment runs. Flour fineness was evaluated after each run, and only runs with fineness similar to the other mills were accepted. The other mills held their settings across experimental runs. The substantially lower efficiency of the WonderMill model, possibly due to its reliance on bushings instead of bearings, led to its exclusion from testing in the prototypes.

Long term durability issues were identified with the remaining mills. Both manufacturers recommend cleaning with brushes and compressed air rather than liquid. However, the mill will be used publicly and cleaned by volunteers, so it was decided that the final design should be cleanable by common foodservice washing and sanitizing practices. The grain feed tube of the GrainMaker model, and the burrs of the Country Living model are manufactured with alloy steel, which will eventually corrode with repeated washing and sanitizing. Both mills also use sealed ball bearings that are not rated for long term liquid resistance.

Both frame systems were adequate during testing with users up to ~130kg, though the rower deflected substantially at the hinge. The rower weighed 28kg, and the cycle 44kg.

Volunteers from ages 11 to ~60 used and evaluated both prototypes. Several adult volunteers were also asked to transport, setup, and clean the mills as well. Comments were generally positive, but some issues were reported with both prototypes.

The ball in tube power indicator on the cycle prototype was a popular feature that drew attention during a public test. Despite the positive reviews, the ball in tube option displayed substantial non-linearity and hysteresis, making it extremely difficult to calibrate. Users also noted several ergonomic and safety problems with the cycle prototype. Through the prototype stage, the chain drive remained exposed and at hand height for small

Table 1: Mill Efficiency and Flour Quality

	Max Torque (Nm)	Efficiency (J/g)	Fineness (%weight)
WonderMill	20.1 ± 0.8	174 ± 15	87
GrainMaker	14.2 ± 0.4	50.8 ± 0.9	89
Country Living	12.2 ± 0.7	70 ± 7	84 ± 2 (avg.)

children. While sitting in a more forward position, the knees of several users made contact with the mill mounting bracket. The long hand grips were comfortable in use, but made getting on and off difficult for some users. Comments indicated a preference for the fewer and lighter large parts of the rowing design.

Some users were uncomfortable with the unfamiliar rowing machine format. One adult user noted that the rowing motion irritated an existing lower back injury. The rowing machine format also relies on a user's hands to transmit force to the grab bar. While not encountered in the focus groups, some users with weak or injured hands or wrists may not be able to pull the grab bar. Several users commented that the power display was small and difficult to read in daylight. The rowing machine prototype was also low to the ground, making it less visible to passers-by.

At the conclusion of the undergraduate capstone design phase, the graduate advisor worked with the project sponsor to identify the issues that impact the overall goals of the project. Once identified, these issues required only minimal work to resolve. Modifications implemented include drive ratio changes to the rower mill, geometry changes to the bike mill, durability improvements, and polycarbonate safety shields for the drive train of both mills.

CONCLUSIONS AND RECOMMENDATIONS

Three commercial flour mills were evaluated for efficiency and flour quality, as well as their suitability for repeated cleaning. The WonderMill model was substantially less efficient. The other two mills had components made from alloy steel that will corrode over time, and bearings that may deteriorate with repeated washing. A hybrid model is recommended for this application. The powder coated aluminum body of the Country Living mill paired with the burrs and adjustment mechanism of the GrainMaker model will substantially increase durability. If it does not substantially impact milling efficiency, replacing the bearings with food safe PTFE or bronze bushings may increase durability as well.

Two undergraduate teams were tasked with developing the prototypes. The two products were evaluated by the students' own criteria, by volunteer testers, and by the project advisor and sponsor. The mills were found to function well. Both mills exceeded the target flour production for the intended age range, and were sufficiently easy to clean and transport. Only minor modifications were necessary to meet the sponsor's requirements.

It is recommended that future versions of this project be based on the reclined bicycle design, largely because user comments indicated that it was more approachable. The ball in tube design was more visible and attracted positive attention. To address negative comments, several changes are recommended. The frame should be fabricated from aluminum instead of steel, and be designed to fold instead of disassembling. The milling unit should be relocated to the top of the front pillar, directly driven by a chain enclosed in a removable cover, and mounted with quick release fasteners. A third mill was requested by a different sponsor after seeing the success of the capstone teams' products. Currently being designed by the project advisor, it is based largely on the bicycle mill prototype, and implements many of the improvements discussed here.

In undertaking this project, the capstone teams were given a chance to take a project from idea to finished product. This was an opportunity to synthesize and practice knowledge and skills acquired during their years of coursework. Starting with their initial interviews with the sponsor, they were required to translate an idea and some rough parameters into testable requirements, then research, design, implement, and test their project in a two quarter span. The research phase included some experimental work to determine design parameters in addition to identifying and evaluating existing systems. Throughout the project, time management, group communication, and supply chain planning were critical to meeting targets. The required written reports gave the students a chance to practice their technical writing. The students received feedback on their writing from both a writing instructor and technical critique from the project advisor. The quality and clarity of the written reports improved greatly during the project.

Part of the success of this project can be attributed to the organizational structure, which was set up to meet the needs of all of the participants. The inclusion of a graduate student advisor assured that the sponsor's final product was acceptable. The advisor gained experience in project design cycle management and customer communication, while serving as a resource to the undergraduate capstone design students. The capstone teams got the experience of taking a project from concept to final product.

Beyond educational uses, this project also has potential applications in rural or developing areas. Many parts of the world still do not have access to reliable electricity or are located far from established mills. Mills similar to these could offer communities inexpensive, efficient, reliable, local milling.

ACKNOWLEDGEMENTS

The following people and organizations have graciously supported this project with time, expertise, funding, and patience: The Moore Family Center for Whole Grain Foods, Nutrition, and Preventative Health, Bob's Red Mill, Capstone Design Teams 42.1 and 42.2, and Rebecca Ostergaard.

REFERENCES

[1] Dencker, M. et. al., 2008, "Maximal Oxygen Uptake Versus Maximal Power Output in Children," J. Sports Sciences, 26(13), pp. 1397-1402.

Proceedings of the ASME 2014 International Mechanical Engineering Congress and Exposition
IMECE2014
November 14-20, 2014, Montreal, Quebec, Canada

IMECE2014-36280

ASSESSMENT DEVELOPMENT FOR ACCREDITATION OF AN INNOVATIVE MECHANICAL AND ENERGY ENGINEERING PROGRAM

Reza A. Mirshams
University of North Texas
Denton, Texas 76207, USA
Reza.Mirshams@unt.edu

Yong X. Tao
University of North Texas
Denton, Texas 76207, USA
Yong.Tao@unt.edu

Xun Yu
University of North Texas
Denton, Texas 76207, USA
Xun.Yu@unt.edu

Azize Akcayoglu
Mersin University
Tarsus, Mersin, Turkey
Azize.Akcayoglu@gmail.com

ABSTRACT

The addition of Energy to the Mechanical Engineering curriculum created a new mechanical engineering model of engineering education in the baccalaureate-level and an opportunity for providing the academic foundation for successful career preparation and lifelong learning for the students. The curriculum has been designed with a system-level approach to traditional mechanical engineering based design, on the fundamentals of undergraduate level engineering within the mechanical engineering discipline, and has provided experiential-oriented approaches for the better understanding of classical mechanical engineering principles. The interdisciplinary nature of energy systems and mechanics requires a cross-cutting education that draws from the synergy of traditional disciplines of mechanical engineering, materials engineering and manufacturing, and computational methods and engineering. We are going to present the outcome based assessment in undergraduate level and discuss components of the program from freshman to senior years, and our successful implementation in developing student learning outcomes assessment, and evaluation approach for ABET accreditation.

INTRODUCTION

Global competition has fueled rapid technology advancements, spurring innovative engineering approaches and ideas as well as placing an increased importance on developing an educated undergraduate work force in Science, Technology, Engineering, and Mathematics (STEM) fields. With increasing scientific and engineering knowledge and the volume of information that should be taught in undergraduate engineering education curriculum, there is a need for a better understanding of the alignment between what we teach in engineering education and what industry seeks in their employees. The success that engineering graduates will have in their careers will depend on their understanding of foundational knowledge, problem solving skills, capability for quantitative reasoning and modeling, communication skills, ability to work in a team environment, and their ethical practices; each of which are considered desired attributes by employers. The graduation rate for STEM fields is lower than other disciplines especially for underrepresented groups (URG). A major barrier in engineering workforce education continues to be recruitment and retention of URGs, in particular female students, to the engineering profession [1-3].

The Mechanical Engineering (ME) discipline has been the catalyst profession for many industries in the past two centuries. Machine design, thermodynamics, combustion, and fluid mechanics have been the basic stem of the ME discipline. In the past, ME contributed significantly to the interdisciplinary promotion of industry and the economy, and today it continues to play that role [4]. The vision of 2030 task force by American Society of Mechanical Engineers (ASME) has reviewed in detail innovation in mechanical engineering and mechanical engineering technology curricula and has recommended components for engineering education [5-6]. The establishment of the College of Engineering at the University of North Texas (UNT) in 2003 with transfer of three departments from the College of Arts and Sciences, as well as the addition of the departments of electrical engineering (EE) and mechanical and energy engineering (MEE), created an excellent opportunity for the development of a modern approach to engineering education. This approach would consist of researching the efficacy of modifying current curriculum content to align with industry need for both existing and newly developed programs. The implementation of this alignment along with project-oriented electrical engineering curriculum coupled with the integration of a "Learning to Learn" component created a unique opportunity to generate an "out of the box" education approach for engineering education; one which keeps pace with the visionary future for "The Engineer of 2020" and alignment with ABET updated requirements for student learning outcomes and evaluations for continuous learning quality feedback.

While experienced engineering educators write about the urgency of reform, and speak of grand challenges in engineering education, most engineering departments have been faced with the challenges of major curricula reform, a process that Robert Maynard Hutchins, president and chancellor of the University of Chicago (1945-1951), likened to moving a cemetery. Even the large engineering coalitions have had little success in institutionalizing national reforms. While NSF programs for reform of engineering education have had some effect, these changes have not resulted in "the [kind of] comprehensive and systemic new models for engineering education reform" that everyone had predicted. This routine resistance to change has continued unabated to the point of bringing into question whether any engineering reform effort can have a national impact.

The movement for quality assurance in engineering education in the United States began in the early 20th century. From this movement came an organization that has been evolved from the Engineers' Council for Professional Development (ECPD) in 1932 to its current name as ABET, Inc. [7]. ABET provides world leadership in assuring quality and stimulating innovation in professional education of Engineering; Engineering Technology; Applied Sciences; and Computing. ABET is a non-governmental organization, and accreditation is a peer-reviewed process by volunteers. ABET has been invited for the review of non-U.S. educational programs worldwide.

Develop effective methodologies for assessing learning improvements: Instructional learning objectives are primarily focused on imparting skills such as problem solving, communication, team work, self-assessment, life-long learning, and problem-based learning. The assessment of these techniques is widely reported in the literature. The assessment methodologies should be tailored for the students in the program in different levels and it could be designed to determine learning in the topics in the course level. Learning objectives at a topic such as Finite Element Methods may include problem-solving skills. At the course level, for example problem-solving and problem-based learning may be assessed. At an engineering program level, such as a ME program, all the skills listed above should be assessed [8,9].

Program level assessment determines if the student could be a productive engineer and valuable member of society. It addresses the fundamental question "Have the students developed the critical skills needed for their professional careers". It is worth looking into the ABET assessment strategy before we go further. ABET criteria defines a minimum set of student outcomes for engineering programs. While the ABET assessment captures the students' learning at topic and course levels, ABET's program-level assessment addresses only a minimum set of requirements. Thus, more efforts are needed to design assessment at the program level. Results of recent research in educational methods are useful to fill this gap.

Methods for assessment suggested in the recent literature include surveys, interviews, conversations, observation, ethnographies, and meta-analysis. These are beneficial only if the instructors are creative in designing activities that will inspire and keep students engaged in the learning processes within the classroom and beyond and if students participate actively in such activities. The measurement of instructional effectiveness is not required by ABET, however, the sharing the results of teaching effectiveness and discussions on improving pedagogical strategies could help to implement better student learning environment.

In this paper, we present an example of an outcome based assessment development process used successfully for the ABET accreditation review of the Mechanical and Energy Engineering BS degree programs at the University of North Texas [10].

ABET REQUIRED ASSESSMENTS

ABET accreditation requires to address quality assurance in general engineering education and specific engineering discipline program requirements. In general, education requirements, the Baccalaureate level degree programs should satisfy eight criteria: Students, Program Educational Objectives, Student Outcomes, Continuous Improvement, Curriculum, Faculty, Facilities, and Institutional Support. Additionally, some specific criteria have been developed by the professional society for specific programs in professional disciplines that must be satisfied in the accreditation reviews.

The most important part of accreditation review is the evaluation of assessment process for quality education assurance, as it is specified in three criteria: (a) Criterion 2: Program Educational Objectives; (b) Criterion 3: Student Outcomes; and (c) Criterion 4: Continuous Improvement.

Criteria 2: Program Educational Objectives

"Program Educational Objectives are broad statements that describe what graduates are expected to attain within a few years of graduation. Program educational objectives are based on the needs of the program's constituencies."

Criterion 3: Student Outcomes

"Student Outcomes describe what students are expected to know and be able to do by the time of graduation. These relate to the knowledge, skills, and behaviors that students acquire as they progress through the program."

Criterion 4 Continuous Improvement

"The program must regularly use appropriate, documented processes for assessing and evaluating the extent to which the student outcomes are being attained. The results of these evaluations must be systematically utilized as input for the continuous improvement of the program. Other available information may also be used to assist in the continuous improvement of the program."

ASSESSMENT PLANNING

1. Establish Program Educational Objectives:

The Program Educational Objectives (PEOs) of Mechanical and Energy Engineering were established by faculty and approved by the Industrial Advisory Committee as follows:

Objective 1: Graduates are successfully employed in mechanical and energy engineering positions and other related fields.

Objective 2: Graduates engage in life-long learning demonstrated by advanced education, professional development activities and/or other career activities.

Objective 3: Graduates successfully demonstrate technical and leadership competence.

An assessment and evaluation process was designed as summarized in Figure 1.

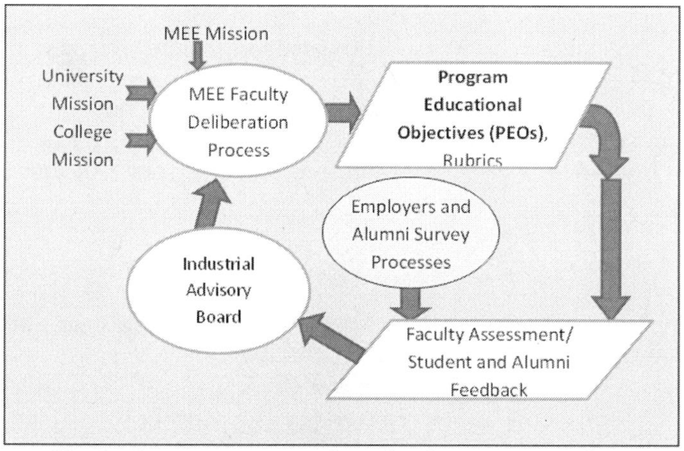

Figure 1: assessment and evaluation process of program educational objectives (PEOs)

Assessment Method and Assessment Team: As illustrated in the above figure, the method used by the department of MEE to assess achievement of program educational objectives mainly includes: (a) Conducting alumni surveys; and (b) Conducting employer surveys. Two separate survey questionnaires are designed to ensure a sustainable process that generates statistically meaningful results, and these questionnaires are posted online. The questionnaires are designed to be concise and only take a few minutes to complete. The faculty ABET coordinator (who is also the ABET committee chair) is responsible for overall coordination. Administrative staff sends notification to alumni and their employers through emails and direct contact. Alumni and employers summit their answers online, and the answers are collected by UNT Co-Op Services.

2. Establish Course Learning Outcomes

Portfolios of MEEN prefix required courses containing the student course work (representative homework, quizzes, exams, projects, etc.) are used for assessment for achievement of course learning outcomes (CLOs). The faculty members who are in charge of instruction of a course propose CLOs for the course to the undergraduate coordinator. Faculty members assigned to those courses are responsible for completing the assessment. The course works data representing those classes are collected using sampling methods each semester.

Surveys: Student surveys evaluating CLO at the end of semesters. Administrative staff is responsible for preparing the survey materials under the guidance of the faculty ABET coordinator (FAC).

The above data provide the sources for annual evaluation of achievements of CLOs and enable the department to triangulate the Student Outcomes and assess overall program efficacy. In the beginning of the 2010-11 academic year, the MEE faculty approved the assessment and evaluation process as shown in Figure 2.

3. Mapping CLOs with Student Outcomes

The initial implementation plan was also approved and is outlined as follows: a) Each instructor develops course learning outcomes (CLOs), for the MEEN course(s) he or she teaches. b) Each instructor in charge of that MEEN course then maps each CLO to ABET a-k Student Outcomes (SOs) [11]. Table 1 illustrates a sample mapping of CLOs to ABET student outcomes of MEEN 3110 – Thermodynamics II. c) Two survey forms were developed: one is for faculty to report his/her assessment results for the course CLOs, and one for receiving students' feedback for the achievement of the CLOs.

4. Mapping SOs with PEOs

The curriculum of the Mechanical and Energy Engineering Program was designed to provide students with course work that meets the overall Program Educational Objectives and gives them the skills to meet the Program Objectives. A student who graduates from the program is prepared to enter the professional workforce as a practicing engineer. The relationship between the SOs and the PEOs is presented in Table 2 below.

ASSESSMENT METHODOLOGY AND FREQUENCY

As shown in Figure 2, the following assessment tools are used collectively to assess the achievement of student outcomes at the time of their graduation:

Course-level assessment

a) Portfolios: Portfolios of MEEN required courses containing the student course works (representative homework, quizzes, exams, projects, etc.) are used for assessment for achievement of CLOs. The faculty members who are in charge of instruction of a course propose CLOs for the course to the undergraduate coordinator. Faculty members assigned to those courses are responsible for completing the assessment. The course work data representing those classes are collected using sampling methods each semester.

b) Surveys: Student surveys evaluating CLO at the end of semesters. Administrative staff is responsible for preparing the survey materials under the guidance of faculty ABET coordinator (FAC).

a. Faculty tools for assessment and determine degree of attainment each CLO.
b. Report to the Undergraduate Curriculum Committee (UGCC) and individual faculty.
c. Analysis of results and matching the established threshold.
d. Feedback from the faculty for improvement and review by the UGCC for approval.
e. Report to the Chair and

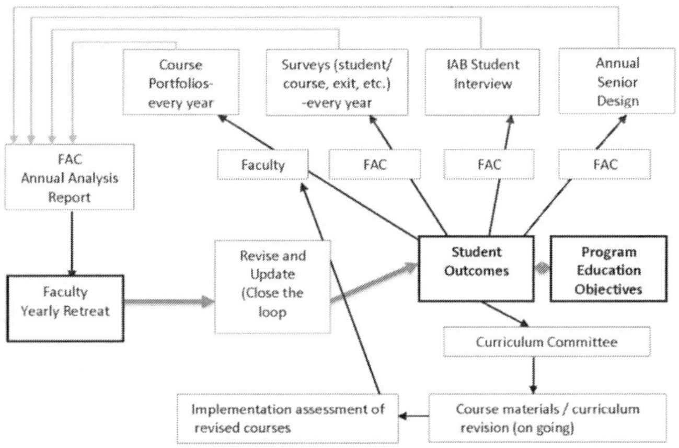

Figure 2: Process for the assessment of the achievement of student outcomes

In the beginning of the academic year, the MEE faculty approved the assessment and evaluation process as shown in Figure 2. The initial implementation plan was also approved and is outlined as follows:

a) Each instructor develops CLOs for the MEEN course(s) he or she teaches.

b) Each instructor in charge of that MEEN course then maps each CLO to ABET a-k Student Outcomes (SOs). Table 1 illustrates a sample mapping of CLOs to ABET student outcomes of MEEN 3110.

c) Two survey forms were developed: one for faculty to report his/her assessment results for the course CLOs, and one for receiving students' feedback for the achievement of the CLOs.

Every semester, the students were asked to evaluate each course they take by indicating the extent to which they believe that each CLO was achieved on a five-level, strongly agree to strongly disagree scale. The survey process is carried out by Scantron® compatible assessment forms, and information is compiled into an electronic report for the instructor and the department. Student surveys were conducted 2-3 weeks before the end of semester. The survey process is such that anonymity of the respondents is assured. Response rates are variable ranging from nearly 100% to as low as about 10 to 15% with an average response rate in the 30 to 40%-range.

The CLOs have been assessed in the prior 2 years for the required ME courses in the BS degree. During an ABET faculty workshop in the beginning of 2011-12 academic year, it was decided the assessment of CLOs to be focused on eight selected core courses in the curriculum, through the 4-year curriculum on the basis of culminating progress from freshman to senior years and coverage of ABET SOs.

The two-semester, senior capstone design courses provide students an opportunity to demonstrate their achievements of SOs. Since 2011, an annual college-wide senior design day in April has been established for all the College of Engineering majors to showcase their projects. MEE students present their design project results in front of Industrial Advisory Board (IAB) members and faculty who evaluate the project and presentation using the evaluation criteria designed by the assessment team. The criteria were grouped into four categories:

(1) Oral communication skills (related to Program Outcomes g and k)
(2) Technical knowledge (related to Program Outcomes a, b, c, e, and k)
(3) Multi-disciplinary teamwork (related to Program Outcomes d and k)
(4) Broader knowledge (related to Program Outcomes f, h, i, and j)

The assessment team for SOs was consisted of the following

- Faculty ABET Coordinator (leader),
- Six core course faculty (course portfolio) – All the faculty members teaching MEEN courses are involved for initial accreditation,
- Staff assistant (exit survey), and

IAB and faculty (senior design evaluation).

EVALUATION OF ASSESSMENTS

Course-level evaluation

Not every course is intended to achieve each SO to the same extent, but it was possible to establish an expectation for the level of attainment by comparing faculty and student opinions of each course. Based on the course portfolio prepared by faculty, expected levels of attainment for student outcomes was set with a performance target of 70% attainment agreed by the faculty for achievement of each CLO. When opinions of the two groups are out of line and below the performance indicator target, changes to the teaching methodology or curriculum may be warranted.

Figure 3 shows the sample of results of evaluation of attainment of CLOs by the instructor for one of required courses, MEEN 3120 – Fluid Mechanics, and the results of students' survey on the attainment of CLOs for the same course during 2010-2012 academic years. Students' data were calculated on Likert scale 1-5 and were converted to percentage for comparison purpose in the charts. The numeric value on Likert scale for each CLO is as follow: 1 for Strongly Disagree, 2 for Disagree, 3 for Uncertain, 4 for Agree, and 5 for Strongly Agree. The faculty provides the assessment of CLOs of the course that he/she was teaching.

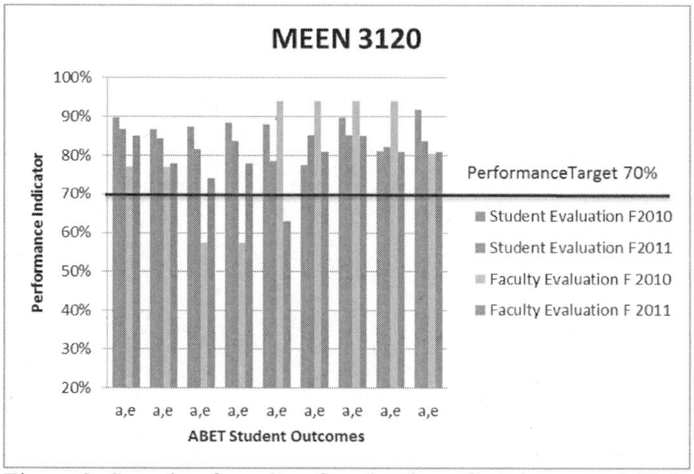

Figure 3: Sample of results of evaluation of attainment of CLOs by the instructor and students

Continuous Improvement

Starting in 2011, the evaluation results have been reviewed by the faculty who are responsible for teaching the courses. A presentation was made by the faculty ABET coordinator at the annual faculty retreat in August of 2011, and the action items for curriculum-wide improvement were identified collectively by faculty for implementation in the academic year 2011-2012. The same process continues and the next faculty retreat is scheduled for August 2014.

The above exercise demonstrates that the UNT MEE program has established and tested its SO assessment and evaluation process. The preliminary data overall showed that the various assessment tools produced a very good benchmark for the achievements of SO by our students. The results also demonstrated that the designed evaluation process for the new MEE program has been established for further improvement. Therefore, it can be concluded that as the program further matures, the quality of the program will improve in a sustainable manner because of the demonstrated evaluation process and its supporting evidences.

CONCLUDING REMARKS

Program level assessment allows us to determine the level of learning and the level of ability of the student(s) to be productive engineers after graduation. They can be valuable members of society by using knowledge skilfully with flexibility. Assessment addresses the fundamental question "have the students developed the critical skills needed for their professional careers". Guidelines have been created by ABET assessment strategy and other related criteria define a minimum set of student outcomes for engineering programs. While the ABET criterion 3 assessment captures the students' learning at topic in the course levels, ABET's program level assessments and evaluation address quality assurance and continuous improvement process for engineering education in undergraduate and master levels. Our efforts to fulfil ABET criteria requirements had successfully was examined and reported with any shortcomings.

REFERENCES

[1] Report on of a Workshop on Science, Technology, Engineering, and Mathematics (STEM) Workforce and National Research Council of the national Academies, 2011, Washington DC, ISBN 978-0-309-25180-8.

[2] "Promising Practices in Undergraduate Science, Technology, Engineering, and Mathematics Education: Summary of Two Workshops," National Research Council, ISBN 978-0-309-18723-7, 2011, Washington, DC.

[3] "Educating the Engineer of 2020," National Academy of Engineering 2005, ISBN 0-309-09649-9.

[4] Sheppard, S.D., Macatangay, K., Colby, A., Sullivan, W.A.: EDUCATING ENGINEERS, Designing for the Future of the Field, The Carnegie Foundation for the Advancement of Teaching, 2009, ISBN 978-07879-7743-6.

[5] Danielson, D., Kirkpatrick, A.T.: AC 2011-2804: VISION 2030 CREATING THE FUTURE OF MECHANICAL ENGINEERING EDUCATION, ASEE Annual Conference, 2011.

[6] Danielson, D., Kirkpatrick, A.T., Ervin, E.: ASME Vision 2030: Helping to Inform Mechanical Engineering Education, 41st ASEE/IEEE Frontiers in Education Conference, October 12 - 15, 2011, Rapid City, SD.

[7] ABET 75th Anniversary Retrospective, ED John W. Prados, ABET, Inc., 2007.

[8] Ambrose, S.A.: Undergraduate Engineering Curriculum - The Ultimate Design Challenge, The Bridge, Volume 43, Number 2, Summer 2013, P. 16.

[9] Baillie, C. and Moore, I.: "Effective Learning and Teaching in Engineering," 2004, ISBN 0-415-33489-6.

[10] ABET Self-Study Report, Mechanical and Energy Engineering Program, University of North Texas, July 30, 2012.

[11] Criteria for Accrediting Engineering Programs, 2011-2012 Accreditation Cycle, ABET, Inc., Baltimore, MD, 21202.

APPENDIX

Table 1: Mapping of CLOs to SOs for MEEN 3110 – Thermodynamics II

Course Leaning Outcomes MEEN 3110	**ABET EAC Student Outcomes**										
	(a)	**(b)**	**(c)**	**(d)**	**(e)**	**(f)**	**(g)**	**(h)**	**(i)**	**(j)**	**(k)**
Conduct energy analysis and improve energy use efficiency	x		x		x			x			
Model and analyze various gas power systems	x				x			x			
Understand gas mixtures, psychometrics, and be able to analyze A/C systems	x		x		x			x			
Analyze reacting mixtures and simple combustion processes	x		x		x			x			
Evaluate and optimize the efficiency of a thermodynamic system	x		x		x			x			
Apply the basic principles of thermodynamics to engineering practices	x				x				x		x
Complete basic level of thermodynamics related projects in engineering practices	x	X		x	x		x	x	x	x	x

Table 2: Relationship between Student Outcomes and Program Educational Objectives for UNT Bachelor of Science Degree in Mechanical and Energy Engineering

ABET Student Outcomes (SOs)	Program Educational Objectives (PEOs)		
	Graduates will be successfully employed in mechanical and energy engineering positions and other related fields in which they apply their engineering knowledge and skills.	Graduates will engage in life-long learning demonstrated by advanced education, professional development activities and/or other career activities.	Graduates will successfully demonstrate technical and leadership competence through ethical conduct, teaming, communication and problem-solving skills teamed in our program.
a	X	X	
b	X		
c	X		X
d	X		X
e	X		X
f	X	X	X
g	X	X	X
h		X	X
i		X	X
j	X	X	X
k	X		

Proceedings of the ASME 2014 International Mechanical Engineering Congress and Exposition
IMECE2014
November 14-20, 2014, Montreal, Quebec, Canada

IMECE2014-36749

DESIGNING A CAPSTONE DESIGN COURSE TO ACHIEVE STUDENT OUTCOMES

Vincent Wilczynski
Yale University
New Haven, CT, USA

Andrew C. Foley
U.S. Coast Guard Academy
New London, CT, USA

ABSTRACT

The assessment of Student Outcomes is an import component for program evaluation and improvement. Though not proposed as the only tool a program should employ to measure the achievement of outcomes, the capstone design course can be a valuable mechanism to measure performance with regards to Student Outcomes. Because of the expansive reach of the engineering design process, capstone design projects present a natural environment to structure assessment activities that directly map to Student Outcomes. This paper presents versions of the Mechanical Engineering capstone design course that have been specifically structured to assess achievement of Student Outcomes commonly included in engineering accreditation criteria. Typically the outcomes are assessed by assignments that guide the engineering design process.

INTRODUCTION

Demonstrating the achievement of Student Outcomes is an important responsibility for engineering educators (1). According to accreditation guidelines (2), these outcomes highlight those aspects of the engineering profession that are expected to be achieved by the time of graduation and include the skills, knowledge and behaviors that define what it means to be an engineer. ABET accredited programs are required to document how students achieve these outcomes, measure student performance with regards to the outcomes, and detail systematic improvements that result from the assessment process.

This paper explains how the Mechanical Engineering capstone design course has been formatted and taught at the U.S. Coast Guard Academy and Yale University to guide students through a design-build-test process while simultaneously assessing the level of performance achieved in the majority of ABET's Student Outcomes.

In this version of the capstone design course, specific assignments ensure that the collective work demonstrates each student's ability to analyze, experiment, design (systems and components), function on teams, practice professional ethics, communicate, apply results in a societal context, and use the techniques, skills and modern engineering tools required of the profession. In addition, the course includes activities that demonstrate the student's ability to incorporate engineering standards and realistic constraints.

Student Outcomes are assessed by addressing each of these concepts explicitly within the course and by requiring the submission of specific reports associated with each of the relevant outcomes. By assembling the design project documentation in this format over the course of the semester, each student team develops a comprehensive collection of reports that details the project's evolution and demonstrates a level of achievement for the Student Outcomes. The course documentation prepared by each design team serves as an artifact to later evaluate the obtainment of outcomes and provides a framework to make program improvements. In addition, the reports are a comprehensive record of learning that may be helpful to external reviewers.

It is noted that the inclusion of topics relevant to Student Outcomes support the course focus, that being an emphasis on practicing three distinct phases of the engineering design process: Conceptual Design, Preliminary/Prototype Design and Detailed Design. Also, the Student Outcomes aligned instruction help emphasize that the design process is broad and encompassing. As such, an outcomes-oriented capstone design course provides a coordinated method for using Student

Outcomes to complement the design process and amplify the need for both technical and non-technical skills in the engineering profession (3, 4, 5).

This explicit method for building Student Outcome evaluation tools into the capstone design course has been applied by the authors at two institutions with similar positive results. The process greatly aids the students' understanding of their design efforts in a larger context, and faculty members - even those not associated with teaching the course - become more aware of how the capstone design course can serve as an assessment tool. It is noted that it is not proposed that an outcomes-oriented capstone design course be used as the only tool for program assessment, but rather that it be one of a number of methods that a program would use to measure the achievement of outcomes and make systematic improvements.

THE OUTCOMES-ORIENTED CAPSTONE DESIGN COURSE METHOD

General Capstone Design Course Structure

An outcomes-oriented capstone design course has been structured and delivered by the authors at the U.S. Coast Guard Academy and at Yale University as single-semester and two-semester courses. The course has a lecture component (typically two hours per week) and a lab component (which may be six hours per week) as described in the course syllabi (posted online at http://seas.yale.edu/capstone-design-syllabi-examples). Projects are client-based where teams of students (maximum of 7 students per team) apply a design-test-build product development process to create a solution to a unique challenge. The projects are derived from problems posed by clients, with the team refining the scope of the presented problem to one that can be solved in the time allocated to the course.

Three axioms of the design process are emphasized, the first being the iterative nature of the design spiral. The best solution is generally not the first solution, but rather the result of a series of iterations and improvements. The second design process axiom that is stressed is that design is best realized as the result of three distinct phases: the Conceptual Design Phase where many possible solutions are presented, the Preliminary (or Prototype) Design Phase where a subset of the concepts are explored, and the Detailed Design Phase where a single (or perhaps a leading few) prototypes are developed and integrated into a complete solution. Along this progression, the scope of the investigation narrows and the attention to design details deepens. The final axiom that is emphasized is that the design process is not complete until a product is built, tested and demonstrated in front of a client or sponsor. These axioms are explicit in the course and reinforce the developmental nature of the engineering design process.

The involvement of external clients for the projects, while not always possible, greatly benefits the students' learning experience. The careful selection of clients and projects is directly correlated with student motivation and performance. Examples of course projects include the design/construction of a solar powered boat, a fuel monitoring system, a bio-fuel test and evaluation platform and a power-assisted equipment dolly. All projects include electromechanical components and the integration of individual subsystems into a composite solution. Generally one instructor leads the course, with other instructors contributing to specific lecture topics. Each project is assigned its own project advisor to provide effective mentorship.

The budget for each project averages $2,500 with the majority of the budget allocated for raw materials. The fabrication, assembly and testing of the designs is accomplished by the students with training and assistance provided by lab staff and faculty. The shop facilities at both institutions are similar and include CNC mills and lathes, electrical and mechanical workshops, and rapid prototyping tools, as well as support from dedicated lab technicians. Provisions exist at each institution to obtain additional manufacturing assistance as needed for specific projects.

Historically, the single-semester option of the course has proven to be the most successful as it provides an intense learn/work period that allows students (and faculty) to focus on the project. The single-semester version of the course also results in a format that integrates instruction and implementation. In addition, the oversight of the projects is much easier when concentrated in a single semester. In those cases when the course is scheduled in the fall term, the single-semester option allows students to continue their work should they be so motivated during the spring semester.

Outcomes-Oriented Framework

The course syllabus is used to detail how the required outcomes for undergraduate engineering majors are addressed in the course. Each ABET Student Outcome is listed in the syllabus along with an indication of how the achievement of that outcome is demonstrated in the course. The syllabus also defines the overall course requirements in the form of a course notebook, progress reports, analysis reports, design reviews, experiments, safety reviews and presentations.

A breakout of individual contributions to the team is used to quantify the otherwise qualitative perspective of measuring an individual's effect on the group project. Assignments are closely integrated with the lecture contents of the course that address an aspect of the design process. For example, a lecture on the use of engineering analysis is presented near the midpoint of the course when the teams progress from the Preliminary/ Prototype Design phase to the Detailed Design phase. The analysis assignment supports this transition to the Detailed Design phase and occurs at a time when enough variables are well defined and the analysis can provide sufficient insight to guide a final design decision.

The format for documenting each assessment activity aligned with a Student Outcome is the submission of an industry standard report. Each report has a primary author and the direct involvement of another team member who reviews the work for its technical content and overall presentation. This industry standard for documentation and review is an important benefit of the formal report structure. Over the course of the semester these individual reports can total up to 40 separate reports for the entire team. As a composite document, the reports provide a history and audit trail of the design choices and serve as a summary document for the semester's work.

The primary mode of documentation is the *Formal Report*. While the majority of the reports are undertaken by individuals there also are a number of team and sub-team *Formal Report* requirements. As to the nature of the reports, there are two distinct types. First, there are the formally assigned reports numbering approximately 12 during a typical semester. These are described and scheduled in the syllabus and have due dates at the approximate time during the semester when a typical project should be undertaking the work required in the report. These reports are typically graded and re-assigned if the quality is not acceptable. The grades are indicative of the entire team's performance and hence all team members, even if not delegated to the specific report, are encouraged to be engaged in each report's production.

The second type of report is the so called *Optional Individual Report*. These reports are self-generated by the student and allow the unique nature of the assorted projects to be documented independent of the syllabus dictated reports. Typically students will undertake 8 to 10 of these reports to document all aspects of the design, fabrication and testing that they have been involved with. The adage "if it isn't written down, it never happened" is used to emphasize that if students wish their contributions to be recognized, then they need to document their efforts in a report format. Ultimately while all the reports contribute to the team's grade, individual grades are distributed around the team's mean grade primarily on the basis of the individual contributions documented in their *Optional Individual Reports*.

As opposed to other documentation techniques (such as requiring a summary report at the end of the semester where all of the work is recalled, presented and framed in a larger context) the rolling *Formal Report* structure has proved to be an effective method for documenting decisions, providing timely feedback, and producing a summary of the entire project. In the end, the assembled reports form a final notebook of the design process in a format that can be examined as part of a program review and by other external reviewers. In addition, the reports serve as artifacts that can be reviewed a-priori by the next generation of students, faculty and project sponsors to readily understand the scope of the work involved in these projects.

STUDENT OUTCOMES ASSIGNMNENT EXAMPLES

Examples of some of the embedded assignments are presented to illustrate how the outcomes-oriented framework of the capstone design course can meet a number of objectives, ranging from serving as a tool for program assessment to realizing effective design solutions.

The "ability to design a system, component or process to meet desired needs within realistic constraints such as economic, environmental, social, political, ethical, health and safety, manufacturability, and sustainability" (ABET Student Outcome (c)) is the primary reason that usually dictates why capstone design courses are often the assessment tool of choice for ABET-accredited programs (1). This Student Outcome is also the reason why the very first, and arguably most important, *Formal Report*, at both Yale and the U.S. Coast Guard Academy, is the *Initial Design Specification Report*. This report must clearly identify the customer, the need statement, the problem to be solved, and a quantified set of design constraints and objectives. The importance of this report cannot be overemphasized as it effectively fixes the playing field on which the project will progress. It protects students from potential mission creep while keeping them focused on the project at hand. As the project progresses there are occasions when the specification needs to be adjusted, but this must always be done with the consensus of the team and in consultation with the customer and/or project supervisor.

The ability to apply knowledge of mathematics, science and engineering (ABET Student Outcome (a)) is, in part, assessed using an "Analysis Assignment" where each student has to apply engineering fundamentals to an aspect of the project such that the results can be used to make a design decision. For example an individual may need to examine the strength of a robot arm using beam theory, accounting for the applied loads and torque required to affect motion. The scope of the analysis need not be substantial, with the needed assumptions made to avoid the all too common "analysis paralysis" experienced by many students. The collective analysis of all team members should ultimately be such that all major components of the design are fully examined and detailed in the individual analysis reports.

Similarly, the ability to design and construct experiments as well as analyze data (ABET Student Outcome (b)) is ensured by tasking sub-teams to design and conduct an experiment that provides information to guide a design decision. While many such experiments are conducted informally as part of the design iteration process, this assignment is more formal in nature. Here, the examined system is controlled, tested, measured and evaluated to the degree that would be performed in a research endeavor. The required report for this review has a format similar to that used in prior experimentation courses and requires a review of the theory, experimental setup/apparatus, results (with error bars) and analysis. As with the analysis report, the experiment needs to be one that will guide design decisions. One example of such an experiment is the measurement of the exhaust volume and temperature produced under varying loads for a turbine engine test stand. While students may initially approach the size of such an exhaust with an arbitrary approximation, the optimal solution for this subcomponent requires information that can be obtained through experimentation. Unlike the analysis reports, the inherent complexities of experimentation require group work for this assignment. To ensure a wide variety of design variables are tested, the team size for this component of the course is limited to two students.

The engineering ethics report (ABET Student Outcome (f)) required in this version of a capstone design course has (surprisingly) been a popular component of the course. While engineering ethics is commonly presented using case studies, having students directly apply engineering ethics to their capstone project has great value. In this assignment students are challenged to find an aspect of their work that has an ethical connection and document that perspective as it relates to the National Society of Professional Engineers Code of Ethics. These reviews frequently channel concerns for public safety and force students to connect their work to the larger population. As an example of one such essay, a student reflected on the inherent risks to the public at large that resulted when a less-than-fully-tested computer code that allowed a machine to work autonomously was deployed and operated on a student designed device.

To ensure students incorporate engineering standards and realistic constraints (ABET Student Outcome (c)) in their project, one course lecture is typically held in the library to access published national standards that guide the design, use, manufacturing and safety of engineered products. These source materials include trade handbooks, government regulations, and industry standards. Following a discussion on engineering standards, students are tasked with finding a standard that is relevant to some aspect of their project. They must then apply that standard to guide an engineering decision. This exercise is often incorporated into the Conceptual Design phase with the results of the investigation then serving as a constraint in the design specification report. It can also be integrated with the experiment assignment, as students discover that experiments to quantify subsystem performance are often prescribed by an oversight body (such as Underwriter's Laboratory).

Safety standards (ABET Student Outcome (c)) receive appropriate coverage in this version of the capstone design course by requiring a report from each student on a safety aspect of their design. Their report must follow one of the safety review methodologies presented in the lecture portion of the course (such as Preliminary Hazard Analysis or Failure Modes and Effect Analysis), with the result of this analysis informing a design decision. For example, a review of the Preliminary Hazard Analysis associated with rotating disks could lead to a design decision on material selection to prevent the rotational speed from causing material separation as well as a need to surround a rotating object with a suitable means to contain material failures. Key to this component of the course is the guided application of safety principles as well as the students' realization that they are directly responsible for the safety of their designs.

The ability to design a system, component or process to meet desired needs (ABET Student Outcome (c)) and communicate effectively (ABET Student Outcome (g)) is assessed using the scheduling component of the course and by having formal design reviews three times during the course at the conclusion of the Conceptual Design Review, Preliminary Design Review and Detailed Design Review phases. These activities reinforce that design is an iterative and planned process. The best solution is not simply discovered in a single step but rather results from the careful review and refinement of a host of potential solutions. Solutions are embedded with details, and these details are only understood through careful analysis. The project schedule assignment is iterative, with the template developed in the first week used as a tool to plan the project's future and to respond to unexpected intermediate developments.

In addition to the three formal reviews, students hold weekly progress meetings within their team. An agenda and previous meeting minutes are distributed two days prior to the meeting and students rotate through the roles of Chair and Secretary. These meetings are intended to be purely administrative in nature and serve as planning sessions for future work. A rolling action list is reviewed, completed actions are closed out, and new actions added and prioritized. Actions are also assigned to specific team members and the requirements for a report that will close out the action may be set during these meetings. Faculty members attending these weekly meetings are present simply in the role of advisors and consultants, and the meetings are always chaired by students. Technical or task-oriented meetings are scheduled separately and generally only include those individuals required to complete the task at hand.

RESULTS

The outcomes-oriented capstone design course has been found to be a very useful approach for covering this important segment of the undergraduate Mechanical Engineering curriculum. This format of the course has been used at the U.S. Coast Guard Academy since 2001 and was introduced at Yale University in 2013. Approximately 25 students were enrolled each year in the U.S. Coast Guard Academy version of the course and 20 at Yale, with the average number of projects each term being 5 and 4, respectively. Both instructors found the format to be a useful tool as it provides structure and clarity for teaching open-ended design. The formal report model to document progress and decision making has been adopted by students in other projects and this use helps validate the value of the report submission format.

The outcomes-oriented capstone design course method structures the course into a sequential process that allows for the delivery of information and demonstration of skills at distinct and well-defined steps. The method also fosters the successful completion of projects by establishing a set of project requirements that must be followed to complete the design project. It has been found that this technique readily identifies teams that are behind their planned schedule early in the semester, thereby offering ample time to recognize deficiencies and implement changes.

The documentation aspect of the outcomes-oriented course is a valuable byproduct of this method of instruction. By standardizing reports from throughout the semester, a detailed record of design decisions (and the supporting reasoning and documentation behind those decisions) is assembled in real time. An alternative method of documentation– assigning a single final report for the entire capstone design project at the conclusion of the semester – runs the risk of missing details due to the passage of time. The report format used in the outcomes-oriented capstone design course creates a comprehensive document that, when combined with the physical design solution, can easily be used to assess the achievement of Student Outcomes. As a record of the design history, the report format also provides a valuable resource to the project sponsor and to future students that pursue additional work on the projects.

The structure of the course presents an opportunity for constant feedback to individuals and the entire team to guide their efforts. Since each step in the design process is documented and reviewed, each assignment and the associated deliverable become a feedback path from the instructor to the team. Faculty comments on the reports can correct potential missteps at a time when corrections can easily be made. The volume of the required reports generates the chance for significant feedback, and this close level of supervision has proved to be an important factor in the successful completion of many challenging projects.

The outcomes-oriented version of the capstone design course is a useful approach that illustrates the iterative nature of engineering design. Further, this approach provides insights that guide decision making at each phase of the design process. Finally, this approach and its emphasis on the design/build/test process for product design and helps promote a visceral understanding of fundamental engineering fundamentals.

The potential for using the outcomes-oriented capstone design course for program assessment is significant. Though not proposed as the only assessment tool a program should use, this method creates a clear track record of student performance for a majority of ABET's specified Student Outcomes. Since a course typically has a number of student projects, a database of student work is created each time the course is offered, and annual versions of the course efficiently establish a large amount of documentation that can be used in during the accreditation review process. It is proposed that the outcomes-oriented capstone design course model can be duplicated at other programs. The enclosed syllabi (Annexes A and B) may serve as a template for such programs.

The outcomes-oriented Mechanical Engineering capstone design course has been successfully implemented at Yale University and the U.S. Coast Guard Academy. Students find the organization of the course and the assignments as useful guides that help structure their design thinking. The format of the course allows students to see how the outcomes are related not only to the design process but also to their responsibilities as engineers. By being very structured with a series of required reports and a sequential timing of assignments, the success of applying an open-ended design process to produce viable solutions is ensured using this method of instruction.

ACKNOWLEDGMENTS

Both authors acknowledge the support of Yale University and the U.S. Coast Guard Academy with regards to this work.

REFERENCES

1. ABET Engineering Accreditation Commission, "Criteria for Accrediting Engineering Programs – 2013-2014 Accreditation Cycle," 2013, retrieved from www.abet.org
2. ABET Engineering Accreditation Commission, "Accreditation Policy and Procedure Manual – 2013-2014," 2013, retrieved from www.abet.org
3. Ullman, D.G., The Mechanical Design Process, McGraw-Hill Higher Education, Feb 2, 2009
4. Wilczynski, V. and Douglas, S.M., "Integrating Design Across the Engineering Curriculum: A Report From the Trenches," Journal of Engineering Education, pages 235–240, July 1995
5. Foley, A.C., "Big Picture, Rational, Engineering Design Methodology," ASEE Annual Conference & Exposition, 2007

This page left blank intentionally.

Proceedings of the ASME 2014 International Mechanical Engineering Congress and Exposition
IMECE2014
November 14-20, 2014, Montreal, Quebec, Canada

IMECE2014-38242

EFFECT OF OFFICE HOUR PARTICIPATION ON STUDENT PERFORMANCE

Michael J. Schertzer
Rochester Institute of Technology
Rochester, NY, USA

Risa Robinson
Rochester Institute of Technology
Rochester, NY, USA

Timothy Landschoot
Rochester Institute of Technology
Rochester, NY, USA

Amitabha Ghosh
Rochester Institute of Technology
Rochester, NY, USA

Alexander Liberson
Rochester Institute of Technology
Rochester, NY, USA

Edward Hensel Jr.
Rochester Institute of Technology
Rochester, NY, USA

ABSTRACT

Office hours are available for students to receive extra help outside of class. Unfortunately, this resource is often underutilized by students despite efforts to schedule convenient and accessible office hour times. Previous survey results from students attending a variety of courses in Mechanical Engineering at the Rochester Institute of Technology (RIT) have shown a positive correlation between low office hour attendance and the following factors: (i) high understanding of course material, (ii) procrastination and lack of time to seek help before deadlines, and (iii) low time studying materials outside of class. Interestingly, the results of this survey did not support the hypothesis that students who attended more office hours performed better.

A new homework grading policy was instituted in Thermodynamics and Fluid Mechanics I in the fall and spring semesters of the 2013 academic year at RIT. Under this policy, students were required to visit office hours to receive credit for completion of assigned weekly problems. Implementation of this policy has provided quantitative information regarding participation and timing of office hour visits. This investigation will examine the effects of attendance and timing of office hour participation on metrics of performance including final class grades and theoretical understanding as measured by performance on multiple choice test questions.

Results presented here suggest that the office hour grading system resulted in high participation rates across a broad range of students. Higher office hour participation rates had a positive impact on student performance in long answer exam problems and low impact on performance in multiple choice questions. While performance was a stronger function office participation at the of end of term than in week five, early semester participation rates can be used as a tool to help identify students at risk of dropping a class or receiving a poor grade.

INTRODUCTION

Frequent, high quality student-faculty interactions have been identified as a key factor in improving student outcomes [1,2]. Regular interactions have been linked to improved outcomes in student recruiting [3], retention [3–8], engagement [9], motivation [10–12], and success [13,14]. These interactions have also been shown to improve skill development in the areas of: (i) problem solving [7], (ii) reflective thinking [15], (iii) group dynamics [16], and (iv) occupational competence and awareness [16]. In one study, the benefit of these interactions was found to be higher among female students [3].

While student-faculty interaction has the potential to positively impact student success, care must be taken to ensure that these interactions are of high quality. In fact, low quality interactions can demotivate students [17–19]. In order to achieve high quality interactions, faculty must be available, approachable, enthusiastic, and demonstrate genuine care for students [1], [12], [19–22].

High quality interactions occur when faculty have a high degree of immediacy, which is most strongly associated with student-faculty rapport [20]. Establishing positive rapport with students can result in increases in studying, class attendance, office hour attendance, and attention in class [23]. One study [19] suggests that faculty can build positive rapport with students by:

1. Encouraging more student-instructor interaction,
2. Inviting students to visit office hours,
3. Using personal examples in class,
4. Calling students by name,
5. Getting to know students and showing genuine concern,
6. Showing enthusiasm for their subject,
7. Making special effort to connect with students who would not normally seek out a relationship with the instructor,
8. And showing respect for all students.

Frequent, high quality student-faculty interactions have often been reported in design competitions [24] and research projects [2], [25–27]. While these types of interactions are available in most Engineering programs, they are most prevalent as students approach graduation. Fostering a culture of high office hour participation has the potential to impact students earlier in their programs, while allowing faculty to connect with students who may not normally seek out these relationships.

Developing a culture where students regularly attend office hours would provide an opportunity to engage in ongoing high quality student-faculty interactions. Unfortunately, students tend to be unaware of the importance of these interactions and have minimal contact with faculty [28]. Efforts to improve office hour performance in the Department of Mechanical Engineering at the Rochester Institute of Technology (RIT) have included: scheduling office hours to minimize student conflicts, increasing availability of faculty office hours throughout the term and before major course events (i.e. exams, project due dates, etc.), and increasing availability of TA office hours [29]. While these measures qualitatively increase office hour attendance, exam performance suggests that students could benefit from increased student-faculty interaction.

A recent survey of students in the Department of Mechanical Engineering at RIT suggests that the main drivers for low office hour participation in the department are (i) high understanding of course material, (ii) procrastination and lack of time to seek help before deadlines, and (iii) low time studying materials outside of class [29]. The same study suggests that students who work in effective study groups, have access to on-line study material or solutions (i.e. solutions manuals, websites, YouTube videos, etc.), or have a strong understanding of the course material will have a low perceived need for office hour visits. The survey data could not conclusively demonstrate that low office hour participation was related to low performance on examinations.

METHODOLOGY

In an effort to foster a culture that encourages frequent, high quality student-faculty interactions, graded homework assignments were replaced with graded office hour visits in Thermodynamics (MECE-110) and Fluid Mechanics (MECE-210). In this system, students were required to complete weekly problems in logbooks and visit a member of the instructional team to present solutions and discuss the issues they encountered. The focus of these meetings was to (i) monitor that students were completing assigned homework problems and (ii) to provide a forum to discuss issues encountered while attempting homework problems. Homework solutions were not graded based on correctness. Instead, a point was awarded to a student provided they made an honest effort to complete the problems. This grading policy was implemented to encourage students to visit office hours to discuss issues and reduce the incentive to seek out and copy correct solutions from peers or on-line resources.

The system adopted here actively (i) encourages more interaction, (ii) invites students to visit office hours, and (iii) allows faculty to connect with all students in the class. The high frequency interactions that result from this system also provide an opportunity to (iv) learn the names of students, (v) get to know and show care for students, and (vi) show respect for all students. As such, this method increases the frequency of student-faculty interaction and has the potential to improve the quality of interactions by involving six of the eight suggestions for improved rapport proposed in [19].

In order to determine the effectiveness of the new grading system, office hour visits were recorded digitally throughout the semester in all offerings. Over the course of a 15 week semester, students could acquire a maximum of 10 logbook points. As such, a student could still receive full credit for visiting office hours two of every three weeks. The goal of this grading scheme is to improve educational outcomes for students by fostering a culture where students actively seek out high frequency interactions with faculty. Implementation of this system also provides quantitative information on office hour participation that can be readily compared to student exam performance.

In order to maximize instructor availability for office hours, the majority of TA hours were allocated to office hour availability. A summary of course enrollment and office hour availability is presented in Table 1. The minimum office hour time per student per week in these offerings was 11.5 minutes. In practice, more face time per students was available as students would often meet with instructors in groups.

Implementation of graded office our participation provides quantitative information that can be readily compared to student exam performance. Exams in both Thermodynamics and Fluid Mechanics are made up of multiple choice and long answer problems. Care is taken to design multiple choice problems to test theoretical knowledge [30], while long answer problems provide an opportunity for students to demonstrate their ability to apply traditional mathematical models to practical problems. Both courses implementing the office hour grading system offer two midterm examinations and a final exam. Midterms were 90 minutes in duration and consisted of 10 multiple choice questions and two long answer problems. Final examinations were 120 minutes in duration and consisted of 20 multiple choice questions and two long answer problems.

Table 1: Office Hour Availability

Course - Semester	Students	Office Hours / Week			Time*
		Faculty	TA	Total	
MECE110-Fall	150	15	18	33	13.2
MECE110 - Spring	53	10	10	20	22.6
MECE 210 - Spring	120	11	12	23	11.5

** Minimum time per student (Minutes per student per week)*

All data that references week five participation rates is relative to performance on the first mid-term examination. Data that references week fifteen participation is relative to performance on all three exams.

RESULTS AND DISCUSSION

The goal of the new homework grading policy introduced in the Mechanical Engineering Department at RIT is to improve student outcomes by developing a culture where students actively seek out frequent high quality interactions with members of the instruction team. In the first full semester of implementation, 1197 office hour visits were recorded by the 150 students who completed thermodynamics. As such, an average week resulted in 80 office hour visits and students averaged eight visits throughout the semester. This result shows that the system presented here is effective in producing frequent office hour visits. Further investigation is required to (i) quantify the quality of office hour interactions, (ii) relate interaction quality to performance, and (iii) examine methods to improve interaction quality along with quantity.

Student office hour participation rates for an entire semester (MECE-110 Fall) and the first five weeks of the course (MECE-110 Fall/Spring, MECE-210 Spring) are shown in Figure 1. Here, overall office hour participation defined as

$$Overall\ Participation = \frac{OH\ Visits}{OH\ Points\ Available}. \quad (1)$$

The overall office participation in week five over all three courses was 55.8%. In the fall section of thermodynamics, the participation rate increased from 66.6% in week five to 79.8% in week 15. This is due to the fact that students could attain full credit by attending office hours in 10 of 15 weeks. As such, even those with 0% participation after week five could attain 100% participation by week 15. By the end of the semester, 55% of students had accumulated 10 office hour visits and over 70% had eight or more visits. Between weeks five and 15, the percentage of students who attended 20% of available office hours decreased from 20% to 10%. The distribution of office hour visits demonstrates that the grading system used here yields high participation rates across a large percentage of the class. However, faculty have an opportunity to improve participation rates by reaching out to students with low office hour participation. Current and future offerings will make more extensive use of available tools for outreach including: personal invitations to students with low participation through the Starfish Early Alert system, updates on global participation rate in lecture, and broadly distributed communications through my-courses.

Of the 120 students enrolled in the spring section in Fluid Mechanics, 107 completed Thermodynamics in the Fall. These students had an average of 3.5 office hour interactions in the first five weeks of the fall semester (participation rate of 70%). Interestingly, the same population had an average of 2.2 office hour interactions through the first five weeks of the spring semester (participation rate of 44%). The cause of this reduction in early participation rate is unknown at this time. The authors will continue to examine the participation rate of this population throughout the semester.

Eight students withdrew from the fall offering of Thermodynamics. These students had a combined office hour participation rate of 29% at week seven. Four of these students had never attended office hours and two of them had only come once. This suggests that (i) students who do not attend office hours early in the course under this system are at risk of withdrawing from the course and that (ii) attending office hours may be correlated to success in the class.

Average student performance on multiple choice problems as a function of office hour participation rate is presented in Figure 2. This data suggest that average student performance tends to increase with office hour participation rate as measured in weeks five and 15. This data suggests that performance on multiple choice problems is a stronger function of participation by the end of the course than it is in week five. While there appears to be a trend in this data, a single factor ANOVA analysis for the week 15 data finds that $F < F_{crit}$ which

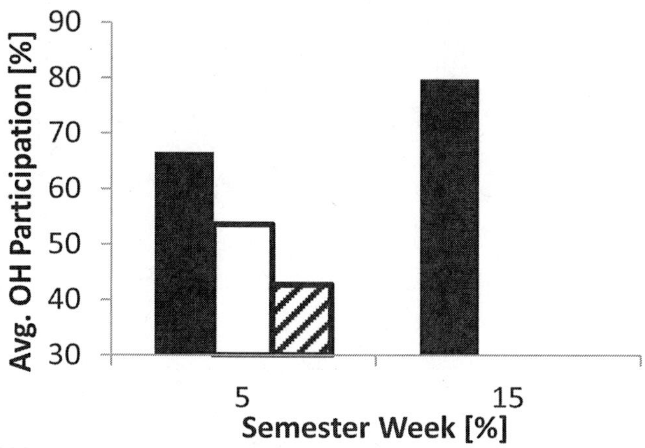

FIGURE 1 Average office hour participation in weeks 5 and 15 for MECE-110 fall *(closed)*, MECE-110 spring *(open)*, and MECE-211 spring *(hatched)*.

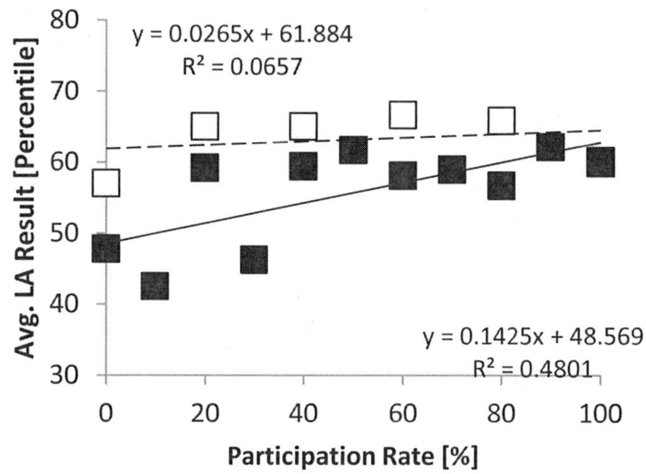

FIGURE 2 Average MC performance as a function of office hour participation in week 5 *(open)* and week 15 *(closed)*.

suggests that the difference in the means of each population is not statistically significant. This was also true for the week five data for the fall semester of Thermodynamics. However, an ANOVA analysis for week five data from all three courses showed that $F > F_{crit}$. While this suggests that the difference in the means is statistically significant, the data shows that multiple choice performance is a weak function of office hour participation early in the course.

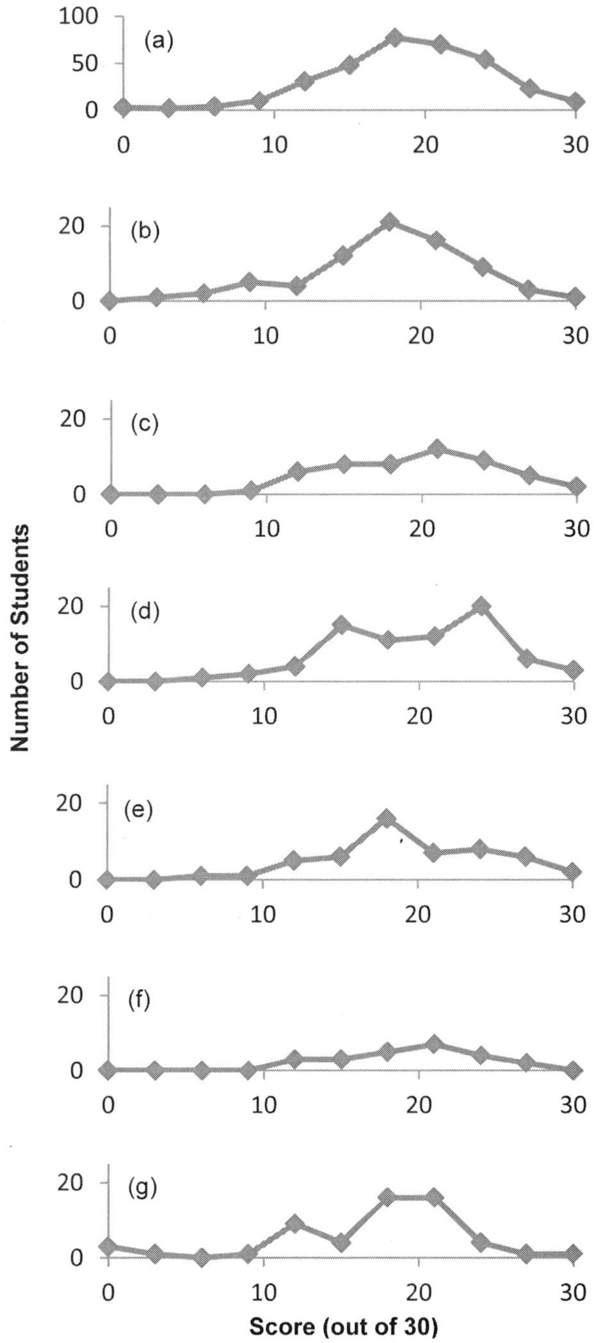

FIGURE 3 Distribution of performance for on multiple choice problems in exam 1 for (a) the entire population, and for office hour participation rates of (b) 100%, (c) 80%, (d) 60%, (e) 40%, (f) 20%, and (g) 0%.

Distribution of multiple choice grades at various levels of office hour participation are presented in Fig. 3. While the grades in the entire population have a distribution that seems normal (Fig. 3a), evidence of bimodality exists for several participation rates (Fig. 3c,d,f,g). This suggests that it is possible to further subdivide each sub-population. This result seems to support the survey data from [29] which suggests that low office hour participation is driven by multiple factors such as procrastination and low perceived need for faculty interaction. If students who do not attend office hours due to procrastination perform worse than those who feel they do not need assistance, bimodal distributions at low participation rates would be expected. The authors are examining methods that may be used to identify these groups within the data.

Average student performance on long answer problems as a function of office hour participation rate is presented in Figure 4. This data again shows an upward trend with participation rate in weeks five and 15. Single factor ANOVA analysis for both data sets shows that $F > F_{crit}$. This suggests that increased office hour participation can improve long answer performance on engineering examinations. While the relationship between performance and participation is weaker in week five, office hour attendance can be used to identify at risk students early in the term.

As office hour participation increases, the percentage of students who fall below the course average tends to decrease (Fig. 5). This decrease was nearly monotonic between 0% participation and 80% participation. The exception to this trend comes at 20% participation. This may be an issue with small sample size as the population in this sample is less than half the size of the next smallest sample. As participation increases from 80-100%, the percentage of students who scored below the average increases from 37% to 47%. Further examination of this group is required to understand this result.

That increased office hour participation shows a statistically significant difference in long answer performance but not multiple choice performances seems reasonable. Anecdotally,

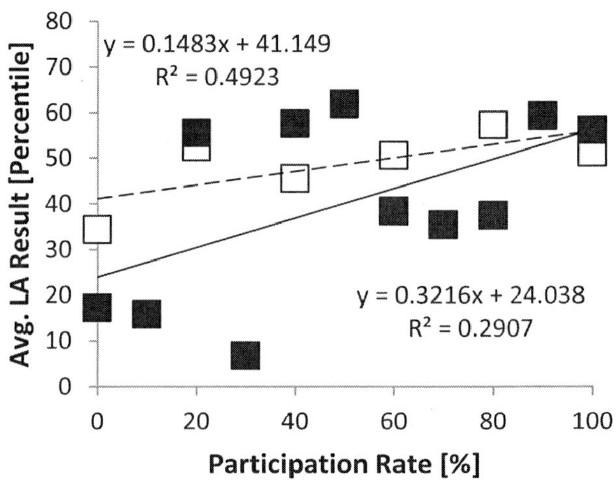

FIGURE 4 Average long answer performance as a function of participation in week 5 *(open)* and 15 *(closed)*.

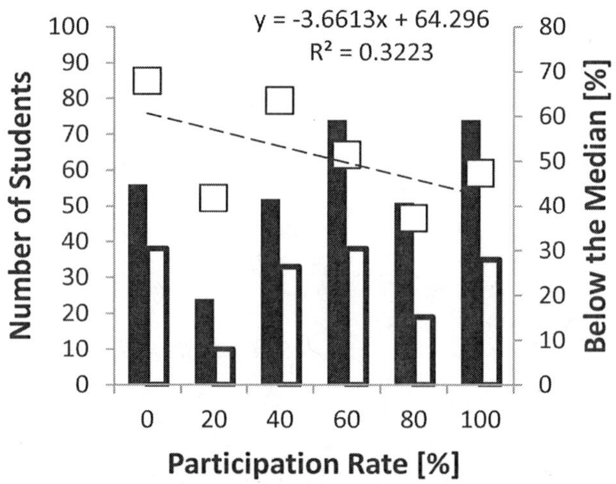

FIGURE 5 Distribution of students above *(closed column)* and below the median *(open column)* for each participation level in week 5. Percentage of students below the median in each group are also presented *(open square)*.

the majority of questions raised in office hour visits focus on issues with assigned homework problems. These problems are more similar to those in the long answer section of an exam than the multiple choice section. As such, student benefit would be expected to be higher in long answer performance.

CONCLUSIONS

Previous studies suggest that frequent high quality student-faculty interactions improve educational outcomes for students. This investigation presents the results of a new homework grading system implemented in three courses offerings. This system requires that students visit office hours to receive credit for completion of assigned weekly problems. The goal of this system is to improve educational outcomes by fostering a culture where students actively seek high frequency interactions with the instruction team.

The grading system examined here produced high frequency interactions with the average student attending office hours eight times in a fifteen week semester. Results from this investigation suggest that office hour participation can be used as an early indicator for at risk students. Interestingly, students who took thermodynamics in the fall and fluid mechanics in the spring had lower average office hour attendance through the first five weeks of the spring semester.

Increased office hour participation tended to increase scores on long answer examination problems. As office hour participation increased, the percentage of students who received below average scores on long answer problems decreased.

While increased office hour participation was statistically significant for performance on long answer results, it appears to have a more modest impact on multiple choice performance. This may be the result of most office hour interaction focusing on solutions to long answer type homework problems. The performance distributions for several office hour participation rates appeared to be bi-modal. This supports previous results that suggest that office hour participation is driven by multiple factors such as procrastination and low perceived need for assistance.

ACKNOWLEDGMENTS

Financial support of this work from the Rochester Institute of Technology is gratefully acknowledged.

REFERENCES

[1] B. E. Cox, K. L. Mcintosh, P. T. Terenzini, R. D. Reason, and B. R. L. Quaye, "Pedagogical Signals of Faculty Approachability: Factors Shaping Faculty-Student Interaction Outside the Classroom," *Res. High. Educ.*, vol. 57, no. 8, pp. 767–788, 2010.

[2] H. L. Chen, L. R. Lattuca, and E. R. Hamilton, "Conceptualizing Engagement: Contributions of Faculty to Student Engagement in Engineering," *J. Eng. Educ.*, vol. 97, no. 3, pp. 339–353, Jul. 2008.

[3] L. Q. Prendergast, "Retention, success, and satisfaction of engineering students based on the first-year experience," Rutgers, The State University of New Jersey, 2013.

[4] G. N. Anderson, E. L. Dey, M. Gray, and G. Thomas, "Mentors and protégés : The influence of faculty mentoring on undergraduate academic achievement," Orlando, FL, 1995.

[5] T. A. Campbell and D. E. Campbell, "Faculty / student mentor program: Effects on Academic Performance and Retention," *Res. High. Educ.*, vol. 38, no. 6, pp. 727–743, 1997.

[6] M. Gasman, "Mentoring Programs for African American College Students in Predominantly White Institutions : Relationships to Academic Success," in *Conference on People of Color in Predominantly White Institutions: Different Perspectives on Majority Rules*, 1997.

[7] E. T. Pascarella and P. T. Terenzini, *How College affects students, Vol. 2: A third decade of research*. San Francisco, CA, United States: Jossey-Bass, 2005.

[8] C. S. Lengsfeld, G. Edelstein, J. Black, N. Hightower, M. Root, K. Stevens, and M. Whitt, "Engineering Concepts and Communication: A Two-Quarter Course Sequence," *J. Eng. Educ.*, vol. 93, no. 1, pp. 79–85, Jan. 2004.

[9] R. S. Heller, C. Beil, K. Dam, and B. Haerum, "Student and Faculty Perceptions of Engagement in Engineering," *J. Eng. Educ.*, vol. 99, no. 3, pp. 253–261, Jul. 2010.

[10] P. R. Pintrich, "A Conceptual Framework for Assessing Motivation and Self-Regulated Learning in College Students," *Educ. Psychol. Rev.*, vol. 16, no. 4, pp. 385–407, Dec. 2004.

[11] C. O. Walker, B. a. Greene, and R. a. Mansell, "Identification with academics, intrinsic/extrinsic motivation, and self-efficacy as predictors of cognitive engagement," *Learn. Individ. Differ.*, vol. 16, no. 1, pp. 1–12, Jan. 2006.

[12] J. I. Rodriguez, T. G. Plax, and P. Kearney, "Clarifying the relationship between teacher nonverbal immediacy and student cognitive learning: Affective learning as the central causal mediator," *Commun. Educ.*, vol. 45, pp. 293–305, 1996.

[13] A. W. Astin, *What matters in college? Four critical years revisited.*, 1st Ed. San Francisco, CA, United States: Jossey-Bass, 1993.

[14] A. W. Chickering and Z. F. Gamson, "Seven Principles For Good Practice in Undergraduate Education," 1987.

[15] K. Kitchener, P. Wood, and L. Jensen, "Curricular, co-curricular, and institutional influence on real-world problem solving," in *American Phychological Association*, 1999.

[16] S. a. Bjorklund, J. M. Parente, and D. Sathianathan, "Effects of Faculty Interaction and Feedback on Gains in Student Skills," *J. Eng. Educ.*, vol. 93, no. 2, pp. 153–160, Apr. 2004.

[17] M. Hoit and M. Ohland, "The Impact of a Discipline-Based Introduction to Engineering Course on improving retention," *J. Eng. Educ.*, no. January, pp. 79–85, 1998.

[18] S. Sheppard and R. Jenison, "Examples of Freshman Design Education," *Int. J. Eng. Educ.*, vol. 13, no. 4, pp. 248–261, 1997.

[19] C. M. Estepp, "The relationships among teacher immediacy, professor / student rapport, and self-regulated learning," University of Floridy, 2012.

[20] J. H. Wilson, R. G. Ryan, and J. L. Pugh, "Professor–Student Rapport Scale Predicts Student Outcomes," *Teach. Psychol.*, vol. 37, no. 4, pp. 246–251, Sep. 2010.

[21] S. a. Meyers, "Do Your Students Care Whether You Care about Them?," *Coll. Teach.*, vol. 57, no. 4, pp. 205–210, Sep. 2009.

[22] J. J. Endo and R. L. Harpel, "The effect of student-faculty interaction on students educational outcomes," *Res. High. Educ.*, vol. 16, no. 2, pp. 115–138, 1982.

[23] T. A. Benson, A. L. Cohen, and W. Buskist, "Rapport : Its Relation to Student Attitudes and Behaviors Toward Teachers and Classes," *Teach. Psychol.*, vol. 32, no. 4, pp. 237–239, 1995.

[24] P. C. Wankat, "Undergraduate Student Competitions," *J. Eng. Educ.*, vol. 94, no. 3, pp. 343–347, Jul. 2005.

[25] A. W. Astin and H. S. Astin, *Undergraduate science education: The impact of different collect environments on the educational pipeline in the sciences*. Los Angeles, CA: Higher Education Research Institute, 1992.

[26] C. M. Kardash, "Evaluation of undergraduate research experience: Perceptions of undergraduate interns and their faculty mentors.," *J. Educ. Psychol.*, vol. 92, no. 1, pp. 191–201, 2000.

[27] G. Kuh, "How are we doing? Tracking the quality of the undergraduate experience, 1960s to present," *Rev. High. Educ.*, vol. 22, pp. 99–119, 1999.

[28] S. R. Cotten and B. Wilson, "Student–faculty Interactions: Dynamics and Determinants," *High. Educ.*, vol. 51, no. 4, pp. 487–519, Jun. 2006.

[29] R. Robinson, D. Culver, M. J. Schertzer, T. Landschoot, and E. Hensel, "Assessing student engagement with faculty outside of class," in *ASME IMECE*, 2014.

[30] A. Ghosh, "Analysis of a feedback assessment loop in engineering sciences core curriculum," in *ASME IMECE*, 2013, pp. 1–11.

Proceedings of the ASME 2014 International Mechanical Engineering Congress and Exposition
IMECE2014
November 14-20, 2014, Montreal, Quebec, Canada

IMECE2014-38698

UNDERSTANDING THE CAUSES FOR LOW STUDENT OFFICE HOUR ATTENDANCE

Risa J. Robinson
Rochester Institute of Technology
Rochester, NY, USA

Dean Culver
Duke University
Durham, NC, USA

Michael J. Schertzer
Rochester Institute of Technology
Rochester, NY, USA

Timothy P. Landschoot
Rochester Institute of Technology
Rochester, NY, USA

Edward C. Hensel, Jr
Rochester Institute of Technology
Rochester, NY, USA

ABSTRACT

It is long understood that many students do not take advantage of faculty assistance outside of class. In an attempt to improve the use of office hours, faculty have made efforts to schedule times that are most convenient to students and are most likely to have high attendance; before homework assignments are due or examinations are being held. Despite these efforts, students rarely take advantage of this support service. As a first attempt to improve student engagement, the number of office hours held by teaching assistants (TAs) was increased, expecting that students would feel more comfortable asking for help from TAs rather than faculty. However, office hour attendance was no better for TAs than for faculty. Yet, exam performance continued to indicate that many students could benefit from help outside the classroom. In an effort to better understand this trend, a survey was conducted to examine reasons why students choose not to attend office hours. In particular, we were looking for the effect of social norms, student's perception of their understanding of the material and their need for extra help, as well as the use of other resources such as on-line solutions to homework problems and cooperative learning with other students. This survey was conducted in six classes (300 students) comprising our engineering science core curriculum, including: Statics, Mechanics of Materials, Dynamics, Thermodynamics, Fluid Mechanics and Heat Transfer. Results indicated that of all the factors tested, the only ones that positively correlated to low office hour attendance were (1) students felt they understood the material well enough and did not need extra help, (2) students procrastinated and therefore did not have time to seek help before homework was due, and (3) students who spent less overall time studying outside of class attended fewer office hours. The data did not support our initial premise that students who attended more office hours performed better. Further study is warranted to explore behaviors that enhance student performance. It is expected that results from these studies will provide information to improve students' efficient use of time outside the classroom.

Key words: student engagement, mechanical engineering, office hour, student performance.

INTRODUCTION

Office hours (OHs) are held by faculty and teaching assistants (TA's) to provide extra help to those students who need reinforcement of material taught inside the classroom with the goal of improving student achievement of course learning outcomes as reflected in examination metrics. In the Mechanical Engineering Department, faculty were concerned that few students were attending office hours. Several approaches have previously been undertaken to improve student participation in office hours. Faculty attempted to increase office hour attendance by setting times that are convenient for students: coordinating office hours with typical student schedules to ensure that their classes do not conflict with faculty office hours, including times before homework assignments are due and adding supplemental OH's prior to examinations. Yet students rarely took advantage of this support service from faculty. In addition, the number of office hours held by teaching assistants was increased, expecting that students would feel more comfortable asking for help from TAs rather than faculty. However, office hour attendance seemed to be no better for TAs than for faculty. Yet exam performance continued to indicate that many students could benefit from help outside the classroom. Anecdotal observations by faculty indicated that the same few students regularly attended OHs.

In this study, we seek to determine the reasons that students choose not to attend OHs. In particular, the faculty are interested in the effect of (a) social norms, (b) student's perception of their understanding of the material and their need for extra help, (c) student use of other resources such as on-line solutions to homework problems, and (d) student practices related to cooperative learning with other students. Surveys were conducted of students in the Mechanical Engineering Department. Office hour attendance trends and student use of resources were analyzed using statistical methods, and results were combined across courses and year levels. We also tested the basic premise that students were doing poorly because they were not seeking office hour help.

LITERATURE REVIEW

A widely used definition of student engagement was coined by Kuh in 2001: Student engagement represents both the time and energy students invest in educationally purposeful activities and the effort institutions devote to using effective educational practices [1]. Kuh was one of the first to develop a framework for student engagement for the National Survey of Student Engagement (NSSE, developed in 1998) based on five benchmarks:

- level of academic challenge
- enriching educational experiences
- active and collaborative learning
- supportive campus environment
- student–faculty interaction

Developing a culture where students regularly interact with faculty outside the classroom has been shown to improve increased motivation [2-4], enhance problem solving skills [5], [5] reflective thinking [6], group dynamics [7], and occupational competence [7]. Students who have regular high quality interactions have and overall student outcomes, [8, 9] and success [10, 11]. Despite the benefits, students tend to have minimal contact with faculty outside of class [12].

Office hours are times set aside to provide students open access to faculty without the barriers of making an appointment. Faculty assume that students see office hours as a low risk welcoming environment, and expect students to take full advantage of the opportunity to ask questions and get clarification on topics that are confusing. However, frequency of office hour attendance is generally low and students do not understand the importance of interacting with faculty [12]. Some attention has been given in the literature to reasons why students do not attend office hours. One study found that establishing positive rapport with students can result in increased office hour attendance [13]. Another studies looked at the relationship between faculty approachability and faculty-student interaction outside of class [8, 14]. The study presented in this paper aimed at determining reasons why students do not attend office hours in an attempt to increase student-faculty interaction outside of class.

METHODS

The survey instrument was administered during class to a total of 301 students. These surveys were conducted in the six classes comprising our engineering science core curriculum: Statics, Mechanics of Materials, Dynamics, Thermodynamics, Fluid Mechanics and Heat Transfer. Students were asked to answer the questions as they pertained to the class in which the survey was taken. Students were asked to self-report their office hour attendance for faculty and teaching assistants (TAs), to report their grade on exam 1 and how many hours, on average per week, they spent studying outside of class (Table 1). Students were then asked to respond to twelve questions related to their study habits and perceptions, on a Likert "agreement scale" from 1 to 5, where 5 was strongly agree and 1 was strongly disagree (Table 2). The questions in Table 2 were designed to test the hypothesis that each of these behaviors and perceptions could be correlated to low office hour attendance.

Table 1. Survey questions related to use of office hours, exam performance and total time commitment outside of class.

q1. How many times have you attended TA-held office hours?
(1) 0 (2) 1-3 (3) 4-6 (4) More than 6

q2. How many times have you attended Faculty-held office hours?
(1) 0 (2) 1-3 (3) 4-6 (4) More than 6

q3. My grade on the first exam was
(1) 100-90 (2) 80-89 (3)70-79 (4) 60-69 (5) Below 60

q4. How many hours do you spend per week outside of class doing homework problems?
(1) More than 10 (2) 6-10 (3) 3-5 (4) 0-2

Table 2. Survey questions related to potential reasons why student are not seeking help outside of class. (Agreement scale: 1=strongly disagree, 2=disagree, 3=neutral, 4=agree, 5=strongly agree)

q5. I work cooperatively with classmates on homework.
q6. I browse the internet/use existing solutions for help solving homework problems.
q7. I ask my professor questions by email.
q8. Office hours/locations should be scheduled for a more convenient time/place.
q9. I understand the material enough and I do not usually require out-of-class help.
q10. I don't have time to ask for help at office hours because I wait too long to start the homework.
q11. My impression is that the TA's are not proficient and helpful.
q12. I am not comfortable asking for help from the TAs outside of class.
q13. I would like more out-of-class help from the TA.

q14. My impression is that the faculty are not proficient and helpful.
q15. I am not comfortable asking for help from the faculty outside of class.

The questions in Table 2 were designed to test the hypothesis that each of these behaviors and perceptions could be correlated to low office hour attendance. The hypotheses associated with each survey question are stated in Table 3.

Table 3. Hypotheses Tested in this Study.

h.5 Students who work cooperatively with classmates on homework are less likely to attend OHs.
h.6 Students who browse the internet/use existing solutions on homework are less likely to attend OHs.
h.7 Students who ask their professor questions by email are less likely to attend OHs.
h.8 Students who perceive OH times or locations as inconvenient are less likely to attend OHs.
h.9 Students who perceive no need for out-of-class help are less likely to attend OHs.
h.10 Students who procrastinate on starting their homework are less likely to attend OHs.
h.12 Students who perceive that the TA's are not proficient and helpful are less likely to attend OHs.
h.13 Students who are not comfortable asking for help from the TAs outside of class are less likely to attend OHs.
h.14 Students who perceive that the faculty are not proficient and helpful are less likely to attend OHs.
h.15 Students who are not comfortable asking for help from the faculty outside of class are less likely to attend OHs.
h.0 Students who attend fewer OHs are less likely to perform well on examinations.

Responses were analyzed first with descriptive statistics to see which behaviors and perceptions were prevalent. Then analysis of variance (ANOVA) was used to assess which behaviors and perceptions might be considered factors in low OH attendance. Finally, we tested our fundamental hypothesis (h0) motivating this study, that low office hour attendance was the reason for low exam performance, or if low performance might be attributed to other factors such as low time commitment on the part of the students outside of class.

RESULTS

Prevalence of Student Behaviors and Perceptions

The histograms of responses for office hour attendance, shown in Figure 1, indicate that overall OH attendance is low for both TA and faculty. Of all students polled, 78% (235 students) reported never attending TA office hours, 45% (134 students) reported never attending faculty OHs, and 41% (123 students) reported never seeking help from either the TA or faculty outside of class.

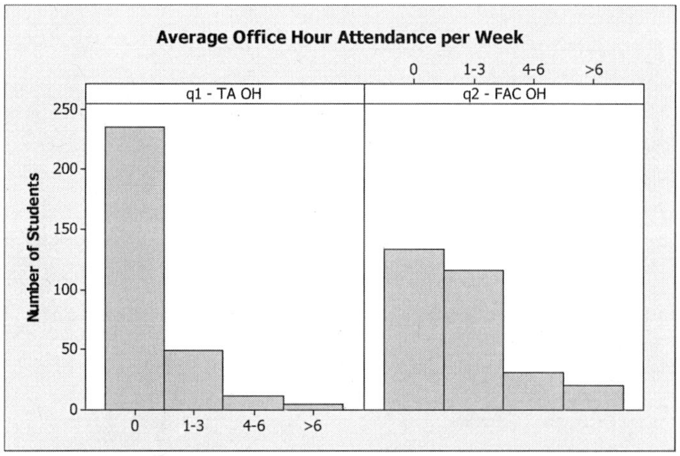

Figure 1. Student reported office hour attendance

Although on average, as indicated by the histograms shown in Figure 2, students did not have negative feelings about either TAs or faculty, there appeared to be a preference for faculty OH over TA OHs. A two sample t-test ($p<0.05$) indicated that more students agreed TAs were not proficient and helpful (mean = 2.9) compared to faculty (mean = 1.7) and more students agreed that they were less comfortable (mean = 2.6) with TAs than with faculty (mean = 1.9). However, a regression analysis indicated ($p<0.05$) that students who frequently attend faculty office hours also frequently attend TA office hours, so apparently students who seek help outside of class are not choosing between faculty and TA but rather are using both.

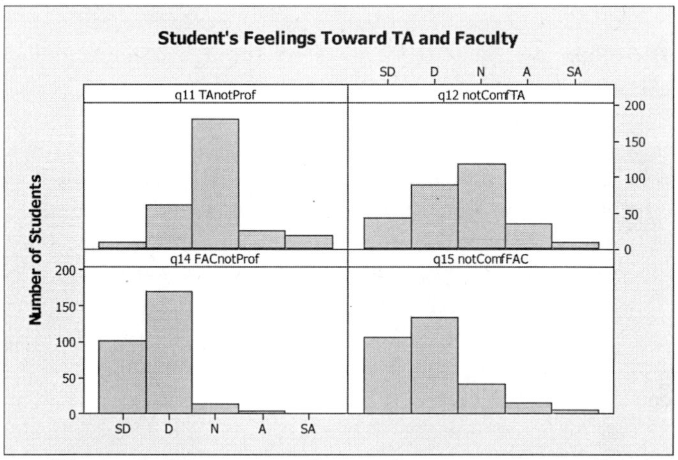

Figure 2. Histograms for student responses to TA (top left) and Faculty (bottom left) Proficiency and Helpfulness and student comfort level with TA (top right) and faculty (bottom right).

Histograms for each of the six questions (q5 through q10) were reviewed to understand which behaviors and perceptions were most prevalent among the students surveyed (Figure 3). Considering whether or not the majority of students either agreed or strongly agreed to each potential factor as an indicator of its prevalence, we found that students are working with peers (q5, 61%), using internet solutions (q6, 54%), and not contacting faculty by email (q7, 14%). The data also showed that 40% of the students perceived that they do not need extra help outside of class (q9) while only 33% of students reported procrastination as a reason for not seeking help (q10). Inconvenient office hour responses were evenly distributed around a neutral response (q8).

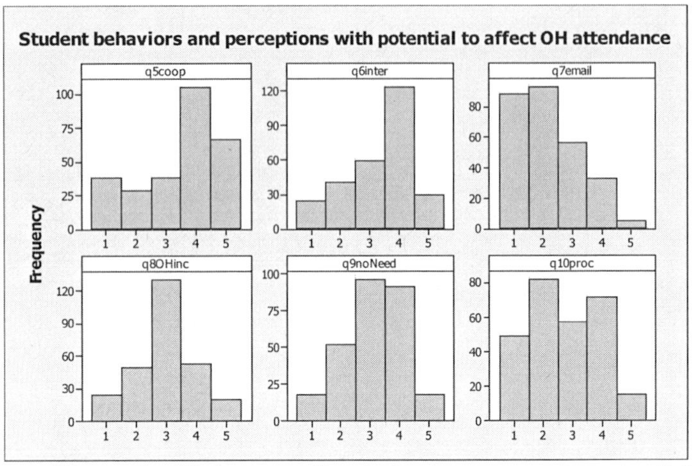

Figure 3. Histograms for student responses to potential reasons for not seeking help outside of class. Cooperation (top left), On-line help (top center), e-mail professor (top right), inconvenient (bottom left), not required (bottom center), and procrastination (bottom right).

The mean response for each factor hypothesized to affect office hour attendance (questions q5 through q16 excluding q13) was ranked from most prevalent (highest average on agreement scale agreement) to the least prevalent (lowest average agreement scale). Means are plotted in Figure 4 for a 95% confidence interval (CI). A response over 3 indicates that the mean of the population agreed to the behavior or perception. The most prevalent behaviors or perceptions, and the only factors to score a mean of 3 or above were cooperative learning (q5), use of internet solutions (q6), and no perceived need for extra help (q9). Inconveniently scheduled OH (q8) scored a neutral and all other factors scored below 3 indicating that the mean population surveyed disagreed. The least prevalent behavior or perception was that students were not comfortable with faculty and that faculty were not proficient and helpful (q14).

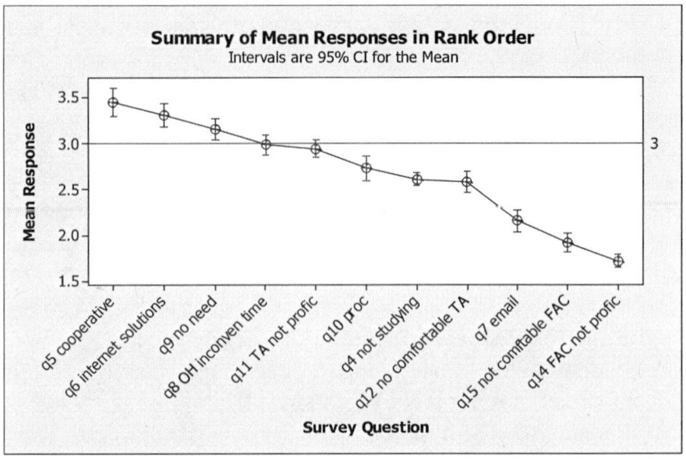

Figure 4. Rank order of mean responses to survey questions. Vertical axis represents the agreement scale where 5 is strongly agree.

Analysis of variance (ANOVA) was used to determine if low office hour attendance could be attributed to any of the student behaviors or perceptions. A main effects plot was created for each question related to student perception of faculty and TAs (Figure 5) and student behaviors and self-perceptions (Figure 6). A negative slope supports our hypothesis that the behavior or perception is decreasing office hour attendance. From the main effects plot and ANOVA it is apparent that potential factors are perceived need (q9) and procrastination (q10). Students who feel that they do not need help outside of class ($p<0.001$) or students who procrastinate ($p<0.01$) attend fewer office hours (Figure 7). We also found that students who were not comfortable with the TA attended more faculty OHs ($p<0.001$). Other factors explored could not be shown to affect office hour attendance with statistical significance.

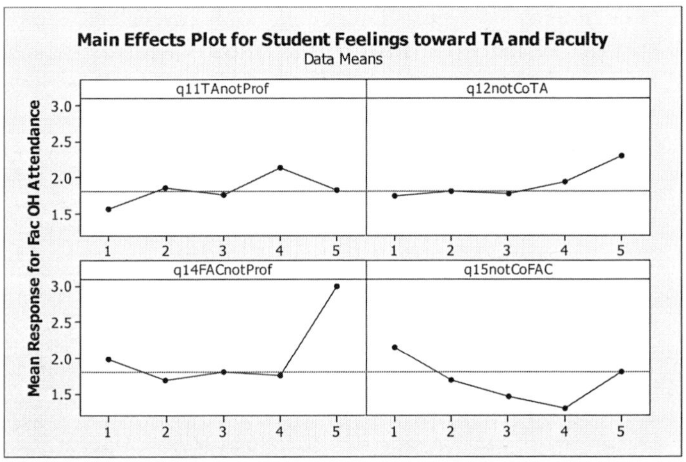

Figure 5. Relationship between student responses to questions regarding feelings toward TA and faculty and student attendance at faculty OH. Higher mean response on the vertical axis indicates more frequent OH attendance.

(1) 0 , (2) 1-3, (3) 4-6, (4) more than 6 times. Horizontal axis represents the agreement scale where 5 is strongly agree.

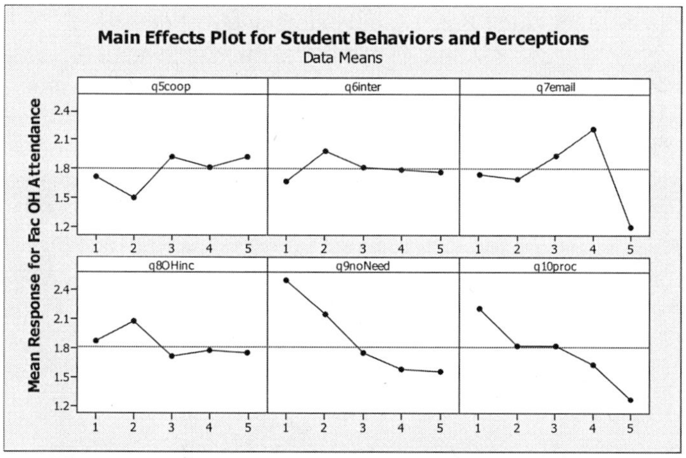

Figure 6. Relationship between student responses to questions regarding behaviors and perceptions and student attendance at faculty OH. Higher mean response on the vertical axis indicates more frequent OH attendance. (1) 0 , (2) 1-3, (3) 4-6, (4) more than 6 times. Horizontal axis represents the agreement scale where 5 is strongly agree.

Effect of OH Attendance on Performance

The original assertion that low OH attendance affected student performance was not supported by the data. An ANOVA showed that there was no significant correlation between either TA (q1) or faculty (q2) OH attendance and exam performance (q3). Even when the responses to q1 and q2 were combined, there was no significant correlation between total office hour attendance and performance.

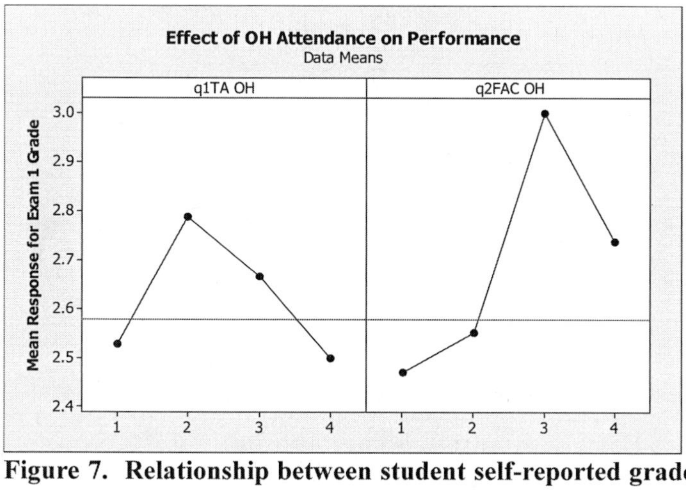

Figure 7. Relationship between student self-reported grade on the first exam and office hour attendance. A higher mean response on the vertical axis indicates a lower mean exam score. (1) 100-90% (2) 80-89% (3) 70-79% (4) 60-69% (5) below 60%. A higher value on the horizontal axis represents more frequent OH attendance (1) 0 (2) 1-3 (3) 4-6 (4) more than 6 times.

CONCLUSION

The hypothesis (h0) that students who attend fewer OHs are less likely to perform well on examinations is not conclusively supported by the student reported data. The most prevalent behaviors or perceptions were (q5) cooperative learning, (q6) use of internet solutions, and (q9) no perceived need for extra help. Student perceptions regarding (q8) convenience of scheduled OHs were neutral. The mean population surveyed disagreed with all other factors studied. The least prevalent student perceptions were (q15) that students were not comfortable with faculty and (q14) that faculty were not proficient and helpful. From the main effects plot, it is apparent that (q9) perceived need and (q10) procrastination are negatively correlated with student attendance at OHs.

RECOMMENDATIONS

`Further study is necessary to correlate student behaviors with examination performance. A follow up study is proposed wherein two behavior metrics including (a) graded weekly participation in OHs, (b) graded weekly participation in on-line quizzes are correlated with student achievement on multiple choice questions on examinations. These behavior metrics are chosen to eliminate the perceived need and (as reflected by grading policy) and procrastination (as reflected by weekly expectations) inhibitors to office hour attendance. In addition, the revised study should perform the assessment using observed performance data, as opposed to student reported survey data. The proposed study is currently under way and will be reported in a future paper.

ACKNOWLEDGMENTS

Financial support of this work from the Rochester Institute of Technology is gratefully acknowledged.

REFERENCES

[1] Kuh, G. D., 2009, "The National Survey of Student Engagement: Conceptual and Empirical Foundations," New Directions for Institutional Research, 2009(141), pp. 5-20.

[2] Pintrich, P. R., 2004, "A Conceptual Framework for Assessing Motivation and Self-Regulated Learning in College Students," Educational Psychology Review, 16(4), pp. 385-407.

[3] Walker, C. O., Greene, B. A., and Mansell, R. A., 2006, "Identification with Academics, Intrinsic/Extrinsic Motivation, and Self-Efficacy as Predictors of Cognitive Engagement," Learning and Individual Differences, 16(1), pp. 1-12.

[4] Rodriguez, J. I., Plax, T. G., and Kearney, P., 1996, "Clarifying the Relationship between Teacher Nonverbal Immediacy and Student Cognitive Learning: Affective Learning as the Central Causal Mediator," Communication Education, 45(pp. 293-305.

[5] Pascarella, E. T., and Terenzini, P. T., 2005, How College Affects Students, Vol. 2: A Third Decade of Research, Jossey-Bass, San Francisco, CA, United States.

[6] Kitchener, K., Wood, P., and Jensen, L., "Curricular, Co-Curricular, and Institutional Influence on Real-World Problem Solving," Boston, MA.
[7] Bjorklund, S. A., Parente, J. M., and Sathianathan, D., 2004, "Effects of Faculty Interaction and Feedback on Gains in Student Skills," Journal of Engineering Education, 93(2), pp. 153-160.
[8] Cox, B. E., Mcintosh, K. L., Terenzini, P. T., Reason, R. D., and Quaye, B. R. L., 2010, "Pedagogical Signals of Faculty Approachability: Factors Shaping Faculty-Student Interaction Outside the Classroom," Research in Higher Education, 57(8), pp. 767-788.
[9] Chen, H. L., Lattuca, L. R., and Hamilton, E. R., 2008, "Conceptualizing Engagement: Contributions of Faculty to Student Engagement in Engineering," Journal of Engineering Education, 97(3), pp. 339-353.
[10] Astin, A. W., 1993, What Matters in College? Four Critical Years Revisited, Jossey-Bass, San Francisco, CA, United States.
[11] Chickering, A. W., and Gamson, Z. F., 1987, "Seven Principles for Good Practice in Undergraduate Education."
[12] Cotten, S. R., and Wilson, B., 2006, "Student–Faculty Interactions: Dynamics and Determinants," Higher Education, 51(4), pp. 487-519.
[13] Benson, T. A., Cohen, A. L., and Buskist, W., 1995, "Rapport : Its Relation to Student Attitudes and Behaviors toward Teachers and Classes," Teaching of Psychology, 32(4), pp. 237-239.
[14] Snow, S. G., 1973, "Correlates of Faculty-Student Interaction," Sociology of Education, 46(4), pp. 489-498.

Proceedings of the ASME 2014 International Mechanical Engineering Congress and Exposition
IMECE2014
November 14-20, 2014, Montreal, Quebec, Canada

IMECE2014-39118

EASILY ADOPTABLE INTERACTIVE TEACHING PRACTICES AND STUDENTS PROGRESS MONITORING STRATEGIES

Pawan Tyagi
Mechanical Engineering
School of Engineering and Applied Science
University of the District of Columbia
Washington, DC, 20008, USA
Email:ptyagi@udc.edu

ABSTRACT

An active class room teaching practice can become highly rewarding for students. An instructor practicing active learning approaches may get significantly higher success in inculcating course materials deeply as compared to a lecture based teaching. However, transitioning from prevailing lecture based instruction to an active learning approach can be hampered by the reservations and prejudices of an engineering educator; a tenure track faculty may find it even more challenging to leave the traditional lecture based teaching approach and adopt an active teaching approach. This paper will describe the active teaching techniques that I, a tenure track faculty, has been practicing to teach mechanical engineering courses; the main discussion will focus on the Fuel Cell Science and Technology course. I have devised my current deep learning and teaching strategies through a yearlong Myrtilla Miner Faculty Fellowship cohort under Dr. Ken Bain and a number of workshops on the group based active teaching and peer interaction based teaching. This paper describes the strategies for developing a teaching and assessment plan for the courses I teach by emphasizing on (i) designing significant learning outcomes before starting a class, (ii) the long term retention of key concepts of a course by fostering student centered deep learning course activities, and (iii) far transfer of the skills students gain from a course. The first topic of this paper is about various strategies to understand students' motivations and inhibitions that may govern their learning curve in a course. The second topic of this paper discusses the crucial aspect of designing a promising syllabus to give students a bigger purpose for learning the course material; a promising syllabus attempt to connect students' long held curiosities and career ambitions with the course to be offered. The third topic delves into the strategies to engage students in self-preparation to assimilate the key concepts to be discussed in a class. This paper will also highlight the approach to design conceptual quizzes to guide student preparation before they come to the class and then use the same conceptual quizzes to conduct peer discussion and define the flow of a class; this strategy is derived from Dr. Eric Mazur's work on peer interaction based teaching. The fourth topic is about the utility of one pager feedback form to be filled by the students after every class. This paper will discuss structure and effectiveness of the feedback form in improving student attention and participation in the class discussions. I have offered two workshops on effective teaching at the University of the District of Columbia to promote active student learning in a wide range of courses. I plan to conduct workshops for the middle and high school teachers to share the effective teaching skills.

INTRODUCTION

An active learning approach is considered a better means to teach course concepts more effectively than traditional lecture based teaching [1]. Active learning is any activity that involves students in doing things and thinking about the things they are doing. Felder & Brent describe active learning as "anything course-related that all students in a class session are called upon to do other than simply watching, listening and taking notes" [2]. There are several strong reasons to advocate the selection of an active learning based class room teaching. An active learning approach can encompass both isolated and highly structured activities to motivate students to take charge of their deep learning [3]. Active learning can be applied to both small [3] and big [4] classes; active learning was found to produce higher and long lasting student learning as compared to the traditional lecture format based teaching [5]. Research studies show that the audience's attention in lectures starts to decline after 10-20 minutes [3, 4, 6, 7]. Incorporating active learning techniques encourage student engagement throughout the class [8]. Active learning also reinforces course contents, concepts, and skills and their long term retention [3]. During active learning students get more frequent and immediate feedback about the depth and accuracy of the material they are focusing [9]. An active learning approach is very effective in addressing students' stereotypes and different student learning styles [10]; a lecture based teaching may give nil to negligible consideration to these crucial factors in the student learning [11]. Active learning can be very effective in creating personal connections between students and the course material, which strongly increases the student's motivation to learn proactively [11]. In addition to the course content, Active learning develops life skills like improving with others feedback, collaboration, and brainstorming to reach the most rational answers. An active learning approach also promotes a sense of community in the classroom by augmenting student-student and instructor-student interaction.

After realizing the aforementioned advantages I started learning various active learning approaches. To learn about various aspects of active learning I attended several workshops by a number of active learning proponents like Dr. Erik Mazur, Dr. Dee Fink, Dr. Prince and Dr. Ken Bain. This paper discusses different approaches I have adopted for my class room teaching. Here I mainly focus on 'MECH 488 Fuel Cell Science and Technology' course which is an elective course and generally has a small enrollment.

TEACHING STRATEGY

Before the commencement of this course in spring 2013, I explored the answers to the following questions: What are my student learning objectives for this course? How to understand student's negative stereotypes and learning style? How to connect students to the course topic early on to motivate them to learn course material? How to ensure students are present in the class? How to continuously motivate students to do their best without blackmailing them for grade? How to make learning fun in the class? How to effectively access and measure depth of their learning? The rest of this paper is organized as answers to these questions and the final outcome.

1.What are my student learning objectives for this course?: After completing this course I expect that students should be able to (1) understand global warming issues and how to address it by using renewable energy technologies like fuel cells, (2) understand the basic components and various types of fuel cells systems and have the ability to decide the suitability of these fuel cells for different applications, (3) understand the thermodynamics concepts behind converting chemical energy into electrical voltage, (4) understand the electrode kinetics concepts behind converting fuel into electrical current, (5) ability to design fuel cell components like electrodes and electrolytes to improve fuel cells, (6) characterize fuel cell systems and electrodes experimentally, and (7) communicate in writing and orally, the science and engineering concepts. After defining the course objectives I focused on the following question.

2. How to understand student's negative stereotypes and learning style?: To understand students mentality and their level of preparedness I conduct a pre-course survey and ask questions, which directly or indirectly gauge students' familiarity with the topic. Other questions focus on how they learn best, what is their primary motivation to attend this course etc. I also asked about the students' hobbies and their most cherished moments; discussing nontechnical points enable students to connect with other students and start becoming comfortable in communication. This survey contains direct or subtle questions to understand the students' motivation and aptitude.

3. How to connect students to the course topic and learning objectives in the beginning of the class?: The Maslow chart is widely acknowledged to describe a hierarchy of needs which motivate humans to do any activity. The first three categories are considered basic needs and the last two categories are considered as higher level needs. Every student is different and presumably influenced by one or more categories of the motivations described in Table 1.

Table 1: Maslow chart of hierarchy of needs providing motivation for human action.

Order of motivation	Category of needs	Action
1	Physiological	Breathing, food, water, sleep,
2	Safety	Security of body, family, employment, health, property
3	Love and belongingness	Friendship, family, community.
4	Esteem	recognition by others Self-esteem, confidence,
5	Self-actualization	Inner bliss from own work, creativity, lack of prejudice

To connect students with the course content I start the first class by telling a motivational story that encompasses a wide range of the motivational categories mentioned in Table 1. This story is also known as 'promising syllabus' [12] and basically designed to boost the student's interest in the subject matter. The sample I used in spring 2013 and 2014 is the following:

Promising syllabus story for the first day of the class:

Question [1] Can there be a single technology producing energy to power small appliances like laptops as well as giant industries; automobile to space ship application?
Steer students' focus on the topic in the course and tell a vivid story.

Let me share one story of a struggling engineer who saw a new hope for his life and society in fuel cell technology that you will learn from this course

Recently I met an engineer named John. After two years of service, John lost his job in a coal power plant in 2008. Actually, this coal plant was a major culprit in polluting air and water and was no longer operational as per the new environmental regulations and economic reasons. John tried hard to get a sustainable and exciting job for six months. According to his credentials, he qualified for the job in a nuclear power plant. In spite of his unwillingness he had to accept the offer to serve as a process engineer in the nuclear power plant. Unfortunately, only after three months of working on the job this nuclear power plant started emitting radioactive pollutants due to a crack in the reactor. John was forced to quit the job with a small compensation.

Question#2: What will you do if you are in John's situation?
After commenting on students' responses I start telling the next phase of the story on what John did.

John started looking for a sustainable job once again. During this uncertain phase, John attended an open house about renewable energy and their social and economic benefits. However, John was suspicious about the fickleness of the renewable energy resources like solar and wind etc. But he decided anyway to explore career opportunities in renewable energybecause this field is good for the environment and is rapidly growing.

Question#3: What are the main renewable energy technologies and what are their advantages?
Like you, John got impressed with the undeniable importance of renewable energy resources and technology.

John realized a new passion for becoming an expert in the renewable energy production. However, John wanted to make sure that his chosen area in the renewable energy field has robust and sustainable career prospects.

Question #4: If you were in John's place, what would you like to know about different renewable technologies before choosing one of them as your career?
Take students' comments and steer their attention towards the unique advantages of Fuel cell technologies-the topic of course.

For some expert advice, he planned to visit the Mechanical Engineering department at UDC. One fine day John took a bus. From an informal discussion with the bus driver he came to know that this bus was using hydrogen as a fuel and emitted water, instead of carbon di-oxide that causes global warming. This water was so clean that he could fill it in a bottle and drink it like spring water from Colorado Springs. While travelling in this special bus he observed that a neighboring female passenger was working on a fancy laptop, which was powered by a fuel cell battery. John could not confine his curiosity about the fuel cell battery and he asked the lady about the advantages of fuel cell battery. John discovered that unlike a conventional battery, you can keep using a fuel cell battery endlessly, and hence they don't produce battery waste and heavy metal pollution. Then, John arrived at the bus stop and walked across the street to catch a connecting bus. Right behind the bus stop was a Wall Mart superstore. Its signboard said that this Wall Mart produces all the electricity within the store by a fuel cell generator. John really got pleasantly surprised that the same fuel cell technology that was powering a laptop was also powering a big superstore. Even before getting to the ME department at UDC, John made up his mind to become an expert of this amazing technology, which is not only full of career prospects, but is also sustainable, and environmentally friendly.

Question #5: What will you do next if you are so excited about the career prospects of the fuel cell technology and its uniqueness?
Steer students' attention towards understanding the fundamentals of this technology tohelp the become familiar with the concepts..

Now dear students you have the opportunity to learn the core principles of the amazing fuel cell technology, which may empower you to solve some of the most pressing problems USA and global community is facing. I invite you to deeply understand the science and technology of Fuel Cells by actively participating in this semester long course.

I currently don't have data to show the effectiveness of the promising syllabus in motivating students. However, according to previous research the 'promising syllabus' activity plays a pivotal role [12].

<u>4.How to ensure students are present in the class?: </u>At UDC many engineering students work full time or part time. It is sometimes a challenging task to ensure their presence in the course. An ven more important task is to ensure that they actively learn when present in the class. To increase students' active presence I ask students to fill out a five minute survey form after the every class; this survey form is given to students within the first 10 minutes of the class and is filled out in the presence of the instructor. This survey has three questions. (i) What were the main concepts we discussed today, (ii) where can you apply these concepts, and (iii) Do you have any questions about the concepts we discussed today. These after-class surveys are graded after each class and account for 10% marks counted toward their grade. Before I started using these,

the average student missed ~10 classes/semester. These surveys helped reduce the absence (Fig. 1). Figure 1 indicates that only one student missed more than two classes. However, one possible reason for the higher attendance may be students' sincerity and self-motivation. More studies are being planned to investigate the effect of end-of the class feedback forms. One possible study may be to study students of the same class for a certain number of classes with and without using a feedback form. In 2014 five students attended the fuel cell course. It was noted that students exhibited low tendency of missing classes without prior notice.

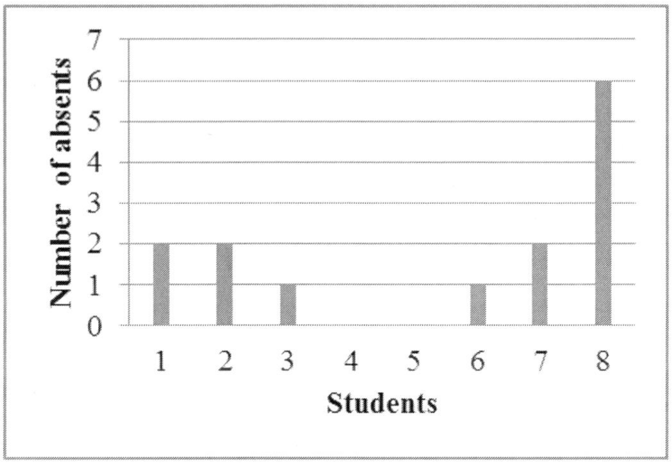

Fig. 1: Number of absents during MECH 488: Fuel Cell Science and Technology course, spring 2013.

5.How to continuously motivate students to do their best to learn the course concepts?: The Major teaching strategy is based on conceptual questions on the major topics. For every chapter or module, major concepts are identified and a number of original conceptual questions are framed. Conceptual questions are generally multiple answers type questions and arranged in terms of a pre-class quiz; however, recently I started asking students to write few lines or sketch about the theme of conceptual questions as necessary. This pre-class quiz and associated reading assignments are prescribed to students before the commencement of the class. Answers to these pre-class quizzes are required to be submitted before the discussion in the class and are also used for grading purposes. These conceptual quizzes ensure that students make themselves prepared for the module to be discussed and become well qualified to grasp the content and concepts during the class discussion.

During the class, conceptual quizzes were used to guide the flow of discussion. Every question on the quiz was first discussed by a small group of students; subsequently students shared their views or argued with each other about the conceptual questions. After the group discussion I invite one representative from each group to share their cumulative or mutually agreed solution and underlying reason with the whole class. I elaborate on student groups' responses and generally add missing reasoning, logic, and information to complete the discussion about the conceptual question of interest.

I also noticed that sometimes a module or math intensive topic may be quite challenging for students to grasp themselves by independent reading. For such modules I ask students to do timed reading and comprehension and share their understanding with the group members. The one representative from the group summarizes the cumulative group understanding. For such difficult topics no gradable pre-class quiz is asked to be completed. However, a reading assignment is given. For such difficult or heavy math based concepts I turn to semi-interactive lecturing about that topic. However, I tried to incorporate multiple interactive strategies to keep students engaged during the class.

6.How to make learning fun in the class?: To ensure that students are actively learning, a number of students activities are set up for them. Since most of the important concepts of this course are dynamic in nature and students are highly likely to benefit by actively participating in it. Occasionally I ask students to design and play small skits about the key course concepts. Wherever possible I also play a role in the skit. I liberally use YouTube videos to aid the comprehension of complicated concepts. To keep the class discussion dynamic I frequently ask students to present whatever they understand about the concept of interest. I also bring commercial components/products and organize industrial visits to industries. For instance, in spring 2013 an industrial visit was organized to visit the fuel cell application. To further deepen the understanding of key concepts, I include a number of experiments to provide experiential learning such as a fuel cell trainer, Versastat electrochemical setup, hydrogen fueling station, fuel cell car etc. To cover a number of important and interesting topics associated with the course theme, students are asked to write a term paper on related topics.

7.How to effectively assess and measure student learning?: In order to assess student learning a number of means were employed. A student's response to the pre-class quizzes before and during the class is one direct measure to do so. The five minute feedback form after the class is another effective means to determine the students' understanding and address any questions they have about the discussion on concept. Traditional home assignments are also assigned to promote collaboration and extensive understanding about the underlying principles of main concepts. I have made significant modifications in the method of giving midterm and final exams. Nearly 60% marks of an exam are based on the quality and depth of the written solutions. In addition, 40% of the marks are based on in class presentation pertaining to the exam questions. For example if a midterm or final exam is 20 marks in total, then 12 marks (60% of total 20 marks) will be assigned for the written answers and 8 marks (40% of total 20 marks) will be assigned for the student presentation and discussion on the exam questions. Typically two randomly selected questions are asked for the in-class discussion.

In addition, I also conducted a survey about the active learning approach. For instance, seven students who participated in the survey during spring 2013 rated active learning approach to be highly desirable (Fig. 2). Similar student response was recorded in 2014. However, student enrollment was only five students. The student rating for the active learning was 4.2±0.83.

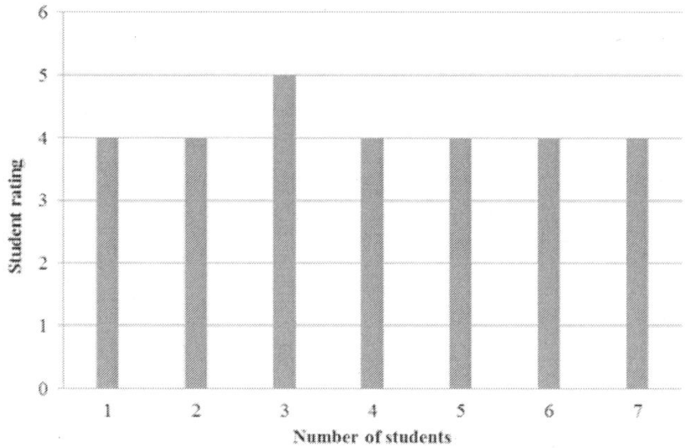

Fig. 2: Student rating for the active learning approach adopted in MECH 488 Fuel cell Science and Technology course (2013).

CONCLUSION

This paper discussed the active learning approaches used in MECH 488 Fuel Cell Science and Technology course at the University of the District of Columbia. Promising syllabus, peer interactions, end-of-class feedback forms, skits, and experiential learning were employed to deepen students' understanding and make learning enjoyable. According to the end of the course survey students enrolled in spring 2013 and 2014 courses gave high ratings to the active learning approach. The author has also organized two workshops about the "Easily adoptable active teaching approaches" at UDC as a part of faculties' professional development event. Both workshops were independently evaluated and found to be highly effective in catalyzing interest in active teaching at UDC.

ACKNOWLEDGMENTS

I acknowledge Dr. Ken Bain for mentoring the author for one year during Myrtilla Miner Faculty Fellowship (2012-2013) and sharing his valuable insights about effective teaching. I also acknowledge National Science Foundation-Research Initiation Award (Contract # HRD-1238802) for supporting the course's buyout time to enable him to practice deep learning approaches.

REFERENCES

[1] Hake, R. R., 1998, "Interactive-engagement versus traditional methods: A six-thousand-student survey of mechanics test data for introductory physics courses," American Journal of Physics, 66(1), pp. 64-74.

[2] Felder, R. M., and Brent, R., 2009, "Active learning: An introduction," ASQ Higher Education Brief, 4, p. 2.

[3] Prince, M., 2004, "Does active learning really work? A review of the research," Journal of Engineering Education. , 93. (3), pp. 223-231.

[4] Crouch, C. H., and Mazur, E., 2001, "Peer Instruction: Ten years of experience and results," American Journal of Physics, 69(9), pp. 970-977.

[5] Thacker, B., Kim, E., Trefz, K., and Lea, S. M., 1994, "COMPARING PROBLEM-SOLVING PERFORMANCE OF PHYSICS STUDENTS IN INQUIRY-BASED AND TRADITIONAL INTRODUCTORY PHYSICS COURSES," American Journal of Physics, 62(7), pp. 627-633.

[6] Johnstone, A. H., and Percival, F., 1976, "Attention breaks in lectures," Education in Chemistry, 13, pp. 49-50.

[7] McDermott, L. C., 1991, "MILLIKAN LECTURE 1990 - WHAT WE TEACH AND WHAT IS LEARNED - CLOSING THE GAP," American Journal of Physics, 59(4), pp. 301-315.

[8] Smith, M. K., Wood, W. B., Krauter, K., and Knight, J. K., 2011, "Combining Peer Discussion with Instructor Explanation Increases Student Learning from In-Class Concept Questions," Cbe-Life Sciences Education, 10(1), pp. 55-63.

[9] Hufnagel, B., 2011, Innovative Strategies for Empowering Your Students to Become Active, Responsible Learners, Astronomical Soc Pacific, San Francisco.

[10] Bain, K., 2012, What the best college students do, Belknap Press

[11] Bain, K., 2004, What the best college teachers do, Harvard University Press.

[12] Bain, K., 2004, What the Best College Teachers Do Harvard University Press.

This page left blank intentionally.

Proceedings of the ASME 2014 International Mechanical Engineering Congress and Exposition
IMECE2014
November 14-20, 2014, Montreal, Quebec, Canada

IMECE2014-36552

INTRODUCING CFD AND WIND TUNNEL TESTING IN AN UNDERGRADUATE FLUID MECHANICS COURSE

Wael Mokhtar and Shirley Fleischmann
Grand Valley State University
Grand Rapids, MI, USA

ABSTRACT

CFD is following the trend of CAD and FEA to undergraduate education especially with recent advances in commercial codes. It will soon take its place as an expected skill for new engineering graduates. CFD was added as a component to an experiment in a junior level fluid mechanics course. The objectives were to introduce CFD, as an analysis tool, to the students and to support the theoretical concepts of the course. The students were asked to complete an experimental two-dimensional study for a wing in a wind tunnel, to use CFD to simulate the flow, and to predict the aerodynamic lift using CFD as well as the experimentally obtained pressure distribution. In addition, they had to compare their results to published data for the studied wing.

Details of the course, the wind tunnel test and the CFD simulations are presented. Samples from the students' work are used in the discussion. The lab activities were successfully completed by the students and the learning objectives were well addressed. One of the valuable outcomes from this lab was the opportunity for the students to integrate multiple fluid mechanics analysis tools and learn about the limits for each tool. CFD also enhanced the learning in the lab activities and increased students' interest in the subject.

INTRODUCTION

In the last two decades, engineering hands-on skills have expanded to include numerical methods. CAD was the first skill to make its way to industry. Later, FEA was added as a powerful design tool. It is now expected that a mechanical engineering graduate will know the basics of CAD and FEA. Most engineering schools offer courses to support these design skills. Software developing companies are competing to develop more user-friendly packages with wider varieties of built-in tools to attract more users. From an educational point of view, it is very important to teach the students the limits of these numerical tools and stress the fact that physical testing is not being totally replaced by numerical simulation. The integration of experimental and numerical methods should be the goal of the users.

Computational Fluid Dynamics (CFD) is currently following the same track to be added to hands-on skills. Starting as a research tool taught at the graduate level, it has recently started to appear in undergraduate level courses. Mokhtar [1] introduced CFD to an undergraduate fluid mechanics course. In his course, he introduced basic concepts of CFD through a set of projects in the lab where the students built their skills by a practice-learning approach. He indicated that the use of CFD supported the introduction of critical thinking and creativity. Later Mokhtar [2] expanded his Project-Based Learning to teach an undergraduate CFD course. Recently, Mokhtar [3] discussed the introduction of a wind energy project in his CFD undergraduate course. In his examples, the author presented assessment data to show the success of introducing CFD in the undergraduate level. He presented a balanced approach between software training and theoretical foundation. He used a commercial CFD package (Star CCM+) in these courses. For more than a decade, instructors have been exploring the introduction of CFD to undergraduate courses; Hailey et al [4] introduced CFD concepts in junior level fluid mechanics course. Their goals were: to support the fluid mechanics topics of the course, to motivate the students to take the advanced CFD course, and to introduce the students to CFD tools used in summer internships. In their courses they used an in-house CFD solver. The authors indicated the success of introducing the CFD at undergraduate level with some minor concerns on the students' responses to the software.

Sert et al [5] used CFD as a teaching tool for a basic fluid mechanics course. In their approach, the students were given virtual flow lab software (FlowLab) where they used a simplified version of CFD to solve basic problems. Details of the numerical methods were not the focus, instead, CFD general concepts such as meshing, boundary condition and flow properties were the main objectives. Blekhman [6] presented further evaluation of using FlowLab to support experiments in a fluid mechanics course. He also indicated the success of using CFD as a teaching tool for undergraduate students with a limited introduction for its concepts.

Cummings et al [7] discussed the development of an undergraduate CFD aerodynamics course in the US Air Force Academy. The focus of the course was on the use of CFD as a practical tool. Topics such as accuracy, stability, meshing and turbulence modeling were covered in the course. Commercial software packages were used for meshing and post-processing such as GridGen, FieldView, Tecplot and MatLab. In-house panel method (Vortex-lattice) software was used as solvers. The authors indicated that advanced topics such as the details numerical algorithms and turbulence modeling were not deeply covered. They also reported that, the course helped the students to better understand fluid mechanics concepts and introduced them well to CFD as a design and analysis tool.

In another example, Guessous et al [8] developed a hands-on undergraduate course in CFD. Through a set of experiments and the use of a commercial code (Fluent), senior students were introduced to CFD as a testing tool without getting into the details of it. The authors reported that the course was successful and the learning objectives were met. The objectives of this approach were to introduce CFD as a "black box" tool and support the fluid mechanics concepts taught in the course. Navaz et al [9] presented two courses for undergraduate students where CFD was introduced. The first was an applied CFD course where basics of CFD including some solver details were covered in the lecture while in the lab the students used CFD commercial codes (Fluent) and a couple of pre- and post-processing tools. In the second course CFD was utilized to teach compressible flow without focusing on the details for CFD methods. Both courses had experimental lab activities where CFD results were verified. They reported that the two courses were successful to introduce the students to CFD and teach them through practice the limits of both computational and experimental tools.

Bullough et al [10] developed modules in fluid mechanics lab where the students studied several designs of diffusers using CFD, experimental and analytical methods. Fluent was used for the CFD part and it was used as an analysis tool without much details about its theory. The objective was to teach the students the importance of using these three methods and to get them to observe the limits for each approach. The authors indicated that their students learned through this hands-on experience some of the challenges and limits to generate accurate results using CFD. As a design tool, Stubley et al [11] introduced CFD to undergraduate students. In their course, a brief overview about CFD concepts was presented in the beginning of the course. Then, a set of pre-computed simulations was given to the students where the focus was to use this CFD tools to analyze complex flow. The simulations were made by a commercial package (CFX) and the students used the post-processing tool that came with the package in their study. The authors indicated that using CFD as teaching tool was helpful to add the realization and visualization aspect of fluid mechanics. Stern et al [12] presented the details of a virtual fluid mechanics lab that was developed by CFD. In this tool, details CFD concepts were not the main focus. The main objective was to provide fluid mechanics instructors with an easy to use teaching tool to support the concepts covered in their courses.

Burban [13] used CFD as a design tool for a team of undergraduate students in the SAE Super-Mileage competition. As he reported, over a year of mentoring, his team was successful to utilize CFD in the design of their vehicle. The focus on this case was the correct use of CFD with limited information about the theoretical concepts. Mazumder [14] mentored undergraduate students in a CFD research study. He reported that the students were successful to use a commercial package (Fluent) to complete a two phase flow study. He indicated that the students utilized CFD with limited knowledge in its theoretical details. Their results compared well with the published experimental data.

PRESENT APPROACH

It can be noticed that over the last two decades CFD has been introduced to undergraduate students. The advances in commercial software could be one of the reasons that helped this transition. Four forms of CFD introduction can be summarizes as follow:

- An applied CFD course with limited background in CFD methods.
- Project-based learning for CFD packages.
- Introducing CFD as a teaching tool in undergraduate courses without focusing on its details.
- CFD as a design and research tool for extracurricular students' activities.

It is clear that CFD, as an applied tool, is ready for undergraduate students. In the present approach, CFD was introduced to support a lab activity in junior level undergraduate fluid mechanics course. Basics of CFD were introduced and the students completed a wind tunnel test for a wing. They then compared results from both the experimental and CFD study to the published data for the studied wing. Details of the lab philosophy are presented follow by the wing wind tunnel test details and the CFD component that was added. Results from the students' work are also presented.

DESIGN IN THE FLUID MECHANICS LAB

In the mechanical engineering program at Grand Valley State University, fluid mechanics course is taught in the end of the junior year as the second required course in the thermo-fluid track. The first course is thermodynamics and the last course in heat transfer. Throughout the thermo-fluid track, design skills are introduced in the form of design projects, lab activities and research topics. Fleischmann et al [15] presented more information about the design philosophy in the thermo-fluid track. Fluid mechanics is the first course to include a lab in the track. The lab is a three-hour session every week and the lecture part is also three hours. The focus in this paper is on the lab portion. The lab activities support the theoretical part of the course and introduce design skills in the form of experimental design and a final design and build project.

Each week, the lab starts with a brief introduction of the experiment objectives and the tools that can be used in the study. Then the students form teams and develop their experimental plan. After the instructor approval, each team preforms the experimental study. Then, each student is individually responsible to report this data and interpret the results.

PREPARATION FOR THE WING EXPERIMENT

The wing experiment occurred at the middle of the semester. In preparation to understand the concept of determining force by integrating a surface pressure distribution on flat as well as curved surfaces, the students completed homework problems and an experiment using a hydrostatic pressure distribution. This experiment was completed by the third week of the semester. Just before the wing experiment they completed a measurement of drag on a circular cylinder using an experimentally obtained surface pressure distribution. In this case the model is very similar to the wing model in that it vertically spans the 1 foot by 1 foot test section using the top and bottom of the test section as end plates to assure 2-D flow. Pressure taps are located around the mid-section of the cylinder at 15° increments just as the surface pressure taps are located around the mid-section of the wing at various percent of chord on the top and bottom of the wing. And also with a 4 inch diameter, the cylinder shows blockage effects just as the wing does at high angles of attack. The models are physically similar.

In our wind tunnel the maximum Re_D is $2.2x10^5$ which, for a smooth cylinder assures a laminar attached boundary layer with a turbulent wake and an early separation point. The speed is close enough that a slight trip (a sand strip on the front) will transition the boundary layer and on half of the cylinder students will observe very low pressures associated with high acceleration in the flow outside the boundary layer affecting the attached flow. Flow separates much later on this half and follows the ideal flow ($C_P=1-4\sin^2(\Theta)$) much longer. This allows students to see what separated flow looks like so that they will recognize stall from the surface pressure distribution. They will recognize regions of favorable and unfavorable pressure gradient and the effects on the surface pressure distribution when flow stays attached longer – as it will on an un-stalled wing.

The blockage effect also clearly shows up as C_P values even less than -3 as predicted by the ideal flow with perfect slip- but only for the attached flow (the turbulent boundary layer side). The regular geometry of the cylinder enables the students to evaluate drag coefficient as an integral of the pressure coefficient times the cosine of the angle relative to forward stagnation. For the wing, with a more complex geometry – especially at various angles of attack – it is clear that the lift can be evaluated by this method of integrating the pressure distribution but the drag would be difficult. This sets the stage for recognizing the benefit of CFD modeling. Because it is also easy to compare the predicted pressure coefficient for ideal flow to the experimental pressure coefficient, students are prepared to expect differences between viscous and inviscid solutions. This is easily obtained with the software used.

WING EXPERIMENT

At this point students were ready to benefit from an experimental study of a two-dimensional wing (airfoil). The goal of the experiment was to calculate the aerodynamic lift by integrating the measured surface pressure. The study showed the effect of wind speed and angle of attack. The experiment was conducted in the open loop closed test section wind tunnel, shown in Figure 1. The model was mounted to a base plate in the floor of the test section that could be turned for a range of angles of attack. Figure 2 shows a Clark-Y wing model that was used in the experiment. As was the case for the circular cylinder, the model vertically spanned the test section using the upper and lower walls as end plates to allow for the two dimensional study. The wing was equipped with 18 pressure openings located at 0, 7.5, 10, 20, 30, 40, 50, 60, and 70 percent chord on both the upper and lower surfaces. Figure 3 shows the wing inside the test section.

The aerodynamic lift was covered in the lecture part of the course before this lab activity. Also the lab manual summarized the basics and the method to calculate the lift from the surface pressure coefficient. Figure 4 shows part of the theoretical concepts that were given to the students in the beginning of the lab. Each team was asked to develop the details of their experiment including the wind speeds, angles of attack and wind tunnel settings.

Figure 1: A photo of the wind tunnel used in the study.

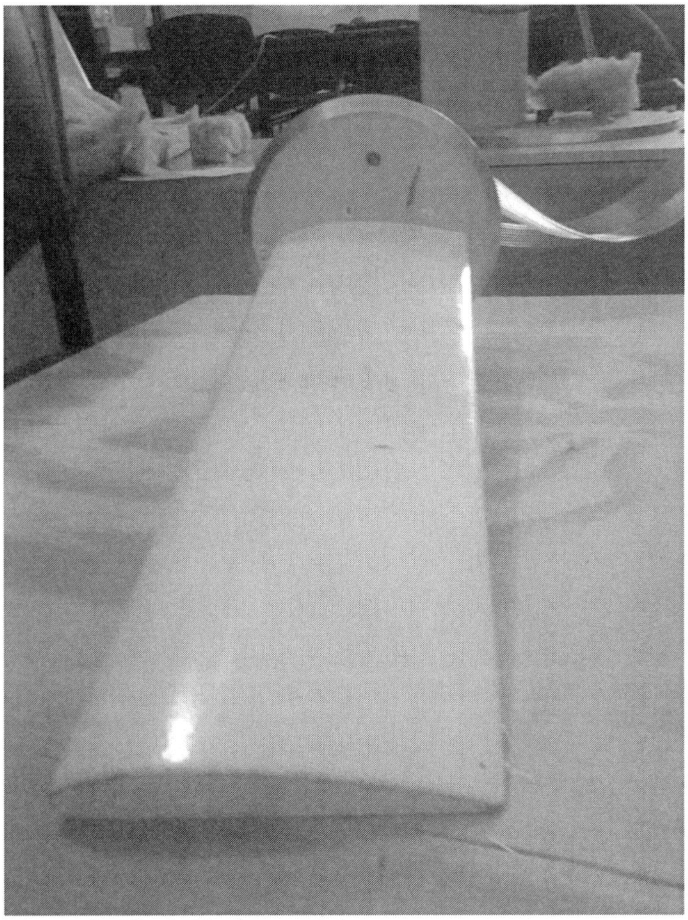

Figure 2: Clark-Y wing model.

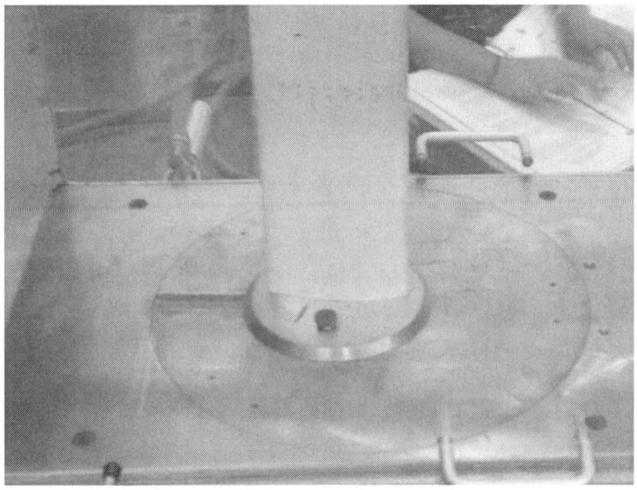

Figure 3: The wing model inside the test section

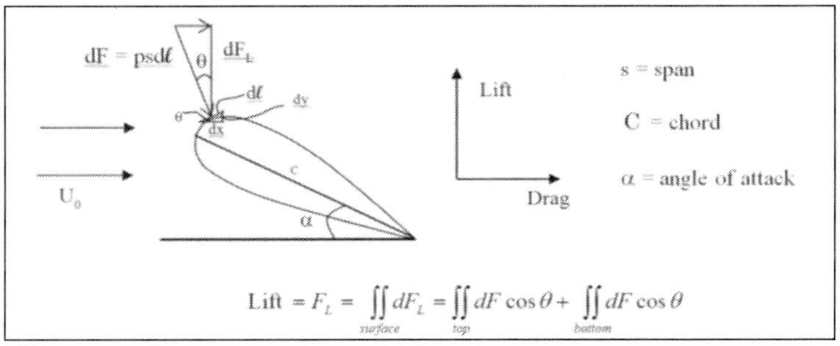

Figure 4: Basics presented to the students.

CFD COMPONENT

The wing experiment had been a part of the lab in fluid mechanics for many years however without a substantial comparison to inviscid flow solutions or other predictive tools such as CFD. The availability of CFD tools as well as free software that could be easily downloaded made this a natural improvement to the lab. An overview of basics of CFD was presented to the students using a commercial code (Star CCM+) that is used in the technical elective taught in the senior year, Mokhtar [2]. The demonstration included the general features of an effective mesh and some of the capabilities and limits of CFD as an analysis tool. At this level, no further details about CFD algorithms and solver schemes could be presented. A case study similar to the wing experiment was used in the demonstration to make the discussion more relevant to the students. Figures 5 and 6 show a zoom in on the two-dimensional mesh and part of the post-processing for the studied airfoil. The demonstration also included some the flow structures such as boundary layer separation, wake formation, stagnation regions and lift force calculation. Having a commercial code allows the use of flow animation and tools such as vector field and pressure and velocity contours to enhance the discussion.

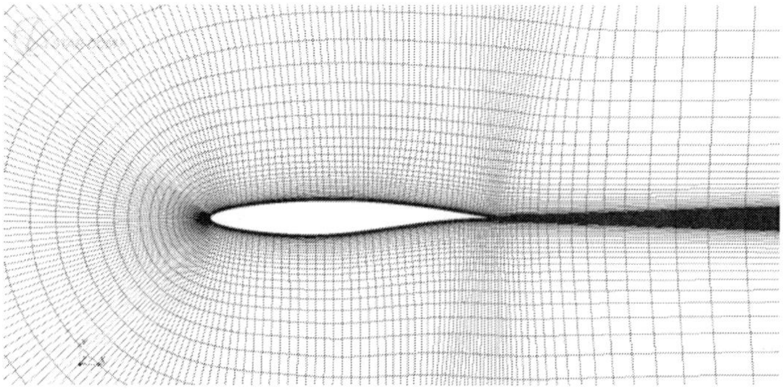

Figure 5: CFD class demonstration, zoom in on mesh.

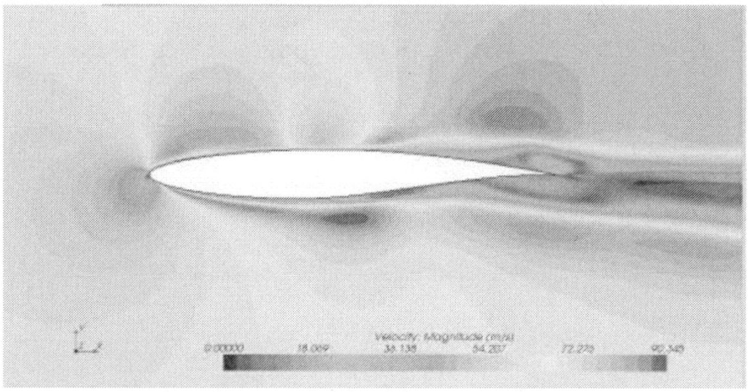

Figure 6: CFD class demonstration, velocity contours.

A commercial code (Star CCM+) was successfully used before to introduce CFD to undergraduate students in fluid mechanics course through a series of demonstrations similar to the one outlined here, Mokhtar [1]. This training needed more time in the class and lab which was not case for this course. To shorten the needed training time and achieve the learning objectives, simplified software was introduced, XFOIL. It is a panel method code that was developed by Drela [16] to study airfoil sections. XFOIL includes several features such as:

- boundary layer formation, Derela [17],
- trailing edge treatment, Derela [18],
- boundary layer tripping,
- meshing control tools,
- aerodynamic loads calculation,
- inviscid and viscous analysis,
- airfoil design tools (inverse method),
- and more features, Ref. [19] through [21].

In the lab, basics of XFOIL use were presented to the students with a focus on the importance of meshing, inviscid and viscous modeling. The profile of the Clark-Y airfoil section was given to the students as well. Figure 7 shows the airfoil section and examples of meshing (paneling) that could be used in XFOIL to add clustering near to the leading and trailing edges.

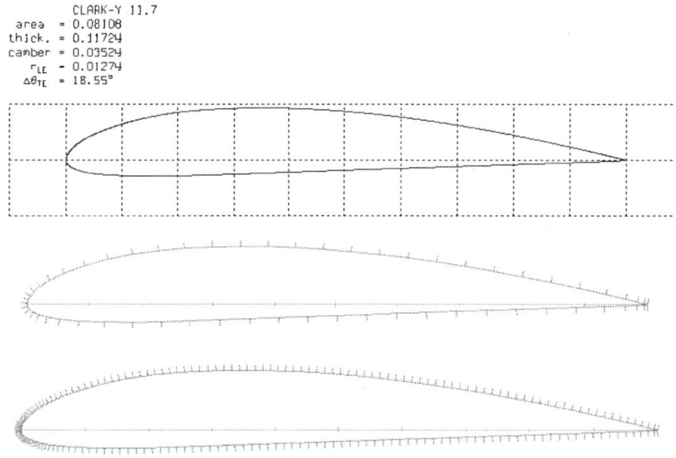

Figure 7:Part of the XFOIL demonstration.

PUBLISHED DATA

Another important part of the lab activities was to get the students to compare their experimental and CFD results to the published data. Clark-Y airfoil section has been used in low speed applications since the 1930's. The students were asked to research and report their references for the polar curves of this airfoil. Among references, NASA [22] reported wind tunnel data in 1930 for a long list of airfoil section including Clark-Y. A more recent source was a book by Lyon et al [23] where they also reported data for many airfoil sections. XFOIL was one of the tools used besides wind tunnel testing in this book.

STUDENTS' WORK

The students were divided into teams where each team developed the experiment plan and complete the wind tunnel test for the wing. Each team was from four to five students. Figure 8 shows one of the teams during the wind tunnel test. Each student was asked to complete the CFD study, research the published data and interpret all the data in a written report. In this section, samples from the students' results will be discussed to observe the level students' responses to the lab activities. Parameters that each team decided were:

- Range of wind speeds.
- Range of angle of attack.
- Wind tunnel accuracy.
- Similarity requirements between CFD and wind tunnel test.
- Sources for published data.

Figures 9 through 14 show the pressure coefficient distribution reported by three students. Student A presented the upper and lower surface in separate graphs for three angles of attack: 0, 10 and -10°. Student B reported separate graphs angles of attack: 0, 7 and 15°. The third student reported angles of attack: 0, 4 and 14°. Student A studied positive and negative values for the angle of attack which is suitable for cambered airfoil section such as Clark-Y. Students B and C ran a high angle of cases to explore the stall conditions. Student C ran two sets of testing through these angles with wind speeds: 80 and 100 mph to report the effect of the wind speed. The first two students did not report that part of the study. Students A and B plotted the pressure coefficient C_p in regular scale while the usual practice in aeronautical application is to have C_p presented in a reverse scale as the way student C reported his data.

To run a CFD simulation using XFOIL, the students had to meet the similarity conditions between the two cases. Reynolds (Re) and Mach (M) numbers were calculated for the tested wing and the same values were used in the simulations. Figures 15 through 20 show samples of the CFD results reported by the students using XFOIL. The software allows both inviscid and viscous modeling. students A and B used viscous modeling and student C used inviscid modeling. All of them used the right modeling condition (Re and M). It can be seen that the measured pressure distribution compared well with the CFD results.

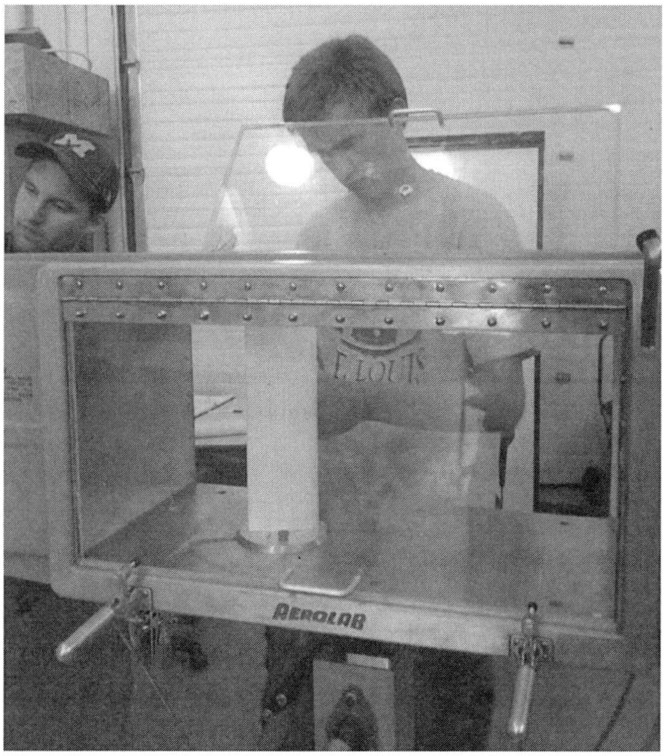

Figure 8: One of the students during the wind tunnel testing.

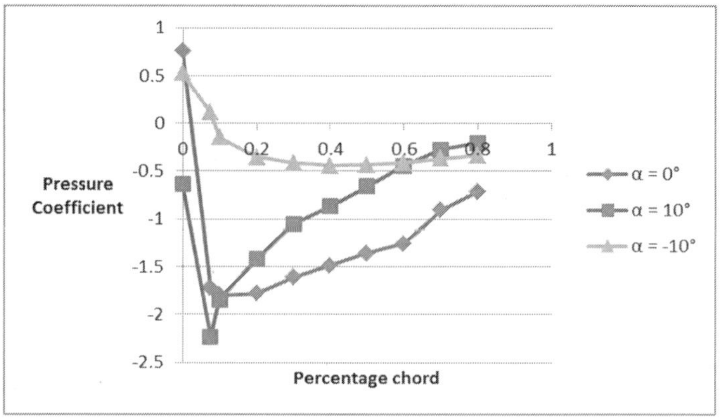

Figure 9: Pressure distribution on upper surface, Student A

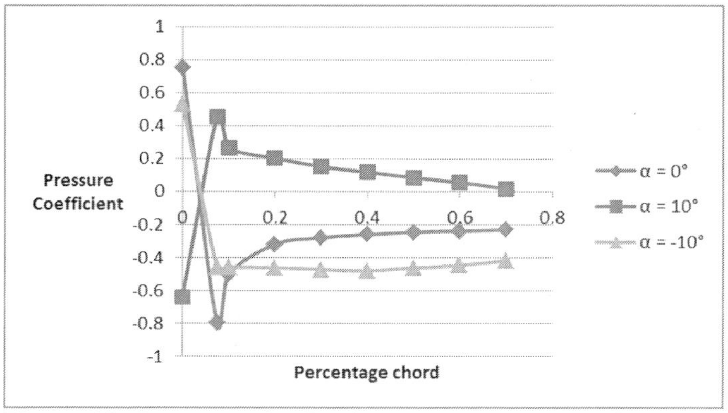

Figure 10: Pressure distribution on lower surface, Student A

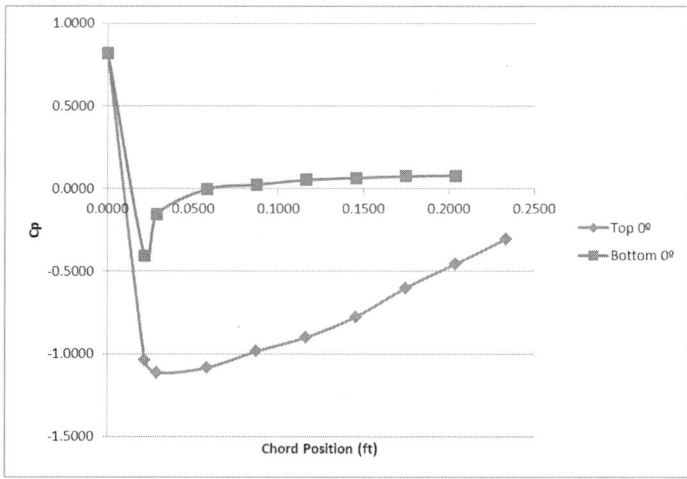

Figure 11 – Pressure Coefficient as Function of Position for α of 0°, Student B.

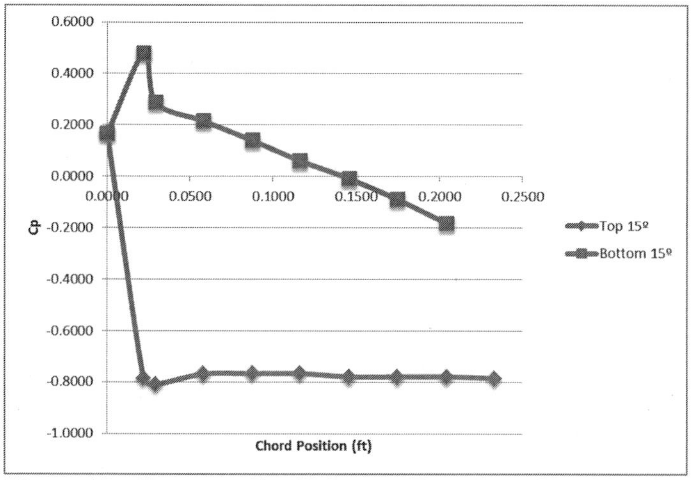

Figure 12 – Pressure Coefficient as Function of Position for α of 15°, Student B.

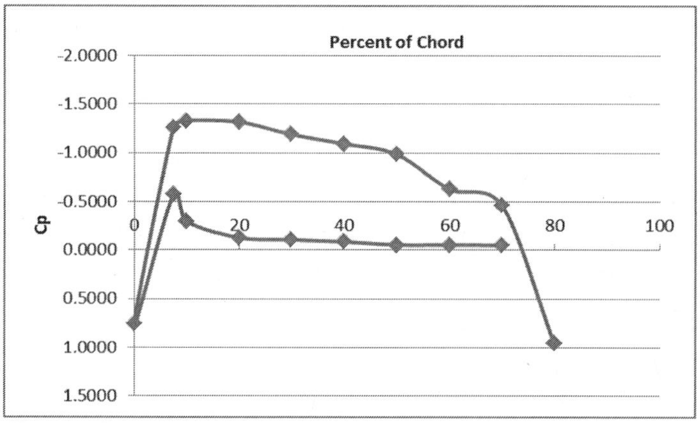

Figure 13 – Pressure distribution for α of 0°, Student C.

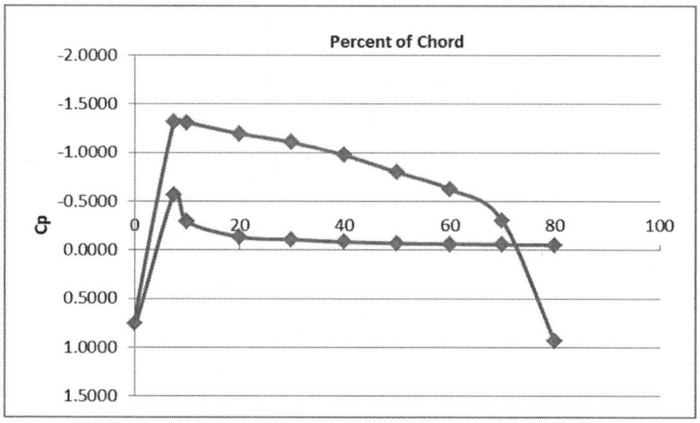

Figure 14 – Pressure distribution for α of 14°, Student C.

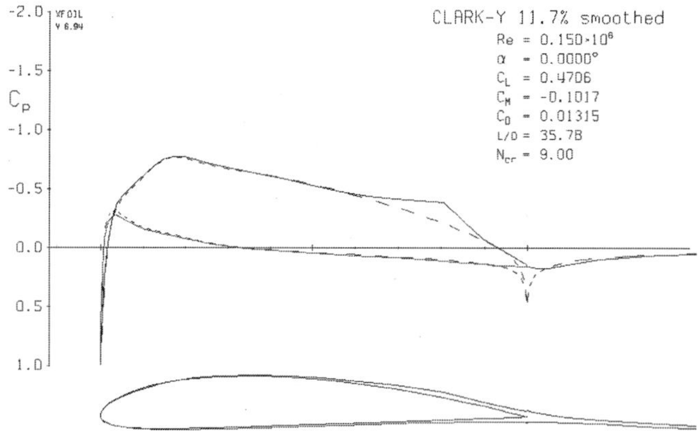

Figure 15: Simulated pressure coefficient for an angle of attack $\alpha = 0^{\circ}$, XFOIL results, Student A

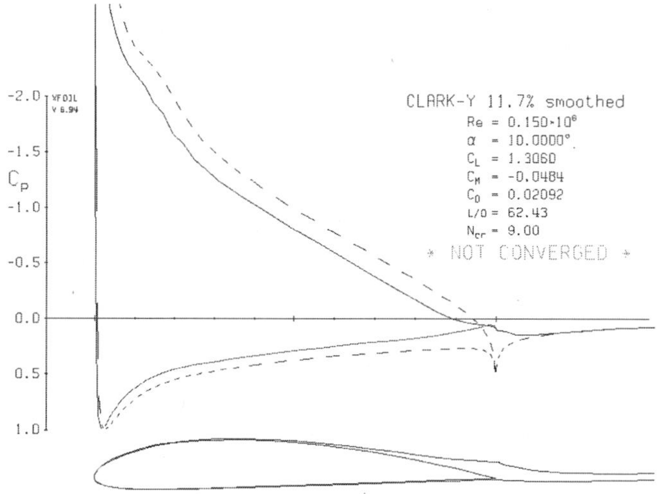

Figure 16: Simulated pressure coefficient for an angle of attack α of 10°, XFOIL results, Student A

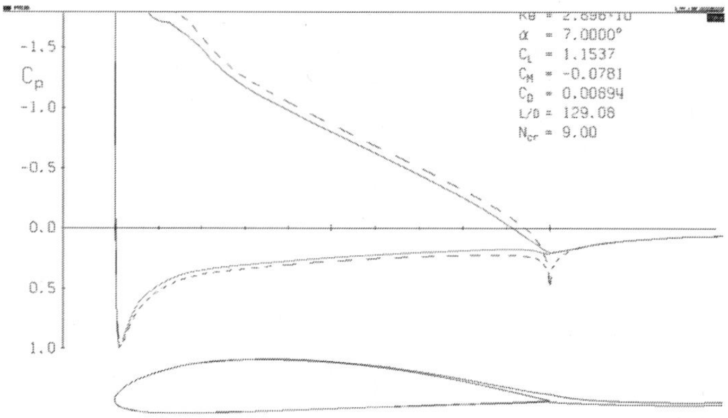

Figure 17: Simulated pressure coefficient for an angle of attack α of 7°, XFOIL results, Student B

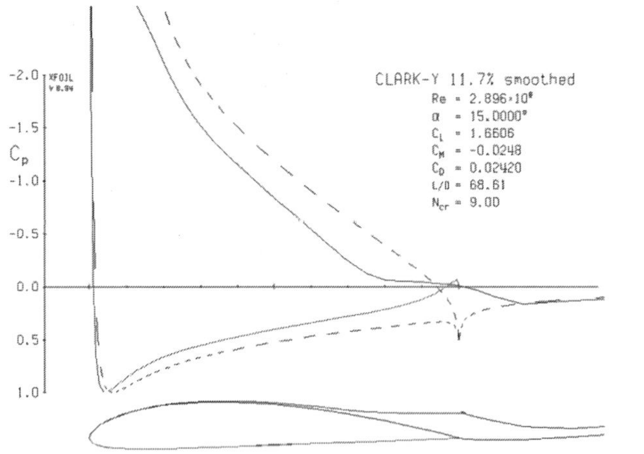

Figure 18: Simulated pressure coefficient for an angle of attack α of 15º, XFOIL results, Student B

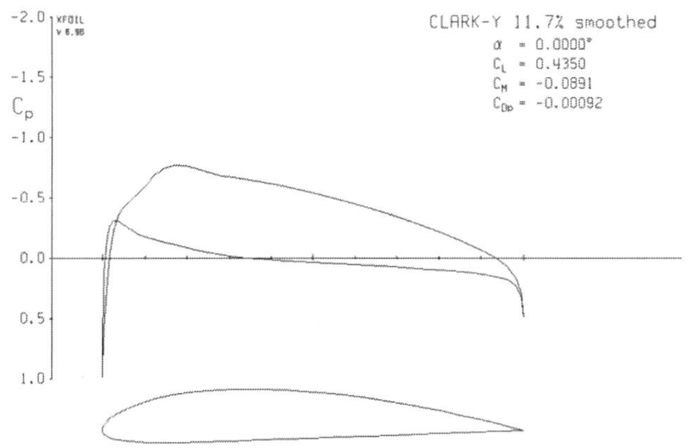

Figure 19: Simulated pressure coefficient for an angle of attack α of 0° , XFOIL results, Student C

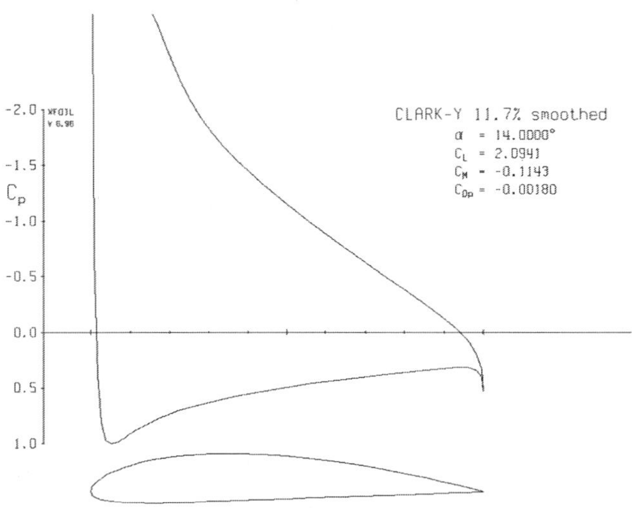

Figure 20: Simulated pressure coefficient for an angle of attack α of 14º, XFOIL results, Student C

The next step in the study was to calculate the lift coefficient C_L. The CFD software, XFOIL, reports the values directly. For the wind tunnel test, the students had to calculate the lift coefficient by integrating the measure pressure distribution. Figure 21 shows the set of equations reported by one of the students for calculating C_L. Tables 1 through 4 show samples of the lift coefficients reports by four students. The first student compared their results to the published data while the rest compared the wind tunnel and the CFD results. All of them commented on the reasons of differences between the data. One of the lessons they learn at this part was that each method had its own sources uncertainty. Below are some the error sources:

- Accuracy of pressure measurements, wind tunnel.
- Blockage interference at high angle of attack, wind tunnel.
- Mesh refinements near to leading and trailing edge, CFD.
- Large separation regions, CFD

$$F_L = \iint_{Surface} dF_L = \int_0^L dF\, cos\theta\ (top) + \int_0^L dF\, cos\theta\ (bottom)$$

$$\int_0^L dF\ cos\theta = \int_0^L p\ s\ dl\ \cos\theta = \int_0^L (p - p_\infty) s\ dl \cos\theta$$

$$\cos\theta = \frac{dd}{dl}\ (Shown\ in\ Figure\ 1)$$

$$\int_0^L (p - p_\infty) s\ dl\ \frac{dd}{dl} = \int_0^L (p - p_\infty) s\ dd$$

$$\mathrm{dd} = \mathrm{dc}\ \cos\alpha\ \ (Shown\ in\ Figure\ 1)$$

$$\int_0^L (p - p_\infty) s\ dc\ cos\alpha$$

$$L = 0.5\rho A U^2 C_L\ = F_L = \int_0^L (p - p_\infty) s\ dc\ cos\alpha$$

$$C_L = \frac{\int_0^L (p - p_\infty) s\ dc\ cos\alpha}{0.5\rho A U^2} = \frac{a\ cos\alpha}{s\ c} * \int_0^L C_p dc$$

Figure 21 –Calculation of the drag coefficient, Student sample.

Table 1: Lift coeficient, Student A

Angle of Attack	0°	10°	-10°
Experimental	0.81	1.34	-0.16
Published	0.42	1.25	-0.50
CFD Analysis	0.4350	1.6285	-0.7721

Table 2: Lift coeficient, Student B

Angle of Attack	Experimental C_L	Theoretical C_L
0°	0.4987	0.4350
4°	1.1209	0.9160
14°	1.8916	2.0941

Table 3: Lift coeficient, Student C

	95 ft/s	105 ft/s
0 degrees	-0.77	-0.71
6 degrees	-0.80	-0.73
10 degrees	-0.80	-0.77

Table 4: Lift coeficient, Student D

α	EXP	XFOIL	Error	Uncertainty
0°	0.8820	0.47	87.7	16.0
10°	1.4900	1.306	14.1	18.4
-10°	0.0834	-7.52	-98.9	13.8

FURTHER DISCUSSION AND ASSESSMENT

Performing a fluid mechanics study using a wind tunnel and CFD and verifying the results against published data were the main goal of this experiment. In the process, the students got to learn about the limits and accuracy of each method. In this section, some of their concluding statements will be discussed. One of the students reported in his conclusion the following statement:

"...Simulations provide an effective tool for evaluating the validity of experimental measurements and vice versa; but regions of separation and unpredictable flow are very difficult to model correctly...."

It is clear that the he learned through the practice, the limits of a CFD study and that there are some flow conditions where wind tunnel testing will be more accurate than CFD. Another student commented about the wind tunnel blockage:

"...The cross-sectional area of the wind tunnel in which the airfoil was placed was 1 foot by 1, and when the angle of attack for the airfoil was adjusted to ±10° it took up a considerable amount of cross-sectional area. This set-up could cause interference between the affected boundary layer around the wing and the sides of the tunnel and in return could affect the pressure at the wing..."

The limits of wind tunnel testing and the size of the tested model is one of the standard constraints for accurate testing. It is clear that this student got this part when comparing his results to CFD and published data. In another student's comment about the accuracy in calculating the lift coefficient by integrating the pressure distribution, he reported:

"...From the previously presented data, the accuracy of the coefficient of lift increased as the angle of attack increased. This was due to the difficulty of numerically integrating C_p versus chord length plots that have outliers, which are present at the stagnation point in situations with a smaller angle of attack..."

Some of the students focused on one source of error while others commented on more than one source of uncertainty. They all realized the value of the integration of CFD and wind tunnel. It is also important to mention that some of the students did not get that deep in the discussion and got busy with reporting the results without focusing on the big picture. However, this group also completed both sides of the lab activities successfully.

The authors ran this lab activities several times in the last couple of years. From their point of view, this integrated lab activities (CFD and wind tunnel) enriched the lab experience for the students and expanded their view for value of both tools. Nearly all the students learned the CFD software after one demonstration with minimal assistance after that. The lab also encouraged many students to take the CFD technical elective in their senior year. One of the students who completed this lab, student A in the previous sections, took the CFD technical elective and another independent study in CFD before his graduation. Another student got even more interested in CFD and he is currently working on his Master degree in CFD applications.

CONCLUSIONS

A lab activity was designed to integrate CFD and wind tunnel testing in an undergraduate fluid mechanics course. Details of the lab were discussed through samples from the students' work. This lab activity was repeated several times in the last couple of years.

The majority of the students were attracted to the CFD tool. Many of them learned the limits of using CFD and the accuracy of running a wind tunnel test to complete a fluid mechanics study. Positive feedback from both the students and the instructors about this lab activity was observed.

As a final remark, CFD can serve as a good tool to support junior level fluid mechanics classes. This work is an example of that use. Usually, students learn this type of software very fast and that saves time for the instructors to focus on theoretical parts of the course. In addition, CFD provides the students with a good visualization tool to enhance the learning environment.

ACKNOWLEDGMENTS

The authors would like to thank the students who participated in the lab activity and provided the samples presented in this paper.

REFERENCES

1. Mokhtar, W., "Using Computational Fluid Dynamics to Introduce Critical Thinking and Creativity in an Undergraduate Engineering Course", The International Journal of Learning, Common Ground Publisher, vol. 17, no. 9, pp: 441-458, 2010.
2. Mokhtar, W., "Project-Based Learning (PBL) – An Effective Tool to Teach an Undergraduate CFD Course", ASEE Annual Conference, ASEE no. 973, 2011.
3. Mokhtar, W., "Introducing Wind Energy to an Undergraduate CFD Course", ASEE North Central Section Conference, ASEE NC no 110, 2014.
4. Hailey, C. and Spall, R., "An Introduction of CFD into the Undergraduate Engineering Program", ASEE Annual Conference, ASEE no. 1556, 2000.
5. Sert, C. and Nakiboglu, G., "Use of Computational Fluid Dynamics (CFD) in Teaching Fluid Mechanics", ASEE Annual Conference, ASEE no. 1560, 2007.
6. Blekhman, D., "Lessons Learned in Adopting a CFD Package", ASEE Annual Conference, ASEE no. 830, 2007.
7. Cummings, R., and Morton, S., "Computational Aerodynamics Goes to School: A Course in CFD for Undergraduate Students", 43rd AIAA Aerospace Sciences Meeting and Exhibit, AIAA 2005-1072, 2005.
8. Guessous, L., Bozinoski, R., Kouba, R., and Woodward, D., "Combining Experiments with Numerical Simulations in the Teaching of Computational Fluid Dynamics". ASEE Annual Conference, ASEE no 2220, 2003.
9. Navaz, H., Henderson, B., Berg, R., and Nekcoei, S., "A new approach to teaching undergraduate thermal/fluid sciences—courses in applied computational fluid dynamics and compressible flow", International Journal of Mechanical Engineering Education Vol 30 No 1, 2000.
10. Bullough, W., Hart, J., and Chin, S., "Comparative studies: CFD, experimental and analytical techniques in the fluids laboratory", International Journal of Mechanical Engineering Education 31(2), 150-159, 2003.
11. Stubley, G. and Hutchinson, B., "The Role of Case Studies in CFD Education" ASME, 2002
12. Stern, F., Xing, T., Yarbrough, D., Rothmayer, A., Rajagopalan, G., Otta, S., Caughey, D., Bhaskaran, R., Smith, S., Hutchings, B., Moeykens, S., "Development of Hands-On CFD Educational Inter face for Undergraduate Engineering Courses and Laboratories", ASEE AC Conference, no. 1526, 2004.
13. Burban, P., Hegna, H., and Zavodney, L., "Role of CFD in Undergraduate Research in the Design of a Supermileage® Competition Vehicle" ASEE NC Conference, 2008.
14. Mazumder, Q., "Integration of Computational Fluid Dynamics Analysis in Undergraduate Research Program" ASEE North Central Regional Conference: Paper no. 73, 2009
15. Fleischmann, S., Sozen, M., and Mokhtar, W., "A Green Heat Transfer Design Project to Introduce Globalization and Society Awareness", ASME 2010 International Mechanical Engineering Congress & Exposition, Paper no: IMECE2010-38285, 2010.
16. Drela, M., "XFOIL: An Analysis and Design System for Low Reynolds Number Airfoils", Conference on Low Reynolds Number Airfoil Aerodynamics, University of Notre Dame, June 1989, also appears as a chapter in: Low Reynolds Number Aerodynamics. T.J. Mueller (Editor), ISBN 3-540-51884-3, ISBN 0-387-51884-3, 1989.
17. Drela, M. and Giles, M.B, "Viscous-Inviscid Analysis of Transonic and Low Reynolds Number Airfoils", AIAA Journal, 25(10), pp.1347-1355, October 1987.
18. Drela, M., "Integral Boundary Layer Formulation for Blunt Trailing Edges", Paper AIAA-89-2166, August 1989.
19. Drela, M., "Elements of Airfoil Design Methodology", Applied Computational Aerodynamics, (P. Henne, editor), AIAA Progress in Aeronautics and Astronautics, Volume 125, 1990.
20. Drela, M., "Low-Reynolds Number Airfoil Design for the MIT Daedalus Prototype: A Case Study", Journal of Aircraft, 25(8), pp.724-732, August 1988.
21. Drela, M., "Pros and Cons of Airfoil Optimization, Chapter in "Frontiers of Computational Fluid Dynamics, 1998", D.A. Caughey, M.M. Hafez, Eds. World Scientific, ISBN 981-02-3707-3
22. Louden, F., "Collection of Wind-Tunnel Data on Commonly Used Wind Sections", NASA Report No. 331, 1930.
23. Lyon, C., Broeren, A, Giguere, P., Gopalarathnam, A., and Selig, M, "Summary of Low-Speed Airfoil Data", vol no 3, SoarTech Publications, ISBN 0-9646747-3-4, 1997.

Proceedings of the ASME 2014 International Mechanical Engineering Congress and Exposition
IMECE2014
November 14-20, 2014, Montreal, Quebec, Canada

IMECE2014-37125

STUDY OF THE CHANGE OF PERFORMANCE OF AN ELASTIC VERTICAL HYDROFOIL WITH DEFORMATION

Alexander Führing
Milwaukee School of Engineering
Milwaukee, WI, United States

Dr. Subha Kumpaty
Milwaukee School of Engineering
Milwaukee, WI, United States

Chris Stack
Milwaukee School of Engineering
Milwaukee, WI, United States

ABSTRACT

In external and internal fluid flow analysis using numerical methods, most attention is paid to the properties of the flow assuming absolute rigidity of the solid bodies involved. However, this is often not the case for water flow or other fluids with high density. The pressure forces cause the geometry to deform which in turn changes the flow properties around it. Thus, a one-way and two-way Fluid-Structure Interaction (FSI) coupling is proposed and compared to a CFD analysis of a windsurfing fin in order to quantify the differences in performance data as well as the properties of the flow. This leads to information about the necessity of the use of FSI in comparison to regular CFD analysis and gives indication of the value of the enhanced results of the deformable analysis applied to water flow around an elastically deformable hydrofoil under different angles of attack. The performance data and flow property evaluation is done in ANSYS Fluent using the k-ω SST and k-ε model with a y^+ of 1 and 35 respectively in order to be able to compare the behavior of both turbulence models. It is found that the overall lift coefficient in general is lower and that the flow is less turbulent because of softer transition due to the deformed geometry reducing drag forces. It is also found that the deformation of the tip of the hydrofoil leads to vertical lift forces. For the FSI analysis, one-way and two-way coupling were incorporated leading to the ability to compare results. It has been found that one-way coupling is sufficient as long as there is no stall present at any time.

INTRODUCTION

The subject of the analysis is a wind surfing wave fin, i.e. a vertical hydrofoil that has the shape of a dolphin fin and resembles the geometry of a tapered swept wing of aircraft. Since fins only differ from wings in their application and the density of the surrounding fluid, theory and findings for tapered swept wings can be applied. In windsurfing, sailing, and other watersports, fins are made of G10 fiberglass as well as other composite materials such as carbon fiber that provide extended strength combined with optimized weight. G10-FR4 fiberglass is used to account for the elastic material in the transient and static structural analyses. Elasticity of the fin is said to be an important factor along with the shape to reach certain performance. Evaluating the deformation and its effect on flow properties and performance could improve the knowledge and lead to a new focus in terms of used materials. Regular CFD comes short due to the assumption of the absloute rigidity and it must be coupled with fluid-structure interaction for improved results. In this research, the one-way and two-way FSI are compared for the prediction of fin tip deflection as a function of different incident angles. A conclusion is drawn concerning the feasibility of one-way FSI in presence of large deformations and stall effects. Two sets of data are obtained for the rigid fin and the deformed fin respectively to predict performance change and allow for a comparison of the realizable k-ε model and the k-ω SST model.

Harper and Maki investigated the stall characteristics of swept wings in 1964 finding that the tip of the geometry stalls already at lower incident angles leading to increasing drag and decreasing efficiency (Harper & Maki, 1964). The maximum section lift is shifted in the tip direction with higher sweep angles (see Figure 1) leading to adverse stall behavior (Schlichting & Truckenbrodt, 2001).

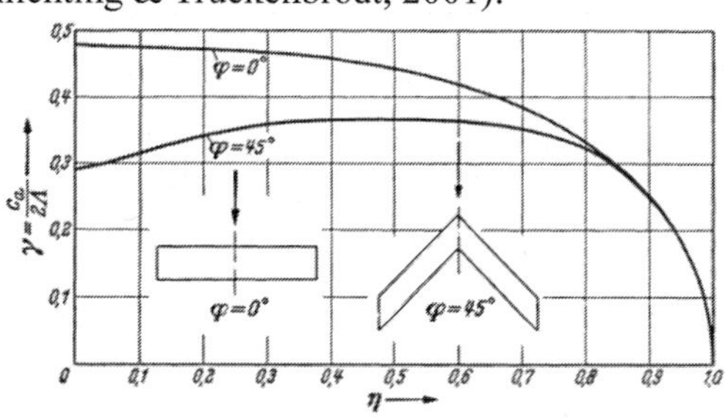

Figure 1: Distribution of Section Lift for Two Wings with Constant Chord and Aspect Ratio

Suzanne Ruth Hutchison from the University of Tasmania conducted a study on the dynamic behavior of hydrofoils using FSI. One of the findings was that one-way coupling yields results that are similar to those of two-way coupled systems. However, she only compared the different approaches for small incident angles so that no conclusion was drawn for stall effects and larger angles (Hutchison, 2012).

Eleni et al. evaluated the realizable k-ε and k-ω SST models along with other turbulence models for the flow over a NACA0012 airfoil concluding that the k-ω SST model was the most accurate model in terms of drag prediction. Also, they found that the k-ε model showed weaknesses in the prediction of post-stall lift coefficients (Eleni, Athanasios, & Dionissios, 2012).

NOMENCLATURE

k – turbulence kinetic energy
ε – scale of turbulence
ω – specific turbulence dissipation
G_b – buoyancy dependent generation of turbulence kinetic energy
G_k – velocity gradient dependent generation of turbulence kinetic energy
Y_M – contribution of the fluctuating dilatation in compressible turbulence to overall dissipation rate
μ – dynamic viscosity
$C_1, C_{1\varepsilon}, C_2, C_{3\varepsilon}$ – turbulence model constants
$\sigma_k, \sigma_\varepsilon, \sigma_{k1}, \sigma_{k2}, \sigma_{\omega 1}, \sigma_{\omega 2}$ – turbulence model constants
$\beta^*, \beta, \beta_1, \beta_2$ – turbulence model constant
γ_1, γ_2 - turbulence model constants
ν_t, μ_t – turbulence viscosity
C_l – lift coefficient
C_d – drag coefficient
ρ_w – density of salt water
α – flow incident angle
v – flow velocity
p – pressure
Ω – absolute value of vorticity
CFD – Computational Fluid Dynamics
FSI – Fluid-Structure Interaction

The Realizable $k-\varepsilon$ turbulence model

Being one of the simplest turbulence models, the $k-\varepsilon$ is one of the most popular two equation models with the earliest development already done in 1945 by Chou. The standard $k-\varepsilon$ model was developed by Launder and Sharma (1974). The $k-\varepsilon$ model includes two extra transport equations to represent the properties of turbulent flow. The two transported variables are the turbulent kinetic energy k, and the scale of the turbulence ε (Wilcox, 1993). Its versatility and economy are the reasons for the wide use in industrial engineering applications. The main application for which the $k-\varepsilon$ model yields the most accurate results are far-field simulations as this two equation model does not resolve the boundary layer and is used with a y^+ that is higher than 30 (Ariff, Salim, & Cheah, 2009).

The Realizable $k-\varepsilon$ model is defined as follows (Eleni, Athanasios, & Dionissios, 2012):

$$\frac{\partial}{\partial t}(\rho k)+\frac{\partial}{\partial x_j}(\rho k u_j) = \frac{\partial}{\partial x_j}\left[\left(\mu+\frac{\mu_t}{\sigma_k}\right)\frac{\partial k}{\partial x_j}\right]+G_k+G_b-\rho\varepsilon - Y_M+S_k \tag{1}$$

And

$$\frac{\partial}{\partial t}(\rho \varepsilon)+\frac{\partial}{\partial x_j}(\rho \varepsilon u_j) = \frac{\partial}{\partial x_j}\left[\left(\mu+\frac{\mu_t}{\sigma_\varepsilon}\right)\frac{\partial \varepsilon}{\partial x_j}\right]+\rho C_1 S\varepsilon - \rho C_2\frac{\varepsilon^2}{k+\sqrt{\nu\varepsilon}}+C_{1\varepsilon}\frac{\varepsilon}{k}C_{3\varepsilon}G_b+S_\varepsilon \tag{2}$$

With

$$C_1=\max\left[0.43,\frac{n}{n+5}\right], n=S\frac{k}{\varepsilon}, S=\sqrt{2S_{ij}S_{ij}} \tag{3}$$

The k-equation describes the production and destruction of turbulent kinetic energy in the flow while the ε-equation models the dissipation rate (Wilcox, 1993). Being valid for calculations in fully turbulent conditions, the standard k-ε model lacks accuracy calculating flows that are separating because of pressure gradients at walls which is why the Realizable k-ε model was developed. The constants of the Realizable k-ε model are:
$C_{1\varepsilon}=1.44$, $C_2=1.9$, $\sigma_k=1.0$ and $\sigma_\varepsilon=1.2$

The $k-\omega$ SST turbulence model

The $k-\omega$ SST, the Menter Shear Stress Transport, turbulence model is a two-layer model combining the standard $k-\omega$ model of Wilcox (1988) and the $k-\varepsilon$ model. Since the $k-\omega$ model performs precisely near walls but excessively predicts separation while the $k-\varepsilon$ model is only valid for fully turbulent flows without wall treatment, the $k-\omega$ SST model combines the advantages of both with a blending function in order to provide accurate calculations of wall bounded flows with separated regions (Georgiadis, Yoder, & Engblom, 2006). The model is very robust and accurate for pressure gradient induced separation which is why the $k-\omega$ SST model has become the new standard turbulence model for industry applications (Lecheler, 2009). The set of equations contains parts of the $k-\varepsilon$ and the $k-\omega$ models as well as additional blending functions and a modification of the eddy viscosity formulation to account for the transport effects of the turbulent shear stress (Menter, 1993). The $k-\omega$ SST model is defined as follows (Eleni, Athanasios, & Dionissios, 2012):

$$\frac{D\rho k}{Dt} = \tau_{ij}\frac{\partial u_i}{\partial x_j} + \beta^* \rho\omega k + \frac{\partial}{\partial x_j}\left[(\mu + \sigma_k \mu_t)\frac{\partial k}{\partial x_j}\right] \quad (4)$$

$$\frac{D\rho\omega}{Dt} = \frac{\gamma}{v_t}\tau_{ij}\frac{\partial u_i}{\partial x_j} - \beta\rho\omega^2 + \frac{\partial}{\partial x_j}\left[(\mu + \sigma_\omega \mu_t)\frac{\partial \omega}{\partial x_j}\right] + 2\rho(1-F_1)\sigma_\omega \frac{1}{\omega}\frac{\partial k}{\partial x_j}\frac{\partial \omega}{\partial x_j} \quad (5)$$

Where

$$\tau_{ij} = -\rho\overline{u_i' u_j'} = \mu_t\left(\frac{\partial u_i}{\partial x_j} + \frac{\partial u_j}{\partial x_i} - \frac{2}{3}\frac{\partial u_k}{\partial x_k}\partial_{ij}\right) - \frac{2}{3}\rho k \partial_{ij} \quad (6)$$

$$F_1 = \tanh\left\{\left[\min\left[\max\left(\frac{\sqrt{k}}{0.09\omega y}, \frac{500\nu}{y^2\omega}\right), \frac{4\rho\sigma_{\omega 2} k}{CD_{k\omega} y^2}\right]\right]^4\right\} \quad (7)$$

$$CD_{k\omega} = \max\left(2\rho\sigma_{\omega 2}\frac{1}{\omega}\frac{\partial k}{\partial x_j}\frac{\partial \omega}{\partial x_j}, 10^{-20}\right) \quad (8)$$

$$F_2 = \tanh\left\{\left[\max\left(\frac{2\sqrt{k}}{0.09\omega y}, \frac{500\nu}{y^2\omega}\right)\right]^2\right\} \quad (9)$$

$$\begin{aligned} &\beta = F_1\beta_1 + (1-F_1)\beta_2\,, \gamma = F_1\gamma_1 + (1-F_1)\gamma_2\,, \\ &\sigma_k = F_1\sigma_{k1} + (1-F_1)\sigma_{k2}\,, \sigma_\omega = F_1\sigma_{\omega 1} + (1-F_1)\sigma_{\omega 2} \\ &\beta^* = \varepsilon/k\omega\,, v_t = a_1 k/\max(a_1\omega, \Omega F_2) \end{aligned} \quad (10)$$

with the empirical constants being:
$\beta^* = 0.09$, $\beta_1 = 0.075$, $\beta_2 = 0.0828$, $\gamma_1 = 0.5532$, $\gamma_2 = 0.4404$, $\sigma_{k1} = 0.85$, $\sigma_{k2} = 1.0$, $\sigma_{\omega 1} = 0.5$, $\sigma_{\omega 2} = 0.856$, and $a_1 = 0.31$.

COMPUTATIONAL METHOD

This paper investigates the flow around a sailboard fin that is set up in two models, one being CFD assuming absolute rigidity of the structure, the other being FSI with material data of G10-FR4 fiberglass. The foil is symmetrical and resembles a NACA 0012 that is highly swept and tapered. For the local Reynolds Numbers, the values for surrounding salt water flow with a velocity of $v_f = 10\frac{m}{s}$, a density of $\rho_w = 1026\frac{kg}{m^3}$, and a dynamic viscosity of $\mu = 0.00114\frac{Ns}{m^2}$ were computed, taking the dimensions of Figure 2 into account:

$$Re_{root} = 1.305 \times 10^6$$

$$Re_{tip} = 3.42 \times 10^5$$

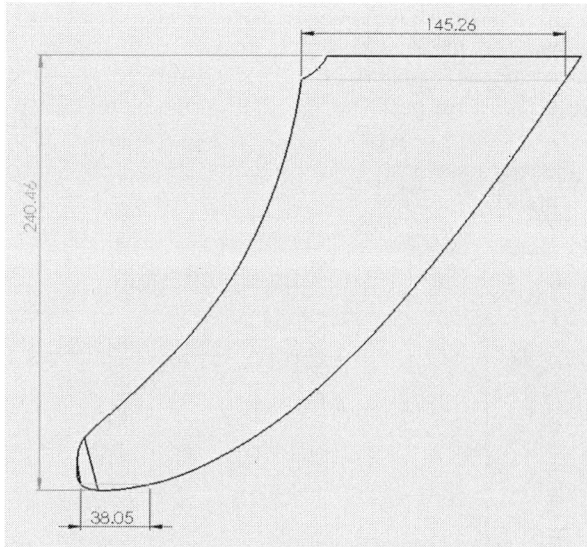

Figure 2: Dimensions of the Used Model [mm]

The commercial code ANSYS Fluent was utilized to execute the CFD calculations while ANSYS Mechanical was used for the structural part of the FSI setup. Calculations were performed for incident angles of $0° \leq \alpha \leq 24°$ with steps of 2° - 4°. In order to deal with lack of computational power, the two-way FSI computations were initially done with a coarser mesh and the Realizable $k-\varepsilon$ model before exporting the static deformed geometry as .stl file and remeshing it with the suitable mesh needed for the $k-\omega$ SST model with a y^+ of 1. Therefore, a boundary layer resolution with an initial inflation layer height of $h_0 = 2.6 \times 10^{-6} m$ was implemented.

The meshes were created in ANSYS Meshing for the CFD and FSI setups whereas refined meshes for the exported and reimported geometries were created in ICEM CFD due to its ability to handle .stl mesh files. Unstructured meshes with boundary layer inflation with up to 30 layers were created with an extended face mesh refinement to account for a smooth transition modeling. Figure 3 shows the used inflation layers with an additional transition refinement of $h_t = 1.4 \times 10^{-3} m$ as well as one half of the symmetric flow volume with the fin located in the center.

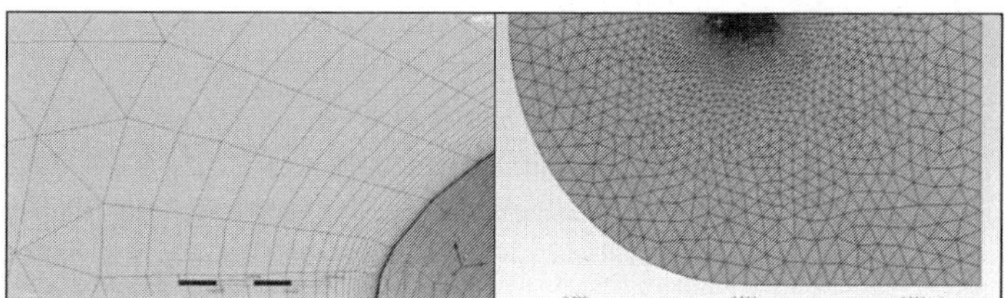

Boundary Layer Resolution	Far-field Modeling

Figure 3: Inflation Layer Setup, Transition, and Far-field of the Mesh

A mesh convergence study was conducted for both turbulence models in order to gain information about the necessary near-field resolution for sufficiently accurate results in terms of fin performance. Due to the significantly lower y^+ value of the $k-\varepsilon$ model, the necessary number of mesh cells could be lowered in comparison to the $k-\omega$ SST model (see Figure 4).

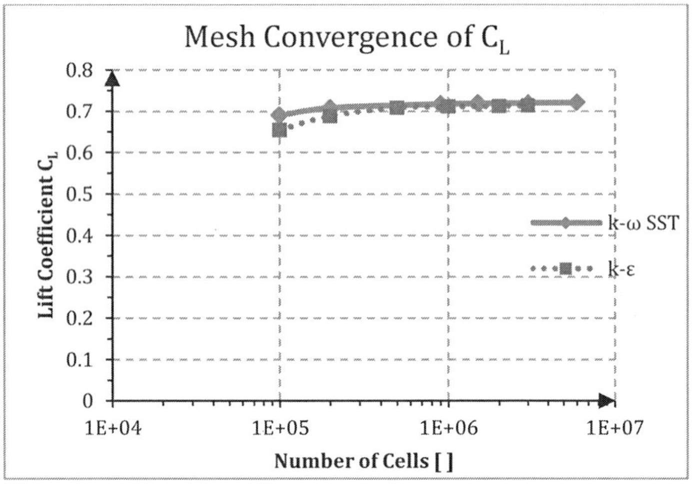

Figure 4: Mesh Convergence Study

For both models, it is noticeable that the values of the lift coefficient C_L converge with a higher number of cells. The $k-\varepsilon$ model is chosen to be tested with a lower maximum number of cells. For both models, it is shown that the eventual results are not dependent on discretization issues and numerical diffusion, but only due to the chosen boundary conditions and model setup. Finally, a mesh resolution of 3×10^6 cells is chosen for the $k-\omega$ SST model whereas the $k-\varepsilon$ computations are utilizing a mesh with 2×10^6 cells.

The accomplishment of the two-way FSI was connected with issues since this kind of simulation contains the coupling of a high-density fluid with a comparably weak material leading to excessive step-deformation of the structure when exposed to the loads of incompressible external fluid flow which in turn causes negative volume elements in many cases. Thus, stability measures were implemented such as a damping ratio of the fiberglass as well as Fluent solution stabilization for dynamic meshes with a volume based stabilization of 0.1. Also, the under-relaxation factors were lowered to 0.3 for pressure and momentum. Finally, a User-Defined Function (UDF) was set up that caused the flow velocity to speed up as a ramp function over 75% of the simulation time of $t_{sim} = 1s$:

$$v_x = 10\frac{m}{s}\left(\frac{t}{0.75s}\right) \quad t \leq 0.75s \text{and}$$
$$v_x = 10\frac{m}{s} \quad 0.75s < t \leq 1s(11)$$

(11)

RESULTS AND DISCUSSION

The investigation of the topic led to three different parts of the discussion:

i. the comparison of rigid and elastic fins for use in high density fluid flow
ii. the comparison of one-way and two-way FSI to sufficiently predict the deformation of an elastic foil in partially or fully turbulent flow
iii. the comparison of the used turbulence models in terms of lift and drag prediction.

Comparison of the Fin Performance

Looking at the computational data, it becomes clear that the performance differences between the rigid and the elastic model computed with CFD and FSI analysis are crucial. As shown in Figure 5, the overall side force - referred to as horizontal lift or lift - performance of the fin is better for the rigid model while both, CFD and FSI, offer maximum side force at an incidence angle of around 16°. This is generally shown by both sets of computations conducted using the $k-\omega$ SST and the $k-\varepsilon$ turbulence models. Between angles of 0° and 8°, the lift coefficient increases linearly before increasing with a smaller slope until reaching the maximum at 16° and decreasing again showing a reduced lift performance for high angles of attack $\alpha > 16°$. The reference area for the coefficients is the surface area of the fin.

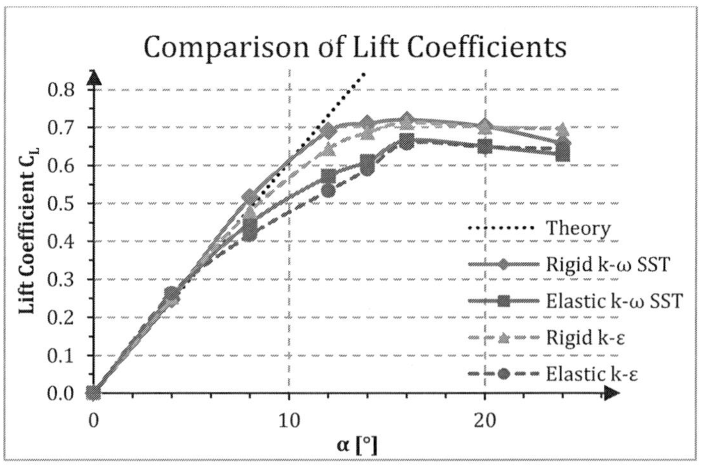

Figure 5: Comparison of Lift Performance

Also, the drag force of the fin changes with elasticity as depicted in Figure 6. Both turbulence models predict a decrease in drag force for the elastic fin model, though the $k-\varepsilon$ predicts a higher drag coefficient for both, the rigid and elastic fin. The decrease in drag force is assumed to be caused by the tip deflection of the fin reducing the drag in the tip section of the fin. With the tip for $\frac{x}{l} > 0.8$ accounting for a relatively small area fraction, the drag reduction is only approximately 8%.

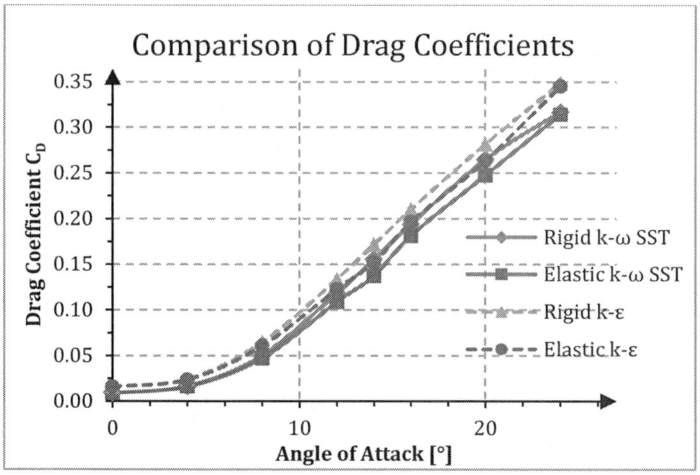

Figure 6: Comparison of Drag Coefficients

From the twist angle β at high root incidence angles α with respect to the fin length (see Figure 7), it can be assumed that the significant changes to the fin shape take place in the tip region as the deformation of the fin at 50% of the overall length is only about one third of the maximum twist that takes place at the fin tip for all three incident angles being 16°, 20°, and 24°. It is concluded that the side force produced by the fin leads to elastic deformation in the tip region lowering the local force coefficient for the tip because of the smaller angle of attack. Also, the relatively small deflection at the middle section of the fin causes a change in the angle of attack which in turn influences the lift and drag behavior.

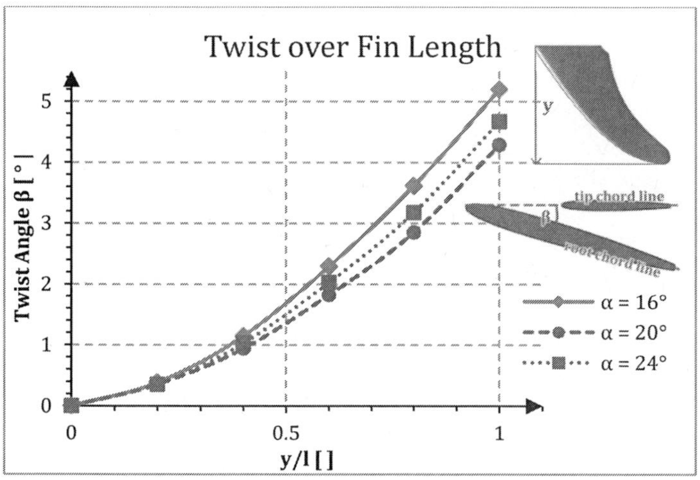

Figure 7: Total Tip Twist over Fin Length

Since the computations were performed for three-dimensional models, there are not only side and drag forces, but also vertical lift forces acting on the fin shape that differ significantly between the considered models. Figure 8 shows the computed vertical lift components for various angles of attack for both, the rigid and the elastic fin model. According to the results, the elastic fin develops increased vertical lift with higher angles of attack while the rigid model produces a suction force in downward direction, pulling the sailboard below the water surface. This could be caused by the associated boundary layer characteristics for swept-back wings (Schlichting & Truckenbrodt, 2001). Skin friction at the foil and spanwise boundary layer movement as well as the three dimensional suction pressure at the backside of the tip of the foil could lead to said downforce. This force is neutralized by the upward pressure due to tip deformation at an incidence angle of about 12° and increases almost linearly up to 24°.

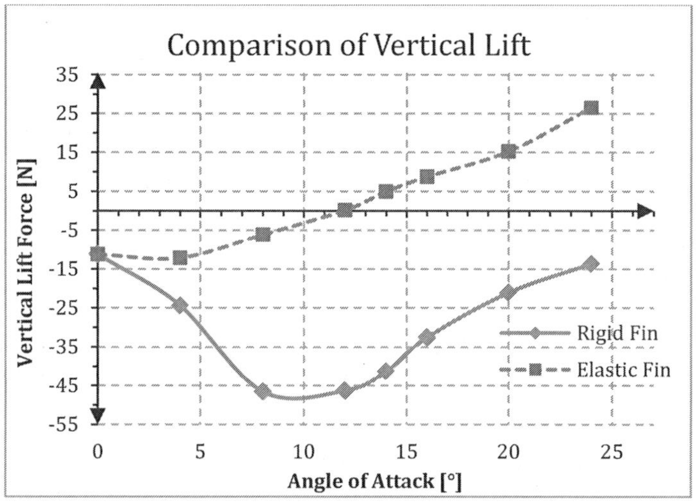

Figure 8: Comparison of Vertical Lift Forces

Additionally, the rigid fin develops an upward vertical pressure with higher incident angles after reaching the minimum of -47N at $\alpha \approx 10°$. It is assumed that this is caused by the fin shape that generally tends to deflect fluid to the downward direction because of its tapered profile thickness as seen on the streamlines in Figure 9.

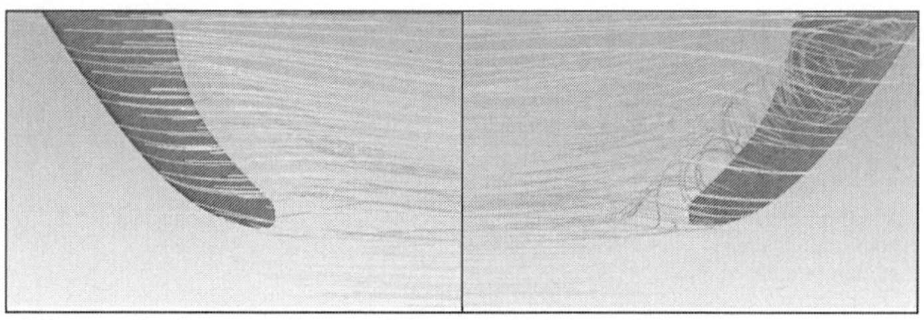

Figure 9: Streamlines of the Rigid Model Flow at $\alpha = 20°$ on the Pressure Side (l) and the Suction Side (r)

The post-stall behavior of the rigid and elastic fin models predicted by the $k-\omega$ SST turbulence model is different as seen on Figure 5. While the lift coefficient of the rigid fin decreases drastically after reaching the stall point at $\alpha = 16°$, the lift of the elastic fin decreases with a smaller slope. This can be caused by the deformation of the fin over length as the local effective incident angle is significantly lowered towards the tip leading to recovery of parts of the fin from stall. As the suction pressure is increased when stopping or decreasing flow separation, this could be causing the smoother decrease in overall lift force of the elastic fin model after initial stall effects.

Comparison of One-Way and Two-Way FSI

One of the secondary outcomes of the computations is the possibility to compare one-way and two-way FSI setups for the use of predicting the deformation of wings or hydrofoils such as windsurfing fins when exposed to high density external flow. Figure 10 shows the magnitudes of the fin tip deformation for both solutions showing significant differences and completely contrary behavior of the techniques for $12° < \alpha < 16°$. This may be explained with the way that both setups work in order to gain information about the tip deflection of the fin. If there was only one way of information needed, both FSI methods would give similar magnitudes of deformation. However, this is not the case so it is assumed that there are secondary effects acting on the fin that are not taken into account by the one-way FSI.

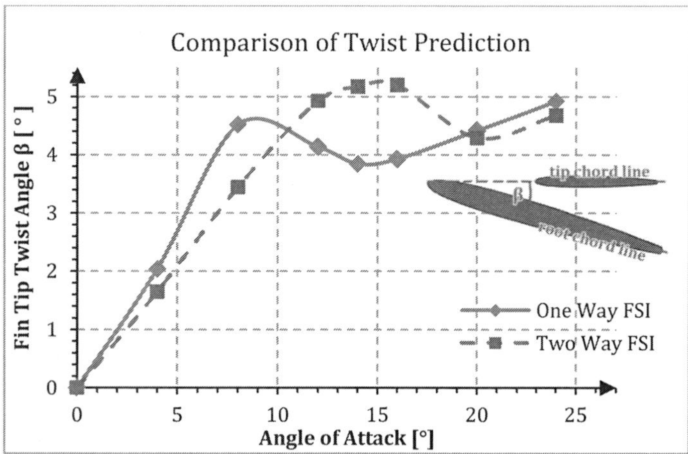

Figure 10: Comparison of Deformation Magnitudes

In order to explain the large differences between the predicted values of the deflection of the fin tip, it is useful to take a look at the forces that cause the deformation of the structure. This can be seen when evaluating the pressure distribution plots (see Figure 11) that show the discretized computed pressures for each node of the mesh. It is noticeable that the fin has an almost homogeneous pressure field at $\alpha = 8°$ while the pressure at $y = 0.2m$ is in the range of approximately $-5 \times 10^4 Pa \leq p_{tip} \leq 1 \times 10^4 Pa$. In contrast to that, the pressure distribution of the fin at $\alpha = 14°$ shows a pressure range at $y = 0.2m$ of $-2.5 \times 10^4 Pa \leq p_{tip} \leq 2.5 \times 10^4 Pa$. Thus, a higher tip pressure is used for the One-Way FSI model at 8° incidence angle leading to a higher initial deflection of the fin.

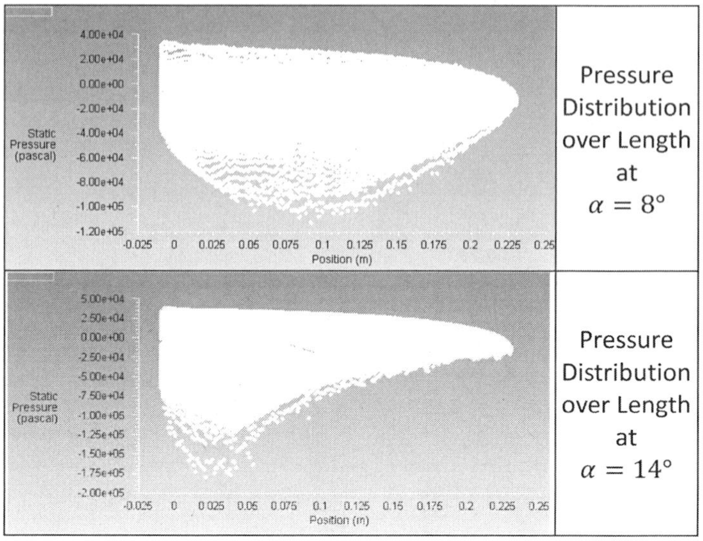

Figure 11: Comparison of Pressure Distribution for $\alpha = 8°$ and $\alpha = 14°$

While the cause of the extended positive tip pressure at $\alpha = 14°$ is due to the higher incidence angle leading to a larger change of momentum of the fluid particles, the lower negative pressure at the suction side of the fin has to be the result of partial stall. The velocity fields of the fin root and tip (see Figure 12) show the nature of the flow around the structure in a linear color scheme with red being high velocity and dark blue being zero velocity of the fluid while y is the distance from the fin root.

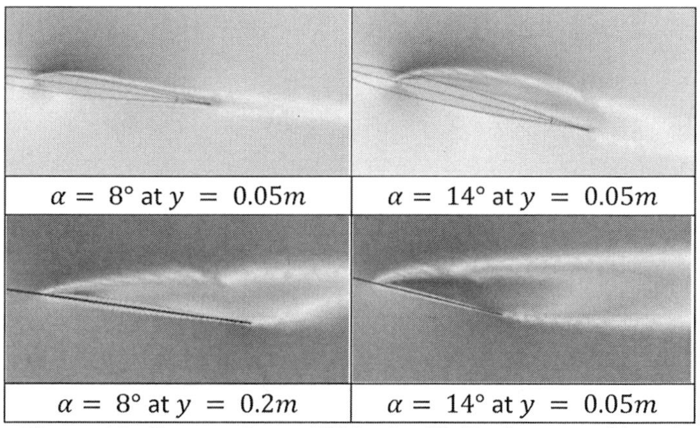

Figure 12: Comparison of Velocity Fields for $\alpha = 8°$ and $\alpha = 14°$

It is apparent that the flow at $y = 0.05m$ is only partially turbulent in both cases and does not differ too much except for the slight flow separation for the fin model angled at 14°. However, the tip flow shown in the two lower pictures differs greatly as the fin tip seems to be stalled at the tip at $\alpha = 14°$ while the fin tip at $\alpha = 8°$ is not completely stalled. This condition leads to the tip pressure being lower for an angle at which the overall fin force is generally higher than for smaller angles. The progression of tip twist in Figure 10 can be explained in a way that the fin tip works efficiently up to an incident angle of 8°, but then suffers from extending stall which is why the twist gets lower until $\alpha = 14°$. The increasing deformation at higher angles can simply be explained with the fin being already completely stalled while the positive pressure at the bottom side increases with higher angles of attack. Ultimately, this leads to the rise in tip deformation.

The cause for the different behavior of both FSI setups have to be secondary effects on the fin tip that also account for the deformation such as damping effects of the surrounding water flow and further lift production of deformed fin parts. At 8° incidence angle, the initial elastic fin deformation is so large, the tip would be exposed to positive pressure from the former suction side of the fin. This leads to damping effects, caused by the water flow. The one-way FSI fails to predict those secondary effects and is only suitable for small incident angles or sufficiently small forces.

Comparison of the Used Turbulence Models

Both turbulence models, the $k-\omega$ SST and the Realizable $k-\varepsilon$ model, predict a qualitatively similar behavior for the rigid and the deformed fin with slight but apparent differences. The $k-\varepsilon$ model computes a lift force of the two models that is approximately 6% lower while predicting a drag force that is up to 20% higher than for the $k-\omega$ SST model. Also, the decrease of the lift force at angles $\alpha > 16°$ is not as clear as for the compared model with shear stress transport. These findings generally correlate with the computational data of Eleni et al. who compared different turbulence models in terms of the prediction of the flow over a NACA 0012 profile (Eleni, Athanasios, & Dionissios, 2012). In their comparison, the lift coefficient of the $k-\varepsilon$ model also showed a slower decrease after the wing stall point. However, the linear part of the lift curve was more similar for both used models which could be due to the use of the two-dimensional foil of the NACA 0012 profile with theoretically infinite length so three-dimensional effects such as spanwise boundary layer movement or tip vortices were not taken into account which are generally better modeled by the $k-\omega$ SST scheme. The drag force comparison showed that the $k-\varepsilon$ model also over-predicted the drag coefficient while the computed values of the $k-\omega$ SST model matched the experimental values of Abbott et al. significantly better (Abbott & Von Doenhoff, 1959). These differences between the used turbulence models are assumed to be caused by their different areas of application. As mentioned above, both turbulence models have their special application and mesh properties that fit best for them. The $k-\varepsilon$ model does not take viscous boundary effects or shear stress into account and models them with a wall function being more or less accurate in predicting the right values to reimburse the missing near-wall treatment. The $k-\omega$ function used in the $k-\omega$ SST model does take those effects into account and is proven to be more accurate in predicting correct lift and drag values for wings.

CONCLUSIONS

This paper showed the investigation of a vertical hydrofoil used for surfing in terms of performance change with deformation as well as comparisons of one-way and two-way FSI approaches and the Realizable $k-\varepsilon$ and $k-\omega$ SST turbulence models. It was found that the fin loses performance with higher elasticity, but is more robust concerning post-stall behavior as the decrease in lift force is damped by the partial recovery through fin twist. Also, a vertical lift force is produced by the elastic model that provides the surfer with a higher chance to stay in planing condition (bottom surface of the board barely touches the water surface) in sharp turns.

It was shown that one-way FSI results match the values of two-way FSI computations only as long as there are laminar flow conditions and no stall, secondary damping or secondary lift effects. This is caused by the early tip stall of swept wings as known from theory which leads to a completely different initial pressure distribution over the length.

The used turbulence models showed similar qualitative behavior with the Realizable $k-\varepsilon$ model being less accurate in predicting the lift curve slope after stall.

FUTURE WORK

An experiment with fins made of materials with different properties could be conducted in a water tunnel in order to validate the obtained results. Also, the fin investigated in this work was of a special geometry used for wave-windsurfing. Other fin styles such as seaweed fins or race/slalom-fins could be evaluated as they work differently as they do not work with such a large amount of twist and tip deflection leading to different findings about deformation effects. As race fins are not shaped like swept wings, but like rectangular wings, they would not provide that much twist and may provide a vertical lift force without losing that much side force performance. Also, they are built with much higher aspect ratios in order to minimize drag. The effect of winglets for deformable fins on their performance could also lead to new insights as they increase the effective aspect ratio leading to better performance.

ACKNOWLEDGMENTS

Special thanks to Dietrich Hanke from Maui Ultra Fins for providing much useful information, and Daniel Aeberli from Choco Fins for helping out with the CAD model data used for the calculations. Also thanks to Dr. Anderson from the Milwaukee School of Engineering for many useful discussions.

This work was accomplished by the first author as a bachelor thesis: *Analysis of a windsurfing fin made of rigid and elastically deformable materials using Computational Fluid Dynamics*, under the advisement of the second author to meet the dual degree requirements of Milwaukee School of Engineering, Milwaukee, WI and the University of Applied Sciences, Fachhochschule Lübeck, Germany.

REFERENCES

Menter, F. R. (1993). Zonal Two Equation k-ω Turbulence Models for Aerodynamic Flows. *24th Fluid Dynamics Conference.* AIAA.

Wilcox, D. C. (1993). *Turbulence Modeling for CFD.* La Cañada: DCW Industries.

Abbott, I. H., & Von Doenhoff, A. E. (1959). *Theory of Wing Sections.* New York: Dover Publications.

Ariff, M., Salim, S. M., & Cheah, S. C. (2009). *Wall y+ Approach for Dealing with Turbulent Flow Over a Surface Mounted Cube: Part 1 - Low Reynolds Number.* Melbourne: CSIRO.

Eleni, D. C., Athanasios, T. I., & Dionissios, M. P. (2012). Evaluation of the Turbulence Models for the Simulation of the Flow over a NACA 0012 Airfoil. *Journal of Mechanical Engineering Research , 4* (3), 100-111.

Georgiadis, N. J., Yoder, D. A., & Engblom, W. A. (2006). Evaluation of Modified Two-Equation Turbulence Models for Jet Flow Predictions. *AIAA Journal* .

Hutchison, S. R. (2012). *Numerical Modelling of Hydrofoil Fluid-Structure Interaction.* University of Tasmania. Tasmania: University of Tasmania.

Harper, C. W., & Maki, R. L. (1964). *A Review of the Stall Characteristics of Swept Wings.* NASA. Washington, D.C.: NATIONAL AERONAUTICS AND SPACE ADMINISTRATION.

Lecheler, S. (2009). *Numerische Strömungsberechnung: Schneller Einstieg durch ausführliche praxisrelevante Beispiele.* Wiesbaden: Vieweg + Teubner.

Schlichting, H., & Truckenbrodt, E. (2001). *Aerodynamik des Flugzeuges* (Bd. 2). Berlin: Springer.

Proceedings of the ASME 2014 International Mechanical Engineering Congress and Exposition
IMECE2014
November 14-20, 2014, Montreal, Quebec, Canada

IMECE2014-37298

TRANSFORMATION OF AN AVIATION TURBOSHAFT ENGINE INTO A EXPERIMENTAL JET ENGINE FOR LABORATORY TESTING UNSTABLE RADIAL COMPRESSOR WORK

Marián Hocko
Technical University of Košice
Department of Aviation engineering
Rampová 7, Košice, 041 21, Slovakia
mail:marian.hocko@tuke.sk

Jiri Polansky
University of West Bohemia
New Technologies Research Centre
Univerzitní 8, Pilsen, 306 14, Czech Republic
mail:polansky@ntc.zcu.cz

ABSTRACT

The article deals with the use of a small aviation turboshaft engine for laboratory purposes. This study describes its transformation into an experimental device for research and education. Various constructional, technological and controlling modifications and settings of the gas turbine test stand were carried out and tested on a stationary configuration. The stationary system can be used as a small backup power generator or as a drive unit for a compressor, pump, etc. New control systems, electronic elements and methods of measuring rotations, pressure and temperature are tested for educational and research purposes. The study includes a schematic description of modelling measurements and subsequent numerical evaluation of the thermodynamic characteristics of the cycle in an experimental gas turbine. The laboratory device presented here is, thanks to technological, material and thermodynamic research, suitable for educating and testing the knowledge of future aviation and mechanical engineers.

The content of the article is a description of the use of transformed small turboshaft engine into small jet engine by means of experimental testing of unstable work of the radial compressor under laboratory conditions.

INTRODUCTION

Unstable work (surge) of compressors is a dangerous phenomenon that occurs in a particular mode of compressors of aviation turbo compressor engine (ATCE) and also gas turbine (GT) that are used in vehicles and stationary applications. That is manifested by sudden changes in pressure and speed of the airflow at the outlet of the compressor and by characteristic sound effects that give rise to intense vibrations of the rotor blades and by changes of the of air flow characteristics in other parts of both ATCE and GT. The consequences of unstable work of the compressor are associated with cessation of activities of ATCE and GT, or their destruction because of mechanical damage of the compressor blades or destruction (burning) of the gas turbine blades. Therefore, the formation of unstable work of the compressor is inadmissible and much attention is devoted to prevention of its starting in ATCE and GT.

Carrying out the research in unstable work of ATCE and GT compressors in a real operational context is associated with potential security risks and significant economic costs due to damage or destruction of expensive engines. Relevant results can be achieved without the mentioned risks under the laboratory conditions using the experimental small turbojet engine “MPM – 20” that was created by transforming the turbine trigger “TS – 20” into a small single jet engine. In experiments conducted at the Department of Aviation Engineering at the Faculty of Aeronautics, Technical University of Kosice, artificially induced unstable work of a centrifugal compressor, virtually confirmed the hypotheses and the relevant consequences for particular parts of the engine [1].

LAUNCH MECHANISM OF UNSTABLE COMPRESSOR WORK

Unstable work of the ATCE compressor is a phenomenon that occurs when a particular mode of work of the compressor is manifested by severe periodic changes in theair flow rate Q_A, the output air pressure from compressor p_2,absolute air speed cand sometimes back-flow of air from the compressor to the input ATCE device.

Concomitants of unstable work of the ATCE compressor are:
- pulsation of pressure p_2 (Fig. 1) and air velocity c_2 at the outlet of the compressor,
- a reduction of the mean pressure at the outlet of the axial compressor $p_{2,m}$.
- characteristic sound effects,
- engine vibration,
- vibration of compressor rotor blades.

The formation of unstable compressor work during the operation of ATCE or ST can occur for various reasons, for exampleas a consequence of:
- unstable work of input ATCE or GT device,
- failures of an engine control system (accelerator machine, high-pressure pump, etc.),
- sucking an object (birds, ice, etc.),
- sudden changes in temperature conditions at the inlet to the

engine (while passing over the fire, active volcano or sucking the off-gasses of another aircraft or launched missiles, etc.),
- sudden changes in active areas of the input device of the engine when aircraft is maneuvering,
- changes in air quality, for example in sucking volcanic ash when flying in the volcanic cloud and the like.

The consequence of unstable work of the ATCE or GT compressors can be:
- an irregular operation of the engine,
- a reduction in thrust (power) owing to reduced flowing amount of air through the engine,
- distortion of the normal steady burning of fuel-air mixture in the main combustion chamber of ATCE,
- the sharp increase in gas total temperature in front of the gas turbine ATCET T_{3t},
- increase of the vibration level of the whole ATCE,
- possibility of breakage of the ATCE compressor rotor blades because of vibration.

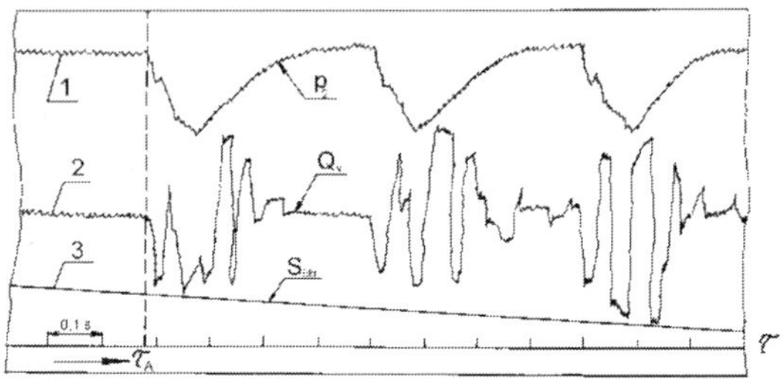

Fig. 1: Recorded parameter changes as the compressor is in surging [2]

1- course of changes in output static pressure p_2,
2- course of changes in air flow Q_A through the compressor,
3- change of the active cross-section flow S_{ACS} through the comp.

Regarding the accompanying phenomena, which may lead to an interruption of the engine, its damage, or destruction of unstable work of the compressor, is admissible [5].

The entire process of creating the unstable work of the ATCE compressor is rather complicated. The physical principle to form unstable compressor work is based on tearing stream lines of airflow at the run around an individual compressor blade when major departure from the calculation mode of operation of the compressor, resulting in a sudden change in the air supply and pressure conditions at the inlet of the compressor ATCE.

LAUNCH OF UNSTABLE WORK OF AN AXIAL COMPRESSOR

In operation, the axial compressor in a computing mode corresponds to he air flow rate $Q_A = Q_{A, cal.}$ of the rotor speed n and the peripheral speed of the rotor u_1 at the absolute inlet air velocity c_1 specifies the initial relative air speed at the rotor blade inlet w_1, which flows in the direction of a tangent towards the central curve profile of a rotor blade to in between rotor blade channel. The outlet absolute velocity of the air from the compressor rotor blade c_2 enters in between a rotor blade channel in the direction of a tangent towards the curve profile of a central stator vane. The airflow through the axial compressor stage is smooth, with out creating turbulence. The axial compressor is stable (Fig. 2a).

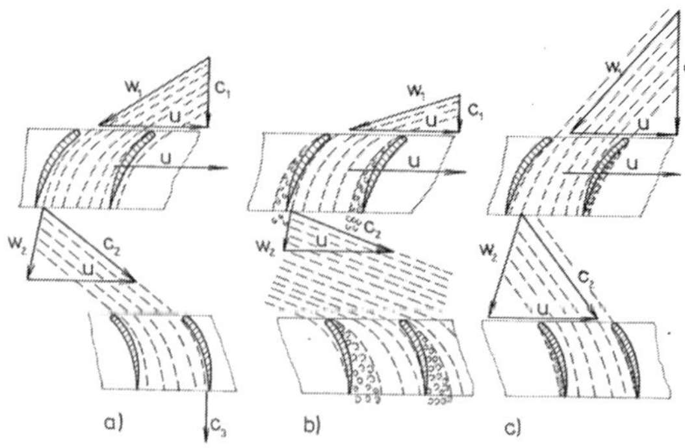

Fig. 2: The airflow character of an axial compressor stage

In reducing the rate of airflow $Q_A < Q_{A, cal.}$ and particular engine speed n, the circuit rotor speed u will create, with reduced absolute speed c_1, the resulting relative speed w_1 the direction of which will deflect from the tangent to the medium curve of a rotor blade profile. This disrupts the smooth rotor blade and on their backs tearing off stream lines will occur and laminar flow to turbulent flow will be converted and spread farther to the stator vanes and other stages of the axial compressor. The roiled air flow in an axial compressor reduces smooth air flow farther into the main ATCE combustion chamber; it will result in backward flow of air back towards the ATCE input device. So the back flow of air in the space in front of the main combustion chamber sharply reduces air pressure and the air starts to move away from the input device to the main ATCE combustion chamber. The unstable air flow (surge) through the ATCE axial compressor results in an abrupt change in the composition of the fuel-air mixture in the main combustion chamber which causes a significant increase in the overall total temperature of the gas in front of the gas turbine T_{Tt} (sparingly rich mixture of fuel and air) and vibration of the rotor blades of the axial compressor with the possibility of their damage due to fatigue fracture (Fig. 2b).

With an increase in the rate of airflow $Q_A > Q_{A, cal.}$, and the particular rotor speed n the circuit rotor speed u will create, with the increased absolute speed c_1, a resulting input relative velocity of the air at the entrance to the rotor blades w_1 the direction of which will deviate from the direction of the tangent to the profile middle curve of the rotor blade. This will cause hindering the smooth laminar airflow in the riverbed of rotor blades. The turbulent flow in the riverbed of blades will decline and there will be no farther spread of the break away streamlines into other stages of the axial compressor. Therefore, in this case there will be no formation of unstable work of the ATCE axial compressor (Fig. 2c).

LAUNCH OF USTABLE WORK OF A RADIAL COMPRESSOR

Unstable work of the ATCE radial compressor results from tearing of the air flow to the impeller blades and the vanes of the diffuser at larger angle of attack changes in the incoming air, which may be caused by changing of the air flow rate, or by changing the rotor speed of the compressor.

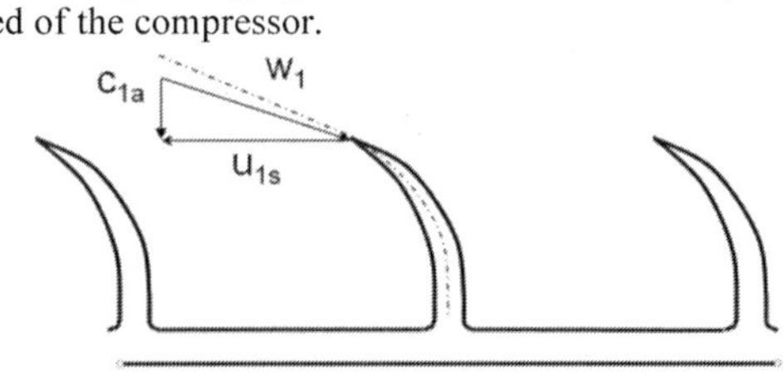

Fig. 3: Stable air flow of radial compressor blades

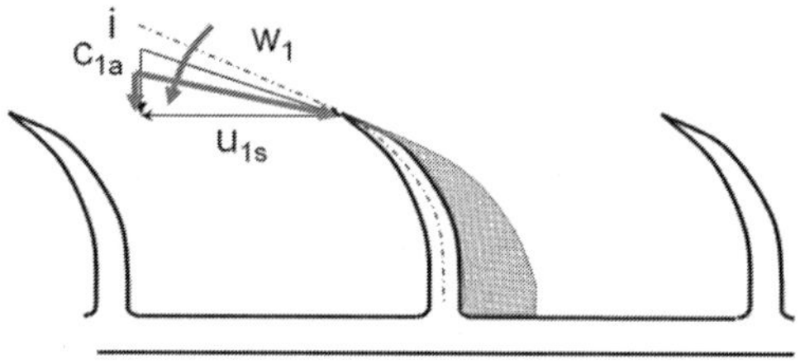

Fig. 4: Unstable air flow of radial compressor blades

The formation of unstable work of the ATCE centrifugal compressor at higher values of calculated speed n c, such as computing speed mode $n_{p,\,cal.}$.

$$n_c = n.\sqrt{\frac{288}{T_{It}}}\ \left[\text{min.}^{-1}\right]$$

$$n_c = n.\sqrt{\frac{288}{T_I.\left(1+\frac{\kappa-1}{2}.M_I^2\right)}}\ \left[\text{min.}^{-1}\right] \qquad (1)$$

Where:

n_c - converted speed [min^{-1}],
n - measured speed of rotation [min^{-1}],
T_{Ic} - the total air temperature at the inlet to the radial compressor [K],
T_I - static air temperature at the inlet to the radial compressor [K],
M - Mach flight number [1].

The equation (1) shows that the calculated speed will increase at low static air temperature T_I and low speed expressed in the figure M_I, corresponding to activities of ATCE in climbing to a great height.

If the air flow rate through a radial compressor equals to the computational rate of air flow $Q_A = Q_{A,\,cal.}$, the air flow of radial compressor blades and diffuser vanes is smooth, the unstable work of a radial compressor will not occur. The radial compressor will work stably (Fig. 3).

If the air flow rate through a radial compressor is greater than the calculated air flow rate $Q_A > Q_{A,\,cal.}$ tearing off the air flow in the riverbed of a radial compressor blades will occur. As the tearing area of the air flow is relatively small, it does not spread. On the diffuser vanes the air flow is tearing off on the back of the blades. Air particles tend to move along the curves that are close to the logarithmic spirals. Roiling of the air flow reduces the efficiency of radial compressor, but the air stream flows through the radial compressor but unstable work of the compressor usually does not occur (Fig. 5).

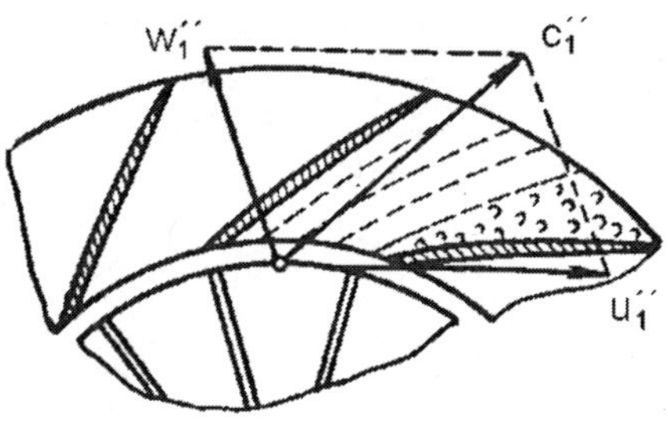

Fig. 5: Tearing off the air stream lines on the back of diffuser vanes at $Q_A > Q_{A,\,cal.}$.

In reducing the rate of air flow through a radial compressor below the calculated amount of air $Q_A < Q_{A,\,cal.}$, tearing off the air flow at the back of the impeller blades (Fig. 4) and diffuser vanes will occur (Fig. 6). The detached air stream loses velocity and reduces the flow area. So the decrease in the kinetic energy tends to increase the static pressure, which leads to the return flow and the consequent restoration of steady flow (surge). This processes again repeated at short intervals [2, 3].

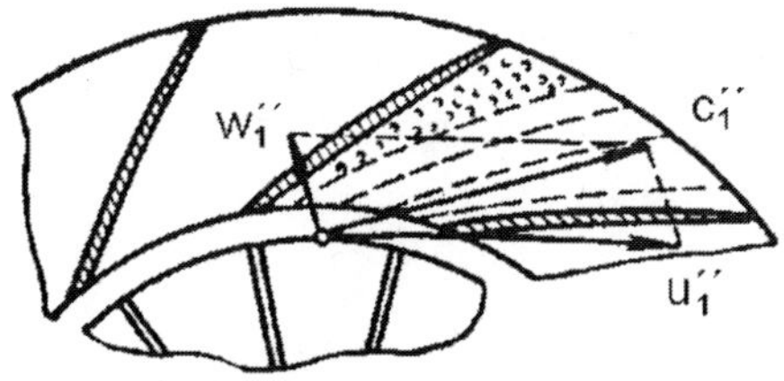

Fig. 6: Tearing off the air stream lines on the back of diffuser vanes at $Q_A < Q_{A,\,cal.}$

EXPERIMENTAL VERIFICATION OF UNSTABLE WORK OF A RADIAL COMPRESSOR

As mentioned in the previous part of this article, one of the causes of ATCE compressor unstable work is a change of air flow through the engine inlet device as a result of throttle down from variety of reasons. In theory, this process is explained in the particular (constant speed) mode by shifting an operating point along the linen = const. in the characteristics of a radial compressor to the left border of the unstable work [4]. In sufficient reduction in air flow rate Q_A via the in put device the operating point of the engine exceeds a thre shold of unstable work and gets in unstable work of the radial compressor (surge area) (Fig. 7).

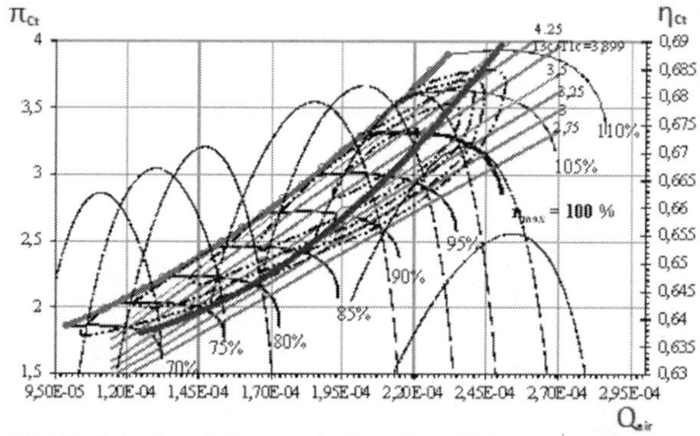

Fig. 7: Calculated characteristics of a radial compressor of an experimental small turbojet engine "MPM-20" [4]

Where:
π_{Ct} – the total degree of air compression on the compressor [1]
η_{Ct} – the total compression efficiency [1]

For practical testing of theoretical hypotheses, the Laboratory for small turbojets at the Department of Aviation Engineering, Technical University of Kosice has been used, in which the experimental small single jet engine is available, it was created by conversion of the turbine trigger "TS-20" [7].

"MPM-20" is a small single jet engine with a diagonal entry system, single-stage centrifugal compressor, an associated combustion chamber, single stage uncooled gas turbine and the fixed exhaust nozzle.

Fig. 8: View of the "MPM-20" at the Laboratory of small jet engines

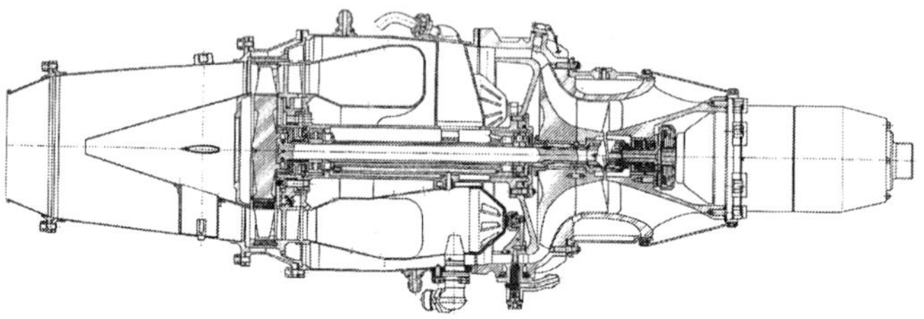

Fig. 9: Section of a small turbojet engine "MPM-20" [7]

THROTTLING OF AIR FLOW IN THE INLET SYSTEM

The practical experimental verification of the impact of throttling the flow of air in the intake device of "MPM – 20" on an unstable work of a radial compressor, the method for gradual increasing the density of nets has been implemented; each was placed on the input system of the engine.

The processed linear mathematical model of "MPM-20" is based on the theory of small changes in [8], described in [1, 10] for the sub-critical conditions of flow in the outlet nozzle, defines the conditions of change in the intake device reflected by changes of all parameters of the working process of single jet ATCE. For practical validation of this theory four sets of experimental measurements were prepared and aimed at gradually throttling the air flow through the input device "MPM - 20th". The measured parameter values for individual experimental measurements are compared with the values obtained in the calibration measurement and the theoretical calculation based on the theory of small changes. Gradually these experimental measurements were realized:

1. Air flow through the input device of "MPM – 20" without the input sieve.
2. Air flow through the input device of "MPM – 20" with the sieve mesh size of 1.5 x 1.5 mm.
3. Air flow through the inlet system of "MPM – 20" with a double sieve (a sieve with a mesh size of 1.5 x 1.5 mm and a sieve with mesh of 2.5 x 2.5 mm).
4. Air flow through the inlet device of "MPM – 20" with a TV viewings creen with openings with a diameter of 0.4 mm.

The change in the independent variable (failures) was calculated according to the formula:

$$\delta X_{si} = \left[\frac{X_{si} - X_0}{X_0}\right].100[\%] \tag{2}$$

Where:
δX_{si} – change in the independent variable,
X_{si} – a value parameter in a simulated malfunction,
X_0 – an original value of the independent variable.
The change inthe dependent variable value was calculated according to the equation:

$$\delta Y_{si} = \left[\frac{Y_{si} - Y_0}{Y_0}\right].100[\%] \tag{3}$$

Where:
δY_{si} – a change of the dependent variable,
Y_{si}– a value parameter in the simulated malfunction,
Y_0 – an original value of the dependent variable.

THE AIR FLOW THROUGH THE INLET DEVICE OF "MPM-20" WITHOUT AN INLET SIEVE

For the conditions of the first experiment the original in put protective sieve was removed from the entrance system of "MPM-20".

Before carrying out the planned experiments, the "MPM-20" was set according to the technical documentation for a calculation scheme [9]. Subsequently 10 measurements of calibration parameters were performed. The mean value of the measured parameters obtained by the calibration measurement served as a bench mark for assessing changes in subsequent experimental measurements.

Fig. 10: Removing the protective sieve from "MPM-20"

A relatively high degree of agreement of measured and calculated thermodynamic parameters is apparent from the results of the calibration measurements processed in "MPM – 20" and its comparison with the results of thermodynamic calculation of heat circulation of "MPM – 20". Minor deviations between measured and calculated values of the thermodynamic parameters are caused by irregularities occurring in the measurement because of the location of sensors in other positions as suggested in the theoretical calculation.

PARAMETER	p_H	T_H	ACTIVITY PERIOD	$p_{2,C}'$	$p_{2t,C}'$	p_{2t}
UNIT	Pa	K	Sec.	Pa	Pa	Pa
MEASURED VALUE	99885,82	290.65	50,8	180787,9	392652,7	351971,5
COUNTED VALUE	101325.2	288	50	181600.13	398549,2	392321,5
PARAMETER	p_3	p_4	T_{2t}	T_{2t}	T_{3t}	T_{4t}
UNIT	Pa	Pa	K	K	K	K
MEASURED VALUE	351326,2	154655,78	458,55	460.25	1170,35	1043.45
COUNTED VALUE	392321.5	162812,95	462,35	462,35	1168,15	1013,27
PARAMETER	Q_{fuel}	G_{fuel}	c_h	F_T	Q_A	c_m
UNIT	cm^3/CYKLUS	kg/CYKLUS	$kg.h^{-1}$	N	$kg.s^{-1}$	$kg.h^{-1}.N^{-1}$
MEASURED VALUE	1362	1,05572	74,87	-	-	-
COUNTED VALUE	-	-	-	698,09	1.2	0,1311

Tab. 1: Comparison of calculated and measured values of the parameters in the calibration measurements of "MPM – 20"

Where:
p_H – atmospheric pressure [Pa],
T_H – atmospheric temperature [K],
$p_{2,C'}$ – static presurebehind compressor impeller [Pa],
$p_{2t,C'}$– total presure behind compressor impeller [Pa],
p_3 – static presure before turbine [Pa],
p_4 – static presure behind compressor impeller [Pa],
$T_{2t'}$–total temperature behind compressor impeller [K],
T_{2t} – total temperature behind compressor [K],
T_{3t} – total temperature before turbine [K],
T_{4t} – total temperature behind turbine [K],
Q_{fuel} – fuel flow [$kg.s^{-1}$],
G_{fuel} – weight of fuel consumption per cyckle [kg. $cycle^{-1}$],
c_h – fuel consumption per hour [kg. h^{-1}],
F_T – thrust [N],
Q_A –air flow [$kg.s^{-1}$],
c_m – specific fuel consumption [$kg.N^{-1}.h^{-1}$].

In the first experiment (Fig. 10) 10 measurements were performed and the same parameters as for the calibration measurements "MPM – 20" in the original configuration (with a protective sieve) were measured and evaluated. The mode of "MPM – 20" operations was set to values that correspond to the calibration measurement (computing) mode; external conditions (p_H, T_H) have not changed.

As is apparent from the data given in [1], that compares the measured values of the parameters in the experimental measurements without the input sieve and the calibration measurements, the differences in individual parameters are minimum and range from -5.4 % (static gas pressure behind the gas turbine p_4) to +2.9 % (total gas temperature behind the gas turbine T_{4t}). Most of the other observed parameters have hardly changed even though the flow area of the inlet to the "MPM – 20", after the removal of the inlet sieve, grew by 35 % (a total area of the inlet sieve with a wire sieve of a mesh size of 2.5 x 2.5 mm and wire thickness of 0.6 mm wire is 0.078593 m^2).

The results obtained by direct measurement of "MPM -20" without input sieve confirm that the choice of the wire diameter and the mesh size has been optimized by the engine manufacturer so that the sieve provides the protection from sucking strange objects with a minimum influence on the parameters of "MPM – 20". Based on the measured parameters, the calculation of thrust F_T and of specific fuel consumption c_m were performed.

THE AIR FLOW THROUGH THE INLET DEVICE OF "MPM-20" WITH THE SOFT INLET SIEVE

In the second experimental measurement, the air flow through the soft sieve of an input device was throttled, which was deployed on the input device of "MPM-20". The total area of the inlet sieve is 0,078593 m^2. When using a wire sieve with a mesh size of 1.5 x 1.5 mm and at hickness of 0.3 mm ire, the occupied area of the input wire sieve is 34.984% of the total area of the input device.

The measured values of parameters of "MPM-20", described in [1], are very close to the measured values in calibration measurements with the original sieve.

AIR FLOW THROUGH THE INLET DEVICE OF MPM-20 WITH A DOUBLE INPUT SIEVE

Third experimental measurements were carried out at throttling off low with the dual sieve (Fig. 11). To the original input device with an original protective sieve of "MPM-20" a soft sieve with the mesh size of 1.5x1.5 mm and a thickness of 0.3 mm wire was deployed, thus reducing the flow area by 58 %. On the basis of the reduction in the flow area of the inlet system the measured thermodynamic parameters have been changed, resulting in a change of the engine-dependent parameters. The total degree of gas expansion significantly changed on the gas turbine π_{Tt} (about 19.194%) and the total temperature of gas behind the gas turbine T_{4t} (by 4,475%).

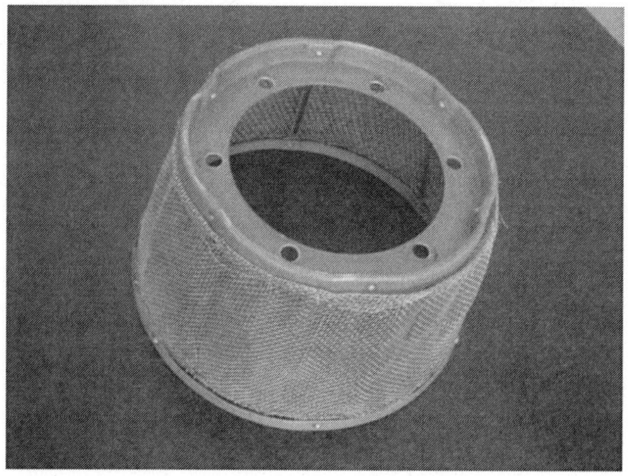

Fig. 11: Dual sieve of "MPM-20"

AIR FLOW THROUGH THE INLET DEVICE OF "MPM-20" WITH THE SOFEST INLET SIEVE

The recent experimental measurement was carried out with strong throttling of flow through the input device, using aperforated metal plate with round holes 0.4 mm (a metal plate was originally used as a color TV viewing screen) (Fig. 12, 13). The perforated plate was placed on the original sieve of an input device, thereby allowing are duction by 83% in the flow area of the inlet system of "MPM-20". This reducing of the flow area of the engine inlet device was substantial degradation of measured parameters compared to measured parameters at the input device without the inlet sieve (Δp_{2t} = -48 978,8 Pa, Δp_3 = 24198,5 Pa, Δp_4 = -24957,82 Pa, T_{2t} = -11,6 K, ΔT_{3t} = -92,4 K, ΔT_{4t} = -177,5 K, c_h = -4,3465 $kg.h^{-1}$, $\Delta\pi_{Tt}$ = 0,6, $\Delta\pi_{Ct}$ = -0,509 a $\Delta\pi_{ENt}$ = -0,257.
Where:
π_{Ct} – the total degree of air compression on the compressor [1],
π_{Tt} – the total degree of gas expansion on the gas turbine π_{Tt} [1],
π_{ENt} – the total degree of gas expansion on the exhaust nozzle [1].

Fig. 12: Perforated plate attached to the original inlet device of "MPM-20"

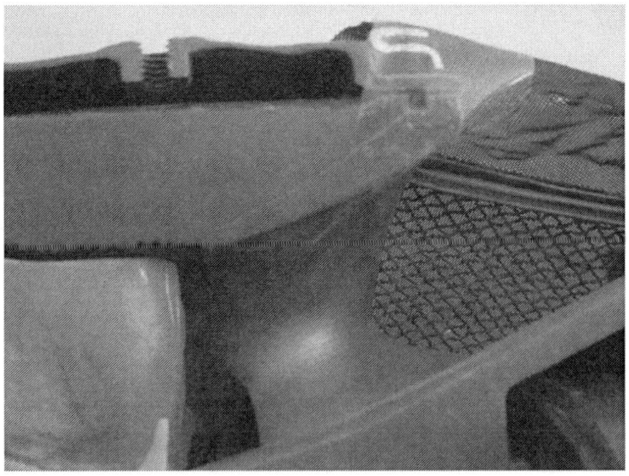

Fig. 13: Perforated plate attached to the original sieve of an input device in cross-section of "MPM-20"

The seventh measurement with 83 % shading of the input device was in an unstable work of a radial compressor of the engine and the destruction of gas turbine blades and to burn the sealant mass in the body of the gas turbine above rotor blades (Fig. 14, 15).

Fig. 14: Destroyed gas turbine rotor blades of "MPM-20" because of unstable work of a radial compressor

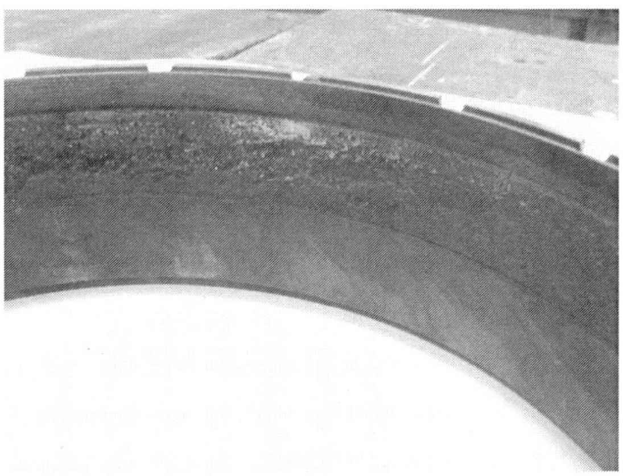

Fig. 15: Burned sealant in the body of the gas turbine rotor blades

PARAMETER	$S_{inlet.}$	p_H	T_H	**Activity period**	$p_{2,C}'$	$p_{2t,C}'$
UNIT	**%**	**Pa**	**K**	**Sec.**	**Pa**	**Pa**
Inlet device without a sieve	100	100525,3	288,75	51,0	178967,1	388941,8
Inlet device with a soft sieve	80	100365,6	290,95	49,7	179136,7	390028,1
Inlet device with a basic sieve	65	99885,82	290,65	51,7	180787,9	391185,7
Inlet device with a dual sieve	58	100125,3	290,65	49,7	198319,4	397633,7
Inlet device with the softest sieve	17	99991,98	290,29	50,0	213852,3	428407,3
PARAMETER	p_{2t}	p_3	p_4	T_{2t}'	T_{2t}	T_{3t}
UNIT	**Pa**	**Pa**	**Pa**	**K**	**K**	**K**
Inlet device without a sieve	351178,4	351524,7	128716,48	458,35	459,55	1169,25
Inlet device with a soft sieve	351571,5	351462,6	128977,1	458,45	459,95	1170,05
Inlet device with a basic sieve	351971,5	351326,2	154655,78	458,55	460,25	1170,35
Inlet device with a dual sieve	366122,2	340125,3	154192,9	460,65	461,95	1172,95
Inlet device with the softest sieve	400157,2	327326,2	153674,3	470,01	471,15	1261,65
PARAMETER	T_{4t}	Q_{fuel}	$G_{pal.}$	c_h	F_T	c_m
UNIT	**K**	**cm^3/cycle**	**kg/cycle**	**$kg.h^{-1}$**	**N**	**$kg.h^{-1}N^{1}$**
Inlet device without a sieve	1039,15	1407	1,09045	77,0065	732,5*	0,1051283*
Inlet device with a soft sieve	1041,15	1405	1,0889	78,8700	731,6*	0,107811*
Inlet device with a basic sieve	1043,45	1397	1,0819	75,2900	698,1*	0,107847*
Inlet device with a dual sieve	1085,65	1330	1,0312	74,7434	598,74*	0,125334*
Inlet device with the softest sieve	1216,65	1293,3	1,0027	72,6600	541,11*	0,13428*

Tab. 2: Comparison of the results obtained in the different experiments on "MPM-20"

CONCLUSION

Carried out experiments on a small jet engine "MPM-20" confirmed the hypothesis.

The throttling of the input device of "MPM – 20" changes parameters of the engine; their character is the same as the change in a coefficient of maintaining the total air pressure in the intake device σ_{input}.

The throttling of the inlet device of "MPM – 20", from the thermodynamic parameters, mostly effects the static pressure of the gas behind the gas turbine p_4 (-19.39 %), the total temperature of the gas behind the gas turbine T_{4t} (-17.081 %), total air pressure behind the compressor p_{2t} (- 13.95 %), total gas pressure the in front of the gas turbine T_{3t} (-7.903 %) and a total air temperature behind the compressor T_{2t} (-2.524 %).

Throttling input device MPM - 20 leads to a higher operating point convergence to the limit in the characteristics of unstable work centrifugal compressor. Despite higher resistance of the radial compressor of "MPM – 20" comparing to against unstable work was the seventh measurement with 83 % throttling of the inlet system to its unstable work and destruction of gas turbine blades and burn of the sealant mass in the body of the gas turbine above the rotor blades.

Changing of the inlet flow cross-section to the ATCE and GT radial compressors has a significant effect on its thermodynamic and performance parameters. When in case of large throttling or sudden reductions in air flow through the input device, unstable flow may affect a specific area or the entire radial compressor, which may cause damage to the ATCE gas turbine (Fig. 14, 15).

Carried out experiments, in the laboratory of small jet engines, for students have demonstrated the negative impact of throttling the input system of the engine on its c parameters and the consequences of unstable work radial compressor MPM-20 for the construction of a gas turbine.

With this test rig we are able to model typical compressor instabilities describe on papers ([11, 12]).

ACKNOWLEDGMENTS

Research was supported by project TAČR CK TE01020036. and MSMT CZ.1.07/2.3.00/20.0139.

REFERENCES

[1] Hocko M., 2003, „Assessment of the ATCE State Based on Changes in Thermodynamic Parameters", thesis, pp. 338

[2] Ruzek, J., 1979, „Theory of Aircraft Engines, Part I (Compressors, Turbines and a Combustion Chamber)", University textbook VA AZ Brno, pp. 373

[3] Ott, A., 1983, „Fundamentals of the Theory and Design of Air Vane Engines, Part I", University Textbook, Military Academy of Aviation, Kosice, pp.140

[4] Med, S.: „Diagnosis of JE - TS -20 in Laboratory Conditions K – 316", master's thesis, Department of Aircraft, Military academy in Brno

[5] Cumpsty, NA, 1989, „Compressor aerodynamics", Department of Engineering University of Cambridge, pp. 687

[6] Hocko, M.,Gaspar,R.,2013, „Impact of rapid thermal fluctuations of the airflow into engine inlet to the gas dynamics stability of the turbojet engine", 12th conference Power System Engineering, Thermodynamics & Fluid Flow, Pilsen. pp. 8

[7] Hocko M., 2003,„The small jet engine MPM – 20", VLA Kosice

[8] Черкез, А., Я., 1975, „Инжинерные расчеты газотурбиных двигателей методом малых одклонений", Издательство Машиностроиение, pp. 379

[9] „The operation and maintenance of the engine AL - 7F - 1 , Book 2 , Book 3,Let -21 -20/2,Let -21 -20/ 3", Praha 1965

[10] Főző , L., Andoga, R., Madarász, L., 2010 , „Mathematical model of asmall Turbojet Engine MPM – 20". In: Studies in Computational Intelligence.Vol . 313, p. 313-322

[11] Polansky J., Kalmar L., Gaspar R.: Aerodynamic performance prediction of centrifugal fan with forward curved blades, Journal of Thermal Science, Vol 22, NO 6, 2013 pp 517-521, ISSN: 1003-2169 , 2013

[12] Cyrus, V.; Polansky, J.: Numerical simulation of the origins of flow pulsations in cascades of the rear blade rows in a gas turbine axial compressor using low calorific fuel. ASME journal 2010, no. 132/3, p. 31012/1-11. ISSN: 0889-504X

This page left blank intentionally.

Proceedings of the ASME 2014 International Mechanical Engineering Congress and Exposition
IMECE2014
November 14-20, 2014, Montreal, Quebec, Canada

IMECE2014-38329

UTILIZING SCHLIEREN IMAGING TO VISUALIZE HEAT TRANSFER STUDIES

Jacob C. Kaessinger
Mechanical Engineering Undergraduate
Donald P. Shiley School of Engineering
University of Portland
Portland, Oregon, 97203
Email: Kaessing14@up.edu

Kramer C. Kors
Mechanical Engineering Undergraduate
Donald P. Shiley School of Engineering
University of Portland
Portland, Oregon, 97203
Email: Kors14@up.edu

Jordan S. Lum
Mechanical Engineering Undergraduate
Donald P. Shiley School of Engineering
University of Portland
Portland, Oregon, 97203
Email: Lum14@up.edu

Heather E. Dillon, Ph.D.
Assistant Professor, Member of ASME
Donald P. Shiley School of Engineering
University of Portland
Portland, Oregon, 97203
Email: Dillon@up.edu

Shannon K. Mayer, Ph.D.
Associate Professor
Department of Physics
University of Portland
Portland, Oregon, 97203
Email: Mayers@up.edu

ABSTRACT

Convective heat transfer beyond explicit solutions to the Navier Stokes equations is often an empirical science. Schlieren imaging is one of the only fluid imaging systems that can directly visualize the density gradients of a fluid using collimated light and refractive properties. The ability to visualize fluid densities is useful in both research and educational fields. A Schlieren imaging device has been constructed by undergraduate students at the University of Portland. The device is used for professorial heat transfer and fluid dynamics research and to help undergraduates visualize and understand natural convection. This paper documents the design decisions, design process, and the final specifications of the Schlieren system. A simple 2-D heated cylindrical model is considered and evaluated using Schlieren imaging, OpenFOAM C.F.D. simulation, and convection analysis using a Nusselt correlation. Results are presented for the three analysis techniques and show excellent verifications between the CFD simulation, Nusselt correlation, and Schlieren imaging system.

1. INTRODUCTION

The scientific topic explored in this research paper is the use of a Schlieren system for imaging and analyzing changing fluid density. The Schlieren system is an important optical technique due to its non-disturbing nature in the fluid flow field [1]. This allows for analysis without interference that could possibly alter results. The Schlieren system uses a collimated beam of light and mirrors to image the density differences in a fluid [2]. Schlieren images can ultimately be used to verify numerical analyses for heat transfer problems.

In mechanical engineering, heat transfer calculations are used to predict the rate that a system will transfer heat to its surroundings. The numerical calculations that predict the transfer of heat must be verified by experimental data. A few examples that would benefit from analyzing natural convection in a fluid are the cooling of small computer chips, cooling of heat fins, and determining of the heat load within a building.

The driving force for natural convection is the density differences present in the fluid, typically due to a temperature difference [2]. The fluid motion is due to the presence of buoyancy forces and the body force being gravitational [3]. It is common for a mass density gradient to arise in a fluid from the presence of a temperature gradient [3]. For example, a heated plate in a fluid would be the source that produces a temperature gradient.

With this knowledge and the imaging from the Schlieren system, there are many different variables for natural convection that can be calculated. For example, the velocity of the fluid and the heat transfer coefficient, h, can be calculated [4-5]. These calculations require the comparison between Schlieren images of the fluid over certain time intervals. This is mainly because the heat transfer coefficient is a function of wall to fluid temperature difference, fluid temperature gradient at the wall, and the fluid thermal conductivity [1].

2. BACKGROUND

Schlieren photography was invented by German physicist August Toepler in 1864 to study supersonic motion [6]. It was used to photograph the flow of air and shock waves from planes in aeronautical engineering [6]. However, with computational fluid dynamics evolving, Schlieren photography is no longer as necessary for airplane designs [6]. Schlieren photography has also been used for analyzing problems involving natural convection, turbulent fluid flow, and even studying disease control by examining the human cough [7,8].

The Schlieren system that captures images of natural convection in a fluid includes a light source, lens or mirror, knife edge or color filter, and a camera [2]. The beam of light passes through a condenser lens before it hits the fluid of interest. The light is dispersed or refracted due to the density differences in the fluid and the knife edge filters out the non-affected light while the refracted light bends around the knife edge and is captured by the camera [2].

Three different types of Schlieren devices were considered for use during this experiment. The first Schlieren device configuration considered was a linear lens Schlieren device (shown in Figure 1). In this type of Schlieren device, a point light source is positioned one focal length away from a convex optical lens; as the diffuse light passes through the lens, it forms a collimated beam of light, or a beam of light in which all "rays" of light are parallel to each other [9]. This collimated beam of light is passed through a test region, containing the test subject [9]. In the test region, some beams of collimated light will be refracted as they pass through the Schlieren around the test subject [9]. After passing through the test region, another optical lens is used to focus the light to a point. At the point (one focal length from the second lens) a razor edge is positioned in such a manner that it will block the collimated light from the original point light source [9]. Light that has passed through density gradients within the test region will not focus back to the original focal point, and will pass around the razor-edge into a camera [10].

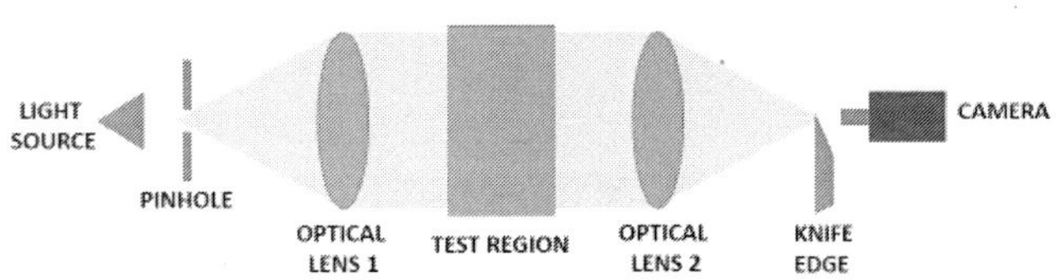

Figure 1. Schematic of a linear lens Schlieren device. Adapted from [9].

The second Schlieren imaging method considered was a v-type Schlieren device. Similar to a linear lens setup, the v-type device uses a small point light source to create a diffuse beam of light; however, in place of the optical lenses, a v-type setup uses a single optical grade concave mirror to focus the diffuse light back to a point, where a razor edge will cut off direct light and allow refracted light to pass around and be captured by an imaging device [9]. V-type Schlieren devices do not use collimated beams of light [9]. Instead, the point light source and razor edge are positioned at twice the mirror's focal length, causing the incoming light to immediately reflect back to a point [9]. The lack of a collimated beam is known to cause issues with image distortion [9]. However, this setup is less costly than the linear lens setup, as optical grade mirrors are currently far less expensive than optical grade lenses [9]. A schematic of a v-type setup is shown in Figure 2.

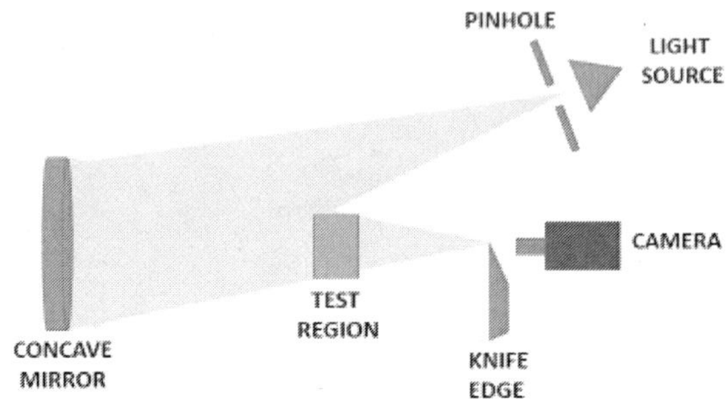

Figure 2. Schematic of a v-type Schlieren device. Adapted from [9].

The third Schlieren imaging method considered was a z-type Schlieren device. A z-type Schlieren device uses two concave mirrors to achieve a result similar to the linear lens setup [9]. In a z-type configuration, a point light is positioned one focal length from the first mirror, which then creates a beam of collimated light [9]. The collimated light passes through the test region, and some light is refracted by the Schlieren present within [9]. The second mirror then focuses the light back to a point, one focal length from the mirror, where a razor edge is used to cut off direct light [9]. Refracted light will pass around the razor edge, and be captured by an imaging device [9]. A schematic of a z-type setup is shown in Figure 3.

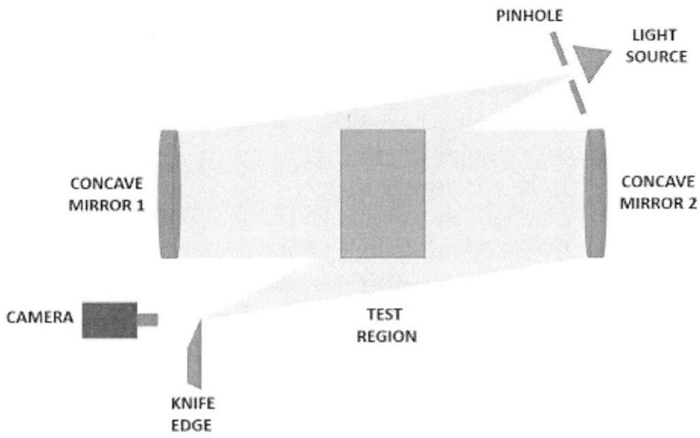

Figure 3. Schematic of a z-type Schlieren device. Adapted from [9].

The typical data gathering technique includes calibrating the Schlieren system based upon the clarity of the images and taking multiple photographs at steady time intervals [5]. These images are usually taken with a quiescent ambient atmosphere to reduce altered results [7]. After images are produced, programs such as MatLAB and specialized software can be used to assist with particle image velocimetry (PIV) or other analysis [5]. MatLAB can be used to interpolate calibration curves when increasing resolution in images, and the other software tools can be used to plot the displacement of particles at certain temperatures [5]. Computational fluid dynamics (CFD) analysis software can also be used to provide simulated data, which the Schlieren imagery can be compared with to validate experimental accuracy [8].

One major challenge for analysis using Schlieren imaging is that there are some natural convection problems that cannot provide consistent experimental data [4]. For example, most natural convection studies do not involve horizontal flat plates because the physical nature of horizontal surfaces is not adequately understood due to inconsistent data [4]. Although there are natural convection problems that do not provide consistent data, Schlieren imaging has made it possible for trends and equations to be made from what data is provided.

3. MATERIALS AND METHODS

There are many different methods of producing Schlieren images by using different designs and components within each design. Once designed and constructed, Schlieren imaging can be used to compare to empirical calculations and computational fluid dynamic modeling.

3.1. Prototyping

In this experiment, three basic Schlieren imaging techniques were examined for use in a final experimental setup, which would then be used for the purposes of comparing Schlieren imagery to CFD and empirical solution results. During prototyping, characteristics of each Schlieren device configuration were examined, including the cost, resulting image quality (measured qualitatively), and amount of space taken up by each configuration.

First, a linear lens Schlieren device configuration was constructed using one inch diameter optical lenses with focal lengths of 10 cm as shown in Figure 4. A small incandescent flashlight and a piece of 1/16 inch acrylic plastic with 0.015, 0.020, and 0.024 inch diameter pinholes were used to create a point light source; the pinholes were positioned at the focal point of the first lens. At the focal point of the second lens, a utility knife razor blade was positioned in the path of the direct light from the point light source. A digital camera was used to capture the remaining image. The linear lens setup offers high image quality when qualitatively compared to the v-type and z-type setups. However, scaling up the test became problematic. Optical lenses are extremely expensive, therefore, while the linear lens setup may be a viable option in situations where economics are not a concern, in this case the linear lens setup would be prohibitively expensive.

The second prototype was a v-type single mirror Schlieren device, constructed using a single 7.75 inch parabolic concave mirror shown in Figure 5. Equipment used in the v-type setup was identical to that used in the linear lens configuration, except for the light source, which was replaced by a 3.0V high brightness LED. The LED was ground flat to remove the diffuser typically added on modern LED's, and the second pinhole (0.020 inch diameter hole in 1/16 inch acrylic plastic) was positioned directly in front of the diode. Results from the v-type Schlieren device configuration were promising; however, "double-imaging" became problematic. Double imaging is prominent in v-type Schlieren devices with "short" focal lengths (such as the 48" focal length in this setup). When a Schlieren occurs in two beams of light at once, the light is refracted twice, producing two Schlieren images which are offset by a small distance.

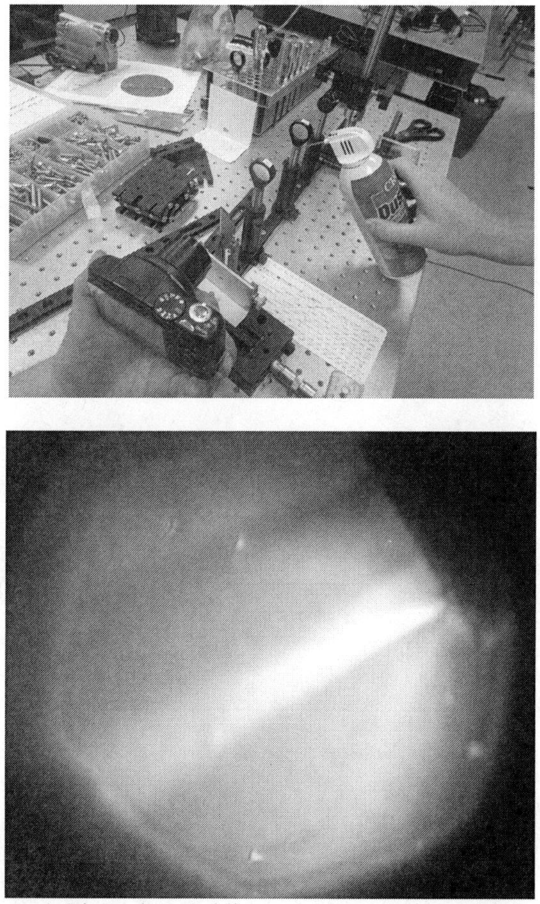

Figure 4. Linear lens Schlieren device prototype (top) with Schlieren of high velocity compressed gas (bottom).

The final Schlieren device configuration prototyped was a z-type Schlieren device. In this configuration, one 7.75 inch mirror parabolic concave mirror (focal length 48 inches) and one 10 inch parabolic concave mirror (focal length 50 inches) were used; otherwise, materials for the z-type prototype Schlieren device were identical to those used on the v-type device (including the LED light source). While the z-type Schlieren device configuration requires two concave parabolic mirrors, and is thus more expensive, it suffers from fewer drawbacks than the v-type setup. So long as the mirrors are positioned sufficiently far apart, double-imaging will not occur, and, like the linear lens setup, a z-type setup utilizes a collimated beam of light in the test region, preventing distortion of the test object.

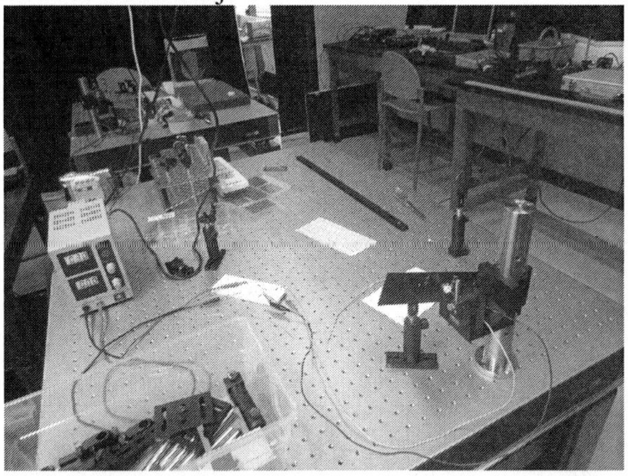

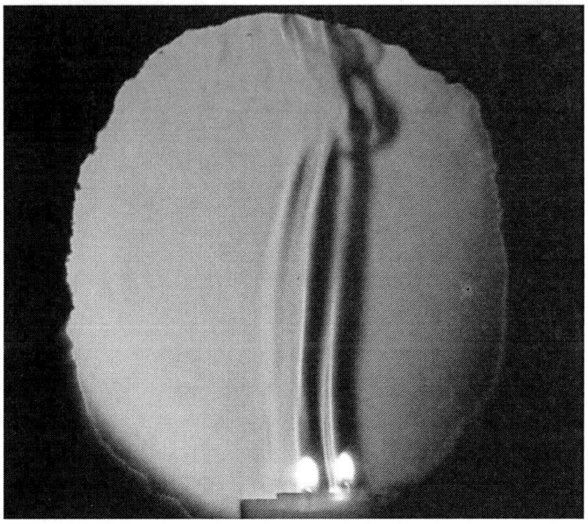

Figure 5. V-type single mirror Schlieren device prototype (top) with double Schlieren imaging of the plume for a single candle (bottom).

3.2. Experimental Schlieren Device

The experimental Schlieren device constructed for the purpose of this project uses a z-type configuration, with two 10-inch diameter parabolic concave mirrors (focal length 50 inches). A z-type configuration was chosen as it provided a middle ground between the low cost of the v-type setup, and the higher image quality of the linear lens setup. A photograph of the experimental device is shown in Figure 7, and a table of critical materials used to construct the device is available in Table 1. A photograph of the heated cylinder used for benchmarking is shown in Figure 9.

The experimental Schlieren device was constructed with an approximate budget of $1000. A sturdy metal cart made from 1" square steel tube with solid rubber wheels was used for the base of the optical table (mobility is important, in this case, as the Schlieren device will double as an educational tool). Since full size optical tables are typically not mobile and are more expensive than the project budget would warrant, the table has been constructed from ¾" composite board with 28 gauge galvanized steel sheet surface.

The concave mirrors selected for the project are 10-inch diameter parabolic mirrors with focal length f/5 (50 inches). Mirror surface quality was of a secondary concern, as Schlieren imaging does not require extremely high quality mirrors. However, the mirrors chosen had above 92% reflectivity and had a wave surface quality of 1/16 RMS. To hold the mirrors, two square cases were constructed which are supported by a simple three point of contact adjustment system, which supports the mirror on a large center fastener and uses two smaller fasteners to adjust the tilt angle of the mirror. The ability to adjust the tilt angle of the mirrors is essential as the mirrors were not secured to an immoveable optical table, and moving the system around would cause misalignment of the system.

Figure 6. Z-type double mirror Schlieren device prototype (top) with Schlieren of air from a hot plate (bottom).

Table 1. Equipment and materials used in the experimental Schlieren device.

No.	Qty.	Item
1	2	GSO Parabolic Primary Mirror, Dia. 10in, Focal Length f/5
2	1	Blue UltraBright® LED (rated for 3.0V at 20mA)
3	1	Canon Powershot ELPH 300 HS
4	1	3.0V DC power supply
5	1	Variable voltage AC power regulator
6	1	Single-edge razor blade
7	1	Omega Medium Temperature Heating Tape (SRT051-040)

Figure 7. Photograph of the experimental z-type Schlieren device with Plexiglass shield cover.

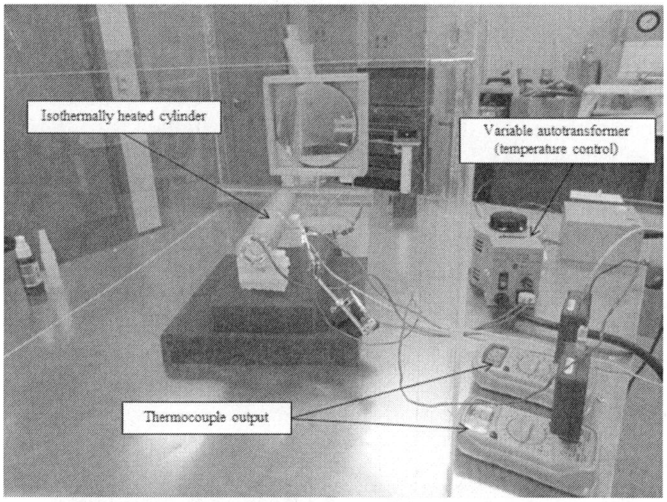

Figure 8 - Photograph of experimental test set-up.

The point light source for the experimental Schlieren device consists of a high brightness blue LED light source, which has been sanded and polished smooth to remove its diffuser. The LED is press-fit into a rapid-prototype ABS plastic frame. The pinhole is also held in the plastic frame, in an adjustable slot. The pinhole is 0.020" diameter, drilled in a small piece of 1/16" inch black acrylic plastic. The razor edge used in the device is a single-edged razor blade. The light source, razor, and digital camera are supported by optical posts and hollow tubes with set-screws, allowing for adequate coarse height adjustment.

A 10 inch long aluminum cylinder, constructed from 2.0 inch diameter, 0.13 inch wall aluminum tube, was used to create a physical model for the benchmark comparison. To heat the cylinder, a 104 Watt heat strip was wrapped around a 1 in. wooden dowel and placed inside the aluminum cylinder; the ends of the cylinder were sealed with aluminum tape. Temperature control over the cylinder is achieved via a manually operated AC power regulator. Temperature control measurements are taken with a type-K thermocouple, which is attached to the surface of the cylinder by aluminum tape. A small piece of non-conductive tape was used to electrically isolate the thermocouple, preventing erroneous readings.

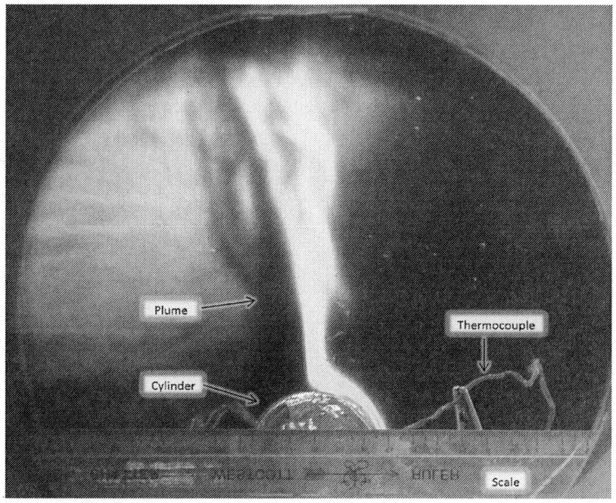

Figure 9. Coarse positioning of the thermocouple with the Schlieren imaging system

3.3. Experimental Verification Using Schlieren Device

To provide a comparison for benchmark testing, the value of h, the heat transfer coefficient, was calculated for a cylinder. In the case of the experimental system, Newton's law of cooling, shown below, can be applied to find the rate of heat transfer:

$$q_s'' = h(T_s - T_\infty)$$

where q_s'' is the local surface heat flux, h is the convection heat transfer coefficient, T_S is the surface temperature of the cylinder, and T_∞ is the ambient temperature at the edge of the boundary layer surrounding the cylinder. For the experimental setup, the surface temperature of the cylinder and the ambient temperature at the boundary layer were both measured using Type K thermocouples. Data acquisition was performed manually, using thermocouple readout attachments for a digital multi-meter. Five temperature samples were taken over a five second period, and averaged to form a data point.

In order to more accurately perform this calculation, the Schlieren device was used to allow for precise placement of the thermocouples. Newton's law of cooling is most accurate when the value for T_∞ is taken directly at the boundary layer; using Schlieren imaging with different levels of contrast, the boundary layer can be both located and measured, allowing for precise temperature measurements and more accurate calculation of the heat transfer coefficient.

As mentioned previously, a 10 in. long, 2.5 in. diameter aluminum cylinder was constructed for use in the experimental analysis. By performing trial-and-error experiments with a variable-voltage AC power regulator, the power required to keep the cylinder at a steady-state condition was determined. In this case, the target surface temperature was determined to be 65°C, since it was easy to replicate with the variable voltage supply, and provided a significant ΔT with respect to room temperature. Large values of ΔT correspond directly to large density gradients in the airflow around the cylinder, which allow for clear, well-defined Schlieren images.

3.4. Empirical Calculation Methodology

Natural convection analysis for various immersed geometries can be completed using Nusselt correlations found in the *Introduction to Heat Transfer* by Incropera, DeWitt, Bergman, and Lavine [3]. The empirical Nusselt correlation for external free convection flows relates the average Nusselt number to the thermal conductivity of the fluid, characteristic length of the geometry, and the average heat transfer coefficient for the object. This correlation is represented by the following equation:

$$Nu = \frac{hL}{k}$$

where Nu is the average Nusselt number, h is the average heat transfer coefficient, L is the characteristic length (diameter for a heated cylinder), and k is the thermal conductivity of the fluid.

For a long, isothermal, horizontal cylinder, a Nusselt correlation for a wide Rayleigh number range was proposed by Churchill and Chu in the *International Journal of Heat Mass Transfer*. This correlation is given in the following equation:

$$Nu = \left\{0.60 + \frac{0.387Ra_D^{1/6}}{\left[1+(0.559/Pr)^{9/16}\right]^{8/27}}\right\}^2 \quad \text{for } Ra_D < 10^{12}$$

where Ra is the Rayleigh number and Pr is the Prandtl number. To calculate the Rayleigh number for a cylinder, the following equation can be used:

$$Ra = \frac{g\beta(T_s - T_\infty)L^3}{\upsilon\,\alpha} = \frac{g\beta\rho^2 c_p(T_s - T_\infty)L^3}{\mu\,k}$$

where g is gravity, β is the volumetric thermal expansion coefficient, T_S is the surface temperature, T_∞ is the ambient temperature, L is the characteristic length, ν is the kinematic viscosity, α is the thermal diffusivity, μ is the dynamic viscosity for the fluid, k is the conduction coefficient for the fluid, ρ is the density of the fluid, and c_p is the heat capacity at constant pressure. Many of these properties were found for the experiment using the Engineering Equation Solver (EES) software to determine thermodynamic properties at the film temperature of the air.

3.5. Computation Fluid Dynamics Analysis Methodology

Computational fluid dynamics (CFD) was used to simulate the test set up. Open source Field Operation and Manipulation (OpenFOAM) version 2.2.2 using Helyx Open Source (HelyxOS) version 2.0 was used as the CFD software. The case considered was natural convection flow created by an isothermal cylinder in a square enclosure. The geometry was simplified to a two dimensional 2 inch (0.0508 m) circle inside a rectangular cavity with an aspect ratio (A_R) of 1.5:

$$A_R = \frac{H}{W}$$

where H is the height of the rectangular cavity and W is the width of the rectangular cavity. The Rayleigh Number calculated for the experiment was 19287 which meets the criteria to be modeled as laminar since $Ra < 10^9$ [3]. The natural convective buoyancies of the air were calculated using the Boussinesq Approximation (ρ_K).

$$\rho_K = 1 - \beta(T - T_{ref})$$

Which is valid when:

$$\frac{\beta(T - T_{ref})}{\rho - \rho_{ref}} << 1$$

where ρ and ρ_{ref} are the fluid density and reference density respectively. The Boussinesq equation is used because the difference in inertia of the fluid at different temperatures is negligible, but gravity is sufficiently strong to make the specific weight different between the fluids at different temperatures [3].

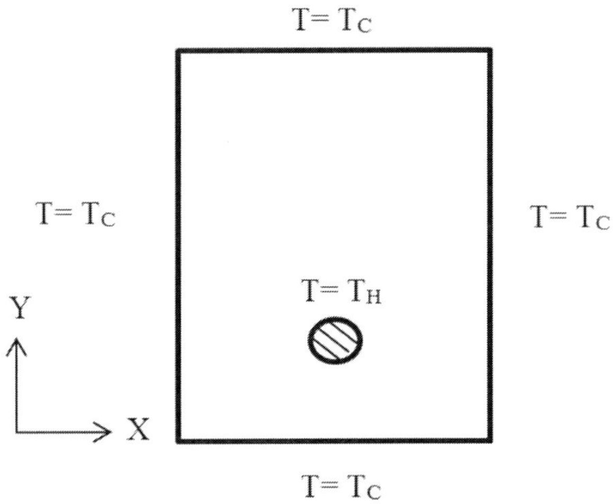

Figure 10. Boundary conditions for CFD model.

The mesh was a 2 meter by 3 meter cavity with 200 elements in each direction. The 2 inch diameter circle had a refinement towards the walls four times smaller than the base mesh. The boundary conditions seen in Figure 10 shows that the four walls of the cavity are maintained at T_c= 294.25 K and the isothermally heated cylinder is maintained at T_H= 338.25 K to correspond with the experimentally observed temperature at steady state. The walls are all fixed without slip, and the front and back face are set to "empty" to make the system 2 dimensional as required by OpenFOAM.

The solver used for all analyses in OpenFOAM was the "buoyantBoussinesqSimpleFoam" solver, which is a steady-state solver for buoyant, turbulent flow of incompressible fluids.

This model uses Reynolds-Averaged Simulation (RAS) turbulence properties set to laminar. The numerical schemes were "Bounded Linear Upwind - 2nd Order" for velocity and "Bounded limited linear – Blending" using a factor of 1 for temperature. The non-orthogonal correction factor was set to 0. The over relaxation setting were set to 0.3 for pressure and 0.7 for temperature and velocity. The field was initialized with a temperature of 294.25 K.

The simulation ran for 50 iterations and the residuals may be seen in Figure 11. The temperature and velocity profiles are shown in Figure 12 and Figure 13 respectively.

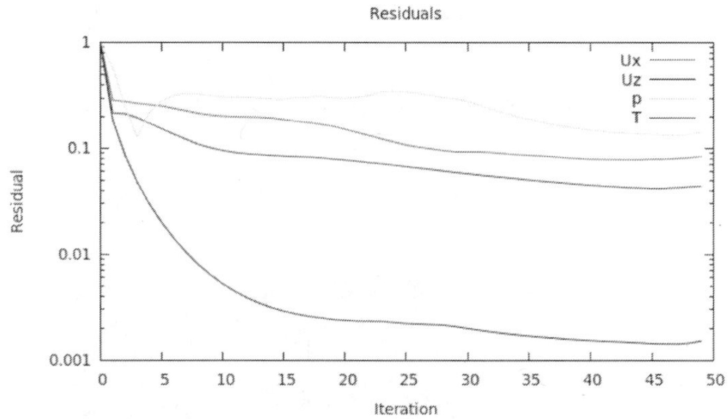

Figure 11. Residuals for 50 iterations of the CFD simulation.

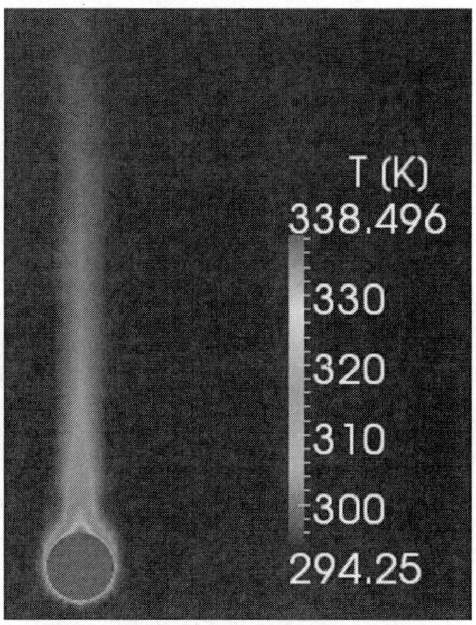

Figure 12. CFD predicted temperatures for the simulated air and cylinder experimental setting.

4. ANALYSIS

4.1. Results of Experimental Analysis

By performing several experimental trials with different levels of contrast in the Schlieren imaging system, the boundary layer for the experimental cylinder was determined to have a minimum width of 6.0 mm and a maximum width of approximately 10 mm. Temperature data points were recorded multiple times, however the temperature values were consistent across multiple tests after the thermocouple at the edge of the boundary layer was repositioned.

Final temperature and power measurements were taken during one single experiment. Average temperature at the surface of the cylinder was constant at 65°C, and the temperature at the surface of the boundary layer averaged approximately 23.3°C. The recorded power into the system was 18 Watts, which accounts for losses in the variable voltage power supply and in the wiring leading up to the heated cylinder. Using these values, Newton's law of cooling returns a convection coefficient of $h = 3.18\ W/m^2K$ when radiation losses from the cylinder are included assuming an emissivity of 0.69.

Finally, the width of the plume given off by the heated cylinder was measured approximately 1.5cm above the outer diameter of the cylinder, and was found to be 2.2cm on average.

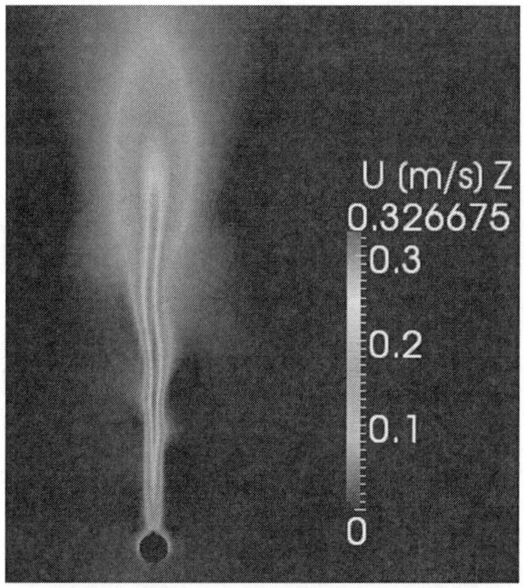

Figure 13. CFD predicted velocities for the simulated air and cylinder experimental setting.

4.2. Results of Theoretical Analysis

The local heat transfer coefficient values that were calculated using Nusselt correlations at different distances away from the surface of the cylinder are shown in Table 2. In the table results, the value for y represents the distance from the outer surface of the cylinder, and the value for h represents the average heat transfer coefficient. Since the boundary layer thickness was found to be 6 mm during experimentation, the value with y = 6 mm was used for comparisons between experimental, theoretical, and CFD analysis.

4.3. Results of Computation Fluid Dynamics Analysis

When an OpenFOAM simulation was run for the long, horizontal cylinder test setup, the following heat transfer coefficients were calculated at different distances from the surface of the cylinder as shown in Table 2.

Again, in the table results, h represents the heat transfer coefficient, and y represents the distance over which this coefficient is calculated. The y = 6 mm value was used for comparison with experimental and theoretical values.

Table 2. Comparison of theoretically predicted heat transfer coefficient with CFD and experimental results from the Schlieren device.

Vertical Distance [mm]	Theoretical Heat Transfer Coefficient [W/m^2 K]	CFD Heat Transfer Coefficient [W/m^2 K]	Experiment Heat Transfer Coefficient [W/m^2 K]
1	20.85	6.45	-
2	11.92	8.85	-
3	8.40	5.77	-
6	**4.34**	**4.10**	**3.18**
10	2.66	2.73	-

Additionally, the width of the plume in the CFD simulation was measured approximately 2 cm above the top of the cylinder. In the CFD simulation, the plume width was found to be 2.0 cm. This compares very well with the experimentally observed 2.2 cm from the Schlieren device.

5. CONCLUSIONS

Comparisons between the experimental, theoretical, and CFD analysis were very well correlated, considering that the experimental setup was created to serve merely as a proof-of-concept device. The initial objective of comparing the heat transfer coefficient and plume thickness of the benchmark model was simply to validate that the experimental setup, theoretical calculations, and CFD analysis were all producing similar results.

The calculated values for the heat transfer coefficient, h, were h = 3.18 W/m^{2}K for the experimental system, h = 4.3 W/m^{2}K for theoretical analysis, and h = 4.1 W/m^{2}K for the CFD analysis were. Comparing these values, all were within the range of h-values that represent natural convection, which is traditionally from h=2-25 W/m^{2}K [3]. Additionally, the results were of the same magnitude, and were within 1.12 W/m^{2}K of each other, so any differences that were present were only large enough to alter the results, and not place them out of the same order of magnitude.

Qualitative comparison of the plume behavior and characteristics also compare well with CFD modeling (Figures 12 and 13). Qualitative images of other experimental systems have also demonstrated the flexibility of the University of Portland Schlieren system as shown in Figures 14-17.

Future work to be done with this Schlieren imaging device will include more experimental testing for various geometries, research involving the efficiency of heat dissipation systems in energy efficient lighting, incorporation with heat transfer and thermodynamic laboratory experiments, and the continuation of improving the current Schlieren imaging device. These experiments and improvements will most likely be carried out by future undergraduate students at the University of Portland.

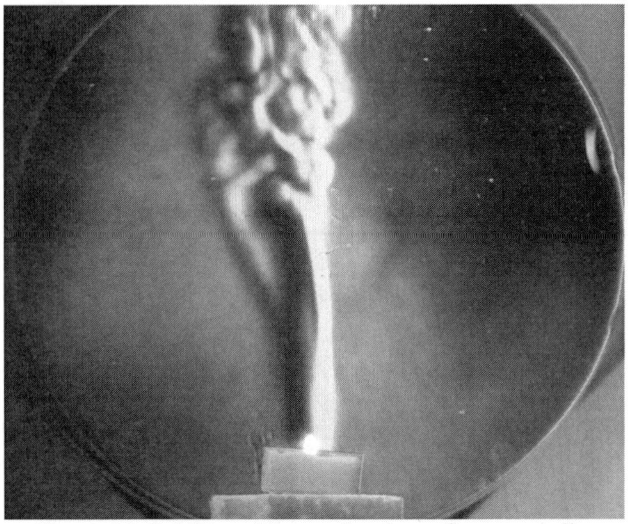

Figure 14. Schlieren image of a plume from a candle.

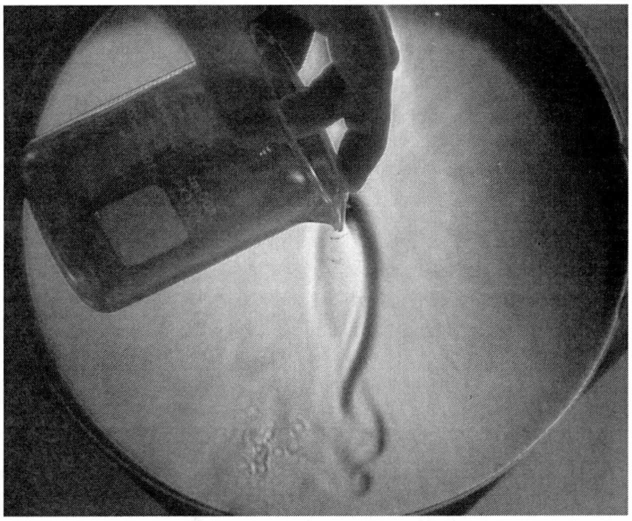

Figure 15. Schlieren image of dense gas poured from a beaker.

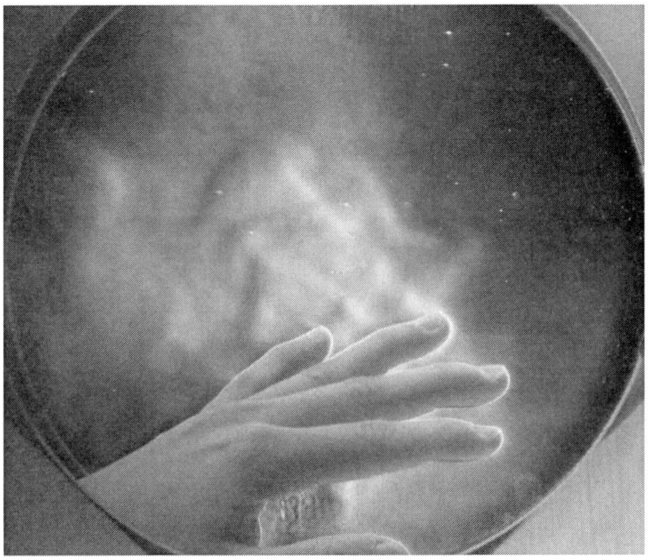

Figure 16. Schlieren image of the plume from a hand.

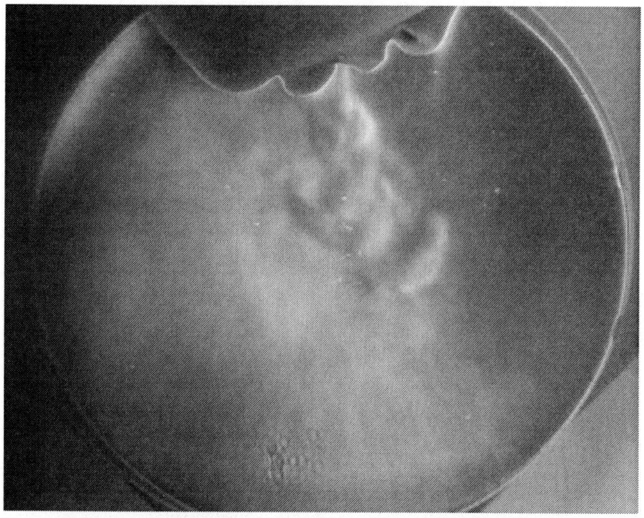

Figure 17. Schlieren image of the human breath.

6. ACKNOWLEDGEMENTS

The authors of this paper would like to acknowledge the University of Portland Donald P. Shiley School of Engineering for providing funding for this research. Also, a special thanks to Dean Sharon Jones, Dr. Kenneth Lulay, Dr. Deborah Munro, Mr. Allen Hanson, Mr. Jacob Amos, Mr. Jared Rees, and Ms. Kimberley Spir for their various forms of assistance in completing this study.

7. REFERENCES

[1] Ambrosini, D., D. Paoletti, and G. Tanda. "Investigation of Natural Convection in Vertial Channels by Schlieren and Optical Holography." *Journal of Flow Visualization and Image Processing* 6th ser. 11.4 (2004): 1-9. Print.

[2] Srivastava, Atul, Katsuo Tsukamoto, and K. Murayama. "Single Lens-Based Schlieren Microscope for Investigation of Three-Dimensional Buoyancy-Induced Convective Flow Fields." *Journal of Flow Visualization & Image Processing* 18.4 (2011): 347-69. Print.

[3] Bergman, Theodore L., Adrienne S. Lavine, Frank P. Incropera, and David P. Dewitt. *Introduction to Heat Transfer*. 6th ed. Vol. 1. Hoboken: John Wiley & Sons, 2011. Print.

[4] Wei, J. J., B. Yu, and Y. Kawaguchi. "Simultaneous Natural-Convection Heat Transfer Above and Below an Isothermal Horizontal Thin Plate." *Numerical Heat Transfer, Part A* 44.1 (2003): 39-58. Print.

[5] Martinez-Gonzalez, Adrian, J. A. Guerrero-Viramontes, and David Moreno-Hernandez. "Temperature and Velocity Measurement Fields of Fluids Using a Schlieren System."*Applied Optics* 51.16 (2012): 3519-525. Print.

[6] Krehl, P., and S. Engemann. "August Toepler-The First Who Visualized Shock Waves." *Shock Waves*. Springer-Verlag, 01 June 1995. Web. 13 Apr. 2014.

[7] Tang, Julian W., Thomas J. Liebner, Brent A. Craven, and Gary S. Settles. "A Schlieren Optical Study of the Human Cough With and Without Wearing Masks for Aerosol Infection Control." *Journal of The Royal Society Interface* 6 (2009): 727-36. *Royal Society Publishing*. Web. 14 Apr. 2013.

[8] Craven, Brent A., and Gary S. Settles. "A Computational and Experimental Investigation of the Human Thermal Plume." *Journal of Fluids Engineering* 128.6 (2006): 1251-2158. *ASME Digital Collection*. Web. 14 Apr. 2013.

[9] Venkatesh Gopal and Julian L Klosowiak and Robert Jaeger and Timur Selimkhanov and Mitra,J.Z.Hartmann, 2008, "Visualizing the Invisible: The Construction of Three Low-Cost Schlieren Imaging Systems for the Undergraduate Laboratory," European Journal of Physics, 29(3) pp. 607.

This page left blank intentionally.

Proceedings of the ASME 2014 International Mechanical Engineering Congress and Exposition
IMECE2014
November 14-20, 2014, Montreal, Quebec, Canada

IMECE2014-39166

USING A FUNDED CAPSTONE PROJECT TO TEACH FLUID POWER AND ADVANCED MECHANICAL DESIGN

James Mynderse, Selin Arslan, and Liping Liu

A. Leon Linton Department of Mechanical Engineering
Lawrence Technological University
21000 W Ten Mile Rd, Southfield, MI 48075

ABSTRACT

The A. Leon Linton Department of Mechanical Engineering at Lawrence Technological University offered a new senior capstone project to a small group of students, funded by a teaching grant from the National Fluid Power Association. All mechanical engineering students at Lawrence Tech must complete a capstone project: either an SAE competition team or a project addressing a particular industry need. The team that worked on the current project consisted of students with various concentrations in mechanical engineering and included an international visiting student from Brazil. Three faculty in Mechanical Engineering, each with different areas of expertise: thermodynamics, heat transfer, fluid mechanics and mechatronics, mentored and worked closely with the students at every step of this project. The objective of this project was to design and fabricate a classroom-scale gantry crane for material handling. The undergraduate students were not only involved in the design of a fluid powered system, but also worked on the modeling of mechanical components and the mechanical system as well as circuit design for an operator interface. The self-guided and real-world design aspect of the project increases the effectiveness of teaching by the faculty and retention of the subject by the students involved in the project.

INTRODUCTION

The A. Leon Linton Department of Mechanical Engineering at Lawrence Technological University requires all students to complete a capstone project in their senior year in order to graduate. The senior capstone projects could involve product design, manufacturing, ergonomics, structures, materials research, mechatronics computer systems and control systems. The department offers two tracks for senior capstone projects: competition projects and industry-sponsored projects (ISPs). Competition projects include Baja SAE, Formula SAE, Formula Hybrid, and SAE Aero Design. These projects are well-known by students with previous vehicles available, strict timelines, and a year-to-year process of continual improvement.

Industry-sponsored projects are more varied, representing time-sensitive real-world problems posed by industry partners. Over the course of two semesters, the students complete 5 credit hours which includes both lectures on the design process and team-based design. For most groups, the first semester involves research into the problem, design of a solution mechanism, numerical validation of the design using appropriate software tools, and a report to the industry partner detailing the proposed design. In the second semester, teams fabricate a working prototype, test the prototype, and present the final design with prototype to the industry sponsor. Each student submits monthly progress reports and the team prepares a comprehensive report each semester. The comprehensive report is presented through with a poster and oral presentation to ME faculty and Industry Advisory Board (IAB) of the department.

Capstone Project course has demonstrated a great success in producing students with skills and experience in project management, oral and written presentation, and the proposing and building of a new design to solve a real-world problem. Studies [1] indicate students finish the senior design courses well-equipped for their role in the rapidly changing world of engineering.

Students involved in capstone projects with support from industry gain a better understanding of the design process, have access to more resources, are more engaged, and ultimately produce a better product [2]. Practical projects, like ISPs, provide the potential of being useful even in graduate level education [3], where students study beyond the characteristic learning objectives of common classroom assignments. Working on real-life design problems has a great impact on students; helping them understand the decision making process, assess impact of design on cost and schedules, giving them a view of real-world engineering [4].

Funded by a teaching grant from the National Fluid Power Association, a new senior capstone project was offered during Fall 2013 and Spring 2014 semesters. Students enrolled in the ISP track were offered the opportunity to join a team designing and building a gantry crane that uses fluid power for material handling.

The motivation for this work was twofold: to facilitate deep learning about fluid power in a design sequence for the enrolled students and to develop a platform for future classroom and laboratory sessions on fluid power and associated topics. In this paper, the results of the student design are detailed and expected future applications of the gantry craned are explained.

The organization of the paper is as follows. First, the project description provided to the students is presented. Next, the resulting mechanical design developed by the students is detailed. The following sections show the pneumatic power system and the electronics for measurement and control. After the student design, the project assessment is presented. Finally, the results are discussed and the paper is concluded.

PROJECT DESCRIPTION AND ORGANIZATION

The objective of this senior project assigned to the students is to design and fabricate a classroom-scale gantry crane for material handling. The crane is to be fluid powered and capable of supporting at least a 50 lb. load while moving across a 10 ft. work space. The crane must be able to lift to a height of at least 2 ft. and move horizontally in a single axis. The support structure will be capable of being disassembled for storage. Operation of the crane will be controlled by a human interface system.

Three full-time faculty members in Mechanical Engineering served as advisors for the project. Each of the three had different areas of expertise, as shown in Table 1. The undergraduate project team was supported during the design, fabrication, and validation by integrating the expertise from all three faculty members. This arrangement was unique among ISPs offered by the department of Mechanical Engineering; all other projects had only a single advisor. While not cost effective, the advisors worked well together and found the arrangement to be convenient for such a complex project.

Table 1 - Faculty advisors and areas of expertise.

Faculty	Expertise
Dr. Mynderse	feedback control theory and mechatronic design
Dr. Liu	fluid mechanics and heat transfer
Dr. Arslan	power generation and thermal sciences

Students enrolled in the ISP course expressed preferences for particular projects during the first week of the Fall term. Working from student preferences, as well as the needs of the project as determined by the faculty advisors, the course coordinator assigned students to project teams. Student skills were informally evaluated from faculty advisor experience, student transcripts, and selection of technical elective courses. Due to the small size of Lawrence Tech, most students considered for the project were easily assessed for skills. The skill assessment was not used for anything beyond selection of students from the available pool. All students enrolled for ISPs were placed into a project.

For this work, the faculty advisors determined that the primary needs were students with skills in design, fluids, and mechatronics. Mechatronics was viewed as valuable for this project for integration of the fluid power, mechanical structure, and electronic control; dynamic modeling of the cart motion; and prototyping of the user interface.

During the Fall 2013 semester, three students were involved in the gantry crane design. During the Spring 2014 semester, four students were involved in fabricating, assembling, and testing the gantry crane. Two students carried over from Fall 2013 to Spring 2014. A third student, who was an original member of the Fall 2013 team, graduated in December 2013. The students and their assessed skills are shown in Table 2.

Table 2 – Students by semester and faculty-assessed skillset.

Student	Fall	Spring	Skills
A	X		Fluids, Mechatronics
B	X	X	Design, Fabrication
C	X	X	Fabrication
D		X	Limited Mechatronics
E		X	Solid Mechanics

The two-semester project allowed the students to split their primary tasks with the semesters. New students took over the remaining tasks in the second semester. During the Fall semester, the students focused on researching into existing gantry cranes and fluid power and used that research to inform their design. During the Spring semester, students fabricated the elements of the design, assembled the elements, and tested the whole project to ensure that it met specifications. For each semester, a Gantt chart was used to schedule the work and deliverables, see Figure 1 and Figure 2. Design differences between the charts arose from turnover in students between semesters.

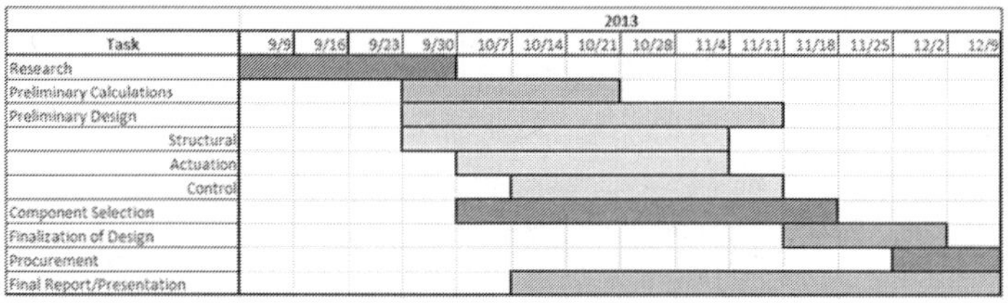

Figure 1 - Gantt chart for Fall semester including research and design.

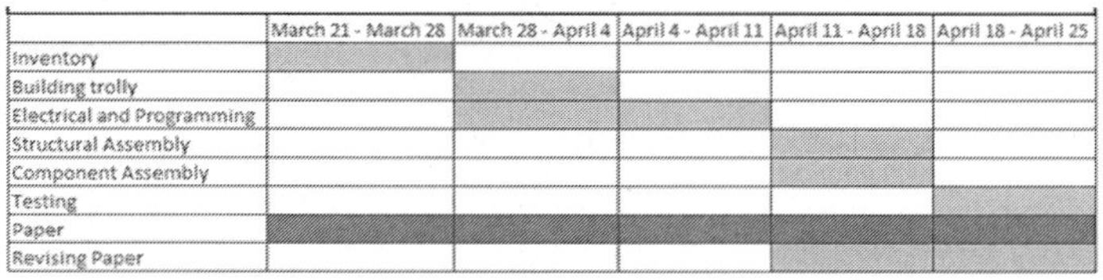

Figure 2 - Gantt chart for Spring semester including fabrication, assembly, and testing.

The students met with advisors weekly who track the progress and offer technical advice as requested by the students. In addition, students were expected to meet weekly as a group to share progress and work collaboratively. These meetings were not tracked by the faculty advisors.

DESIGN

A gantry crane is a machine for lifting and lowering a load and moving it horizontally where the gantry mechanism, trolley, is rigidly supported on two or more legs running on fixed rails. Hoist, the lifting component of the crane is an integral part of the equipment. In industrial applications, gantry cranes, such as the example shown in Figure 13, are an important part of the fabrication and production process as they aid in the transportation of materials to and from work cells. The use of overhead gantry cranes can reduce labor costs and can improve working conditions and safety.

Figure 3 - An industrial gantry crane [5]

The objective of this project was to design and fabricate a small, classroom-scaled fluid powered gantry crane. The gantry crane was to be used in material handling and was required to be capable of supporting a 50 lb. load. Additionally, the crane was required to be able to lift the load to a height of at least 2 ft. and move it horizontally in a single axis across a 10 ft. work space. The structure must easily be disassembled for storage and transportation purposes. Motion of the crane was intended to be controlled by a human operator using an unspecified human-machine interface system.

The students used the first semester of the two-semester project to research existing gantry crane designs, determine additional specifications such as movement speed and type of operator interface, and design the crane. The final gantry crane design incorporates a mechanical design for the structure, a pneumatic design for the actuation, and an electronic design for the control and operator interface. Each portion of the design is detailed in the sequel.

Mechanical Design

The structure design of this project consisted of two T-frame sides that support the bridge, the beam on which the trolley is positioned, see Figure 4a. In order to provide easy movement and the positioning of the gantry crane, the frame structure has locking swivel caster rollers at the base. The frame structure is fabricated from 2 in. × 4 in. rectangular aluminum tube. The top beam is attached with four bolts and is removeable from the sides for easy storage. The columns are 2 in. wide by 6 ft. tall and support the top beam which measures 9 ft. by 4 in. Air compressor was supported by a shelf, attached to one of the T-frames.

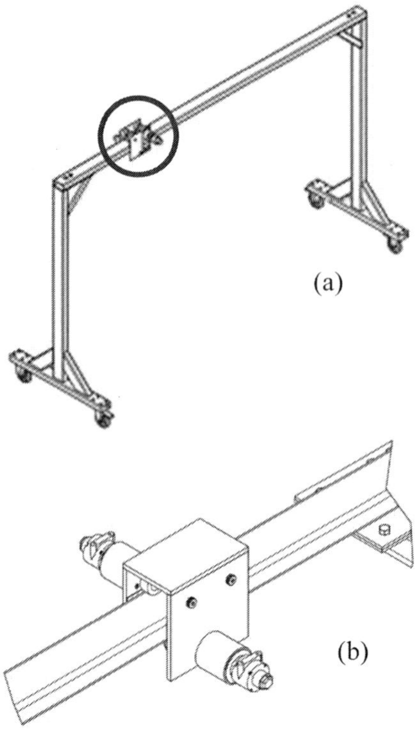

Figure 4 - (a) Gantry crane structure; (b) Magnified trolley section

Figure 5 – Fabricated Trolley

The trolley, shown in Figure 4b and Figure 5, was assembled on the top beam and housed two low speed pneumatic motors, shown in Figure 6. Pneumatic motors were responsible of the vertical and horizontal movement of the 50 lb. load. The motor that was responsible for the horizontal movement was attached to two 3-in wheels, which drives the trolley horizontally.

Figure 6 – Low Speed Pneumatic Motor [6]

Pneumatic Design

In order to control the speed and the direction of both the trolley and the crane hook, two pneumatic H-bridges were designed by the students. It was determined that only one motor would be actuated at a time providing either lateral or vertical motion of the load, but not both.

One pneumatic H-bridge is shown schematically in Figure 7. An off-the-shelf air compressor with a tank provides a supply of compressed air. A proportional regulator valve, mounted between the air supply and H-bridge controls the pressure supply to be delivered to the motors through the H-bridge. This proportional regulator allows the operator to control the speed of the gear motor which in turn controls the speed of the load in either the lateral or vertical direction.

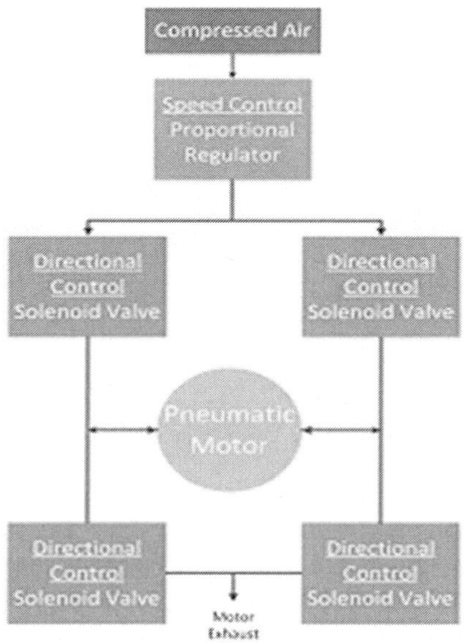

Figure 7 – Pneumatic H-Bridge

The H-bridge itself is based on an electrical H-bridge, as might be found controlling the direction of a DC motor. Four solenoid valves for directional control bracket the motor, as shown in Figure 7. By enabling two solenoid valves at a time, the air flow through the pneumatic motor may be reversed, which reverses the direction of the motor. For example, enabling the upper left and lower right valves will cause the motor to turn clockwise.

The pneumatic H-bridge system was demonstrated to provide bi-directional motor control during prototyping on a Festo pneumatic training station in Lawrence Tech's Mechatronics Lab. This prototyping may be seen in the video report to NFPA [7]. The pneumatic H-bridge is shown after final assembly in Figure 8.

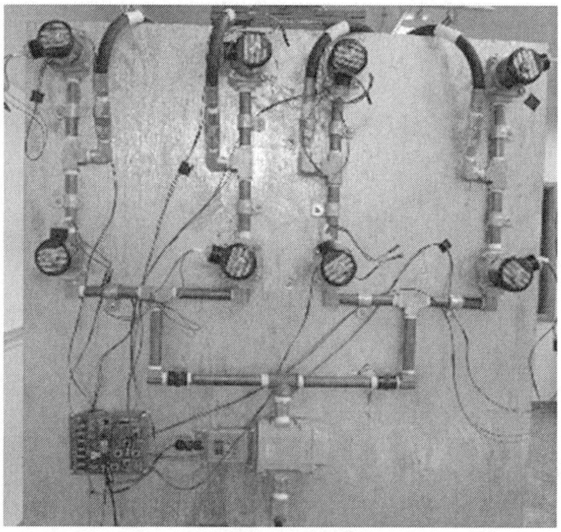

Figure 8 – Pneumatic H-bridge after installation on gantry crane.

Electronic Design

Students were also responsible for the design of a human-machine interface for the operation of the system. Based on research into existing methods, the students selected a pushbutton pendant for the control of the trolley and load. Four pushbuttons controlled the direction: left and right to control motion of the trolley across the beam and up and down to control the vertical position of the load relative to trolley. For safety, an enable pushbutton was also incorporated. The enable button was required to be pressed to allow any control action. The proportional regulator valve adjustment to control the speed of the motors is also done by the operator using the potentiometer located on the tethered remote.

The control logic was implemented on an Arduino Micro microcontroller board. The Arduino Micro uses an ATmega32u4 microcontroller with 20 digital I/O pins and 12 analog input channels. Like other Arduino products, the Micro enjoys a free, easy-to-use integrated development environment with library files. A large ecosystem supports the Arduino products with online documentation, forums, and many sample projects. The Arduino Micro selection was based in part on prior experience of the responsible group member in a mechatronic design course taught by Dr. Mynderse.

Utilizing methods from the mechatronic design course, the control logic software was written as a finite state machine with states for stop, left, right, up, and down. Releasing the enable button always transitioned the system to the stop state. The behavior of the finite state machine is shown in Figure 9. The digital outputs, shown in the blocks, are: trolley motion enabled (M), move hook up (A), move hook down (B), move trolley left (C), and move trolley right (F). The digital inputs from the user, shown on the transition arrows, are: enable button (E), up button (U), down button (D), left button (L), and right button (R). The 'x' digits indicate that either a true (1) or false (0) is acceptable.

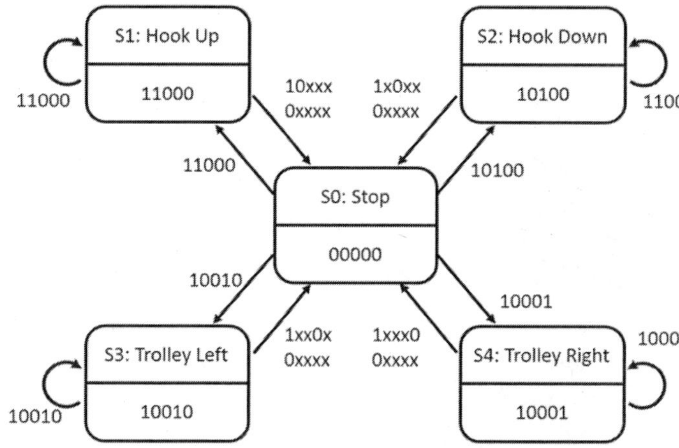

Figure 9 - Human interface control software designed as finite state machine with digital inputs in sequence EUDLR and digital outputs in sequence MABCF.

The electronic design was prototyped on a breadboard as shown in Figure 10 and implemented with a custom-designed PCB as shown in Figure 11. While clearly not a complicated design, due to the limited electronics content within the BSME curriculum at Lawrence Tech, this proved to be a challenging task. The PCB design was self-taught using Eagle PCB Design software with some limited guidance from the faculty advisors. Following fabrication of the PCB, several minor modifications were necessary to address errors in the design.

Rotary optical encoders with quadrature outputs were incorporated into the trolley design to allow for the measurement of position and speed of each pneumatic gear motor. The quadrature signals were decoded by a National Instruments Compact DAQ system for validation of the gantry crane performance. No attempt was made to implement input shaping or feedback control at this stage. Future work by graduate students enrolled in control systems will use the rotary encoders for feedback in a position control experiment.

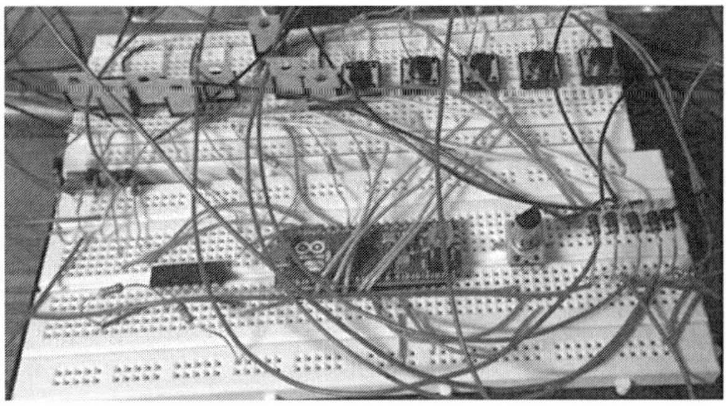

Figure 10 - Student designed electronics for human interface and control during prototyping.

Figure 11 - Student designed PCB for human interface as implemented on gantry crane.

ASSESEMENT

During each semester, student progress is assessed through a series of four progress reports. Every student has to submit his/her progress report individually. The progress report followed a strict template with several fields to be addressed by the student, and each field is assigned a weighting (see Table 3). For consistency all progress reports submitted for a single deadline were evaluated by one faculty advisor.

At the end of each semester, the students are required to present four items for evaluation: individual logbooks, a group written final report, a group poster, and a group oral presentation. Individual logbooks and the group poster are required as part of the courses EME4252 Senior Project Fundamentals and EME4253 Senior Capstone Project.

Table 3 – Progress report format and rubrics

Report Item	Points
Summarize your own accomplishments since the last progress report.	15
What expenditures did you have since the last progress report? Were these within the budget?	10
What work do you expect to do during the next month? Are you on schedule?	10
Are there any issues, problems or budget concerns? What tasks are behind schedule?	10
How many hours have you worked on this project since the last report? What is the average hours/week worked on this project?	15

They are evaluated by the course coordinator, not the faculty advisors. The group report is graded using a rubric shown in Table 4. Even though the students receive a group score on their report, they receive individual scores on their logbooks. There is also a team member evaluation to allow students rate each other based on their perception of others' conduct and contribution toward this group effort.

Table 4 – Project final report grading rubrics (summary)

Report Dimension	Weight
TECHNICAL PERFORMANCE, 50%	
Project Overview and Description	5%
Design Approach Summary	15%
Cost and Schedule	10%
Sustainability and Environmental Aspects	10%
Report Synthesis	5%
Previous Work / Future Work Plan and Schedule	5%
WRITING PERFORMANCE, 50%	
Structure	10%
Graphics	10%
Figures, Tables and Equations	5%
Formatting	5%
Mechanics	15%
Documentation and References	5%

The group presentation was made to the faculty advisors and members of the department Industrial Advisory Board (IAB). The students presented their problem, design, analysis, and continuing work. Following the presentation, audience members were invited to ask questions to the group. Faculty members as well as IAB members evaluated each student in the team using rubrics (see Table 5). The course coordinator compiled faculty and IAB evaluation into presentation scores for each student in the group.

Table 5 – Student oral presentation rubrics (summary only)

Oral Dimension	Weight
Delivery	15%
Language	15%
Content, structure, organization	20%
Visual Aids	20%
Audience Questions	20%
References	10%

In order to improve the student learning and collect valuable suggestions for future senior design projects, an informal survey was distributed to students to acquire their feedback and feeling about the work. As shown in Figure 12, all the students felt encouraged to participate in the group work, and most of them believed that working on this project significantly improved their understanding on mechanical design and control theories.

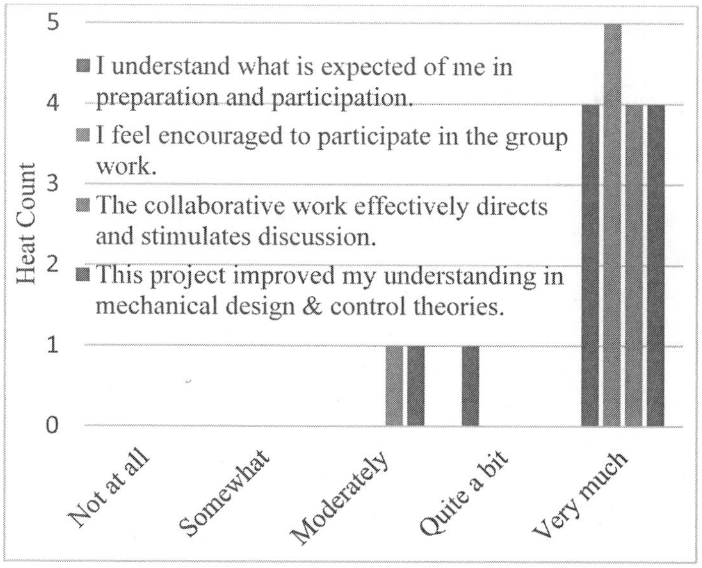

Figure 12 – Student feedback on the project

Below are the answers from the team when they were asked that "*What hands-on experience have you acquired through this project?*" It can be clearly seen that students successfully integrated their individual expertise to accomplish this design and fabrication task.

- "In this project, so far I have had the opportunity to work hands-on with electrical circuit connections and bread boarding of a circuit system."

- "I was able to design and prototype/test both electrical and pneumatic circuits in the lab to validate them for use on the final project."

- "While working on this project I was able to gain more experience with handling electronics and programming electrical systems."

- "The hands-on experience that I have acquired includes: fabrication of the trolley, assembling the piping and valves for the pneumatics, painting, and assembling the trolley."

The team expressed that they "liked most about this project is the ability to apply knowledge in practice." "It provides an opportunity to take a real world problem and allow me to use my educational knowledge that I have been developing for many years." "We had freedom to discuss and choose the parts and components of the project", and very much enjoyed "the variety of disciplines covered from fluid dynamics to electro-mechanical systems."

STUDENT OUTCOMES AND DISCUSSION

While the deliverable for this project was the fluid-powered gantry crane, the goal was a deeper understanding of fluid power and the design process for all students involved. Students indicated that this goal was achieved,

> "I was able to design and prototype/test both electrical and pneumatic circuits in the lab to validate them for use on the final project."

In addition, several students showed growth in the areas of mechatronics, fluid power and leadership. Many students in the BSME program at Lawrence Tech wait until their last semester to take the only required course in mechatronics. Thus, existing skill in mechatronics is rare during the senior capstone project. The electronic portion of the design was developed by a student who had taken the required mechatronics course and was taking an elective mechatronic design course simultaneously. The student demonstrated transfer of skills from the previous coursework to a new application.

In contrast, the electronic portion of the design was prototyped, updated, fabricated, and assembled by students who had not taken the elective mechatronic design course. These students relied on a much more limited background but seemed to learn relatively more regarding electronic integration, as informally assessed by faculty advisors. For example, these students enjoyed opportunities to learn about rotary encoders, practice debugging circuits, and construct electromechanical systems. These were advantages not usually explored by students enrolled in ISPs.

Although EME3123 Fluid Mechanics course is a core course in the Mechanical Engineering curriculum at Lawrence Tech., there is no specialized course on fluid power, hydraulics and pneumatics. The students took the Fluid Mechanics class before joining the project Faculty members guided the students during fluid flow, energy calculations and flow measurements. Students conducted extra research into fluid power, especially pneumatics. This was necessary to address design specifications such as deciding between pneumatic and hydraulic actuation. Students were able to apply what they learned in the previous year to a more specialized topic in the same field successfully.

Students enrolled in senior capstone projects at Lawrence Tech are usually also enrolled in a required course on leadership. These students must complete a leadership portfolio which includes examples of their work which shows growth in leadership. The final assignment which is included is a letter from their senior capstone advisor describing their demonstrations of leadership over the course of the project. Both the students' request for the letter and the writing of the letter itself made for valuable teaching moments concerning time management, different aspects of leadership, and evaluating their own growth in leadership skills.

SUMMARY AND CONCLUSION

Mechanical Engineering students at Lawrence Tech that were assigned to the described project participated in a collaborative capstone design project to design, fabricate, and test a classroom-scale gantry crane for material handling. The students gained hands-on experience in building a mechanical system, acquired deeper understanding on fluid power, explored mechatronic design, and practiced project and time management.

The gantry crane will provide a test bed for future educational activities at both the undergraduate level and the graduate level. Future undergraduate students will benefit from a laboratory showing a practical application of fluid power combined with instrumentation and classical control theory. Graduate students in the Mechatronic Systems Engineering program will benefit from a test bed for implementation of more advanced feedback control designs incorporating nonlinearities and control implementation strategies. Other expected applications of the gantry crane include the study of statics (the structure), dynamics (motion of the cart), modeling of dynamic systems (motion of the cart and load), and fluid flow (from the fluid power).

ACKNOWLEDGEMENT

The authors thank the National Fluid Power Association for support of this project through a 2013 Fluid Power Teaching Grant.

REFERENCES

[1] J. C. Grundy, "Experiences with Facilitating Student Learning in a Group Information Systems Project Course", *Proc. SE:E&P,* Dunedin, New Zealand, January 1996.

[2] R. C. McVay, R. Floersheim, A. Ayyar, K. Hollander, A. Boehler, M. Holgate, "The Benefits of Involving Industry in Engineer Capstone Courses: A Case Study", *Proc. of Capstone Design Conference,* Champaign-Urbana, IL, USA, May 2012.

[3] A. Hess, D. Rombach, R. Carbon, D. F. Murphy, M. Hoeh, C. Bartolen, "The Role of Collaborative Capstone Project-Experiences from Education, Research and Industry", *International Journal of Engineering Education,* Vol. 29, No. 5, pp. 1088-1099, 2013.

[4] I. G. Guardiola, C. Dagli, "Using University-Funded Research Topics to Teach System Design Process and Tools", *IEEE Transaction of Education,* Vol. 56, No. 4, November 2013.

[5] http://eotcrane.org/

[6] http://gopro.pro/gast_gear_motor_1up-nrv-11-gr11.htm

[7] Video Report to NFPA. https://www.youtube.com/watch?v=SZzGdzq2giY

Proceedings of the ASME 2014 International Mechanical Engineering Congress and Exposition
IMECE2014
November 14-20, 2014, Montreal, Quebec, Canada

IMECE2014-36635

INTEGRATING THE CASE METHOD AND DESIGN PROJECTS IN THE INDUSTRY-SPONSORED ACADEMIC EDUCATION

Manuel A. Nunez
University of Connecticut
Storrs, Connecticut, U.S.A.

Zbigniew M. Bzymek
University of Connecticut
Storrs, Connecticut, U.S.A.

ABSTRACT

This is a companion paper to IMECE 2013 - 63278. The paper describes a course in which practical designing of industrial products and processes is supported by the analysis of operations management cases taken from actual manufacturing companies. Through the case method, students assume the role of decision-makers who have to use their engineering and business knowledge to deal with real-life problems. Such an approach helps to support and complement the students' senior design experience and cover those subjects left out from their sponsored design projects. The cases emphasize operations management concepts; economic analysis of manufacturing processes; process analysis, design, and improvement; integration of experimental analysis and research methodologies in diverse manufacturing industries; as well as the interaction between manufacturing technologies and the competitive strategy of the firm. This way, students not only practice solving manufacturing problems, but also develop a framework for dealing with practical situations they are likely to face in their career development. We provide teaching recommendations and practical examples of the case method in this context.

INTRODUCTION

The Management and Engineering for Manufacturing (MEM) program consists of a unique curriculum to teach students modern manufacturing practice from two different but complementary perspectives. The program's goal is to develop manufacturing professionals with a solid foundation in engineering and business, who will have a total enterprise vision, and an intimate knowledge of the most advanced manufacturing methods [10]. The curriculum is built upon a powerful philosophy: be the most effective in transferring cutting edge innovations in technology to industrial environments involving design, manufacturing, production, and operations management. The capstone course of the program is entitled *Advanced Manufacturing Systems* (AMS) and has two major components: senior design projects and analysis of company cases. By becoming familiar with case studies of selected industries, students develop a thorough understanding and knowledge of manufacturing and production markets. Then they can apply that knowledge to complement their senior design project experience and later, in their professional practice. An integrated engineering and business education provides the most competitive advantage in design, operations management, production, and other areas of the manufacturing.

NOMENCLATURE

ABET: Accreditation Board for Engineering and Technology
AIC: Advanced Integrated Circuitry, Inc.
AMS: Advanced Manufacturing Systems course.
HBS: Harvard Business School.
MEM: Management and Engineering for Manufacturing program.

ADVANCED MANUFACTURING SYSTEMS

AMS is a one-semester (fourteen weeks) four-credit team-taught course offered during the last semester of the MEM program. In the course students have the opportunity to analyze and design products and processes in actual production systems through a company-sponsored team project, integrating the engineering and business knowledge they have gained through their undergraduate program. The projects in the class require students to: (1) work with sponsors, perform analysis, and conduct research to develop requirements and performance metrics for a manufacturing problem; (2) design the system or solution process; (3) implement a prototype, model, method or process according to the project requirements; (4) operate the prototype, model, method or process; (5) evaluate the prototype, model, method or process with respect to the performance requirements using the defined metrics; (6) document the process, products and results; and (7) manage the project, contact with the sponsor and the course instructor [6].

In addition, students analyze smaller projects (industry case studies) to complement their senior design experience and extend the application of engineering and business knowledge to areas not covered by the company sponsored projects. These smaller projects are taught by using the case method and by requiring the students to write an in-depth report in the context of the studied companies. These case studies emphasize: (1) operations management concepts; (2) economic analysis of manufacturing processes; (3) process analysis, design, and improvement; (4) integration of experimental analysis and research methodologies in manufacturing such as simulation, sensitivity analysis, spreadsheet modeling, optimization, and statistical analysis; (5) diverse manufacturing industries; (6) the interaction between manufacturing technologies and the competitive strategy of the firm; and (7) cultural, ethical, societal, and global issues associated with the company under study. In this way, students will not only have practice in solving problems, but also have a framework for determining which problems are most critical to the success of the firm and what analysis methodologies to apply [6].

AMS is taught by one faculty member from the School of Business and one or two faculty members from the School of Engineering. The students have to complete a design project and analyze six industry cases, where the project receives the same weight as the six cases in the final grading.

CASE STUDY METHOD

The case analysis section of the AMS course is implemented by a using the case method for teaching. In this method, students are placed in a more strategic role, as operations managers or decision makers, in the context of real-life situations. The cases borrow materials from realistic situations: real companies facing real issues [3]. Being very popular in business and law schools across the world, the case method has been effectively and successfully used to teach students since early in the twentieth century [7].

In AMS the case method is implemented by first having the instructor or a team of students present an overview of the company under study. This overview includes aspects such as chronology of events, main characters, product description, market and industry conditions and competition, operations strategy, production process, and any other relevant background information. Next, it follows a discussion concerning the main problems faced by the company and preliminary solution proposals (brainstorming). In this discussion, the full class is encouraged to participate with questions and answers; in some occasions even by using random cold calls. Finally, the discussion moves into a more detailed and technical analysis phase, where the case issues are further analyzed using more data and rigorous methodologies. In some situations, the discussion may include side topics involving cultural, ethical, or global issues related to the manufacturing processes at the company. The analysis of a given company spans a period of two to three weeks, which includes two fifty-minute sessions per week.

A major difference between the case method approach in AMS and the more traditional case method used in places such as the Harvard Business School (HBS) lies in the technical phase of the analysis. In the HBS and similar institutions, the case method is often targeted to operation management audiences with wider backgrounds than MEM. Therefore, in many occasions, the traditional analysis includes simplifying assumptions to make the discussion more amenable to the audience. The analysis in those institutions also tends to focus more on strategic issues with simple mathematical analysis involving a few calculator-type equations. The analysis is also more visual, relying more on graphs, charts, tables, and similar displays.

In contrast, the case method in AMS during the technical phase analysis uses more sophisticated mathematical tools to look for insights on how to solve the case problems, depending on how much data is provided in each situation. These tools include optimization (linear and non-linear), simulation, statistical analysis (regression, hypothesis testing, time series, forecasting), calculus (derivatives, limits), queuing theory, stochastic processes, and so on. The analysis usually starts with a discussion on what are the most important elements to incorporate into a model of each scenario. Then students are asked to suggest ways to translate the tradeoffs in the problem into mathematical expressions representing objective functions and constraints. Finally, students are asked to use mathematical tools to find solutions and perform sensitivity analysis on key parameters.

Additionally, the cases used in AMS contain many additional topics related to the companies under study that mesh well with general accreditation metrics (e.g., ABET program outcomes [1]) such as:

- *An ability to apply knowledge of mathematics, science, management, and engineering.*
- *An ability to design and conduct experiments, as well as to analyze and interpret data.*
- *An ability to design a system, component, or process to meet desired needs within realistic constraints such as economic, managerial, environmental, social, political, ethical, health and safety, manufacturability, and sustainability.*
- *An ability to identify, formulate, and solve engineering and management problems.*
- *The broad education necessary to understand the impact of engineering and management solutions in a global, economic, environmental, and societal context.*
- *An ability to use the techniques, skills, and modern engineering and management tools necessary for engineering practice.*

Table 1 in Annex A provides a sample of eighteen cases used in AMS. The cases cover a wide range of industries, from food industry to advanced electronics and aeronautics. They span a period of thirteen years from 2002 to 2013. Also, they encompass different production systems such as job shops, multiple assembly lines, and manufacturing cells. Notice that some cases were eventually replaced as time passed; this is because some cases are based on technologies that have become dated or because some of the issues do not reflect more current trends in manufacturing.

A tool commonly used in the analysis of many cases is process flow simulation. In particular, AMS has used two main simulation software environments: DELMIA/QUEST and Rockwell's Arena. QUEST is a 3D factory simulation environment for process flow analysis [8], whereas Arena is a 2D process flow simulation environment based on the SIMAN programming language [2]. Even though QUEST offers a greater range of capabilities and graphical design, Arena is a more preferred tool in classroom settings because of its fast development environment and simpler user interface.

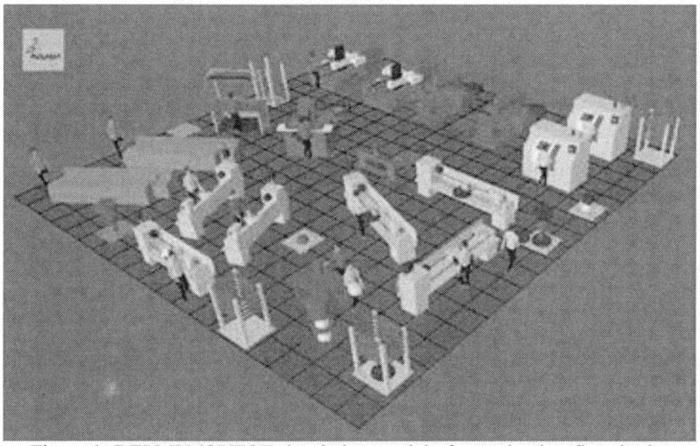

Figure 1. DELMIA/QUEST simulation model of a production floor in the aeronautics industry.

Figure 1 shows a QUEST simulation model in the context of process for making turbines for jet engines. This model was used in a senior design project involving process design and batch size optimization at the production floor [5].

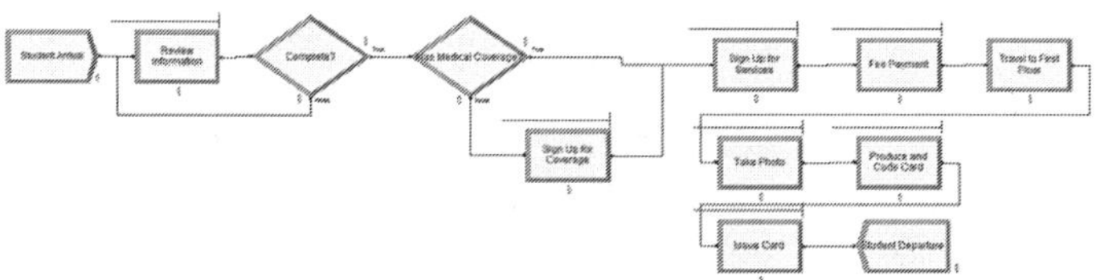

Figure 2. Rockwell's Arena simulation model of a college registration process.

Figure 2 shows an Arena model of the registration process of students at a small college. This is part of a business case used in AMS to teach simulation modeling.

EXAMPLE: AIC NETBOOKS

This case, developed by Wheelwright and Yong, 2011 [11], is about a consumer electronics firm named Advanced Integrated Circuitry, Inc. (AIC). The case focuses on the production of certain type of netbooks by using four assembly lines with ten operations per line. Even though the planned cycle time of each line is supposed to be forty one seconds per netbook, the actual lines are very unbalanced and some of them reach cycle times of fifty seconds per netbook. The main issue in the case is how to balance the lines and improve the cycle times to obtain closer values to the planned performance.

The traditional analysis of the case for a general audience, containing students with non-engineering background, would entail a quick assessment of the output rate of the current process in terms of units per month and a comparison to the planned output. Next, it would perform an assessment of theoretical efficiency, that is, computing the ratio of direct labor content to available labor. Then, it would incorporate an economic analysis of the potential impact of a fully efficient line, and conclude with a set of recommendations for improvement; some of them based on the numerical analysis and others based on a more general managerial approach. All of the computations mentioned above only involve simple arithmetical operations that can be done with a standard calculator.

In the AMS approach, we extend the traditional analysis by first creating a flowchart of one of the assembly lines at AIC. Based on the flowchart and the data given in the case, a simulation model of the process is created. Figure 3 presents an Arena model of one assembly line at AIC.

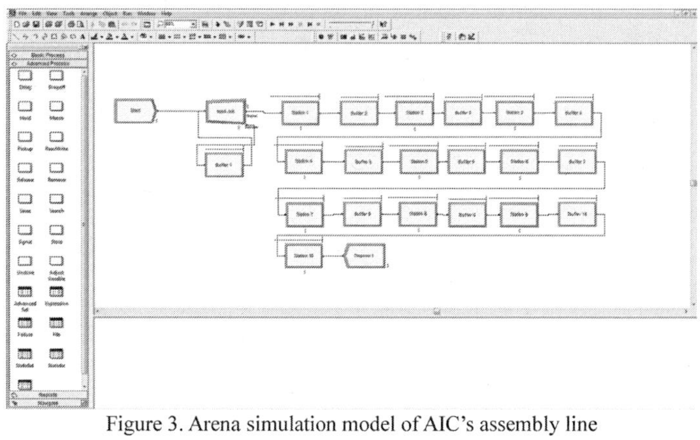

Figure 3. Arena simulation model of AIC's assembly line

The simulation model is used to better understand the assembly process, to determine the constraints in the production process, to assess current and potential production, and to perform sensitivity analysis of key parameters.

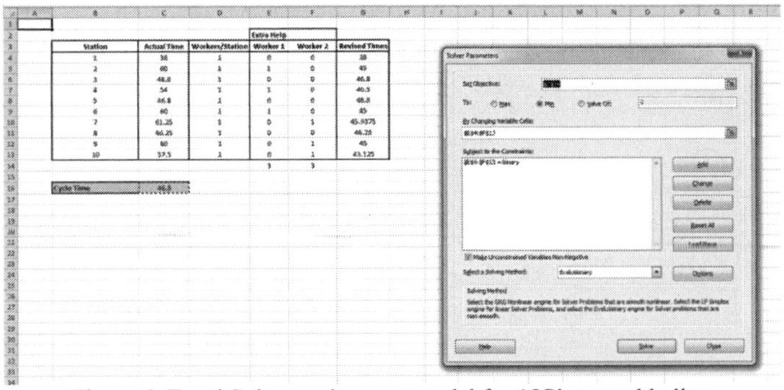

Figure 4. Excel Solver assignment model for AIC's assembly line.

Next, the students are required to create a model for personnel assignment to an assembly line in order to improve cycle time and balance the line. Figure 4 shows an Excel spreadsheet optimization model to determine the optimal assignment of two additional cross-trained workers to the current assembly line. Once the model is formulated, a solution is found by using Solver, which is an Add-in software tool that is part of Excel. The optimization model in this case is not trivial to implement because it contains a non-linear objective function and binary decision variables. Figure 5 illustrates this model, where U_i and V_i are the decision variable and T_i is the current cycle time at station i in the assembly line. Variable U_i is interpreted as equal to 1 is additional worker 1 is assigned to station i and equal to 0 otherwise; while variable V_i is similarly interpreted only that for additional worker 2. Therefore, it is necessary to use a heuristic algorithm provided by Solver to find a solution. This heuristic is based on an evolutionary algorithm [9]. Using this optimization model, students are able to better understand the impact of worker distribution along the line and the dependency of cycle time on labor. Students are also asked to modify the optimization model and combine it with the simulation model to study other scenarios where more workers are added or there are learning curve effects (workers become more efficient as time progresses).

$$MIN\ Z = \underset{i=1,\dots,10}{MAX}\left\{\frac{T_i}{1+\dfrac{U_i}{\sum_j U_j}+\dfrac{V_i}{\sum_j V_j}}\right\}$$

subject to

$$U_i, V_i \in \{0,1\}$$

Figure 5. Optimization problem for the AIC case.

Finally, the case analysis moves into other more general topics, such as the global impact of the company (AIC is a large Taiwanese company with sales all over the world, and with production plants in China and other countries), cultural effects (cheaper labor cost in Asian countries), and societal (the impact on society from the frequent introduction of new computer technologies).

CASE METHOD AND SENIOR DESIGN

In our previous paper [4] we fully presented and discussed the implementation of the senior design project section in AMS. Summarizing, the senior design experience consists of team projects developed by students and sponsored by local manufacturing companies in the New England region. The subject of a typical project can be classified into two types: product/process design and operations management. Product/process design projects deal with engineering design (or redesign) of products or manufacturing solutions to production problems. Approximately, 40% of the design projects are product/process design, whereas 60% are operations management. Senior design projects have been sponsored by well-known companies such as Pratt & Whitney, Sikorsky, Frito Lay, Dymax, MorphoTrust, and General Electric [4].

In addition to complement the course coverage in the AMS course, the case method is also used to support senior design projects. This is done by requesting the student teams to incorporate similar elements used in the case analysis such as drawing a flow chart of the process used by the sponsor company, using simulation (either Arena or QUEST) to assess the effectiveness of the proposed solutions, providing a cost analysis of the proposed solutions, and using mathematical modeling tools in formulating solutions.

TEACHING RECOMMENDATIONS

Based on our experience, the following is a list of learned lessons and recommendations for people who would like to implement a similar methodology in their courses:

1. Case selection is crucial. The operations management section of the Harvard Business Publishing website (http://hbsp.harvard.edu/) is an excellent source of manufacturing cases and also includes cases from other sources (e.g., Stanford University, Ivey, Indian Institute of Management, etc.) Look for cases that involve supply chain management, process analysis, capacity planning, and so on. The best cases are those that include detailed descriptions of the production processes, numerical data in terms of cycle times and production constraints, and incorporate cost analysis. Also, students relate well to cases that involve interesting narratives having a 'hero' (decision maker) facing difficult problems in the case.
2. Having a team of students (two or three people) present a case overview and lead the discussion is very helpful in motivating student participation. We also found very valuable letting the audience provide constructive feedback about the students' job as presenters and moderators.

3. Ask students to find connections between cases or between similar situations in other industries. For example, typical questions are: How does the job shop process at company X differ from the process at company Y? How does the supply chain of the company under study compare to the supply chain in the automobile manufacturing industry?
4. Integrate cost analysis to the analysis of the issues in a case. Challenge students to assess the economic benefits allowing the company to solve their production cost problems and improve effectiveness.
5. When creating mathematical models in a case, start by introducing an oversimplified, easy to understand, model and then, with the help of the students, continue refining the model by adding more realistic features. How far to go in this process would be determined by the tradeoff between obtaining an increasingly more complex and difficult to solve model versus how much is gained from adding more details.
6. As previously mentioned, Arena is a good tool in the classroom to quickly produce running simulation models of a given situation as the case discussion progresses.

CONCLUSIONS

Integrating the case method and a design experience in the senior-level capstone course AMS has proven to be an excellent approach to channel students' knowledge acquired during their years of study and foster professional development early in the students' career. It is also very useful for students to gain experience working on real practical problems with sponsoring companies and real-life companies taken from industry cases. The case method has been of tremendous help in senior design in the MEM program, by complementing and supporting the bread of knowledge of the students. The authors' experience developed by teaching the AMS course can be very useful for other instructors who already teach or would like to create similar courses in related manufacturing programs.

ACKNOWLEDGMENTS

The authors would like to acknowledge the support of Prof. Robert G. Jeffers, who served as Co-Director of the MEM Program until 2011 and as the Mechanical Engineering Department Head for thirteen years. Dr. Jeffers was one of the key developers of the MEM program and a fervent supporter of the senior design activities through his tenure at the University of Connecticut. As a Senior Vice President of ASME International and later a member of the Board of Governors of ASME, he stressed the importance of engineering education and strongly supported it through ASME activities.

REFERENCES

[1] ABET. 2014-2015 Accreditation Policy and Procedure. Available at http://www.abet.org/, 2014.
[2] Altiok, T., and Melamed, B. Simulation Modeling and Analysis with ARENA. Academic Press, 2007.
[3] Barnes, L.B., Christensen, C.R., and Hansen, A.J. Teaching and the Case Method. Third Edition. Harvard Business School Press, Boston, MA, 1994.
[4] Bzymek, Z.M., and Nunez, M. Integrating Design and Production Problem-Solving in Industry-Sponsored Academic Education. Proceedings of the International Mechanical Engineering Congress and Exhibition, IMECE 2013, San Diego, CA, November, 2013.
[5] Bzymek, Z.M., Nunez, M., Li, M., and Powers, S. Simulation of a Machining Sequence Using Delmia/Quest Software. Computer-Aided Design and Applications, Vol. 5, No. 1-4, pages 401-411, 2008.
[6] Bzymek, Z.M., Nunez, M., and Van Scoter, D. Advanced Manufacturing Systems Syllabus. University of Connecticut, Spring 2014.
[7] Corey, R. Case Method Teaching. Harvard Business School 9-581-058, Rev. November 6, 1998.
[8] DELMIA. QUEST: The Systems Integration, Process Flow Design and Visualization Solution. Available at http://www.cardix.com/pdf/delmia/DELMIA_QUEST.pdf, 2014.
[9] Michalewicz, Z. Genetic Algorithms + Data Structures = Evolution Programs. Third Edition. Springer, 1999.
[10] University of Connecticut. MEM Mission Statement. http://www.business.uconn.edu/cms/p343, 2014.
[11] Wheelwright, S.C., and Yong, S. AIC Netbooks: Optimizing Product Assembly. Harvard Business School 4245, January 21, 2011.

ANNEX A

SAMPLE CASE STUDIES

Table 1. Sample of HBS industry cases used in AMS through the years.

Company	Industry/Product	Main Subjects	Years Taught
AIC Netbooks	*Consumer electronics*	*Assembly line analysis, line balancing, labor distribution, simulation*	*2014*
Alara	*Food industry*	*Assembly line analysis, demand uncertainty, risk management*	*2011, 2013*
Baria	*Manufacturing consulting*	*Capacity planning, resource utilization*	*2012*
Bayonne Packaging, Inc.	*Paper industry*	*Process design, job shop analysis, simulation*	*2014*
Bergerac	*Veterinary diagnostic instruments*	*Vertical integration, make or buy decisions, supply chain management*	*2012*
Corning	*Glass products*	*Adapting new technology, quality control, data analysis in manufacturing*	*2002-2010*
Donner	*Printed circuit boards*	*Scheduling, job shop design, simulation*	*2002-2007*
Dore-Dore	*Fashion industry*	*Quick response systems, manufacturing cells, demand uncertainty*	*2004-2011*
Flextronics	*Smartphones*	*Manufacturing process design, manufacturing cells, simulation*	*2014*
FoldRite	*Stackable chairs and foldable tables*	*Production planning, metal fabrication, risk management*	*2012*
Hindustan Aeronautics	*Aircrafts, aircrafts parts*	*Supply chain management, outsourcing decisions, inventory management*	*2014*
Maria Yee, Inc.	*Furniture*	*Innovation, outsourcing decisions, supply chain management*	*2010, 2012*
NoBull Burger	*Food industry*	*Process and capacity analysis, simulation*	*2014*
Providence Seals	*Mechanical seals*	*Group technology, job shop analysis, manufacturing cells*	*2002-2005*
Sandvik	*Cutting tools*	*Recycling, reverse supply chain, logistics*	*2011, 2013*
Scharffen Berger	*Chocolate Maker*	*Process design, quality control, simulation*	*2006-2012*
Sof-Optics	*Contact lenses*	*Yield analysis, quality control, new technology analysis*	*2002-2009*
Taiwan Semiconductors	*Semiconductors, chips*	*Production scheduling under uncertain demand*	*2014*

Proceedings of the ASME 2014 International Mechanical Engineering Congress and Exposition
IMECE2014
November 14-20, 2014, Montreal, Quebec, Canada

IMECE2014-36755

RAISING GLOBAL LEADERS IN SCIENCE AND ENGINEERING UNDER TRILATERAL COLLABORATION

Dr. Hiroshi Honda, P. E.
Industry Frontier Research & Development Committtee,
#1-301, Higashi Narashino 2-17, Chiba 275-0001 Japan

Dr. Hephzibah Kumpaty
University of Wisconsin- Whitewater
Whitewater, WI, United States

ABSTRACT

This paper discusses on how globalization affects industry, business and engineering practice, and what kind of education is considered and attempted at selected high schools and colleges to raise global leaders from the United States, India and Japan. Case studies for selected schools in the United States, India and Japan are also presented. In particular, details on the participation of undergraduate students in an integrated, global research culminating in global leadership and outlook with specific examples from the ongoing collaboration of the University of Wisconsin-Whitewater and Indian Institute of Chemical Technology, Hyderabad, India are presented to corroborate the beneficial effects of globalization. With the goal of effectively raising global leaders in science and engineering fields, the authors propose a scheme for the trilateral collaboration between the U. S., India and Japan, with a focus on difference in education, characters of the peoples, and preferred models of global leaders among these nations.

INTRODUCTION

As human activities became increasingly bound to the global economy, each nation came to explore the ways to raise global leaders in science, engineering and technology among other fields. Crucial to the realization of such a goal, collaborative endeavor is a must among various stakeholders. A good example is a trilateral collaboration among three universities from different nations. *Stoichiometric and polymorphic variations in cocrystal synthesis* is a trilateral project among UK-USA-India (University of Bradford, University of Minnesota and University of Hyderabad) aimed at overcoming the challenges in manufacture of pure and stable pharmaceutical cocrystals[1]. This is a current work funded by the UK-India Education and Research Initiative established in 2006. Similarly another successful collaborative endeavor, the U.S. National Science Foundation (NSF)-sponsored International Research Experiences for Students (IRES) Site at University of Wisconsin-Whitewater (UWW): *Collaborative Research in Complex Natural Product Synthesis with Bio, Nano, and Green Applications at Indian Institute of Chemical Technology (IICT), Hyderabad* (US-India Collaborative Research, Grant No. OISE-1065543)[2] is presented to provide compelling argument for the benefits of globalization.

As a Japanese model for global leaders in science and engineering, astronauts such as Dr. Koichi Wakata[3], the first Japanese commander for International Space Station (ISS), may be pointed, as a person who was born and raised in Japan, worked at Japan Airline after receiving Master's degree from Kyushu University, married to a German woman, and educated and trained by the National Space Development Agency of Japan (NASDA) and National Aeronautics and Space Administration of the United States (NASA). He was in space, as we spoke during the course of writing this paper.

As an Indian model for global leaders in science and engineering, recently appointed Chief Executive Officer of Microsoft[3], Satya Nadella who was born and raised in Hyderabad, India comes to mind. He received his bachelor degree in engineering in India and furthered his education at the University of Wisconsin- Milwaukee and University of Chicago in the United States. He rose in ranks in his 22-year career at Microsoft. Based on these models, trilateral collaboration schemes are discussed and proposed in this paper.

OBJECTIVES OF *IRES*

UWW accomplished the following objectives through the IRES site:

1) Provided *global, integrated research opportunities* for undergraduate science students to gain hands-on experiences with the cutting-edge research in bio/nano/green applications of chemical syntheses;

2) Developed *interdisciplinary approaches (and entrepreneurial approach) to research and education* through planned synergistic visits to Avra Laboratories, Center for Cellular and Molecular Biology, Evolva Biotech India and Indian Drugs and Pharmaceuticals Limited (all located in Hyderabad) resulting in students' broadened view of collaborative research and entrepreneurship and

3) Facilitated the success of IRES participants in gaining critical thinking skills, intercultural adaptability, global competence, presentation and communication techniques and data analysis skills, resulting in *published work* in journals/

conference proceedings and their *continued research participation* beyond the program.

IRES PROGRAM ELEMENTS

The NSF-IRES Site (2011-2013) at UWW has provided successful global research opportunities to 12 students and involved Principal Investigator (PI) Dr. Kumpaty, senior personnel Dr. Catherine Chan, and several research mentors at IICT, Dr. Yadav, Dr. S. Chandrasekhar, Dr. Likhar, Dr. Lakshmi Kantham and multiple research scholars. In addition to the use of an ***interdisciplinary*** research environment and direct involvement with the advisors, the communication skills of IRES participants have been enhanced through the development of project abstracts in the beginning, and PowerPoint presentations by the end of program at IICT, participation in poster sessions, talks and writing of final research papers in the subsequent academic year at UWW, and by direct participation at the annual ***National Conference on Undergraduate Research*** in April following the IRES participation.

The demographics of the 12 participants shows our commitment to diversity: Minorities and others (**25%**) and Women (**75%**). There were a total of 8 female and 4 male participants. Two African-American and one Hispanic constituted the 25% of the participation by minorities and others. More minority students were offered the opportunity than those who availed. What is unique about this IRES site is in each cohort, three out of four (**75%**) were first-generation students to enter college from their families. One is a transfer student from a 2-year college, UW-Rock County to UWW. This kind of involvement in research in a global setting is a testament to the NSF support in encouraging diversity and equal opportunity, and the PI, Kumpaty's commitment to ensure that IRES is making a difference in the trajectory of these participants and their future generations.

Each year, the participants and Dr. Kumpaty were warmly received by the IICT, which provided excellent accommodation on its campus at the Abhinandan Guest house, and 24/7 transportation facility within a reasonable budget afforded by the grant. A dedicated driver was available for the group and all the tranportation to the synergistic visits was covered by the IICT. The IICT has provided free doctor visits to its on-campus medical clinic to receive first aid when needed. Generally, the participants are introduced to various facilities and scientists on campus and the first week is about learning, adjusting to the new culture and weather; and there are a series of meetings to come up with project descriptions accounting for some leeway for what is feasible within the project areas that have already been selected. For example in Year 1, ***Michelle Storage*** was provided the particulars to work on a project in biopesticides (isolation of natural product and its application as biopesticides). ***Margaret Legan*** would work on a nanocatalyst, NAP-MgO stablized Pd(0). ***Melissa Mueller*** and ***Natalie Huss*** were given projects in core organic synthesis-benzimidazole and chrysin. Similarly in Year 2, ***Milia Mims'*** work on a biopesticides project included testing for insecticidal activity of *Nerium odorum*. ***Brittny Henderson*** took on the synthesis of the UREAS as inhibitors of the protein ACE. ***Adam Bucane*** continued work on the nanocatalyst-NAP-MgO stablized Pd(0). ***Dustin Olson*** researched on natural product multistep synthesis of intermediate toward C14-C26 building block of Eribulin Mesylate. Year 3 projects described under conference papers involved four participants, namely, ***Amanda Loch, Cassandra Heller, Alex Steeno*** and ***Casey Mueller***. Year 1 participants served as good role models for the aspirants of Year 2 participation as they continued research in the final year at UWW and similarly the Year 2 participants fulfilled that role for the Year 3 aspirants. Year 1 participants presented their research work at the ***National Conference on Undergraduate Research*** (March 28-31, 2012, Weber State University, Ogden, UT). Year 2 participants likewise presented their research work at the ***National Conference on Undergraduate Research*** (April 9-11, 2013, University of Wisconsin-La Crosse, La Crosse, WI). Year 3 participants have presented their research work at the ***National Conference on Undergraduate Research*** (April 3-5, 2014, University of Kentucky, Lexington, KY). Before the attendance at a national conference, the students had already presented their research findings before several audiences- Department, UWW fall and spring Undergraduate Research Days (September and March), chemistry department, Chem Club and tri-beta biology club. Melissa Mueller joined Dr. Kumpaty in an invited talk at the Whitewater Senior Citizens home-called Fairhaven lecture series on February 20, 2012. In October 2011 following the Year 1 participation, Dr. S. Chandrasekhar of IICT visited the UWW campus (travel supported by IICT) and had special meeting with the Chancellor, the Provost and the faculty of the Department of Chemistry.

In Figure 1, Brittny (IRES 2012) is in her lab working and with her post-doctoral mentor; and in Figure 2 it can be seen that the four UWW students were catching up with their research progress documentation in a dedicated air-conditioned room with internet facility which was specifically set aside for us as a workplace to work on the reports and presentations while being away from the laboratory.

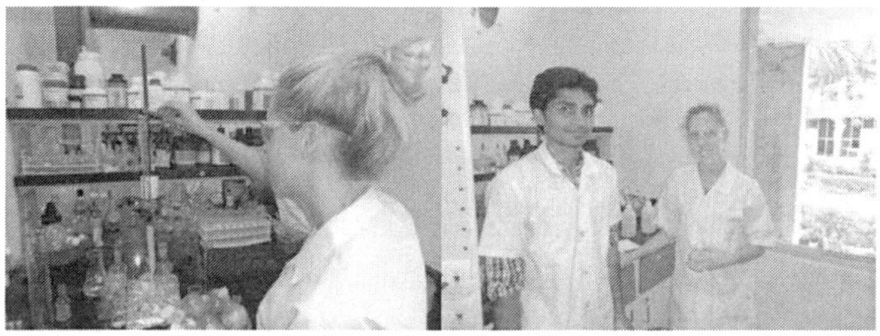

Figure 1: Brittny in her lab and with her post-doc mentor

At this room the participants had several meetings with scientists and other junior mentors who directly assisted our students in the labs. The students have throughly enjoyed the lab experience and most of them preferred to spend more time in lab, preferring to skip even dinner meetings, so they could concentrate on research work. During break times students also took advantage of visiting other labs, attending seminar talks, interacting and learning with other staff at IICT to gain the whole IRES experience. The IRES students primarily worked in the Indo-French organic laboratory, nano chemistry laboratory and the biopesticide laboratory all of which are closely located to each other on IICT campus.

Figure 2: Research Progress Documentation

Figure 3 presents the pictures of final presentations. One 2011 participant wrote about the final research presentation, *"The final research presentation was a wonderful, yet slightly intimidating experience. Many of the Ph.D. students, along with some senior scientist attended our presentations. Many asked questions and challenged our critical thinking. It was a great practice for our future careers."* One 2012 participant expressed, *"My most shinning accomplishment in the program was getting a chance to present my research in front of my advisors and faculty of Indian Institute of Chemical technology. It was a chance to show off the hard work I did and the time people around me invested in me."*

Figure 3: (top) Dr. S. Chandrasekhar introducing the occasion of Final Presentations by 2011 IRES participants; (bottom) Final Presentations at IICT: Melissa Mueller on combinatorial synthesis of Benzimidazoles and Margaret Legan on use of a heterogeneous nanocatalyst NAP-MgO-Pd(0) in selective reduction reactions

The already existing international collaborations at IICT is a testament to its commitment to global research and an ***International Symposium on Chemistry and Chemical Biology of Natural Products*** was held during 1st week of August 2-2012. Very many dignitaries attend this conference representing various countries. Dr. C.N. Rao of Indian Institute of Sciences, keynote speaker was extremely pleased with IICT's collaboration with UWW through this NSF grant and desired that other CSIR Laboratories will also establish such memorandum of understanding that exists between IICT and UWW. Figure 4 presents the picture taken during the dinner reception which shows cultural assimilation: the ladies from UWW are dressed in sarees and the picture includes a foreign delegate from Puerto Rico.

Figure 4: Cultural Assimilation

There were several synergistic center visits such as ICICI Knowledge Park, Biotechnology Incubation Center, Alexandria Technology Park and High Tech City (Microsoft and Infosys). The IRES participants had a good share of the cultural exposure during weekends. The visits included Salarjung Museum, Charminar, Golconda Fort, Birla Mandir, Ramoji film city to name a few historical landmarks. The participants had wonderful culinary experiences as well as shopping. Dr. Subha Kumpaty from the Milwaukee School of Engineering brought another IRES team for research in materials and biomedical applications at the Non-Ferrous Materials Technology Development Centre, also in Hyderabad. There were occasions both sets of participants met and interacted with one another. In Figure 5, both IRES teams (8 students of 2012) along with the hosts are seen in front of Salarjung Museum.

Figure 5: IRES Participants at Salarjung Museum

PAPERS

The following is the list of papers (publications/ conference presentations) by the twelve IRES participants.

1. Amanda Y. Loch (Hephzibah Kumpaty, Catherine Chan, Praveen Likhar), Nano-Nickel Hydroxide Catalyst Synthesis & Industrial Applications with Oxidation Reaction, *Proceedings of the National Conference on Undergraduate Research*, University of Kentucky, Lexington, KY, April 3-5, 2014 and UWW-Fall Undergraduate Research Day, Sept 2013.

2. Cassandra Heller (Hephzibah Kumpaty, Srivari Chandrasekhar), Ireland-Claisen Sigmatrophic Rearrangements, *Proceedings of the National Conference on Undergraduate Research*, University of Kentucky, Lexington, KY, April 3-5, 2014 and UWW-Fall Undergraduate Research Day, Sept 2013.

3. Casey Mueller (Hephzibah Kumpaty, Srivari Chandrasekhar), Green Asymmetric Organocatalyzed Michael Reactions, *Proceedings of the National Conference on Undergraduate Research*, University of Kentucky, Lexington, KY, April 3-5, 2014 and UWW-Fall Undergraduate Research Day, Sept 2013.

4. Alexander Steeno (Hephzibah Kumpatyar), Environmentally Friendly Suzuki Cross-Coupling Reactions through Eco-Friendly Solvents, *Proceedings of the National Conference on Undergraduate Research*, University of Kentucky, Lexington, KY, April 3-5, 2014 and UWW-Fall Undergraduate Research Day, Sept 2013.

5. Dustin Olson (Srivari Chandrasekhar, Hephzibah Kumpaty), Multistep Synthesis of Intermediate Toward C14-C26 Building Block of Eribulin Mesylate, *Proceedings of the National Conference on Undergraduate Research*, University of Wisconsin-La Crosse, La Crosse, WI, April 9-11, 2013. Also presented at the UW System Chemistry faculties meeting, Oct 2012 and UWW-Fall Undergraduate Research Day, Sept 2012.

6. Brittny Henderson (S. Chandrasekhar, Hephzibah Kumpaty), The Synthesis of UREAS as ACE Inhibitors, *Proceedings of the National Conference on Undergraduate Research*, University of Wisconsin-La Crosse, La Crosse, WI, April 9-11, 2013.

7. Adam Bucane (Pravin Likhar, Hephzibah Kumpaty), Nanocrystalline Magnesium Oxide-stabilized Palladium (0): A heterogeneous catalyst and cyclization reactions of indole and benzofurans, *Proceedings of the National Conference on Undergraduate Research*, University of Wisconsin-La Crosse, La Crosse, WI, April 9-11, 2013 and UWW-Fall Undergraduate Research Day, Sept 2012.

8. Milia Mims (Catherine Chan, Hephzibah Kumpaty, S. Chandrasekhar), Testing for insecticidal Activity of *Nerium odorum*, *Proceedings of the National Conference on Undergraduate Research*, University of Wisconsin-La Crosse, La Crosse, WI, April 9-11, 2013 and and UWW-Fall Undergraduate Research Day, Sept 2012.

9. Margaret Legan (Hephzibah Kumpaty, Pravin Likhar, Lakshmi Kantham, S. Chandrasekhar, J.S. Yadav), Nanocrystalline Magnesium Oxide-Stabilized Palladium (0): Heterogeneous Catalyst for Selective Reduction of Nitro Compounds in Aqueous Media, *Proceedings of the National Conference on Undergraduate Research*, Weber State University, Ogden, UT, March 28-31, 2012 and UWW-Fall Undergraduate Research Day, Sept 2011.

10. Michelle Storage (Hephzibah Kumpaty, Catherine Chan, K. Suresh Babu, S. Chandrasekhar), Isolated Compounds from Plant Extracts Screened Against Insects as Biopesticides, *Proceedings of the National Conference on Undergraduate Research*, Weber State University, Ogden, UT, March 28-31, 2012 and UWW-Fall Undergraduate Research Day, Sept 2011.

11. Natalie Huss (Hephzibah Kumpaty, J.S. Yadav, S. Chandrasekhar, K. Suresh Babu), Isolation and Synthesis of Chrysin and Coumarin Alkaloids, *Proceedings of the National Conference on Undergraduate Research*, Weber State University, Ogden, UT, March 28-31, 2012 and UWW-Fall Undergraduate Research Day, Sept 2011.

12. Melissa Mueller (Hephzibah J. Kumpaty, S. Chandrasekhar, Prathama, Mrunal Pendske, Radikha), Parallel Synthesis of Benzimidazoles, *Proceedings of the National Conference on Undergraduate Research*, Weber State University, Ogden, UT, March 28-31, 2012 and UWW-Fall Undergraduate Research Day, Sept 2011.

EVALUATION RESULTS

In terms of evaluation, at IICT, students completed a participant evaluation individually and then they had the same instrument completed by the advisors - an advisor evaluation of the participant, to help the participant assess his/her performance and discuss strengths and weaknesses. There are significant things learned from this process for both students and advisors. Then, upon return, the participants completed ***IRES program evaluation*** and Dr. Kumpaty used the responses for continuous improvement of the program. The students thoroughly enjoyed research at the world-class institute and comments indicate appreciation for the interactions, people and facilities at IICT. One wrote, "*I liked the ability this program afforded to network and meet other scientists from other countries and to be able to say that I worked in an international, world-renowned chemical institute. This experience is one of the best in my life so far and I hope it opens doors for others to come.*" Another said, "*The equipment and facilities were very user friendly and suited all of our research needs. Some of the techniques used were environmentally friendly.*" Another said, "*I found the skills that I was taught as a UWW student prepared me extremely well for the project. I also enjoyed learning the minor skills that I had not already acquired, from my Ph.D. student mentor.*" The participants had varied degrees of their accomplishments based on the time needed for their projects. One participant expressed, "*I would have liked to get further along with my synthesis but time did not allow this; this would have allowed me to conduct various reactions and allowed for more experience.*" Another noted as her most shining accomplishment, the completion of 15 reactions and added, "***Because my reactions were multistep and time-consuming, completing 15 reactions was a great amount.***" A six-week period, however intense, is definitely a very short time for organic synthesis projects. The students continued their research upon return to UWW campus in their final year which helped them to accomplish a significant piece of work and enabled them to present at fall and spring UWW undergraduate symposiums and at the *National Conference on Undergraduate Research.* As anticipated, more coordination of projects and clarification of project assignments were targeted for 2013 participation as students indicated they would like more familiarity with the projects and more practice prior to research at IICT. The program evaluation has indicated that students have had a very positive impact and their worldview has enlarged and learning to do research with global scientists is a major accomplishment of this IRES Site. Typical comments include, "*I believe this research program included everything that I wanted to do. I was able to experience India in both the cultural experience and scientific experience; it was very well-balanced.*", "*This research program exceeded my expectations in more ways than one and I came out of it not only a better scientist but also a **global leader**.*", "*My favorite part was not only learning about my favorite thing in the world (science) but also learning about a culture that was so different from my own. The whole experience was so hands-on.*", "*This program provided us with unimaginable opportunities that I didn't think anyone could ever provide me with. It was a great life experience and I'm very proud I got chosen. So the life experience was the best part about this IRES.*", and "*I really like that this research experience was in a whole different country. It's nice to get an internship before you graduate, but to have it in a completely different culture was the best part.*" Presenting before IICT scientists and post-graduate researchers, though intimidating and daunting, was seen as a career building experience by the participants. The thrill of discovery is interwoven in statements such as, "*I was able to discover a successful reaction using a nano catalyst that has never been tested before*".

PROGRAM IMPACT: BENEFITS OF GLOBALIZATION

The IRES program has made a huge impact in the career choices of the 12 participants as they are gainfully employed or in graduate school. It has been learned that the IRES trip seems to be the central theme during the placement interviews as the employers found the ***global citizenship and adaptability*,** especially travelling and living in a country which is very different from home**,** as a pivotal strength the participants brought to the table. Melissa Mueller (2011) has been employed at a chemical company, Trostel, which is dedicated to expanding business globally and plans to pursue Ph.D. in chemistry in the near future. Dustin Olson (2012) is in Ph.D. in Chemistry at University of Wisconsin- Milwaukee. Michelle Storage (2011) is awarded scholarship to pursue M.D., Ph.D. at Medical College of Wisconsin. Brittny Henderson (2012), while employed at S.C Johnson in Racine, WI was accepted into three of the top 10 pharmacy schools in the nation including UW-Madison and she chose to join the University of Texas at Austin for D. Pharm. She highlighted IRES experience as a strong factor in her selection at industry and future study. In her program evaluation, Brittny described how participation in the IRES program has affected her career goals by stating, "***The participation in this program was the one of the main reasons that I was hired as a fragrance lab technician at SC Johnson. This experience also made me realize that I was to go into more of the research side of Pharmacy***". Regarding how participation in the IRES program has affected her attending graduate school, Brittny wrote, "***Schools were extremely impressed with my participation in this program and I believe that it was one of the main reasons that I was accepted into three of the top ten pharmacy schools in the nation. Many schools asked about this program in my interview.***" The experience is similarly true for all other participants. Natalie Huss (2011), joined MBA program at UWW after a successful one-year career in industry. Her ambition is to work for a global chemical company upon graduation. Milia Mims (2012) is pursuing MBA as well and Adam Bucane has advanced in position at a company, Air Products in Janesville, WI. Year 3 participants are currently in the final year of their undergraduate programs with bright future ahead of them. Margaret Legan (2011), upon graduation went to Loma Linda University, Loma Linda, CA and worked as a Research assistant while pursuing MS in Physician Assistant program. She has been accepted into several Ph. D programs starting in Fall 2014. Based on her work and subsequent research, a journal publication was submitted: Margaret Legan, Hephzibah Kumpaty, Pravin Likhar, Lakshmi Kantham, S. Chandrasekhar, Nanocrystalline Magnesium Oxide-Stabilized Palladium (0): Environmental Friendly Catalyst for Selective Reduction of Nitro Compounds in Aqueous Media, *Manuscript submitted to Green Chemistry Journal.*

As far as the impact on career goals, the comments were very affirming, for example, "*I have always wanted to continue my education in chemistry. The program has made me think of possibly working outside the U.S.*" and "*It has made me more ambitious for a terminal degree that I might be able to move forward with international research.*"

The students liked interactions with people both at work and outside. The top three picks for one participant are, "My research with NAP-MgO-Pd(0), the culture and connections with IICT and faculty" and to another, "The people's kindness and generosity, the food, and the lab experience". The participants consider IRES a lifetime opportunity and an amazing experience. Typical comments along this thought are: "*If given the opportunity, I would return to India to conduct research again. This program has impacted my life and I will never forget it.*", "*I was able to learn a lot from the experience and it gave me the opportunity to travel outside the United States which was an amazing experience.*", "*I would like to thank everyone associated with making this possible. It was an experience that I always remember.*" and "*I wish more students would be able to get this experience as it is really a life changing experience and I enjoyed every minute of it.*" Due to such a significant impact, despite personal sacrifices, the PI Kumpaty is committed to providing global experiences for UWW students by seeking renewal grants with expanded scope and collaboration.

TRILATERAL COLLABORATION SCHEME

Based on the successful bilateral collaboration between UWW and IICT and similar such collaborations that include institutions in Japan, the authors propose trilateral collaboration among U.S.-Japan-India. A few trilateral collaborations were considered in the brainstorming of various scenarios. China-Japan-US Trilateral Symposium on Lifeline Earthquake Engineering[5] had its sixth meeting in October 2012. This collaboration started in 1990 continues to address earthquakes and seismic management of lifelines. The trilateral symposium *Architectured biomaterials, Medical and Tissue Engineering*[6] to be held in December 2014 is organized by the Science & Technology Department of the French Embassy in Berlin, in collaboration with the Max Planck Institute of Colloids and Interfaces in Potsdam (Germany), the University of Grenoble (France), and the Warsaw University of Technology (Poland). Officially launched in 2006, U.S.-Ireland R&D partnership[7] enables collaborative research in the four areas of sensor technology, nanotechnology, cystic fibrosis and diabetes among U.S.-Northern Ireland-Ireland universities. It has been quite successful in making advances in research in these areas and contributing to economic development in Ireland. Organized by European Commission (EC), U.S. Department of Energy (DOE), Japan's Ministry of Economy, Trade and Industry (METI), and the New Energy and Industrial Technology Development Organization (NEDO), conferences on Critical Materials are held annually since 2011 under the EU-US-Japan Trilateral Critical Materials Initiative[8] to improve extraction, find substitutes, improve use efficiency and promote recycling. Such collaborations require commitment and give-and-take by the stakeholders for successful implementation.

Table 1 shows comparative gives and takes of US, India and Japan in the fields of materials science, machine element engineering and technology, culture and sports, among others.

Table 2 shows candidate partners for the subject trilateral collaboration. Many combinations of exchange are possible from the presented choices for universities, institutions and high schools. Globalization has no bounds but brings many beneficial effects on all participants.

Table 1. Comparative Gives and Takes of US, India and Japan in the Fields of Materials Science, Culture, and Sports.

Nation	USA	India	Japan
What the Nation Can Give	First Class Basic Science and Technology, American and Diversified Cultures, American Spirits, Experiencing and/or Enjoying American and European Sports (Football, Basketball, Baseball, Golf, Tennis etc.)	Fruits of IICT Research, Human Resource in Mathematics, Computer Science and IT, Hindu and Islamic Cultures, Indian Spirits, Experiencing and Enjoying Indian and British Sports (Cricket, Hockey, etc.)	First-Class Science and Technology, Industrial Application, including Machine Element, Transportation and Energy Engineering and Technologies, and Training Senior Expertise, Japanese Culture, Food and Tea, Japanese Spirits, Experiencing and/or Enjoying Japanese Sports (Judo, Kendo, Karate, and Sumo Wrestling)
What the Nation Want to Take	Industrial Application Technology, Communication Skills in Indian and Oriental Languages and Cultures, American Students' Experiences in India and Japan	Industrial Application Technology, Communication Skills in American and Oriental Languages and Cultures, Indian Students' Experiences in USA and Japan	Enhanced Diversity of Talents by Gaining Those from USA and India, Communication Skills in American and Indian Languages and Cultures, Japanese students' Experiences in USA and India
Possible Addition	Research and Other Activities Utilizing Space, Student Formula SAE Competition, Robotics Competition, Regulatory Engineering, Relevant Policies	Research and Other Activities Utilizing Space, Student Formula SAE Competition, Robotics Competition, Regulatory Engineering, Relevant Policies	Research and Other Activities Utilizing Space, Student Formula SAE Competition, Robotics Competition, Regulatory Engineering, Relevant Policies

Table 2. Possible Schemes for Exchange Programs of Universities, Associated Institutes and High Schools between and/or among USA, India and Japan

Nation	USA	India	Japan
Universities	Selecting from Public/ Private Universities, including UWW and the PSU Schreyer Honors College	Selecting from IITs, and IISc., Bangalore	Selecting from Public Ivies (Tokyo, Kyoto, Osaka, Tohoku, Kyushu, Nagoya, Hokkaido), Tokyo Institute of Technology, among others
Institutes	Argonne National Lab, Sandia National Lab, Oak Ridge National Lab, Pittsburgh Energy Technology Center	Council of Scientific and Industrial Research labs : IICT, Materials Research Lab, ARCI (Powder Metallurgy), Non-Ferrous Materials Technology Development Center	Selecting from NIMS, National Institute of Advanced Industrial Science and Technology (AIST), and Japan Science and Technology Agency (JST)
High Schools	Schools in the Midwest (WI, IL, IN, MN, MI) and PA	Schools in Hyderabad, Bangalore and Chennai, among others	Schools in Tokyo, Chiba, Kyoto, Osaka, Tohoku and Kyushu, among others

Note: High school exchange students can select one or two universities from any of the three nations for possible joint degrees.

Table 3 shows citations rankings of Top 10 in the world in the field of materials science[9]. The proposed US-Japan-India trilateral collaboration has valid support through this data presented.

Tables 4 to 7 show QS 2014 World Rankings of universities in mechanical, aeronautical and manufacturing engineering, chemistry, chemical engineering and civil and structural engineering[10], which can be a useful reference when the trilateral collaboration considers materials science, chemistry and relevant engineering fields.

Honors colleges, such as the Schreyer Honors College shown in Table 2, are beneficial in raising students with a broader horizon, during the course of being educated simultaneously as a global leader and as an expert in scientific and engineering disciplines; therefore, these colleges can be integrated into the trilateral scheme for the students so wish.

Education for raising future leaders in science and engineering can be started at a high school level or even earlier in infancy. In Japan, there are featured high schools such as the super science high schools and the global leaders high schools to raise elites in science and engineering and in global leadership. Organizations such as Consortium for Renovating Education of the Future (CoREF)[11] are helping these high schools to provide them with effective and efficient programs.

Table 3. Citation Rankings in the Field of Materials Science[9]
(Source: NIMS (Compiled from the ESI database by Thomas Rueter))

World (January 2006 – April 2011)		
Rank	**Institution**	**Citations**
1	Chinese Academy of Sciences	45,196
2	Max Planck Society, Germany	16,271
3	Massachusetts Inst. Technology, USA	11,481
4	National Inst. Materials Science, Japan	11,161
5	National University of Singapore	11,120
6	Tohoku University, Japan	10,289
7	Tsing Hua University, China	10,264
8	Indian Institute of Technology, India	9,467
9	Georgia Tech., USA	9,412
10	Univ of California-Berkeley, USA	8,946

Table 4. 2013/14 QS World Rankings in the Field of Mechanical, Aeronautical and Manufacturing Engineering[10]

Rank	Institution
1	Massachusetts Inst. of Technology, USA
2	Stanford Univ., USA
3	Univ. of Cambridge, U.K.
4	Imperial College London, U.K.
4	Univ. of California-Berkeley, USA
6	National Univ. of Singapore, Singapore
6	Harvard Univ., USA
8	Univ. of Tokyo, Japan
9	Georgia Inst. of Technology, USA
10	Univ. of Michigan, USA
33	Kyoto University, Japan
41	Tokyo Inst. of Technology, Japan
43	Indian Institute of Technology, India

Table 5. 2013/14 QS World Rankings in the Field of Chemistry[10]

Rank	Institution
1	Massachusetts Inst. of Technology, USA
2	Univ. of California-Berkeley, USA
3	Univ. of Cambridge, U.K.
4	Harvard Univ., USA
5	Stanford Univ., USA
6	Oxford University, U. K.
7	ETH Zurich, Switzerland
8	California Institute of Technology, USA
9	Univ. of Tokyo, Japan USA
10	Univ. of California-Los Angeles, USA
13	Kyoto Univ., Japan
20	Tokyo Inst. of Technology, Japan
28	Osaka University, Japan
34	Tohoku Univ., Japan
49	Nagoya Univ., Japan

Table 6. 2013/14 QS World Rankings in the Field of Chemical Engineering[10]

Rank	Institution
1	Massachusetts Inst. of Technology, USA
2	Univ. of California-Berkeley, USA
3	Stanford Univ., USA
4	Univ. of Cambridge, U.K.
5	National University of Singapore
6	Imperial College London, U. K.
7	Univ. of Tokyo, Japan
8	Delft Univ. of Technology, Netherlands
8	Kyoto Univ., Japan
10	California Institute of Technology, USA
28	Tohoku Univ., Japan
31	Osaka Univ., Japan
41	Tokyo Inst. of Technology, Japan

Table 7. 2013/14 QS World Rankings in the Field of Civil and Structural Engineering[10]

Rank	Institution
1	Massachusetts Inst. of Technology, USA
2	Univ. of Illinois-Urbana Champaign, USA
3	Univ. of California-Berkeley, USA
4	Univ. of Tokyo, Japan
4	Univ. of Cambridge, U.K.
6	Univ. of Texas-Austin, USA
7	Naional University of Singapore
8	Kyoto Univ., Japan
9	Imperial College London, U. K.
10	Univ. of Hong Kong
29	Tokyo Inst. of Technology, Japan
49	Indian Institute of Technology-Madras, India
50	Indian Institute of Technology-Bombay, India

Therefore, enhancement of exchange programs at high school levels will give opportunities for ambitious, high school graduates to enter into colleges and universities in nations of their choice at both undergraduate and graduate levels, leading to joint degrees, possibly in multiple languages, from those in the U.S., India and Japan.

The lead author has constructed a scheme of collaboration for his alma mater Ibaraki High School, one of the ten global leaders high schools, designated by Osaka Prefecture.[12] The high schools in both the U. S. and India may also be benefitted from such a scheme.

CONCLUDING THOUGHTS

This paper showed it is possible to raise global leaders in science and engineering via collaboration. Emphasizing the success of a bilateral international experience for students site at UWW (a U.S. university) in collaboration with IICT (a foreign research institute based in India), the paper projects a trilateral collaboration scenario among U.S.-Japan-India wherein students could visit research institutes in other countries or attend universities across the globe. Enlarged world view, learning to do research with global scientists, and cultural adaptability are some of the characteristics gained through global participation by these students. Under the theme of materials science, the citations from Essential Science Indicators by Thomas Reuter affirm the value of trilateral collaboration among the three countries. The authors predict the continual interest of policy makers to encourage global citizenry so students will take advantage of opportunities made available to them. Whether pursuing graduate school or a career move, the global experiences will be valued highly in the interviews and selections. Collaboration among the nations would entail forming work culture that includes cultural and style differences and staying aligned to the purpose of achieving the intended goals. Cooperation and open communication are keys to successful collaboration besides the technical expertise and competence. The authors are grateful that the trilateral collaboration to raise leaders such as Wakata and Nadella is indeed possible.

REFERENCES

1. http://www.bradford.ac.uk/research/our-research/rkt-centres/pharmaceutical-engineering/news-and-events/news/ukieri-uk-us-india-trilateral-research-in-partnership/Date Visited: April 22, 2014.

2.http://en.wikioffuture.org/IRES:_US-India_Collaborative_Research_in_Complex_Natural_Product_Synthesis_with_Bio,_Nano,_and_Green_Applications_at_the_Indian_Institute_of_Chemical_Technology,_Hyderabad. Date Visited: April 22, 2014.

3. http://www.astronautix.com/astros/wakata.htm NASA Official Biography. Date Visited: April 22, 2014.

4. http://www.microsoft.com/en-us/news/ceo/index.html New CEO Announced. Date Visited: April 22, 2014.

5. http://www.asce.org/chinajapanussymposium2012/ Date Visited: April 22, 2014.

6. http://www.idw-online.de/en/news583417 Date Visited: April 22, 2014.

7. http://www.sciencediplomacy.org/perspective/2014/ trilateral-partnership-for-supporting-research-and-relationships Date Visited: April 22, 2014.

8. http://ec.europa.eu/research/industrial_technologies/event-13_en.html Date Visited: April 22, 2014.

9. http://www.rvo.nl/sites/default/files/bijlagen/NanoinNIMS.pdf Date Visited: April 22, 2014.

10. http://www.topuniversities.com/university-rankings/university-subject-rankings/2013/engineering-mechanical Date Visited: April 22, 2014.

11. http://coref.u-tokyo.ac.jp/en

12. Honda, H., Globalization of Industry, Business and Engineering Practice and Role of Education (15), to be published by Science of Machine Journal (Yokendo Co., Ltd. November or December, 2014).

Proceedings of the ASME 2014 International Mechanical Engineering Congress and Exposition
IMECE2014
November 14-20, 2014, Montreal, Quebec, Canada

IMECE2014-39952

PRIVATE ENGINEERING EDUCATION SCENARIO IN INDIA

Sandip Kale
Trinity College of Engineering and Research
Pune, Maharashtra, India
sakale2050@gmail.com

S. N. Sapali
College of Engineering, Pune
Pune, Maharashtra, India
k_sapali@yahoo.co.in

ABSTRACT

In India, government aided and private engineering institutes provide engineering education. Government aided institutes include Indian Institutes of Technology (IITs), National Institutes of Technology (NITs), Regional Engineering Colleges (RECs) and government engineering colleges. Ten percent of the total students get education in government-aided institutes and are globally accepted too. Remaining ninety percent of the total students get education in private self-financed engineering institutes. To meet the increasing demand of engineers from various industrial sectors, a quantitative growth of private engineering institutes took place with an average annual intake capacity of four hundred to five hundred students. With increasing annual intake capacity, the trend of vacant seats in private engineering institutes is also increasing rapidly year wise. Indian industry demands many engineers, but only a few students passed out from private institutes are employable. There is a challenge to build the gap between what industries are looking for the engineers and the education provided in the institutes. In this article, the authors have tried to frame the strengths, weaknesses, opportunities and threats (SWOT) analysis and recommend some remedial actions needed for private engineering institutes in India.

INTRODUCTION

Indian policy makers recognized the importance of engineering education since independence and established five IITs and twenty RECs in the country immediately. In the 21st century the knowledge based economy and remarkable technological growth around the globe, demanded a focused view towards engineering education. Until 1980s the engineering education, was available through government-aided institutes such as IITs and NITs only. To cater the ever-increasing need of skilled engineers, government liberalized its policy of technical education and some of the private institutes started their engineering colleges in the year 1983. In the subsequent years, the private engineering education sector witnessed progressive growth.

Indian government runs IITs, NITs, IIITs, and RECs. In additions, state governments run few engineering colleges in respective states. These governments provide finance to start and develop government or government aided engineering institutes. Self-finance private colleges, can be started by any institute after fulfilling regulations of All India Council for Technical Education (AICTE). In self-finance private colleges, the fee collected from students is used to fulfill all financial needs (building development, laboratory development, salaries of faculties and staff). The fee structures of private institutes are regulated by AICTE based on development and facilities provided to students and revised annually.

In the year 1947, when India became independent, there were 36 engineering institutes with an annual intake of 2500 students [1]. In the year 1997 number of engineering institutes increased up to 416 with an annual intake of 0.131 million [2,3]. In the year 2005 number of engineering colleges increased up to 1451 with an annual intake of 0.52 million [4,13]. The number of institutes reached 1511 with annual intake 0.55 million in the year 2006. From the year 2006 to the year 2012 there is more than twofold increase in the number of institutes of 3495 and threefold increase in annual intake of 1.8 million as shown in Fig. 1[5].

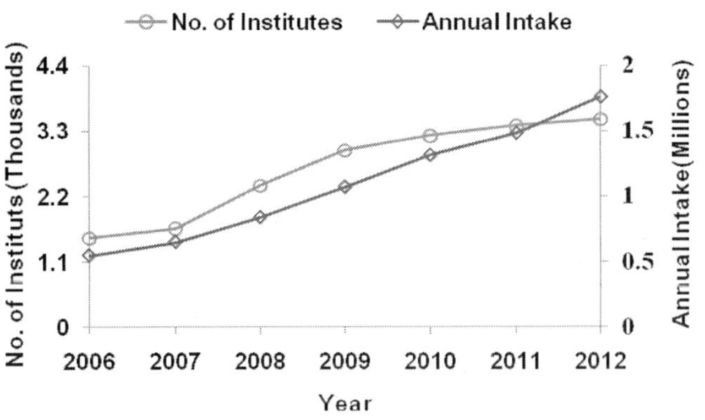

Fig. 1: Increase in No. of Institutes and the Annual Intake in India

In India, a statutory body AICTE governs all regulations related to engineering education. During the tenth plan (2002-2007), the number of AICTE approved engineering institutes has been raised from 1057 to 1522 and the annual intake from 0.29 million to 0.58 million. Though the tenth plan outlay for the technical education sector was INR 47 billion, an expenditure of only INR 34.16 billion was incurred. The eleventh plan (2007-2012) envisages setting up of eight new IITs, ten new NITs, twenty Indian Institutes of Information Technology (IIITs). Till date, there are more than 3500 engineering institutes in India.

The annual intake of students in all IITs is 7500, NITs 0.035 million and the rest, more than 1.5 million (about 97 % of the total) is accounted by the private institutions. The drastic increase in intake capacity through increased number of engineering institutes faces many problems such as lack of infrastructure, insufficient teaching faculties, less manpower in universities to which they are affiliated, lower merit student admissions, etc. has brought down the quality of engineering education to a very low level. Employability of students graduated from these institutes raises a big question. In order to enhance the quality of engineering education the National Board of Accreditation (NBA) has started evaluating the quality of education imparted in various engineering institutes. In spite of this quantitative growth, many recent studies and surveys have posed some serious questions regarding the quality and sustainability of the education imparted in these institutes. Hence, in this paper, the authors have tried to work out the SWOT analysis, issues and challenges and present some suggestions for the private engineering education sector in India.

SWOT ANALYSIS

Strengths

- Many institutes have the ability of capital investment for starting and developing engineering institutes as per rules and regulation laid by AICTE.
- The ability of the higher middle class, now, to pay for engineering education in private self-financed institutes is increasing.
- The government is providing scholarship for the specified caste students and economically backward students to promote engineering education. The amount of scholarship is offered to both, government and self finance private institutes.
- Because of AICTE regulations, engineering institutes are equipped with modern infrastructure.
- Geographically engineering institutes are within reach of the students and in the radius of 50 km from any location.
- Potential of becoming finishing schools,
- Nurture the youth, typically from the rural areas.
- The Quality Improvement Program (QIP) schemes from AICTE have contributed significantly to the upgrading of qualifications of faculty in some of engineering institutions
- AICTE also promotes enhancement of quality engineering education through different schemes such as Research Promotion Scheme (RPS), Research Park, Faculty Development Program (FDP), Technical Education Quality Development Program (TEQIP), Nationally Coordinated Project Schemes, etc.
- Increasing interest of industry associations (such as CII, FICCI) and of professional societies to collaborate with academic institutions.
- Some of the companies prefer the engineers from local institutes, due to easily available manpower
- Few private un-aided institutes, doing phenomenally well have received the autonomous status, that will boost the progress of the institute as well as society.

Weaknesses

- Deficiency of competent teaching faculties affects pedagogy. The minimum qualification for faculty is post graduation in respective branch. The number of institutes offering postgraduate programs is not increased proportionally to that of graduate programs.
- Average performing students
- Lack of enough industry-institute interaction [6, 14],
- Curriculums are outdated and not compatible with the latest developments and industry requirement.
- Less attention is given to research projects and less interest shown towards post graduation and research degrees; many faculties are pursuing these degrees because of compulsion made by AICTE.
- Numbers of students vacant seats are increasing every year in private engineering institutes.
- Lack of availability of qualified and experienced faculty
- Mismatch between education and training (knowledge and skills) received by students and the industry requirements
- The widely prevalent affiliating system in our universities precludes timely curriculum updating and the introduction of innovative reforms.

Opportunities

- International collaboration and student exchange programs
- Value added courses could be included in curriculum or conducted additionally e.g. soft skills, aptitude, etc.
- Open sources of education,
- Raising standard through accreditation,
- Global competence in improved written and verbal communication [10, 11],
- The upcoming World Bank project will provide the necessary resources for up gradation of technical education in the country.
- Distance education possibilities for continuing education as per recommendation from the National Knowledge Centre (NKC) [7],
- Networking of technical institutions with research and development labs and industries,
- Strengthening the alumni- alma mater for help and support from them
- As the software engineering field, many untapped areas in the field of engineering, where Indian engineers can be in great demand
- Foreign universities are setting up their campuses in India and raise competition to Indian engineering institutes.

Threats

- The number of engineering graduates is increasing and face tough competition - "survival of the fittest"
- Devaluation of the Ph.D. and postgraduate degrees is a serious threat. Many faculties of engineering institutes are pursuing these degrees for the sake of promotion.
- Disproportionate spread over of the institutes causing regional imbalance and interstate migration of students
- The proportion of vacant seats, increasing from last few years
- Polytechnic/ Diploma courses are getting hampered, as admissions to engineering degree course are easily available,
- Bright engineering graduates are not interested pursuing teaching careers.
- The scientific foundation in the country is getting weaker, which will have an adverse impact on our capacity for technological development

ISSUES AND CHALLENGES

Though Indian engineering education has expanded rapidly to meet the growing demand of engineers, the quality of private engineering institutes is falling day by day. Therefore, this section addresses different issues and challenges such as vacant seats, employability of students, quality of students admitted, facilities available in institutes, etc.

Vacant Seats

About 19 percent of seats in engineering colleges in the country remained unoccupied in the academic year 2011-12, i.e. 0.28 million seats remained vacant out of 1.485 million. The trend continued in the academic year 2012-13 with around 20-30 % of vacant seats. During the year 2013-14 this trend drastically changing as shown in Table. 1 [15-18]. Uttar Pradesh, one of the Indian states, has around 306 engineering colleges and 58 colleges could not fill a single seat [19].

Table. 1: Vacant Seats in Private Institutes in Major States of India

Indian State	Intake capacity (million)	Number of vacant seats (million)	% vacant seats
Andhra Pradesh	0.340	0.076	29.23
Tamilnadu	0.236	0.080	33.89
Maharashtra	0.155	0.043	27.74
Uttar Pradesh	0.120	0.100	83.42
Madhya Pradesh	0.109	0.060	55.04
Karnataka	0.092	0.024	26.08
Gujarat	0.063	0.010	15.87
Rajasthan	0.058	0.014	24.13
Odissha	0.046	0.029	63.04
West Bengal	0.028	0.006	21.43

Fig. 2 shows trend of vacant seats in the state of Tamilnadu from year 2009 to 2013. The percentage vacant seats are around 30 %.

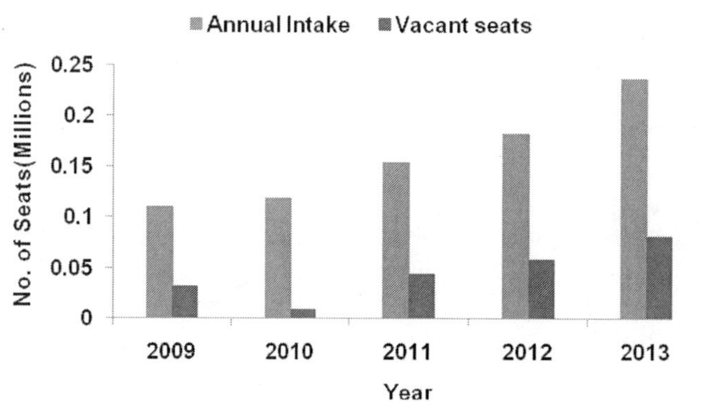

Fig. 2: Trend of Vacant Seats : Tamilnadu State

Employability of the Students

A National Association of Software and Services Companies (NASSCOM) study says, engineering services are a $40 billion opportunity for India by 2020, but only 17 % engineering graduates were employable as per 2012 report. Employability is the function of the application oriented technical knowledge, soft skills and all round progress of the students. Most of the private institutes fail to prove themselves as finishing schools resulting in poor employability.

Quality of Students Admitted

In India, students who have completed twelfth class are admitted to first year degree courses and students who have completed polytechnic are admitted to the second year degree course. There is a successive lowering down of the eligibility criteria for admission to the engineering courses from the authorities. The criteria are a non-zero Common Entrance Test (CET) score and aggregate twelfth class, 45% marks (40 % for caste wise reserved category) last year and still many students were not eligible to take admissions to the engineering courses. In some states like Andhra Pradesh, the criterion was more relaxed and students with zero score in the entrance examinations were declared as eligible. The fact is, the toppers in the entrance examination always prefer the premier institutes (most probably government engineering institutes) and the rest having no other option, get enrolled in the private engineering institutes. This is a serious issue letting the "garbage in" and the key reason for the poor performance of these institutes. Each state conducts its own CET. In Maharashtra state, objective type of CET is conducted with maximum 200 marks. Table. 2 shows result analysis of CET conducted in Maharashtra state from 2010 to 2013.

Table. 2: Result Analysis - Maharashtra State CET

Academic Year	2010	2011	2012	2013
Total Student Appeared	249052	254208	280042	256466
Score above 190	183	37	39	31
Score above 100	36992	21807	27025	23582
% Student above 100 out of 200	14.85	8.57	9.65	9.19

Other Issues

Thus, setting up of a state of the art infrastructure, conformance to the standards, attracting students and faculties, pedagogy and developing the technocrats through excellent teaching-learning process are the major issues and challenges in front of the private engineering institutes across India.

REMEDIES

It is required to improve quality of private engineering education to make more engineers employable and readily acceptable by the industries. Various stakeholders such as governments and regulatory bodies, engineering institutes, industries, alumni, primary and secondary education institutes, etc. play an important role to decide the quality of education. This section focuses on different remedies to improve education quality by the active role of these stakeholders.

Role of Government and Regulatory Authorities

This section discusses the role of government and regulatory authorities to improve quality of education by strengthening the primary and secondary education, accreditation, improved governance and a way of appointing mentors from premium institutes.

Strengthening the Primary and Secondary Education

Most of the students enrolled for engineering in the private institutes are not having at par academic excellence. A good performance cannot be expected from an engineering student without a sound base of science and mathematics. Therefore, strengthening the primary and secondary education and needful reforms are highly essential for quality improvement of the incoming students.

Mandate of Accreditation

In India, AICTE plays an important role for proper planning and coordinating the qualitative and quantitative growth of technical education. The NBA is the autonomous institute of AICTE, which is the key authority for the accreditation of technical institutes. The NBA accreditation of a particular program is run by the institute, through the self - assessment score and scoring against a well defined set of attributes. Therefore, like the International Organization for Standardization (ISO) certification NBA accreditation ensures the quality and set standards for the program. However, due to lack of awareness, proper work systems, non- maintenance of records many private engineering institutes avoids going for accreditation. So, for the renewal of the institute's approval or for sanctioning additional intake for the course, the NBA accreditation with a benchmark score is to be set as criteria.

Improved Governance

In NKC report, the need for reform of the governing authorities in the engineering education sector is highlighted. At present, the private institutes, are affiliated with state universities, and governed by AICTE through the respective state wise Directorate of Technical Education (DTE). The NKC promotes full autonomy to the institutes, wherever possible and under the control of two authorities- AICTE for academics, faculty development and curriculum and for the governance, it would be under control of Independent regulatory Authority of Higher Education (IRAHE). This structure will help the institutes in bringing transparency and professionalism in their governance and it will bring the uniformity in the curriculum and pedagogy throughout the country.

Premium Institutes as Mentor

In India, 45 technical institutes are of national importance, which include IITs, NITs, RECs and Government engineering colleges. The products of these few institutes are globally accepted. In the private sector, only a few institutes are ranked as premier technical institutes and for the rest of the institutes, quality of education imparted and their functioning is a key issue. To change this situation, the private institutes must idealized the premiere institutes and improve their functioning. The other way is the elite institutes can play a mentoring role for the private colleges in their respective region to bring about a qualitative transformation in the engineering education in India.

Role of the Institutes

Only regulations made by government authorities are not useful to improve quality. Institutes should actively work by taking forward steps such as faculty attraction and retention, encouraging research, conducting finishing schools, increasing industry institute interaction, maintaining strong alumni relationships, collaborating with international recourses, etc.

Faculty Attraction and Retention

Most of the private engineering colleges have acute short of qualified and experienced faculty. The attraction of the IT / corporate world because of the handsome salary packages, unwillingness for higher education are the two prime reasons, why the young graduates do not wish to choose teaching profession. In India, only 5% of the graduates opt for post graduation while in the USA it is up to 50%. Better salary, incentives, modern infrastructure, better working and living environment in the private institutes will attract the cream towards the teaching profession. As well as safe and secure environment, faculty development programs, authority and due responsibility, promotions and incentives, performance-based assessment are highly essential for retention of good faculties. In association with Technical Teacher's Training Institute (TTTI), along with the FDP, an induction program for the teachers, will also serve the purpose. AICTE, ISTE are always promoting one or two weeks FDP and provide a platform for engineering faculties along with the appropriate funds.

Encouraging Research

Encouraging and promoting original research work in the areas of fundamental as well as advanced interdisciplinary engineering areas and its measures in terms of papers published, patents filed, commercialization of a product developed, books written will help in establishing the image of the institute as a research hub and will result in many tangible and intangible benefits. The private institutes have huge potential of research and technology transfer, and they should work out a plan for the same, consisting of fully funded vibrant Ph.D. programs.

Finishing Schools

In many private engineering institutes, though the students are technically strong, they lack in written and oral English communication, soft skills and professionalism, i.e. the key indicators influencing their employability. There is a strong need for mapping the skill sets of students in this area and developing an activity based training in the curriculum for all the eight semesters. Thus, the private engineering schools should take positive efforts towards becoming the competitive finishing schools.

Industry Institute Interaction

Often there is a complaint from the Industry that only 10 to 20% of their engineers are competent for the job and they have to train the students for about 6 months to a couple of years. This is because of the widening gap between Industry and academia. To bridge this gap and to enhance the industry-institute interaction the institute must have a road map. Arranging frequent industrial visits, seminars and guest lecturers from industry experts, sponsorship from industry for setting up a research lab, summer and winter internship of students, undertaking the industrial projects, consultancies and training programs are some of the measures taken by the Institutes through memoranda of understanding (MoU).

Strong Alumni- Alma Matter Relationship

This parent-child like relationship must reciprocate the feel good factor, pride and unfailing communication between the two. Nurturing this relationship will produce both the tangible and intangible benefits for the institutes.

International Collaborations

Tying up with a global competent player in the field of education and research will be an insignia of excellence for an institute. As an example IIT Madras, has a tie up with 36 international universities. The student exchange programs, deputing students and faculties for the higher studies/ research work, sharing the knowledge bases, undertaking the joint ventures are the forms of such collaboration. Therefore, the private institutes must strive to earn the credentials to have such link ups and opening doors of international education to its students.

Open recourses

The concept of a 20th century engineering teacher, standing on a stage in front of the class and delivering the notes from his course file is becoming obsolete worldwide. To keep the students on the global track of learning, the institutes must take the initiative in the most advanced method of pedagogy and the effective use of audio-visual tools available. There are many open sources to facilitate the e-learning process, like eduX or Coursera, from which the institutes may benefit. The government of India recently has announced the opening of 200 e-learning schools under Edusat, an Indo- US project, through which the Indian students will benefit from the expertise of professors from US technology universities. The private institutes must take the efforts to keep up with the competition.

Public Private Partnership (PPP)

Like the practice of taking the projects on PPP basis, as in infrastructures, industries, the engineering education sector can also enjoy the benefits of the PPP model. The engineering colleges set up by government, working in its regulatory framework and run by private institutes could be a better option for the qualitative growth of this sector.

CONCLUSIONS

The private engineering institutes have huge potential to produce quality engineers as per requirement of industry. The government policies are always supportive to make engineering education better through private engineering institutes. Because of lack of competent teaching faculties, research and development facilities, student's interest in learning, overall skills, poor education at the primary and secondary level raise the question of employability of students passed from private engineering institutes. These factors affect admissions in private engineering institutes and the result is observed in the form of increasing vacant seats in engineering colleges.

To overcome the challenges in the near future, private engineering institutes in India must adopt the excellent practices of the accreditation complying with the set norms and procedures. The other prime steps taken in this direction must be faculty attraction and retention, enhanced industry-institute interaction, getting good mentors from premier institutes, use of open resources of learning, curriculum reforms and adopting PPP models. The government and concern authorities should also review the modes of directing and controlling the engineering education with a special attention on the strengthening of primary and secondary education.

NOMENCLATURE

AICTE	:	All India Council for Technical Education
CII	:	Confederation of Indian Industry
DTE	:	Directorate of Technical Education
FDP	:	Faculty development program
FICCI	:	Federation of Indian Chambers of Commerce and Industry
IIM	:	Indian Institute of Management
IIT	:	Indian Institute of Technology
IIIT	:	Indian Institutes of Information Technology
INR	:	Indian National Rupees
IRAHE	:	Independent regulatory Authority of Higher Education
ISO	:	International Organization for Standardization
ISTE	:	Indian Society for Technical Education
NASSCOM	:	National Association of Software and Services Companies
NBA	:	National Board of Accreditation
NIT	:	National Institute of Technology
NKC	:	National Knowledge Centre
REC	:	Regional Engineering College
SWOT	:	Strength, weakness, opportunity and threat
TTTI	:	Technical Teacher's Training Institute

REFERENCES

[1] UEC, 1983, "Report of the University Education Commission (Dec. '48 – Aug '49)," Vol. I. Second Reprint Edition, Publication No. 1396, Ministry of Education and Culture, Government of India, New Delhi.

[2] AICTE, 1999, "Technical education in independent India, 1947-1997: a compendium to commemorate the 50th anniversary of independence," AICTE, New Delhi.

[3] Prabhu, S. S., 2000, "Engineering Education in a Flux : The IIT Kanpur Story," Cover Story, volume 3, No. 5, IIT Kanpur.

[4] AICTE, 2006, "Annual Report: 2005-2006," AICTE, New Delhi, pp. 92.

[5] AICTE, 2013, Approval process handbook: 2013-14, AICTE, New Delhi, pp. 14-15

[6] R. Natarajan, "Emerging Trends in Engineering Education –Indian Perspectives", 16th Australian International Education Conference 30 September – 4 October 2002, Hobart, pp. 1-10

[7] NKC, 2009, "National knowledge commission, Report to the nation, 2006-2009," Government of India http://www.aicte-india.org/miskcr.htm

[8] RUSA, 2013, "Rashtriya Ucchatar Shiksha Abhiyan Report," Ministry of Human Resourse Department, Government of India. http://www.aicte-india.org/misrush.html

[9] Garygereffi, Vivekwadhwa, Ben Rissing, Ryanong, 2008, "Getting the Numbers right: International engineering Education in the United States, China, and India," Journal of Engineering Education, January pp .12-25

[10] Khurana, S., Shivani, 2012, " Technical Education System in India: Present Scenario," International Journal of Research in Economics and Social Sciences, Volume 2, Issue 10 , pp. 39-49

[11] Subbarao, E. C., 2013, "India's higher engineering education: opportunities and tough choices," Current Science, 104(1), pp. 55-66

[12] 2012, "Indian Higher Education Sector Opportunities aplenty, growth Unlimited," Deloitte Touche Tohmatsu India Private Limited

[13] Banerjee, R., Muley, V. P., 2009 'Engineering Education In India", Presentation at INAE Workshop at Bangalore , pp.1-30

[14] Veeranna, D. K, Joshi, A. K. Sahadeva, G. N, 2011, "Quality Issues In Engineering Education: An Indian Perspective", International Journal of Engineering Science and Technology, Vol. 3 No. 3, pp. 2044-2051

[15] http://www.business-standard.com/article/management/problem-of-plenty engineering-seats-lying-vacant-in-west-bengal-110083000071_1.html

[16] http://www.thehindu.com/news/national/karnataka/over-24000-engineering-seats-in-karnataka-vacant-this-year/article4017142.ece

[17] http://www.successcds.net/engineeringentranceexam/many-vacant-seats-engineering-colleges-country.html

[18] http://www.telegraphindia.com/1120909/jsp/7days/story_15955325.jsp

[19] http://www.engineering.careers360.com/articles/58-engineering-colleges-in-get-zero-students-where-are-we-heading.

This page left blank intentionally.

Proceedings of the ASME 2014 International Mechanical Engineering Congress and Exposition
IMECE2014
November 14-20, 2014, Montreal, Quebec, Canada

IMECE2014-37374

PROPOSAL OF "FUTURE APPLIED CONVENTIONAL TECHNOLOGY"

Chieko Narita
Kyoto Institute of Technology
Matsugasaki, Sakyo-ku, Kyoto, Japan, 606-8585
chieko.narita@gmail.com

Yutaro Shimode
Future-Applied Conventional Technology Center,
Kyoto Institute of Technology
Matsugasaki, Sakyo-ku, Kyoto, Japan, 606-8585
shimode-yutaro@xa2.so-net.ne.jp

Iwako Yamamoto
Kyoto Institute of Technology
Matsugasaki, Sakyo-ku, Kyoto,
Japan, 606-8585
ewhay@kit.ac.jp

Noriyuki Kida
Kyoto Institute of Technology
Matsugasaki, Sakyo-ku, Kyoto,
Japan, 606-8585
kida@kit.ac.jp

Hiroyuki Hamada
Kyoto Institute of Technology
Matsugasaki, Sakyo-ku, Kyoto,
Japan, 606-8585
hhamada@kit.ac.jp

ABSTRACT

Japanese traditional industry has evolved depending on demand, so that it has been contributed to Japanese manufacturing for a long time. Therefore, it is considered that learning and understanding the basis of that skill connects to learning the culture of one's own country and supporting future manufacturing. Consequently, we propose the development of human resources who can contribute not only to Japanese manufacturing, but also to world manufacturing through Future-Applied Conventional Technology. That is consistent subject—not just a unit—from elementary to high school. Its aim is to deepen a person's familiarity with and understanding of the culture and manufacturing in his or her own country. In Japan, an individual accomplishes this by learning Japanese traditional manufacturing—practical and indwelling knowledge of the predecessor in traditional industry—and developing the ability to absorb the acquired knowledge and skill through inventive ideas based on the "visiting old, learning new" concept.

1. INTRODUCTION

The preamble of the Japanese Fundamental Law of Education states the following: "we shall esteem individual dignity, and endeavor to bring up people who long for truth and justice, honor the public spirit, and are rich in humanity and creativity, while promoting education that transmits tradition and aims to create a new culture." Section V in Article 2 also states that "we shall foster respect for our traditions and culture, love for the country and region that nurtured them, respect for other countries, and a desire to contribute to world peace and the development of the international community" [1]. The value of tradition and culture in education is deemed to have increased because these sentences were added when the Japanese Fundamental Law of Education was revised in December 2006.

The National Institute for Educational Policy Research put into force the "practical study of traditional and cultural education" as a research project on the curriculum in 60 schools of the country from 2010 to 2011. Practical studies on educational guidance regarding the traditions and culture of the country were done on each subject. The point of this project was to make school children experience the traditions and culture of their own country; deepen their interest and understanding; encourage them to think a great deal about tradition and culture, and strive for wealth; and get data that can be used as reference in improving the standard of the curriculum. The key practice contents were the tea ceremony, ikebana (flower arrangement), Japanese dance, kimono (Japanese robe), and traditional crafts [2].

Furthermore, for the past few years, Japan had been moving to register Japanese food as an intangible cultural heritage in the United Nations Educational, Scientific and Cultural Organization (UNESCO); it was recognized in 2013. Japanese food is the fifth intangible cultural heritage about food; the others are French gastronomy, Mediterranean food, Turkish keskek, and the traditional food of Mexico. It is obvious from this trend that people have great interest in a particular country's tradition and culture [3].

Many practical case studies in Japan bring the local tradition industry to the classroom. [4, 5]. However, the contents of this study depend on regions, and it does not cover the whole country. Besides, Y. Tsujii pointed out the limitation imposed by time and cost. Other problems have been reported, among them the difficulty of collecting teaching materials about traditional crafts as current state and future agenda in relation to practical study on traditional crafts [6, 7].

This is surely not a simple problem, but it is thought that the Japanese identity is in traditional culture, which has been inherited from and developed through generations. Japanese can reflect on their traditional culture and transmit it to the world by solving this problem and basing traditional culture on education. In addition, learning traditional manufacturing and applying it to advanced technology are thought to contribute to contemporary and future manufacturing. What needs to be done is not only to make traditional culture part of education, but also to create an environment wherein students can learn traditional culture practically, understand skill and knowledge mutually, and improve their ability to apply what they learned to the occasion.

2. PROBLEM IDENTIFICATION

2.1 CURRENT CONDITIONS OF THE ART EDUCATION IN THE SCHOOL

A. Fujii pointed out recently that children hardly show attachment to their work in class, such as drawing and handicrafts. She also indicated that since attachment to their work is self-identification, it means they do not accept themselves and cannot accept others, as argued by V. Lowenfeld [8]. That is to say, of the number children who can become attached to the work and art of others is decreasing. From this point, A. Fujii continues the research on classes that aim to encourage children to have an attachment to regional traditional crafts. She reported that she experimentally adopted the appreciation and creation of regional traditional crafts into the class of social studies in elementary school [9] for three school hours.

This study proposes not a short-term program for a particular grade or school, but a long-term program—from elementary to high school. In the case of Japan, this spans primary school to high school; in other words, the student is 7 years old at the start and 18 years old at the end. The reason the long term is preferred is that contents would be limited and partial in a short-term program, and it may not get the interest of many children.

2.2 PROBLEMS OF JAPANESE SCHOOL SYSTEM

The problems of the Japanese school style are described as follows. Firstly, subjects from primary to high school that are related to the practical skill of manufacturing in the current curriculum are:

Elementary school:	Drawing and handicrafts *Homemaking course*
Junior high school:	Technical and homemaking course *Art*
High school:	Elective subject (music, art, crafts, and calligraphy) *Homemaking course*

Secondly, social studies or history include traditional industry, but the ratio of manufacturing and traditional industry is too low, and their relationship with other subjects is weak. Further, the contents are concluded in one or two classes, so that they have little potential for development later. Thirdly, the classification of Japanese high schools and their learning contents should be reconsidered. Japanese high school are classified into three kinds:

a.	General course:	General education (mainly)
b.	Special course:	Special education (mainly)
c.	Integrated course:	General and special (comprehensively, by elective subject)

Subjects are subdivided and most high schools are of either the general- or special-course type. In contrast, high schools in the United States or Britain have many integrated courses [10]. Moreover, manufacturing, such as drawing and handicrafts, or technical courses are not included in the general course, so that junior high school students must take special courses to learn such things.

Nevertheless, it turns out to be a big conflict when the university entrance rate and employment rate are considered. In the 2012, the Japanese university entrance rate for general-course schools was 62.8% and for special-course schools, 21%. As for the rate of employment, general-course schools had 7.8% and special-course schools, 49.9% [11]. What is also evident here is that most students who go to university or college to acquire high-level expertise are from general-course schools. In brief, this shows that most students go to the department which majors in manufacturing, such as the department of technology, without enough knowledge about manufacturing. In contrast, most students who learn manufacturing in special-course schools would rather seek employment than go to a university. The insufficiency of a long-term career education appears to have a bad influence on Japanese education.

Ministry of Economy, Trade and Industry shows a direction of training and securing core human resources at "Monodzukuri (Manufacturing)" sites by the cooperation such as manufacturing industry and universities [12]. However the existing school system would have bad influence on it.

2.3 TECHNICAL AND HOMEMAKING COURSE AND THE PERIOD OF INTEGRATED STUDY

Specifically, technical courses are excluded from the high school curriculum; only homemaking courses are included. That is to say, it is too difficult to learn practical manufacturing skills at general-course high schools today. Certainly, learning basic subjects such as mathematics, physics, and chemistry contributes to the basic knowledge of industry. However, theoretical knowledge of basic subjects with no practical application to manufacturing narrows the scope of education.

The "period of integrated study" was introduced gradually for from elementary school to high school beginning 2000, though this subject was not a direct solution to the problem. The meaning of this subject is "each school has to perform educational activity with inventive ideas based on students' interests or cross-sectional and integrated study, depending on actual the conditions of the regions, schools, and students." The decline in academic ability was addressed with the introduction of this subject, and the existence of this study was questioned, although some successful examples were reported [13]. The capacity to utilize knowledge acquired in other subjects in the period of integrated study has also been questioned.

2.4 PURPOSE OF THIS STUDY

Therefore, this study proposes "Future-Applied Conventional Technology" as consistent subject—not just a unit—from elementary to high school. However, it is not an independent and complete subject by itself; in fact, it teaches tradition and culture with manufacturing and connects them with other subjects. This subject aims to remove the problems pertaining to how current subjects relate to manufacturing and the period of integrated study.

In this study, the example of the class "Applied Conventional Technology" of H. Hamada, a professor, is introduced. Afterward, Future-Applied Conventional Technology is proposed, based on H. Hamada's class.

3. APPLIED CONVENTIONAL TECHNOLOGY

H. Hamada has been researching about experts' skill in traditional industry. He interprets it as tacit knowledge and tries to translate tacit knowledge into explicit knowledge by scientific methods. In addition, he tries to apply them to advanced technology [14-16]. Since 2010, has been conducting "Applied Conventional Technology," a class at a Japanese university, based on the results of his studies.

The aim of this class is to consider new manufacturing by understanding the methods that make tacit knowledge, which is inherent in traditional industry, into explicit knowledge, and by understanding the skill that was made into explicit knowledge. The syllabus of Applied Conventional Technology is shown in Table 1 and the manufacturing schema of Applied Conventional Technology is shown in Figure 1.

Table 1 Syllabus of Applied Conventional Technology.

Applied Conventional Technology	
Instructor	Prof.Hiroyuki HAMADA
Term	Spring
Day of the week, Period	Saturday , 1 (9:00-10:30)
Credit	2
Course Outline The aim of this course is to consider new manufacturing by understanding the method to change the tacit knowledge, which is immanent in traditional industry, into explicit knowledge by scientific methods.	
Lecture Plan	
1	The definition of "Applied Conventional Technology"
2-3	The secret of "Kyoto Bow"
4-5	The secret of "Kyoto style earthen wall"
6-7	The secret of "Kyo-gawara"
8-9	The secret of "Japannese Urushi lacquer work"
10-11	The secret of "hexagonal pattern"
12-13	Learning from traditional knowledge
14-15	Application of traditional knowledge

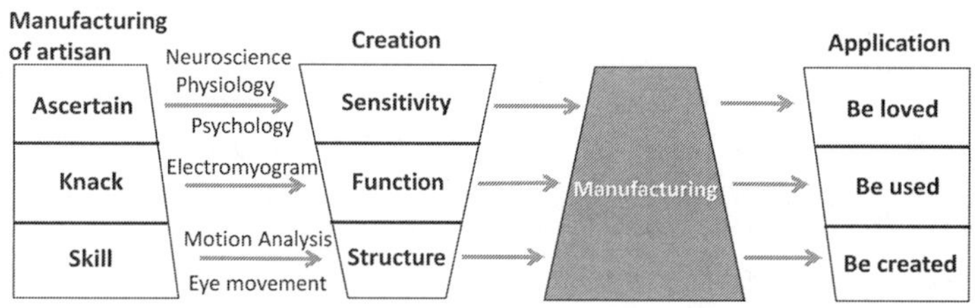

Figure 1 Manufacturing Schema of Applied Conventional Technology

The contents of the class introduce the tacit knowledge of traditional industry, such as "Kyoto bow", "Kyoto-style earthen wall", "Kyo-gawara", "Japanese *urushi* (*urushi* natural lacquer) crafts", and "*kana-ami*"; and the process of making explicit knowledge out of them. His class not only explains the outline of traditional industry, but also makes students practice the manufacturing of traditional industry. In addition, he teaches students the pleasure and difficulty of manufacturing by making them understand quantitatively the difference between their work and that of expert. This class makes students learn and understand skill and knowledge practically, so that they not only learn the traditional industry, but also acquire various ways of thinking. An example of the *kana-ami* lecture is shown in Figure 2.

Furthermore, H. Hamada explains how to apply explicit knowledge to current industry. He connects tradition to the present by showing how basic handicrafts connect to advanced technology. Merely knowing the application of manufacturing is not tantamount to understanding its essence. Talented people who understand the manufacturing of their own country and can spread their knowledge to society must be nurtured.

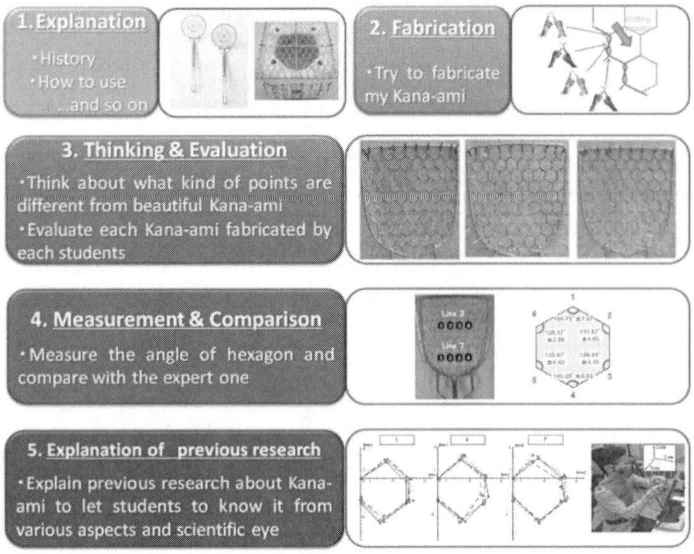

Figure 2 Lecture on Kana-ami

The number of participants in Applied Conventional Technology increases every year. This trend means that university students are highly interested in traditional industry, and it underscores the importance of thinking over how a class should be conducted. Meanwhile, a questionnaire was conducted in 2013 on 106 students regarding their familiarity with traditional crafts. About 60% said they knew the term "traditional craft" and about 30% said they had used traditional crafts in the past. The results of the survey (see Table 2) show that the nation's educational goal regarding tradition and culture has not been sufficiently achieved. Early and long-term education, such as Applied Conventional Technology, which starts at elementary school, may help solve this problem. Through it, the scope of children's interests and career-path selection in the future would be widened.

Table 2 Results of the Questionnaire on Traditional Crafts for Students in the Class of Applied Conventional Technology 2013

N = 106 (Men:76, Women:30)

	Questions	Yes	No
1	Do you know the word "traditional craft"?	66 (0.62%)	40 (0.38%)
2	Have you ever used "traditional craft" ?	31 (0.29%)	75 (0.71%)

4.FUTURE-APPLIED CONVENTIONAL TECHNOLOGY

The contents of Future-Applied Conventional Technology are described below. Its aim is to deepen a person's familiarity with and understanding of the culture and manufacturing in his or her own country. Its aim is to deepen a person's familiarity with Japan's culture and manufacturing by learning the country's traditional manufacturing in a practical manner and acquiring the indwelling knowledge of the predecessor in traditional industry. Moreover, this subject seeks to develop a student's ability to absorb the acquired knowledge and skills through inventive ideas based on the "visiting old, learning new" concept.The scheme of Future-Applied Conventional Technology is shown in Figure 3.

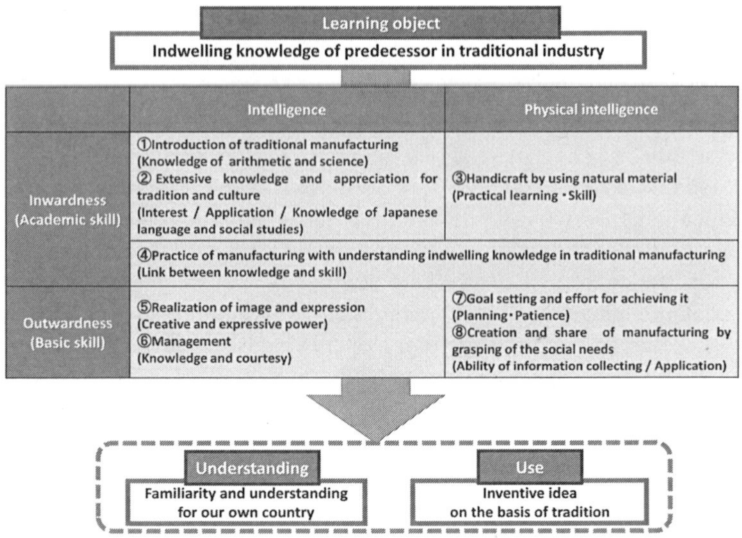

Figure 3 Scheme of Future-Applied Conventional Technology.

In this new subject, a student learns not only tradition and culture through manufacturing, but also how to relate them to other subjects. Afterward, this subject needs to establish a mutual link between intelligence and physical intelligence.

This point is considered in comparison with the problems that are described in PROBLEM IDENTIFICATION below. For example, the aim of a technical course in junior high school is to acquire the basic knowledge and techniques related to materials and processing, energy conversion, raising organisms and information through practical and experiential learning activity such as manufacturing; and to develop the students' ability and attitude so that they can evaluate and utilize technique suitably by deepening their understanding of the relationship between technique, society, and environment [17]. For instance, students can easily make woodwork by learning how to use a handsaw in class. However, as it stands, weight is attached to the mere fact that something is made. Few cases discuss the reason certain materials and tools are used for a certain type of manufacturing, how to use them properly and efficiently, or why a saw cut depends on not only the material's properties, but also how the person handles it. Additionally, even if the same materials are used, the completed works are not same, as they depend on the students' ability or sensitivity. The reason behind it should be expressed in words, and the difference should be investigated with a scientific eye based on knowledge of mathematics or science. The following sentences are described in guidelines of the technical course [18].

> Teaching methods should have a scientific basis in this learning; for example, students discuss the possible processing with the properties of materials, consider how to use them based on the structure of

tools or machines, process the materials, and so on. Teachers have to make students understand that the elements of work movement, such as posture and line of sight, how to use tools, and the distribution of power are related to the safe use tools and machines.

The school education act sets the school hours of technical and homemaking courses at 70 hours for the first and second grades, and 35 hours for the third grade. However, the limited number of school hours is to be divided between technical and homemaking courses. It is difficult to cover such content in detail when time is limited. Moreover, the technical and homemaking course is not a main subject in Japan's high school entrance examination. Thus, today, the environment in which to utilize this subject is not attained because much emphasis is put on the educational background. This appears to be related to the lack of long-term career education mentioned above.

The annual number of hours of the period of integrated study is about 70 for elementary school students in the third to sixth grades. For junior high school students, the first grade can use 50 hours and the second and third grades, 70 hours. High school students have 105–210 segments of unit time until graduation; a unit time segment equals 50 minutes [19-20]. The contents of this subject vary by school, although the same number of school hours is distributed as technical and homemaking courses. Academic ability declined when this subject was introduced and the time allotted for other subjects was reduced. Therefore, if the new subject's cooperation with other subjects is considered, the improvement of basic scholastic achievement is essential. Furthermore, the aim of the period of integrated study needs to be clarified because it has been shown that its original gist and principle has not been sufficiently achieved.

Future-Applied Conventional Technology tries to solve this problem by combining the technical and homemaking course and the period of integrated study. Strengthening manufacturing education in the period of integrated study has been proposed because there are no technical and homemaking courses in elementary and high schools. The important point is how this subject contributes to the improvement of basic scholastic achievement in its cooperation with other subjects. First, the example for elementary school students is shown in Figure 4.

The theme of this class is Japanese paper, and the students can learn about the materials and process used in making it. They experience making paper as practical learning and learn the importance of this resource from various viewpoints; in fact, they are unaware that they use it in their daily lives. It is difficult for the current period of integrated study to contribute to the increase of basic scholastic achievement because the position of the other subjects is not defined in it. The aims based on the contents that students learned in each subject are thus set in this subject. With this method, students can practice the contents of what they learned by thinking for themselves, so that it does not remain empty theory.

The example for junior high school students is shown in Figure 5.

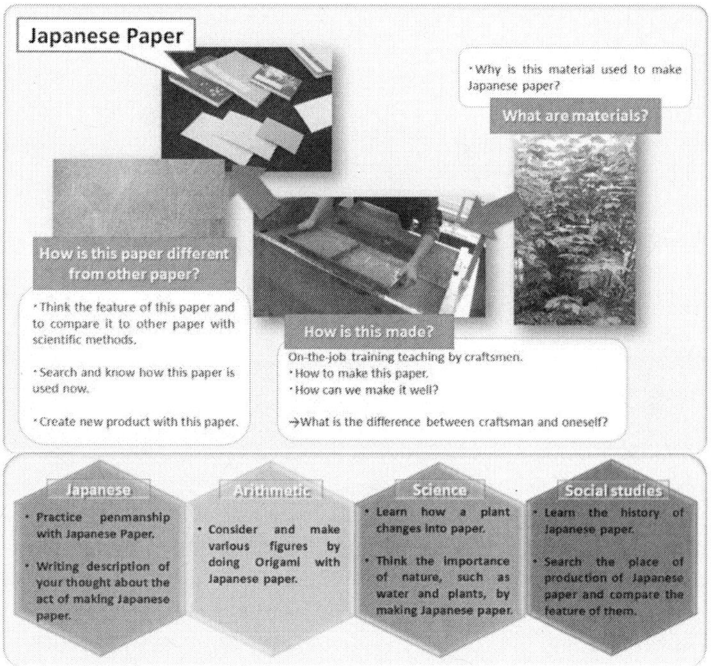

Figure 4 Scheme of the Lecture that Featured the Theme of Japanese Paper for Elementary Students

Figure 5 Scheme of the Lecture on the Kitchen Knife for Junior High School Students

The problems of technical and homemaking courses, as previously described, may be solved by combining them with the period of integrated study and making them specialize in manufacturing. The class contents in junior high school are more subdivided and become harder than those of elementary school. Besides, there are two kinds of Japanese high schools, and the lives of the students are affected by the kind that they choose, as mentioned in PROBLEM IDENTIFICATION. For these reasons, Future-Applied Conventional Technology for high school students must not only increase the basic scholastic achievement, but also expand the choices and possibilities of students. In junior high school, the aims of several subjects are to develop the students' ability to challenge themes by applying the content that they learned.

In high school, the contents of the class are determined by the kind of high school, as the students have to choose between going to the university and finding employment.

The teacher in charge of each subject and extramural lectures conducts this class. The burden for teachers, such as preparation time for the class, can be suggested as problems of the period of integrated study. In particular, the teacher in charge is set for each subject in junior high and high school, but the expert for the period of integrated study is not set. Traditional industry workers, such as craftspeople, are invited to the class and the experts' know-how of regions is included to solve this problem. It can reduce the burden for teachers and make a connection between schools and regions. In addition, teachers rarely learn it by themselves, so their career enhancement is supported through the class. Finally, this subject is held only once a month because it may progress faster than the other subjects if it were held once a week. As such, therefore, a larger amount of time can be allotted for reviewing the other subjects.

5. CONCLUSIONS

In this study, we proposed Future-applied conventional technology as a positive disappearance method to remove existing problem of Japanese schools system and the current curriculum which related to the practical skill of manufacturing. Future-applied conventional technology is not only to make traditional culture part of education, but also to create an environment wherein students can learn traditional culture practically, understand skill and knowledge mutually, and improve their ability to apply what they learned to the occasion.

This subject was proposed that it is introduced gradually for from elementary school to high school. Ericsson et al. mentioned that "deliberate practice" of more than 10,000 hours is needed to achieve expertise in various fields, such as sports [21]. He also mentioned that experts in any particular field have commonality, as they have been in an environment where they could experience the task enough from childhood. Furthermore, Nakazawa et al. researched the childhood environments of experts also found that their physical and personal environment for artistic activities was arranged in a better way than that of nonexperts [22]. Thus, a well-arranged physical and personal environment and long-term teaching are needed to raise specialty in any field.

Traditional industry has evolved depending on demand, so that it has been contributing to Japanese manufacturing for a long time. Therefore, it is considered that learning and understanding the basis of that skill connects to learning the culture of one's own country and supporting future manufacturing. Raising human resources who have skill and knowledge of manufacturing is important, instead of cramming the mind with knowledge and merely mastering things theoretically. Consequently, we propose the development of human resources who can contribute not only to Japanese manufacturing, but also to world manufacturing through Future-Applied Conventional Technology.

REFERENCES

[1] Japanese Fundamental Law of Education (Revised ver. 2006).

[2] National Institute for Educational Policy Research, (in Japan), 2011, http://www.nier.go.jp/index.html.

[3] Agency for Cultural Affairs, Press release from The Agency for Cultural Affairs of Japan, 2013, http://www.bunka.go.jp/ima/press_release/pdf/unesco_wasyoku_131205.pdf.

[4] H. Shimizu, K. Sasaki, and N. Kamiyama, 2010, Teaching of traditional culture, Utsunomiya University, 33, pp. 259-266.

[5] N. Takimoto, 2007, The Trial of the Training Lesson by Cooperation of a Traditional Industrial Studio: Work Training of the Original Kitchen Knife by Forge, (54), pp. 60-61.

[6] Y. Tsuji and S. Ishihara, 2010, The Practical Study of Handicraft Education II–Learning through Experience of Traditional Craft in the Region, Gifu University, 12, pp. 97-110.

[7] Y. Tsuji, 2007, The Practical Study of Handicraft Education I–The Introduction of Traditional Craft into the Classrooms, Gifu University, 56(1), pp. 43-52.

[8] V. Lowenfeld, 1987, Creative and Mental Growth, 8th edition, The Macmillan Company (New York), p.510.

[9] A. Fujii, 2008, Bijutsu Kyouiku, No. 291, pp. 86-93.

[10] M. Sato, 1999, Kyoiku kaikaku wo dezain suru (in Japanese), Iwanami, 5th edition, p. 201.

[11] Ministry of Education, Culture, Sports, Science and Technology, (in Japan), 2012, http://www.mext.go.jp/a_menu/shotou/shinkou/genjyo/021202.htm.

[12] Ministry of Economy, Trade and Industry, 2006, White Paper on Manufacturing Industries (Monodzukuri) , p.230.

[13] T. Fujioka, 2007, Development of Environmental Education from the Viewpoint of Period of Integrated Study, The Japanese Society of Environmental Education, 17(2), pp. 26-37.

[14] Y. Shimode, Y. Takahashi, A. Endo, C. Narita, Y. Takai, H. Nishimoto, K. Yamada, A. Goto, H. Yasunaga, and H. Hamada, 2011, Wide Angle Reflection Properties of Usushi (Japanese Lacquer) Products and Black Injection Molded Article, Proceeding of 12th Japan International SAMPE Symposium–JISSE12, TRA-2.

[15] A. Odawara, H. Sato, T. Sugei, K. Morisako, T. Shiono, S. Kitajima, A. Nakai, and H. Hamada, 2009, Kyokabe as a

Composite Material, Journal of the Japan Society for Composite Materials, 35(1), pp. 27-32.
[16] M. Tada and H. Hamada, 2002, Determining the Structure and Manufacturing Parameters of Coupled Square Braids, Journal of the Society of Fiber Science and Technology, 58(2), pp. 40-45.
[17] Ministry of Education, Culture, Sports, Science and Technology, 2008, Explanation of Courses of Study for technology and home economics, Kyoiku-tosho Co., Ltd., Tokyo, pp. 14-22.
[18] Ministry of Education, Culture, Sports, Science and Technology, 2008, Explanation of Courses of Study for elementary school – General rules – Kyoiku-tosho Co., Ltd., Tokyo, p. 43.
[19] Ministry of Education, Culture, Sports, Science and Technology, 2008, Explanation of Courses of Study for junior high school–General rules–GYOSEI Corporation, Tokyo, p. 45.
[20] Ministry of Education, Culture, Sports, Science and Technology, 2009, Explanation of Courses of Study for high school–General rules–Higashiyamashobo Co., Ltd., Kyoto, p. 29.
[21] K. A. Ericsson, R. T. Krampe, and C. Tesch-Romer, 1993, The role of deliberate practice in the acquisition of expert performance, Psychological Review, pp. 363-406.
[22] J. Nakazawa, K. Nakamichi, M. Asahina, and A. Koga, 2008, Differences between environments in young childhood of artistic expert and novice, Bulletin of the Faculty of Education, Chiba University, 56, pp. 125-130.

This page left blank intentionally.

Proceedings of the ASME 2014 International Mechanical Engineering Congress and Exposition
IMECE2014
November 14-20, 2014, Montreal, Quebec, Canada

IMECE2014-40039

CROWD-FUNDING FOR INSPIRING GRADUATE STUDENTS TO EDUCATE K12 STUDENTS ABOUT STEM

Christine Taylor
Georgia Institute of Technology
Atlanta, GA, USA

Suresh K. Sitaraman
Georgia Institute of Technology
Atlanta, GA, USA

ABSTRACT

Often when people who are not in the field hear about electronic packaging, they immediately presume that it is exclusive to electrical engineering; however, electronic packaging has opportunities for many different Science, Technology, Engineering, and Mathematics (STEM) areas. Many projects in micro- and nanotechnology are interdisciplinary in nature, and thus, a broad background of various disciplines is needed to conduct research and development in these areas. At the Georgia Institute of Technology, an initiative called the Meindl Legacy project has been created to use crowd funding to help graduate students in the nanotechnology area to create "teachable moments." The intention of the teachable moment is to broaden the research to younger audiences, so that they are inspired to take the necessary background classes needed to pursue a STEM career path. The use of crowd-funding allows for industry partners and the general public to become involved with research that is currently ongoing at the Georgia Institute of Technology and to educate K-12 students. The "teachable moment" outlined in this paper was created to demonstrate how different materials' coefficients of thermal expansion can affect the interfaces and potentially lead to cracking damage in an electronic package.

1.0 INTRODUCTION

In the late 1950s Russia launched Sputnik, and one reaction of the United States was to provide a push in mathematics and science education. This reaction can be seen in today's education system where the federal government continues to fund the standardized testing which resulted as part of the National Defense Education Act of 1958. The Act also prompted school counselors to identify talented math and science students, and encouraged the idea of "gifted" students. However, as time has passed there has been less focus on maintaining STEM advancement in school education. Now, as the generation of America's STEM workers reach retirement, another STEM worker shortage is emerging [1].

At the Georgia Institute of Technology graduate students are encouraged to do STEM outreach at local schools. In particular, the "teachable moments" that were created by the Meindl Legacy project used crowd funding to help fund graduate students in the area of microelectronics as STEM ambassadors. The "teachable moment project" allowed graduate students to have a hands-on experience in developing STEM curriculum for an audience within the ages for the levels K through 12 (K-12). The main criterion for the "teachable moment" is that it should be related to the graduate student's research experience. The "teachable moment" discussed in this paper is currently titled "Don't crack under stress" and relates to current electronic packaging research. While creating this "teachable moment," several factors were needed to be considered including:

1. Target Audience
2. Available Resources
3. Teacher and Audience Involvement
4. Practice
5. Follow up

Currently, "Don't crack under stress" is under development. It is part of the first group of ten projects to be sponsored at the Georgia Institute of Technology.

2.0 TARGET AUDIENCE

The target audience will define several key parameters including the level of the terminology, the funding involved, and the goal of the message. K-12 is a broad range of audience age, potential background knowledge, and developmental level.

2.1 "DON'T CRACK UNDER STRESS" AUDIENCE

For this project one of the intentions of the teachable moment is to expose younger audiences to various aspects of science and technology research, so that they are inspired to take the necessary background classes needed to pursue a STEM career path. The target audience was late elementary because by middle school varying levels of math are available.

A secondary-level of audience includes parents, teachers and mentors of an individual student who are shown the teachable moment module.

Lastly, there is the audience of the donors, who may be industry partners, peers, friends, or anyone who wants to support STEM outreach. The main support for this project is through crowd-funding. At the Georgia Institute of Technology, there is a crowd-funding website, *starter.gatech.edu*, which allows faculty and students post ideas to be funded by the public after a legal review process. For the website, a short video clip explains the project and motivation behind the project for the general public. There is also a link to let people fund the project and incentives for different payment amounts. Advertisement is done through media online.

The secondary audience and donors may or may not have a STEM background. However, they still have expectation about the student educational outcomes.

3.0 RESOURCES

There were three levels of resources areas to be considered:

1. Presenter Resources
2. Primary Audience Resources
3. Secondary Audience Resources

All three levels of resource areas consist of available technology, feasibility of supplies and scalability.

3.1 "DON'T CRACK UNDER STRESS" RESOURCES

For this "teachable moment," the idea was to create a hands-on project that the primary and the secondary audiences could carry out on their own. This means that the objects used in the study should be easily available and are reasonably priced that the objects are affordable to most of the students in a class or so that the school can afford to buy enough objects to reach all students in a class.

4.0 INVOLVEMENT

There are different aspects of involvement. The involvement of the presenter is the primary focus, and the presenter's relation to the audiences is the secondary focus.

4.1 "DON'T CRACK UNDER STRESS" INVOLVEMENT

The main criterion for the "teachable moment" is that it should be related to the graduate student's research experience. The lead author of this paper, a PhD student, is developing a way to monitor stresses which develop in an electronic package as shown in FIG. 1. Thermo-mechanical stresses are induced in a microelectronic package because different materials expand at different rates under thermal excursions. The material property called the coefficient of thermal expansion (CTE) governs how much a material changes in length per unit length of the material and per °C change in temperature. These thermo-mechanical stresses could result in cracking of the components and thus, the failure of the microelectronic package. Showing how different the CTE of materials can affect the interfaces in an electronic packaging and potentially lead to cracking damage was the educational goal of this "teachable moment." In addition to microelectronic packaging, thermo-mechanical stresses due to CTE mismatch are present in composite materials which are commonly used in various applications.

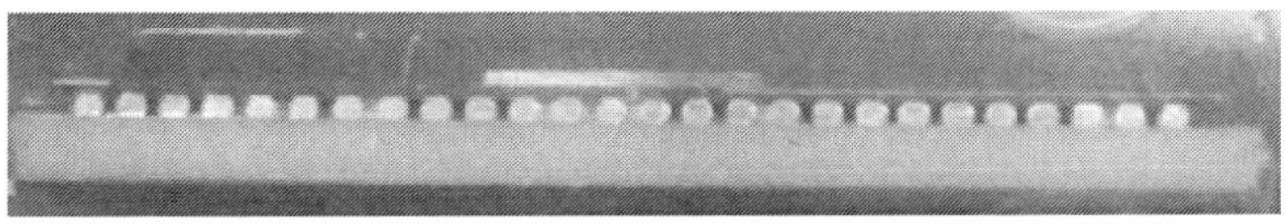

FIG. 1: Cross section of an electronic package

For the secondary audience in connection with the target student, a target area of project location is a kitchen or a place for a microwave to be plugged in. To lower the cost of supplies, an experiment was centered on a snack presentation that was analogous to an electronic packaging with several material components.

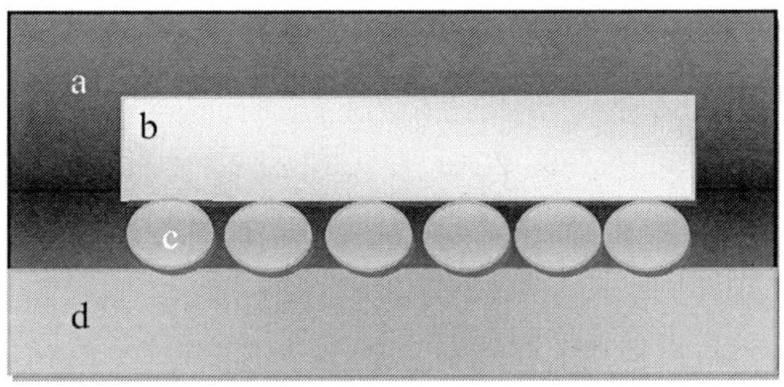

FIG. 2: Schematic of a basic electronic package a) mold compound, b) silicon, c) solder, and d) substrate

As electronic packaging technology develops newer interfaces/interconnects to get denser input/output interconnections to get better electrical performance, the material properties affect how the chip(s) perform. To simplify the model, the electronic package consists of silicon, solder, substrate, and mold compound as shown in the schematic in FIG. 2.

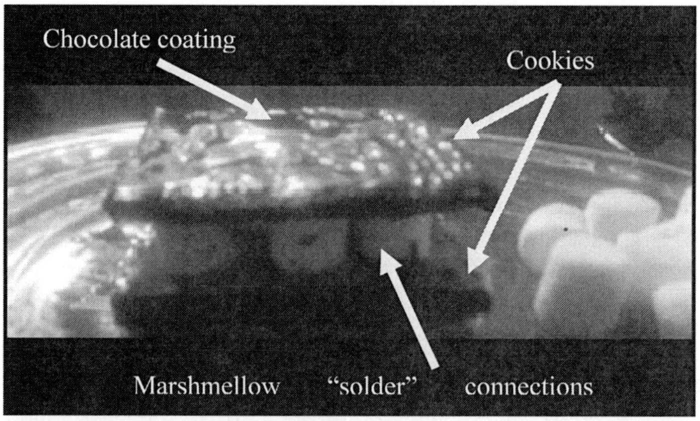

FIG. 3: "Cookie" Package before heating

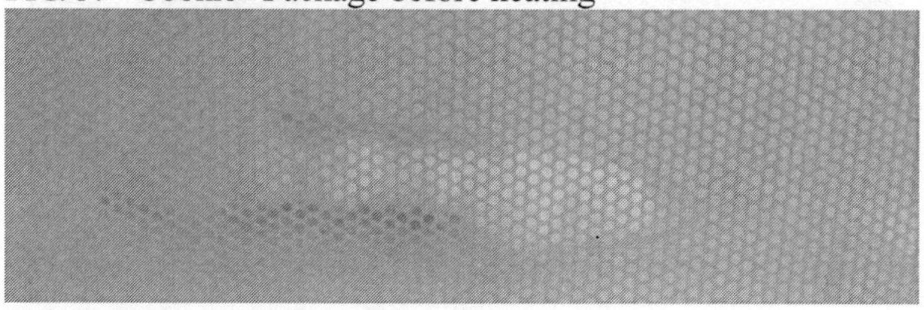
FIG. 4: Package under heat in microwave

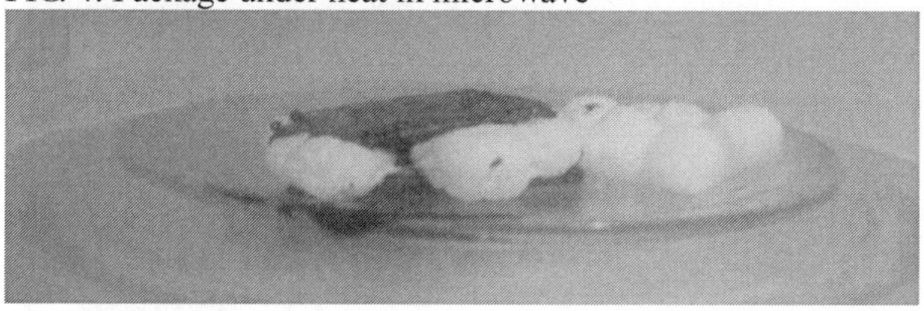
FIG. 5: Marshmallow still can be seen expanded before coming out of the microwave

In the hands-on experiment, a cookie "package" as shown in FIG. 3 could be analogous to the simplified model of an electronic package shown in FIG.2. The "outer" cookies resemble the substrate as well as the silicon layers. The marshmallows resemble the solder bumps. The chocolate coating on top of the cookie is the mold compound for the cookie package. In FIG. 2 the chocolate coating is thinly applied to the top of the cookie; however, it could be applied thicker.

Upon heating, the marshmallows expand more than the cookies as shown in FIG. 4 and FIG. 5. This could lead to the top cookie sliding off and/or potentially crack the chocolate coating if the coating has not melted. Depending on the heat source, the chocolate may or may not melt as the marshmallow expands with the heat. Thus, we will explain to the audience that the solder bump or silicon die or mold compound could crack in a microelectronic package due to the CTE mismatch. Thus, the failure induced by the CTE mismatch could be demonstrate with an in-house simple experiment using common household food items.

The snack experiment can also demonstrate general electronic packaging terminology like 2.5D/3D stacking, ball grid array, peripheral array or PoP (package on package). Students can try different configurations like stacking multiple layers of cookies and marshmallows or a package set on top of another package to see how the CTE mismatch can cause failures.

When the cost of the experiment is beyond the reach of certain students or schools, the concept of donation from public at large can be explored. To be able to attract donations, incentives, however small, can be employed. Such incentives are also intended to show appreciation to the donors. For this project, scanning electron microscope (SEM) artwork in a .pdf file format was used as an incentive for $25 or more donations. The SEM picture in FIG. 6 is an electroplated solder on a slide without a scale bar. Without the scale bar, it resembles an image taken from the space. It is particularly interesting for a "lay" person to realize that the image shows a solder ball with a diameter that is of the thickness of a typical hair. Depending on the level of donation, different incentives or encouragements may be explored. These incentives could include, for example, a visit to our lab to observe the experiments, a view the solder ball in an SEM, a tour of the cleanroom, a discussion of the project with the research students, and/or demonstration and discussion of the project to a local school of the donor's choice.

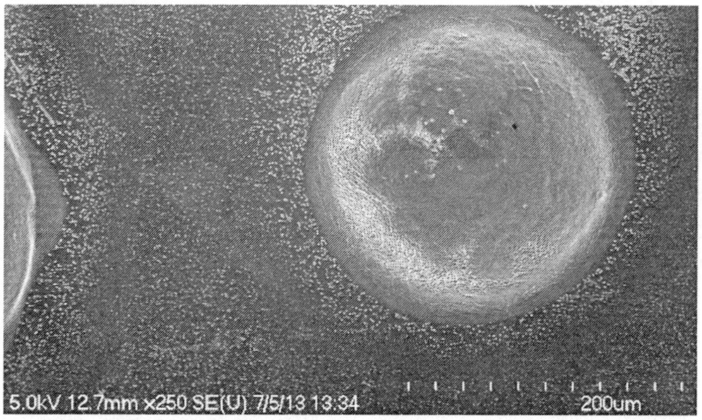

FIG. 6: Solder Solar System example of an incentive art

5.0 PRACTICE

Presentation of a technical material needs practice so that it is tailored at the appropriate level for the intended audience group. A presentation that is well-made reaches a large group of students and exposes them to STEM aspects of education. If the experiment is not put together at the appropriate level and is viewed too difficult or too abstract to grasp, the presentation could have opposite effect of discouraging students to pursue STEM education.

5.1 "DON'T CRACK UNDER STRESS" PRACTICE

Exposure to the intended audience is done through STEM outreach at Georgia Tech with already developed hands-on presentations. The STEM ambassadors attended a workshop to help with learning previously-developed demonstrations and improve their presentation skills for younger audiences. Thus far, the STEM ambassadors have reached over 900 K-12 students in the first four months with demonstrations done at local area schools' STEM nights, field trips to the Georgia Institute of Technology, local camps, and student leadership conferences.

For the "teachable moment," several videos were developed to inspire potential donors. With video editing, the best message was created to reach the donors. In the same way, the final demonstration video will be created to obtain an enjoyable learning experience clip to reach the students.

6.0 FOLLOW-UP

There are two follow-ups needed: the primary audience and the donors. For the primary audience, it is essential to make sure that the STEM message of the teachable moment experiment is understood by the K-12 students. With the donor, it is important to ensure that they are satisfied with the outcome so that future endeavors may be funded. A short survey may be conducted to assess the satisfaction of the donor.

6.1 "DON'T CRACK UNDER STRESS" FOLLOW-UP

For the primary audience it is important that the student understands the concept demonstrated: how CTE mismatch can lead to damage if a package is placed under thermal excursion. Also a student should recognize the multi-disciplinary aspect of microelectronic packaging. Many projects today are multi-disciplinary, and many different backgrounds are needed to understand a problem.

The other aspect of follow-up is with the donors. This will be done through the reward incentives given on the website. Many of the higher-level incentives are to encourage contact with the presenter either with an interview onsite or over Skype to discuss the project results.

7.0 DISCUSSION

Once the "Don't crack due to stress 'teachable moment'" is finished, there will be an online video demonstration along with instructions. The intention of the "teachable moment" is to expose younger audiences to various aspects of science and technology research, and inspire students to take the necessary background classes needed to pursue a STEM career path. The use of crowd-funding allows for industry partners and the public to become involved with the ongoing research at a university to inspire K-12 students toward STEM career path, and at the same time, helps a graduate student to be involved in creative teaching in addition to financial incentive. Efforts such as these are essential to bring together K-12 students, university students, and the public in imparting STEM education to our future engineers and scientists.

ACKNOWLEDGMENTS

The authors would like to acknowledge the Meindl legacy project for the opportunity and Georgia Tech Starter for the help with the videos and crowd funding. James Meindl is a Professor Emeritus of the School of Electrical and Computer Engineering of the Georgia Institute of Technology.

REFERENCES

[1] Jolly, J.L., 2009, "The National Defense Education act, Current STEM Initiative, and the Gifted." *Gifted Child Today*. Vol. 32, no. 2, pp 50 – 53.

Proceedings of the ASME 2014 International Mechanical Engineering Congress and Exposition
IMECE2014
November 14-20, 2014, Montreal, Quebec, Canada

IMECE2014-37376

PROBLEM SOLVING TECHNIQUES TAUGHT THROUGH VALIDATION OF AN INSTANTANEOUS RIGID FORCE MODEL

Richard B. Mindek, Jr, Assoc. Prof.
Western New England University
Mechanical Engineering Department
Western New England University
Springfield, MA 01119

Joseph M. Guerrera, Mech. Eng.
Colt's Manufacturing Company
West Hartford, CT 06110

ABSTRACT
Educating engineering students in the appropriate methods for analyzing and problem solving fundamental manufacturing processes is a challenge in undergraduate engineering education, given the increasingly limited room in the curriculum as well as the limited time and resources. Although junior and senior level laboratory courses have traditionally been used as a pedagogical platform for conveying this type of knowledge to undergraduate students, the broad range of manufacturing topics that can be covered along with the limited time within a laboratory course structure has sometimes limited the effectiveness of this approach. At the same time, some undergraduate students require a much deeper knowledge of certain manufacturing topics, practices or research techniques, especially those who may already be working in a manufacturing environment as part of a summer internship or part-time employment. The current work shows how modeling, actual machining tests and problem solving techniques were recently used to analyze a manufacturing process within a senior design project course. Specifically, an Instantaneous Rigid Force Model, originally put forward by Tlusty (1,2) was validated and used to assess cutting forces and the ability to detect tool defects during milling operations. Results from the tests showed that the model accurately predicts cutting forces during milling, but have some variation due to cutter vibration and deflection, which were not considered in the model. It was also confirmed that a defect as small as 0.050 inches by 0.025 inches was consistently detectable at multiple test conditions for a 0.5-inch diameter, 4-flute helical end mill. Based on the results, it is suggested that a force cutting model that includes the effect of cutter vibration be used in future work. The results presented demonstrate a level of knowledge in milling operations analysis beyond what can typically be taught in most undergraduate engineering laboratory courses.

INTRODUCTION
Manufacturing is a very precise process of converting raw materials components or parts into a finish product that meets customers' expectations and specifications. Manufacturing commonly employs a machine setup with a division of labor in a large scale production setup. These processes, such as milling, turning, refinishing, etc., have gone from a hands-on job requiring a large amount of skill to an automated process, which can reproduce results to tolerances of 0.001 inches or less. Although these processes have been automated, maintenance and oversight of the operations are still needed to ensure that the machines are running properly and remain within the set tolerances for the particular job being executed.

With any process of converting raw materials into a finished product there is wear that occurs on machine components and specifically the cutting tools that are used to shape the material. This slow degradation of the tool can cause the tolerances on the machining process to become larger than required and can also cause damage to the machined part if the tool becomes damaged or broken in any way. In a manufacturing process where automation is a key role of the process, machines need to be able to sense tool degradation and damage so that defectively machined parts will not be produced, wasting material, time, and money.

The objective of this work is to continue the work that a former engineering student, John P. Fontaine, pursued in the spring of 2008. Fontaine's work validated, in part, milling theory for end mills (1,2) with helical cutting teeth by measuring cutting forces using a force dynamometer. He was also able to detect defects that had purposely been placed on a cutting tool (3). The current work uses conventional milling and force measurement techniques to further validate the milling theory developed by Tlusty (1,2). Subsequent to this, additional work is performed to identify cutting force signal differences in defect-free versus defective cutting tools, with a specific goal of determining what size defect can be detected. The results of this work could be used, for example, to alert a machine operator that a cutting tool needs to be changed before part damage occurs. Although not novel, this work demonstrates a pedagogical approach, that is, using an experimentally-based senior design project, to teach machining concepts, beyond what a typical undergraduate engineering lab course could effectively teach, for educating students in the area of machine process analysis and monitoring.

INSTANTANEOUS RIGID FORCE MODEL

A theoretical approach developed by Tlusty (1,2) that describes how forces on end mills with helical teeth are calculated is known as the "Instantaneous Rigid Force Model" (IRFM) because it does not take into account cutter deflection or vibration. From this theory a plot of the cutting force versus the angle of cutter rotation can be made, as shown in Figure 1.

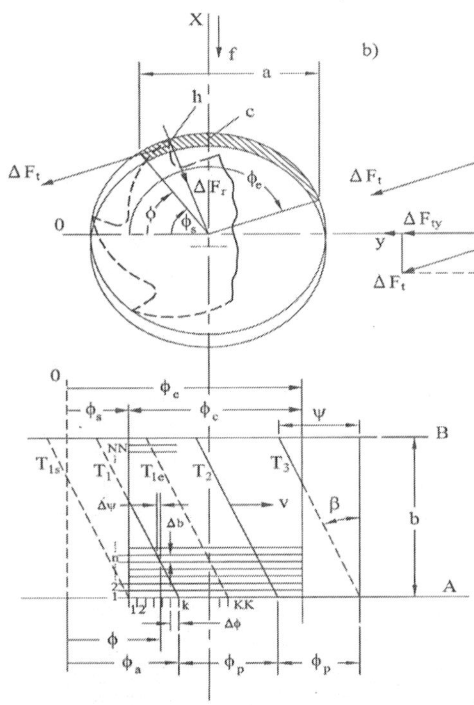

Figure 1: Cutting Model Diagram (1)

Considering Figure 1 in more detail, an element is shown of an end milling cutter with a radial depth "a" and an angle of tooth engagement "φ" measured from a "zero" reference position. The axial depth of cut is "b" and the feed direction of the work piece into the cutter is "f." The engagement of the cutter in the work piece, ranges from φ_s to φ_e. The cutting angle is defined as:

$$\phi c = \phi e - \phi s \tag{1}$$

where:

ϕe= Ending Angle Of Tooth Engagement, radians

ϕs= Starting Angle Of Tooth Engagement, radians

The image at the bottom of Figure shows an unrolled view of the cutter and the helix angle of the cutter is depicted as β. The engagement of a tooth can be defined by the following equation:

$$\psi = \frac{b * \tan(\beta)}{Cutter\ Radius} \tag{2}$$

where:

ψ= Tooth Engagement Angle, Radians

Since the actual cutting engagement of a tooth exists only between φ_s to φ_e there may be multiple teeth cutting simultaneously depending on the cutting conditions. The number of teeth that may be cutting at one time is defined by the following equation.

$$MM = round\left(\frac{\phi c + \psi}{\phi p}\right) + 1 \tag{3}$$

where:

ϕp= The Pitch Angle

$$\phi p = \frac{2\pi}{Number\ of\ Teeth} \tag{4}$$

The cutting forces are periodic within a "tooth period." This corresponds to a rotation by the pitch angle. Because of this the leading point of tooth T_1 at level A is just entering the cut at $\varphi_a = \varphi_s$ and the motion of this tooth can be examined until φ_a to $\varphi_s + \varphi_p$. Tooth T_1 will move through this range from the position shown as T_{1s} to the position T_{1e}. The teeth T_2 and T_3 move in a similar manner, but are always φ_p and 2 φ_p ahead, respectively. The motion of the teeth through the range of φ_p is incremented by Δϕ, which is defined as:

$$\Delta\phi = \frac{\phi e - \phi s}{1000} \tag{5}$$

Using the flowchart from Tlusty (1), Δϕ is incremented with the count K for KK steps, where KK is defined as:

$$KK = round\left(\frac{\phi p}{\Delta\phi}\right) \tag{6}$$

This flowchart considers every edge incrementally with the count N in a total of NN elements where NN is selected based on the resolution desired. For the purposes of this work, NN = 1000 was used. With these elements spread over $\Delta\psi$ and having a height of Δb, these parameters are defined as:

$$\Delta\psi = \frac{\psi}{NN} \quad (7)$$

$$\Delta b = \frac{b}{NN} \quad (8)$$

Using this approach, a computer program was constructed in MATLAB (v R2007a) to simulate the forces on an end mill with helical teeth. This program is shown in its entirety by Guerrera (4).

During the calibration of the force dynamometer the following equation was used to determine the voltage gains and amplifications.

$$V = F * G * V_{exc} * Sens * 10^{-6} \quad (9)$$

where:
V=Output Voltage (Volts, 10V Max)
F= Load to be Measured (pounds)
G=Gain (1k, 2k, 3k, 4k)
V_{exc}= Excitation Voltage (2.5V, 5V, 10V)
Sens= Sensitivity (micro-volts/V_{exc}-lb)

The cutting tool (end mill) was modeled as a cantilevered beam in order to determine its stiffness, given as equations (10) and (11) below. However, Tlusty (1,2) assumes the force portion of the equation and the deflection portion be treated as constants.

$$\delta = x = \frac{PL^3}{3EI} = \frac{P}{k} \quad (10)$$

$$k = \frac{3EI}{L^3} \quad (11)$$

where:
E= Modulus of Elasticity (psi)
I= Moment of Inertia (in^4)
L= Length of the tool protruding from the machine collet (in)
k= Stiffness of the cutter (lb/in)

The natural frequencies of a cantilevered beam for a uniformly distributed mass along the beam for the first, second, and third harmonic frequencies, are given by:

$$fn_1 = 0.56\sqrt{\frac{g_c EI}{\rho A L^4}} \text{ (First Harmonic)} \quad (12)$$

$$fn_2 = 3.507\sqrt{\frac{g_c EI}{\rho A L^4}} \text{ (Second Harmonic)} \quad (13)$$

$$fn_3 = 9.895\sqrt{\frac{g_c EI}{\rho A L^4}} \text{ (Third Harmonic)} \quad (14)$$

where:
E= Modulus of Elasticity (psi)
I= Moment of Inertia (in^4)
L= Length of the tool protruding from the machine collet (in)
ρ= Density
gc= gravity constant (in/s^2)
fn= Natural frequency (Hz)

VALIDATION OF THE CUTTING FORCE MODEL

The next portion of the project involved repeating cutting tests that Fontaine (3) performed to verify both the current work, as well as the Instantaneous Rigid Force Model itself. For these cutting tests a 4 flute, ½ inch diameter, helical, high speed steel end mill was used. The test matrix in Table 1 shows the conditions tested. It consists of six total tests with two different machining conditions. These machining conditions are representative of the machining conditions used by Tlusty when validating his machining (milling) model and a portion of the tests previously run by Fontaine.

Table 1: Test Matrix For Rerun of Previous Work

Cutting Parameters					
Test	**Radial Immersion (inches)**	**Depth of Cut (inches)**	**Cutter RPM**	**Feed (ipm)**	**Feed (ipt)**
1	0.125	0.25	2000	30	0.0038
2	0.125	0.25	2000	30	0.0038
3	0.125	0.25	2000	30	0.0038
4	0.25	0.25	2000	30	0.0038
5	0.25	0.25	2000	30	0.0038
6	0.25	0.25	2000	30	0.0038

This testing was started by first securing the force dynamometer to the bed of the machine, as shown in Figure 2.

Figure 2: Work Piece Mounted to Dynamometer

An aluminum work piece (6061-T6), previously cut with steps on either side, was then secured to the dynamometer and aligned to the longitudinal path of the cutter (x-axis), as shown in Figure 8. The dynamometer was then connected to an amplifier and the amplifier output was connected to a data acquisition system. During cutting tests data was recorded on three channels (F_x, F_y, and F_z). Since a cutter speed of 2000 rpm was used during testing, a sampling rate of 5 kHz was chosen when collecting data in order that the passing of each flute cutting force could be recorded. Data was taken for six trials using the same cutter with two different test parameters used. During these cutting tests coolant was applied to the work area. In order to minimize the force that the coolant might impart to the dynamometer, the flow rate was kept to a minimum and was directed on the cutter as opposed to directly hitting the work piece attached to the dynamometer, which might have given a false vertical (z-axis) cutting force reading.

Once the tests were completed, the cutting force data was plotted for each set of conditions given in Table 1. Figure 3 is a plot of the cutting force data versus one full cutter rotation for tests 1 through 3 from Table 1. Overlaid on this data, at the same test conditions shown in Table 1, are previous cutting force data taken from Fontaine (3), as well as Tlusty's IRFM, described in the previous section. Note that the force values for the IRFM were actually obtained using a MATLAB program of the IRFM developed by Fontaine (3). The plot itself shows reasonable qualitative agreement between Tlusty's model and both sets of experimental cutting force data, including the four peaks in force magnitude (one for each flute pass through the work piece) and the general shape expected from the theoretical model.

Figure 3: Tests 1-3, 0.25" DOC, 0.125" Rad Immer., 0.5 dia., 4 flute end mill

Since the experimental data taken does show some disagreement with the cutting force model in terms of force amplitude variation, several rotations of the cutting force were plotted to assess whether or not this variation had an identifiable pattern. Figure 4 shows an apparent once per cutter revolution peak in cutting force. Since Tlusty's model does not consider cutter deflection or vibration, additional testing with higher stiffness end mills was conducted in an attempt to identify the source of the cutting force variation.

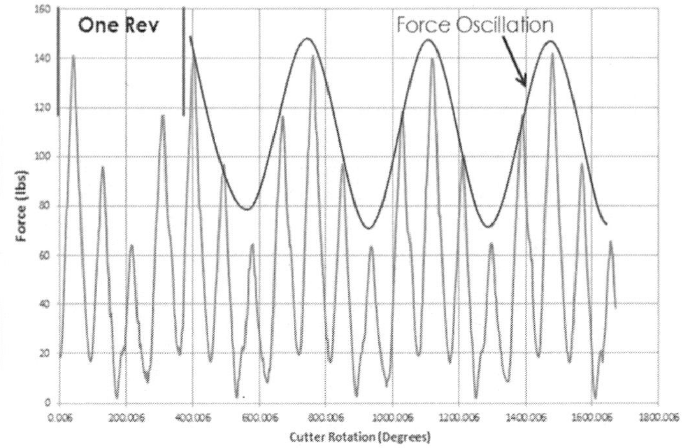

Figure 4: Force Oscillation Plot, 0.25" DOC, 0.125" Rad Immer., ½ inch dia., 4 flute end mill

One Inch Diameter End Mill Testing

During the next phase of testing using a stiffer (1 inch diameter) end mill, a "Kurt" Vise was mounted on top of the dynamometer to allow the work piece to be taken on and off of the machine quickly and without repeated referencing to the machine. A photo of the new set-up used is shown in Figure 5.

With the change in the setup, and an unknown source of cutting force variation, a frequency response test was performed to identify any resonances that might be occurring from the machine vise or force dynamometer. This was done by imparting an excitation to the center of a mounted work piece with a rubber hammer, while using the data acquisition system to record the force output signal from the hammer impact. Figure 6 shows a Fast Fourier Transform (FFT) plot of the force output data from this test. Note that the fundamental frequency of the dynamometer, vise, work piece assembly occurs at about 275 Hz, well above the cutter rotational frequencies used during all machining tests. Therefore, fixturing is not the likely cause of the cutting force variation observed previously.

Next a second set of milling tests was performed using a 1 inch diameter, 4 flute, helical end mill. In these tests variation of radial depth of cut (immersion) and depths of cut, along with different sampling rates, were investigated. The conditions tested are shown in Table 2. Note that data was taken at sampling rates of 5, 10 and 25 kHz for each test condition, to ensure that any high frequency characteristics of the signal were not aliased. Changes in radial immersion and depth of cut were included so that a variety of comparisons could be made to the theoretical model.

Figure 5: Revised Cutting Test Setup

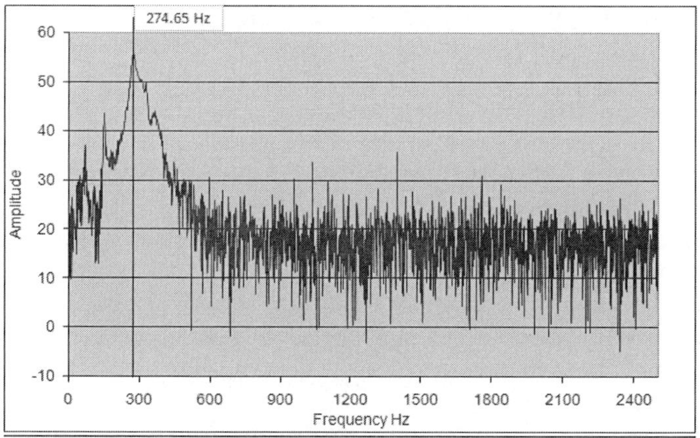

Figure 6: Amplitude vs. Frequency Plot for Test Setup

Once the tests were completed, the cutting force data was plotted for each set of conditions in Table 2. Figure 7 is a plot of the cutting force data versus one full cutter rotation for tests 7 through 9 from Table 2. Overlaid on this data is the cutting force predicted by Tlusty's cutting force model at the same test conditions as shown in Table 2. Figure 7 shows general agreement with Tlusty's model in terms of frequency of the force peaks. However, agreement with the predicted cutting force magnitude is marginal. In Figure 8, where more cutter

Table 2: Cutting Parameters for Test Set #2

Cutting Parameters Test Set #2				
Test	**Cutter Dia. (inches)**	**Radial Immersion (inches)**	**Depth of Cut (inches)**	**Feed (ipt)**
7	1	0.125	0.125	0.0038
8	1	0.125	0.125	0.0038
9	1	0.125	0.125	0.0038
10	1	0.125	0.25	0.0038
11	1	0.125	0.25	0.0038
12	1	0.125	0.25	0.0038
13	1	0.25	0.25	0.0038
14	1	0.25	0.25	0.0038
15	1	0.25	0.25	0.0038

rotations are plotted, shows that peaks that should represent the passing of one flute through the work piece are instead modulated by a higher frequency. One possible explanation for this modulation is that the higher stiffness of the cutter, relative to the first set of tests (that is, one inch diameter cutter versus 0.5 inch diameter cutter) is causing the end mill to rebound almost completely off of the work piece due to the initial impact of the cutter, and then recover within the time that the same flute is passing through the work, causing what is essentially an oscillating force deflection in the cutter, or chatter. If this is true, cutter stiffness (varied here using cutter diameter), should change the shapes of the cutting force peaks. In comparing Figures 3 and 7, it is evident that the shape of the cutting force peaks has changed, and is likely due to the change in cutter stiffness. It should also be noted that no significant change in the shape of the cutting force signal was observed when comparing data taken at the same cutting conditions for

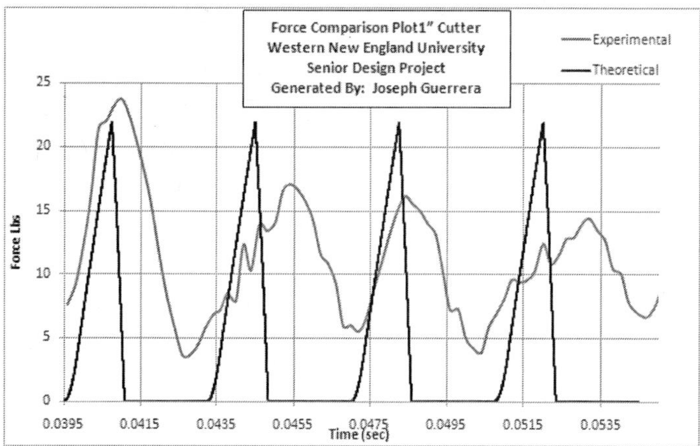

Figure 7: Tests 7-9, 0.125" DOC, 0.125" Rad Immer., 1 inch dia. 4 flute, end mill, sampling rate of 5 kHz

sampling rates of 5, 10 and 25 kHz. Therefore the sampling rate was kept to a minimum at 5 kHz for subsequent testing.

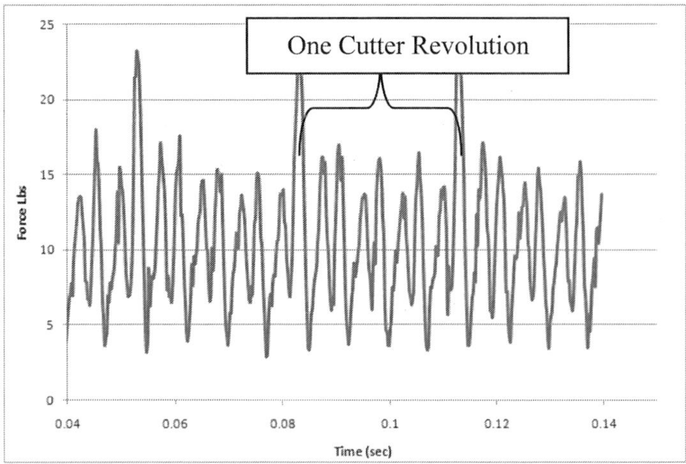

Figure 8: Tests 7-9, 0.125" DOC, 0.125" Rad immer., 1 inch dia., 4 flute, end mill, sampling rate of 5 kHz

Again, since the Tlusty model used does not account for cutter deflection or vibration, it is possible that the end mill was chattering while cutting through the work piece. In order to investigate this issue further, another set of cutting tests was performed using a 0.25 inch diameter, 4-flute, end mill. Stiffness calculations of each end mill diameter were also made, and then compared to their respective cutting force magnitudes to determine whether or not cutter stiffness was, in fact, the cause for the observed changes in cutting force magnitude. This comparison is shown later in this paper.

0.25 Inch Diameter End Mill Testing

The third set of testing used a 0.25 inch diameter, 4-flute, helical end mill. In this set of tests, radial immersion and depths of cut were varied, as shown in Table 3, so that different comparisons could be made to the Tlusty force model. The sampling rate used was again 5 kHz for each test.

Table 3: Cutting Parameters for Test Set #3

Cutting Parameters Test Set #3				
Test	**Cutter Dia. (inches)**	**Radial Immersion (inches)**	**Depth of Cut (inches)**	**Feed (ipt)**
16	0.25	0.0625	0.125	0.0038
17	0.25	0.125	0.125	0.0038
18	0.25	0.0625	0.25	0.0038
19	0.25	0.125	0.25	0.0038

Once the tests were completed, the cutting force data was plotted for each set of conditions in Table 3. Figure 9 is a plot of the cutting force data versus one full cutter rotation for Test 18 from Table 3. Overlaid on this data is the cutting force predicted by Tlusty's cutting force model at the same test conditions as shown in Table 3. Figure 9 shows reasonable agreement with Tlusty's model in terms of both frequency of force peaks and cutting force magnitude. However, some variation, that is, high frequency modulation of the force within each peak, still exists. Why this is present is not completely clear, although it is apparent that changing cutter stiffness, by using a change in cutter diameter, does significantly change the shape of the cutter force.

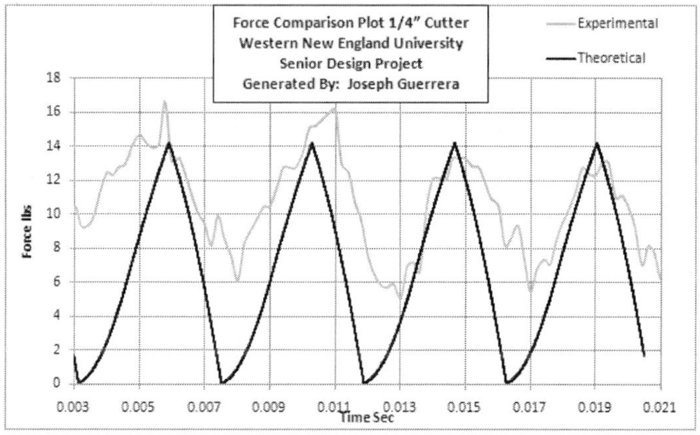

Figure 9: Test 18, 0.25" DOC, 0.0625" Radial Immer. 0.25 inch dia., 4 flute, end mill, 5 kHz

Natural Frequency and Stiffness Comparisons

Due to the frequency modulation present in the force signals of the 1 inch and 0.25 inch end mills it was decided to investigate the natural frequency of the cutters in relation to cutter stiffness, as previously discussed. First, the frequencies for each of the three cutter diameters was calculated in terms of both the once per revolution passing frequency for each tooth and the frequency associated with the passing of four flutes per revolution. Also calculated were the 1st, 2nd, and 3rd harmonic frequencies of the end mills, modeling each as a cantilevered beam with a distributed mass equal to the respective cutter mass. These frequencies were then compared to the natural frequency of the entire setup (Figure 6), from the frequency response test, and the natural frequency of the dynamometer given by the manufacturer. A summary of these results is shown in Table 4. Note that none of the frequencies associated with end mill rotation are within the bandwidth of the first, second, or third harmonic frequency of the end mill, or the natural frequency of the dynamometer setup itself. Although these calculations did not show a direct correlation to variations in the force data presented previously, they did confirm that there is no obvious resonance occurring. Further analysis of the setup should be done to confirm these results and find other frequencies that may be modulating with the machining forces.

Table 4: Natural Frequencies and Stiffness Data

	1/4" End Mill	1/2" End Mill	1" End mill
One tooth rev (Hz)	53.33	33.33	16.66
4 teeth rev (Hz)	213	133	66.5
fn1 (Hz)	4918	3004	8580
fn2 (Hz)	30804	18815	53733
fn3 (Hz)	86913	53087	151608
Stiffness k (kip/in)	7.167	19.238	2192.5

Also note the cutter stiffness calculated for each cutter in Table 4. From these calculations it can be seen that the stiffness of the one inch end mill is extremely high, approximately 2 million lbs/inch, relative to the very low stiffness of the 0.25 inch cutter of only about 7,000 lb/inch. It's clear from these values that cutter stiffness should be considered in the force model, since it likely has a significant impact on cutting forces, as suggested by the previous experimental data. Further testing with the use of a visual aid, like a high speed camera, might reveal more detailed results.

IDENTIFICATION OF TOOL DEFECTS

A number of tests were conducted in order to determine how small of a defect can be detected using a cutting force signal. First, baseline tests were conducted with a cutter having no defects so that a baseline force curve could be plotted and later compared to the force signal obtained from tools with cutter defects. For these baseline tests, a vertical depth of cut was fixed at 0.125 inches, which guaranteed that defect sizes less than that value would remain immersed as the flute passes through the arc of cut within the work piece. Two levels of radial immersion were tested to better show the effect of force magnitude on defect detection. The cutting parameters used for the baseline tests are shown in Table 5 below.

Table 5: 0.5 Inch End Mill Baseline Tests

Cutting Parameters Defect Comparison Tests				
Test	**Radial Immer. (inches)**	**Depth of Cut (inches)**	**Feed (ipt)**	**Defect Size**
28	0.125	0.125	0.0038	no defect
29	0.25	0.125	0.0038	no defect

A base line cutting force plot for the 0.5 inch, 4-flute, defect-free end mill is shown in Figure 10 below. Note the relatively constant peak force for each of the 4 flutes during the single cutter rotation shown in Figure 10. Also note that each peak is essentially smooth and free of any discontinuities in force.

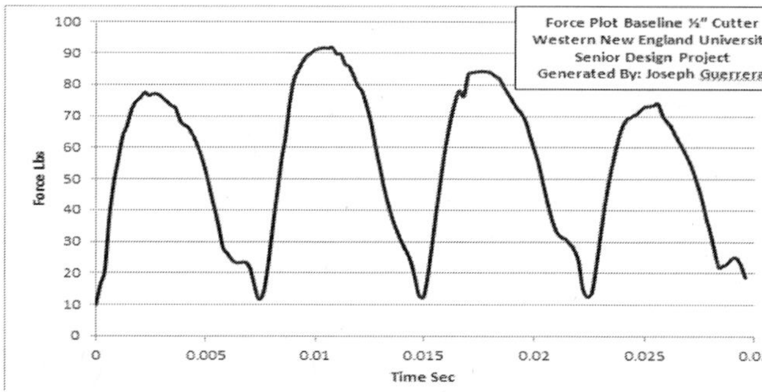

Figure 10: Test 29, 0.125" DOC, 0.25" Radial Immer. 0.5 inch dia., 4 Flute, end mill (No Defect)

Single Flute Defect Testing

For testing with tools having defects on the cutting flutes, two different sets of tests were performed. The initial testing used a single defect on a single flute, beginning at the tip of the end mill, to best simulate a defective end mill that had been chipped during milling operations or had been dropped by the operator and chipped from the impact. Two different size defects, 0.05 and 0.1 inches by 0.025 inches, were used, each tested at the same radial immersion as the baseline tests. This allowed for comparison of the cutting force signals to show any difference in the signal dependent on the presence and size of the defect. The defects were ground into one of the flutes of the cutting tools using a small grinding wheel mounted on a hand-held tool. Figure 11 shows the end mills with no defect (left) and with a ground 0.1 inch defect (right), before machining. Table 6 shows the test parameters used for testing with the single defect tools. Note from Table 6 that four milling trials were conducted for each of the single defect tools, including two tests each at radial immersions of 0.125 inches and 0.25 inches.

Figure 11: End Mill Showing No Defect (left) and 0.1 inch Defect (right), respectively, 20X

Table 6: Single Defect Testing Parameters

Cutting Parameters Single Defect Test Set #1					
Test	**Rad Immer. (inches)**	**Depth of Cut (inches)**	**Cutter RPM**	**Feed (ipt)**	**Defect Size**
20	0.125	0.125	2000	0.0038	0.050
21	0.125	0.125	2000	0.0038	0.050
22	0.125	0.125	2000	0.0038	0.1
23	0.125	0.125	2000	0.0038	0.1
24	0.25	0.125	2000	0.0038	0.05
25	0.25	0.125	2000	0.0038	0.05
26	0.25	0.125	2000	0.0038	0.1
27	0.25	0.125	2000	0.0038	0.1

Figure 12 shows the cutting force results of the tool with no defect, overlaid onto the cutting force results of the tool with a 0.1 inch defect. From these results it is apparent that each peak force represents a flute passing through the work piece. Note that the tool without the defect shows the peak force remains essentially constant as each flute passes through the work piece. On the other hand, the tool with the defect shows a significant reduction in cutting force once per revolution (once every four peaks), immediately followed by a significant increase in cutting force from the flute right after the defective flute. This portion of the signal was present in the 3rd and 4th peak of every cycle, demonstrating that as the defective flute made its path through the work piece there was a force reduction on the defective tooth, and a subsequent reduction in material removed, which was then removed by the flute following the defective flute, thus increasing the force seen by that flute by up to 60%.

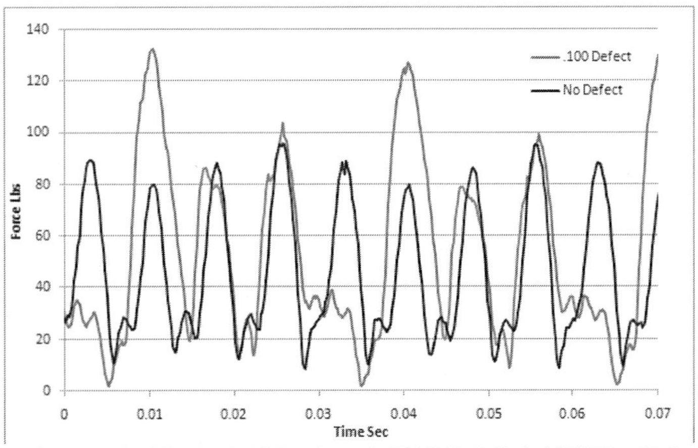

Figure 12: Single 0.1" Defect, 0.125" DOC, 0.125" Radial Immersion

Tests with Adjacent and Opposing Flute Defects

The second set of defective tests were done to further substantiate the results found in the single flute defective testing and to demonstrate the effects on cutting force when the defects exist on adjacent or opposing flutes of a four-fluted cutting tool. For these tests the depth of cut was held constant at 0.125 to assure the entire defect was within the depth of cut. Two different radial depths of cut (radial immersions) were used for both the 0.05 inch and 0.1 inch sized defective tools. All other parameters were kept the same, relative to the baseline tests with no defect, for comparison purposes. Table 7 shows the testing parameters used for this set of machining trials.

Table 7: Adjacent and Opposing Defect Testing Parameters

Cutting Parameters Opposing/Adj. Defect Test Set #1					
Test	**Rad Immer. (inches)**	**Depth of Cut (inches)**	**Feed (ipt)**	**Defect Size**	**Defect Orient.**
30	0.125	0.125	0.0038	0.050	Opp.
31	0.125	0.125	0.0038	0.050	Adj.
32	0.25	0.125	0.0038	0.050	Opp.
33	0.25	0.125	0.0038	0.050	Adj.
34	0.125	0.125	0.0038	0.1	Opp.
35	0.125	0.125	0.0038	0.1	Adj.
36	0.25	0.125	0.0038	0.1	Opp.
37	0.25	0.125	0.0038	0.1	Adj.

The first tests performed were with defects ground onto two adjacent (successive) flutes of a 4-flute cutter. From the test results shown in Figure 13, it is apparent that the adjacent defects are prominent. Specifically, during the 4-peaks per revolution, corresponding to the 4-flutes, there are two missing

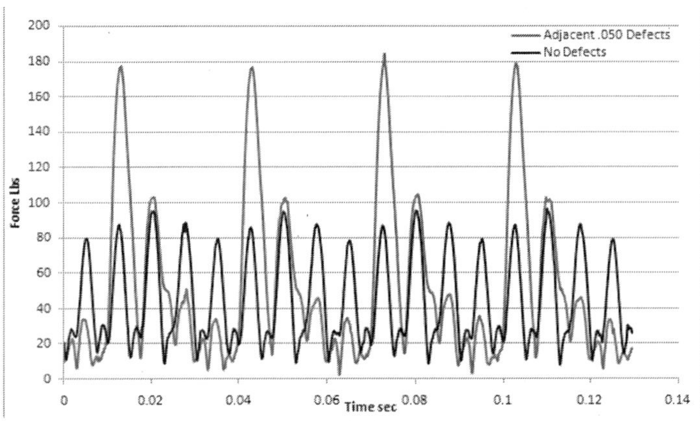

Figure 13: 0.05" Adjacent Defects, Test 31, 0.125" DOC, 0.125" Radial Immersion

peaks, where the defective flutes enter and leave the work piece. After the two missing peaks is a single peak of approximately 55% greater amplitude than nominal, followed by a peak with close to the expected nominal force value. The reason for the large increase in force in the third peak, following the two missing force peaks in each cycle, is likely due to the excess material removed by the third flute, which the first two flutes in each cycle missed.

The last set of machining tests used a tool with a single defect ground on opposing flutes of the end mill. This set of machining tests was done and compared to the single defect tests. The test results, shown in Figure 14, exhibit the same effects observed in both the single defect and adjacent defect tests. Specifically, this figure shows the defective flute experiences a greatly reduced cutting force as it passes through the material, but is immediately followed by a flute without a defect that experiences a significantly increased load relative to the no defect tool. Again, this is likely because the flute immediately following the defective flute is removing extra material left from the defective flute. The flute that follows the defective flute experiences approximately 64% more cutting force relative to the no defect tool cutting force, confirming that additional material is removed by the flute that follows the defective flute.

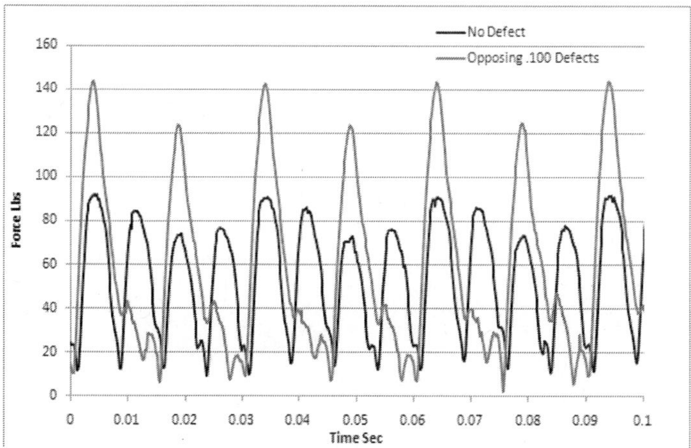

Figure 14: 0.1" Opposing Defects, Test 36, 0.125" DOC, 0.25" Radial Immersion

CONCLUSIONS

The following conclusions can be made from the results of this work:

(1) The Tlusty Instantaneous Rigid Force Cutting Model can be used to predict cutting forces during milling operations. However, the accuracy with which the forces can be predicted is significantly affected by effects of tool deflection and vibration.

(2) Tool defects as small as 0.050 inches by 0.025 inches can consistently be detected by measuring or monitoring the cutting force during machining.

(3) During machining tests conducted with defective tools, the peak cutting force signal generally increases following the flute with the defect. This likely occurs due to the additional material removed by the flute following the flute with a defect.

(4) Using senior design projects as a platform for teaching more advanced machining concepts can be an effective pedagogical tool.

ACKNOWLEDGMENTS

The authors would like to thank Mr. Richard DeCelle for assisting with parts ordering and Mr. Peter Bennett for helping to set-up and run the machining trials.

REFERENCES

(1) Smith, S., and J. Tlusty. "**An Overview of Modeling and Simulation of the Milling Process.**"*Journal of Engineering for Industry* 113.2 (1991).

(2) **Tlusty, George.** *Manufactuing Processes And Equipment.* New Jersey : Prentice Hall, 2000.

(3) **Fontaine, John.** "The Use of a Force Dynamometer to Validate Cutting Theory in Milling and Identify Tool Defects." Senior Design Project Report, Western New England University. Spring 2009.

(4) **Guerrera, Joseph M..** "Identification of Tool Defects Using a Force Dynamometer During Machining Operations." Senior Design Project Report, Western New England University. Spring 2013.

(5) **Kalpakjian, Serope and Schmidt, Steven R.** Manufacturing *Engineering and Technology*. New Jersey: Prentice Hall, 2001.

(6) Kim, Kug W. "**Predicting Cutting Forces in Face Milling with the Orthogonal Machining Theory.**" *INTERNATIONAL JOURNAL OF PRECISION ENGINEERING AND MANUFACTURING* 6.3 (2005).

This page left blank intentionally.

Proceedings of the ASME 2014 International Mechanical Engineering Congress and Exposition
IMECE2014
November 14-20, 2014, Montreal, Quebec, Canada

IMECE2014-38516

THE USE OF PATENTS DOCUMENTATION INFORMATION IN TEACHING, ACADEMIC RESEARCH AND DEVELOPMENT OF ENGINEERING PROJECTS

Jeziel Nunes
National Institute of Industrial Property Academy
Rua Mayrink Veiga, nº 9, sala 1803
Rio de Janeiro, RJ, Brazil
jezielnunes@gmail.com

Marcia Batalha
Federal University of Rio de Janeiro – COPPE
Av. Pedro Calmon s/n – Ilha do Fundão
Rio de Janeiro, RJ, Brazil
marciabatalha.nunes@gmail.com

ABSTRACT

The present work aims to demonstrate the use of information contained in patent systems for teaching and academic research in mechanical engineering and related fields. The approach proposed herein is focused on academic research and the engineering disciplines to provide a more complete training of students and researchers. In the proposed approach, the Patent System is used in three different ways: As a source of technical information; as a source of inspiration for designing solutions; and for patent protection of academic research and their eventual commercialization. The patent system also enables teachers to keep up with technological trends and check their possible impacts on the training of engineers. The methodology was applied in practice, with results exceeding expectations.

INTRODUCTION

With the introduction in 2004 of a new law in Brazil, academic institutions of higher education were required to use patents to protect results of their research and development. The law, known as the "Innovation Law", forced public universities to create their own technological innovation offices (Núcleos de Inovação Tecnológica - NITs). As such, Brazilian universities had to adapt to a new context, different from the traditional way of performance evaluation based only on scientific publications. Actually, the Brazilian academic system performed very well in this regard by occupying the 19th position in the world ranking of scientific publications in 2013.

After about eight years of establishing this new context, a significant increase had occurred in the number of patents filed by universities, but only a very small fraction of these patents has been used by Brazilian society. The immediate question arises is of why this happens. The answer is relatively simple for those who know the academic system, the dynamics of the inventive process, and the patent system: The generation of inventions that could be protected by patents is not a casual activity. Quite the contrary: it is an activity that must be one of the ultimate goals of research.

The patent system remains largely unknown in the Brazilian academic system, which does not use it either for protection of knowledge or as a source of technological information, as observed by Nunes and Goulart [1], Nunes et al. [2] and Nunes at al. [3], although about two thirds of all technical knowledge is revealed through the Patent System as shown by WIPO [4].

The number of patents filed worldwide per year has already surpassed more than 2 million. This is undoubtedly an indication of the importance of patents in furthering the development of new technical areas, as well as unprecedented provider of technical information. The information contained in a patent system also prevents possible duplications in technological development, thus saving time and economic resources.

How then to overcome this difficulty without losing the fundamental purpose of the university, which is to do basic research to broaden the knowledge base and to provide high-level training for professionals formed by their courses? Although patenting and research or education at first sight seem to be antagonistic activities, in practice they are not. In this work we show one approach of addressing this problem, which has already been applied in practice with excellent results.

In the present paper we demonstrate the feasibility of using information contained in a patent system for engineering teaching purposes, based on a methodology that strengthens the focus on engineering design independently of tools or methodologies used in the project itself.

USE OF THE TECHNOLOGICAL CONTENT OF PATENTS IN EDUCATION AND ACADEMIC RESEARCH

The proposed methodology enables the introduction of technological information contained in patent documentations into engineering educational activities through the implementation of selected extracurricular courses that introduce a new approach for dealing with engineering problems and project development.

At first, the courses are given during holiday periods so as not to overwhelm students in their last engineering course years; as well as for postgraduate students. Students in graduation or in these MSc or PhD engineering courses are instructed to solve a specific engineering problem in their final course project; or as part of their research. This proposed methodology does not interfere in the project theme defined by the students with their advisors.

In our study the patent system is implemented using three different steps or stages. The first step is studying various techniques of the Invention Process so that students learn how to identify the problem that needs to be solved. Students in this step make the first contact with the Patent System to learn its basic features. They learn how to recover technological information and how to identify the state of the art, as well as to verify solutions that are already protected by the patent system so that avoid duplication of efforts.

The second step is the Invention Design stage in which students seek the most appropriate solution to the invention problem, and the main sources of information available for its solution. Among several possible sources, the patent system stands out as being very unique one. It must be shown to the students that the patent system can be used as a source of inspiration for solutions, since they can see how other researchers, in different places and with different technologies, have addressed similar problems.

The third stage relates to the realization or embodiment of the invention, that is, its physical construction. This stage involves the transformation of the characteristic problem solution of the invention into an engineering project for construction. At this stage, several feasibility analyses are carried out, including development of a detailed design, rapid prototyping, and preliminary tests. At this stage, the patent system is used to verify the patentability of the envisioned solution and various aspects dealing with patent protection.

Inventive Process

Possible sources of inspiration for ideas for an invention or for solving an inventive problem are many and varied. Ideas can come from a sudden personal perception or as result of a systematic study, which is the motivation of this study.

In the inventive process it is studied how ideas come up. Studying the inventive process can generate specific ideas, or sourced for ideas, that will lead to inventions. It will also show how the inventive process works and what its features are. The inventive process additionally involves the identification of the characteristic problem of an invention that needs to be resolved.

Engineering students throughout their academic training are usually taught how to solve problems that are presented in the form of a technical question. They are seldom asked to identify a novel problem and to propose solution without any direction. Most of the time, students are driven by the problem statement proposed by teachers or advisors to search for a solution within the context of the problem statement. This practice restrains student's creativity when searching possible novel solutions to engineering problems.

Identifying the existence of a given problem is the first step of the creative process, since the vast majority of inventions are born of an idea to solve a problem that has not yet been realized or settled.

In the present case the problems are already identified since the projects of final course work or research topic of the MSc or PhD students are used in the proposed methodology. Even then the mechanisms and alternatives for generation ideas must be taught to the students.

Once the problem is identified, one needs to know how to solve it. While the search for a solution is a common activity in engineering, students still need guidance where and how to find the solution. The first tool they have is the technical knowledge acquired in engineering courses, but interestingly they often do not even realize that these skills are important and how they can help in the inventive process.

The patent system can be extremely helpful in finding solutions to a problem. A search in the patent system provides access to a large amount of information, which can be very useful to the creative process. Also important is using the patent system to determine the state of the art (i.e., "What does exist so far?", and "Who did it?") to avoid wasting time if solutions of the particular problem being investigated already exists.

In Brazil, the use of patent databases in technical information surveys for use in engineering projects, particularly in universities, is virtually nonexistent. The use of the patent system by Brazilian universities represents 0.6% of the total resident applications, despite being an important source of technological information widely used by industry.

Most patent databases at present have free access through web, including providing copies of documents. Access to these databases is not restricted, facilitating recovery of patents filed worldwide, since most offices provide robust search engines that enable more advanced searches. As an example of searching sites, Google launched the page "www.google.com/patents", which enables full-text searching of U.S. patents and bibliographic data from various patent collections in several countries. The site also allows advanced searches using many other criteria. Other databases such as the European Patent Office; the World Intellectual Property Organization; the U.S. Patent and Trademark Office; and the German Patent and Trade Mark Office are very important also since they allow exporting the search results for later analysis and selection.

To perform the search it is necessary to clearly define the subject to be searched, in this case the problem of the invention or the idea of the solution. Several issues are taught: Search preparation techniques; how the subject is classified; keyword selection; the search strategy; choice of databases; and the coverage period; among others. It is important to use the most appropriate words and phrases that describe the problem or the invention since success of the search depends on the use of words and phrases contained in the patent.

In our study, we used the software PERISCOPE [5], developed in Brazil by the Federal University of Mato Grosso. This software was used for analysis of the search results since it can process documents in both English and Portuguese, among many other features. This software imports the databases of the patent documents cited above (or any other source of information), allowing for an analysis and selection of relevant information to be used for solving problems.

This step is illustrated in Example 1, which concerns a survey of patent information for development of a modular micro-reactor for biodiesel production from ethanol and soybean oil without generating glycerin. The example is part of the PhD thesis of one of the students of the course. A diagram of the micro-reactor is shown in Figure 1.

The main problem related to biodiesel production is the amount of glycerin that is created during its production. The proposed idea is to avoid the creation of glycerin during biodiesel manufacture, rather than looking for possible use of the glycerin being produced

To perform the search for a solution in the patent system, one need to formulate a specific search strategy that includes a set of classifications and keywords, which describe the problem functionally and structurally, in such a way that these "search parameters" allow selecting documents from the patent system as a whole. That is, the search is not limited to mechanical fields since in this case we chose to use the Altshuller's Inventive Principles directly. This topic will be addressed in the next section.

The results achieved were very relevant because it identified already some initiatives to produce biodiesel without producing glycerin. The recovered patents did not show glycerin elimination at an acceptable cost. No solution of the problem being solved by the student was found in the patent system.

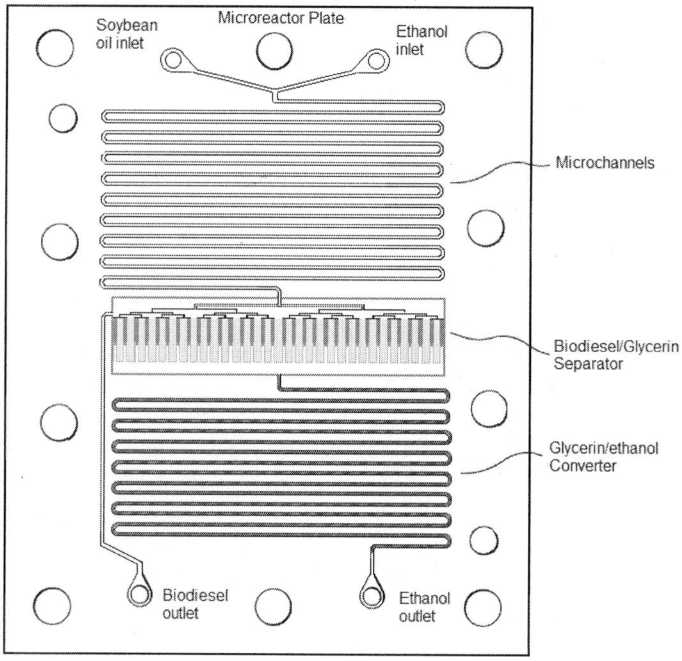

Figure 1 – Diagram of the biodiesel micro-reactor with a module for converting glycerin to ethanol

Moreover, the search was important because it allowed the student to verify the existence of other useful solutions that could improve his original research, thus avoiding mistakes already addressed by other researchers.

A direct result obtained from the patent information was the replacement of reactor material by a non-precious metal that functions as a catalyzer in the glycerin to ethanol conversion process. Figure 2 illustrates this process.

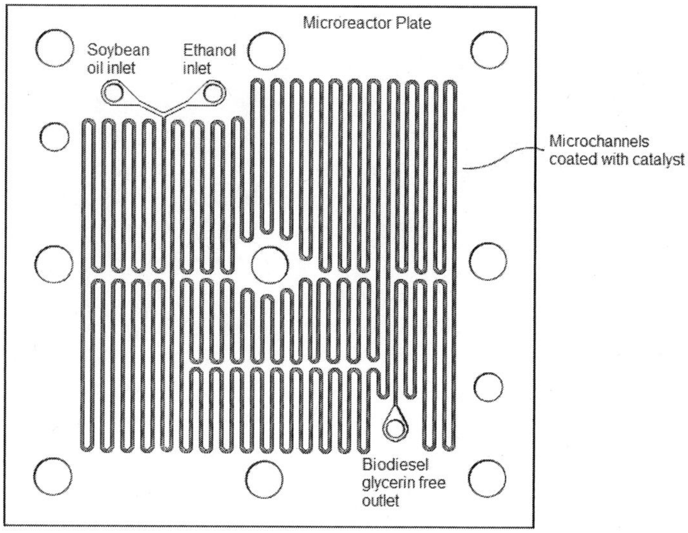

Figure 2 – Biodiesel micro-reactor diagram with full conversion

It is important to note that for this example case we choose to use directly the Inventive Principles without a need of using the Altshuller's Contradiction Matrix. Since the number of retrieved documents was small, it was possible to make a direct analysis of the technical content of the documents selected by the PERISCOPE software [5].

Finally, using information of the patent system from the very beginning of the inventive process lets users know if there is already a solution for the identified problem, or if there are elements that may be used for the possible solution.

Invention Design

After the characteristic problem of the invention is identified, a viable solution to the problem should be sought. The first resource to be used for this is application of the engineering principles the students have learned. The use of basic concepts that serve as building blocks for inventions is a classic way to invent. The more one delves into physics, chemistry, biology and engineering fundamentals, the greater the likelihood of identifying the most appropriate principles for solving the problem of the invention.

The patent system should always be used, regardless of whether or not this provides a solution to the characteristic problem of the invention. Information contained in the patent system has served as a valuable resource of technical information for all branches of knowledge, and sometimes even provided an immediate solution to the problem being investigated. The patent system often also indicates solutions to a closely related technical problem, thus serving as inspiration for new solutions or showing applications of the same principle to different problems. It is the invention conception step where one can make the search for solution methodologies or techniques far more systematic.

Fortunately, a methodology exists of combining the application of engineering principles with information contained in patents. This methodology is contained in the Theory of Inventive Problem Solving - TRIZ, which was developed by Altshuller [6], [7], [8], [9], [10], based on patent system information. Besides being a methodology, TRIZ is a set of tools based on knowledge, and as such favors the use of the patent system, leading to one of the most complete sources of technical information as shown in Marmor [11] and WIPO [4]. These studies demonstrate that information provided by patent systems represents two thirds of all information published in the technological world, which currently publishes more than 2 million documents per year, and that this most of this information is revealed only by the Industrial Property System and no other means. This information itself would justify its inclusion in engineering teaching.

The Theory of Inventive Problem Solving, TRIZ, was developed as a result of an analysis of more than 200,000 successful patents worldwide. In the analysis, standards of creativity and inventiveness of patent content were observed, leading to conclusions that the inventive activity of these patents correspond to the solution of one or more technical contradictions (i.e., cases where the improvement of one particular parameter depreciated another parameter) and also that technical systems obey certain evolutionary patterns or predictable evolutionary lines.

Altshuller in his research found that more than 90% of engineering problems had already been solved and followed the same fundamental principles. He also found that some solutions used in a specific area had already been used for inventions of other technical areas. One great advantage of TRIZ is the fact that it provides a systematic process to define and solve any kind of problem. This is very different from the traditional approach using trial and error as a method of solution, which is not feasible for very complex problems.

Use of the TRIZ methodology requires that specific training is done in an introductory module where students learn the TRIZ fundamentals. In this course students learn such topics as the inventive principles and Altshuller matrix; modeling and analysis of the problem in subfields; an introduction to inventive patterns; technical contradictions and examples for their solution; physical contradictions and the principle of separation; the ideal final result; and the Algorithm for Inventive Problem Solving - ARIZ. To assist in the use of TRIZ methodology and follow its steps, we used the Trisolver 2.1 Idea Generator & Manager software - Professional Edition [12] by TriSolver Consulting.

Using the TRIZ methodology has proven exceptional in finding inventive solutions to problems, even when students had difficulty when the solutions were outside of their immediate discipline. This difficulty can be addressed through implementation of a module in the PERISCOPE [5] software (referenced above) that reclassifies patent documents classified in the Cooperative Patent Classification (CPC) and the International Patent Classification (IPC) based on the Altshuller's 40 Inventive Principles, and in the 39 Engineering Parameters. This feature greatly enhanced the search for solutions in the patent system since solutions to the contradictions can be found directly in the patent documentation.

The Example 2 illustrates this step was the search to solve the contradictions of a problem concerning the optimization of the liquid cooling system of a set of photovoltaic cells in a solar concentrator panel (HCPV), using micro heat exchangers with micro channels. This problem was the MSc dissertation theme of a student; its diagram is shown in Figure 3.

This problem showed several contradictions, of which we cite two to illustrate application of the method. One of the contradictions that needed to be resolved was the possible reduction in pumping power versus an increase in the working fluid flow. Another contradiction was to maintain the cell temperature within some optimum range versus the loss of one or more micro-exchangers (for example due to clogging).

These contradictions could be solved by applying the corresponding principles with the aid of Trisolver 2.1, which suggests Inventive Principles that could solve the contradictions.

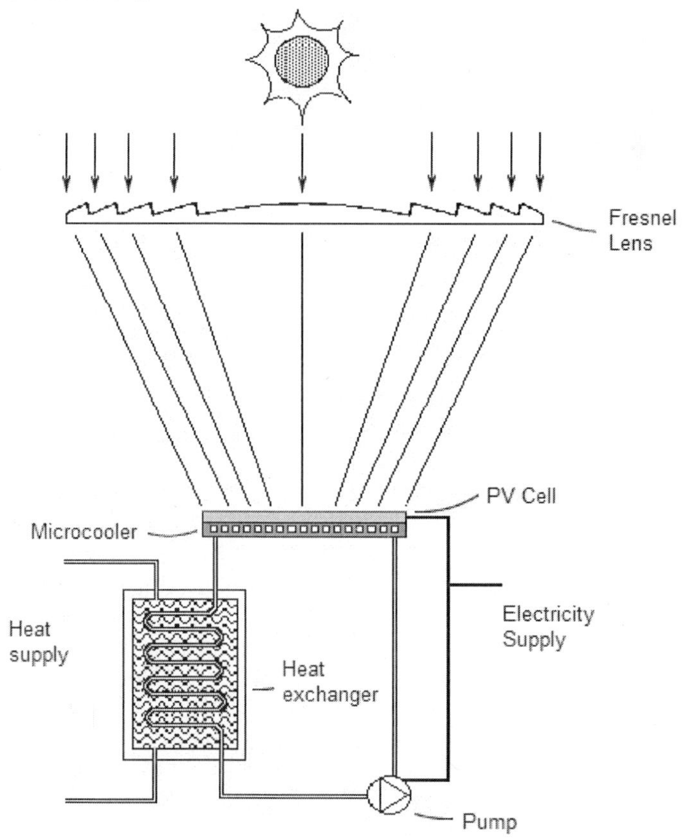

Figure 3 – Diagram of an HCPV cooling system

To reduce the pumping power (parameter 19 – Energy spent by moving object) and to keep constant flow (parameter 7 – Volume of stationary object), the Altshuller Contradictions Matrix suggests using the following Inventive Principles: 35 – Change physical or chemical properties; 13 – Other way around; and 18 – Mechanical vibration.

To keep thermal stability of the system (parameter 17 – Harmful side effects) when there is a stoppage of operation of one or more exchangers (parameter 31), the Contradictions Matrix suggests: 22 – Blessing in disguise; 35 – Physical or chemical properties; 2 – Separation or extraction; and 24 – Intermediary.

Once the principles provided by the Altshuller Matrix for the stated contradictions are identified, the patent system should be searched for possible solutions. For this purpose we elaborated a search strategy that enabled searches in the various patent databases as well as in several CPC and IPC classification sections, using selected keywords for the text search.

The search strategy in this case was much more complex as compared to the straightforward application of principles used in the first example. It is not trivial to elaborate a search strategy in patent documents describing the Inventive Principles suggested by the Contradictions Matrix since these principles cover all knowledge fields.

The critical aspect that helped this task was the use the PERISCOPE [5] software which, besides enabling patent documents reclassification through data mining and artificial intelligence, incorporated each identified solution into a problem solutions bank and as such increases the likelihood of finding new solutions through its use.

The search in the patent system to find solutions for the contradictions were important in that they provided information that helped in the solution. Although no document was found that provided a clear-cut solution, it highlighted the novelty requirement as well as showed that it was possible to search directly in patent system for solutions of the contradictions.

The contradiction related to reducing pumping power while keeping fluid flow was solved using ideas disclosed in the patent of a system of a water cooler used in personal computers. The main finding was to change the properties of the working fluid; in conjunction with filing a patent for an engine fuel feed system that would provide a solution for the head loss.

On the other hand, the contradiction related to thermal stability of the system with stoppage of one or more exchangers was not directly solved. However, it inspired a change in the originally designed system, thanks to information acquired from a patent of a general use voltage regulator system. This example shows, as suggested by Altshuller, that many solutions are available outside the original invention field of the invention one wants to develop.

If not found using the patent system, possible solutions could be investigated using other sources of ideas, such as in Nature. Learning from and imitating Nature has been a major source of inspiration for inventions and for problem solving. A classic example is the invention of the Velcro hook-and-loop fastener, conceived based on the characteristics of a plant that inspired the Swiss electrical engineer Georges de Mestral to invent it.

Another source that is worth investigating is information from experienced professionals working in the sector or in the context of operation of the invention, particularly for areas of knowledge different from that of the inventor.

In this part of the training, students must suggest a solution and explain why it will work by proposing an idea of how to solve the problem. Factors such as size, security, and time must be observed, which may require the use of a model and simulation. The model must be able to demonstrate that the suggested solution works, and the project is complete with all components.

The proposed methodology includes searching of the patent system, to be done by the student, to verify if another inventor has already revealed the solution that he or she has proposed. If affirmative, the student is encouraged to deviate from what is protected or to make an improvement of what is patented, thereby avoiding abandonment of the project.

Invention Development

To have an idea and to implement it, through a sketch or a prototype, is the first step to realize an invention. For this purpose, the problem needs to be defined in detail and be specific, in order to create a clear picture of what needs to be solved. The more specific and well-defined the problem, the sharper is the solution that will directly impact on the invention.

The physical embodiment of an invention still requires one to transform the idealized invention into a real object – hardware. Another important aspect, which must be clear, is that the invention depends on the technology involved. This transformation is not an easy task, since it requires several other attributes of the inventor. Although there is no magic formula to do it, some relevant procedures facilitate the task.

The first is to develop the idealized solution together with some measure of how the invention will change things within the context of its use. Measurement is essential to good science and engineering since only in this way one can present evidence that the predicted changes in the invention will occur.

If possible, one should always follow a "top-down" approach when constructing the proposed solution to a problem. The top-down approach looks for large systems or functional elements of the invention prior to focus on smaller details. This is an iterative process in which each step is constructed from the former. This is a crucial stage of the invention project, which requires time and economic resources. The invention process must then be tested if the proposed solution relies on a prototype or can be modeled/simulated via software. If possible, the later approach is preferred because it is faster and cheaper. Figure 4 shows a flowchart of the design of an invention to the final detailed project. The steps related to manufacturing design and life cycle of the product are not included in this approach.

Once a basic representation of the idea is complete, the hardest part is over since it has overcome the barrier of moving from an idea to something physically functioning. After this it is necessary to conduct an invention functionality test. In this test procedure, if possible, each part of the project must be tested one at a time, to ensure basic functionality. One must verify if each component is working according to what was designed. Once this is ensured, the system must work as a whole, i.e., all components must perform correctly the functions for which they were designed to work as a whole.

The development of the invention description containing the construction drawings, the parts list, operation of the parts, and a full functioning description, is fundamental. This should be checked several times before assembling.

Functional testing and the corresponding collection of measured data should be performed next. Data acquisition is essential because it is through these data that it will be possible to know whether the invention works correctly. Without data, there is no way to demonstrate that the invention works. A data analysis using graphs, tables and measured data must be carried out to see if they correspond to the theoretical parameters of the project, or by using any other means to assist verification.

One technical report with a detailed description of the invention should be prepared to record the entire assembling process by identifying key problems found and adopted solutions. This document will be useful for the preparation of a patent application. One should be very careful with this information and never share it before deciding what will be the form of protection and eventual commercialization of his or her invention.

Economic Exploitation of the Invention

First, when talking about commercial exploitation of the invention, it is paramount to address the issue of protection of the invention. The decision to seek protection of a new product or process must be handled with great care. A careful assessment must be made, similarly as with other commercial transactions.

Just because a new product or process was invented does not mean that it will be eligible for patent protection. In particular, patent protection does not necessarily mean that the costs of obtaining protecting and conserving are justified from an economic standpoint. Patent protection should be seen as part of marketing. Before obtaining patent protection, one must consider the economics as well as other legal requirements and strategic points.

There are basically three forms of protection: Directed publication; trade secret and patent protection. Here we focus only on patent protection.

Patents provide protection for inventions related to an apparatus, method or process that is new, useful and not obvious. The patent is a contract between the inventor and the government, whereby the inventor agrees to disclose his or her invention to the public, while in return he or she receives the right to exclude others about his invention. This exchange has been viewed as being useful in the development of a technological knowledge base available to the public, as well as for promotion of science and technology.

The patent does not grant the owner the right to exploit the invention, but it allows him or her to exclude others from making, using, selling, offering for sale or importing the object of the invention.

The question of how to make money through an invention always seems to come in any discussion of inventions. As soon as someone mentions working on an invention, the question of its potential market emerges.

There are different ways to profit from the invention. As in almost all financially driven ventures, wealth is proportional to risk. It is important to be aware of several alternatives on how best to economically exploit an invention. The simplest is to work as an inventor in an institution, while receiving a salary to invent. Another way is to invent as an autonomous person and obtain the rights to invention. Yet another approach is to invent and license the invention for others to explore, and to start a company to produce the invention and exploit it commercially.

It is recommended that students take other courses on industrial property where the basic concepts for submission of the patent application are introduced. This subject is outside the scope of this work.

If the option for patenting an invention is followed, it is recommended to the students seek professional help to prepare the patent application and to monitor the proceedings before the patent offices.

The use of information contained in the patent system helps identifying the origin and evolution of a particular technology; possible information sources are technical reference textbook for use in engineer training. Through this information one can also check the directions of the various technological sectors, which make it possible to assess the possible impact of these changes within the content of engineering courses, thus assisting in change planning.

The use of patent information facilitates a new approach to addressing technical problems and complementing the theoretical framework of teaching, with the analysis of possible solutions adopted by major players in the market place since it is possible to have access to detailed descriptions of these solutions, how the techniques were handled, and identification of technological pathways used. Patent information will also assist the identification of radical and incremental innovations and their differences, thereby qualifying a student to identify the so-called "inventive activity".

Another important consequence of the use of patent system information is the identification of significant changes in areas of research and development by the main players internationally, and the possible implications of these changes in engineering education. Regarding academic research, it is possible to identify which technologies are protected and by whom, thus avoiding duplications and exposing the unpublished field for further research.

CONCLUSIONS

The use of the information contained in a patent system as supplementary information allows enhancement of the final work of undergraduate students and the research of post-graduate students (MSc and PhD levels) in mechanical engineering and related disciplines. Given the results achieved, we strongly recommend using these resources in student education.

The proposed method is being implemented through Summer Courses. The first one, entitled "Curso Prático de Invenção I" (Invention Practical Course I), was held in January 2014. The course contained the following topics:

1. The Inventive Process: Discussion about how ideas and inventions arise; identification of the problem to be solved; introduction to the patent system; patent information recovery forms;

2. The Invention Design: Discussion of project assumptions of an invention; solutions to the invention characteristic problem; introduction to TRIZ;

3. The Invention Development: Embodiment of invention; rapid prototyping; modeling or simulation through software; tests and data acquisition, assessment and verification; documentation;

4. Economic Exploitation of the Invention: Ways of protecting inventions; ways of commercial exploitation;

5. Presentation of results and evaluation.

The first course was attended by 10 engineering students: two undergraduate students in mechanical engineering; three MSc students in mechanical engineering; one PhD student in civil engineering; one PhD student in physics; and three PhD students in mechanical engineering.

The course was taught in a 40h/class including exercises. The knowledge acquired enables students to solve relatively simple inventive problems using TRIZ. For more complex problems we found that further development of some topics and complementing the class with practical sections were necessary.

Patent information will be part also of other courses, notably Invention Practical Course II, scheduled for the summer of 2015. This course will use already specific literature describing the methodology, including several examples.

When students applied the methodology and realized that they were trying to discover unique solutions, two students invented two devices and products related to the themes of their academic research, while a third student invented a safety device for swimming pools. After confirming the patentability of their inventions, they are now preparing documents to deposit their patents with the Brazilian Patent Office.

We observed much interest in the use of information of the patent system by the students as they realized that such aggregated information is useful for developing and implementing engineering projects. The possibility to file a patent application was highly motivating as well. A review of the course was done through questionnaires. Results demonstrated that the course was very successful and achieved its purpose.

It was very evident to the students that those who graduated from the course had a great advantage over others who know little about the subject. The skills obtained with the course will constitute an important part of their future professional life as engineers. This perspective reflects total acceptance of the topic by students in engineering courses.

REFERENCES

[1] Nunes, J. and Goulart, L., 2007, "Universidades Brasileiras - Utilização do Sistema de Patentes de 2000 a 2004", INPI, Rio de Janeiro

[2] Nunes, J., Oliveira L., Barcelos S., 2008, "The use of Patents Documentation information in the Engineering Teaching and Academic Research",, ENCIT 2008

[3] Nunes, J., Batalha, M.S., Rodrigues, R.C., 2009, "Proposal for the use of the information contained in the Patents system in the engineering teaching" Proceedings of COBEM 2009, November 15-20, 2009, Gramado, RS, Brazil.

[4] WIPO - Wold International of Intelectual Property, 2013, "World Intellectual Property Report 2013", available from <http://www.wipo.int/econ_stat/en/economics/wipr/>
[5] PERISCOPE available from <www.ic.ufmt.br/periscope>
[6] Altshuller, G. S., 1984, "Creativity as An Exact Science - The Theory of The Solution of Inventive Problems". 1st. ed. Luxemburg: Gordon & Breach, NL.
[7] Altshuller, G. S., 2004, "And Suddenly the Inventor Appeared". Worcester: Technical Innovation Center, 4th Edition, MA
[8] Altshuller, G.S., 2005a, "40 Principles – Triz Keys to Technical Innovation", Worcester Technical Innovation Center, 4th Edition, MA
[9] Altshuller G.S., Zhuravlyova, V., 2005b, "Balad of the Stars", Worcester Technical Innovation Center, 1st Edition, MA
[10]Altshuller, G.S, 2007, "The Innovation Algorithm", Worcester Technical Innovation Center, 2nd, Edition MA.
[11]Marmor, A.C., Lawson, W.S., Terapane, J.F., 1979, "The Technology assessment and forecast program of the United States Patent and Trademark Office", World Patent Information, Munchem, vol.1, n.1, pp. 15-23

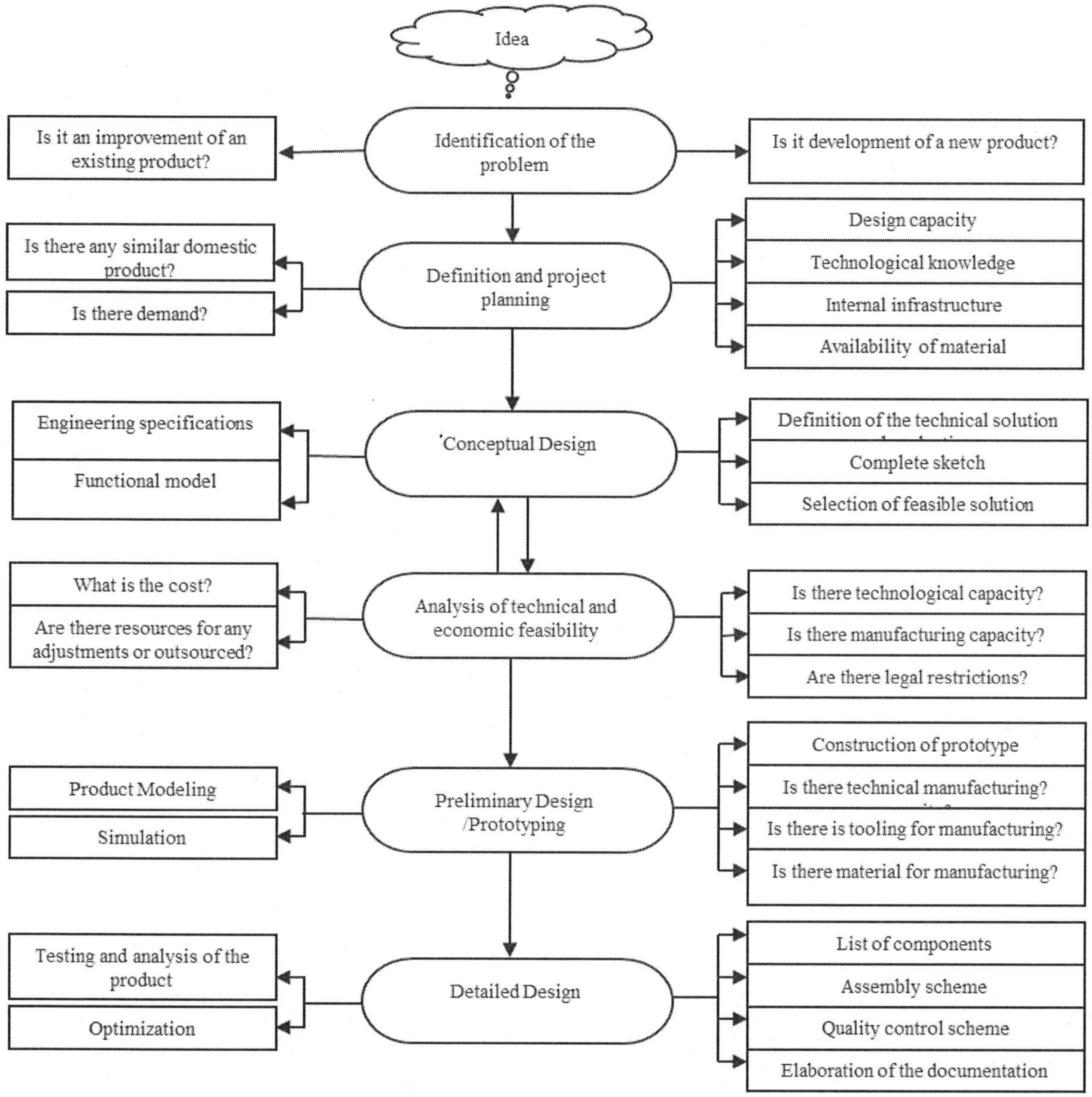

FIGURE 4 - FLOWCHART OF THE DESIGN OF AN INVENTION

This page left blank intentionally.

Proceedings of the ASME 2014 International Mechanical Engineering Congress and Exposition
IMECE2014
November 14-20, 2014, Montreal, Quebec, Canada

IMECE2014-39296

ORGANIZATION AND MANAGEMENT OF AN INDUSTRY SPONSORED CAPSTONE SENIOR DESIGN COURSE

Vito Moreno
University of Connecticut
Storrs, Connecticut, USA

Thomas J. Barber
University of Connecticut
Storrs, Connecticut, USA

ABSTRACT

Senior level capstone design courses are run in many different ways in the academic community. A growing number of institutions strive to promote immersion into the real world of engineering through industrially sponsored projects. While this approach offers many immediate benefits to near-graduating seniors, it introduces many unique problems to the academic community. Developing and sustaining an industrially-sponsored capstone design program requires an understanding of the synergies and differences between academia and industry. Key issues discussed in this paper are program management, company sponsorship, diversity of projects, level of oversight required to make a successful project and legal implications of sponsoring a meaningful project.

INTRODUCTION

Senior level capstone design courses are run in many different ways in the academic community. Many institutions have programs based on developing a single product entity, like an SAE or mini-Baja vehicle[1,2]. Student teams work on individual vehicle components, but the goal is to complete a single entity for a series of trials at the end of the program. Other institutions do not require the development of a working model; rather their goal is a paper design study, e.g. an airplane or a model manufacturing plant. Some institutions have their students work at developing new laboratory experiments to support their academic programs. In contrast to these models, a growing number of institutions strive to promote immersion into the real world of engineering through industry sponsored projects. This offers many immediate benefits to near-graduating seniors but introduces several challenges to the academic community[3].

The University of Connecticut (UCONN) Department of Mechanical Engineering has developed, within southern New England, an industry recognized Senior Design Capstone program. Consistent with the ABET learning objectives, this 2 semester program provides fourth year students the opportunity for a "major **design** experience in which they apply the principles of engineering, basic science, and mathematics; to model, analyze, design, and realize physical systems, components or processes; and **prepare students to work professionally**". The number of sponsored projects, within Mechanical Engineering, has grown consistent with the increase in class size (figure 1). Currently, the program manages between 40 to 50 projects each year This paper discusses the program management, project solicitation and selection, as well as, the formation of student teams for the execution of these projects.

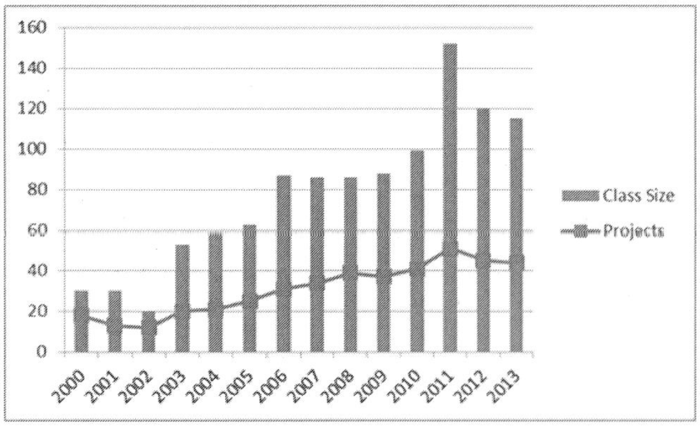

FIGURE 1: SENIOR CLASS SIZE AND PROJECTS

SOLICITING INDUSTRY SPONSORSHIP

Within Connecticut, UCONN benefits from an industrial base that includes large diverse corporations, such as United Technologies and General Electric, as well as, many other companies of various sizes engaged in a wide range of technologies, manufacturing processes and design disciplines. Recognition of the Mechanical Engineering Senior Design (Capstone) program has grown through continuous yearly sponsorship by companies, a growing and active alumni base and efforts of the UCONN Foundation and other organizations to reach out throughout the State and Alumni network. An additional significant contributor to the awareness of the program is the Senior Design Demo Day held in May at the conclusion of the Spring semester. This highly publicized event (figure 2) provides an opportunity for the industrial community to review all projects and interact with the students. A panel of industry judges selects the top 3 projects, based on innovation, engineering judgment and presentation, which receive cash awards from the Department.

FIGURE 2 2014 DEMO DAY

Solicitation of projects for the upcoming fall semester begins in June. Each potential sponsor provides a one page Statement of Work including problem description, scope, expected deliverables and any restrictions (US citizen, export, etc.). The Senior Design Program Manger works with each company to develop the SoWs which become the basis for the students to identify their candidate projects.

In 2013-2014, a total of 47 projects were sponsored by 33 companies and Government agencies. Table 1, shown in the Appendix, illustrates the diversity and scope of the sponsorship and projects.

PROGRAM MANAGEMENT

The tactical and logistical issues associated with a program of this size require a strong management structure and oversight process. This poses a significant time challenge to a tenure tract faculty member engaged in research and teaching. Over the last 14 years, the Mechanical Engineering Senior Design program has been managed as a non-tenure tract position (Professor in Residence) with individuals having significant previous industry experience. The Manager is the focal point for the program and solicits all projects, works with sponsors to develop Statements of Work, forms the student teams, mentors a number of teams, presents the lecture portion of the program and works with the Department staff to ensure accurate financial tracking of each project. A Faculty Advisor is also selected for each team, based on his/her area of expertise, to provide technical guidance and meet regularly to monitor progress. Each Industry sponsor provides a focal point project mentor that the students interact with on a regular basis. Each team is required to provide written and oral progress reports to the Faculty Advisor and Industry Sponsor and meet at the company multiple times during the course of each semester. Overall responsibility and information flow structure for a project is shown in Figure 3.

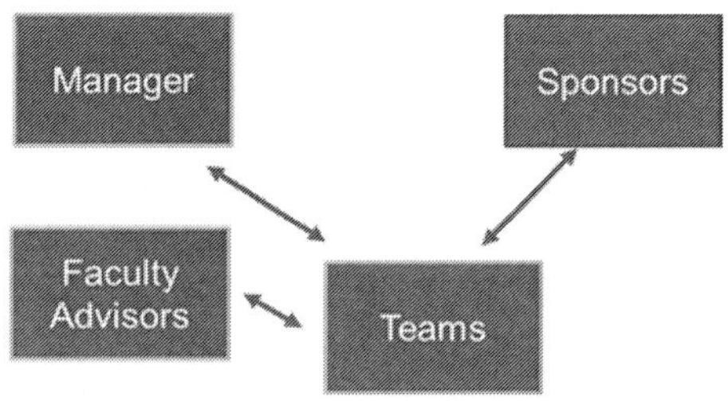

FIGURE 3 PROGRAM MANAGEMENT

STUDENT TEAM FORMATION AND PROJECT EXECUTION

The industry sponsored capstone project is intended to provide students with experiences that they will encounter in the work force. These are generally not the same issues and learning they experience in the class room. During the first week of the fall semester, they individually review the Statements of Work for each project and select their top five projects. They take a Myers Briggs or similar personality test, define their general area of interest (thermo/fluids, structures, etc...) and identify preferred (and not preferred) partners. Using this data, the Manager assigns students to 2 to 3 person teams depending on project scope. Their first team assignment is to meet with the Faculty Advisor and Company Sponsor to further develop the project statement, clearly define expected deliverables and define a technical approach (research, analysis, test, etc..) They also establish which commercially available software will be used in the analysis of the project problem (heat transfer, fluid flow, structural dynamics, electromagnetic, etc...). The team is responsible to define individual roles and responsibilities for each member (communication lead, analysis, rig design, testing, etc.) and the team "Mode of Operation" that defines frequency and location of meeting times, decision making process, etc.

As potential solutions are identified, the initial evaluation can pose a problem as the students may not have had any exposure to the required analysis codes during previous course work. This requires a review of available tutorials and analysis of basic unit problems (laminar flow in a pipe, beam analysis, etc…) to develop some level of expertise before they work on the actual project problem. Generally, an issue they hadn't anticipated.

The importance of organization structure and roles and responsibilities become obvious as a project schedule is developed for timing of analysis completion, experiment design, hardware ordering, testing, data analysis, etc.

Within the classroom portion of the course, each team makes four 12 minute oral presentations (2 per semester). The audience includes their peers, the Manager, Faculty Advisors and Sponsors. Each team is assessed on their presentation of status, results to date and schedule. They are introduced to the concept of risk analysis and required to present their top risks and mitigation plans. A representative risk summary from one of this year's teams is shown in figure 4.

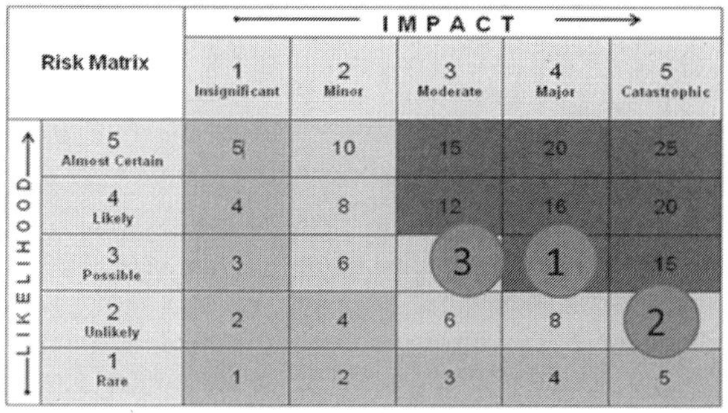

FIGURE 4 TEAM 48 RISK ASSESSMENT

At the conclusion of the fall semester the teams publish an Interim Project Report to the Sponsor and Faculty Advisor. This is followed by a Final Project Report in May which includes all analytical results, CAD models, experimental design, data, and instruction manuals generated as a product of executing the project.

TEAM ASSESSMENT

Teams receive and give assessments of their performance on the projects during the 2 semester course. Feedback on individual student performance is provided to the Manager by the Faculty Advisor and Industry Sponsor each semester. This data is provided back to the students (in summarized form). A representative sponsor feedback scorecard and comment statements are shown in figure 5.

Knowledge Area	Student #1	Student #2	Student #3	Student #4
1. Technical Performance				
1a. Analysis [4]	B	C	D	
1b. Design [2]				
1c. Problem Solving ** [1]	A	B	C	
2. Communication				
2a. Technical Discussions (team meetings [5]	A	B	C	
2b. Oral Reports/presentation skills [5]	A	A	A	
2c. Written Report [5]				
3.Professional Performance				
3a. Effort [3]	A	B	D	
3b. Attending meetings [3]	A	A	D	
3c. Team Work [3]	C	C	C	
3d. Planning and Scheduling [3]	A	C	C	

Areas for improvement:

They seem to have a difficult time giving us straightforward updates on what they are doing and the progress they have been making given the project….. This makes meetings take longer than they otherwise would need to and gives the impression that progress is coming too slowly or the plan is not fully understood,……So communication is definitely an area for improvement.

FIGURE 5 FEEDBACK ASSESSMENT FORM AND SPONSOR COMMENTS

The students also complete a self-assessment of their peers and overall team performance using the same feedback form show in figure 5. The comments generally reflect the challenge of balancing priorities, learning to work together and working to a schedule as shown in following comments.

Areas for improvement:
"We need to improve our motivation. We seem to be moving very slowly and are just getting to the due dates on that day. We need more motivation to go past our client's expectations. We keep getting caught up with work from other classes, which distracts us………"

Areas for improvement:
"Although we have great communication via email and cell phone, we should have more team meetings and everyone should try harder to attend."

SPONSORSHIP AND OTHER LEGAL ISSUES

For each project, a Sponsor provides a grant to the Department and covers any additional expenses of incidental hardware associated with constructing and running experiments. Students are required to take a shop safety class before they can do any fabrication or testing. During the course of discussions, a Sponsor may share proprietary or other intellectual property with students and require them to sign NDAs (Non-Disclosure Agreements) and Intellectual Property Agreements. Project content may have export restrictions and be governed by commercial EAR (Export Administration Regulations) or military ITAR (International Traffic in Arms Regulations) regulations. The students are required to take University provided export training and comply with all regulations in their written and oral communications and presentations. Under the current structure of the Senior Design Program, the Sponsor solely owns the IP generated from a project. Historically, if a company submits a patent application as a result of the project, the students are generally included on the application.

CLASSROOM LECTURE TOPICS

In addition to spending a significant amount of time on the projects (estimated at 500 hrs. each student), the Senior Design course is scheduled to meet twice a week for discussions, selected topic presentations and feedback from the Program Manager. Invited lectures presented by Industry and University experts are intended to, not only support the project work, but expose the students to relevant topics that they will encounter as they prepare to become working professionals. The lectures included in the 2013 -2014 Senior Design course are shown in Table 2.

TABLE 2: 2013-2014 SENIOR DESIGN LECTURES

Mr. Stephen Heath, Pratt & Whitney, (Retired)	*'Project Management'*
Mr. Mark Austin, CT Society of Professional Engineers	*'Professionalism and Licensing'*
Dr. Mary Caravella, UCONN	*'Product Marketing'*
Dr. Tom Barber, UCONN (Retired)	*'Engineering Ethics'*
Mr. Robert Shemenski, Pratt & Whitney	*'Design of Experiments'*
Mr. Thomas Meyer	*'Fatigue in the Real World'*
Mr. Frederic T. Tenney, Pratt & Whitney	*'Patent Law and Intellectual Property'*
Mr. Brian Montanari, Habco Inc.	*'Lean Manufacturing'*
Mr. Chris D'Angelo, Zachry Engineering	*'Energy Market Overview'*
Mr. William Collins, United Technologies (Retired)	*'Codes and Standards'*
Mr. David Ware, UConn	*'Product Liability & Warrantee'*

CONCLUSIONS

The Senior Design (Capstone) program of the Mechanical Engineering Department at the University of Connecticut provides students the opportunity to apply their collective academic experience to a real world problem that directly benefits an industry sponsor. They experience, not only the application of engineering knowledge and principles to develop a solution, but the challenges of teamwork, clear communication, changing requirements, schedules, budgets and other issues. This course through the capstone project and lecture/discussion sessions is intended to aid them as they transition to working professionals.

REFERENCES

[1] Todd, Robert D., Magleby, Spencer P., Sorenson, Carl D., et al., "A Survey of Capstone Engineering Courses in North America," ASEE Journal of Engineering Education, Vol. 84, No. 2, April 1995, pp.162-174.

[2] Barber, Thomas J., Mechanical Engineering Department Senior Design Program, Universityof Connecticut, www.uconn.edu/me/seniordesign, November 2006.

[3] Barber, Thomas J., "Critical Issues Associated With an Industrially Supported Capstone Design Program," ASEE 2007.

APPENDIX A

TABLE 1 2013-2014 MECHANICAL ENGINEERING SENIOR DESIGN PROJECTS

Team No.	Sponsor	Project Title
ME1	Agrivolution	Cloud Based Monitoring and Management of a Residential Vertical Farming Unit
ME2	Alstom Power Co	Fossil Fuel Boiler Backpass Cavity Utilization
ME3	Alstom	Seismic Isolator for Tower Top Solar Receiver
ME4	ASML	Advanced 3D Printer Material Characterization
ME5	ASML	Design for Manufacturability and Assembly Redesign of an Advanced Water Channel
ME6	Barnes Aerospace	Design of Shot Peen Nozzle Apparatus for Internal Surfaces of Drum Rotors
ME7	Belcan	Bidirectional One-way Clutch
ME8	Capewell	Next Generation Capewell Reliability Rig
ME9	Covidien	Polymer Reactor Sampling System
ME10	CT Corsair	F4U Corsair Simulator
ME11	Emhart Glass	Energy Analysis on the Forming of Glass Containers
ME12	2Fly Wines/ Spotwine	High Capacity Oxygen-Depleted Filling Environment
ME13	GE	Contact Erosion Test Apparatus Development
ME14	GKN Aerospace	Interlaminar Stresses In Thick Walled Carbon Fiber Tubes
ME15	HABCO	RFID Tool Tagging Machine
ME16	UTAS	Continuous Fluid Flow Measurment
ME17	UTAS	Additive Manufacturing Cold Plate Design
ME18	Jacobs Vehicle Systems	Optimizing the Flow of a Solenoid Valve
ME19	Koffee Karousel	Designing a Cashless K-Cup Vending Machine
ME20	Henkel Loctite	Peel Resistance of Adhesive Bonds
ME21	MSH1 Bicycle Works	The Design, Build and Calibration of a Bicycle Frame Stress and Fatigue Life Test Facility
ME22	OTIS	Emergengy Hydraulic Brake Release Mechcanism
ME23	Parker Hannifin Corporation	Incorporating Position Sensing Into a Solenoid Coil
ME24	Pioneer Aerospace	Aerodynamic Deceleration System Test Rig
ME25	Pratt & Whitney	Design for Variation on Wear and Tolerance of Compressor Kinematic Systems
ME26	Pratt & Whitney	Turbine Tip Clearance Design for Variation Tool Development
ME27	Pratt & Whitney	Spiral Wound Gasket Load Characterization
ME28	Pratt & Whitney	Designing a Robust Air Inlet for Combustion
ME29	Pratt & Whitney	CMAS Testing Rig
ME30	Schick	Design for Improved Flow Control of Gel Products
ME31	Sikorsky Aircraft	Additive Manufacturing of Non-Metallic Tooling Design for Complex Composite Panel Repair
ME32	Sikorsky Aircraft	Wireless Instrumentation for Rotating Parts
ME33	Sikorsky Aircraft	Fatigue Test of Electon Beam Welded Titanium
ME34	Sikorsky Aircraft	Power Available vs. Required: Stand Alone Low Cost Flight Display System
ME35	Sikorsky Aircraft	Laser Assisted Manufacturing for Honeycomb Shelves
ME36	Stanley Access	Sliding Door Seal
ME37	UCONN	Passive Solar House
ME38	Uconn	Development of a Quiet Gas turbine Test Facility
ME40	Westinghouse	Tornado Missile
ME41	Zachry Nuclear	Determination of Transition Heat Transfer Coefficients
ME42	Windham Dental Laboratory	Immediate Functional Loading of Dental Implants
ME43	NASA	Design, Build and Test of a Cold Gas Reaction Control System Thruster for the Nanolaunch 1200
ME44	Carlyle Johnson	Non-Contact Centrifugal Pressure Seal
ME45	Center for Clean Energy	CO mitigation in PEM fuel cells
ME46	U.S Dept of Energy	Compact Chemical Kinetic Mechanisms for Tranporations Fuels
ME47	National Science Foundation	The Role of Chemical Explosive Modes as Flame Diagnostic
ME48	U.S Dept of Energy	Retrofitting a Rapid Compression Machine with Crevice Containment

Proceedings of the ASME 2014 International Mechanical Engineering Congress and Exposition
IMECE2014
November 14-20, 2014, Montreal, Quebec, Canada

IMECE2014-39692

SOLVING CONFLICTING ENGINEERING PROBLEMS IN EDUCATION, RESEARCH AND PRACTICE - ENHANCED APPROACH

Zbigniew M. Bzymek
University of Connecticut
Storrs, Connecticut, USA

ABSTRACT

The nature of engineering is problem solving. The challenge of ongoing design research is to develop a tool that would support the most difficult phase of design – solving problems with contradictions and finding the best possible idea for conceptual design of products. The Brief Theory of Inventive Problem Solving (BTIPS) is a prospective tool for performing such a task. Derived from TRIZ, TSIP and TIPS, BTIPS slightly differs from those methods. Principles, Effects and Prediction modules in BTIPS are enhanced to meet the newest challenges of engineering pedagogy and technology development. To meet those challenges principles of Size Reduction, Miniaturization, Nanotechnology and Biotechnology were added. Design principles and technological effects were enriched with new developments based on nanotechnology and biotechnology. Furthermore the procedure of the Virtual Element approach was added to the Prediction module. The tests of functions' separation and minimum information contents to evaluate the derived end solution are also the new additions. BTIPS is living and developing; it is taught and used, and, thus, constantly improved. This paper points out the enhancements and shows some ways of BTIPS application in solving problems with conflicting constraints in conceptual design.
Key words: Problem solving, Conflicting constrains, Conceptual design, Ideal Solution.

NOMENCLATURE

Acronyms, names and abbreviations:

- BTIPS: Brief Theory of Inventive Problem Solving;
- BS: Brain Storming; a method of idea generation introduced by Allex Fraikney Osborn in 1963 and based on the theory of the subconscious;
- Invention Machine: A problem solving software developed in the late 1980s and the 1990s by the Invention Machine Corporation based in Minsk, Belarus and in Boston, Massachusetts;
- IHS Inc.: Information Handling Services (IHS); the US company based in Douglas County, Colorado that acquired Invention Machine Inc. on 7th of November 2012;
- TechOptimizer: Improved edition of Invention Machine;
- GoldFire: A problem solving software that optimizes the innovation processes and enhances the productivity; developed on the basis of Invention Machine, presently being distributed by HIS;
- Ideal Solution: The solution that satisfies a design problem statement, requires minimum information to be described, and takes minimum energy to function;
- TRIZ: Teoria Reshenia Isobratatielskich Zadatsh (Russian: теория решения изобретательских задач – pronounce: *teoriya resheniya izobretatielskih zadatch)* - Theory of Inventive Problem Solving;
- TSIP: Theory of Solving of Inventive Problems, the name for TRIZ introduced in the book: *Creativity as an Exact Science* by G. S. Altshuller - translated from Russian by Anthony Williams (1984);
- TIPS: Theory of Inventive Problem Solving, the method derived from TRIZ introduced by the Invention Machine ™ Corporation and the team under leadership of Valery Tsourikov.

INTRODUCTION

In the fast growing world's economy of the current century the most important factor for maintaining a leading technological and economic is to invest properly into engineering research and education. This may help the US economy to retain its position - one that according to some predictions, may be lost by 2035 [1]. One of the ways of keeping the technological leadership is to develop problem-solving methods in engineering education, research, and practice to the degree in which it could be most successfully used. In this paper some aspects of the problem solving education, research and practice using an algorithmic method called BTIPS [2] are discussed.

The method introduced by G. S. Altshuller [3] as TRIZ was formalized as TIPS [4] and developed further as a program package by Invention Machine Corporation [5, 6 and 12] and V. Tsourikov [4, 7]. It was further changed, abbreviated and equipped with additions at the University of Connecticut and named BTIPS [8]. While BTIPS was created mainly for teaching purposes, it still maintains its ability to be effectively used in practical industrial design. Modules of BTIPS have been already re-defined, described and published [10]. To make the method more effective the vocabulary of key words, a glossary of basic terms and definitions, algorithms and formulas have been developed and formalized. This paper describes some developments of BTIPS's procedures and its application in practical solution of engineering problems.

At first, problems with contradictions may appear impossible to solve, but after careful study and appropriate effort they may become solvable. The key to solve "impossible" problems is in the use of knowledge about the world, in which problems exist. Such world is composed out of the matter that could be present in four phases. Engineers should be prepared to understand this and apply in design. Solving "impossible" problems might determine the success of the product and give the companies and organizations that launched the product significant advantage over their competitors. This paper introduces the main characteristics of BTIPS describing its use. The new developments of BTIPS – new principles, new attributes, and new approaches in the analysis of the designed system, its components and their interactions - will be briefly presented.

THINKING AND PROBLEM SOLVING

In some way the problem solving depends on thinking process (Figure 1) [11]. BTIPS was developed from TRIZ [3, 14] and TIPS [4]. It is in a group of algorithmic methods targeted towards the possibility of finding the solution that is unknown and it cannot be derived by adjusting or accepting another solutions. Additionally, out of all algorithmic methods, BTIPS is the one that is mainly oriented towards education. With the application of matrices and tables, general algorithms of problem solving (Figure 2), and formalization of the theory by description of keywords, definitions, theorems and corollaries (Table A1 in Annex) it became a handy teaching tool that still maintains its viability and usefulness in engineering practice.

BTIPS AND THE QUEST FOR THE IDEAL SOLUTION

At all times, problem-solving procedures are the main focus of engineering activities. After introduction of so-called "exact solutions," it became apparent that even with assumptions as close to reality as possible and the most precise analysis, a product may be not as successful as expected. It was understood that the attractiveness of the product's concept determines the success. The degree of precision in the analysis assures its safety, but not necessarily its practicality. Thus,

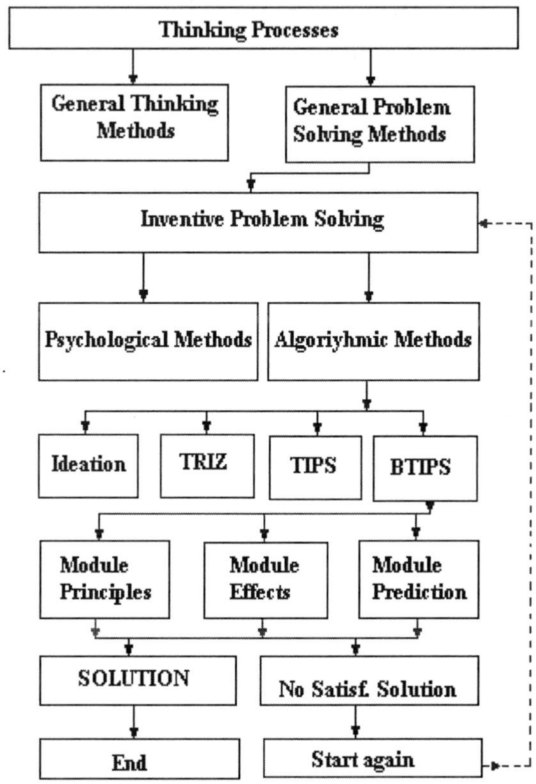

Figure 1. The place of BTIPS in the Thinking Process.

the problem solving process is decisive. It should be chosen and applied wisely. The method described in this paper is derived from Altshuller's works [3] and IM corrections [4], which were abbreviated, and slightly enhanced by the author of this paper and called BTIPS [8, 9, and 10]. BTIPS contains 44 principles, is more formalized than TRIZ [14], TSIP [3] and TIPS [4], and introduces several new design categories [10]. BTIPS theory, matrices of principles and algorithms together with Invention Machine software such as IM® [5, 6, 17, 18, 23 and 24], TechOptimizer [12], and Goldfire [26], turned out to be very effective in teaching conceptual design and still maintaining its practical capability. Other methods, such as the

Brain Storming (BS) [25], generate a diversity of concepts in different directions and therefore carry a danger of slowing down or even defocusing the solution generation process.

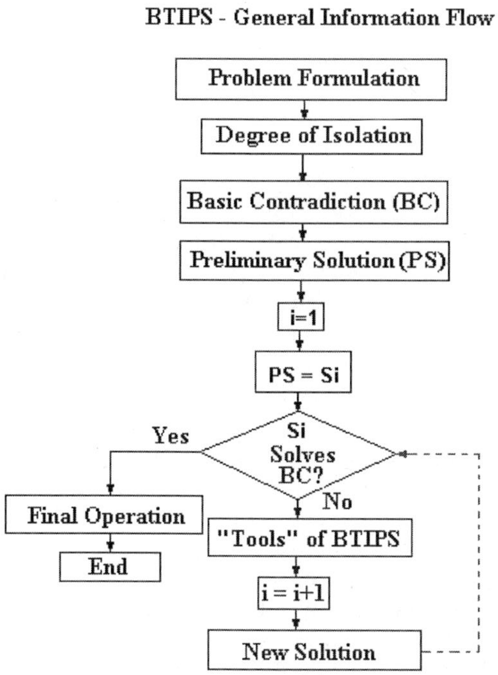

Figure 2. General flow diagram of BTIPS [8, 9].

BTIPS helps to avoid those dangers. In fact, in many cases it immediately assists the user in generating the key solution that, without BTIPS, would probably never be found.

FACING THE CHALLENGE OF ENGINEERING EDUCATION

Engineering education is a global and national asset on which the economic growth and well-being of a nation depends. To meet this challenge, the engineers need to generate creative ideas, cutting edge innovations, and revolutionary high quality products. There are many ways of thinking that are called problem solving methods, but they do not always lead in the direction of the Ideal Solution. They are mostly based on a psychological approach [25]. Algorithmic problem solving methods, however, have a much greater chance to lead in the direction of the Ideal Solution. There are several published techniques and methods of thinking and problem solving; however, because they were not developed for educational applications in mind they are somewhat ineffective, vague, and unsuccessful in the engineering classroom. BTIPS, on the other hand, proved to be effective when used in classes, in engineering practice, and in research. This paper underlines its user-friendliness, ease to learn, and high degree of effectiveness in reaching desired problem-solving results.

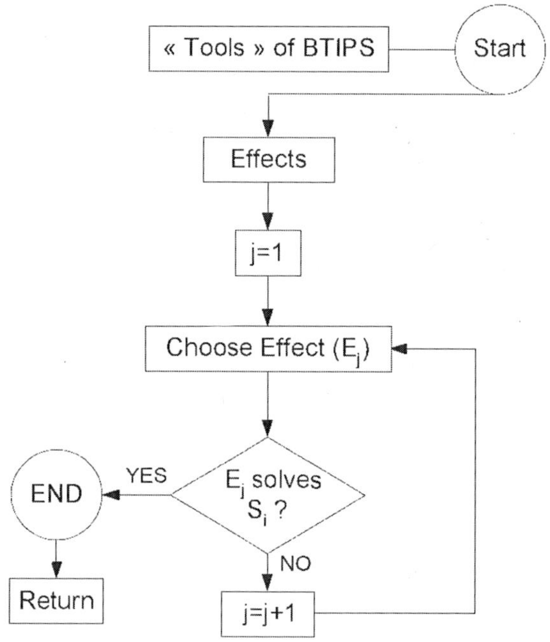

Figure 3. General flow diagram of BTIPS: Effects [8, 9]

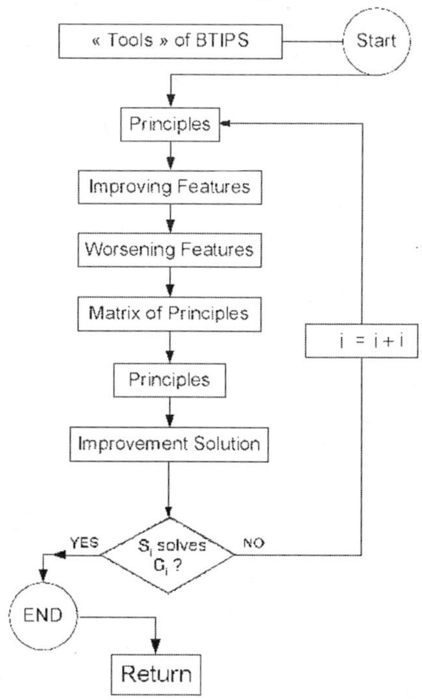

Figure 4. General flow diagram of BTIPS: Principles [8, 9].

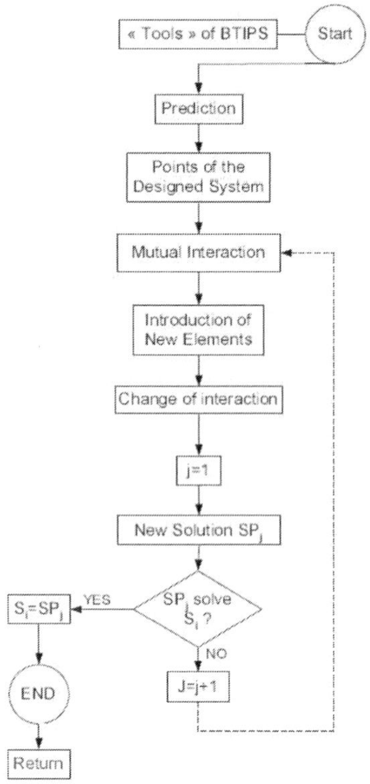

Figure 5.General flow diagram of BTIPS: Prediction [8, 9].

PROBLEM SOLVING FOCUS

Problem solving is the key to the engineering design. BTIPS meets all the challenges of engineering problem solving methods (Table 1). As a successful engineering problem solving method, BTIPS could have a major influence on improving the education of engineers, raising technology levels, and producing useful, high quality of products.

Table 1. The Challenges of Engineering Conceptual Design

To meet the challenges of engineering Conceptual design one has to:

(i) Introduce most effective method of engineering problem solving in teaching and practice
(ii) Equip this method with all the necessary teaching tools that will allow the instructor to teach the method and the students to gain experience in its application
(iii) Test and prove the effectiveness of the method and tools
(iv) Develop a procedure for evaluating learning effectiveness
(v) Make the teaching tools available in the public domain for all those involved in engineering education.

FORMALIZATION OF BTIPS

Clear formalization is the most important criterion of effectiveness of every engineering theoretical tool. BTIPS is such a tool for successful teaching and learning of the conceptual design process. As the first step of BTIPS formalization, the key words were listed, basic terms and algorithms were defined, and vectors of features and matrices of principles were described, both numerically and in English (Table A1 – Annex). As the second step the theorems were introduced and the corollaries were derived; algorithms and flow charts were also constructed (Table A1 – Annex). As the last step the basic algorithm was described and flow diagrams (Figures 2, 3, 4, and 5) were developed to help students to understand the fundamentals, learn the theory, and apply procedures. The BTIPS theory is based on modern science and, thus, it is constantly being upgraded and developed as the technology progresses [20, 21 and 22]. Several actualizations have already been done at the University of Connecticut and some new additions are under development. In addition, the list of principles and features was slightly enlarged to reflect these developments. A greater number of effects, simplified prediction techniques and a new method of applying of virtual elements in designed system were introduced. This makes BTIPS an even more unique tool for problem solving in engineering research, design, practice and education. Furthermore, the examples showing the method's application in engineering make BTIPS even more useful, applicable and understandable [19]. BTIPS is taught not only as a problem solving method but also as a problem solving philosophy that can be used with or without computers. In practical applications however the use of computers, especially with the IM software such as IM v.2.1 for Windows [5], TechOptimizer [12] and

Table 2. Basic Concepts of BTIPS Application

1.	Hierarchical system approach to the designed system
2.	Degree of design problem separation
3.	Concept of ideal solution
4.	Concept of basic contradiction
5-7.	Approaches to contradiction solving: ▪ Using problem-solving principles ▪ Using physical or chemical effects of the applied material substances or mathematical/geometrical attributes of the model structure • Causing, changing or stopping subsystems' intervention
8.	Concept of solution trimming
9.	Concept of recurrent application of BTIPS to the designed system on different levels
10.	Concept of the end solution adaptation and testing
11.	Concept of end solution evaluation

Goldfire [26] packages, makes application of BTIPS much more effective.

METHODOLOGY OF PROBLEM SOLVING

Before starting the process of design, there are some necessary actions one has to perform. In order that time and efforts are not wasted, one should carefully study the Problem Statement to insure that the design does not concern the known solutions. The following dangers have to be overcome in the solution process: psychological inertia; shortage of knowledge (re-inventing); wrong objects (solving wrong, badly formulated problems); and avoiding solving the contradictions that may lead to low quality products. There are several solution search algorithms recommended by TRIZ and TIPS: an Algorithm for Formulation of an Engineering Problem; an Algorithm for Solution of Inventive Problems (ASIP) [3]; an Algorithm for Solving Especially Difficult Problems [4]; and an Algorithm for Transposing the Advanced Solutions from Costly to Economical Systems [4]. The last one is used when the solution is known, but the system that carries it is too expensive to apply to the design in question. In such a case, the algorithm lists operations necessary to isolate the solution from the expensive system and apply it to the economical one. The basic concepts of solution search operations used in BTIPS [2, 8] are presented in Table 2. Solution procedure requires formulating the problem, defining the contradictions and applying the BTIPS tools (Figure 2) [9].

Table 3. Algorithm for Formulation of Engineering Contradiction

1. Define the problem clearly.
2. Define the desired ideal solution.
3. Point out the basic contradiction.
4. Define the design system to include its surroundings, and the advantages of the desired solution.
5. Define the disadvantages that should be eliminated.
6. Define which of the following kinds of situations the disadvantage belongs to:
   a. Does some harmful action exists which must be eliminated?
   b. Is there a useful action that must be improved?
7. If disadvantage belongs to a harmful action:
   a. Define an obvious or well-known way of reducing the harmful action.
   b. Define what useful action would prevent the harmful action.
   c. Formulate the Engineering Contradiction pointing out how to reduce the harmful action in the obvious or well known way.
   d. Formulate the action to eliminate or minimize the harmful action or intensify the useful one.
7. If the disadvantage belongs to the weakness of the useful action:
   a. Define the useful action precisely.
   b. Define the obvious or well-known way of improving the useful action
   c. Formulate the Engineering Solution, pointing out how to reinforce the useful action.

Table 4. Eliminating Contradictions

Contradictions can be eliminated by:

1. Deriving the solution as close to the Ideal Solution as possible.
2. Representing engineering contradiction as physical contradiction.
3. Formulating and analyzing a system model for solving physical contradiction.
4. Applying the algorithms to solve the contradictions.
5. Applying the tools of Principles and Effects to solve the contradiction and applying the Prediction to minimize the harmful action and/or maximize the useful action.
6. Reducing the designed system in such a way that the function will be fulfilled satisfactorily, but the redundant system's elements will be reduced or eliminated.

GENERAL ASPECTS OF BTIPS APPLICATION

Upon hearing the words "product design" one understands the concept to entail a process leading to the description of a product to be manufactured. Inventive design refers to new, previously unknown characteristics of a product. Copying designs with slight (or even big) alterations is not an example of inventive design. Using the catalog of solutions and changing them are designs, but not inventive designs that would bring new ideas into existence. In such designs, there are no conflicts to resolve. Inventive design takes place in the case where there are strong conflicts that the designer has to resolve. It often happens that the conflict is that the object should be both small and large. This is a paradox; but, analyzed in time, the conflict can be solved.

The applications of BTIPS are described using instructions, procedures and examples. Some of them come from TIPS [4, 5]. The theory itself is introduced in a concise way, which is adequate for the students to learn, for designers to begin to work with, and for the managers to understand and evaluate its usefulness. A number of basic concepts (Table 2) can be used in reaching a solution with BTIPS. In particular one should pay a special attention to: defining the problem; describing the basic contradiction; description of ideal solution for the problem; system approach and defining the degree of separation (or isolation) of the designed system, its relations to the super-system, other systems and subsystems as well as systems' interaction; defining and examining the surrounding systems and examining resources; describing possible solutions and defining their closeness to the ideal solution; changing the system in such a way that the closest solution to the ideal solution can be reached; reducing or trimming the found solution by eliminating functions from one subsystem and delegating them to the another; and finally analyzing the possibility for reducing the entire system and its components in such a way that the function will remain but the system will be greatly reduced or eliminated leaving its function to be performed by something else, as for example by surrounding systems. Such a transformation should not change the desired ideal solution its function separation characteristics and ability to be described with the minimum information contents.

PROBLEM DEFINITION

A proper problem definition is essential in application of BTIPS. The definition should point out the existing situation, the contradictions that had to be eliminated and the characteristics of desired solution.

DEGREE OF ISOLATION

The proper Degree of Isolation is essential for successful design idea generation. The art of determining the boundaries around the designed system is called isolating the system from the environment. The description of "how much" the system is isolated is called the degree of isolation of the designed system. The degree of isolation has to be assumed in such a way that it is neither too large nor too small. If the problem is too isolated one might not have enough resources to develop the sought-after solution. If it is too large the problem may be lost in the area of the broad surroundings; for example if it includes too large a number of systems and subsystems, the solution generation is complicated and the design goals are unclear. Thus, the degree of isolation has to be manageable. Getting the degree of isolation "just right" depends on the designer's intuition, practice and knowledge. While BTIPS gives guidance, it does not formulate explicit rules on the degree of isolation; that is left up to the designer. In order to isolate the system properly, one has to realize how it is built, what its components are, what their functions are, what resources they need and provide, how they are interrelated, and what their relations are to the super-system and the subsystems.

.

BASIC CONTRADICTIONS

In defining the problem, the main conflicting property of a designed system should be described in the form of antonyms: stationary and moving, big and small, open and closed, deep and shallow, fast and slow, far and near, etc. Such a property is called the basic contradiction. At first, such a contradiction seems insoluble; surprisingly, however, in many cases there are possible ways to solve it. Two steps are employed in solving contradictions: first, apply the rule of separation; second, use one of the three main BTIPS modules. In most cases, the separation rule says that contradiction phenomena can be separated in time or space or that they are associated with situation, activity or with changing fields; i.e. external influences. The same object may have different characteristics in different times or locations and under different circumstances (i.e. under pressure or in a different temperature). This can be discovered using the Principles module [9, 22].

SOLVING CONTRADICTIONS

As it was pointed out the contradictions can often be solved if separated in time or space [8]. For example, the tailor's needle eye should be large at the time of thread insertion and small at the time of sewing [4]. When threading, the needle could be held close to a light, which is not needed during sewing. Light could influence the material of the objects to expand making the hole larger. If one can find a physical effect that will cause such enlargement, it may be the key for solving the problem. Knowing this, one can look for physical effects, which could be applied in this case. So, applying time separation and material behavior effects one may solve this contradiction.

If one considers an example taken from civil engineering, such as building foundation on piles, one can define another contradiction. In this case the foundation pile should be sharp at the time of being driven into the ground and dull when it is already in place [4]. Time and place separations are the keys to the solution of the problem. The pile can be preliminarily designed in such a way that it would change its properties in different moments and different places; for example being sharp upon initialization of the driving process and becoming dull upon reaching its final destination. It is an example of separation in time.

USING BTIPS MODULES

Following TIPS [4], BTIPS introduces three modules that can be used in problem solving. The user should first apply the Effects module [8] and try to find the effect that will solve the problem. For example the effects of gravity can solve the downhill transportation problem, while the effect of temperature can open windows of the green house when the temperature raises and close them when the temperature drops [21].

The second module that should be applied is the Principles module [4, 5, 6 and 9]. Using this module one employs the use of principles to solve the contradiction. BTIPS Principles puts the principles of the newest technology (i.e. Nanotechnology, Biotechnology, Miniaturization) for the disposal of the user.

The third module is Prediction [4, 5, and 8]. In using this module the user must study the designed system, predict the interaction of its modules and make the desired changes. He or she has to change any negative actions into neutral ones, make positive the neutral actions and retain the positive actions.

During this study the designer may use the virtual elements in order to obtain the required properties of the system. All these should lead to the Ideal Solution.

FINDING THE IDEAL SOLUTION

At first, the concept of an Ideal Solution could seem to be the irrational. Surprisingly, however, this apparent irrationality often turns into reality when one applied a BTIPS approach, as the following examples illustrate. For instance, one could consider the design of an automatic window that opens according to the weather conditions outside (opening when it is warm and closing when it is cold) as an illustrating example of a technological effects application. By applying materials with memory that would change shape depending on temperature, it is possible to design and build such a window that would function by itself [10].

The next example can illustrate the application of design principles. In this example the design of the system that would prevent fire in steel quenching in oil is considered [4, 5]. The hot red steel part when submersed in oil will cause oil to burn. To prevent burning the oil container has to be covered with lid that will cut the oxygen supply. While the lid will stop oxygen supply, at the same time would hinder the operation of the part submersing. The contradiction here is that while *the lid should be there* to stop the oxygen supply, *the lid should not be there*

Table 5. The Steps Changing the Design System using the Virtual Element procedure

One:	Identifying the components and structural hierarchy of the system. such as the system, sub-systems and elements.
Two:	Describing the interaction of the system component and the mutual influence on each other – usual such interaction is positive, negative and/or neutral.
Three:	Changing the system by alternating its existing contents, adding new components or eliminating some existing ones.
Four:	Changing the negative actions into neutral or positive and the neutral into positive.
Five:	Eliminating components causing the negative action and adding ones that may improve the system.
Six:	If the component that is needed is unknown, introducing virtual component that wouldn't exists physically but would perform needed action.
Seven:	Checking the function of the modified design system.
Eight:	If the function of the modified system is satisfactory - substituting the virtual components with physical ones and testing the system.

for free part submersing. However, this contradiction (the lid *is* and *isn't* there) can be solved by realizing that matter can exist in four phases and therefore, instead of a solid lid, applying the lid made out of gas – let's say a layer of CO_2 - on the top of oil. Such a lid both prevents the supply of oxygen and allows for undisturbed access to the oil container. This way one can say that *the lid is there* and the *lid is not there.* Thus, the contradiction is solved and the designed system may be ideal and function at a minimum of effort and operation costs.

The case of the sewing needle can be considered as another example of technological effects application [4, 5]. Such an application introduces a "friendly" needle with a large eye for threading and a small one when sewing. The discovery of new materials has made this dream possible, and now such a needle can be constructed using material that enlarges considerably when brought close to light and shrinks when taken away.

Another modern application of BTIPS is the design of a screen onto which movies can be projected. The ideal solution would be a screen located in the cinema theater in space, without any support resisting the gravity field thanks to the fact that the weight of the screen would be less than or same as the weight of air. Such a screen could be located in any place in the theater with the angle and tilt convenient for viewers. One can hypothesize even further that perhaps the physical screen can be eliminated, but its function could be saved. This function could be passed on to a projector that could generate light rays of a regulated length at the end of which the picture would appear in the air in front of the audience. As this example demonstrates, in the problem solving process there is no limitation in choosing an Ideal Solution; BTIPS encourages the designer to go as far as his or her imagination can reach. in the problem solving process. While limitations will be imposed by the real world due to the limited accessibility of materials, energy and knowledge of physical phenomena, i.e., by the limits of technological development. Since technology is developing very rapidly, it is wise to check from time to time to determine whether yesterday's dream solution has not already become a reality today.++

APPLICATION OF THE PREDICTION MODULE AND THE DESIGNED SYSTEM

The Prediction module should be used in case the attempts to use Effects and Principles modules did not bring about the desired solution. In such a case the designed system should be analyzed and the relations between its components should be studied. The precise definition of relations between the super-system, the designed system, the subsystems as well as the elements and objects should be described in order to solve successfully the basic contradiction using Prediction. The elements of the designed should be listed and depicted. Their definitions should be described as simply as possible. The interactions can assume three forms. The elements can be "hostile" or "friendly" towards each other, or they can be "indifferent" ("neutral"). Two elements of the same or different levels, if not indifferent, can have positive or harmful influences on each other. Sometimes the relations of two elements are simultaneously friendly and hostile, depending on time and situation. The designed system should be changed (see Table 5) to satisfy the problem statement. If necessary, the change of the problem statement should be considered. A virtual element may be used if there are no physical elements available.

In all cases, a clear statement of relations between the super-system, the system and sub-subsystems, along with the elements and objects, opens the path to a proper solution to a problem. In seeking a solution, the designer should try to reinforce the positive actions and weaken the negative ones. For example, the positive action of air in controlling the speed of a falling body can be reinforced by adding additional surfaces to the body; i.e. a special "flying bat suit" or a parachute. This way, the interaction is used positively. If one is interested in fast falling, the air action is negative and can be reduced by assuming an appropriately aerodynamic shape to the body. Thus, a good understanding of the interactions of the elements of a system is the key to inventive design.

TESTING THE IDEAL SOLUTION

There are several tests for the Ideal Solution. If the end solution satisfies the expectations, cannot be reduced (trimmed) in its physical and functional aspects, and requires minimum information for description and minimum energy to exist and function, then it is an Ideal Solution. There is no limitation in choosing an appropriate Ideal Solution in the problem solving process; in fact, it is recommended to go as far as designer's imagination can reach. The current level of technology development, not the designer's imagination, will determine any limitations. If it is not possible to find the Ideal Solution using Principles and Effects, the Prediction module should be applied.

REDUCING -TRIMMING THE SYSTEM

After coming to the end solution, one must answer the question: Is it possible to reduce or eliminate any element of the designed system or sub-system while still retaining its function? Or: Is it possible to delegate its function to other element and eliminate the element in question completely? If possible, this should be done. One should keep in mind however, that the Ideal System derived from the Ideal Solution to the problem is the one that can be described the minimal information, has its functions separated [13], and uses minimum energy to function and exist - all while providing the optimal service possible [8]. For this reason, it cannot be reduced or trimmed.

NO SOLUTION CASES

In spite of the power of BTIPS, there can be cases in which, even after applying all procedures and expending great effort, an ideal solution for the problem cannot be found. In such cases, the designer will have to propose the solution which is less-than- Ideal Solution and put the problem aside for the time being. One can always come back to the process of finding a solution once the technology relating to the problem has made further advances. For example Samuel Colt [15, 16] had to wait to make improvement to his firearm, which he originally designed as a powder gun, until the invention of rounds facilitated the loading operation and he could create his pistol.

CONCLUSIONS

Brief Theory of Inventive Problem Solving (BTIPS) is a compact, enhanced version of TIPS created especially for engineering education; however it is also suitable for research and practice as a useful tool in inventive engineering design. Thanks to its advantages over psychological methods, BTIPS is faster and surer in obtaining solutions. Thus is also born a serviceable tool for industry that will be able to support the most uncontrollable phase of design: concept generation. This method helps in the innovation of existing products and in the design of new ones. Indeed, a broad application of the method may mean a real technical revolution in design and production. Therefore, it is important that this method, along with the software tools and manuals, and notes, and examples, be made commonly available in both academia and industry. Together, with the IM software and its database, BTIPS can be of invaluable assistance to teachers, students and professional designers.

ACKNOWLEDGEMENTS

Some suggestions proposed in student reports and course materials from the University of Connecticut ME 3221 fall 2012 and 2013 courses were used in this article. The author appreciates the students' input and is grateful for the attitude and cooperation of the class. Special thanks go to Katie Kornacki of Willimantic, CT for her suggestions of text improvements and for editing this paper. The author is grateful to the former Invention Machine Corporation for the grant of IM v. 2 for Widows and TechOptimizer software and especially to Dr. Valery Tsourikov for his encouragement and help with beginning both the teaching and research on developing BTIPS.

This paper has been submitted to the IMECE 2014 session devoted to Prof. Robert G. Jeffers, Co-Director of the MEM Program at UConn until 2011 and Mechanical Engineering Department Head at the University of Connecticut for thirteen years. Dr. Jeffers was one of the key developers of the Management and Engineering for Manufacturing Program at UConn. As a Senior Vice President of ASME International, and later a member of the Board of Governors of ASME, he stressed the importance of engineering education and supported it through his ASME activities, as well as, academic and professional work.

REFERENCES

[1] The Carnage Policy Outlook entitled "The World Order in 2050." Projected G20 Economic Growth 2009-2050. http://www.Carnagieendowment.org/publications/index.cfm?fa=view&id=24195, visited on 8/13/2010.

[2] Bzymek, Z. M.: New Principles in Conceptual Product Design Using BTIPS, Proceedings of the Fifth International Workshop on Advanced Manufacturing Technologies, London, Canada, (L. Wang and X.Y. Liu, eds.), Sponsored by National Research Council of Canada—Integrated Manufacturing Technologies Institute, pp. 105–110, 2005.

[3] Altshuller, G.S. "Creativity as an Exact Science. Theory of the Solution of Inventive Problems," 2nd ed., Gordon and Breach Publishers Inc., New York, 1984.

[4] Tsourikov, V. M. "IM Course Notes," Invention Machine, Cambridge, MA, 1994.

[5] Invention Machine™ v.2.1 for Windows. Principles, Effects, Prediction. User Guide, August 1995, Invention Machine Corporation, 200 Portland Street, Boston, MA 02114-1722 USA.

[6] Invention Machine, Getting Started with Goldfire, Training for New Goldfire Users, Document Last Updated August 9, 2010 for v.6.0; Invention Machine Corporation, 28 State Street, Suite 3600, Boston, MA 02109, USA, tel. 1-617-603-5290; www.invention-machine.com.

[7] Petrovich, N. and Tsourikov, V. "*The Way to Invention*," (in Russian), Moscow,, 1980.

[8] Bzymek Z., 2006, Problem Solving using BTIPS – Learning and Applying in Conceptual Design, Research in Interactive Design: Proceedings of Virtual Concepts 2006, Fisher X., Couterllier D. - editors, Springer, p. 17-22.

[9] Bzymek, Z. M., 2005, "Brief Theory of Inventive Problem Solving—Enhanced Approach Using Principles," Innovations 2005—World Innovations in Engineering Education and Research, (A. Wing et al. eds.), International Network for Engineering Education and Research (iNEER), Begell House, pp. 503–516, 2005.

[10] Bzymek, Z. M., 2000, Conceptual Product Design Using BTIPS, DE-Vol.109, recent Advances in Design for Manufacturing (DFM), ASME 2000, pp. 139-147, (DFM), ASME, New York, NY.

[11] Quality Engineering and BT Thinking, Proceedings of the Fifth Conference, Culver City, CA. 1995.

[12] Invention Machine Corporation, 1998, TechOptimizer, Version 3.0, Invention Machine Corporation, 28 State Street, Suite 3600, Boston, MA 02109, USA, tel. 1-617-603-5290; www.invention-machine.com.+++

[13] Nam P. Suh, 1990, The Principles of Design, Oxford University Press, 1990.

[14] Kosse, V. "Some Limitations of TRIZ Tools and Possible Ways of Improvement," DE-Vol.103, "Conceptual and Innovative Design for Manufacturing,"ASME 1999. ASME IMEC and E, Nashville, Nov.14-19, pp. 111 -115, 1999.

[15] Keating, B. "The Flamboyant Mr. Colt and His Deadly Six-Shooter," Doubleday & Company Inc., Garden City, NY, 1978.

[16] The 'Russian' Colds. From Colonel Samuel Cold to the Russian Imperial Court."Design by Scott Beinfeld, phot. by Sid Latham, published on the occasion of the presentation of Russian Cold revolvers to the Museum of Connecticut History, Beinfeld Publishing, 1976, 1979.

[17] "Invention Machine ™ v2.1 for Windows; Principles, Effects, Prediction; UseGuide," Invention Machine Corporation, 200 Portland Street, Boston, MA 02114-1722 USA., August 1995.

[18] "TechOptimizer™ 3.0. Professional Edition for Windows® 95, Windows 98 and Windows® NT; Training & Reference Manual," Invention Machine Corporation, 200 Portland Street, Boston, MA 02114-1722 USA., 1998 Edition.

[19] Bzymek Z., 2006, The Quest for the Ideal Solution, Research in Interactive Design: Proceedings of Virtual Concepts 2006, Fisher X., Couterllier D. - editors , Springer, p. 50;

[20] Bzymek Z., 2007, Computer-Aided Problem Solving in Conceptual Design – Formalization and Development, CAD 2007, Hawaii, June 2007.

[21] Bzymek Z., Vahidi S., Spottiswoode H., 2007, Solutions of the 21 Century –Teaching Computer-Aided Conceptual Design, CAD 2007, Hawaii, June 2007.

[22] Bzymek Z.M., Innovations 2011, World Innovations in Engineering Education and Research, Ch. 40, iNEER 2011.

[23] Innovative Technology™ and Invention Machine™; January 1995, Invention Machine Corporation P. O. Box 38-0794 Cambridge, MA 02238, USA.

[24] Invention Machine™, 1995, v.2.1 for Windows; Principles, Effects, Prediction; Software and User Guide; August Edition, Inventor Machine Corporation, 200 Portland Street, Boston, MA 02114-17222, USA.

[25] Alex Faickney Osborn, Applied Imagination, 1963, ISBN-13: 978-0023895203

[26] Goldfire - https://gfi.goldfire.com/

ANNEX

Table A1. Summary of definitions, theorems and corollaries in BTIPS

<table>
<tr><th>Definition</th><th>Abbreviation</th><th>Theorems</th><th>Corollaries</th></tr>
<tr><td>1. Problem Description</td><td>PD</td><td rowspan="8">Theorem One*)
The Minimum Energy Theorem:
The Ideal Solution is such a solution that creates the simplest possible system that solves the problem and performs the desired function in the most satisfactory manner at the minimum energy expense.</td><td rowspan="4">Corollary 1.1:
The Ideal Solution solves the basic contradiction.</td></tr>
<tr><td>2. Problem Statement</td><td>PS</td></tr>
<tr><td>3. Problem Solving Process</td><td>PSP</td></tr>
<tr><td>4. Conflict Problem</td><td>CP</td></tr>
<tr><td>5. Basic Contradiction</td><td>BC</td><td rowspan="4">Corollary 1.2:
The Ideal Solution cannot be trimmed.</td></tr>
<tr><td>6. Degree of Isolation</td><td>DoI</td></tr>
<tr><td>7. Solution Model</td><td>SM</td></tr>
<tr><td>8. Solution Procedure</td><td>SP</td></tr>
<tr><td>9. Preliminary Solution</td><td>PS</td><td rowspan="8">Theorem Two *)
The Minimum Information Theorem:
The complete set of information describing the Ideal Solution is the least possible amount of non redundant data required to describe solution of the problem.</td><td rowspan="8">Corollary 2.1:
The complete Description of The Idea Solution Containing non redundant information cannot be simplified.</td></tr>
<tr><td>10. Current Solution</td><td>CS</td></tr>
<tr><td>11. Final Solution</td><td>FS</td></tr>
<tr><td>12. Trimming</td><td>T</td></tr>
<tr><td>13. Test of the Final Solution</td><td>TFS</td></tr>
<tr><td>14. Customer Requirements</td><td>CR</td></tr>
<tr><td>15. Specifications</td><td>Ss</td></tr>
<tr><td>16. Ideal Solution</td><td>IS</td></tr>
</table>

*) Adapted with some changes from [13]

Proceedings of the ASME 2014 International Mechanical Engineering Congress and Exposition
IMECE2014
November 14-20, 2014, Montreal, Quebec, Canada

IMECE2014-36196

EFFECT OF CHUCKING MOVEMENT WITH THE INDENTATION ON THE WORK-PIECE SURFACE IN CHUCK JAWS GRIPPING OF A LATHE BETWEEN AN EXPERT AND A NON-EXPERT

Porakoch Sirisuwan
Advanced Fibro Science,
Kyoto Institute of Technology,
Kyoto, Japan

Masayuki Nakamura
Mechanical Engineering,
Niihama National College of Technology
Ehime, Japan

Takashi Yoshikawa
Mechanical Engineering,
Niihama National College of Technology
Ehime, Japan

ABSTRACT

Analysis of depth and the roughness from the chuck jaws indentation investigated by the Olympus LEXT-OLS4000 Laser microscope. Strain gauges were used for measuring the gripping force of the jaws. The three-characteristic movements of an expert who ensured that the work-piece was kept steady and balance did not measurably affect the surface indentation of the work-pieces. The characteristic movement of the non-expert often straddled the left body to the left side while he was twisting the chuck-key has appeared the surface indentations. The depth inspection of them found the inside of surface indentation deeper than outside. Moreover, the results of a strain gauge measurement of all movements both an expert and the non-expert indicated the inside of the jaw had higher the strain than outside. Nevertheless, the results showed the most strain on work-piece surface occurred with the body movement of the non-expert.

NOMENCLATURE

The Standing Posture and the Hand Holding Position

AC Stand quite aligns with the chuck.

RC Stand on the right side of the chuck

LC Stand on the left side of the chuck

LR Both the left hand and the right hand held on the handle of chuck key

SR Only the right hand held on the handle and the left hand held on the shank of chuck key

INTRODUCTION

Chuck

Fig. 1: The characteristic of universal scroll chuck, the independent chuck and the chuck key

In the most common method of work holding, the chuck has either three or four jaws and mounts on the end of the main spindle (Figure.1). The universal scroll chuck usually has three jaws, which move in unison as an adjusting pinion rotated. The advantage of the universal scroll chuck is its ease of operation in centering the work for concentric turning. This chuck is not as accurate as the independent chuck but, when in good condition, it will centre the work automatically to within 0.003 of an inch. In addition, the universal scroll chuck is capable of holding and automatically centering round or hexagonal work-pieces. However, this chuck is unable to effectively to hold square, octagonal, or irregular shapes.

The independent chuck includes four jaws that are adjusted individually on the chuck face by means of adjusting screws. The jaws of the independent chuck may reverse so that the steps face in the opposite direction; thus, work-pieces capable grip either externally or internally. The independent chuck can uses to hold square, round, octagonal, or irregular shape work-pieces in either a concentric or an eccentric position due to the independent operation of each jaw. Because of its versatility and capacity for fine adjustment, the independent chuck is commonly to use for mounting work-pieces that require extreme accuracy. The jaws of the universal scroll chuck and the independent chuck have a series of teeth that mesh with spiral grooves on a circular plate within the chuck. This plate is rotated by the chuck-key inserted in the square socket [1].

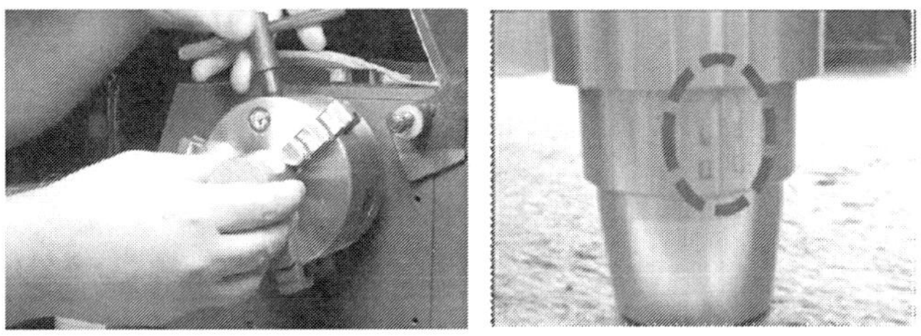

Fig. 2: The work-piece indentation by jaws gripping

Occasionally, the mistake of the indentation on the work-piece surface has occurred by chuck jaws gripping (Fig.2). The aim of the current study was investigated the differential factors between an expert and the non-expert such as the characteristic of standing posture, the hand holding on chuck-key, the balance movement of the body and the compression force that influenced of the surface indentations on the work-pieces which were arisen from chuck jaws gripping faces.

EXPERIMENTAL APPARATUS

The research on the difference of the behavior for chucking movement on the lathe processing between an expert worker and a non-expert worker was taken by a motion analysis system (sampling rate: 100Hz). An expert worker was 85 years old and has worked 70 years in lathe processing meanwhile the non-expert was 16 years old and has worked only 1 year. Fifty-six minutes assessment of an expert operating on the lathe machine found that there was a tightened characteristic of chucking movement on group I around 3 times, 4 times on group II and 3 times on group III meanwhile thirty-eight minutes observation of the non-expert on group IV had a 4 times (Table 1). The characteristic of an expert worker group I (Figure 3) was the standing quite alignment with the chuck (AC), both left and right hand held on the handle of chuck key (LR). The second one (Figure 5), an expert stood on right side of the chuck (RC) when the positions of hands were like the first one (LR). The standing posture on the third one of expert worker (Figure 7) was RC and only the right hand held on the handle but the left hand held on the shank of the chuck key (SR). Nevertheless, group IV of the non-expert (Figure 8), he always stood on the left side of the chuck (LC) but his hands position was like an expert on group I and II (LR).

Tab. 1: The groups of characteristic movements, which separated by the standing posture and the hand holding position

Type	Working hour	Frequency for tightly chucking	Standing Posture	Hand holding Position	Group
Expert	56 minutes	3	AC	LR	I
		4	RC	LR	II
		3	RC	SR	III
Non-expert	38 minutes	4	LC	LR	IV

Waist and Shoulder Movements [2]

Group I

The comparisons on the 3 times of chucking movement on bilateral of waist and shoulder movement on YZ plane of Group I (AC and LR) on Figure 3 and 4, the virtually identical in their indicated that most trend line of all movements in expert worker were fairly constant. The expert pretty less changed the Y position while he was turning the chuck key. He always kept the position on left and right body along Y-axis. The positions between left and right shoulder or left and right waist were quite similarly position on Y- axis. The distance movements both left and right side were quite equally. It showed that his body did not move up and down.

Fig. 3: Group I of an expert worker, his standing quite aligned with the chuck (AC), both left and right hand held on the handle of chuck key (LR).

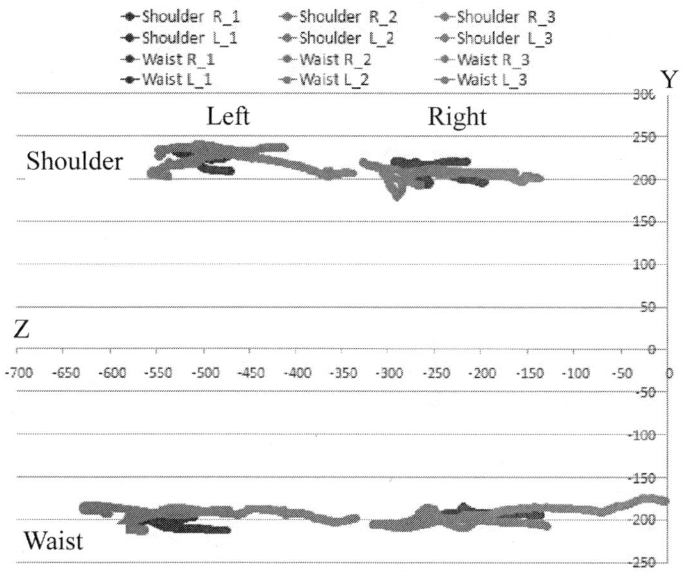

Fig. 4: Comparison the body movement between waist and shoulder on Group I (AC and LR)

Group II

Four-times chucking of the expert worker who stood on the right side of the chuck (RC) while both left and right hand held on the handle of the chuck key (LR) had a four-time for chucking (Figure 4). In this case, Figure 6 showed that all characteristic of body movement had difference the movements of group I. The direction movement of right waist quite sloped up when the left waist sloped down. However, the direction movement of waist and shoulder were quite similar, they had same the pattern movement on each side because when the waist sloped up or down the shoulder followed in them. Meanwhile, the distance movements of waist were shorter than the group I then the expert had a less movement. However, this characteristic movement still kept the position between left and right side of waist or shoulder equally along the Y-axis.

Fig. 5: Group II of an expert worker; he stood on right side of the chuck (RC), both left and right hand held on the handle of chuck key (LR).

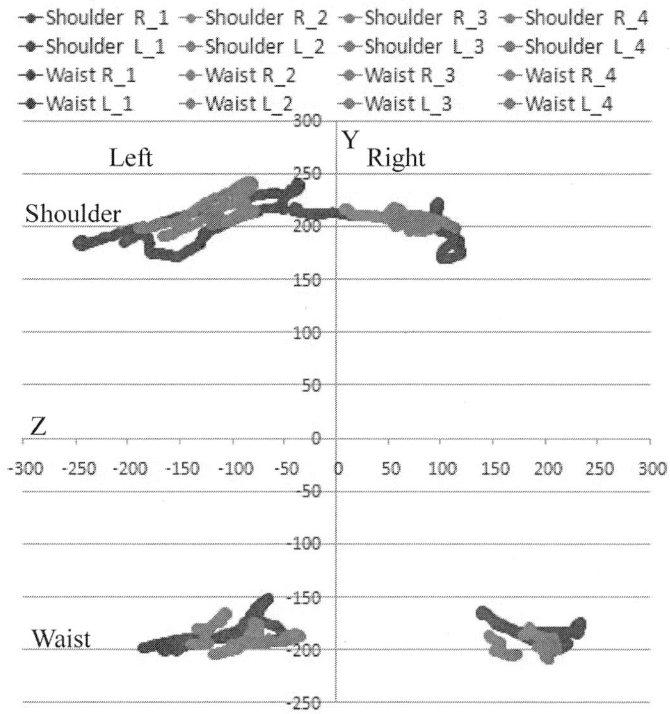

Fig. 6: Comparison the body movement between waist and shoulder on Group II (RC and LR)

Group III

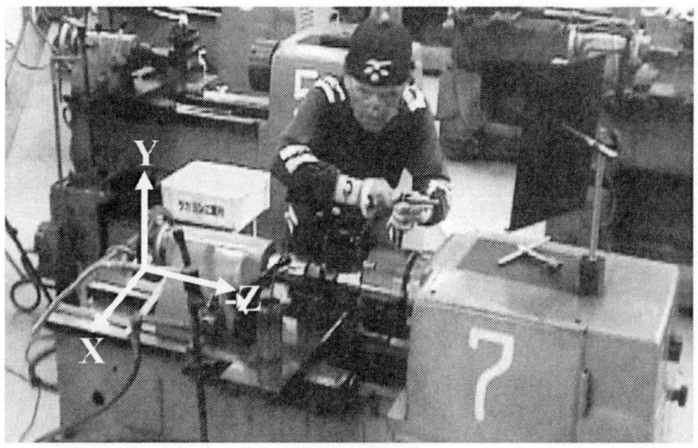

Fig. 7: Group III of an expert worker; he stood on right side of the chuck (RC), only the right hand held on the handle but the left hand held on the shank of the chuck key (SR).

Although the standing position of group II (Figure 5) was like group III (Figure 7) but the movement characteristics, both bilateral waist and shoulder quite differed. The several pattern movements of waist were too short as if the expert fixed the waist and moved only his shoulder. Even if the pattern movement of shoulder on this group is like the group II, shoulder right sloped up but shoulder left sloped down, but the several distance movements on group III were shorter than group II furthermore the line graphs were more smooth than group II (Figure 8).

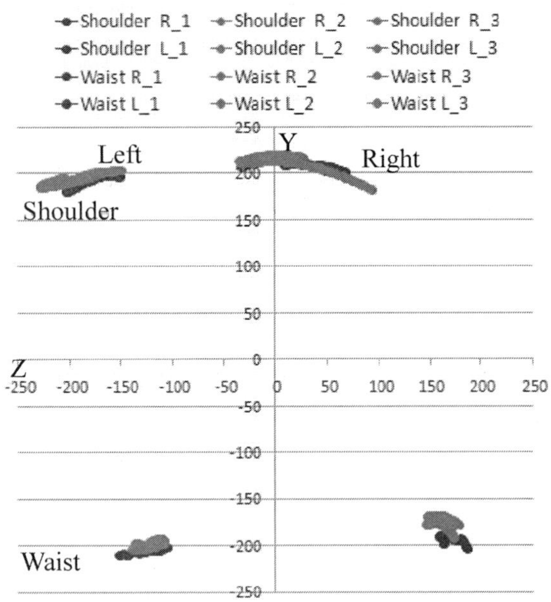

Fig. 8: Comparison the body movement between waist and shoulder on Group III (RC and SR)

Group IV

Fig. 9: Group IV of the non-expert worker; he stood on left side of the chuck (LC), both left and right hand held on the handle of chuck key (LR).

Figure 9 showed the analysis chucking behavior of the non-expert in Group IV. Even though the position both left and right hands of the non-expert held on the handle as though an expert group I and II but his pattern movement was not. This graphs on Figure 10 indicated the uncommon movement of the non-expert worker. During chucking performance, the non-expert has inclined the body left down all the time. In addition, he did not keep the balance movement between left and right body. Sometimes found that the position data of left waist or shoulder was higher than the right side sometimes was contrary. Furthermore, all of distance motion of this one was longer than an expert movement.

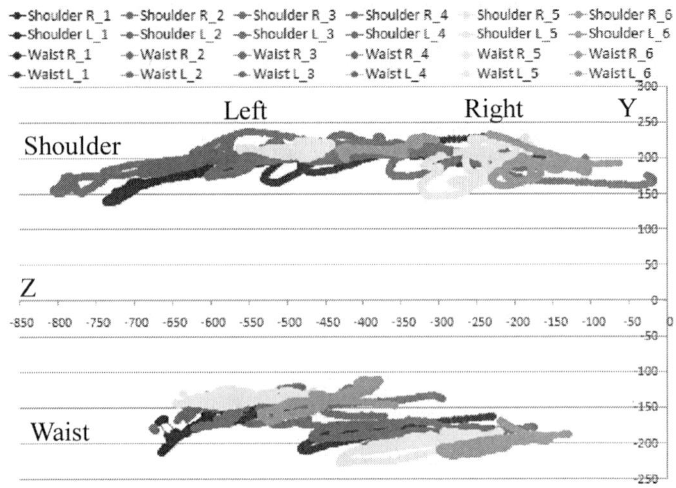

Fig. 10: Comparison the body movement between waist and shoulder on Group IV (LC and SR)

Center of Gravity movement

Group I

Figure 11 as the expert worker began twisting the chuck-key, both the waist and the shoulder would turn together meantime his elbows swung to the left side. The performance of waists, shoulder and elbow on the top plane (XZ plane), which are like the curves line could confirm that during the expert was chucking he used the center of the body as the line of backbone for center of rotational movement. Furthermore, the expert always move the shoulder forward the waist in XY plane.

Group II

Figure 12 the jaws tighten up by this procedure often used his waists for center gravity for twisting. The expert worker still moved the shoulder forward the waist and still bent his elbow both left and right side stretched it out of his body. He would fix the waist and jerk the elbow for pulling the shoulder to the left side together when he got starting. Both shoulders and hands would follow up when the elbows moved. On top plane (XZ), the body of right side moved less than the left side. The expert still keep steady straight between the position of hands and elbows of right side. However, he a few leant the shoulder, hand and elbow of left side. This motion group did not appear the leaning between the bilateral waists.

Group III

Figure 13, this movement was quite similar pattern of group II but he had a few leaning the left shoulder. On the top plane (XZ) the left body movement such as the waist, shoulder, elbow and hand movement hardly changed the position movement when was compared with the movement of right body and compared with the left body movement of group II. The expert would fix the left body as like the center of turning.

Because the maximum distance movement of group III was belonged to the hand and the elbow of right side hence, he used only the right body to exert the force when he was turning the chuck-key.

Group IV

Conversely, on group IV of the non-expert in Figure 14 indicated that he could not keep the balance movement when he was chucking. The YZ plane showed that he often leaned the left body. On XZ plane, the non-expert kept standing upright along the movement of chucking. The positions of each waist and shoulder on each side were quite concentrically. It showed that the non-expert did not move the shoulder forward.

Beside he rotated both shoulder right and waist right to backward for exerting force to twist the chuck-key. At the same time, both left shoulder and waist have moved to the left side as a rotational pattern. The trend line of body movement on top plane showed that the non-expert always has rotated the body movement together with swung the body to the left side by used the left hand as a center of rotation. Therefore, the distance movement of left side has been less than the right side. The non-expert inclined the body left down all the time.

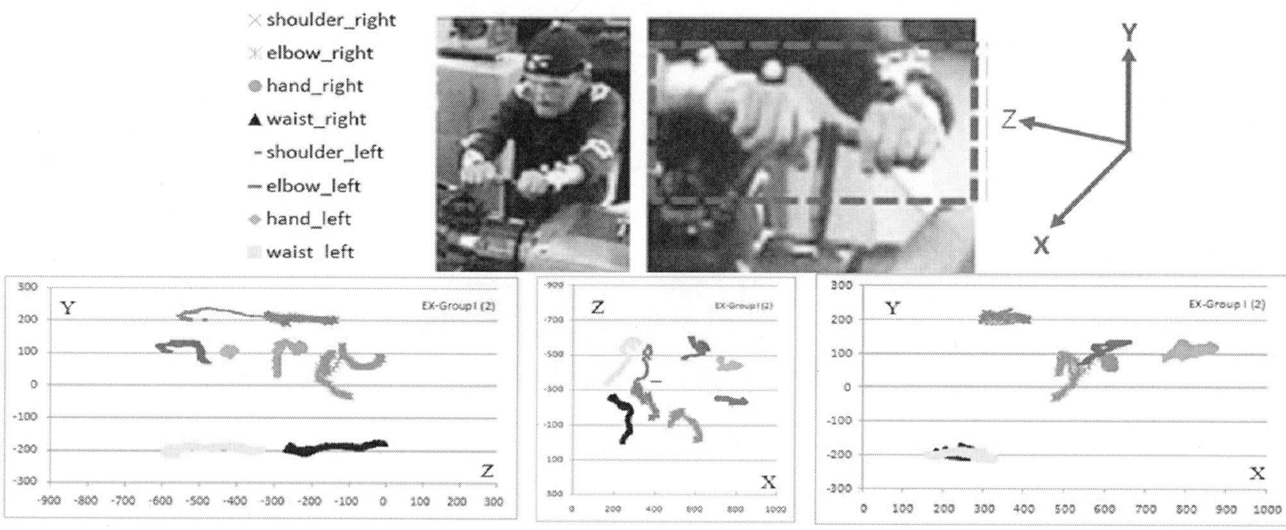

Fig. 11: The Trend line of the body movement of Group I; Standing posture: AC, Hands Position: LR

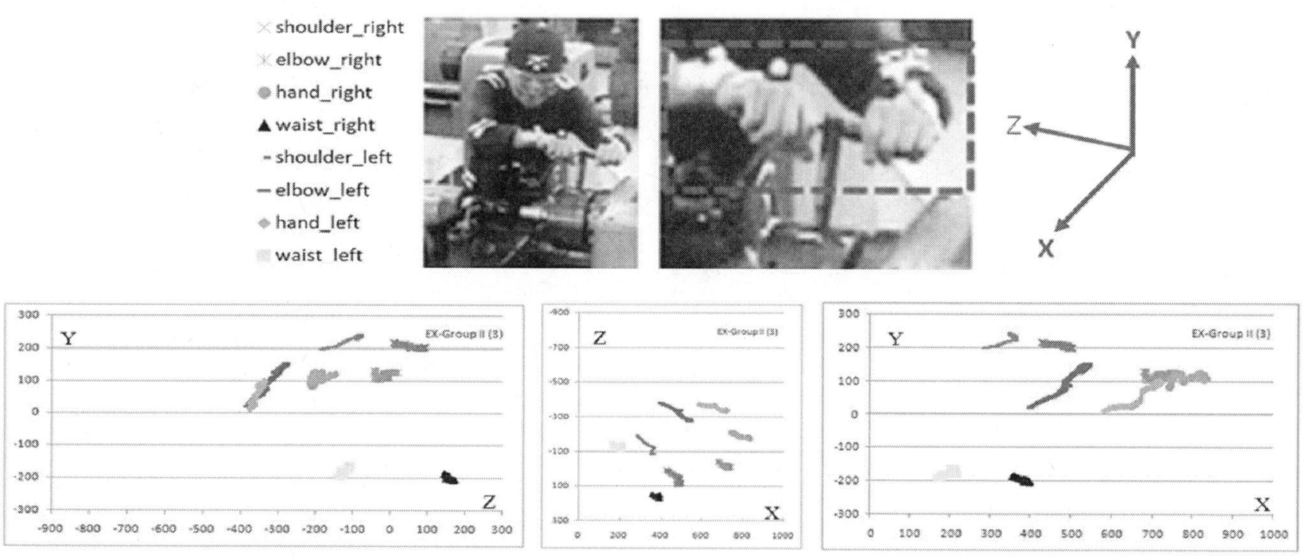

Fig. 12: The Trend line of the body movement of Group II; Standing posture: RC, Hands Position: LR

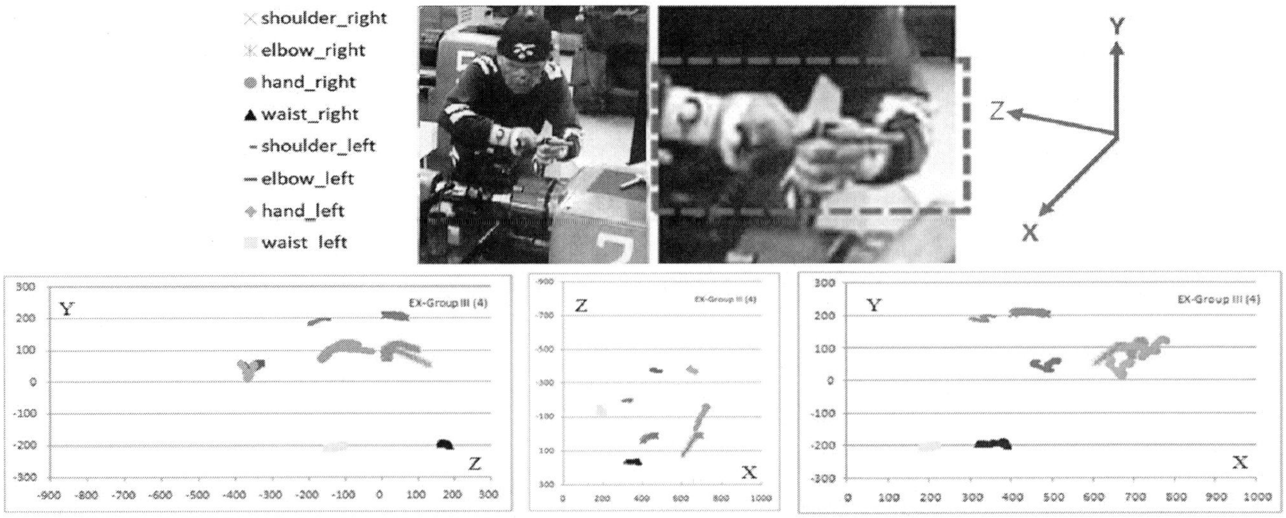

Fig. 13: The Trend line of the body movement of Group III; Standing posture: RC, Hands Position: SR

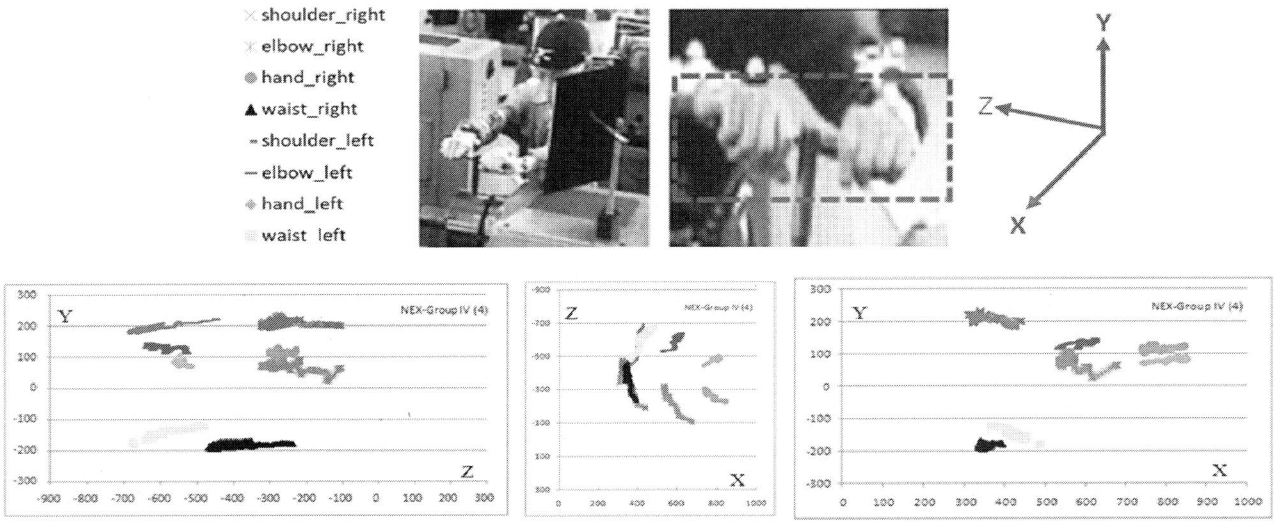

Fig. 14: The Trend line of the body movement of Group IV; Standing posture: LC, Hands Position: LR

Observation on Work-piece Surface Indentation

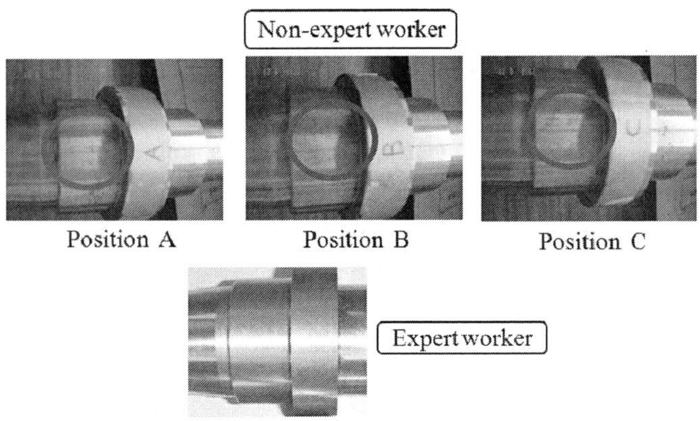

Fig. 15: The work-piece indentation by jaws gripping between an expert and the non-expert worker

The observation with the indentation of work-piece surface that occurred in the four-categories of chucking movement of an expert and the non-expert worker found that although the expert worker had a three-type of posture movement while chucking but disappearance the surface indentation on his work-piece. Only the work-piece surface of the non-expert had the indentation from the jaws (Figure.15). The work-pieces material was used a S45C.

The Olympus LEXT-OLS4000 Laser microscope was used to inspect the accuracy in roughness and depth of these surface indentations. Position A, B and C on the work-piece surface on Figure.15 indicated to be the chuck jaw number (Figure.16).

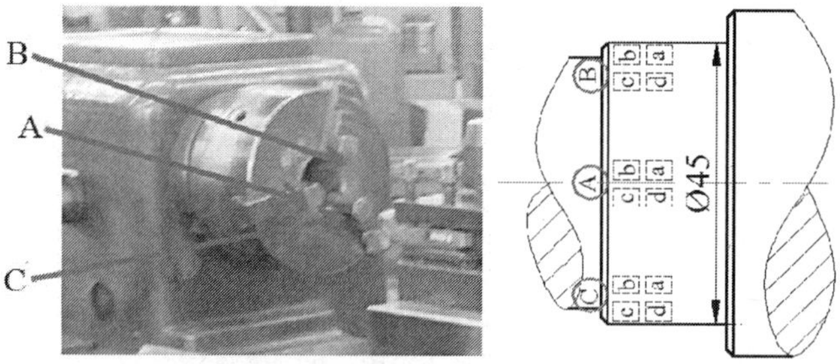

Fig. 16: The connection of position A, B and C with the jaws and the surface indentation

Furthermore, on Figure.8 each tooth of jaw indentation was indicated by a small letter such as a, b, c and d. However, the corner of jaws indentation used the number 1, 2, 3 and 4 for easy determination.

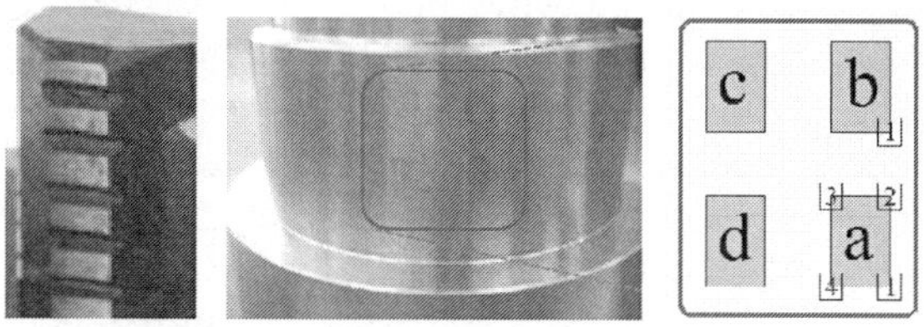

Fig. 17: The quality of jaw teeth, jaw indentation and the indication numbers of position on work-piece surface

The investigation by the Olympus LEXT-OLS4000 Laser microscope found that the deepest position of jaw numbers A, B and C consists of position C-c-2, A-b-3 and B-b-1 on work-piece surface respectively (Figure.18). All deepest positions were the position at left side contacted inside of jaw teeth (Figure 19).

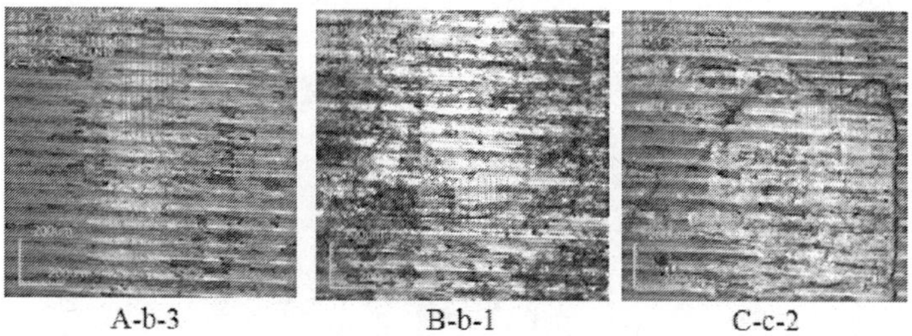

Fig. 18: The deepest of jaws indentations

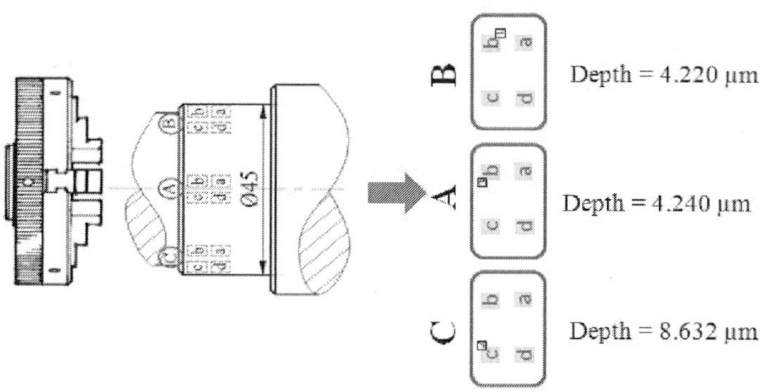

Fig. 19: The deepest of surface indentation by jaws clamping

PROCEDURE AND INSTRUMENTS

The aim of this study would like to examine how only the chucking behaviors of the non-expert worker has influenced the surface indentation [3]. This experiment compared the compression of jaws after chuck-key using for tight gripping of the jaw between an expert and non-expert worker behavior.

Subjects

Three males of non-experts who were an under graduate student in Niihama national college JAPAN which had the ages between 20 – 21 years with known the lathe experience around 30 hours to serve as subjects. They imitated the chucking behavior of an expert group I, II, III and non-expert group IV.

Strain gauge

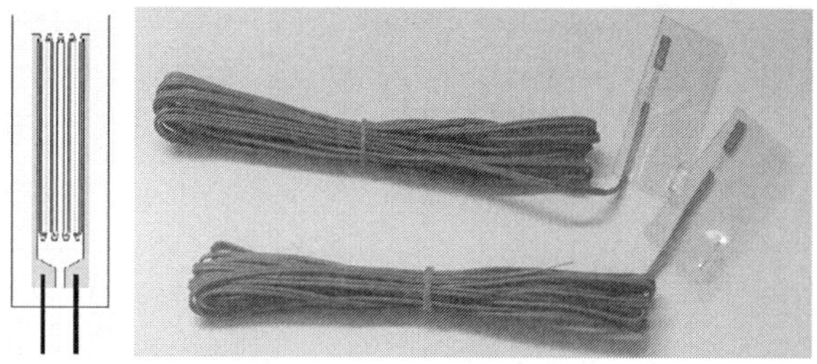

Fig. 20: The Single-axis strain gauges

The single strain gauges on Figure.20 were a device used to measure strain of the work-pieces into the jaws. Most commercial strain gauges had a sensing element made from a thin metal wire of a foil pattern with an electrical resistance that changes when the element was compressed, expanded of bent. While the work-pieces are deforming the metal wire or foil was deformed and transmits to electrical resistance to change. The specimens were a steel-pipe which 60 centimeters outside diameter and 56.5 centimeters inside diameter. Four-strain gauges were fixed inside the steel pipe and used the capital letters A', B', C' and D'. The position of each strain gauge showed that they were attached directly to the position of jaw gripping (Figure.21).

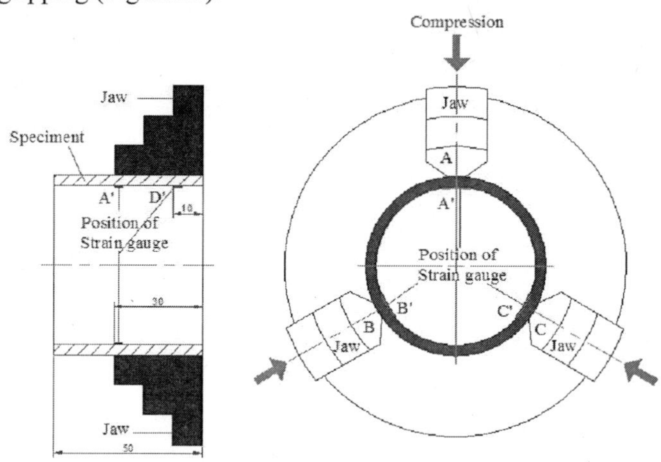

Fig. 21: Position of strain gauges inside the steel pipe

A Beam Torque wrench

A torque wrench in this experiment (Figure.22) was used for measurement how tightly a jaw. A beam wrench worked by the beam bending in response to the torque applied. The unit of this one was Kilogram x Centimeter.

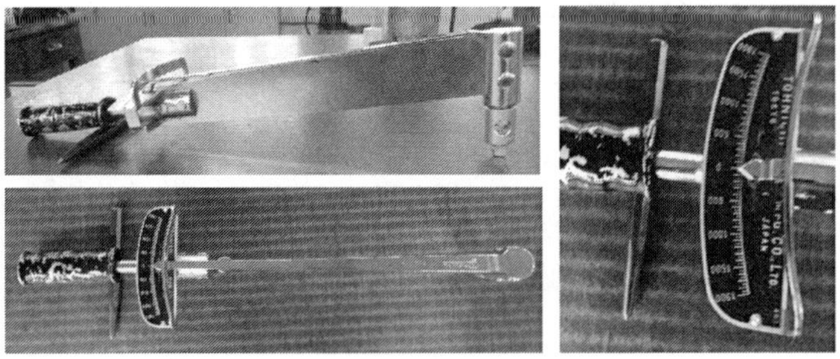

Fig. 22: Using the beam torque wrench

RESULTS AND DISCUSSION

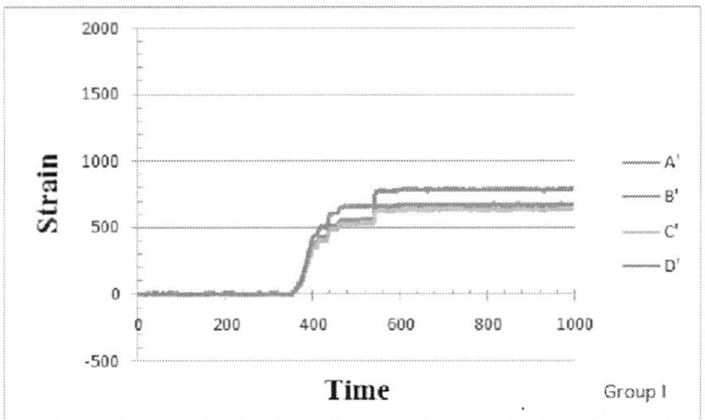

Fig. 23: The trend line of strain on work-piece surface which clamped on the chucking movement

Recovery lines pattern were plotted for each movement showing the strain value on four positions of strain gauges such as position A', B', C' and D' were located on work-piece specimens (Fig.23).

Consistent with the indiscrete pattern changes demonstrated by the strain gauges measurement, the strain also not changed pattern at all performance. This was clear from the bar graph that for every chucking performance (Group I – IV) on Three non-experience of lathe operation. In fact the peak number of strain value on completely behavior was achieved in position D' which located deeply inside.

Figure 24 showed the strain value changes amongst four movement categories of three non-experts in chucking behavior. There were the trends of strain sequence similarly pattern although the subjects had a differently personality. It was clear that the imitated chucking mobility of group IV on the non-expert number 1, 2 and 3 are the highest in terms of strain value. The average of this one was around 1960, 1174 and 874 respectively. The next group of the imitated chucking behavior of group II showed almost significantly higher strain value than the group I and III. Interestingly, the bar graph on group III of the behavior simulation of an expert from the non-expert altogether indentify that this pattern movement got the lowest strain value during turned the chuck-key for transference the power to the jaws for clamping the work-piece.

The strain ratio comparison among the behavior on group I, II, III and IV by the three non-expert subjects in Table 2 found that almost evaluation of strain ratio has quite resembled. The data analysis showed that the power exertion by the chucking behavior group IV as well as around the second half, double and four times of the Group I, II and III respectively.

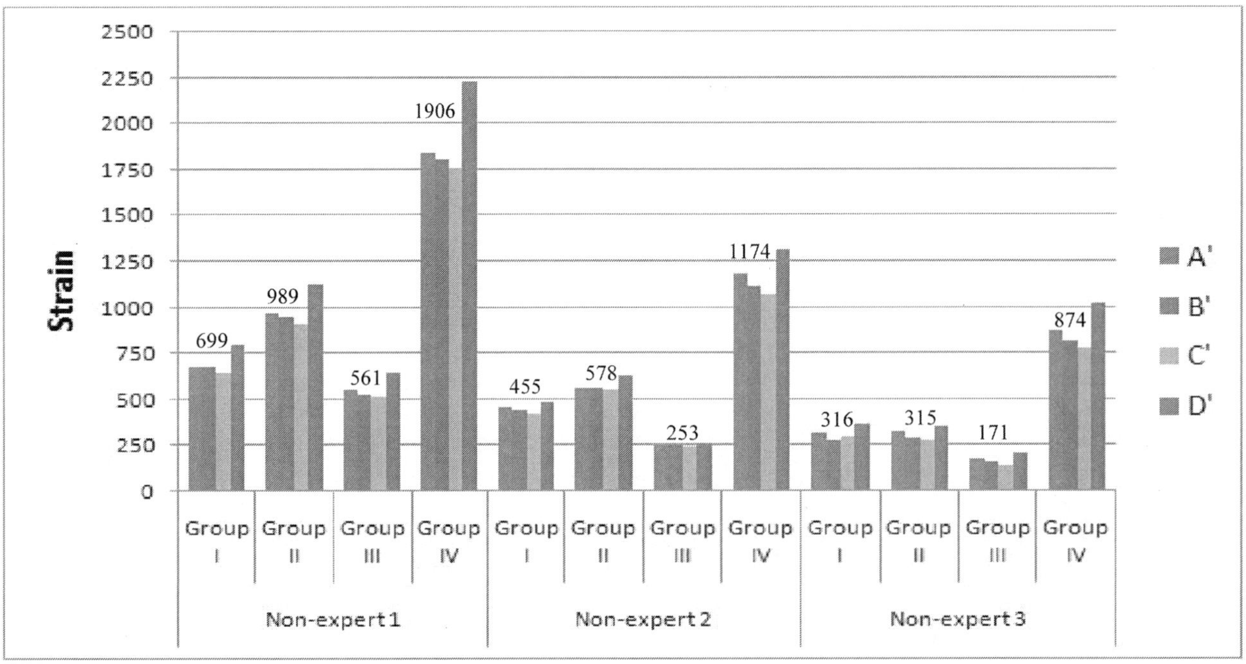

Fig. 24: Comparison of strain value depend on the type of chucking behavior

Tab. 2: The strain ratio among the chucking performance on group I, II, III and IV by the behavior simulation

Subject	Group I	Group II	Group III	Group IV
Non-EX1	0.37	0.52	0.29	1
Non-EX2	0.39	0.49	0.22	1
Non-EX3	0.36	0.36	0.20	1

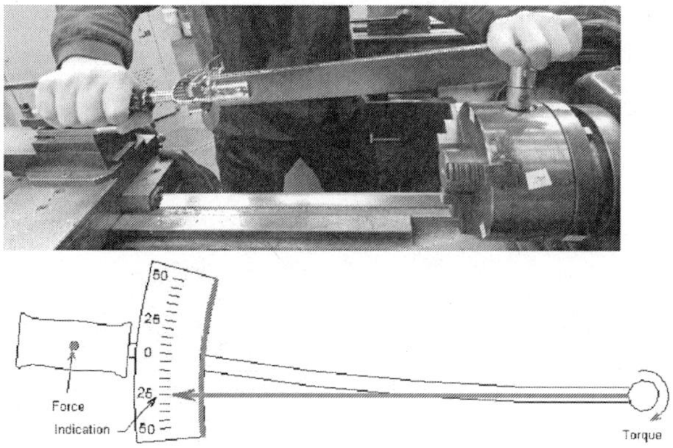

Fig. 25: Using the beam torque wrench for measurement the torque of chucking

The results of strain values were transferred to the torque value by using the beam torque wrench[4-5]. At first, put the square wrench on it into the square socket of the chuck (Figure.25). After that, turned the beam torque wrench for tighten the work-piece. Play an attention on the value of torque (Kg.cm) that indicated on the torque scale stayed in the range 100, 200, 300 and 500 Kg.cm together with assessing the strain values on the strain gauges position A'. In order of torque scale, the strain included 173, 393, 695 and 937 respectively.

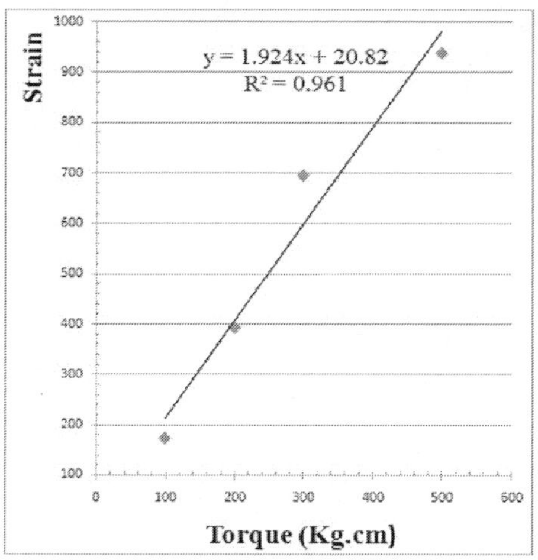

Fig. 26: The correlation of strain and torque of chucking movement

Afterwards, the strain values and the torque values were brought to make the graph of Fig.26 for transformation the tendency strain values of position A', B', C' and D' on four groups of chucking movement to the actual torque (Table 3).

Tab. 3: The actual torque on chucking movement depend on the chucking behaviors

Subject	Group	Strain ($x10^{-7}$)	Torque (Kg.cm)
Non-EX1	I	699	353
	II	989	503
	III	561	281
	IV	1906	980
Non-EX2	I	455	226
	II	578	290
	III	253	121
	IV	1174	599
Non-EX3	I	316	153
	II	315	153
	III	171	78
	IV	874	443

CONCLUSION

The process of surface indentation analysis, which performance with the chucking movement of an expert and non-expert workers found that;

1. The strain and torque of position D' was the greatest without the irrespective of the characteristic of the chucking movement. Even if the chucking behaviors have changed the standing postures; AC, RC, LC; and the hand holding position on the chuck key; LR and SR; but they have not affected the highest strain and torque still belong to the position D'.

2. The deepest of indentations by jaw tighten have always occurred at the interior of jaw teeth therefore, the results analysis of strain and torque were affirmed the more reliability. The more the strain has increased the torque and the depth of surface indentations have become the better.

3. All resultants analysis have confirmed that the chucking behavior of the non-expert worker whom could not keep the balance movement by often leaned the left side of body while he was beginning chucking cause of the surface indentation. The torque on the chucking behavior of the non-expert was excessive and affect to the compression of jaw teeth.

REFERENCES

[1] Army Institute for Professional Development, 1988, "Lathe Operation", Edition 8, pp. 30-33.

[2] Sirisuwan, P., Koshino, T., Narita, C. and Yoshikawa, T., 2013, "Difference among an Expert and a Non-expert in the Behavior for Chucking in the Lathe," *Proceeding, the*

ASME2013 International Mechanical Engineering Congress & Exposition., San Diego, California, USA, pp. 245.

[3] Yoshikawa, T., Ito, R. and Tsujinaka, T., 2012, "Stance of chucking by the Takumi on Lathe Working," *Proceeding, The Japan Society of Mechanical Engineering*., Japan, pp. 517.

[4] Ascanio G, Castro B. and Galindo E., 2004, "Measurement of power consumption in stirred vessels-a review," *Trans IChemE, Part A, Chemical Engineering Research and Design,* 82(A9): pp.1282–1290.

[5] Anna L.B., Karin H and Paulien B., 2004, "The effect of using a laptopstation compared to using a standard laptop PC on the cervical spine torque, perceived strain and productivity", *Applied Ergonomics*, Volume 35, Issue 2, pp.147-152.

Proceedings of the ASME 2014 International Mechanical Engineering Congress and Exposition
IMECE2014
November 14-20, 2014, Montreal, Quebec, Canada

IMECE2014-36631

COMPARISON OF DIFFERENT GRINDING SOUND DURING METALLOGRAPHIC PREPARATION

Takuya Sugimoto
KOYO Netsuren Corporation
Kyoto, Japan
t-sugi@koyo-kinzoku.com

Eriko Aiba
The University of Electro-Communications
Tokyo, Japan

Akihiko Goto
Osaka Sangyo University
Daito, Osaka, Japan

ABSTRACT

The carburizing process requires metallurgical inspection by means of ground metallurgical mounts. Grinding process for a metallurgical mount is an important process. In this study, we investigate the difference in the outcome of the sound during the grinding process between an expert and a non-expert execution. We aim to identify the evaluation criteria in grinding technique based on the sound information, in order to establish more efficient training method for acquiring the grinding techniques for non-expert inspectors. As a result, we found the factor in the sound that are essential for the efficient grinding.

INTRODUCTION

Carburizing is one of the most common surface hardening techniques used for gears and mechanical parts. In addition to measuring surface hardness, test specimens were evaluated by case depth measurement and metallographic examination. The metallographic analysis included austenite quantity, carbide networks, inter-granular oxidation depth; and grain size measurements. To accurately examine the sample test specimens requires a near-surface microstructure. The preparation process for test specimens is shown in Figure 1. First, a section is cut perpendicular to the surface at the measurement location. Second, the obtained specimen is hot-mounted with epoxy resin and then filed using two different types of coarse-grained papers. Stepwise grinding is then performed to produce a flat surface, followed by refined abrasive grinding to obtain a mirror finish. During the preparation process, if the sample is damaged or its edges become rounded, accurate metallographic information cannot be obtained. The resin used is weaker than the sample, and therefore may be ground away more quickly. In this case, the surface of the sample becomes rounded, preventing a microscopic, near surface microstructure analysis. Difficulty arises when pin-pointing the exact cause of process failure, especially when evaluating the quality of the heat-treatment process.

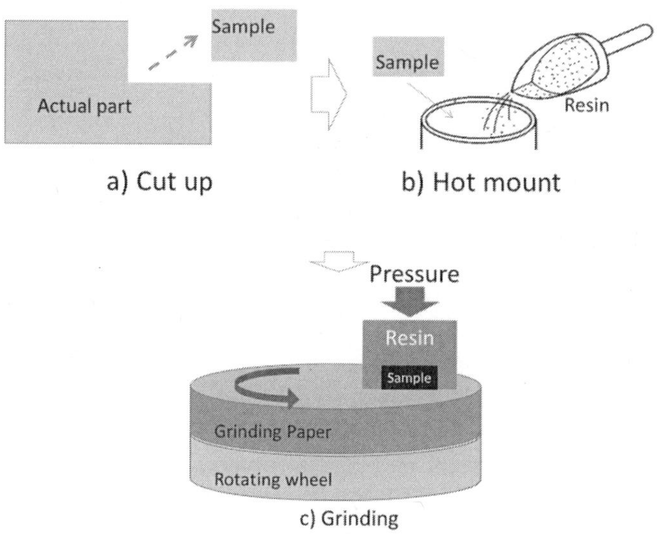

Fig. 1 The preparation process for test specimen

This preparation process is standardized in ASTM E3[1] as semi-automated. However, the outcome may differ depending on the skill of the technician. The hardness and shape characteristics of the sample vary depending on their components. Expert technicians thus change their polishing and grinding processes accordingly. In this study, carburized and hardened gear parts were polished by expert and non-expert technicians, and the effective case depth was measured in four different positions on a pitch circle. The results, shown in Figure 2, indicated σ: 0.017mm for the expert's specimens, and σ: 0.027mm for the non-expert's specimens. Evidence suggests that the expert had less variance than the non-expert between the positions measured. However, proper grinding technique is a sort of intuition that is acquired through long experience. This technique has been conveyed across generations in the form of tacit knowledge, without formalization. Thus, training new technicians is a difficult and inefficient process. L. E. Samuels established the theory behind the preparation process for metallographic test specimens [2], and M. F. Smith compared thermal sprayed specimens created by different abrasives [3]. However, no study has yet been conducted on the skills acquired when handling a specimen, which experts obtain through long-term experience. We have researched the expert's skill on grinding process and ground surface [4-5]. In this study, we compared the grinding sounds generated by expert and non-expert technicians, and verified that it is possible to evaluate grinding technique using acoustic information. By utilizing acoustic information, we were able to evaluate technique without installing large-scale equipment or attaching measuring devices to the technicians' bodies. The ultimate goal of this study is to use the results of our acoustic evaluation to try to establish a more efficient and reliable system to train new technicians in proper grinding techniques.

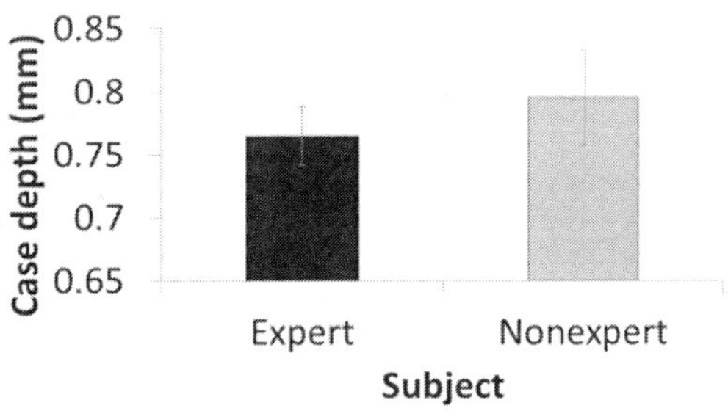

Fig. 2 Case depth between subjects

EXPERIMENT METHOD

One subject was an expert technician with 20 years of experience, (height 170cm, dominant hand; right, age 48 years, duration of employment, 28 years). The other is a non-expert technician, with 2.5 years of experience, (height 165cm, dominant hand; right, age 32 years). Both subjects were engaged in this type of work on a daily basis at the time of this study.

The grinding sample used were gears, specifically a 9310 (AMS6265,) cut into a quarter of a gear consisting of four teeth, (24.5mm pitch diameter). Each specimen was finished with a "hot mounting" treatment before it was ground and polished.

Experiment equipment

The experimental situation is shown in figure 3. Grinding sound was recorded using linear PCM (Pulse Code Modulation) and a digital audio recorder (LS-10, Olympus), equipped with a linear PCM microphone. The audio was recorded with a sampling frequency of 44.1 kHz at 16-bit.

Grinding and polishing procedures were performed by the technicians, utilizing the same semi-automatic grinder (Refine Tech, STO-228K) they normally use. The grinder controls used are a lever to adjust grinding pressure, a rotation regulator for the adjustment of the rotational speed of the grinding disc, and a snap switch to supply water to the grinding disc. Grinding was carried out by pressing a test specimen to the rotating disc attached with abrasive paper. The polishing motions at each step in the process were recorded by video camera. We evaluated the difference between technicians at each step in the polishing process.

a) Grinding machine

b) Location of digital audio recorder

Fig. 3 The experimental situation

Experiment Procedure

The subjects performed grinding in standard conditions and procedures, using P120 abrasive paper. We recorded the grinding sound during the operations. The conditions for the number of grindings, grinding pressure, disc RPM, and water supply were set to mimic the conditions of the technicians' typical grinding. Each subject repeated the grinding operation five times for a total of ten recordings.

RESULTS

Grinding time:

Figure 4 shows the five grinding times of each subject. We conducted a t-test based on this result and found out that the grinding time of the expert's trials was significantly longer ($P<.05$), with a larger standard deviation than the non-expert's trials. This suggests the possibility that the expert changed the grinding time to suit his sense of the specimen.

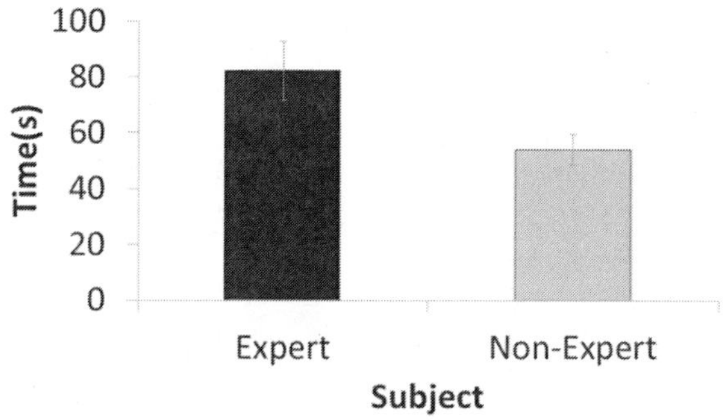

Fig. 4 Average grinding time between subjects

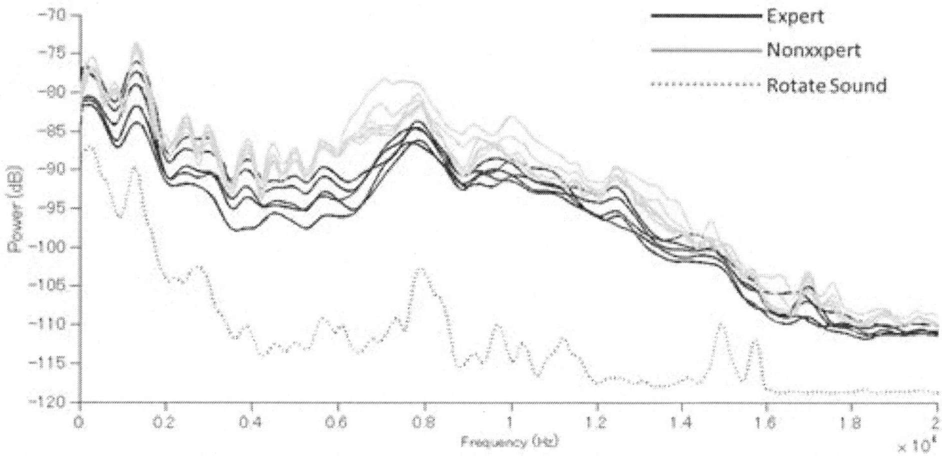

Fig. 5 Spectrum between subjects

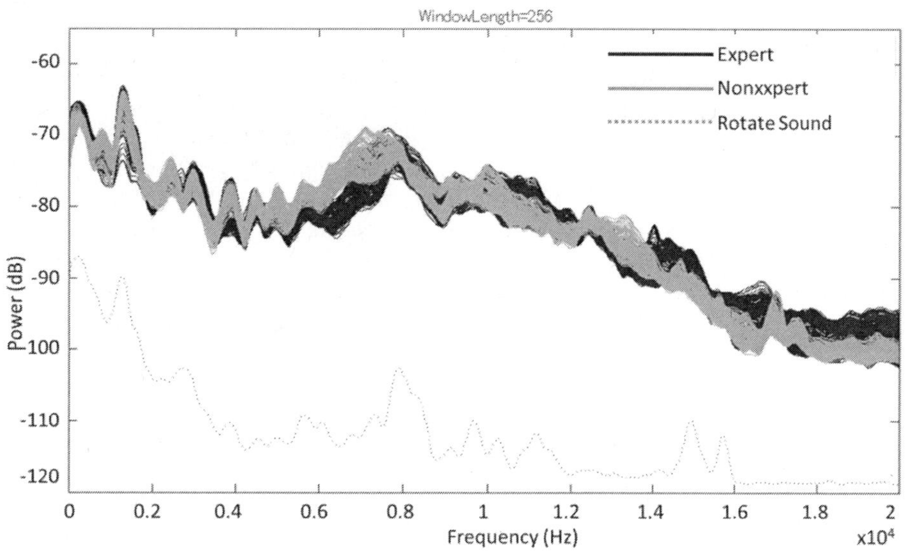

Fig. 6 Normalized spectrum between subjects

Spectrum

Figure 5 shows the frequency components contained in the grinding sound made by each subject. The dotted line indicates the state in which only the rotation disc of the grinding machine was moving. The spectrum during grinding operation is indicated in black for the expert technician and in gray for the non-expert technician. The non-expert consistently showed greater power overall. The shapes of the frequency component are almost the same. However, sound in the frequency around 4000-8000Hz was only detected in the recordings of the non-expert.

Normalized spectrum

The sound generated during the non-expert technician's trials was louder than the recordings of the expert's trials, and the spectrum shown in Figure 5 generally indicates higher levels of sound across all frequency components for the non-expert. Thus, in order to make a fair comparison between subjects, we normalized the sound pressure level. We extracted 10 second intervals from the waveform every second (ie. 0 to 10 seconds, 1 to 11 seconds, 2 to 12 seconds) and adjusted all effective values at the same rate. Then, we acquired the spectrum based on the extracted 10-second waveform.

The normalized spectrum is shown in Figure 6. The dotted line shows only the rotating sound, the gray indicates the frequency components in the recording of the expert, and the black indicates the frequency in the recordings of the non-expert.

This result indicates that the recording of the grinding sound during the expert technician's operation includes more relatively high frequency components, above 14 kHz. On the other hand, the non-expert technician's operation included more frequency components between 6 and 8 kHz.

Amplitude Transition

The amplitude transition waveforms of the 5 trials of each subject are shown in Figure 7. Rapid collision noise (peak) is more common in the amplitude of the non-expert's recordings. We also detected peaks 30dB louder than the baseline acoustic pressure level of the disc rotation, and compared the number of peaks per second. The result of number of peaks per second is shown in Figure 8. The non-expert's recordings showed significantly more peaks.

DISCUSSIONS

P120 abrasive paper is used for rough polishing and can create deep scratches in specimens. If a deep scratch or edge rounding occurs at this stage, the ground surface obtained will hardly meet quality assurance in the following fine polishing process.

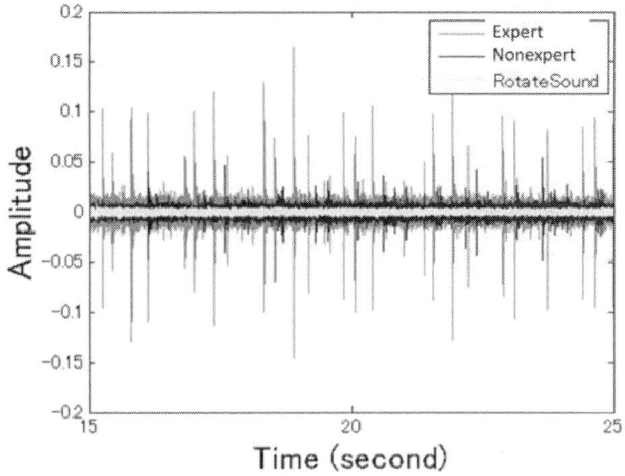

Fig. 7 Amplitude waveform between subjects

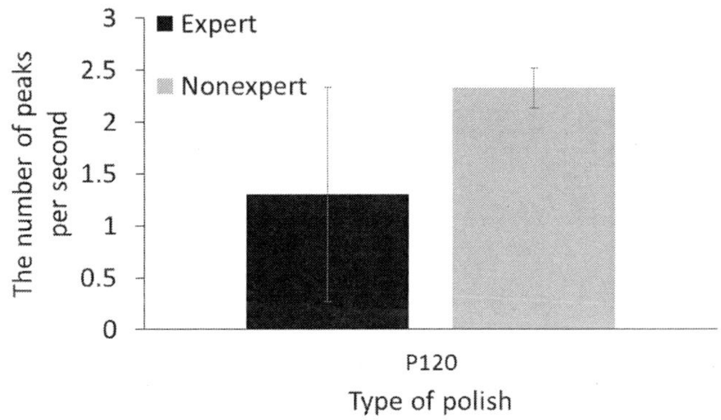

Fig. 8 The number of peaks per second

Due to the frequency components, we can see that the non-expert used comparatively more power and applied more grinding pressure than the expert, in order to eliminate the scars caused by cutting. Thus, the grinding disc with the abrasive paper was pressed more firmly against the specimen, and more collision noise was created, as shown by the amplitude transition.

Ideally, the material is cut away solely by the grains of the abrasive paper. If, during grinding operation, the rotating disc is pressed too strongly to the specimen, plowing may occur, and large scratches will remain. This damage will prevent a flat ground surface from being obtained. Expert technicians know this from experience, and the expert technician in this trial avoided pressing the specimen to the disc too strongly. This allowed him to grind the surface evenly and decrease the standard deviation of the effective case depth, as mentioned at the beginning of this paper. We consider that the average grinding time of the expert was longer because the expert technician kept the grinding pressure lower, grinding more slowly and carefully than the non-expert.

CONCLUSION

In this study, we compared the sounds generated during the grinding process for expert and non-expert technicians and verified that it is possible to evaluate grinding technique using acoustic information. The results showed that many peaks were included in the amplitude transition of the recordings created by the non-expert technician. We also found that more power was detected in the frequency components in the non-expert's recordings. In comparison, the recordings of the expert technician contained more relatively high frequency components of 14kHz or above. These results suggest the possibility of evaluation of grinding pressure and technique from the amplitude transition and frequency components.

References

[1] ASTM International, 2011, "Standard Guide for Preparation of Metallographic Specimens," E3-11

[2] Samuels L. E., 2003, "Metallographic Polishing by Mechanical Methods," *ASM International*.

[3] Smith M. F., 1993, "A Comparison of Techniques for the metallographic Preparation of Thermal Sprayed Samples," *Journal of Thermal Spray Technology, pp. 287-294*

[4] T. Sugimoto, T Sakamoto, Y. Kuwahara, T. Koshino, Y. Takai, A. Goto, "Comparison of different preparation processes and final surfaces of metallographic specimen", 2013, Proceedings of the ASME 2013 International Mechanical Engineering Congress & Exposition

[5] T. Sugimoto, T Sakamoto, Y. Kuwahara, T. Koshino, Y. Takai, A. Goto, "Stepwise comparison of Expert and Non-expert preparation of a polished metallographic surface", 2013, Proceedings of the ASME 2013 International Mechanical Engineering Congress & Exposition

This page left blank intentionally.

Proceedings of the ASME 2014 International Mechanical Engineering Congress and Exposition
IMECE2014
November 14-20, 2014, Montreal, Quebec, Canada

IMECE2014-37732

VISUALIZATION OF BRUSHING IN "SHIBO" PRODUCTION OF HAND-MADE JAPANESE PAPER

Shinji NOJIMA
Niihama National College of Technology
Niihama, Ehime, JAPAN

Takashi YOSHIKAWA
Niihama National College of Technology
Niihama, Ehime, JAPAN

Akihiko IMAJO
Niihama National College of Technology
Niihama, Ehime, JAPAN

Takeshi TSUJINAKA
Saijo city
Saijo, Ehime, JAPAN

ABSTRACT

We deal with "Danshi" that has a special wrinkle called "Shibo" on the surface structure of the Japanese paper. Three sheets of wet papers are superimposed to make the Shibo. Wet papers are extended the slack by brushing. Superimposed wet papers are made the Shibo by especial pulling movement. This time, we investigate about the movement of brushing for extending the slack.

Subjects are two people which are wearing 18 reflecting markers on the body. An expert has manufactured Danshi for 23 years of experience and a non-expert has done less than 1 year of experience. The movement of brushing process is measured as sampling time of 100 [Hz]. We will investigate differences of the physical movement between the expert and the non-expert.

We make a hearing about the differences and work on visualization about the skill of the expert.

The method of making Shibo structure has five procedures. First, one piece of wet paper is placed on the inclined work table. Second, wrinkles of the paper are extended by brushing. Third, next wet paper is put on the first sheet, it is extended the slack again. Fourth, moreover third wet paper is put on the second sheet, it is extended the slack by brushing again. Fifth, the Shibo structures are produced by pulling in a state of three superimposed sheets.

We compared the number of brushing in each sheets. There was not difference in the number of brushing in each sheet in a non-expert. In other hand, number of brushing by the expert was two times that of first and second sheets in third sheet. The number of brushing in first and second sheets did not change by the expert. In consequence, he was a notion that important things for making the clear Shibo structure are not only extending the slack, but also pulling out the bubbles.

We divided brushing movement into eight directions. In third sheet of the expert, the number of brushing to body direction has increased over 10%. It was found that the expert removed air bubbles from the corner of the top right. Furthermore, we investigated the change of angle of the right wrist in the expert and the non-expert. It became clear that angle of the wrist is reduced when the expert is brushing forward.

INTRODUCTION

A special Japanese called "Danshi", that has many peculiar wrinkles called "Shibo" its surface structure, is shown in Figure1. Danshi was a good quality paper made in the bark of "Mayumi" (Defoliation small trees of celastraceae) in old days. In late Heian, Michinoku paper was made from mulberry. Also since the middle Ages, a good-quality and thick paper was made from the same raw materials in the Btsutyuu. Danshi is the longest paper name in lifespan. Danshi is made of over 100 years in Ehime Prefecture Toyo region. Production of Danshi was the golden age in around 1950. Suppliers of more than 100 houses were present at the time, but the only suppliers of 6 houses are guarding the tradition now. In Japan there are three problems in the hand-made Japanese paper industry. First is the aging of workers. Second is the slump of Danshi demand. Third is upbringing of successors is difficult. Therefore, we think so that inheritance of traditional technology threatened. In other words, the establishment of technology inheritance of the expert in existence is sought.

The method of making Shibo structure has five procedures. (1): One piece of wet sheet is placed on the inclined work table. (2): Extend the wrinkled sheet by brushing. (3): Second sheet is put on the first sheet, and it is extended the slack again. (4): Moreover third wet sheet is put on the second sheet, it is extended the slack by brushing again. (5): The Shibo structures are produced by pulling in a state of three piled sheets.

Fig.1 Surface of Danshi

Process of superimposing three wet papers is difficult for the non-expert because wet papers can tear easily. And this process takes a long time in production Danshi. This time, we find out about brushing in the step of extending the slack in the brush after wet papers have been laid on a workbench as show Fig.2.

Fig.2 Brushing on piled wet sheet

EXPERIMENTAL APPARATUS

Subjects are two people. An expert has manufactured Danshi for 23 years of experience and a non-expert has done less than 1 year of experience. They are wearing 18 reflective markers on the body as shows Fig.3. We collect data such as the location coordinates of the reflective markers by using six infrared cameras to measure changes in the time in the body part of the subjects. The movement of brushing process is measured as sampling time of 100 [Hz] by mac 3D system made Nac.

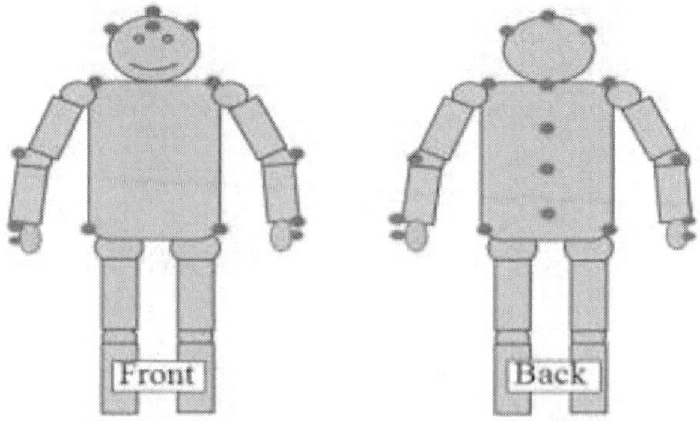

Fig.3 Location of reflecting markers

RESULTS AND DISCUSSION

We defined times of touching to brush on the paper's surface as the brushing time. One set is defined a process of piled three sheets. The data of an expert was obtained from 6 sets. On the other hand, non-expert one was done from 7 sets. As a result, an expert had brushing times of 79 per set, and a non-expert had brushing times of 101 per set.

Fig.4 shows the average number of brushing in each sheet of three piled sheets by the expert and the non-expert.

In the expert, brushing time of first sheet is similar to second sheet. The value of third sheet is approximately twice of first and second sheets. In first and second sheets, the brushing times of the expert are less than that of non-expert one.

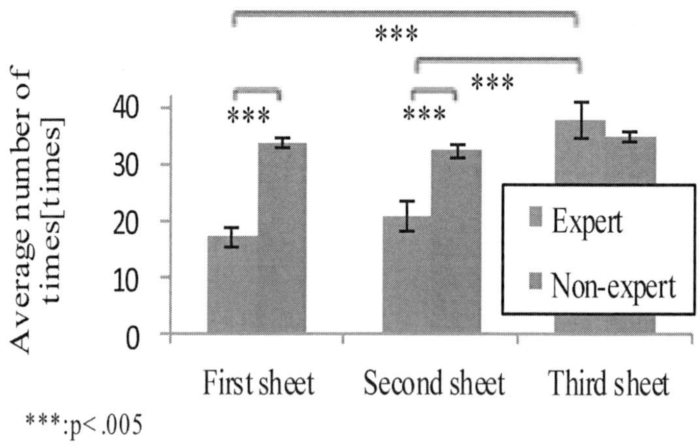

Fig.4 Average number of brushing

The other hand, in the non-expert, values of three sheets are like each other. It was not different by significance test in the brushing time of each sheet in the non-expert. It means that in the non-expert, brushing times are substantially constant regardless of order.

Initially, the expert was saying, "There is nothing to be aware in the brushing process". Thus, she could not been conscious of difference. We showed her this graph, and then asked again. Consequently, she said, "In third sheet, I might be carefully, because it became a bad quality of Shibo if it included a few air bubbles".

We divided brushing movement into eight directions as Fig .5. Fig.6 shows the average of sweeping by Expert in each direction in each sheet. Percentage of "b" direction has the highest in first and second sheet. However, in the third sheet, the proportion of "e" direction is highest and has increased over 10% than one in first and second sheet. At this time, the percentage of "e" direction showed 30% of the total in the third sheet. When we asked about this result to the expert, She said "I remove air bubbles from the corner of the top right, So "b" direction may be maximum". Furthermore, she mentioned that in the third sheet "e" direction might be maximum because she must remove small bubbles in final brushing.

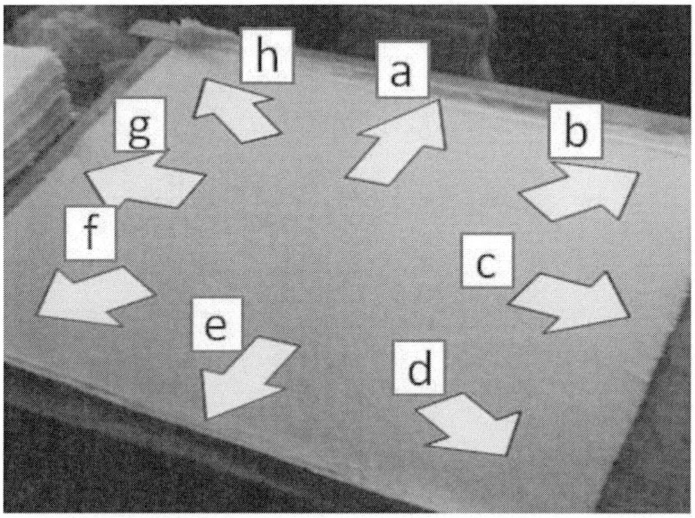

Fig.5 shows 8 directions of brushing

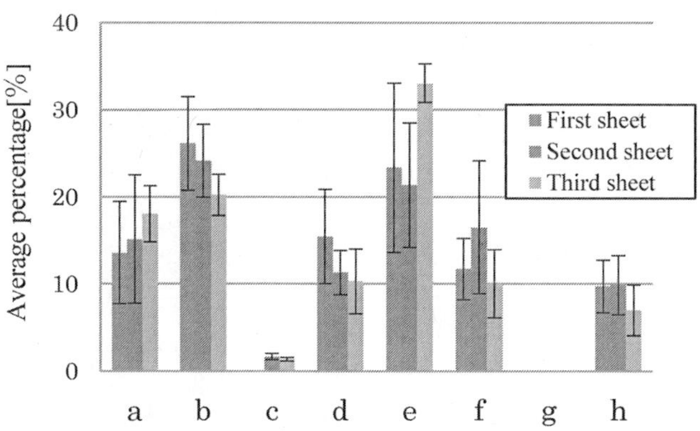

Fig.6 Average of brushing times in each direction

We investigated the change of angle of the right wrist in the expert and the non-expert. Fig.7 and Fig.8 show the change of angle of the right wrist in "a" direction by the expert and the non-expert, respectively. In Fig.7 initial angles range from 145 degree to 175 degree. It was found that right wrist angle of initial brushing of expert is always bent. There are three types of angle shift. First type is that decreases from 170 degrees to 160 degrees. Second type is that keeps its angle constant of 160 degrees. Third type is that are swaying from 145 to 170 degrees. And it is clear that final angles of brushing became about 160 degrees.

In Fig.8 the non-expert ranges in initial angle of the right wrist in the beginning of brushing movement from 155 degree to 180 degree. Almost angle shifts kept 180 degrees constant. Others shifted to final 180 degrees.

As time passes, angle of right wrist is increase in the non-expert. But in the expert, an angle of right wrist is reduced.

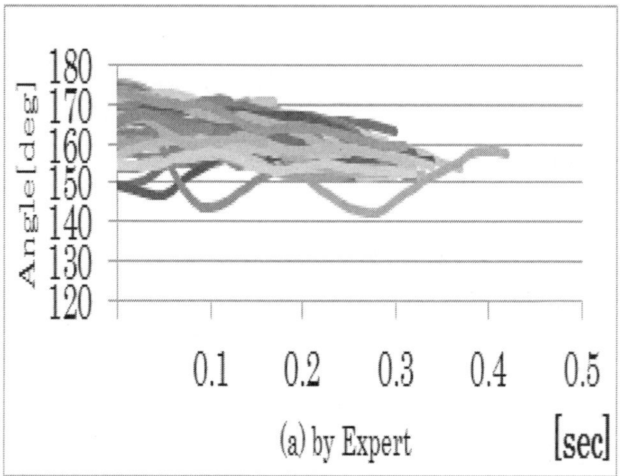

Fig.7 Changes angle of right wrist in Expert

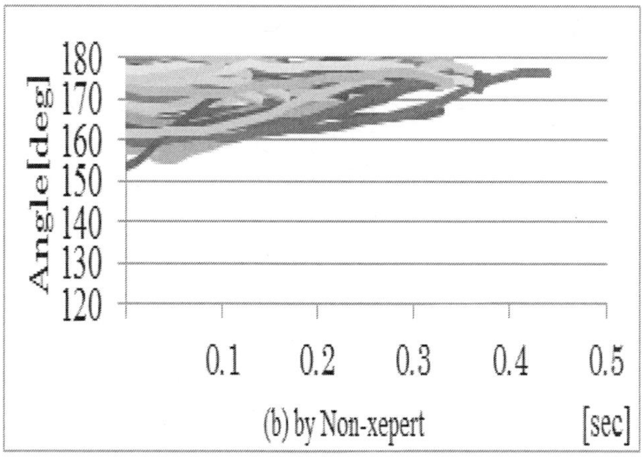

Fig.8 Changes angle of right wrist in Non-expert

Conclusion

We found that by analyzing the right wrist motion of an expert in the brushing of Danshi manufacturing, as following:

1. Important things for making the clear and fine Shibo structure are not only extending the slack of wet sheet, but also removing the bubbles. As the result the brushing times are many in the third sheet.
2. The expert removed air bubbles from the corner of the top right. Thus brushing times of "b" direction are mainly a lot.
3. Expert fixes the upper right portion of the piled paper and removes slacks from it.

ACKNOWLEDGMENTS

We are grateful for the assistance given by Mrs. Asano in production of Danshi. In provided handmade Japanese paper, we are deeply grateful to Mr. Hiromi SUGINO of Ehime traditional craftsman who belong to Toyo Handmade Paper Promotion. In data collection as well as 3D motion analysis system, I have received a cooperation by Professor Dr. Hiroyuki HAMADA of Kyoto Institute of Technology, Professor Dr. Akihiko GOTO of Osaka Sangyo University, the university assistant Dr. Yuka TAKAI, and students of Hamada laboratory.

Proceedings of the ASME 2014 International Mechanical Engineering Congress and Exposition
IMECE2014
November 14-20, 2014, Montreal, Quebec, Canada

IMECE2014-37771

ANALYSIS OF THE RHYTHM OF THE NECK IN HANDMADE JAPANESE PAPER MANUFACTURING MOVEMENT

Akihiko Imajo
National Institute of Technology, NIIHAMA College
Niihama, Ehime, Japan

Takashi Yoshikawa
National Institute of Technology, NIIHAMA College
Niihama, Ehime, Japan

Nobutaka Saeki
Saijo City
Saijo, Ehime, Japan

ABSTRACT

In Japan, the production of traditional handmade Japanese paper using the spring water has been performed in the Shuso area of Shikoku Island. However, the method of production was been handed down by trial and error and observation of the expert movement until now. The produced papers are counted one unit that is five hundred sheets of Japanese paper. Five hundred sheets of Japanese paper are called 1 [lot]. The paper has two type of thickness and its size is 600mm x 1500mm. Thick paper is 9[kgf] per lot. The other hand, thin paper is 8[kgf] per lot. In other words, the expert is making paper which is the difference of 2[gf] per sheet. This time, we have been able to obtain cooperation with traditional craftsman for the digitizing of his skills. Therefore, we visualize the tacit knowledge of the expert's skill.

The expert subject has 34 years of experience of traditional hand-made Japanese paper. We have digitized his manufacturing movement by using motion capture. His movements are analyzed by attaching infrared reflective markers of 20 on each parts of the body.

In this study, we found that constant rhythm of the neck in handmade Japanese paper manufacturing movement. Furthermore, the first scoop is performed at the time much shorter than in the case of other scoops.

INTRODUCTION

Handmade Japanese paper is one of the traditional crafts in Japan, and has long history. It had prospered since the nineteenth century [1]. However in Japan, there are three problems in the handmade Japanese paper industry. First is the worker of advanced age. Second is the slump of Japanese paper demand. Third is upbringing of successors is difficult. Therefore, we think so that inheritance of traditional technology threatened. In addition, the establishment of technology inheritance of the expert in existence has been sought.

In our previous studies, the ocular movement of the expert in process of the handmade Japanese paper was concluded as follows: [2][3]

(1) The expert mainly focuses on edges of a frame than the central region.
(2) Glancing at the whole of a frame, an expert confirms that fibers uniformly pile up.
(3) An expert judges that thickness of paper is uniformity by glancing about 0.3[sec].

Purpose

We investigated about "the expert are doing random direction of the fiber length of 3~5[mm] during manufacturing and, he makes a paper for becoming uniformly the thickness". And therefore, we work on the visualization of the tacit knowledge of the expert's skill in of the handmade Japanese paper. This study focused the movement of each body part of the expert in process of the handmade Japanese paper. To examine them, we investigated the motion by attaching Infrared reflective markers to the expert. And then, we think aim that we will be able to help people to learn handmade Japanese paper by revealing the skill of the expert.

EXPERIMENTAL APPARATUS

Subject

The subject is a traditional craftsman who has worked 34 years for the making Japanese paper. He will strive to save traditional crafts of the locality-specific, and promote conveyed to generations of posterity.

Fig.1 shows the producing of handmade Japanese papers by an expert attached infrared reflective markers.

Figure 1: The expert is making Japanese paper

Figure 2 shows the points that the expert was affixed with 20 infrared reflective markers. Analysis equipment of the manufacture movement in process of the handmade Japanese paper is using a motion capture consisted of 6 infrared cameras and 2 video cameras. We recorded the motion with sampling time of 100Hz.

Figure 2: Infrared reflective markers were attached to each body part in total 20.

Handmade Japanese paper processing

There are 8 procedures for produce the handmade Japanese paper. First of all, the expert pours water in a vessel. Figure 1 shows vessel called a "Suki-fune" in Japan. Second of procedure, the fibers of pulp as raw material for making paper are put into the vessel. Figure 3 shows the fibers in a bucket. The fibers are made from a pulpwood. Third, he stirs them in order to not create a mass of fibers by a left hand. Fourth, after stirring them, he pours the natural sap as glue in the vessel. The natural sap called "tazu" for connecting the fibers of the Japanese paper. Fifth, he stirs them again by using a plate. A raw material of fiber solution is prepared for handmade paper by these 5 procedures. Sixth, he scoops the solution by the filter, and shakes the filter in order to become uniformity thickness of the paper. Figure 4 shows the filter. Seventh, he finishes up 1 sheet of paper by scooping up several times. Eighth, the sheet is put to dry behind him after he scooped up a wet Japanese paper on the filter.

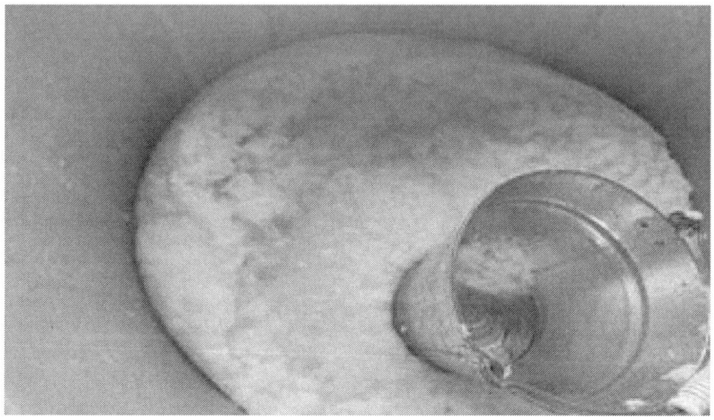

Figure 3: The fiber solution as a raw material in a bucket

Figure 4: Filter

The filter is used putting on the frame that is called "Suki-keta". In handmade Japanese paper production process, the filter is used to swing the frame in all the directions including back and forth, right, left and diagonally. Figure 5 is enlarged view of figure 4. The filter is made of uniform bamboo of 0.5 millimeter in diameter. The characteristic of this filter is shifted dare combined points. Because, in order to increase the strength of the filter. In addition, the expert are using valuable tool that are bamboo of uniform diameter which is the raw material of the filter is few. And, there are not a craftsman weaving it.

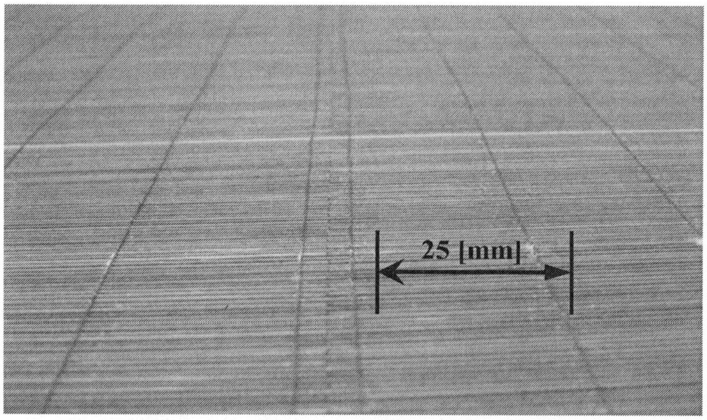

Figure 5: Enlarged view of the filter

Scoop movement & Shake movement

In order to pile up fiber on the filter, a behavior is divided two processes of "scoop movement" and "shake movement". The scoop movement is behavior that is he scoops up the solution of raw material for paper on the filter. Figure 6 shows the moment of the scoop movement. Next, the shake movement is behavior that swings the filter. Figure 7 shows the shake movement swinging the filter.

Figure 6: Scoop movement

Figure 7: The shake movement rippled a surface of the solution

RESULTS & DISCUSSIONS

The time of shake movement

In this time, six sheets of Japanese paper were made from the fiber solution of a vessel.

In order to make a sheet, several scoop and shake movements were necessary. The number of scoops of six sheets was different. These number of scoops at first sheet, second sheet, third sheet, fourth sheet, fifth sheet and six sheet were 4, 5, 5, 5, 5 and 6 times respectively. A number of scoop and shake movement are same.

We focused on shake movement and analyzed. Figure 8 shows the shaking time of each sheet. In the figure it found that the number of shake movement increased according to a number of making sheets. The times of first shake movement of every sheet were very short and constant. The times of shake movement varied widely according to a number of shake after second shake.

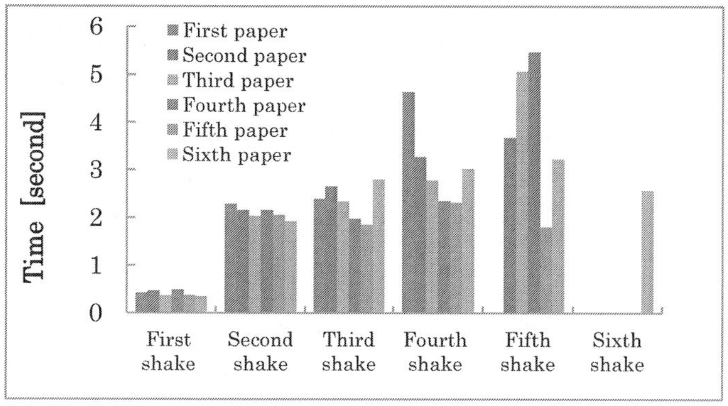

Figure 8: The time of shake movement

When the solution be low concentration in the vessel, it may be to make a constant thickness of paper by many shake movement.

Figure 9 shows the average time of each shake movement of Fig.8.

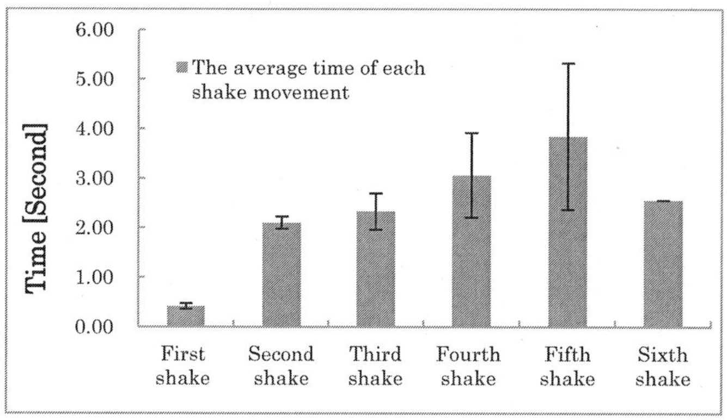

Figure 9: Average time of each shake movement.

We found that the average time of the first shake is the shortest time at 0.42± 0.06 [sec]. This time is very short and little deviation. It looks like to mean that the shake is very important because the first scoop and shake are foundation for thickness uniformly. The deviations of time of shake in second shake, third shake and fourth shake are respectively ±0.12, ±0.37 and ±0.86. These deviations increased.

Movements of the top of head and the neck

In the process to analyze the handmade Japanese paper manufacturing operations, we noticed that the expert have taken the rhythm at parietal and the neck. Therefore, we focus on movements of the top of head and the neck. We analyzed the movements of the top of head and the neck in the vertical direction in process of the finish of one sheet.

Figure 10 shows displacement of the top of head that contains scoop- and shake- movement in the vertical direction during the making fourth sheet. The lines of 4-1 to 4-5 indicate displacements respectively in the first, second, third, fourth and fifth shake of fourth sheet. The initial height of top of head in first shake starts from the position of the 400 [mm] shorter than in other shakes. The highest initial height of top of head is 550 [mm] in the second shake. The initial height of one decreases according to proceeding after second shake. Heights of top of head were low or constant during fluctuation.

We measured the fluctuation in the top of head of the second shake. As a result, It has been found that the cycle of the top of head is 2.38±0.30 [Hz].

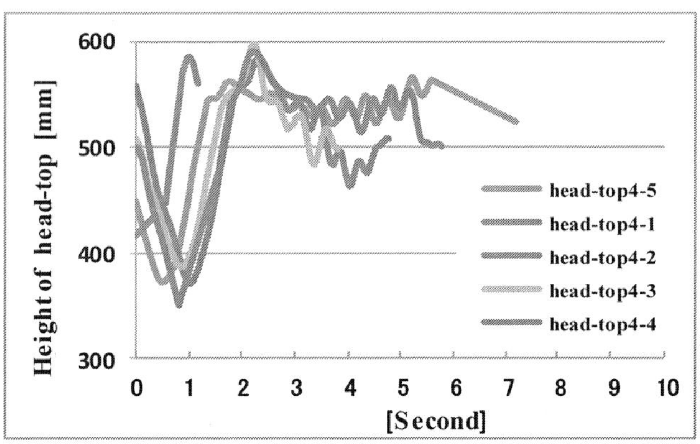

Figure 10: Displacement of the top of head in the vertical direction

Figure 11 shows the displacement of the neck in the vertical direction at 4-1 to 4-5. The initial height of neck in first shake starts from the position of the 320 [mm] shorter than in other shakes.

We can see that the parietal region and the neck's trajectory are similar. However, initial heights of neck were same after second shake. Heights of neck were up during fluctuation.

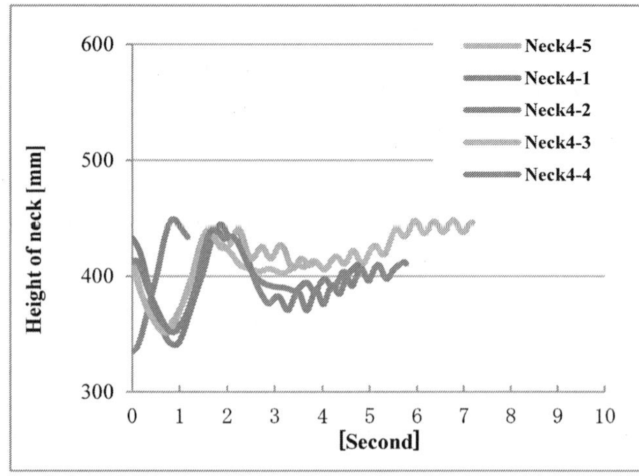

Figure 11: Displacement of the neck in the vertical direction

The velocity of neck in scoop movement

In this section, we focused on the height of neck that is the lowest position in the scoop movement. Figure 12 shows the relationship between the height of the neck and the number of scoop. The horizontal axis indicates the number of scoop. And, the vertical axis indicates height of neck. We found that the height of neck that became the lowest position, rise with the number of times of scoop increases during making one paper. And then, correlation between the number of scoop times and the height of neck that became the lowest position were researched. As a result, the number of scoop times and the lowest height of neck, were found to have average R=0.843±0.124 correlation.

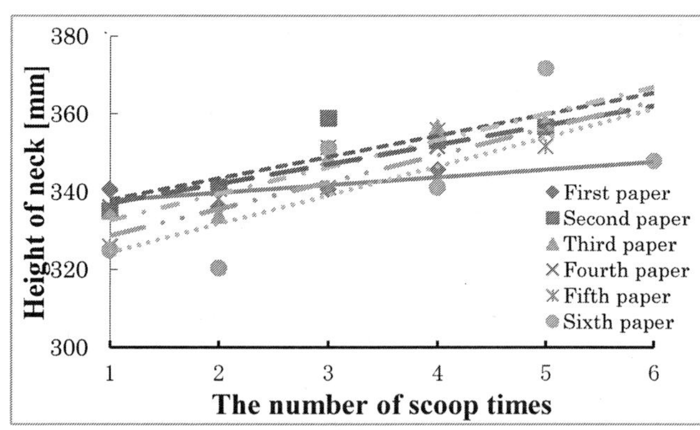

Figure 12: Relationship between the height of the neck and the number of scoops

We researched the velocity of radial direction of neck in scoop movement. Figure 13 shows the average of velocity of each scoop. We found that the average velocity of all phase are decreased from 284 [mm/s] to 205 [mm/s].

In the figure, the data of first, second and third sheets is called the first phase. The other hand, the data of Fourth, fifth and sixth sheets is called the last phase.

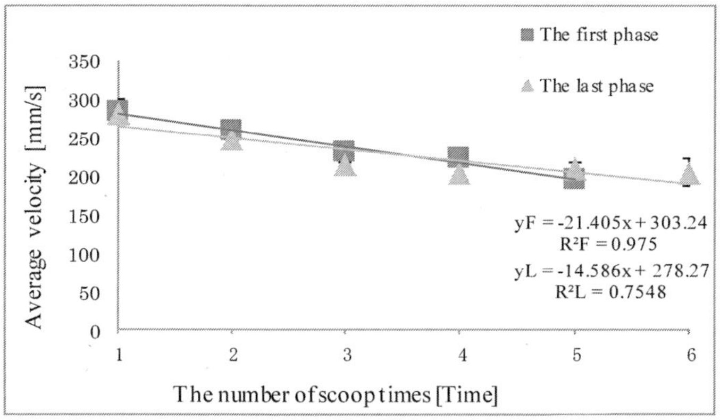

Figure13: Average velocity of the neck in scoop movement

Correlation coefficients of both phases were 0.99 and 0.87. In other words, the velocities of neck decrease according to increasing of number of scoop.

In a result of the interview, the expert said that he scoop up fiber most deeply in the first scoop. After that, he raise the height of the neck gradually. This is because the fiber that were piled up until now, don't scatter by scoop up much fiber. We found that the expert piles up the fibers without scattering the fibers stacked on the filter by raising the height of the neck at the scoop movement and slowing the velocity.

The velocity and acceleration of the neck in shake movement

The velocity and acceleration of the neck in shake movement were measured. Figure 14 and 15 show the velocity and acceleration in shake movement. In order to make a sheet of paper, several shake movements are necessary. In all processes, it seems to us that the velocities of the neck become slowly, when the shake movement advances. We estimated correlation coefficients in both phases that were 1.00 and 0.76. In other words, the velocities of neck decrease according to increasing of number of shake. In last phase until fourth shake, the velocities had strongly the relationship of correlation.

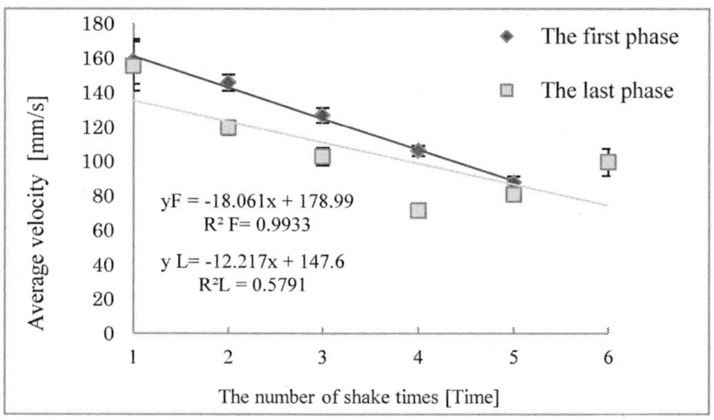

Figure14: Average velocity of the neck in shake movement

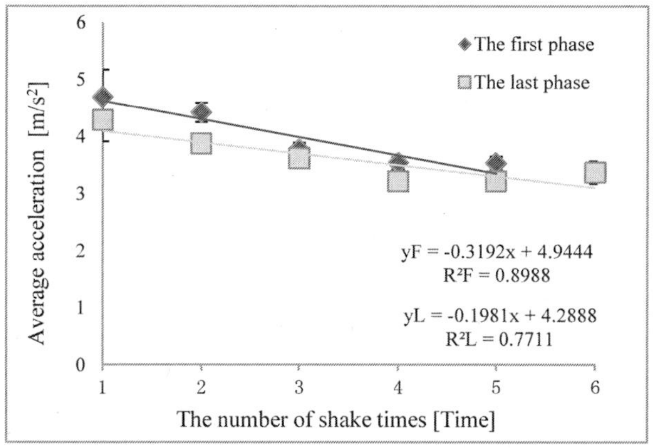

Figure15: Average acceleration of the neck in shake movement

As is the case with velocity, First shake was seen that he have done manufacture movement by an equal acceleration in the range of standard error. Furthermore, we seem that as the process advances, the average acceleration converge to the same value.

In results of the interview, the expert said that he is shaking quickly only first shake in order to pile up fiber uniformly on frame. In addition, when the fibers piled on the filter uniformly, the fibers don't leak from the filter. Therefore, it is possible to slow the velocity of shaking. Accordingly, as the shake movement advances, the averages of velocity become slower. We found that the expert performs an operation with the most careful in the first shake movement. Consequently, the expert could make a sheet of paper of uniform thickness.

It was confirmed that "The number of years required for the acquisition of technology, neck swing is three years and waist swing is five years" by the expert answered in interview.

CONCLUSIONS

In this study, analyze the neck movement of the expert in handmade Japanese paper manufacturing movement was concluded as follows:

(1) We found that constant rhythm of the neck in handmade Japanese paper manufacturing movement.
(2) The first scoop is performed at the time much shorter than in the case of other scoops.
(3) After second shake, the fluctuation at the top of head is taking the rhythm of 2.38 ± 0.30 [Hz].
(4) In the first shake of handmade Japanese paper, initial height of top of head is 400 [mm] that is shorter than other shakes.
(5) We found that the expert piled up the fibers without scattering the stacked fibers on the filter by raising the height of the neck at the scoop movement, and slowing the velocity.

(6) We found that the expert performed an operation with the most careful in first shake. After that, he could make a paper of uniform thickness.

REFERENCES

[1] Kato M. "History and culture of Saijo City handmade Japanese paper" Okadakoubundou, 2008

[2] Yoshikawa, T. Sogame F., Ito I., Saeki N., Tsujinaka T., "About Ocular Movement of the Expert in the process of Manufacture by Hand making Japanese paper as Shuuso", The Japan Society of Mechanical Engineers, M&P 2012 No.525, 2012.

[3] Yoshikawa, T., Imajo, A., Goto, A,. Saeki, N., "Regarding the ocular movement of the expert in process of the handmade Japanese paper", JAPAN International SAMPE symposium & exhibition, ID 2422, 2013.

Proceedings of the ASME 2014 International Mechanical Engineering Congress and Exposition
IMECE2014
November 14-20, 2014, Montreal, Quebec, Canada

IMECE2014-37886

RESEARCH OF ADHESIVE EFFECT ENHANCED BY POUNDING BRUSH ON SECOND LINING POUNDING PROCEDURE FOR JAPANESE SCROLLS

Yasuhiro Oka
Oka Bokkodo Co., Ltd.
113 Fukunagacho Nakagyo-ku Kyoto, Japan

Akihiko Goto
Osaka Sangyo University
Osaka, Japan

ABSTRACT

Hanging scrolls are a traditional Japanese ornamental art, which allow paintings and calligraphy to be unrolled and hung on a wall or in an alcove for display, and rolled up and stored in a box. Hanging scrolls should hang straight when unrolled, and roll smoothly for proper storage, without damaging the artwork beneath. For this purpose, scrolls are lined with several layers of Japanese paper, and adhered together with a weak, aged paste made from wheat starch, which gives the paper the flexibility required when the scrolls are rolled up for storage. While this old paste facilitates winding a scroll because it does not become hard even when dried, it does not have sufficient adhesive effect to grip Japanese paper. In order to increase the adhesive power of this aged paste, craftsmen employ a traditional technique of pounding the paper with a special "pounding brush."

This pounding technique is an important part of the fabrication process of hanging scrolls, but it is a difficult task for each generation to pass down the proper pounding technique. This study was intended to verify the effects of the pounding technique on aged paste and Japanese paper. We prepared samples with the pounding technique and investigated their adhesive properties of samples by peel text. In order to verify the importance of this traditional technique and the traditional materials, we compared and analyzed the differences in adhesion between craftsmen of different skill and differences introduced by paste concentration and backing paper quality.

INTRODUCTION

Hanging scrolls are one of the traditional forms of ornamental art common in Japan. Displayed in alcoves or from beams, these scrolls allow paintings or works of calligraphy to be rolled up for convenient storage. As in Figure 1, scrolls hang flat when unrolled, but roll tightly for storage as in Figure 2. What makes hanging scrolls special is their flexibility. Storing the scrolls rolled up protects the artwork inside from exposure to light and air, making hanging scrolls a superior way to ensure the ornamental art will last. This unique property is the result of traditional Japanese paper, starch paste, and a technique called lining pounding. Typically, the lining pounding is carried out starting with the second layer. For this procedure, a large brush, called a pounding brush is used. The surface of the paper is repeatedly pounded with the bristles of the brush to promote the desired adhesive properties.

Fig 1. When the scroll is hung in an alcove or a beam, it has to be hung straight.

During pounding, the tips of the bristles pass through the lining paper to the next layer. As pounding continues, the fibers of each layer become entangled with the layer below, helping to increase the adhesion between layers. Figure 3 shows an illustration of the lining pounding. Figure 4 shows the work being performed. The technique this research focuses on has been passed down for generations. Our goal is to analyze the benefits of pounding and the use of aged paste by conducting peel tests of papers prepared at the second stage of lining pounding. Also, the pounding technique requires tremendous skill, and is thought to be extremely difficult. For our research, we will be using samples prepared by both an expert and a non-expert craftsman, and comparing the adhesive properties of each using a peel test.

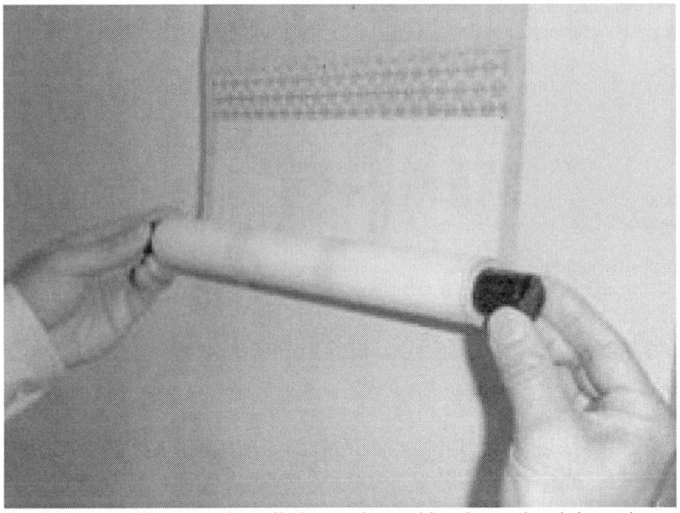

Fig 2. The scroll is generally rolled up and stored in a box, when it is not in use.

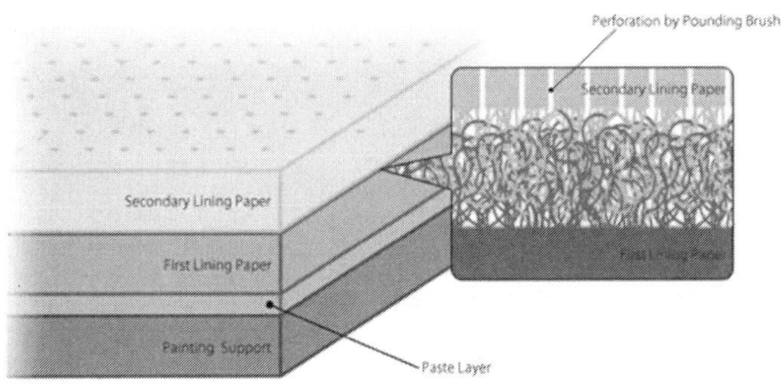

Fig 3. Illustration of the adhered papers after lining pounding process

MATERIALS USED IN THE FABRICATION OF HANING SCROLLS

A hanging scroll consists of an underlying painting or work of calligraphy, on paper or silk, backed by four layers of lining paper. These layers are known as the first, second, third, and final lining papers, in order of proximity to the artwork, and are adhered using a paste made from wheat starch. The artwork used in a hanging scroll is typically surrounded by silk, gold brocade, or damask.

First, the first and second lining papers are applied to the artwork and silk respectively. Once the lined silk has been allowed to dry completely, it is applied to the border of the artwork. Then, the third and final lining papers are each applied and pounded to ensure a complete join. The artisan performing the lining confirm the age of the artwork by adjust the thickness of the paper used, as well as other variables. In the end, the scroll will hang straight, with no bends or curves in the supporting papers, and roll easily for storage without needing to be folded.

Fig.4 Image of the pounding process

The order of the lining papers is important, as three different types of paper, made with three different types of paper, made with three different paper making techniques, are used. The first lining paper is a thin *Mino* paper, made from the fiber of the paper mulberry grown in *Nasu*, Tochigi prefecture. It is thin and lightly springy. This strong, light paper provides direct support to the artwork. A different type of paper, *Misu* paper, is used for both the second and third lining papers. *Misu* paper is made mainly from the fiber of paper mulberries grown in the *Tosa* region of Japan, and also contains a white pigment commonly found in Japanese painting, from the calcium carbonate in ground oyster shells. Because *Misu* paper does not undergo the compression found inmost types of paper making, it is unusually pliable. This is the defining characteristic of *Misu* paper. The last layer to be applied, the final lining paper, is made using *Uda* paper. *Uda* paper is made using fiber from paper mulberries grown in the Yoshino district of Nara prefecture, and a type of white clay found specifically in that region.

The first lining paper is applied using a starch paste made from wheat starch that has been heated for a time. However, the other layers are affixed using a starch paste that has been aged for ten years to allow it time to ferment and cure. This is the aged paste traditionally used as an adhesive for hanging scrolls. During the aging process, the molecular weight of the starch paste decreases, weakening its adhesive properties. This weaker paste does not stiffen as it dries, allowing the scrolls to be rolled and stored, and is the main source of the hanging scroll's flexibility. However, because the adhesive properties of the aged paste are so weak, the second, third, and final lining papers are pounded with the pounding brush, in the lining pounding, in order to strengthen the adhesive.

DESCRIPTION OF THE POUNDING PROCESS

The second lining pounding process, used to affix the second lining paper using *Misu* paper and aged paste, is described by the following steps.

1. Placing one sheet of Misu paper on a plain wooden table, carefully coat the paper with the specially prepared aged paste to avoid wrinkles.

2. Take the dried result of the first lining pounding, place it on the same table, and apply an appropriate amount of water evenly to the surface with a brush. Then, carefully lay the Misu paper prepared in step 1 on top, without causing any wrinkles to form.

3. After visual confirmation that no wrinkles or bubbles have formed, firmly stroke the paper with the brush.

4. Pound the points of adhesion with the bristles of the pounding brush, to help promote adhesion.

5. Confirm visually that the paper has been evenly and thoroughly pounded, and allow it to dry naturally.

These steps provide a rough description of the process by which Misu paper is adhered using aged paste in the second lining pounding.

DETAILS OF THE PEEL TEST

For this experiment, our subjects were an expert artisan (male) with 21 years of experience, and non-expert artisan (female) with 2 years of experience. We used plain pieces of silk to hold the place of artwork, and produced samples of peel apart. Each sample was 70 cm long and 40 cm wide. Subjects were given specimens lined with the first lining paper and dried, and sheets of *Misu* paper with aged paste applied. To help us examine the effects of the aged paste and the pounding brush, the expert artisan produced the three samples types described below.

Ⅰ. After applying the aged paste to the *Misu* paper, pound the samples as usual with the pounding brush.

Ⅱ. After applying the aged paste to the *Misu* paper, do not pound the sample.

Ⅲ. Use water instead of aged paste to apply the *Misu* paper to the sample and does not pound with the pounding brush.

We also had the non-expert prepare a fourth sample, Ⅳ, following the same steps as the expert artisan's sample Ⅰ. The aged paste used in samples was diluted by the expert craftsman as per his usual work process, with the concentration measured before fabrication. The concentration of aged paste used was 3.74%.

All samples were cut into specimens 3 cm wide longitudinally, with 5 specimens taken from each sample, and placed into a desiccator set to 100% humidity for 24 hours. After this, we conducted the peel tests on each specimen in a room with the humidity stable at 80%. Peel tests were conducted using a *Shimazu* AUTOGRAPH AGS-X, with speed set to 5 mm/minute, and stroke length set to 100 mm.

RESULTS OF THE PEEL TEST

First, let us examine the force required to peel apart the samples made by the expert artisan. The averages, maximum force, and minimum force for each sample are shown in Table 1. Note that all numbers are expressed in Newtons (N).

SAMPLE	MAX (N)	MIN (N)	AVERAGE (N)
Ⅰ	2.03	0.03	1.14
Ⅱ	0.54	0.03	0.42
Ⅲ	0.41	0.03	0.30

Table 1. Peel test results for Sample Ⅰ, Ⅱ, Ⅲ

There is not difference seen between samples in the minimum value at which peeling began, but the maximum and average values are largest in the case off sample Ⅰ, where the pounding brush was used. The difference between these and the values for samples Ⅱ and Ⅲ is significant. Compared to Sample Ⅱ, in which aged paste was used, but the pounding process did not take the place to strengthen the adhesion, the maximum force for Sample Ⅰ is 3.76 times higher, while the average is 2.71 times higher. It is plain to see that the lining pounding process has a large effect on adhesive strength.

To see the effects of the aged paste, we look at Sample Ⅱ. The aged paste used was diluted, and did not feel stick to the touch, but the peel test confirms that it is more effective than using plain water. Sample Ⅱ specimens required, on average, 1.4 times the force of specimens from Sample Ⅲ. This proves that aged paste does have an adhesive effect.

Next, let us consider the non-expert's imitation of the expert artisan's Sample Ⅰ, the maximum, minimum, and average results for which are show in Table2. As in Table1, all numbers are expressed in Nowtons (N).

SAMPLE	MAX (N)	MIN (N)	AVERAGE (N)
Ⅳ	0.87	0.03	0.65

Table 2. Peel Test Results for the sample Created by the Non-Expert with using aged paste and pounding brush.

Despite using the same materials, tools, and processes, the non-expert's Sample Ⅳ was unable to attain the same strength as the expert's Sample Ⅰ. The maximum force for Sample Ⅰ is 2.33 times that of Sample Ⅳ, and the average is 1.75 times that of Sample Ⅳ.

Examining the surfaces of the samples, after peeling test, the difference is visible. The expert's sample had small, visible holes in the second lining paper, from the pounding brush. It appeared that the holes penetrated into the first lining paper.

Figure 5 shows an image of microphotograph (X50) of the surface of Sample Ⅰ. By comparison, the surface of Sample Ⅳ is shown in Figure 6 , but we could not confirm whether the same conditions were present. Few small holes by the bristles of the brush could be confirmed in the non-expert's sample, however, the numbers of the holes are fewer than in the expert's sample. While the values for Sample Ⅳare slightly higher than those for Sample Ⅱ, where no pounding took place to enhance the aged paste's adhesive properties, it seems clear that achieving the maximum effect requires the correct pounding movements, and pounding the correct locations.

Fig 5. The microscopic image of the surface of the lining paper which was pounded by the expert. Small holes of the bristles of the pounding brush can be confirmed.

CONCLUSION

The results of these peel tests have allowed us to confirm many aspects of the use of the traditional pounding brush during the second lining pounding. Comparing the three samples produced by the expert artisan, we can confirm that pounding the paper with the pounding brush helps promote adhesive properties between layers of paper. Also, even with the aged paste diluted to a low concentration, it was evident that the aged paste had stronger adhesive properties than water. Relying only on the strong paste would not produce the characteristic pliability and softness typical of the Japanese hanging scroll. These results essential in promoting the adhesive properties necessary to fabricate hanging scrolls. Even using the same tools and materials, the craftsman must be suitably skilled for the adhesive properties to be strengthened, as can be seen in the difference between the expert's and non-expert's samples.

Fig.6 The microscopic image of the surface of the lining paper which was pounded by the non-expert. The numbers of the holes by the bristles of the brush are fewer then the expert's sample.

ACKNOWLEDGMENTS

I should like to thank all colleagues at Oka Bokkodo Co., Ltd. for preparing all samples of the peeling tests. Also thanks to Ms. Noriko Hayakawa, Independent Administrative Institution, National Research Institute for Cultural Properties, Tokyo, for supporting the peeling test.

REFERENCES

(1) Kyotaro Nishikawa; "Science of Japanese Painting Frame-Special Research "Scientific Research for Preservation and Restoration of Japanese Hanging Scroll", National Research Institute for Cultural Properties, Tokyo (1977)

(2) Noriko Hayakawa, Rika Kigawa, Wataru Kawanobe, Hisashi Higuchi, Yasuhiro Oka, Iwataro Oka, "Basic Research of "Aged Paste" (Furunori)-Traditional Japanese restoration Material – by GPC and HPLC for Organic Acid Analysis", Current Awareness Portal No.41 (2002)

(3) Noriko Hayakawa, Takayuki Kimijima, Kyoko Kusunoki, Yasuhiro Oka, "The adhesive Effect of Uchibake (Beating Brush) in Japanese Paper Conservation", Current Awareness Portal No. 43 (2004)

This page left blank intentionally.

Proceedings of the ASME 2014 International Mechanical Engineering Congress and Exposition
IMECE2014
November 14-20, 2014, Montreal, Quebec, Canada

IMECE2014-37927

ASSESSING REMOTE PHYSIOLOGICAL SIGNALS ACQUISITION EXPERIMENTS

Carla Barros
R&D Centro Algoritmi, Department of Production and Systems, School of Engineering, University of Minho
Guimarães, Portugal

Celina P. Leão
R&D Centro Algoritmi, Department of Production and Systems, School of Engineering, University of Minho
Guimarães, Portugal

Filipe Pereira
R&D Centro Algoritmi, Department of Industrial Electronics, School of Engineering, University of Minho
Guimarães, Portugal

Filomena Soares
R&D Centro Algoritmi, Department of Industrial Electronics, School of Engineering, University of Minho
Guimarães, Portugal

José Machado
R&D CT2M Centre, Department of Mechanical Engineering, School of Engineering, University of Minho
Guimarães, Portugal

Demétrio Matos
R&D Centro Algoritmi, Department of Industrial Electronics, Engineering School, University of Minho,
Guimarães, Portugal
IPCA-EST, Barcelos, Portugal

Vitor Carvalho
R&D Centro Algoritmi, Department of Industrial Electronics, Engineering School, University of Minho,
Guimarães, Portugal
IPCA-EST, Barcelos, Portugal

ABSTRACT

A great number of remote laboratories has been implemented in the engineering field. Nevertheless, there are few approaching the bioengineering area. The present paper will describe not only an innovative remote laboratory developed for biomedical engineering education, but also its assessment based on the target public's feedback. The remote laboratory developed by the research team intends to provide the physiological signals remote acquisition from human body, supported by theory to a greater understanding of learned concepts. This tool is geared towards the undergraduate biomedical engineering students. Therefore, a sample of twelve students took part in a limited study conducted to quantitatively and qualitatively assess the remote laboratory. The study was undertaken using two questionnaires, one distributed before and other after the performance of a remote experiment. Moreover, the information about the learning style/method, employed by each student, was collected in order to devise strategies for future applications development and to make the remote laboratory suitable for the target public.

***Keywords**:* Biomedical engineering education; Learning styles; Perceptions; Physiological acquisition systems; Remote Labs.

INTRODUCTION

One of the greatest, if not the greatest, remote laboratories' goals is to meet the same educational goals as the physical experiments or, in other words, as the hands-on practice [1–3]. With the development of the technology and with even greater use of computers and computing capabilities available, new challenges, new areas and opportunities emerge [4]. Remote laboratories are seen as important tools to complement learning/teaching traditional classes, from which students can remotely control a real device and get real data, acquiring practical skills according to its main purpose [5]. This kind of pedagogical platforms should be suitable for the target public, optimizing the advantages that come from this approach. In this respect, it is possible to use many different lines of reasoning to adjust the available experiments as well as the way that students access the platform. Attention should be given to the academic degree, students' knowledge about specific theoretical concepts, and their awareness of a given experimental proceedings. On another level, the students learning styles can be used to perceive their preferred methods of study. The combination of these strands should provide an improvement of the students' learning process.

Learning styles

Information about each student's learning method/style might be highly significant to identify. This knowledge may be applied to design and implement new learning strategies, capable of achieving that aim. The learning style's concept describes the individual learning, taking into account the learner's preferences.

The Kolb's experiential learning theory is one of the best-known educational theories in higher education. According to this theory, new knowledge, skills or attitudes are achieved through confrontation among some modes of experiential learning [6]. The model works on two levels or two learning activities, perceiving and processing, and Kolb's emphasize four distinct learning styles (accommodating, diverging, assimilating and converging) based on four learning stages. In this respect, Kolb's model is particularly elegant, providing an understanding of individual learning styles and the explanation of a circle of experiential learning, which applies to all of us [6, 7]. A typical representation of the two Kolb's learning activities is based on two axes (Figure 1): the horizontal axis corresponds to the measurement of the processing activity, the way how we approach a situation, ranging from the active experimentation to the reflective observation; and the vertical axis to the measurement of the perceiving activity by distinguishing between the concrete experience and the abstract conceptualization, quantifying our emotional response or our thinking way. The learning styles are the combination of two lines of the axes [7].

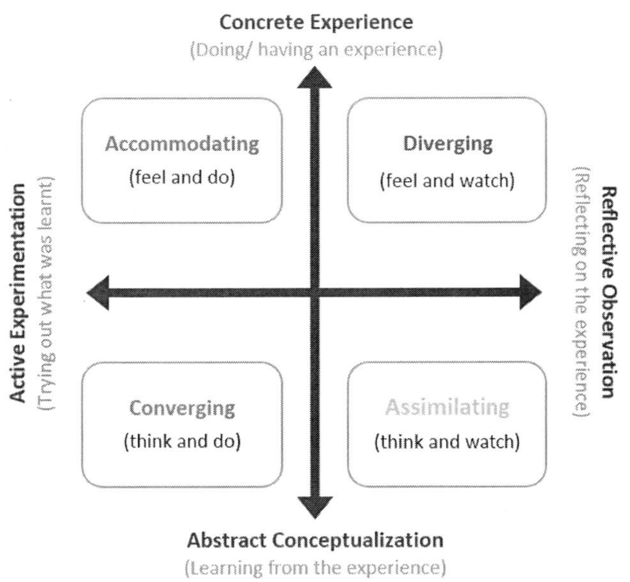

FIGURE 1. DIAGRAM ILLUSTRATING THE KOLB'S LEARNING STYLES (ADAPTED FROM [6]).

The proposal drawn up by Felder and Soloman to index the learning styles is based on the assessment of students' preferences according to Felder and Silverman model [8]. This model describes the learning style of a learner in more detail, distinguishing between preferences on four dimensions. Another main issue is based on tendencies, saying that learners with a high preference for certain behaviour can also act sometimes differently [9]. According to Carver et al. [10], "the Felder model is most appropriate for hypermedia courseware" and it is often used in technology-enhanced learning and that is designed for traditional learning [9].

The first dimension of Felder model distinguishes between an active and a reflective way of processing information. Active learners learn best by working actively with the learning material, by applying the material, and by trying things out; they prefer to learn by working in groups where they can discuss (Figure 2). In contrast, reflective learners prefer to think about and reflect on the material; they prefer to work alone or maybe in a small group. The second dimension covers sensing versus intuitive learning. Learners who prefer a sensing learning style like to learn facts and concrete learning material; they like to solve problems with standard approaches. The sensing learners tend to be more practical than intuitive ones. In contrast, intuitive learners prefer to learn abstract learning material, such as theories and their underlying meanings. They are more able to discover possibilities tend to be more innovative and creative than sensing learners. The third, visual-verbal dimension differentiates learners who remember best and therefore prefer to learn from what they have seen (e.g., pictures, diagrams and flow-charts), and learners who get more out of textual representations, regardless of whether they are written or spoken. In the fourth dimension, the learners are characterized according to their understanding. Sequential learners learn in small incremental steps and therefore have a linear learning progress. They tend to follow logical stepwise paths in finding solutions. In contrast, global learners use a holistic thinking process and learn in large leaps. They tend to absorb learning material almost randomly without seeing connections but after they have learned enough material they suddenly get the whole picture. Then they are able to solve complex problems, find connections between different areas, and put things together in novel ways but they have difficulties in explaining how they did it. Because the whole picture is important for global learners, they tend to be more interested in overviews and in a broad knowledge whereas sequential learners are more interested in details [11].

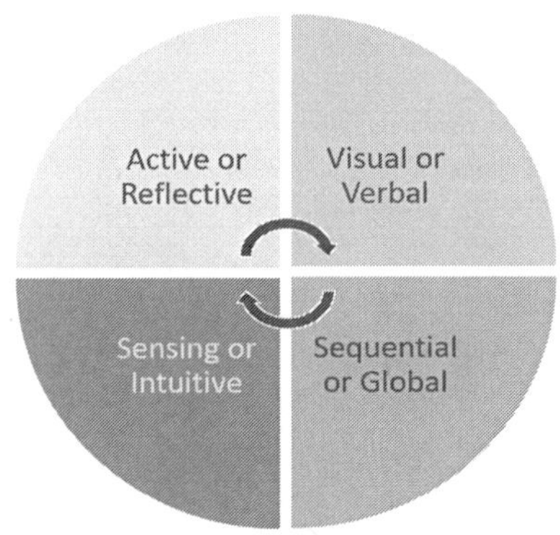

FIGURE 2. FOUR DIMENSIONS OF THE MODEL PROPOSED BY FELDER AND SOLOMAN [17].

Teachers should ensure that activities are designed and conducted to offer each learner the opportunity to engage appropriately by the identification of the students learning style. Furthermore, learners can learn more effectively by the application of the learning cycle [6, 7]. The awareness of the learning style would provide better engineering educational experiences for students and may help instructors to better understand their preferences [12].

Motivation

A wide range of remote learning applications has been developed in engineering and sciences areas such as physics, biology, electronics, and mechanics [2]. Although a large number of remote laboratories are associated with engineering, the number is fairly small when searching the bioengineering area. This paper describes an innovative remote laboratory for biomedical engineering education, based on physiological systems' study. This new remote lab intends to, in an intuitive way, provide the physiological signal remote acquisition from the human body, being supported by theory for the fully concepts understanding. It is designed for undergraduate students and intends to be an auxiliary learning tool.

The paper also presents a study conducted for the assessment of the remote laboratory based on the target public's feedback about its using. In addition, this study also identifies the students learning style and it determines if the design strategy used to develop this remote lab is suitable for them and if it achieves the educational goals purposed.

THE REMOTE LABORATORY

The RePhyS (Remote Physiological Systems) Laboratory is an innovative remote laboratory based on physiological data acquisition [13]. This platform is aimed towards to undergraduate biomedical engineering students. Since this course requires a huge knowledge of physiological systems functioning and of the biosignals acquisition methods [14], this tool intends to provide the remote acquisition of physiological signals, supported by theory for the full understanding of the important concepts and the experimental proceedings [13].

The remote acquisition of physiological signals is conducted using a small device, designed and built by Shimmer Research™, called Bio Starter Kit®. This equipment allows the wireless data capture from human body and is composed for ECG (electrocardiogram), EMG (electromyogram), GSR (galvanic skin resistance) and body acceleration sensors. Another device, the e-Health Sensor Platform is used to capture more biometric data. This Arduino kit allows acquiring signals such as the blood pressure, body temperature (BT), heart rate, or the body position. Integrating both equipment the remote laboratory was implemented with LabVIEW software, making use of the Web Publishing tool, which makes possible to deploy the user interface on a web browser page. Through the Internet connection, the access to the user interface enables the remote control and monitoring of experiments [15]. Figure 3 shows the interface during an experiment of ECG acquisition.

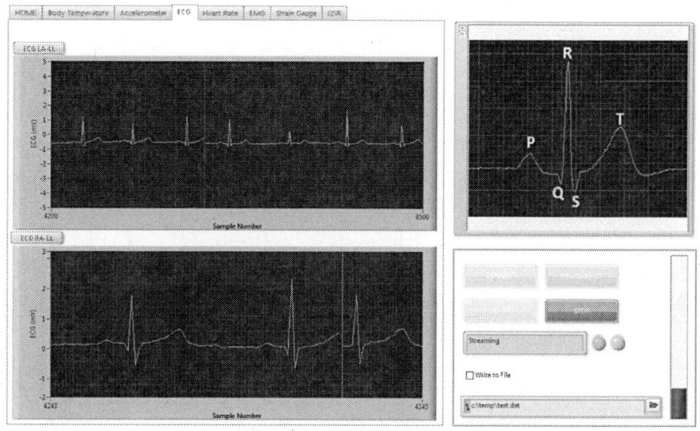

FIGURE 3. REPHYS USER INTERFACE DURING AN ECG ACQUISITION.

ASSESSMENT STUDY

The study on the RePhyS was performed in order to validate some educational and technical features. The study will show the importance of the remote laboratory developed for physiological data acquisition, enabling to conclude about issues such as its efficiency and effectiveness as a new learning tool.

Since this is a recent approach in biomedical engineering area, especially in the physiological signals acquisition, it is crucial to know how receptive and motivated are the students in relation to the usage of a remote laboratory to assist their learning process. The students' satisfaction was assessed by their comfort and acceptance to the remote access experiment.

This assessment study complements the one performed before for the RePhyS first version evaluation [16]. The procedure of the first study is similar to that used in the present work. The participants performed a remote experiment through the RePhyS platform, following a protocol.

The class of 12 students was divided into groups of two. Each group, following the protocol designed for remote physiological signals acquisition, performed the experiment. This experiment took about 15 minutes, during which the students were able to explore the online platform and some of its functionalities.

The protocol, similar to that used in [16], aimed the access to the online platform and the remote connection to the real lab. The laboratory procedure involved ECG, body acceleration and BT values acquisition from a subject, who agreed to realize the experience, at a distant place. Students could visualize and identify the different signal's components and to understand the behaviour of these physiological signals in response to various situations, in particular, to the body's movement. It was also possible to change some device's parameters.

To obtain students' feedback on the platform and procedures, two questionnaires were developed. One was answered before the experiment (named Pre-Questionnaire) and the other delivered after the accomplishment of the experiment (named Post-Questionnaire). Close, open, multiple-choice responses, and five-level Likert-scale was used to rank the students' motivation, satisfaction and knowledge about some important educational issues. Both questionnaires had the same questions of those answered under the first study conducted [16]. Nevertheless, some additional questions were proposed in the post-questionnaire to assess the new experiment available (BT values acquisition) and to identify the students' learning styles, according to the Felder and Silverman model [17].

STUDY RESULTS

In this section, the quantitative and qualitative analyses of the results of both questionnaires are discussed.

Participants' characterization

The participants in this study (n=12) are students who attend, for the first time, the fourth year of the Integrated Master in Biomedical Engineering, of School of Engineering, University of Minho, Portugal, for the 2013/2014 academic year. The average age is 21 years and 58.3% (7/12) of students are male. Before their first contact with the RePhyS, 83.3% of them asserted that they have never heard about this type of educational approach, while the remaining mentioned they have heard about remote laboratories at the university. No students held a remote experiment before this study.

Participants' learning styles

To infer about the learning styles, the participants were asked to identify their action, based on Kolb's theory activities (processing and perceiving). Below is exemplified the question exposed to the students.

Processing:

Active	○	○	○	○	○	○	Reflective
	1		←	→		6	

Perceiving:

Abstract	○	○	○	○	○	○	Concrete
	1		←	→		6	

The participants' responses show that about 41.7% of the students have an accommodating learning style (Figure 4, second quadrant). This result indicates that, in processing field, the students do the things in an active way and they have a more concrete perception of the situations. As can be seen from Figure 4, the majority of the students (66.7%) prefer to process information more actively. The same percentage of students prefers to understand the concepts through concrete experiences.

Based on Kolb's theory, this behaviour is characteristic of engineers: they are motivated to investigate how situations are processed, their strength is the practical application of ideas [7]. The instructional methods, such as the interaction, the hands-on, the lab's work, or the assisted computer training, are suitable for the students who employ this kind of learning style [7]. This is an essential aspect to take into account because it allows perceiving if the tool is designed for the students subjected to the test.

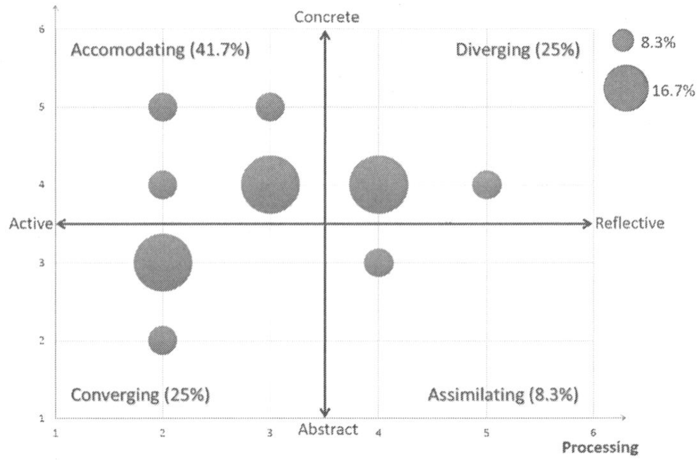

FIGURE 4. LEARNING STYLES DIAGRAM OBTAINED (BASED ON KOLB'S THEORY): 1ST QUADRANT – DIVERGING STYLE; 2ND QUADRANT – ACCOMMODATING STYLE; 3RD QUADRANT – CONVERGING STYLE; 4TH QUADRANT – ASSIMILATING STYLE.

Through the answers of six questions presented in the post-questionnaire, it is possible to determine the participants' learning style according to the Felder and Soloman model [17]. These issues worded allow to know how important is, for each student:

A1 - to perform the experiment step by step (sequential learner);
A2 - to exploit freely the platform functionalities (global learner);
A3 - to reflect about the experience before running it (reflective learner);
A4 - to conduct the experience, reflecting on it following (active learner);
A5/A6 - the instructions/protocol to be presented in written form or through a video demonstration (verbal/visual learner).

Without taking the questionnaire proposed by Felder and Soloman [17], and through simple questions about the students' experience in this remote laboratory, it is possible to have insight about their learning preferences, according the four dimensions model. The result is presented in Figure 5, where the importance of each question is rated in ascending order (from 1 - the least important to 5 - the most important). This outcome shows the average values.

Through the results obtained (Figure 5) it is possible to make up the differences between the importance levels given to each question. The issues are grouped according the learning styles. From the analysis of question group A1/A2, it is concluded that the students consider more important to do the things in large steps, therefore they are more global learners. Likewise, these students are more visual and active learners:

they like to work in-group and they learn from what they see, preferring information provided in the form of charts, figures or demonstrations.

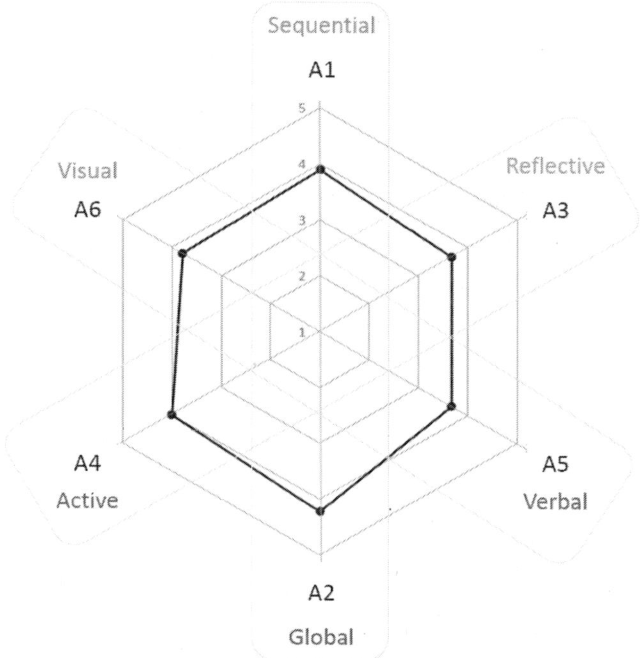

FIGURE 5. LEARNING STYLES DIAGRAM OBTAINED (BASED ON FELDER AND SOLOMAN MODEL [17]).

Although these differences, all groups of questions have very similar values between each one, they considered them as important. This could mean that their preferences are not so strong and are liable to undergo changes.

Participants Insights about Remote Labs

Once the protocol and the laboratory intended the physiological signals acquisition, it was examined the group's experience in data acquisition. All students have already acquired physiological signals from the human body: 63.7% of them indicated that they performed it without difficulty, while 36.3% considered the acquisition difficult.

When asked about the possible existence of differences between traditional and remote acquisition, 91.7% asserted that the remote acquisition was different and 66.7% asserted that the remote experiment could become less real. Traditional should be understood as acquisition in person, as performed into clinical environmental. When asked about the reality of the experiment, the students answer if they find or not they are acquiring real physiological data.

After their first experience with remote labs, the participants have changed their point of view and 44.4% realized that no differences between the both methods exist; this change of opinion is statistically significant ($W=10$, $p<.05$). Concerning to the experiment's sense of reality, the change in the opinion was not statistically significant after the experiment ($W=10$, $p>.05$) and 77.8% considered the experience was not less real than the traditional one, justifying with the achievement of similar results.

The students' motivation to control and interest to conduct a remote experiment deserved a special highlight in this study. Before the experiment, the analysis of students' answers showed that the majority (58.3%) of them were not so driven to deal with remote laboratories.

Experiments assessment

The experimental design was a crucial step to develop a learning strategy, as mentioned before, to capture the students' attention and to make a positive impact on them. Above all, the experiment was designed in order to evaluate the educational goals achievement. The remote experiments involved an ECG, body acceleration values and the body temperature acquisition. The main goal was the full understanding of each experiment, the visualization of the signals and the identification of the signals' components. Moreover, another objective was the understanding of the electrodes placing in the human body, i.e., the acquisition method. Figure 6 shows the results of the experiments assessment, reflecting the students understanding of each experiment.

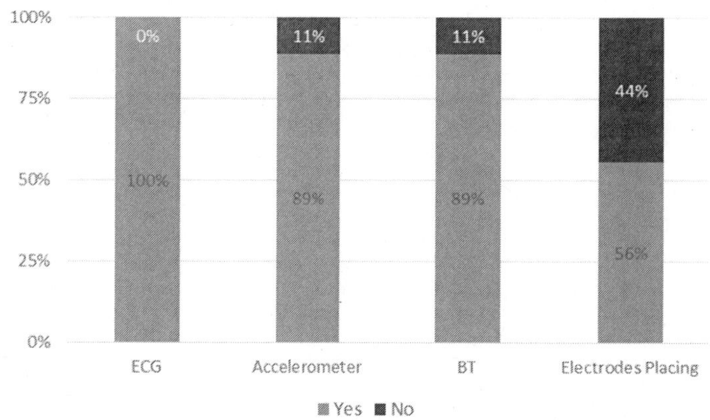

FIGURE 6. UNDERSTANDING OF THE EXPERIMENTS PERFORMED DURING THE STUDY CONDUCTED WITH STUDENTS

The ECG acquisition was valuable in so far as they ensured the assessment of the understanding, the utility and the visualization of the signals. All students (100%) understood the ECG acquisition and they were able to identify the constituent waves of the signal, during the data real-time visualization. The electrocardiogram was supported by theoretical concepts, capable of being explored before or after the acquisition. When asked about the accelerometer experiment, the result was not so consistent, 89% of the students have been able to understand and to visualize the acceleration values' capture around the three orthogonal planes XYZ. Indeed, the fact that not all students have understood this experiment was an expected result because it was not available a theoretical support for it and this signal' acquisition is not commonly taught in this biomedical engineering course's disciplines.

Concerning to BT acquisition, it was verified the same result as the accelerometer one. About 89% of the students understood the BT acquisition as well as its utility for the physiological systems study. A less positive result was obtained in the electrodes placing evaluation. Only 56% of the students understood the signals acquisition method, against the expectations once enough theoretical information was available in the platform. However, and due to a technical problem throughout the experience, the students could not remotely visualize the electrodes placing in real time (the webcam did not function). That can justify their negative perception.

The visualization of the entire process is important for students' acquisition of knowledge and for the experiment's similarity with the hands-on one. For instance, in this case, the electrodes placing or the movement detection of the person in the real lab are interesting aspects to take into account when the users are analysing the signals acquisition' s results.

Technical features were also evaluated by students, attributing a classification in a 1 to 5 rating scale, where 1 is "poor", 2 is "insufficient", 3 is "sufficient", 4 is "good" and 5 is "very good".; and the remote configuration of the Shimmer kit's parameters. The majority of the students, about 60%, rated them as "good" and "very good". General evaluation of the platform involved the intuitiveness, the user friendliness, the effectiveness of the results' monitoring, and the platform's graphic environment rating. Other factors were evaluated, such as the easiness to access to the theoretical support concepts, the real time acquisition and the remote configuration of devices' parameters.

DISCUSSION

Comparing the results of the first study [16] with the present work's outcomes, it can be seen that they are very similar, although the understanding of the experiments had been better in the second one. In general, they are very positive. About 77.7% of students considered the remote lab a useful tool. Their receptiveness to this new educational tool is remarkable and is a great motivation for the development of new remote applications in this area of knowledge and learning. In general, the students' opinion about this new laboratory was very interesting:

"*useful tool*",
"*rewarding experience*",
"*interesting*",
"*technology with potential application in many areas*".

The students were asked for comments about all experience in order to identify some aspects capable of being improved in a near future. The majority of them identify the missing of the remote real-time visualization of the physical experiment as the main fail. The electrodes placing in the person/patient, who underwent the experience in the real lab, was also the target of this evaluation. Due to fact that the network's velocity has not allowed the image data transferring over the Internet, the students were not able to real-time monitoring the experiment, making it difficult to see the electrodes placing. To overcome this difficulty, the online platform contained theoretical support explaining the electrodes application. As future work, a small video for explanation will be prepared.

The knowledge of the students' learning styles allows to recognize and to identify the learners' preferences. A Kolb's model-based analysis shows that the biomedical engineering students present an accommodating learning style, which was expected. An active processing and a concrete perceiving activities characterize this style. Taking into account the Felder and Soloman model, it is evident that these students are more global, active and visual learners, although they have not shown strong preferences. Anyway, these facts should be used to adjust the platform to the students' needs, in order to enhance their learning process. For instance, it could be provided the instructions in video form to meet their visual capabilities. The main characteristic, common to both models, is the active way of processing the information. The future developments will require greater emphasis on the implementation of forums or a kind of social network, which enables the information sharing between students and where they can discuss and explain it.

Given that the study was conducted on a small sample of individuals, it does not present the level of statistical significance that the results must attain. However, the principal findings and conclusions drawn from it are of key concern for the research team, providing them mechanisms to drive future RePhyS' improvements.

FINAL REMARKS

Data acquisition of physiological signals has been a very important issue since, depending on their values different health treatment will be given. With recent developments in technology, new medical devices and new ways of monitoring and acquisition of physiological information emerge. From the point of view of higher education level, same opportunities must be given preparing the engineering students before entering the labour market. Following this drift, this paper described an assessment study of a new remote lab, RePhyS, for biomedical engineering students in undergraduate level. The remote acquisition of an ECG, signal, BT values and body acceleration data was the experiments performed to evaluate the platform. A group of 12 students of the fourth year of biomedical engineer course, for 2013/2014 academic years, participated in this study, as target public.

Besides the great results from assessment of RePhyS and the experiments available, this study allowed the identification of the students learning style. This information is important in order to address preferences, but balanced perception and processing activities, and to create a platform suitable for biomedical engineering learners.

In essence, the purpose of this pilot study was accomplished. The remote lab was regarded as a useful learning tool, promoting the students' motivation and interest. The positive results achieved are very encouraging for the development of future applications, concerning to remote laboratories for biomedical engineering education. The study's outcomes prove that this innovative tool provides enriching experiences for biomedical engineering students, able to engage them and fulfilling the remote laboratory pedagogical goals.

It is authors purpose to extend RePhyS remote lab to different curricular subject and different higher education institutions. At the present time, the team is drafting a cooperation protocol in order to overcoming some possible limits.

NOMENCLATURE

BT – body temperature
ECG – electrocardiogram
EMG – electromyogram
GSR – galvanic skin resistance
RePhyS – Remote Physiological Systems Laboratory

ACKNOWLEDGMENTS

The authors are grateful to the FEDER funds through the Operational Programme for Competitiveness Factors - COMPETE and National Funds through FCT- Foundation for Science and Technology within the project FCOMP-01-0124-FEDER-022601 (ref. FCT PTDC/CPE-PEC/122329/2010) and to all the students for their voluntary cooperation and help in the evaluation of the remote lab experiments.

REFERENCES

[1] Zubía, J.G., and Alves, G.R. (2012) Using Remote Labs in Education: Two Little Ducks in Remote Experimentation, Universidad de Deusto.

[2] Ma, J., and Nickerson, J.V. (2006) Hands-on, simulated, and remote laboratories: A comparative literature review, ACM Comput. Surv., **38**(3).

[3] Feisel, L. D., and Rosa, A. J. (2005) The Role of the Laboratory in Undergraduate Engineering Education, J. Eng. Educ., **94**(1), pp. 121–130.

[4] Neuman, M., and Kim, Y. (2006) The Undergraduate Biomedical Engineering Curriculum: Devices and Instruments," Ann. Biomed. Eng., **34**(2), pp. 226–231.

[5] Popescu, D., and Odbert, B. The Advantages Of Remote Labs In Engineering Education, Educ. Corner.

[6] Kolb, A. (2005) The Kolb learning style inventory–version 3.1 2005 technical specifications, Boston, MA Hay Resour. Direct, pp. 1–72.

[7] Kolb, D. (1984) Experiential learning: Experience as the source of learning and development.

[8] Litzinger, T.A., Lee, S.H., and Wise, J.C. (2005) A study of the reliability and validity of the Felder-Soloman Index of Learning Styles, Education, **113**, p. 77.

[9] Graf, S., Viola, S.R., and Kinshuk, T.L. (2006) "Representative characteristics of Felder-Silverman learning styles: An empirical model, Proceedings of the IADIS International Conference on Cognition and Exploratory Learning in Digital Age (CELDA 2006), Barcelona, Spain, pp. 235–242.

[10] Carver Jr, C. A., Howard, R.A., and Lane, W.D. (1999) Enhancing student learning through hypermedia courseware and incorporation of student learning styles, Educ. IEEE Trans., **42**(1), pp. 33–38.

[11] Felder, R.M., and Silverman, L.K. (1988), Learning and teaching styles in engineering education, Eng. Educ., **78**(7), pp. 674–681.

[12] Tokdemir, G., and Cagiltay, N.E. (2007) Using Learning Style Theory in Remote Laboratory Applications, Personal, Indoor and Mobile Radio Communications, 2007. PIMRC 2007. IEEE 18th International Symposium on, pp. 1–3.

[13] Barros, C., Leão, C.P., Soares, F., Minas, G., and Machado, J. (2013) RePhyS: A Multidisciplinary Experience in Remote Physiological Systems Laboratory, Int. J. Online Eng., **9**(S5), p. 21.

[14] Bronzino, J.D., and Peterson, D.R. (2013) The Biomedical Engineering Handbook, Taylor & Francis Group.

[15] Barros, C., Leão, C.P., Soares, F., Minas, G., and Machado, J. (2013) Issues in remote laboratory developments for biomedical engineering education, Interactive Collaborative Learning (ICL), 2013 International Conference on, pp. 290–295.

[16] Barros, C., Leão, C.P., Soares, F., Minas, G., and Machado, J. (2013) Students' perspectives on remote physiological signals acquisition experiments, Engineering Education (CISPEE), 2013 1st International Conference of the Portuguese Society for, pp. 1–8.

[17] Soloman, B.A., and Felder, R.M. (2004) Index of learning styles, www2. ncsu. edu/unity/lockers/users/f/felder/ public/ILSdir/ILS-a.htm

This page left blank intentionally.

Proceedings of the ASME 2014 International Mechanical Engineering Congress and Exposition
IMECE2014
November 14-20, 2014, Montreal, Quebec, Canada

IMECE2014-38355

DESIGN, MACHINING AND PRODUCTION INTEGRATION PROBLEMS IN MANUFACTURING AUTOMATION

Zbigniew M. Bzymek
University of Connecticut
Storrs, Connecticut, USA

Alicia Benjamin
University of Connecticut
Storrs, Connecticut, USA

ABSTRACT

This paper describes the process of integrating engineering design, manufacturing, and production in the area of manufacturing automation. The work was done within the scope of a Mechanical Engineering senior course that's objective was to introduce students to the processes of advanced manufacturing and to solving practical engineering problems in manufacturing automation. The students' efforts at integration covered automation of conceptual and geometric designs, automation of machining process, and machine sequence optimization. The CAD/CAM software, CAMM3 Micromodeler, G-code, NX8, Solid Works, DELMIA/QUEST, and Mastercam were used successfully in a sequence. A survey of the students' opinions about the effectiveness and user friendliness of the software was summarized at the end of the semester. The elements of the course were integrated in the Final Project. Full automation of integrated design and manufacturing data exchange were found to be too difficult to accomplish. However, the use of the automation software in a sequence, together with data export and import, marks a significant step forward towards integrated manufacturing automation. The research to accomplish this will continue and the results will be applied in order to reinforce the teaching and practice of Manufacturing Automation.

Key words: engineering education, manufacturing automation, problem solving, course syllabus, design/manufacturing integration.

NOMENCLATURE

Acronyms and abbreviations

DELMIA – Digital Enterprise Lean Manufacturing Interactive Application; a digital manufacturing and production software developed by Dassault Systèmes that enables manufacturers to virtually define, plan, create, monitor, and control all production processes

GoldFire - Design software created to optimize innovation processes and enhance productivity; developed on the basis of Invention Machine

G Code – a milling machine code

IHS Inc. - a US company based in Douglas County, Colorado; acquired Invention Machine Inc. on 7 November 2012

Invention Machine - A problem-solving software package developed in the 1990s

Mastercam - NC programming software for multi milling; used for design, modeling, and machining of mechanical parts

Micromodeler – CAMM3 desktop milling machine code

Unigraphics NX8 - CAD software for geometric design

Tormach PCNC – Personal 1100 Model 2010 automatic machine; Personal Computer Numerical Control by Tormach Company

QUEST – DELMIA/QUEST - a 3D digital factory environment for process flow simulation and analysis, combined with object-based, discrete event simulation environment, visualization, and robust import/export capabilities

TechOptimizer – an improved version of Invention Machine

Symbols

Z^u -- upper tolerance limit

Z^l -- lower tolerance limit

INTRODUCTION

The Manufacturing Automation course described in this paper was based on two foundations: the textbook, with its theoretical material, as well as exercises and assignment problems; and laboratory projects. The first foundation was the theory of

Manufacturing Automation. It was covered through lectures, reading assignments, homework assignments working with problem solving, and working through the textbook examples. The second foundation was composed of laboratory projects based on problem solving together with material gathered from the university library, Internet websites, and other sources outside of the class. These two streams were at the base of teaching the problem solving that is described in this paper. There are not many textbooks available for Manufacturing Automation. The course textbook [1] was used as the main reading source, supplemented with articles (such as [2, 3, 4, 5, 6, 7]) and Internet references (as, for example, [8, 9, 10]) as recommended material.

During various stages of the course, different software was used: Invention Machine™ [11, 12, 13], TechOptimizer [14], and IM™ GoldFire [16] (now supported by HIS [15]) for conceptual and preliminary design. Furthermore, Unigraphics - NX8 [17, 18], Solid Edge [19], SolidWorks [20], and Pro/ENGINEER (Creo) [21] were used for geometric design, while Micromodeler [22], G Code [23], and Mastercam [24] were used for model and part cutting, and DELMIA/ QUEST [25] was used for machining sequence optimization. In an ideal situation, the results generated by the software used in one stage of manufacturing automation should serve as data to be used in the subsequent stage. This was not completely possible, and the data had to be transferred in a semi-automatic or sometimes conventional way. Using the software in a sequence however, in which the results of one stage can be passed to and used in the subsequent stage, was already a partial success. First using Invention Machine, TechOptimizer, and GoldFire for problem solving and conceptual design, then using Unigraphics NX8 and other geometric design software, and finally applying G code, Micromodeler, and Mastercam for part cutting, represented a partial software sequencing success of the process. The machining cue optimization using DELMIA/QUEST was done outside of the sequence.

SOLVING DESIGN, MACHINING, AND MANUFACTURING PROBLEMS – LECTURES AND TEXTBOOK USE

Serial and Concurrent Engineering Approach

In the first section of the course, solving design problems was considered in two approaches: a Serial Engineering approach [26] and a Concurrent Engineering approach. This was done with intensive use of the textbook [1]. These approaches were illustrated using the case of choosing tolerances in a shaft design. The shaft was specified with a dimension and tolerance of 1 ± 0.003 inch. The process average and the standard deviation were estimated to be 1.00 and 0.003 inches, respectively. Other relevant data for the raw material, unit salvage, and unit-processing fee (Table 1) were used by students following recommendations in the reference [1].

Table 1. Data for Serial Engineering Calculations

1	Unit cost of raw material	$12.00
2	Unit salvage value	$2.00
3	Unit processing fee	$7.00

Unit Cost and Scrap Calculations

As a turret lathe was the first manufacturing technology option, $j = 1$; assuming a tolerance of 1 ± 0.003 inch as the first design option, $k = 1$. All parts above and below the tolerance limits are scrapped. For the given data,

$$Z_{11}^{u} = 1.00 \text{ and } Z_{11}^{l} = -1.00.$$

The unit output cost per part was calculated as $26.90.

In Concurrent Engineering, marketers, designers, manufacturing engineers, and all other participants in the product development process are brought together to discuss integrating issues [1]. This multifunctional team is responsible for addressing various issues of conceptual and detail design, initial problem solving, machining, manufacturing, production, and marketing. In class exercises, this process led to a sequence of interactions through which the students arrived at a series of decreasing unit processing costs from $21.87, to $24.06, to $24.00, and finally to $21.00. Students also worked this sample problem using different sets of data. It definitely proved the advantage of a Concurrent Engineering approach.

Machine Programming and Part Fabrication Using NC Programs in G Code

The students had to write an NC program to machine a ½-in wide capital letter shape slot (the U-shape shown in Figure 1) in a mild aluminum work piece with dimensions of 3 × 4 × 0.80 inches (see Figure 1 and Table 2).

The machining was done with the assumption that the cut would start and finish on the top or the side of the plate (see Figure 1). The program prepared by one of the student teams is shown in Table 2 with a brief explanation of the effect of each command. The program was run on the Tormach PCNC Personal 1100 Model 2010 automatic machine with 12 individual alterations and 12 aluminum plates of different shape of cut (see Figure 2).

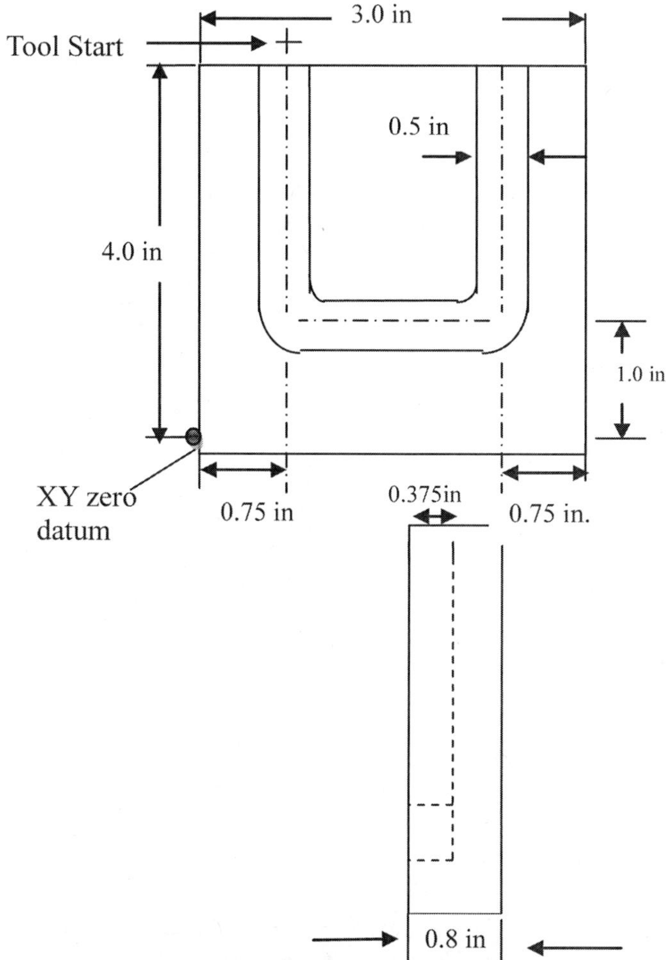

Figure 1. Top and Side Views of the Part

Table 2. A Program in G- Code

%	Indicates start of program
N005 G90 G70	
N010 G97 G94 T01	Specifies absolute dimensions, English units (inch)
N015G00 X0750 Y4500 Z250 F0	
	Specifies units for speed and feed rate; load first tool
N020 G01 Z-375 M03 S500 F10	Rapid positioning of tool to start point turns on spindle, feeds tool to required depth
N025 Y1000	Machines the first vertical portion of the U
N30 X2500	Machines the horizontal portion of the U
N35 X2250 Y4500	Machines the other vertical part of thc U
N040 Z250	Retracts the tool to 0.25 in. above part surface
N045 X-1000 Y-1000 F0	Moves to a safe location at rapid rate
N045 M30	Turns off all machine functions

In total, the students completed twelve homework assignments, one per week, as indicated on the Syllabus (see Annex). The problems, which covered all main areas of Manufacturing Automation (see Table 3), were solved using data supplied by the instructor.

Figure 2. An Aluminum Plate Cut by Students

Machine Programming and Part Fabrication Using NC Programs in CAMM-3 Code

Following the G-code example (Table 2), another two programs were created using CAMM-3 internal code. The internal code was entered into the CAMM-3 machine and run. First, a letter cube (see Figure 3) was created and then Lego®-like blocks were designed and entered. Both the letter cube and the blocks (see Figure 4) were cut out of wax. Students' initials were engraved on each block. After the Lego®-like blocks were produced, the space structure was built using the blocks. If the tolerances were correct and the blocks were cut with the required accuracy, the structure was solid and stable. The defective blocks that did not fit were put aside. From the initials cut on the walls of the blocks, the names of students who produced scrubs were known, assisting the instructor in making comments, evaluating the jobs, and assigning grades.

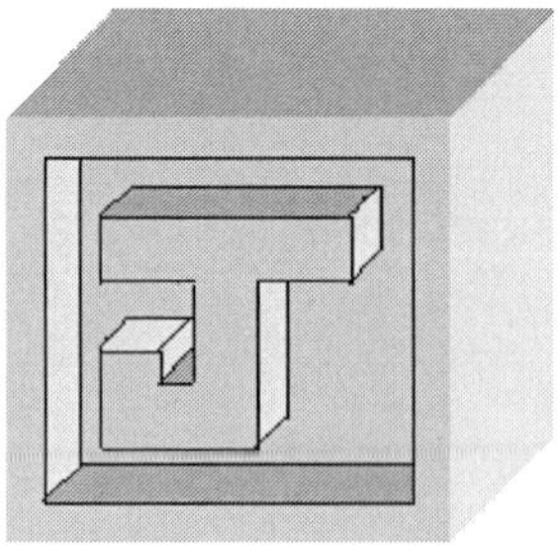

Figure 3. A Letter Cube Cut from Wax

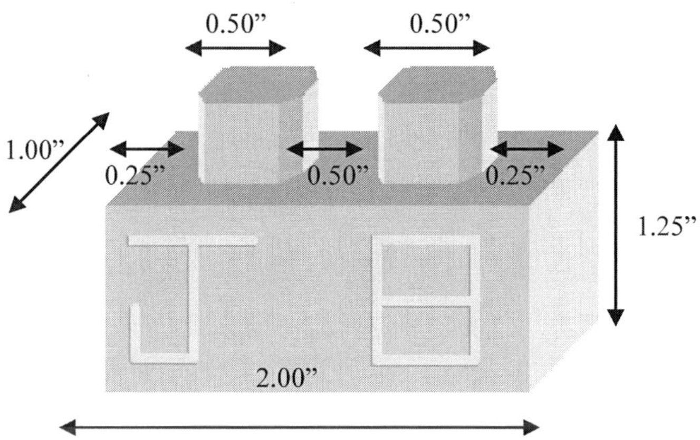

Figure 4. Lego®-Like Block (top) with the Initials “J B” Cut on the Side

Introduction to Manufacturing Automation Areas and Concepts

In this course, students were introduced to the fourteen main areas of Manufacturing Automation listed in Table 3. The examples described above represent areas three and five, respectively (Table 3). However, problems pertaining to all thirteen areas were covered in the course (Table 3, Annex).

Table 3. Thirteen Main Areas of System Approach to Computer-Integrated Design and Manufacturing

---

1. Computer-Aided Design Hardware, Software, and Firmware
2. Geometric Modeling
3. Concurrent Engineering
4. Computer-Aided Process Planning
5. Computer Control of Manufacturing Systems
6. Automated Material Handling and Storage Systems
7. Robotics System
8. Quality Engineering and Automated Inspection
9. Manufacturing Planning and Control
10. JUST-IN-TIME Manufacturing Systems
11. Group Technology
12. Flexible Manufacturing Systems
13. Enterprise Integration

Table 4. The Generic Steps for Geometric Design of a Nozzle

---

1. Create a **New** file:
   a. Select file extension '.prt'
   b. Define file name and folder extension for saving
2. Use **Sketch** command:
   a. Click the **Sketch** command icon from the toolbar
   b. Select a plane for the sketch (typically the x-y plane is used)
   c. Select the **Line** command and draw the desired shape
      i. For this example, draw the top half of a cross section of the nozzle
   d. Use the **Inferred Dimensions** command to define each line
      i. Once sketch is defined, click **Finish Sketch**
3. Use **Revolve** command:
   a. Click the **Revolve** command from the toolbar
   b. When prompted to **Select Curve**, choose the 2D polygon from **Sketch 1**
   c. When prompted to **Specify Vector**, choose the inferred vector
   d. When prompted to **Specify Axis**, choose the axis of rotation
      i. For this example, choose the bottom line segment
   e. Click **OK/Apply**
4. Use **Hole** command:
   a. Click the **Hole** command from the toolbar
   b. Click **General Hole**
   c. Click OK
   d. When prompted for **Position**, click the center of the hole and the front face of the part and the hole will be cut through
   e. When prompted for **Direction**, click **Normal to Face**

f. When prompted for **Dimension**, specify the desired diameter and select **Through Body** (since the hole will be going though the entire part)

g. Click **OK/Apply**

5. Insert **Thread**:

Click **Insert** from the top pane, scroll to **Design Feature** and select **Thread**

i. Select **Detailed** for the thread

ii. Generate the thread

SOLVING DESIGN, MACHINING, AND MANUFACTURING PROBLEMS - THE LABORATORY PROJECTS

Nozzle Design

For the laboratory projects, students were asked to design a nozzle with a variety of software tools including Unigraphics NX8, SolidEdge, SolidWorks and Pro/ENGINEER. Students designed the generic procedure (Table 4) and then followed it, using the specific capabilities of each software package. Generated on a PC computer, the final images were satisfactory and looked good in two dimensions (see Figures 5 and 6), but they were not as convincing as the stereo images [27] generated on Silicon Graphics during previous years in which the course ran. Table 4 summarizes the generic steps the students followed in their nozzle design projects.

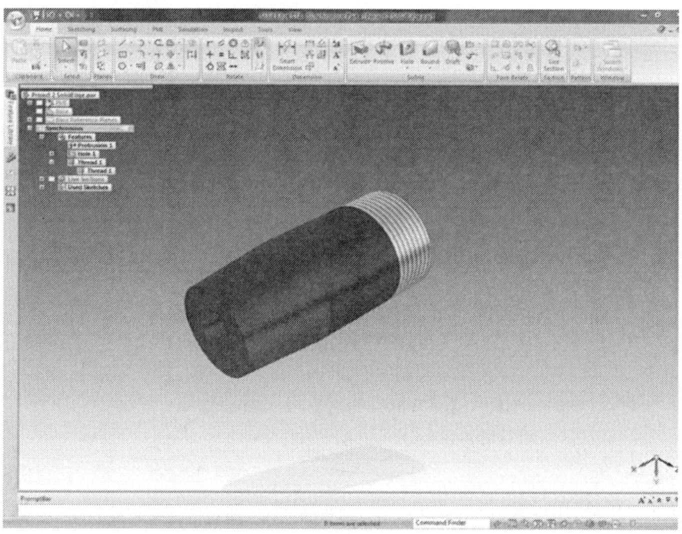

Figure 5. A Nozzle Created with Unigraphics NX8

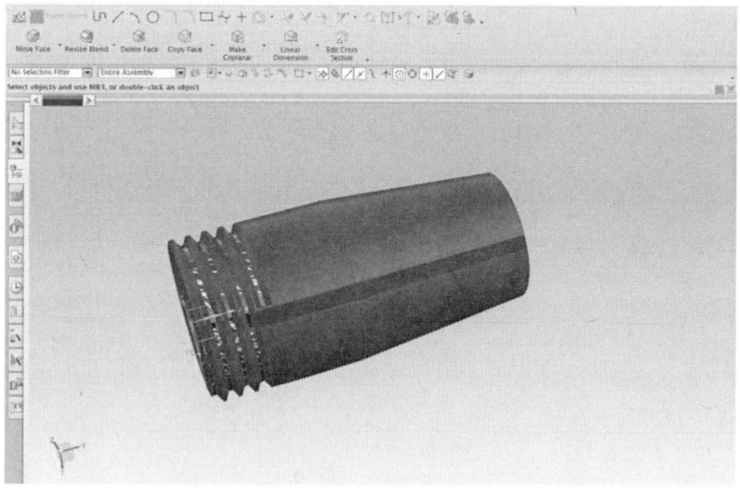

Figure 6. Nozzle Designed with SolidWorks

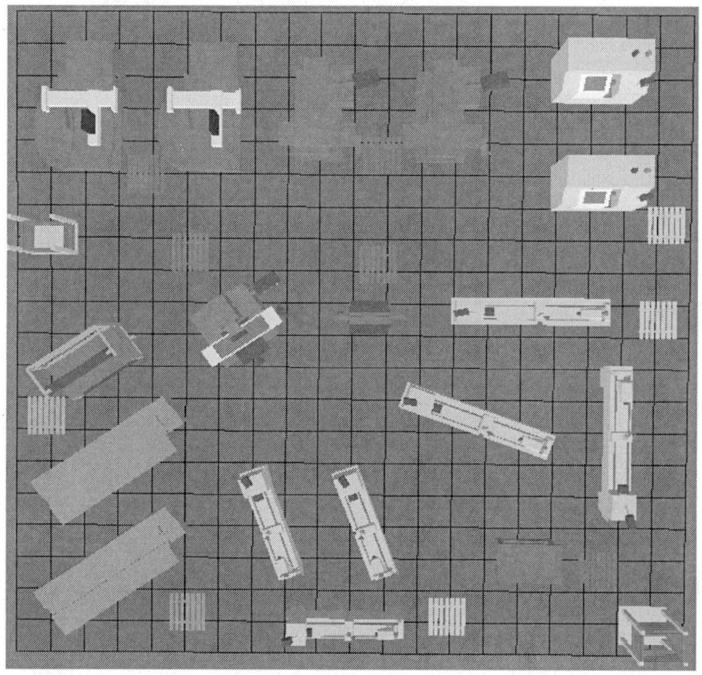

Figure 7. DELMIA/QUEST Manufacturing Floor Simulation

The process used to cut Lego®-like blocks was similar to rapid prototyping techniques – known as stereo-lithography. However, an automated desktop milling machine, as was used in this class, is far simpler - the process is shorter, faster, more educational, and doesn't involve complex chemistry as several of the stereo-lithographic techniques do. The nozzles generated by different CAD packages are shown in Figures 5 and 6.

Computer-Aided Machine Sequence Optimization

Computer-aided machine sequence optimization was done using DELMIA /QUEST software [25]. Students investigated nozzle machining using two to six different machines (lathes and millers). Using two machines, it took students 14 seconds to produce the nozzle. With six machines, on the other hand, the production time was reduced by 7 seconds per nozzle. The machine configuration generated by QUEST is shown in Figure 7.

Virtual Machining Using Mastercam

The last project before the Final Project was devoted to virtual machining. Using Mastercam software [24], the students simulated fabrication of a Lego®-like block. The results are shown in Figure 8.

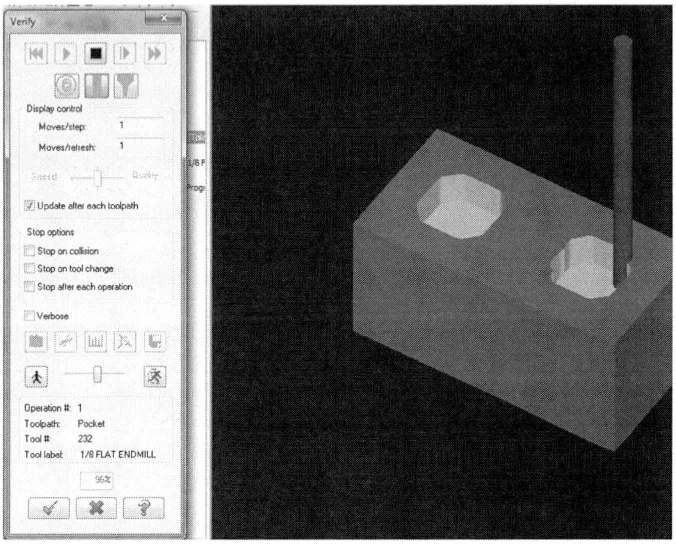

Figure 8. Mastercam Output

CONCLUSIONS

The course included lecture material based on the textbook [1], twelve homework assignments, two exams, and five projects. The final exam and the final project required students to demonstrate that they had acquired the fundamental knowledge and were able to apply it to real manufacturing automation problems in the industry. All the software used elicited positive evaluation from students. The automatic transfer of data from conceptual design to manufacturing was not fully achieved. Some data transport had to be done in a conventional way.

ACKNOWLEDGMENTS

This paper is a result of a team effort between the authors, along with the participation of some ME and MEM 2012/2013 students. Andrew Walton prepared the theoretical examples and Kayla Kamerer submitted student project materials. They were both students in the ME 3221 Fall 2013 class. The authors would also like to acknowledge Lauren Bradley, as well as other students whose work was used for reference examples in this paper. Special thanks go to Katie Kornacki of Willimantic, CT for her suggestions for text improvements and for editing this paper. Thanks to Invention Machine and CNC Mastercam for the software grants. Finally, thank you to John Wiley and Sons for supplying the textbook for the course.

REFERENCES

[1] Singh, N.: System Approach to Computer-Integrated Design and Manufacturing, Wiley & Sons, INC., New York, NY, 1996.

[2] Bzymek, Z.M., 2005, "Brief Theory of Inventive Problem Solving—Enhanced Approach Using Principles," Innovations 2005—World Innovations in Engineering Education and Research, (A. Wing et al., eds.), International Network for Engineering Education and Research (iNEER), Begell House, pp. 503–516, 2005.

[3] Bzymek, Z. M., 2000, Conceptual product Design Using BTIPS, DE-Vol.109, recent Advances in Design for Manufacturing (DFM), ASME 2000, pp. 139-147.

[4] Bzymek, Z. -M.: New Principles in Conceptual Product Design Using BTIPS, Proceedings of the Fifth International Workshop on Advanced Manufacturing Technologies, London, Canada, (L. Wang and X.Y. Liu, eds.), Sponsored by National Research Council of Canada—Integrated Manufacturing Technologies Institute, pp. 105–110, 2005.

[5] Bzymek, Z.M., Nunez, M., Li, M., Powers, S.: Simulation of a Machining Sequence Using DELMIA/Quest Software, Computer-Aided Design and Applications, Vol. 5, Nos. 1- 4, 2008, p. 401 - 411.

[6] Bzymek, Z. M., 2012, Engineering Problem Solving – Learning and Practice, Paper #88527, Proceedings of the International Engineering Congress & Exposition IMECE 2012, Houston Texas, USA..

[7] Bzymek, Z. M., Yang Xu, 2012, Effectiveness of Problem Solving Software in Engineering Conceptual Design Paper #86495, Proceedings of the International Engineering Congress & Exposition IMECE 2012, Houston Texas, USA.

[8] Customized Automated Manufacturing Equipment: Assemble, Test, Inspection and Accuracy; http://www.setpointusa.com/?gclid=CK-hPic37YCFYPd4Aod_jwA9A (visited 8/22/2012).

[9] Biz Plenty, Why is electronic manufacturing not fully

automated? http://www.bizplenty.com/Manufacturing_ _Automation?o=3986&qsrc=999 (visited 4/02/2014).

[10] MAS - A company which specializes in the custom design and development of automation systems for manufacturing http://www.mfgautosys.com (visited 4/22/2012).

[11] Invention Machine ™ v.2.1 for Windows. Principles, Effects, Prediction. User Guide, August 1995,

[12] Invention Machine: A Problem Solving software, 1996, https://invention-machine.com/, visited: 4/14/2014 www.invention-machine.com , visited 4/14/14.

[13] Tsourikov V. M., 1994, IM Course Notes, Invention Machine, Cambridge, Massachusetts.

[14] TechOptimizer™ 3.0; Professional Edition for Windows R 95, 98 and Windows NT R ; Training and Reference Manual, Invention Machine Corporation, 200 Portland Street, Boston Ma 02114 – 1722 USA, 1998 Edition.

[15] IHS Inc. - a company based in Douglas County, Colorado that acquired Invention Machine, https://www.google.com/#q=IHS, visited on 4/15/2014.

[16] Goldfire - https://gfi.goldfire.com/;

[17] NX: plm.automation.Unigraphics: http://www.plm.automation.siemens.com/en_us/products/nx (visited on 4/14/2014).

[18] Unigraphics NX: http://www.brighthub.com/engineering/mechanical/articles/34681.aspx (Visited on 4/14/2014).

[19] SolidEdge, http://www.plm.automation.siemens.com/en_us/products/velocity/solidedge/index.shtml, visited, 4/14/14.

[20] SolidWorks, http://pages.solidworks.com/lp/solidworks/Online_Tour.html?&utm_ keyword=solidworks.com, visited 4/14/14.

[21] Pro/ENGINEER, (Creo), PTC Product and Service advantage, http://www.ptc.com/products/proengineer. visited on 4/14/14.

[22] Rolland CAMM3, Three-Axis Device PNC-2500, California Digital, Inc., http://www.cadigital.com/camm3.htm, visited 4/14.14.

[23] Warfied R., CNC G-Code Programming Tutorial, Course, Examples ,*www.cnccookbook.com/CCCNC**GCode**Course.htm*, visited on 4/14/14.

[24] MASTERCAM, CNC Software, Inc., http://www.mastercammill.com/MastercamX3_Web_Brochure.pdf , visited 4/14/14.; *www.**mastercam**.com/*, visited 4/14/14.

[25] DELMIA/QUEST User Manual

[26] Deutschman, A. D., Michels Walter J., Wilson Charles E.: Machine Design. Theory and Practice. Macmillan, New York, 1975.

[27] Stereo Graphics Cristal Eyes 3, Real 3D, http://reald-corporate.com/scientific/crystaleyes.asp, visited: 4/14/14

ANNEX

SYLLABUS OF THE ME 3221 FALL 2013 MANUFACTURING AUTOMATION COURSE

MECHANICAL ENGINEERING DEPARTMENT

Course No. ME 3221 Instr.: ………..
Course Name: Manufacturing Automation Office hours: ………..
Classes: ……………………….

Fall ………………..

Textbook: Systems Approach to Computer-Integrated Design and Manufacturing, Nanua Singh, John Wiley & Son Inc., 1996

References: Manufacturing Automation, Morris, A. Cohen and Uday M. Apte, McGraw-Hill, 1997
Modern Manufacturing Process Engineering, Benjamin W. Niebel, Alan B. Draper, and Richard A. Wysk, McGraw Hill Publishing Company, 1989.
List of other references will be distributed in the class.

Week	Topic	Reading	Project Due (Collected Wednesdays)
Week 1	Instruction/Overview of Manufacturing Historical Perspective, Design Process & CAD, Characteristics of a CAD package, Transformation, Scaling/Scalar operations.	Chapter 1-2 pages 1-42	
Week 2	Geometric Modeling Serial Engineering Discuss reading 1-2	Chapter 3-4 pages 43-151	Proj.1: Comp-Aided Probl. Solv. For Design.
Week 3	Product Design Discuss reading 3-4		
Week 4	Process Design, CA Machining NC Part Programming Process Control	Chapter 5 pages 153-192	
Week 5	Machining Processes and CNC	Chapter 6 pages 193-255	Proj.2: Geom. Des.
Week 6	Machining Processes and CNC Introduction to CAM-3 CAMM -3 Demonstration (EII, room 108) Discuss reading 5-6	Manuals	
Week 7	Automated Material Handling and Storage	Chapter 7 pages 257-301	Proj.3A CAMM 3 (Cutting programs)
Week 8	MIDTERM EXAMINATION		
Week 9	Robotics Systems	Chapter 8 pages 303-348	
Week 10	Quality Engineering, Planning and Control Discuss reading 7—8	Chapter 9, 10 pages 349 - 434	Proj. 4 Quest Model*) (System Description)
Week 11	Just in Time and Kanban Systems, Group Technology	Chapter 11, 12 pages 435-527	Final Project Defin
Week 12	Flexible Manufacturing Systems Enterprise Integration.	Chapter 13, 14 (Pages 529-634)	Proj. 3B Mastercam Program
Week 13	THANKS GIVING BREAK		
Week 14	Final Project Presentations (selected teams)		
Week 15	Final Project Presentations - General Review		Final Project Due
Week 16	Final Examinations begin		

*) Thanks to the sponsorship of the Connecticut Center for Advanced Technologies (CCAT)
Students' chapter presentations – next class period after the lecture based on the chapter.

4.1 Homework (collected Mondays):

Ch. 2 Ex.: 2.1 (1, 4), (3, 6), (5, 4.5);
Ex.: 2.2 (three times), Ex.: 2.3 (30deg.), Ex.: 2.4 (newx = x+2); Problem 2.7 (simple and homogeneous)
Ch. 3 Ex.: 3.6: vo=(0,0,0) ; v1=(0,3,0),v2=(6,3,0), v3=(6,0,0), Problem 3.16
Ch. 4 Ex.: 4.1 (Unit cost of raw material $12.00); Problem 4.1
Ch. 5 Ex.: 5.1 (unit raw m.c. $11., unit salvage $2.4); Problem 5.11
Ch. 6 Ex.: 6.1 (work piece with dimensions 3x4x0.80); Problem 6.23
Ch. 7 Ex.: 7.1 (vehicle speed =220 ft/min), Ex.: 7.5 (the average cycle time 0.5 min); Problem 7.15
Ch. 8 Ex.: 8.6 (link lengths are 2m and 4m), Ex.: 8.9 (distances of 6, 12, 14); Problem 8.16
Ch. 9 Ex.: 9.1 (it costs manufacturer $250. and it may cost only $17.40 for labor charges); Problem 9.6
Ch. 10 Ex.: 10.2 (alpha=0.30), Ex. 10.3 (study and write your comments), Problem 10.4
Ch. 11 Ex.: 11.1 (alpha =0.50); Ex. 11.2 (study and write your comments), Problem 11.6
Ch.: 12 Ex.: 12.1 (study and write your comments), Ex.: 12.2 (study and write your comments); Problem 12.4
Ch.13 Ex.: 13.1 (study and write your comments), 13.2(study and write your comments), and 13.3 (study and write your comments)
Ch.14 Ex.: 14.6 (40 messages per second, 40 workstations per LAN);

Grading:
Four projects 35%, Final Project: 10%, Exams: Midterm 15%, Final (cumulative) 20%; Homework: 10%, Students' chapters presentations (oral and slides): 10%.

This page left blank intentionally.

Proceedings of the ASME 2014 International Mechanical Engineering Congress and Exposition
IMECE2014
November 14-20, 2014, Montreal, Quebec, Canada

IMECE2014-40014

CONNECTING FINITE ELEMENT MODELING WITH STRENGTHS OF MATERIALS AND VIBRATIONS USING BEAM EXPERIMENTS

Julian L. Davis*
Department of Engineering
University of Southern Indiana
Evansville, Indiana 47712
Email: julian.ly.davis@usi.edu

Natasha L. Smith
Department of Engineering
University of Southern Indiana
Evansville, Indiana 47712
Email: nlsmith@usi.edu

ABSTRACT

Students tend to compartmentalize individual classes throughout their time in school; associating that which is taught in one class with only that class. A finite element class offers a unique opportunity to illustrate the connection between several engineering topics through modeling and experiments since it is used to solve many different types of problems (e.g. force-deflection, stress-strain, heat transfer, fluid transport, and vibrations). In addition, providing hands-on experiments is an excellent way to illustrate concepts. Particularly, experiments can help students visualize the additional abstractions present in a finite element model while demonstrating the connections between several prerequisite engineering courses. Here we present experiments that connect finite element modeling with two classes: Strength of Materials and Vibrations. These experiments highlight the effects of finite element modeling choices and illustrate errors in the approximations.

INTRODUCTION

Often throughout undergraduate engineering programs students compartmentalize their engineering courses and therefore any of the connections that might exist between them [1–4].

Finite element analysis (FEA) offers an opportunity to connect many different courses in an engineering curriculum: Solid Mechanics, Fluid Mechanics, Heat-Transfer and Vibrations for example. Many faculty enhance courses to using FEA solutions to visualize principles (or deformations) that may often be too small to measure or see. These examples often connect solid mechanics and FEA [2, 5–7]. Some use finite element modeling to validate theory learned in respective classes - illustrating the power of FEA to solve large complex problems [2, 5, 6, 8]. There are few [9] using FEA to make connections to theory through experiments.

Here we present two simple experiments that provide tangible connections between a Strength of Materials course and a Vibrations course through finite element analysis. These experiments are excellent examples illustrating our approximations in finite element analysis and connecting both static (Strength of Materials) and dynamic (Vibrations) solutions. These experiments are also unique in that they provide realizations of concepts discussed in the finite element course: distributed load allocation and errors arising from symmetry approximations. Students are introduced to the respective topics for 2 to 3 lecture hours. Students are then introduced to the experiments and given 3 to 6 lab hours to complete the experiments. They are asked to plot and compare their experimental results to theoretical results based on beam theory.

The Experiments

We present two beam experiments that offer students an opportunity to connect finite element modeling to several engineering classes. These are hands-on experiments that are then modeled and compared with finite element results. Each of these experiments provide physical illustrations of mathematical con-

*Address all correspondence to this author.

cepts and illustrate to students the horizontal integration of mechanics in the engineering curriculum.

Distributed Loads & Force Allocation

The first experiment connects strength of materials and finite element modeling, while illustrating two concepts: 1) errors in the finite element solution both at and between nodes, and 2) distributed load allocation to finite element nodes. In Strengths of Materials, students learn beam theory to determine the analytical solutions for the transverse deflection of a beam under various loading conditions. These results stem from the solution to Eqn. (1) [10], where x represents axial location, y is the transverse deflection, and w is loading. A stiffness factor, EI comes from the material elastic modulus, E, and area moment of inertia, I. Both may vary with respect to x.

$$\frac{d^2}{dx^2}\left(EI(x)\frac{d^2\left[y(x)\right]}{dx^2}\right) = w(x) \tag{1}$$

For a constant distributed load ($w = constant$) on a beam with uniform cross-sectional shape and material ($EI = constant$), the exact solution to Eqn. (1) is a fourth order deflection curve. The first three derivatives of deflection are also of interest as they provide beam slope, moment, and shear respectively. In fact, many Strengths of Materials texts offer "double integration" of the moment curve (rather than $4X$ integration of Eqn. (1)) as the primary method of solving beam deflection problem. When the topic is introduced in finite element analysis, students learn to approximate the solution using cubic shape functions. As a consequence, errors in the deflection curve resulting from the difference between 4th order and cubic functions are magnified for moment and shear (and thus for beam stresses). These errors may be reduced by decreasing the element size ($h - level\ refinement$) or by increasing the order of the shape functions ($p - level\ refinement$). Understanding the source of error in a finite element solution is a critical concept for anyone using finite element analysis.

The cubic FE model is developed from the solution to Eqn. (1) when $w(x) = 0$, thus in essence, a beam with a distributed load must be modeled as one with no distributed loading. Any forces and moments must be applied as nodal boundary conditions, so distributed loads are allocated to the nodes as point forces and moments. Nodal force allocation for an evenly distributed load is calculated using Eqn. (2) [10]. In this equation, e represents the element number and n represents the node of the element for a two node beam element; essentially half of the total load is allocated to each node of each element. In addition, a nodal bending moment (see Eqn. (3) [10]) is needed to improve the accuracy of the FE model. Again, understanding the abstraction from the analytical model (with $w = constant$) to the FE model (with $w = 0$) is fundamental to the students appreciation of the finite element method.

$$f_n^e = -\frac{1}{2}w\,dx \tag{2}$$

$$m_n^e = (-1)^n\frac{1}{12}w^2\,dx \tag{3}$$

For this project, students developed both an analytical solution and finite element solutions using two different discretizations for a simply supported beam with a constant distributed load. They were required to predict the error at various locations either at or in-between the nodes of the FE models, and then to develop plots to illustrate errors in deflection, slope, internal bending moment, and internal shear force. Finally, students performed an experiment designed to illustrate the differences in the models.

The experimental set-up included a 32 inch (813 mm) long, simply supported, Balsa wood beam with a flexural stiffness $EI(x)$ of $20,900\ lbf \cdot in^2$ ($60N \cdot m^2$), and a $15.625 \times 10^{-3}\frac{lbf}{in}$ ($27.4\frac{N}{m}$) evenly distributed load. The set up is illustrated in Fig. 1A. Students were asked to measure deflections at the mispan, quarter-span and one-eighth of the total length of the beam from the simple support. Next students re-distributed the load to the beam mimicking 2 and 4 element finite element models. The concept of load allocation (Eqns. 2 and 3) is enhanced though a physical allocation of the distributed load to element nodes. Realization of the force re-distribution is illustrated in Figs. 1B & 1C.

To complete the experiment, students discussed why their finite element results may be different from experimental. This leaves an opportunity for students to discuss how the experiment did not provide moments at the ends as required by Eqn. 3. The moments from adjacent elements cancel one another for interior nodes but remain at the ends as seen in Figs. 1B and 1C. Again from Eqn. 3, the magnitude of the moment is proportional to the step size dx. Thus from the experiment, students observed that the four element model was a better predictor than the two-element model to the distributed load, in large part due to the reduction in the magnitude of the moments which were missing from the experiment. This experiment could be expanded to include different shaped distributed loads and, with some work, a moment applied at the nodes.

Vibration Modes and Nodes of Vibration

The second experiment expands on the previous, drawing from Strength of Materials by borrowing Eqn. (1). This experiment also draws from Vibrations by allowing the transverse

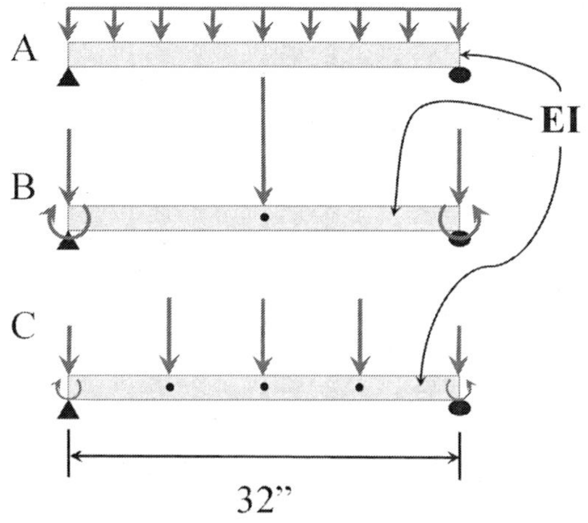

FIGURE 1. FINITE ELEMENT APPROXIMATION OF EVENLY DISTRIBUTED LOAD: **A**, illustrates an load distributed evenly across a beam. Distributed loads are allocated to nodes as forces and moments. **B**, & **C** illustrate two and four element approximations of **A** with three and five nodes as forces (linear arrows) and moments (curved arrows).

FIGURE 2. STATIC BEAM EXPERIMENTAL SET UP: A balsawood beam is simply supported near its ends with small weights (black rectangular prisms) applied to the upper surface (Shown in **A**). Deflection measurements are collected using small dial calipers placed under the beam at three locations: $\frac{L}{2}, \frac{L}{4}, \& \frac{L}{8}$. **B** & **C** illustrate an approximation of nodal forces for a two and four element beam (respectively).

displacement to change with respect to time and by replacing the right hand side of Eqn. (1) with mass and acceleration (See Eqn. (4) below). Equations (4) though (7) are extracted from *Elements of Vibrations Analysis* [11].

$$\frac{d^2}{dx^2}\left(EI(x)\frac{d^2\left[y(x,t)\right]}{dx^2}\right) = m(x)\frac{d^2\left[y(x,t)\right]}{dt^2} \tag{4}$$

FIGURE 3. VIBRATION BEAM EXPERIMENTAL SET UP: In the foreground - the aluminum beam is suspended from string using 4 chemistry stands. Three accelerometers are placed on the beam at 3 specific locations: 1) at the center along the length and width, 2) at the center of the far edge and 3) at the far corner of the beam

Using separation of variables, a trigonometric solution for the spatial differential equation is developed, shown below.

$$Y(x) = a_1\, sin(\beta\, x) + a_2\, cos(\beta\, x) + a_3\, sinh(\beta\, x) + a_4\, cosh(\beta\, x) = 0 \tag{5}$$

A solution for the time domain equation is also developed.

$$T(t) = c_1\, sin(\omega\, t) + c_2\, cos(\omega\, t) \tag{6}$$

Coefficients for Eqn. (5) are determined using boundary conditions. Coefficients for Eqn. (6) are determined from initial conditions. Here, β is a function of the flexural stiffness of a beam (EI), the density (ρ), and the natural frequency (ω_n).

$$\beta^4 = \frac{\rho\,\omega_n^2}{EI} \tag{7}$$

This experiment connects vibrations and finite element modeling while 1) illustrating the concept of symmetry errors associated with two dimensional approximations and 2) providing an audible realization of nodes of vibration. Students developed the analytical solution using a two dimensional model based on Eqn. (4). The FEA solution was developed using using a commercial code (SolidWorks) with three dimensional elements. Natural frequencies and locations of node lines were compared to the analytical solution.

Using the results from the three dimensional FEA, the experimental set up for modal analysis was discussed. The experiment consisted of three uni-directional accelerometers and a small modal shaker. Students are asked to determine the best locations to place accelerometers in order to determine the most number of natural frequencies. Discussion is steered towards the

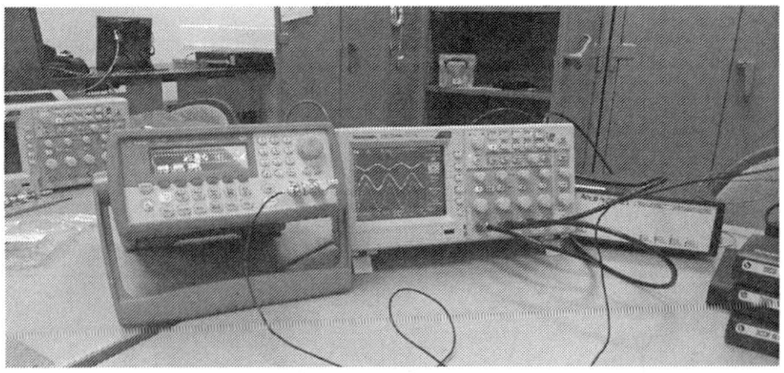

FIGURE 4. VIBRATION BEAM EQUIPMENT: Output from the accelerometers is sent thtrough the charge amplifier (right) and into the oscilloscope (middle). The function generator (left) is used to generate a signal for the modal shaker (not shown).

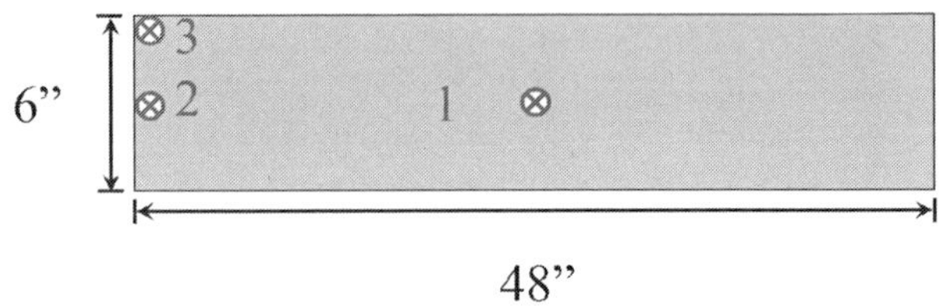

FIGURE 5. ACCELEROMETER PLACEMENT AND BEAM DIMENSIONS: Dimensions of the aluminum beam used in the vibration experiment are shown. Accelerometer placement (1, 2 & 3) to determine resonant frequencies and nodes of vibration is also shown. Please see experimental locations in Fig. 3

locations shown in Fig. 5. Students determined betwen 5 to 15 resonant frequencies based on the output from the accelerometers and a changing excitation frequency.

Students reported and compared natural frequencies from their experimental, analytical and FEA results. In addition, students compared the locations of each node (or line) of vibration for each mode. This aspect of the experiment provided an *audible* connection to a physical interpretation of nodes of vibration. Students placed a short and thin piece of metal on the edge of the vibrating beam to hear the vibration created by the beam. They ran this short piece of metal along the length or width of the beam until the node of vibration is identified (no/minimal noise due to vibration) at each natural frequency. Measured natural frequencies and locations of these nodes of vibration are compared with theoretical data. We expected students to be able to closely match the frequencies and node lines associated with the three dimensional FE model with the experiment. We also expected students to identify the natural frequencies and mode shapes missing from the two dimensional analytical solution. Students were able to predict frequencies and locations of node lines to within 15 % of the analytical results. This experiment could easily be expanded to handle different boundary conditions applied to the beam.

Summary

Both of the experiments discussed here illustrate the connections between several classes within the engineering curriculum using finite element modeling. We illustrate the connection though the study of differential equations as it applies to static and dynamic problems. These examples require students to take measurements, illustrating the idea that finite element models are only approximations. And finally these examples both provide physical realizations (visual and audible) to concepts discussed in the finite element class.

An informal survey of our students illustrates that the finite element class in conjunction with these hands-on experiments are helpful in understanding the connections to their previous coursework. More experiments like these should be developed and introduced in early course work to demonstrate the horizontal (as opposed to vertical) connections to other courses.

ACKNOWLEDGMENT

Thanks to our Spring 2014 Finite Element Class!

REFERENCES

[1] Avitabile, P., 2007. "An integrated undergraduate dynamics systems teaching methodology utilizing analytical and experimental approaches". *American Society for Engineering Education*.

[2] Milton-Benoit, J., Grosse, I. R., Poli, C., and Wolf, B. P., 1998. "The Multimedia Finite Element Modeling". *Journal of Engineering Education*.

[3] Felder, R. M., 2012. "Engineering Education : A Tale of Two Paradigms". *Shaking the Foundations of Geo-Engineering Education*, pp. 9–14.

[4] Roede, R., El-Ghazaly, S., and Aberle, J., 1998. "An integrated upper division course in electronic materials and electromagnetic engineering - Wave Phenomena for Electrical Engineers ". *28th Annual Frontiers in Education.*, pp. 921 – 926.

[5] Coyle, M. F., Ph, D., and Keel, C. G., 2001. "Teaching Finite Element Analysis to Second Year Students".

[6] Shakerin, S., and Jensen, D., 2001. "Enhancement of Mechanics Education by Means of Photoelasticity and the Finite Element Method". *International Jounral of Mechanical Engineering Education,* ***29***(4).

[7] Steif, P., and Gallagher, E., 2004. "Transitioning students to finite element analysis and improving learning in basic courses". *34th Annual Frontiers in Education.*, pp. 1258–1262.

[8] Abumahaimed, M., and Rencis, J., 2011. "Mechanics of Materials, Machine Design, and Vibrations Finite Element Learning Modules for Undergraduate Courses". *American Society for Engineering Education*, pp. 1–17.

[9] Ural, A., and Yost, J., 2010. "Integration of Finite Element Modeling and Experimental Evaluation in a Freshman Project". *Proceedings of the ASEE Mid-Atlantic Annual Conference*.

[10] Logan, D. L., 2012. *A First Course in the Finite Element Method*, 5th ed. Cengage Learning, Australia.

[11] Meirovitch, L., 1986. *Elements of Vibration Analysis*, 2nd ed. McGraw Hill, New York.

This page left blank intentionally.

Proceedings of the ASME 2014 International Mechanical Engineering Congress and Exposition
IMECE2014
November 14-20, 2014, Montreal, Quebec, Canada

IMECE2014-40099

SOFTWARE FOR THE KINEMATIC ANALYSIS OF A SERIAL MANIPULATOR FOR ACADEMIC PURPOSES

Julio C. Correa, Juan A. Ramírez-Macías, David Rozo
Facultad de Ingeniería Mecánica
Universidad Pontificia Bolivariana
P.O. Box 56006
Medellín, Colombia
julio.correa@upb.edu.co
juan.ramirez@upb.edu.co
david.rozo@alfa.upb.edu.co

ABSTRACT

This paper addresses the development of software created at the Universidad Pontificia Bolivariana, located in Medellin, Colombia, for the direct and reverse analyses of a serial robot manipulator. Results provided for the software are used for the simulation of the manipulator and also provides data for the control of a serial manipulator prototype. Both, the prototype and its control system were also developed at the University as part of a project to establish a robotics lab. Since joint angles control is affected by the complex relationships presented in the gear transmission mechanisms, the kinematics of gear transmissions is also studied. In this way appropriate mappings between actuators, links and the Cartesian space are established. Software requirements, architecture, and its implementation are addressed. The software includes user-friendly graphical environments that allow one to navigate through all the capabilities of the program. The software has been tested intensively in the prototype and can be extended to serial manipulators with the same structure but different dimensions. Authors expect that virtual tools as the presented in this paper, can help to reduce the time to understand concepts related to motion analysis in the fields of robotics.

INTRODUCTION

The Universidad Pontificia Bolivariana (UPB) at Medellin, Colombia, has been developing a laboratory for robotics education. This laboratory has both, industrial devices and self developments. Some of them have been reported in [1–3]. Although developing a prototype of a serial manipulator is a demanding task, it is highly beneficial in terms of knowledge in manufacturing and components integration. It is also a good opportunity to present to the students all the details involved in a mechanical design. With this goal, the Automation and Design Research Group (A+D Group) of the UPB, developed a prototype of an academic six-degree-of-freedom (6-dof) serial manipulator.

It is known that in the academic environment, robots are being used in research projects to explore their capabilities to perform new tasks. They are currently used in industry for tasks such as welding, pick and place, painting, among others; but at the academic level, is easy to identify other environments in which these robots could also participate, e.g., in partnership with computer vision system or different sensor sets to improve performance [4–6]. Using real equipment helps to create the best scenario to provide undergraduate and graduate students appropriate tools to acquire advanced knowledge in robotics. However, not all universities have access to such systems due to high related costs. Considering this situation, simulation software for multi-dof machines becomes a priority. Using programming languages and high level programming environments, simulation software can be developed in order to simulate robots motion, and facilitate the understanting of kinematics and dynamics.

In addition to the development of mechanical and electrical

components for the academic 6-dof serial manipulator, a software tool is needed. Hence, it is required to develop a user-friendly software environment that allows the user to perform the direct and reverse kinematic analyses, off-line trajectory design, simulate motion in a 3D virtual environment, and generate information to execute trajectories in the real prototype. The development of such a tool, requires not only the knowledge of the mechanism's parameters and dimensions, but also the kinematic model of mechanisms required to transmit the motion from actuators to the robot's links. Although these mechanisms are not required to perform the kinematic analysis, knowing their behavior is essential for controlling the device. Therefore, the software must contain the additional kinematic models of transmissions that include velocity ratios, in order to adjust data provided by the regular kinematic analyses before sending them to the robot controller.

A good option to develop a robot simulation software is Matlab^{TM}, which is a popular tool used to program applications with a high content of mathematical functions. For instance, Kumar et al. [7] developed a software that allows one to solve and simulate the forward and reverse kinematics of the Mitsubishi RV-M1 serial manipulator. Other developments use different high level programming environments such as LabVIEW, see for instance [8], in which the analysis of a variety of manipulators with different degrees of freedom (up to 5-dof) is allowed. It is possible to find also developments that combine Visual Basic and CAD packages, such as the work presented by Neto et al. [9].

This work addresses the development of a software written in Matlab^{TM}, and used to simulate the behavior of an academic serial robot created at the Universidad Pontificia Bolivariana, in Medellín, Colombia. The paper is organized as follows: the robot prototype is described, considering the parameters that define the model. Then, the mathematical model for direct and reverse kinematic analyses is shown. Next, the software requirements and the main interfaces are presented. Finally, tests for independent validation of the results are explained.

PROTOTYPE

El Robot Azul (that stands for *The Blue Robot* in Spanish), shown in Fig. 1, is an academic 6-degree-of-freedom serial manipulator developed at the Universidad Pontificia Bolivariana, in Medellín (Colombia), as a part of the robotics lab. The development included design, manufacturing, electronics, control, software, and testing. The prototype was developed in two main stages: 1) mechanical design, and 2) final development. The first stage included the concept, degrees of freedom, geometry, mechanisms, mechanical parts, and costs. The second stage comprised manufacturing, assembly, design and implementation of hardware and software required for open-loop motion control, development of kinematic models and simulation platforms, and testing. Because this work focuses on kinematic modeling and simulation, dimensions and other geometric parameters are described in detail.

The robot has six moving serial links, coupled through revolute joints (6R manipulator). All six joints are named A1, A2, A3, A4, A5, and A6. A1–A3 are used for positioning and A4–A6 are used for orientation. The motion ranges are shown in Table 1, the direction of rotation of all joints are indicated in Fig. 2 a), and the general dimensions are depicted in Fig. 2 b).

TABLE 1. Joint extreme angles

Joint	Range [°]
A1	−68/+72
A2	−23/−107
A3	+60/+121
A4	−175/+172
A5	±85
A6	±175

Each joint is actuated with a stepper motor, and all of them have power transmission mechanisms. All links, joints and power transmission mechanisms are illustrated in Fig. 3. In detail, A1 uses a 10:1 reducer and a 44:19 belt & pulley stage; A2 uses a 15:1 reducer; A3 uses a 10:1 reducer and a 4R planar-parallelogram rocker-rocker stage; A4 uses one 62:21 spur-gear stage; A5 has a 42:21 spur-gear stage, a 36:18 bevel-gear stage, and a 44:15 belt & pulley stage; finally, A6 has two 40:20 bevel-gear stages and a 48:24 belt & pulley stage. As noticed, A4–A6 use a spherical wrist configuration.

FIGURE 1. El Robot Azul (*The Blue Robot*)

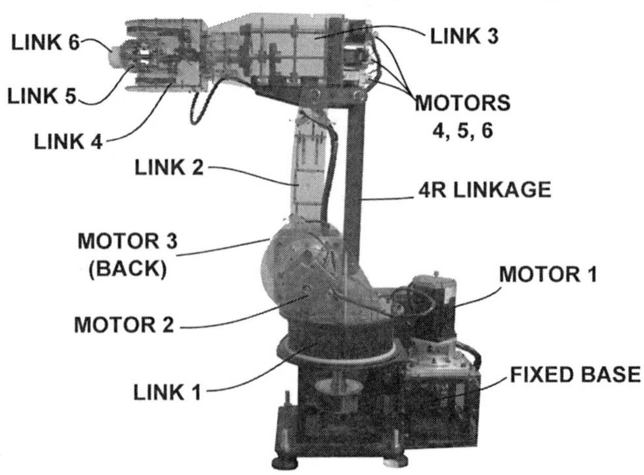

FIGURE 2. Joints, directions and general dimensions

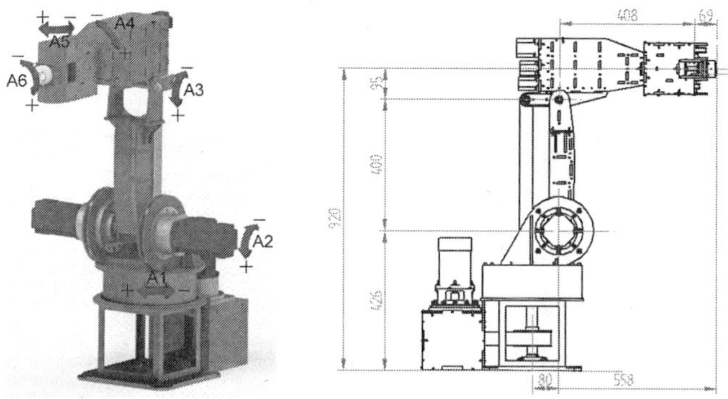

a) Directions of rotation b) General dimensions

FIGURE 3. Links, joints and power transmissions

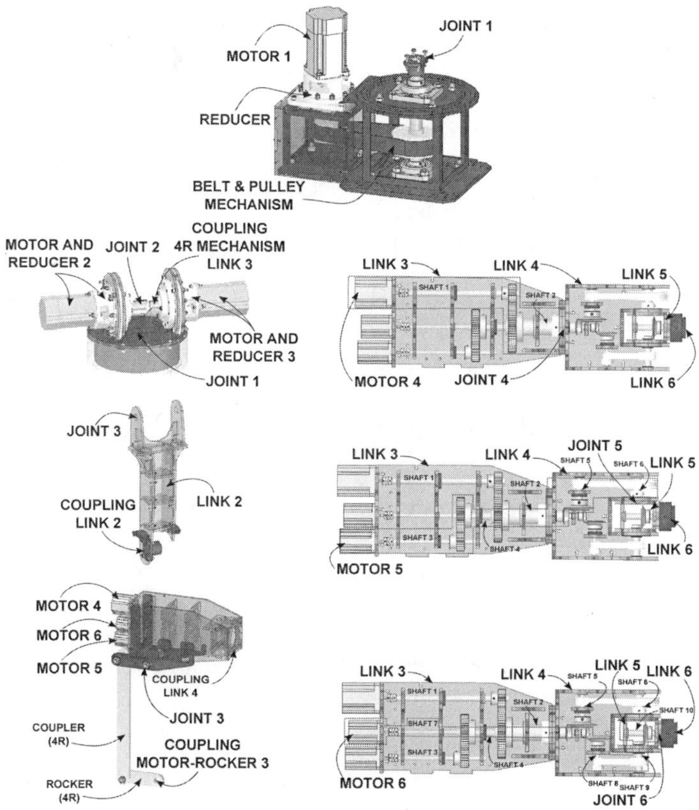

MATHEMATICAL MODEL

The mathematical model, used in the software for simulation purposes and trajectory planning, requires the definition of mappings between three spaces: the Cartesian space, the links space, and the actuators space (Fig. 4).

The direct kinematics considers two mappings: from the actuators space to the links space and from the links space to the Cartesian space, i.e., if the actuators' motion is known the links' motion can be calculated, and if the links' motion is known the

FIGURE 4. Mappings between spaces

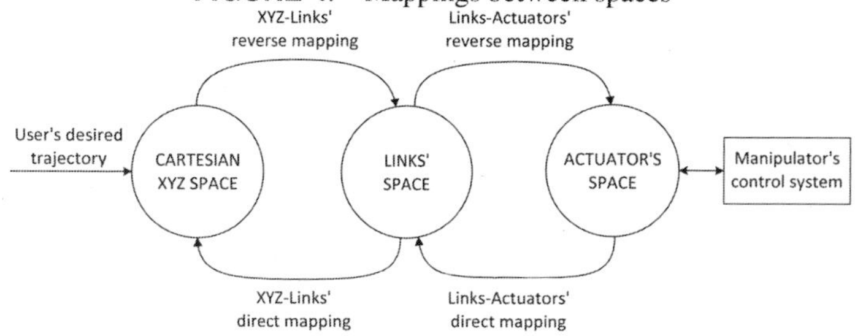

end-effector's motion can be calculated also. Those mappings are used for simulation and visualization purposes, because they predict the motion of the entire robot knowing the motion of the individual actuators. In Fig. 4 these mappings are depicted from right to left.

The reverse kinematics considers two mappings also: from the Cartesian space to the links space and from the links space to the actuators space, i.e., if the desired motion of the end-effector is known the links motion can be calculated, and if the links motion is known the actuators motion can be calculated. Those mappings are used for trajectory planning, because they allow knowing how each actuator has to move in order to maintain a desired trajectory of the end-effector. This information is useful for the control system that moves the robot's actuators. In Fig. 4 these mappings are depicted from left to right.

In the space of the actuators each motor is assigned a ϕ_i variable, with $i = 1,\ldots,6$; each ϕ_i represents the angle of rotation of each motor. The space is given by a vector $\Phi = \left[\phi_1 \cdots \phi_6\right]^T$. The behavior of this set of variables is highly important for control tasks.

In the space of the links each joint is assigned a θ_i variable, with $i = 1,\ldots,6$; each θ_i represents the relative angle between links i and $i-1$. The space is given by a vector $\Theta = \left[\theta_1 \cdots \theta_6\right]^T$.

The Cartesian XYZ space is rather different and require the definition of a fixed-frame $\{f\}$ attached to the base link (or link 0). Once this frame is defined, the space is represented by vectors ${}^f\mathbf{P}_{tool}$, the location of the tool tip, and ${}^f\mathbf{S}_6, {}^f\mathbf{a}_{67}$, the orientation of the robot end effector in $\{f\}$ [10].

Actuators-links mappings

The mappings involving the actuators' and links' spaces depend on the geometric constraints imposed by gears and transmissions, as described in the previous section. Because all velocity relations are considered constant, i.e., all elastic effects are neglected, the mappings can be described by the linear transformations $\Phi = Z\Theta$ and $\Theta = Z^{-1}\Phi$, where Z is a block diagonal

6×6 constant matrix given by

$$Z = \begin{bmatrix} Z_1 & 0_{3\times3} \\ 0_{3\times3} & Z_2 \end{bmatrix},$$

where

$$Z_1 = \begin{bmatrix} -R_1 N_{1,1'} & 0 & 0 \\ 0 & -R_2 & 0 \\ 0 & R_3 & -R_3 \end{bmatrix} \tag{1}$$

and

$$Z_2 = \begin{bmatrix} -N_{2',2} & 0 & 0 \\ N_{3',3} & -N_{3',3}N_{5,4}N_{6',6} & 0 \\ -1 & -N_{8,7}N_{9',9} & N_{8,7}N_{9',9}N_{11,10} \end{bmatrix}. \tag{2}$$

As it may be noticed, Z_1 includes velocity ratios for A1–A3 and Z_2 include ratios for A4–A6 (the spherical wrist mechanism). In (1) and (2) the R_i represent reducer ratios, the $N_{i',i} = T_{i'}/T_i$ represent spur-gear ratios, and the $N_{j,i} = T_j/T_i$ represent bevel-gear ratios. Fig. 5 shows a simplified diagram which illustrates how the more complicated relations of Z_2 in (2) are obtained, this diagram was developed following the convention in [10,11]. The ratios of A1–A3 in (1) are obtained from information given in the previous section.

FIGURE 5. Spherical wrist diagram

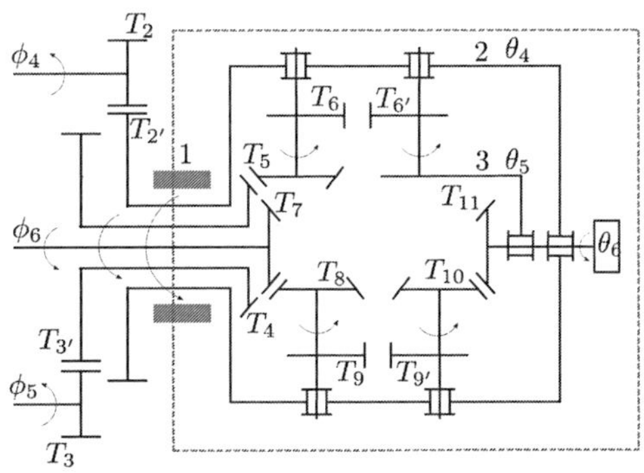

Links-XYZ mappings

The mappings involving the actuators' and links' spaces are described using common serial manipulator kinematics. The model relies on the DH notation and in the proper definition of a series of frames attached to each link, i.e., $\{f\}$, $\{1\}$, …, $\{6\}$,

FIGURE 6. Kinematic diagram

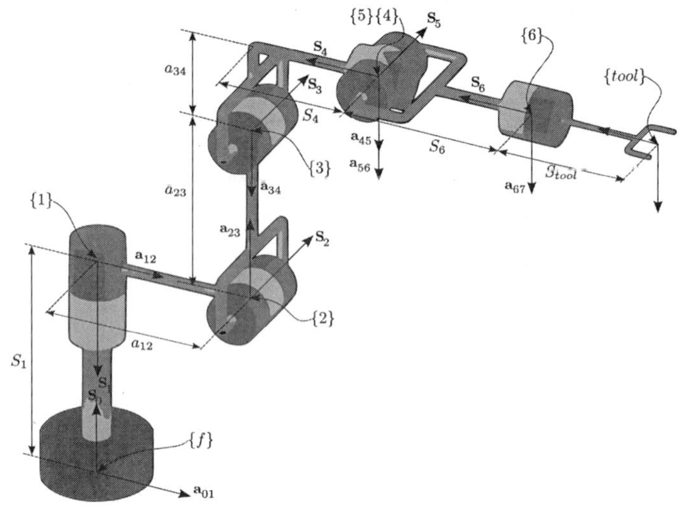

and $\{tool\}$ [10]. Fig. 6 shows how every frame and parameter in the DH notation is defined. In order to make it explicit, the DH parameters are defined using the following guidelines [12]:

- the *axis vector* $\mathbf{S}_{i+1}$ is parallel to the $(i+1)$th axis and its direction is arbitrary;
- the *link vector* $\mathbf{a}_{i,i+1}$ is perpendicular to vectors $\mathbf{S}_{i+1}$ and $\mathbf{S}_{i+2}$ and its direction is arbitrary;
- the *link distance* $a_{i,i+1}$, is the distance between the lines defined by axes $\mathbf{S}_{i+1}$ and $\mathbf{S}_{i+2}$;
- the *joint offset* S_{i+1}, is the distance between the lines defined by vectors $\mathbf{a}_{i,i+1}$ and $\mathbf{a}_{i+1,i+2}$;
- the *twist angle* $\alpha_{i,i+1}$, is the angle between vectors $\mathbf{S}_{i+1}$ and $\mathbf{S}_{i+2}$, around $\mathbf{a}_{i,i+1}$ using the right-hand rule; and
- the *joint angle* θ_{i+1}, is the angle between vectors $\mathbf{a}_{i,i+1}$ y $\mathbf{a}_{i+1,i+2}$, around $\mathbf{S}_{i+1}$ using the right-hand rule.

The numerical values of each parameter in the DH notation are shown in Table 2. These parameters are used in the definition of

TABLE 2. DH parameters

i	$a_{i,i+1}$ (mm)	$\alpha_{i,i+1}$ (deg)	S_{i+1} (mm)	θ_{i+1} (deg)
0 (f)	0	180	−426	θ_1
1	80	90	0	θ_2
2	400	0	0	θ_3
3	−95	−90	−408	θ_4
4	0	90	0	θ_5
5	0	90	−69	θ_6

transformation matrices

$$T_{i+1}^{i} =$$
$$\begin{bmatrix} c\theta_{i+1} & -s\theta_{i+1} & 0 & a_{i,i+1} \\ c\alpha_{i,i+1}s\theta_{i+1} & c\alpha_{i,i+1}c\theta_{i+1} & -s\alpha_{i,i+1} & -S_{i+1}s\alpha_{i,i+1} \\ s\alpha_{i,i+1}s\theta_{i+1} & c\theta_{i+1}s\alpha_{i,i+1} & c\alpha_{i,i+1} & S_{i+1}c\alpha_{i,i+1} \\ 0 & 0 & 0 & 1 \end{bmatrix}, \quad (3)$$

which represent the homogeneous transformation from frame $\{i+1\}$ to $\{i\}$, i.e., any ${}^{i+1}\mathbf{P}$ in $\{i+1\}$ could be obtained in $\{i\}$ through

$$\begin{bmatrix} {}^{i}\mathbf{P} \\ 1 \end{bmatrix} = T_{i+1}^{i} \begin{bmatrix} {}^{i+1}\mathbf{P} \\ 1 \end{bmatrix}.$$

More specifically, any ${}^{6}\mathbf{P}$ in $\{6\}$ could be obtained in $\{f\}$ through

$$\begin{bmatrix} {}^{f}\mathbf{P} \\ 1 \end{bmatrix} = T_1^f T_2^1 T_3^2 T_4^3 T_5^4 T_6^5 \begin{bmatrix} {}^{6}\mathbf{P} \\ 1 \end{bmatrix}. \quad (4)$$

Direct kinematics The direct mapping between the actuator space and the links' space can be formulated as follows:

given $\theta_1, \theta_2, \theta_3, \theta_4, \theta_5, \theta_6$; the joint angle variables that define the relative orientation between links, and
S_6 and the direction of $\mathbf{a}_{67}$ relative to $\mathbf{S}_6$ in order to establish the sixth coordinate system $\{tool\}$;
find ${}^{f}\mathbf{P}_{tool}$: the location of the tool tip in the fixed coordinate system, and
${}^{f}\mathbf{S}_6$ and ${}^{f}\mathbf{a}_{67}$: the orientation of the robot's end effector.

This mapping is obtained taking (4) one step further. Since ${}^{f}\mathbf{S}_6$ and ${}^{f}\mathbf{a}_{67}$ are known, $a_{6,7}$, $\alpha_{6,7}$, S_{tool}, and θ_{tool} can be used to define T_{tool}^{6}. Once T_{tool}^{6} is known the mapping from $\{tool\}$ to $\{f\}$ is

$$\begin{bmatrix} {}^{f}\mathbf{P} \\ 1 \end{bmatrix} = T_1^f T_2^1 T_3^2 T_4^3 T_5^4 T_6^5 T_{tool}^6 \begin{bmatrix} {}^{tool}\mathbf{P} \\ 1 \end{bmatrix}. \quad (5)$$

Equation (5) represents the regular mapping from the links to the XYZ space (namely the tool's space). If the model is required for simulation and 3D modeling purposes, not only the mapping from $\{tool\}$ to frame $\{f\}$ is needed, but also the mapping between any frame $\{i\}$ and $\{f\}$. Any of these mappings take the general form

$$\begin{bmatrix} {}^{f}\mathbf{P} \\ 1 \end{bmatrix} = T_1^f \cdots T_i^{i-1} \begin{bmatrix} {}^{i}\mathbf{P} \\ 1 \end{bmatrix}. \quad (6)$$

Reverse kinematics The reverse mapping between the actuator space and the links' space can be formulated as follows [12]:

given S_6 and the direction of a_{67} relative to S_6 in order to establish the sixth coordinate system,
${}^{6}\mathbf{P}_{tool}$: the location of the tool tip in the sixth coordinate system,
${}^{f}\mathbf{P}_{tool}$: the desired location of the tool tip in the fixed coordinate system, and
${}^{f}\mathbf{S}_6$ and ${}^{f}\mathbf{a}_{67}$: the desired orientation of the robot end-effector;
find $\theta_1, \theta_2, \theta_3, \theta_4, \theta_5, \theta_6$: the joint angle parameters that positions the end effector as desired.

In the reverse kinematic analysis, mathematically for a unique position and orientation of the end-effector, up to eight possible solutions of the joint angles can be obtained. How the eight solutions result is illustrated in Fig. 7, this diagram follows the notation in [12]. The diagram shows that starting from ${}^{6}\mathbf{P}_{tool}$, ${}^{f}\mathbf{P}_{tool}$, ${}^{f}\mathbf{S}_6$, and ${}^{f}\mathbf{a}_{67}$ as known data and usign the equations shown later in this section, more than one solution appear en each step, e.g., solving the first equation two solutions for θ_1 appear ($\theta_1^{(a)}$ and $\theta_1^{(b)}$), and solving the second equation two solutions for θ_3 appear for each θ_1 ($\theta_3^{(a)}$ and $\theta_3^{(b)}$, and $\theta_3^{(c)}$ and $\theta_3^{(d)}$). Fig. 7 is a guideline of how the reverse kinematic model is solved.

FIGURE 7. Reverse kinematics solution diagram

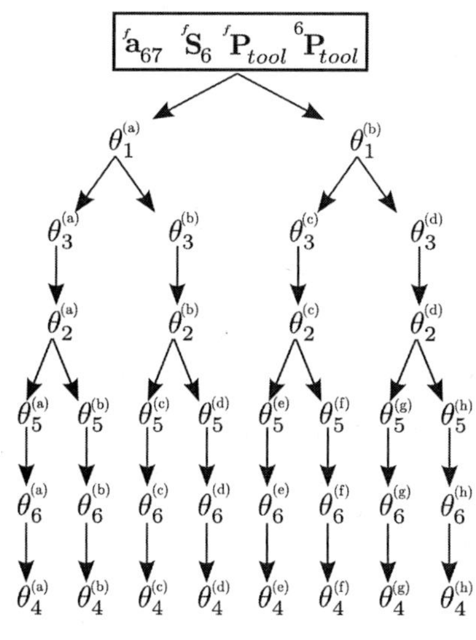

From these eight solutions usually four are complex (which are discarded), and from the four real remaining solutions possibly two of them do not satisfy the allowable motion ranges of the joints. Hence, after analyzing the results, two possible configurations are obtained, so the robot correctly positions the end-effector. Further analysis is required to choose which of the remaining feasible solutions is the most appropriate.

In order to carry on the solution, it must be considered that the first three joints of a serial manipulator are responsible for positioning the end effector in the Cartesian space, and the last three are responsible for the orientation; therefore, those two problems can be independently analyzed. This means that it is possible to use the kinematic detaching methodology to divide the equations into two semi-dependent groups [10]. In this methodology, a series of explicit equations that allows one to determine the numerical values for each joint angle can be obtained. The mathematical detail of this equations is addressed in [10, 12–14].

Considering that XYZ variables p_x, p_y and p_z are known. The first joint angle is

$$\theta_1 = \arctan(-p_y/p_x), \tag{7}$$

with two possible solutions

$$\theta_1 = \theta_1 \pm 180^\circ. \tag{8}$$

The third joint angle results from

$$Ac(\theta_3) + Bs(\theta_3) + D = 0, \tag{9}$$

where $A = -2a_{2,3}a_{3,4}$, $B = 2S_4a_{2,3}$, and

$$\begin{aligned} D = {} & (p_y c(\theta_1) + p_x s(\theta_1))^2 + \\ & (a_{1,2} - p_x c(\theta_1) + p_y s(\theta_1))^2 + \\ & (S_1 + p_z)^2 - S_4^2 - a_{2,3}^2 - a_{3,4}^2. \end{aligned}$$

The second joint angle results from solving the system

$$AX = B, \tag{10}$$

where

$$A = \begin{bmatrix} -a_{3,4}s(\theta_3) - S_4c(\theta_3) & a_{3,4}c(\theta_3) - S_4s(\theta_3) + a_{2,3} \\ a_{3,4} + c(\theta_3) - S_4s(\theta_3) + a_{2,3} & a_{3,4}s(\theta_3) + S_4c(\theta_3) \end{bmatrix},$$

$$X = \begin{bmatrix} s(\theta_2) \\ c(\theta_2) \end{bmatrix}, \text{ and}$$

$$B = \begin{bmatrix} p_x c(\theta_1) - p_y s(\theta_1) - a_{1,2} \\ -p_z - S_1 \end{bmatrix};$$

and finally solving

$$\theta_2 = \arctan 2(s(\theta_2), c(\theta_2)). \tag{11}$$

The fourth to sixth joint angles are found with the expression

$$\begin{bmatrix} kc(\theta_4) - hs(\theta_4) & bc(\theta_4) - ms(\theta_4) & dc(\theta_4) - ns(\theta_4) \\ -hc(\theta_4) - ks(\theta_4) & -mc(\theta_4) - bs(\theta_4) & -nc(\theta_4) - ds(\theta_4) \\ e & f & g \end{bmatrix}$$
$$= \begin{bmatrix} c(\theta_5)c(\theta_6) & -c(\theta_5)s(\theta_6) & -s(\theta_5) \\ s(\theta_6) & c(\theta_6) & 0 \\ c(\theta_6)s(\theta_5) & -s(\theta_5)s(\theta_6) & c(\theta_5) \end{bmatrix}.$$

Then, it results for the fifth joint

$$\begin{aligned} \theta_5 &= \arccos(g), \text{ and} \\ \theta_5 &= -\theta_5; \end{aligned} \tag{12}$$

for the sixth joint

$$\begin{aligned} c(\theta_6) &= \frac{e}{s(\theta_5)}, \\ s(\theta_6) &= \frac{-f}{s(\theta_5)}, \\ \theta_6 &= \arctan 2(s(\theta_6, c(\theta_6))); \end{aligned} \tag{13}$$

and finally for the fourth joint

$$\begin{aligned} &\begin{bmatrix} -k & -h \\ -b & -m \end{bmatrix} \begin{bmatrix} s(\theta_4) \\ c(\theta_4) \end{bmatrix} = \begin{bmatrix} s(\theta_6) \\ c(\theta_6) \end{bmatrix} \text{ and} \\ &\theta_4 = \arctan 2(s(\theta_4), c(\theta_4)). \end{aligned} \tag{14}$$

The solution of (7)–(14), completes the reverse kinematics analysis.

SOFTWARE

Due to its capabilities and its broad range of ready-to-use functions, Matlab™ is the programming platform selected for the development of the software environment. This includes the Graphical User Interface (GUI), all auxiliary functions that solve the kinematic models, the 3D model visualization and animation, and generation of data-exchange files.

Design specifications

The design parameters used for the development of the application were:

- Provide a Graphical User Interface (GUI), that facilitates interaction with code and commands.
- Allow the user to enter data either manually or by means of a file using a prescribed format.
- Allow the user to view an animated simulation of the robot, so it can be analyzed.
- Allow the user to export the results in an exchange data format for further analysis (.csv, .txt).

Architecture

The kinematic analysis software is divided into two independent programs: the first one allows one to solve the direct kinematics and the second one allows one to solve reverse kinematics. Each program has a three-layer architecture and runs on a single computer. The first layer contains the GUI. The second layer contains the set of functions that perform mathematical calculations of the robot's kinematics. The third layer is dedicated to data, i.e., required geometrical parameters. Figs. 8 and 9 show how information flows for the direct and reverse analysis programs, respectively.

Figure 8 shows a simplified diagram of how information flows in the direct kinematic analysis software. The user has information of a sequence of positions for all six joints A1–A6, namely waypoints. This sequence is entered into the software through the GUI. The information is computed in order to calculate the detailed motion of all robot's links, including waypoints and points in-between. Then a 3D graphical representation of the motion is shown in the GUI. Once the user considers motion is feasible, data exchange files may be generated and used to generate motion of the real prototype.

FIGURE 8. Function diagram for the direct analysis

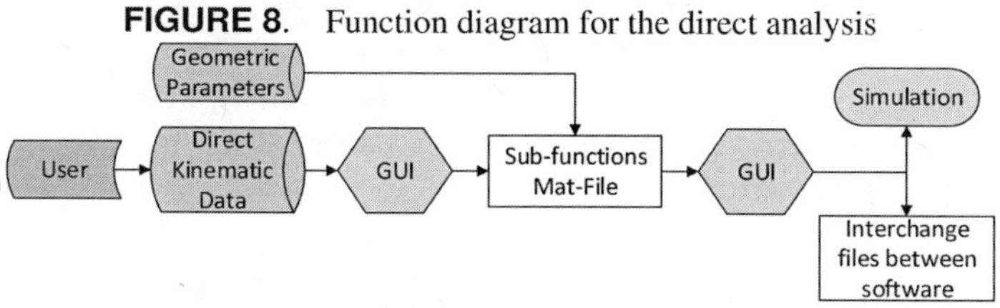

Conversely, Fig. 9 shows a simplified diagram of how information flows in the reverse kinematic analysis software. The User has information of a sequence of positions of the end-effector, namely waypoints for the robot's tool. This sequence enters the software through the GUI. The information is computed in order to calculate the reverse kinematics and find proper link's motion. Then a 3D graphical representation of the motion is shown in the GUI. Finally, once the user considers motion is feasible, data exchange files may be generated and used to generate motion of the real prototype.

FIGURE 9. Function diagram for the reverse analysis

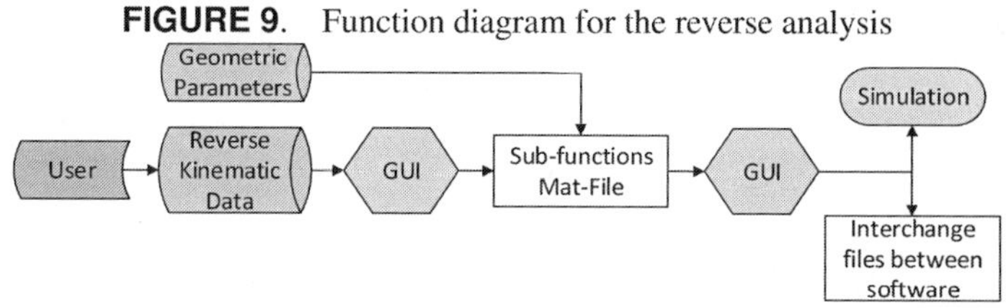

3D model and link positioning

Figure 10 illustrates the assembly of all the graphical objects that make up the virtual manipulator in the developed software. This representation is based on the CAD model of the robot. This means that starting from a stardard CAD file a series of data manipulations are required in order to generate a Matlab^{TM} 3D object representation.

FIGURE 10. Complete 3D model

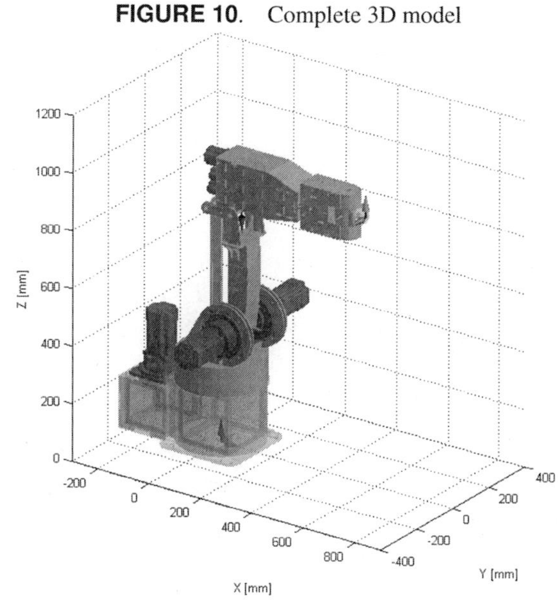

Figure 11 shows the sequence used to convert a CAD file to a Matlab^{TM} 3D object representation of the robot. First, each robot link must be modeled using CAD software, and drawn with respect to the appropriate frame of reference $\{i\}$. Then, the

CAD representation has to be exported to the generic .stl format. This format defines each part's surface as a mesh. The .stl file is then converted to a 3D object, namely a patch object, using the open source program `cad2matdemo.m` [15]. This representation contains the coordinates of all link's geometry. Using transformation matrices, defined in (5)–(6), it is possible to rotate and move these objects to the appropriate position in the Cartesian XYZ space (frame $\{f\}$). The use of Matlab™ objects in combination with transformation matrices makes posible the animation of the robot's movement.

GUI: Graphical User Interface

In the implementation of a Graphical User Interface (GUI) the broad Matlab™ library of user interface controls was used, including: pushbuttons, edit texts, static texts, pop-up menus, among others [16, 17]. Fig. 12 shows the GUI developed for the direct kinematic analysis. The interface has a group of different controls, and the graphical object that represents the serial manipulator. In order to illustrate the GUI's capabilities, the following is a list of controls present in the software:

① Menu Bar.
② Table of angles.
③ Table Solution Forward Kinematics.
④ Virtual manipulator.
⑤ Control Panel Manual.
⑥ Control Save Position.
⑦ Control Animation.

Figure 13 shows the GUI developed for the reverse kinematic analysis. This interface also has a group of different controls and the graphical object that represents the serial manipulator. In order to illustrate the GUI's capabilities, the following is a list of controls present in the software:

① Menu Bar.
② Table Position and Orientation of the End Effector.
③ Table Reverse Kinematic Solution.
④ Manual Control.
⑤ Control Load/Calculate.
⑥ Control Save/Calculate.
⑦ Control plotted angles.
⑧ Pop-up menu of kinematic Solution.
⑨ Virtual manipulator.
⑩ Animation Control.

FIGURE 11. CAD to Matlab™ conversion sequence

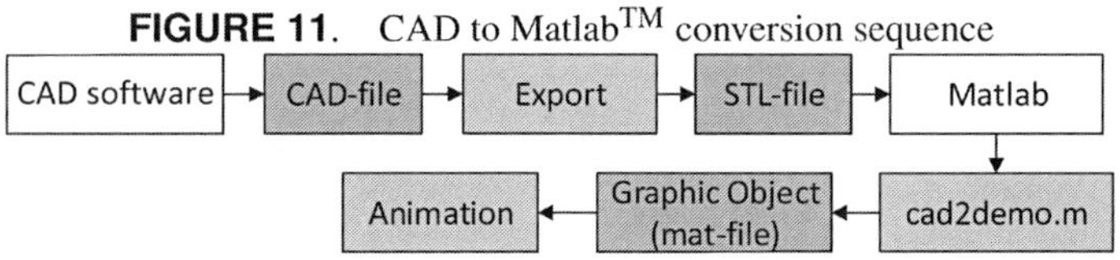

FIGURE 12. GUI for the direct analysis

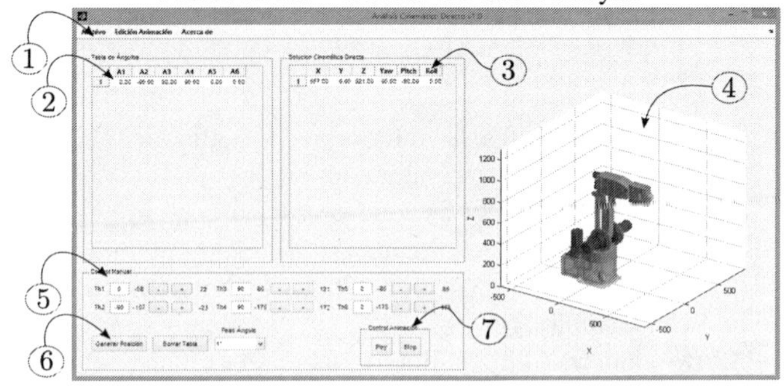

FIGURE 13. GUI for the reverse analysis

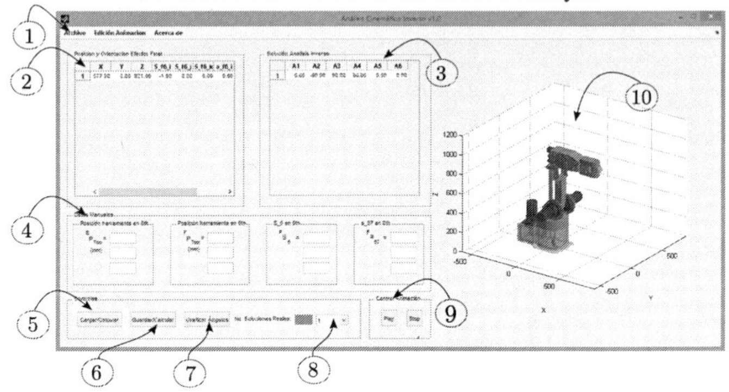

Data-exchange file generation

In order to expand the capabilities of the software, generic .txt and .csv files can be generated with information useful to the prototype's control system, starting from trajectory information available. That makes this software a high-level trajectory planning tool that uses user data and generates control-system level data. It is an off-line programming tool in the sense that it allows the user to design trajectories to be executed later in the real prototype.

The trajectories are always represented by a table with a sequence of waypoints in the space of the links. This table has two header lines: the first one sets velocity and the second one sets acceleration. Then, each following line has a different waypoint and each column represent a different datum. The meaning of each column is respectively:

1. Angular position of A1 (θ_1) in degrees.
2. Angular position of A2 (θ_2) in degrees.
3. Angular position of A3 (θ_3) in degrees.
4. Angular position of A4 (θ_4) in degrees.
5. Angular position of A5 (θ_5) in degrees.
6. Angular position of A6 (θ_6) in degrees.
7. Required command: 1, movement; 4, speed definition; 5, acceleration definition; and 8, pause.

This information may be easily arranged either in a .txt or a .csv file, using the order shown. Fig. 14 shows an example of how information is arranged in a .txt file.

FIGURE 14. Example and explanation of a .txt exchange file

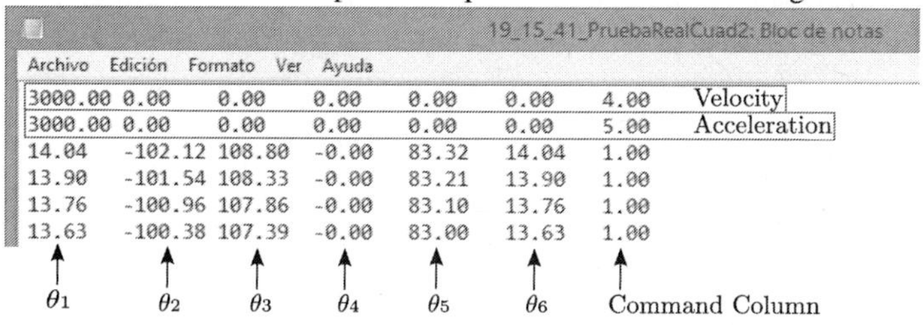

19_15_41_PruebaRealCuad2: Bloc de notas

Archivo Edición Formato Ver Ayuda

θ_1	θ_2	θ_3	θ_4	θ_5	θ_6	Command Column	
3000.00	0.00	0.00	0.00	0.00	0.00	4.00	Velocity
3000.00	0.00	0.00	0.00	0.00	0.00	5.00	Acceleration
14.04	-102.12	108.80	-0.00	83.32	14.04	1.00	
13.90	-101.54	108.33	-0.00	83.21	13.90	1.00	
13.76	-100.96	107.86	-0.00	83.10	13.76	1.00	
13.63	-100.38	107.39	-0.00	83.00	13.63	1.00	

Further use of the software

Even though the developed software was centered in the shown prototype, the methodologies explained here were also tested to simulate the kinematics of a KUKA-KR6 industrial robot. Fig. 15 shows the GUI developed to interact with a virtual 3D representation of this robot. This software is to be used in the training of early users of this equipment and students learning robot kinematics.

FIGURE 15. Software for the simulation of known KUKA-KR6 kinematics

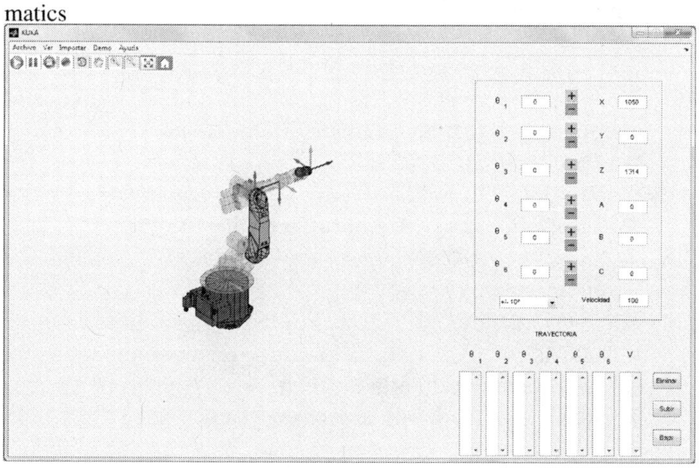

TESTS

A procedure for checking the serial manipulator kinematics derived from data in the Cartesian space was developed. The procedure was as follows: 1) define a circle of radius $r = 60$ mm located at the origin of a Cartesian coordinate system (x, y, z), 2) perform a coordinate transformation, so the circle is in the workspace of the end effector of the serial manipulator, i.e., $(x = 600, y = 0, z = 500)$, as shown in Fig. 16.

FIGURE 16. Trajectory transformation

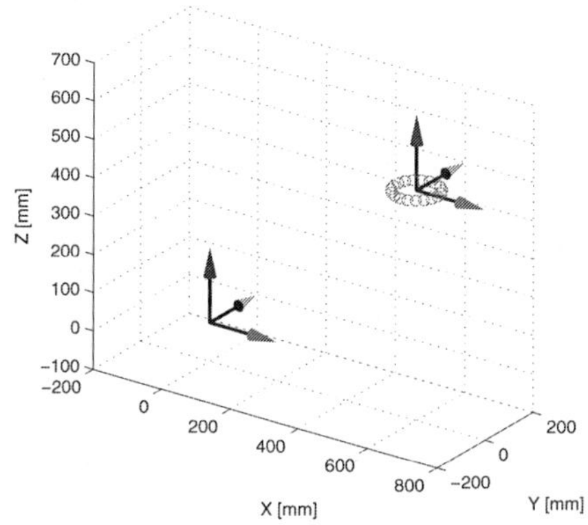

Using the reverse kinematic analysis software, the trajectory in the XYZ space is transformed to the links' space, as shown in Fig. 17. The software allows an analysis of the results: it

FIGURE 17. Software Reverse Kinematic Analysis

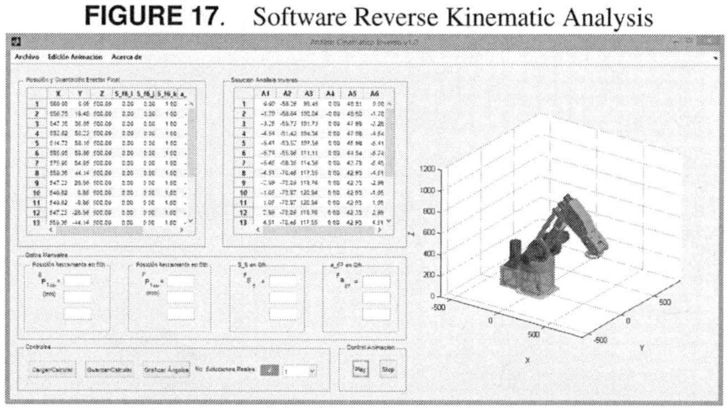

shows all feasible configurations for the manipulator to follow the trajectory, and using the virtual model of the manipulator it is possible to visualize any selected solution, in order to identify if the end-effector is carrying out the assigned task and if the links-joints are behaving well, this displays also the trajectory described by the final effector in the Cartesian space by using the direct kinematic model.

In order to verify if at the software level the direct and reverse kinematic models are correct, one of the solutions obtained in the inverse kinematic analysis was selected (for this problem two solutions were obtained). This solution was transformed using the direct kinematic analysis software and compared to the results obtained using the reverse kinematic analysis software.

As illustrated in Fig. 18 the original path and the transformed one match very well.

FIGURE 18. Comparison between original trajectory vs. transformation trajectory.

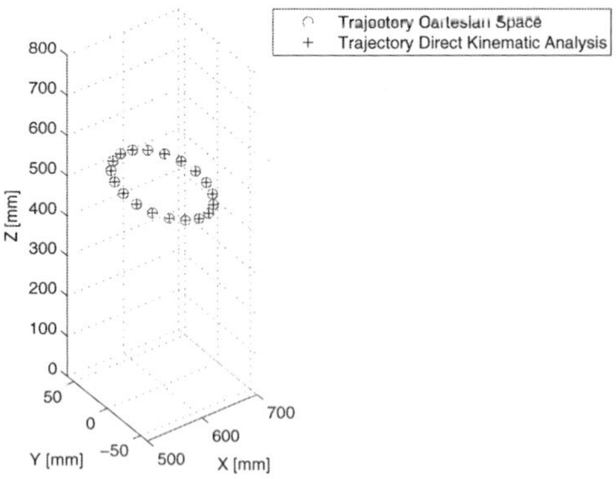

Moreover the dispersion of the absolute value of the difference between each point of the trajectory is in the order of 1×10^{-12} mm. From these results, we can conclude that the two analyses successfully solved the kinematics of the six-degree-of-freedom serial manipulator at the software level.

CONCLUSIONS

This work presented the process that was followed to develop a simulation software for a serial manipulator, created for academic purposes within the robotics lab of the Universidad Pontificia Bolivariana, Medellín, Colombia.

The software provides user-friendly interfaces, performs the kinematic analyses, and exchanges data in different formats. Several independent methods to verify accurate results were used. The developed software tool provides data required to command the actuators of the real system, so the user can easily go from a simulation environment to a real robotic system. Further work require experimental validation of the results in order to obtain data such as precision and repeatability.

The methodology followed for the software development, can be adapted to other robotic devices. The use of this virtual environment can help to accelerate the understanding of advanced concepts that are usually found in the motion analysis of robotic systems.

REFERENCES

[1] ASME-IMECE, 2010. Implementation of a Laboratory for the Study of Robot Manipulators.

[2] CIMM, 2013. Software de Simulación para un Robot Manipulador Serial KUKA KR6.

[3] ASME-DETC, 2011. Software for Programming an Industrial Serial Manipulator.

[4] G. Michalos, S. Makris, A. Eytan, S. Matthaiakis and G. Chryssolouris, 2012. "Robot path correction using stereo vision system". In 45th CIRP Conference on Manufacturing Systems 2012.

[5] M.A. Montironi, P. Castellini, L. Stroppa and N. Paone, 2014. "Adaptive autonomous positioning of a robot vision system: Application to quality control on production lines". *Robot and Computer-Integrated Manufacturing*, ***30***, pp. 489–498.

[6] Zhongxu Hu, Chris Marshall, Robert Bicker and Paul Taylor, 2007. "Automatic surface roughing with 3d machine vision and cooperative robot control". *Robotics and Autonomous System*, ***55***, pp. 522–560.

[7] Rajesh Kumar, Parveen Kalra and Neelam R. Prakash, 2011. "A virtual rv-m1 robot system". *Robotics and Computer-Integrated Manufacturing*.

[8] T. J. Mateo Saguino and J. M. Andújar Márquez, 2011. "3d-ras: A new eductional simulation tool for kinematics analysis of anthropomorphic robotic arms". *International Journal of Engineerieng Education*.

[9] Pedro Neto and Nuno Mendes, 2013. "Direct off-line robot programming via a common cad package". *Robotics and Autonomous System*.

[10] Tsai, L.-W., 1999. *Robot analysis: the mechanics of serial and parallel manipulators*. Wiley-Interscience.

[11] Tsai, L.-W., 1988. "The kinematics and of spatial and robotic bevel-gear and trains". *IEEE Journal Of RobRobot And Automation*, ***4***, p. 7.

[12] Crane III, C. D., and Duffy, J., 1998. *Kinematic Analysis of Robot Manipulators*. Cambridge University Press.

[13] Jazar, R. N., 2010. *Theory of applied robotics: kinematics, dynamics, and control*, 2 ed. Springer.

[14] Rozo Osorio, D., 2013. Desarrollo de una aplicación para el análisis cinemático de un prototipo de manipulador serial de seis grados de libertad. Tech. rep., Trabajo de Grado, Facultad de Ingeniería Mecánica, Universidad Pontificia Bolivariana.

[15] Riley, D., 2003. File exchange-import stl file to matlab. Available Online 09/04/2013, `http://www.mathworks.com/matlabcentral/fileexchange/3642-cad2matdemo-m`.

[16] MathWorks, 2013. Uicontrol documentation. Available Online 09/04/2013, `http://www.mathworks.com/help/matlab/ref/uicontrol.html`.

[17] MathWorks, 2013. Uicontrol properties documentation. Available Online 09/04/2013, `http://www.mathworks.com/help/matlab/ref/uicontrol\_props.html`.

Proceedings of the ASME 2014 International Mechanical Engineering Congress and Exposition
IMECE2014
November 14-20, 2014, Montreal, Quebec, Canada

IMECE2014-40110

GREEN ENERGY MANUFACTURING LABORATORY DEVELOPMENT FOR STUDENT LEARNING EXPERIENCE ON SUSTAINABILITY

Richard Y. Chiou, Radian Belu, and Michael Mauk
Department of Engineering Technology
College of Engineering
Drexel University
Philadelphia, PA 19104

Tzu-Liang (Bill) Tseng
Department of Industrial, Manufacturing and Systems Engineering
The University of Texas at El Paso
500 West University Avenue
El Paso, TX 79968

ABSTRACT

This paper discusses the development of a green energy manufacturing laboratory for student learning experience in the emerging fields of renewable energy and green manufacturing. The development involves a creation of a series of experiments to stimulate discoveries and developments that promise to sustain a wave of new technological innovations on energy and manufacturing throughout the world. The needs for engineering students and practicing engineers to understand sustainability concepts and concerns have been noted by educators, scientists or engineers and all engineering students need to become versed in sustainability ideas. This paper describes key factors in enhancing the ability of future engineering graduates to better contribute to a more sustainable future, preserving natural resources and advancing technological development. Two main components are used to incorporate sustainability into the green energy manufacturing laboratory, including: (1) renewable energy and (2) manufacturing energy efficiency. The efforts presented in the paper also include life-cycle assessment, development of innovative thinking skills, better understanding of sustainability issues, and increasing students' interests in the engineering and technology programs. A concluding section discusses laboratory development for student hands-on learning experience within the context of a project. The paper will present the how it establishes its long-term sustainability and support through a variety of mechanisms including the energy mission, the award of federal grants, program projects, private foundation support, partnerships, and university-based investments. The GEM/Institute/Community College research model and the supporting the hardware, software and middleware are being installed, developed and enabled by the joint project between two universities in the nation.

INTRODUCTION

This paper discusses a joint educational effort that incorporates sustainability in engineering and technology laboratory development at Drexel University (DU) and University of Texas at El Paso (UTEP). A critical component of a national "green industries/green jobs" effort is to motivate our citizenry to become proficient in STEM and associated manufacturing fields and societies, thus ensuring we have a 21st century workforce. Sustainable engineering is about design that recognizes the constraints applied by natural resources and the environmental system. The needs for engineering students and practicing engineers to understand sustainability concepts and concerns have been noted by many educators, scientists or engineers, and it is the philosophy of the authors that all engineering students need to become versed in sustainability ideas. This paper describes key factors in enhancing the ability of future engineering graduates to better contribute to a more sustainable future, preserving natural resources and advancing technological and societal development.

The field of renewable energy technologies has changed dramatically over the last two decades due to significant progress and advancement in wind turbines, power electronics, controls, solar energy, photovoltaic (PV) and fuel cell technologies, instrumentation and electric machines. People at all levels must be educated regarding the need of emerging energy technologies, the uses for these technologies, and their roles in the alleviation of growth in energy demand for the future. In addition, it will be imperative to create a highly educated workforce that can contribute to deal with challenges associated with renewable energy sources. One method of supporting workforces development in future energy solutions is to incorporate new and emerging energy technology directly into required undergraduate coursework. Engineering and technology educators must develop new curriculum solutions in advanced energy technologies to fill the gaps in existing coursework and prepare the next generation of students to support renewable energy, energy efficiency, and sustainability [1-2].

In addition to the increase in population, developing countries are becoming more industrialized. The increased manufacturing activities have led to per capita increases in the resource use and economic activity, contributing to serious environmental problems as well. The economic activity is dependent on the environment and its underlying resources. When the environment and the resources are polluted, degraded, dissipated, and irretrievably compromised, then the economy goes bankrupt with it because the economy is just a subset within the ecological system. Recently, there has been an intense, global focus on the rising costs of non-renewable energy sources and the negative impacts that our societies have on the global environment. There are many reasons for this increased focus. As fuel costs rise, newer technologies are driving down the costs of many of the more expensive renewable sources. These two factors are combining to make the renewable sources economically feasible. Furthermore, with the current economic and military situations, there is an understanding that increased use of some non-renewable fuels is increasing national trade deficits and supporting undesired entities.

Sustainability issues are increasingly important among governments, consumers, researchers, and corporations. Many companies are directing their resources to reduce the environmental impact of their products and services, to remain competitive in the global economy. The companies must recruit employees who understand and possess the knowledge of the impacts of their decisions on the environment and society. It is the mission of universities to prepare these future employees to meet this need. Green manufacturing and design for sustainability are emerging fields in recent years and is also the sustainable development model for modern manufacturing industries. Sustainable green manufacturing encompasses the concept of combining technical issues of design and manufacturing, energy conservation, pollution prevention, health and safety of communities and consumers. Many industries are directing their resources to reduce the environmental impact of their produced products and services. To remain competitive in the global economy, these industries need to train engineering and technology professionals to understand the impact of their decisions on the environment and society. It is important for universities to prepare these future engineering technologists to meet this need. Many technology programs do not offer this type of information to their undergraduate students. However, the Accreditation Board for Engineering and Technology (ABET) requires that graduates be able "to design a system, component, or process to meet desired needs within realistic constraints such as economic, environmental, social, political, ethical, health and safety, manufacturability, and sustainability." [3-6]

The topic of sustainability has become ubiquitous. It is part of corporate strategy, consumer choice processes, university initiatives, engineering, and technology programs within the business discipline. This moves toward more sustainable business practices and education is a direct result of an increasing awareness of the significant green manufacturing covers a broad spectrum of manufacturing, from development of green technology products, implementation of advanced manufacturing and production technologies, and introduction of energy efficient and environmentally friendly manufacturing processes and systems, from the plant floor to the enterprise level, and the whole supply chain.

SUSTAINABLE LABORATORY DEVELOPMENT

The EPA defines sustainability as "meeting the needs of the present without compromising the ability of future generations to meet their own needs." The principles of sustainability can stimulate technological innovation, advance competitiveness, and improve our quality of life. Industrial designers make their living developing designs that meet the needs of the present, but are not always bound to consider future generations. They are integral components of a manufacturing system that is bound and determined to maintain product flow without interruption or foreseeable end in the name of economic development. Sustainability in engineering has its roots in two engineering sub-disciplines: green engineering and environmental engineering. In general, green engineering has focused on design that is in greater long-term harmony with the environment while environmental engineering has addressed the deleterious effects engineering design (or, sadly, lack of design) has had on the environment [7].

Sustainable engineering is, in many ways, more related to green engineering as it focuses on design that requires fewer natural resources, produces less (or no) waste, and reduces, reuses, or recycles waste products. Several approaches have been proposed clear educational objectives for incorporating sustainable development concepts into the undergraduate curriculum [8-14]. The ABET accreditation criteria for engineering and technology programs in the United States calls on programs to directly address issues of sustainability and environment. However, in an already crowded engineering curriculum, it is difficult to envision where and how these concepts can fit. In this paper, we refine the recommendations of others into yearly goals and objectives for the undergraduate curriculum. Given the increasing importance of sustainability, it is incumbent upon academia to educate future engineers and other decision makers on sustainability topics, i.e., incorporate sustainability into engineering and technology curricula.

The sustainable development, green manufacturing and renewable option incorporates energy efficiency and conservation, energy sustainability, renewable energy, and carbon management of systems will be related to industrial processes, processes of commercial, and light commercial and residential facilities. Students learn a combination of technical and management theories designed to prepare them for careers in the green economy sector associated with facilities management, energy efficiency, energy conservation, renewable energy generation, green manufacturing, and sustainable construction. This program offers practice based theoretical knowledge in the classroom, experimental design in the laboratory, and hands-on experience in the field. Ideally, graduating engineers would be able to create a process flow and a life-cycle inventory for competing product (or process) designs with the aid of a computational software tool. From the inventory, they would be able to assess the relative merits of the competing designs based on ecological and resource indicators. These advanced skills draw on the students' knowledge of their discipline. But they also require a set of analytical approaches that encourage holistic, systems thinking. In systems thinking, one pays attention to the inputs and outputs of the system and their effects on the surroundings. These approaches can be integrated into the framework of existing engineering laboratory.

Promoting sustainability through laboratory design involves articulating what students will take away from a learning experience and connecting this outcome to some kind of assessment. Laboratory developers use systematic, consistent frameworks to organize learning outcomes, activities, and assessment tasks. The laboratory components include learning modules with associated assessment tasks. These deliverables are the vehicles by which sustainability is evidenced and communicated to learners and the public. In order to incorporate sustainability and green manufacturing principles and methods into green energy manufacturing laboratory, we must understand the barriers that need to be overcome. In the past, barriers to laboratory development have included: i) accreditation focus on "bean counting" as opposed to process improvement, ii) conventional thinking of some faculty members, and iii) company expectations and recruiting trends. While progress has been made in surmounting or reducing some of these barriers, much improvement is still needed.

The Accreditation Board for Engineering and Technology (ABET) is charged with the task of "Quality assurance in higher education" for programs in applied science, computing, engineering, and technology. Institutions pursuing accreditation must demonstrate that the program meets a set of general criteria [5]. Of particular interest are the requirements of Criteria #2, #3, and #5, which are focused on Program Educational Objectives, Program Outcomes and Assessment, and Faculty. These requirements include:

1. A process based on the needs of the program's various constituencies in which the objectives are determined and periodically evaluated (Criterion #2);
2. The students in the program must attain "an ability to design a system, component, or process to meet desired needs within realistic constraints such as economic, environmental, social, political, ethical, health and safety, manufacturability, and sustainability" (Criterion #3); and
3. The overall competence of the faculty may be judged by such factors as education, diversity of backgrounds, engineering experience, teaching experience, ability to communicate, enthusiasm for developing more effective programs, level of scholarship, participation in professional societies, and licensure as professional engineers (Criterion #5).

In addition, there are specific requirements for an institution that is seeking accreditation of each engineering or engineering technology program. The criteria and requirements act only as guidelines to ensure a certain degree of uniformity among different programs. However, the ABET Criterion #5 requires faculty to show enthusiasm for developing more effective programs. Laboratory innovation, however, is not specifically identified as a goal. It is believed that at present, most programs too often channel student thinking in traditional directions when solving a problem. For example, meeting personal mobility needs using an automobile, as opposed to telecommuting. New approaches are needed to foster faculty enthusiasm for developing innovative instructional techniques, promote creative problem solving, and nurture unconventional approaches to address societal needs.

One potential concern associated with our undergraduate curricula is the lack of knowledge integration and limited emphasis on higher orders of thinking. Very few courses actually require students to integrate the knowledge acquired in previous courses. Laboratory experiments are structured to simply transfer knowledge from the faculty member to the student, with little thought given to helping the student contextualize the material. Of course, expectations vary from instructor to instructor, and those instructors who foster knowledge integration and higher thought should be applauded.

DESCRIPTION OF LABORATORY DEVELOPMENT

1. Renewable Energy Laboratory Experiments

The course EET 320 - Renewable Energy Systems was introduced in the Department of Engineering Technology at Drexel University in summer 2013, for those who wish to learn the basics of renewable energy - including where it is found, how we can harvest it for use in our homes and how it can help ease pressures on the environment. By the end of the course students know renewable energy in its many forms and which form of energy is the right choice for an application. The course consisted of eight in-class lectures and two laboratory sessions. In the ten weeks of the term, students were given a broad idea of renewable energy manufacturing and then upcoming technologies in green energy manufacturing.

The lab sessions were divided in two sessions covering three wind energy experiments and two solar experiments. Students performed experiments on Lab-Volt kits. A short demonstration of the De Lorenzo GREEN KIT was also conducted. The following table shows the lab schedule.

Week 1 : Wind Experiments	Week 2 : Solar Experiments
Voltage vs Speed Characteristics	The Solar Panel
Torque vs Current Characteristics	Effect of Temperature on Solar Panels
Power vs Wind Speed	Demo of DL Lorenzo Green Kit

The Lab-Volt wind turbine equipment consisted of a Dynamometer and a Wind Turbine generator that is coupled through a pulley. The turbine speeds can be altered by simulating the effect of different wind speeds. The Figure 1 shows the system. The Solar system equipment consisted of a mono-crystalline solar cell and a bright light source. The light source replicates the solar energy. The De Lorenzo GREEN KIT consisted smaller solar panels and wind turbines. It also had a small DC fan and motor. Generated energy can be directly coupled with these equipment and observe renewable energy being put to use. This system communicated through a USB cable with the LVDAC-EMS software that runs on the Computer. This software can be used to obtain real-time data from the system and save/export data for further analysis.

During the first week, the Voltage v Speed Characteristics of a Wind Turbine was the first one to be performed. The objective of this experiment was to understand the effect of generator speed on the amplitude and frequency of the generated voltage. Experiment 2 gave students an understanding of the Torque-current characteristics of the wind turbine generator. Once the effect of velocity and torque was studied, the students took the next step in experiment 3 by studying the relationship between the output power and wind speeds.

Figure 1: The Lab-Volt Wind turbine system

During week two, the students performed experiments related to solar energy using solar cells. Figure 2 shows students performing experiments on the solar kit. This kit consisted of a bright source of light that simulated solar energy. The mono-crystalline solar panel is excited with this light source. In the first experiment students studied about fundamentals of a solar cell. The experiment introduced students to the E-I curve of a PV cell. In the successive experiment, students study the effect of temperature on solar cells.

Figure 2: Students performing solar experiments.

Figure 3 shows the DeLorenzo kit, which was demonstrated to the students. Different experiments such as effect of tilt; shade on solar panels was performed. A representative demonstration of these effects was shown by coupling the power output to a

DC-motor. This direct supply of power to instruments helped students understand this concept to greater depth.

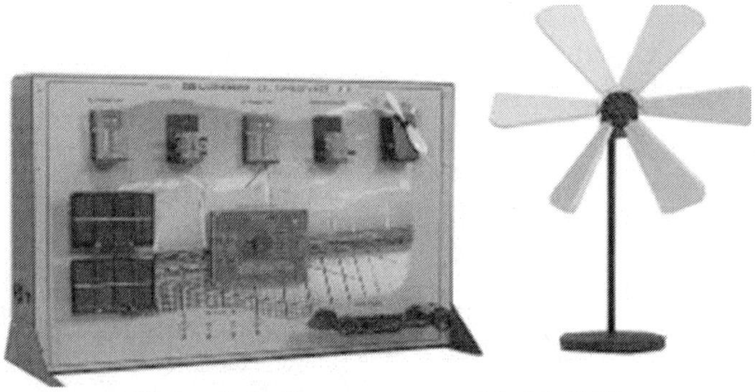

Figure 3: DeLorenzo renewable kit [15]

The laboratory sessions were followed by a feedback evaluation. A questionnaire was circulated amongst the students who answered them while maintaining anonymity. The general feedback received was overwhelming. Students could follow the manuals very well and understand the goal of the experiment. The lab experiments functioned as a good supplement for the EET 320 - Renewable Energy Systems course. The Lab-Volt experiments was well structured in that they started off with experiments that explained the basic concept in harnessing renewable energy and developed into experiments that derived important characteristics of the device to harness maximum energy.

2. Microfluidic Heat Exchanger Energy Efficiency

A representitive experiment uses an infrared camera to capture the two-dimensional temperature profile of a microfluidic heat exchanger for energy efficiency analysis. This experiment introduces students to the techniques of infrared imaging. Its subject of analysis is a microsystem instrumented with sensors and hosting microfluidic 'circuits' carrying heated liquids. The fluid flow can also be imaged with a visible imaging camera to make videos of the flow, and quantify flow characteristics. This shows the complementary and supplemental information provided by imaging in two spectral regions. Such microsystems are of technical interest for *lab-on-a-chip* technologies which seek to provide miniature, sustainable implementations of fluidic processes that are traditionally implemented on laboratory benches, or pilot plants at much larger scale. These systems will reduce consumption of materials, generation of waste, and use of energy. Further, they are amenable rapid prototyping and so can speed product development times. These systems also simulate conventional fluid and thermal systems but on a microscale. They will allow educational institutions to substitute these desktop microscale systems for tradtional laboratory equipment that normally is a factor of 10 or more greater in floor space, capital and operating costs, and create certain safety and enviromental issues. Clearly such miniaturization is a dominant theme of sustainable green manufacturing and product development, and our imaging experiments will give students much experience in many aspects of microscale engineering. Further, their low impact in cost, space, and other resources will facilitate their dissemination to many educational institutions.

The microfluidic chips are designed with AutoCAD™ or SolidWorks™. The chips are fabricated as laminates. A middle layer is machined with through cuts to define the fluidic circuit in a thin (1 mm) sheet of plastic using a laser cutter or CNC machine (Figure 4). The channels and chambers of the middle layer are enclosed by attaching top and bottom plastic sheet layers using thermal bonding, solvent bonding, adhesives, or special double-sided tapes. Samples and reagents can be pipetted directly into the chip or extracted from the chip through inlet/outlet ports. Alternatively, fluidic interconnections are provided by small-diameter (Teflon) tubing inserted through sidewall ports of the chip. We also found disposable pipette tips (e.g., Denville Sharp™ precision barrier tips), shorted by cutting off the tip and with fibrous plug removed, made good transitions between tubing and chip ports (drilled holes). The tubing and sideport hole diameters are selected to make a snug, leak-tight fit. Using tubing interconnections, the liquids can be delivered to the fluidic circuit with a pipette or a syringe, actuated either manually or with a programmable syringe pump.

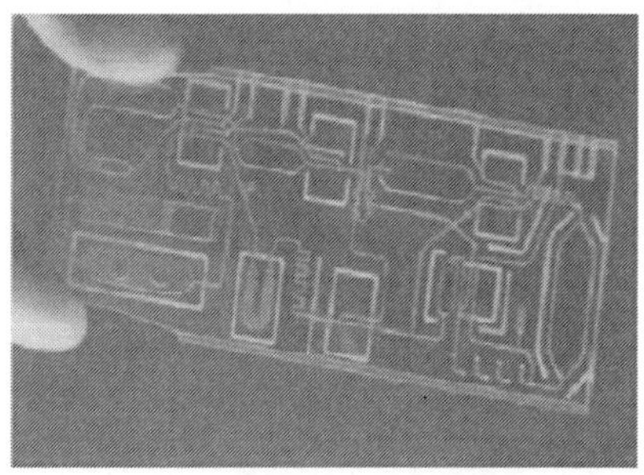

Figure 4: SolidWorks desgin of chip and fabricated chip in acrylic

As shown in Figure 5, the experimental setup consists of the microfluidic chip wherein a fluid enters at an inlet temperature and exits at an outlet temperature. A syringe pump causes the fluid to flow through the chip by pulling fluid from the fluid reservoir that is located on a hot plate at a constant temperature 175°C (although the temperature of the fluid in the reservoir was only measured to be approximately 75°C). A thermoelectric Peltier cooling device pumps heat from the cold side to the hot side where the fan circulates the ambient air between the heat sink's fins to absorb the collected heat. A DC regulated power supply provided power to the cooling fan and current to the thermoelectric cooler.

Figure 5: Experimental setup for infrared thermography of microfluidic chip

The experiments are carried out in two modes. In the first mode, the DC power supply is turned off to observe the heat transfer caused by syringe pump induced fluid flow through the chip. The DC power supply is turned on in the second mode to observe the heat transfer associated with the thermoelectric Peltier cooling device, heat sink, and cooling fan. For each mode, the syringe pump is set to three volumetric flow rates to be called low, medium, and high flow rate. The low, medium and high flow rates are 4,000 μL/min, 5,000 μL/min, and 6,000 μL/min, respectively. Although these are the values given by the syringe pump, they are not the actual flow rates due to disconformities between the syringe pump and the experimental system.

Heat Transfer Analysis

Transfer of heat from a load to the environment is a function of conduction and convection. Conduction is the transfer of heat through matter and is a function of the temperature difference across the material, physical dimensions, and thermal conductivity of the material. Convection is heat transfer across the boundary of air at the surface of the material and is a function of the temperature difference across the boundary layer and the rate of air movement at the surface.

When a moving fluid comes into contact with a surface at some temperature difference the fluid will transfer heat to or from the surface and advect heat away from the point of contact by its motion. Newton's Law of Cooling provides a simple expression for this rate of convective heat transfer.

$$Q = hA\,(T_s - T_f) \qquad (1)$$

Where Q is the rate of heat convected in watts; T_s is the temperature of the solid in C°; T_f if the temperature of the fluid in C°; A is the area of the surface in contact with the fluid in square meters; and h is the convective heat transfer coefficient in watts/ square meter-C°.

Many factors affect convective heat transfer such as its geometry, type of flow, boundary conditions, type of fluid used and its properties. In this experiment, forced convection occurs when fluid flow, induced by the syringe pump, moves through the chip. The flow of fluid through the chip can be classified as laminar or turbulent. The heat transfer coefficient, h, used for calculating forced convective heat transfer is determined through a correlation of the dimensionless numbers: Nusselt number, Nu, Reynolds number, Re, and Prandtl number, Pr.

$$Nu = hD/k \quad Re = DV\rho/\mu \quad Pr = \mu C/k \qquad (2)$$

Where D is tube diameter in meters; V is characteristic fluid velocity in m/s; k is thermal conductivity of fluid in kJ/hr-m-K; ρ is fluid density in kg/m^3; C is the constant pressure specific heat of the flowing fluid in J / kg-K; and μ is fluid viscosity in N-s /m^s.

Thermographic Acquisition

As shown in Figure 6, Infrared thermography can determine the temperature at the walls of the fluidic circuit and in the fluidic flow without disturbing the system. The IR camera was connected to a computer via a USB cable and recorded video of the fluid flow through the chip using the thermographic data acquisition and analysis application, ExaminIR. The recorded video can then be enhanced by adjusting the temperature scale limits and changing how the color palette is mapped to the data. Temporal plots can be obtained from the recorded video by the placement of ROI (region of interest) cursors that record the average temperature value of a 3x3 box of pixels over a time period. This data can then be exported to Excel for further manipulation.

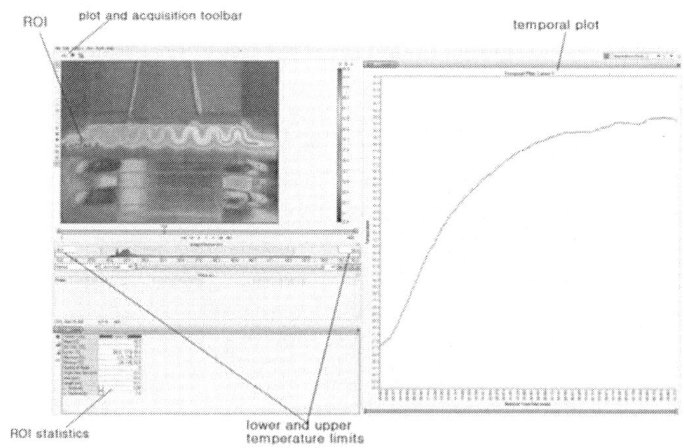

Figure 6: Screenshot of ExaminIR

The infrared camera is attached to a tripod and positioned directly above the microfluidic chip to record the fluid flow through the chip. The water flowed from right to left and inlet and outlet temperature values are recorded by placing ROI cursors in the middle of these ports as shown in Figure 7.

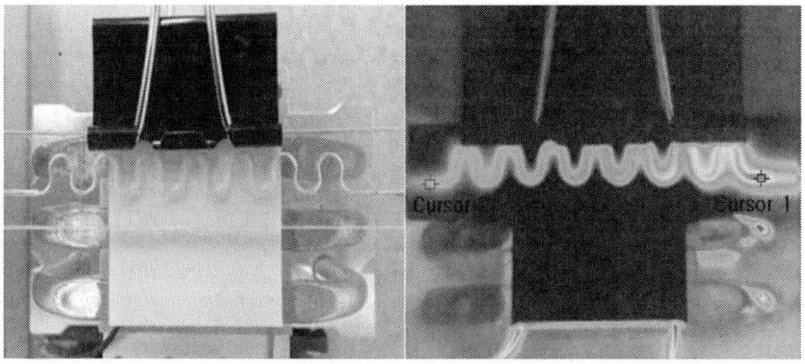

Figure 7: Normal IR image of fluid flow through the microfluidic chip on the Peltier cooling device

3. Solar Cell Surface Roughness Measurement System

Solar energy is one of the most promising renewable energies. Cell efficiency depends on the silicon dopant, light intensity and wavelength, optical thickness and surface texture. The efficiency of current solar cells is around 20% but some specialized surface texturing processes can help boost efficiency up to 25%. As with any new technological paradigm, new challenges have emerged which are unique to the renewable energy manufacturing. The manufacturing difficulty of such processes implies a higher cost, making them prohibitive. One of these challenges in particular is the development of effective monitoring technologies to compensate for the decentralized nature of remote solar cell surface quality monitoring and control.

There may be electrical issues such as breaks in the contact lines. In the manufacturing process, there may be issues where the titanium dioxide anti-reflection coating (ARC) on the top of the cell is not uniform. The surface of the solar cell is textured (i.e., roughened) to reduce reflectivity and increase optical coupling. The roughness is optimized in combination with the ARC. Increased texturing improves optical coupling but may lead to breaks or discontinuities in the ARC and/or metallization lines. As shown in Figure 8, if the surface of the cell is too smooth then too much light will be reflected rather than absorbed. If the surface is too rough, then reflection may also be excessive.

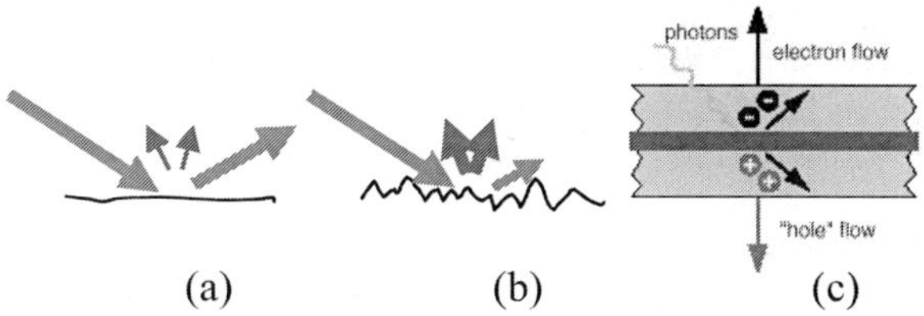

Figure 8: Reflection dependent on surface roughness (a) Smooth surface – weak electron flow and (b) rough surface – strong electron flow, and (c) Photovoltaic effect.

Remote Robotic Workcell

Utilizing robots in the process of performing surface roughness scanning allows for both full automation as well as precise and repeatable measurements. The two robots used in this cell are the Yamaha YK250X SCARA four-axis robot and the Yamaha F1010-700 one-axis robot. The measurement head of the Laser Check system is mounted to the tool arm of the SCARA robot. The piece to be scanned is mounted to the one-axis robot. The scan is performed by incrementing the one-axis robot in what shall be designated as the y-axis i number of times. Once completed, the one-axis robot moves back to its initial position and the SCARA robot increments in what shall be designated as the x-axis, and the one-axis robot begins its incrementing cycle again. This process repeats $j+1$ times, j being the number of increments the SCARA robot makes. This produces a 2D array or matrix of data: $M_{i+1,j+1}$, which can later be used to plot a 3D graph. The robotic workcell can be seen in Figure 9.

Figure 9: Remote robotic workcell for surface roughness monitoring

To make the system fully automated there must be a way for the robots to send and receive signals from each other, as well as to and from LabVIEW and the Laser Check control unit. To accomplish these connections a PLC is used. The type of PLC used for this automation system is an Allen Bradley Control Logix unit. A simple PLC ladder logic diagram is employed to coordinate certain inputs to in turn trigger specific outputs. The ladder logic is used in this system. This ladder logic basically functions as a form of virtual wiring between components. The first rung connects an output from the one-axis robot to the laser trigger of the Laser Check control unit. Whenever the robot's program sends that output, it turns on the laser. The second and third rungs connect an output and input from one robot to the input and output of the other. This allows the robot's to signal when, for instance, they are both in the initial position and ready to start a scan. The ladder logic is not the only way that the PLC is used in the automation system. The LabVIEW program can access and flip bits in the PLC via an Ethernet connection. These connections are used to initiate the robots' programs remotely as well as control the timing of data flow into the LabVIEW program.

There is more than one way to gather data from the Laser Check system, but by using a LabVIEW program it allows for not only data gathering and processing but also for a degree of system control. The LabVIEW block diagram breaks down into two main sections, which can be seen visually by noting the two large while loops that are outside of each other. The top part of the program is responsible for communicating with the Laser Check control unit and gathering the data from it. Once the data has been gathered, it is then processed in subsequent case structures, which produces the floating point data that is needed. This data is then built into an array and graphed on a 3D plot in the LabVIEW program.

To initiate the connection between either the serial port or serial server a configuration block, specific to the type of connection, is used. Once the connection is established, TCP/IP or Serial Read/Write blocks are used to issue a call for data to the Laser Check system and then read the data that comes in. To control the flow of data in, such that data is not erroneously read twice or skipped over, a case structure is employed, which is triggered by the falling edge of the laser trigger in the PLC. This essentially allows the program to read the data only right after the laser turns off. The data that flows in comes in the form of an ASCII string. To convert this into a floating point value, the string is parsed and header and footer values are removed. The remaining value is then converted to a float value using a type cast function. This data is then sent along and placed into a 2D array that is a set of 1D arrays that represent each scan along the face of the material. This array is then sent to the 3D graph and table on the front panel, as well as to an external spreadsheet file.

The second part of the program is responsible for initiating the two robot's programs. The majority of the program is inside the case structure, which is inside the first larger while loop. This case structure has four different cases which it switches through based on button presses on the front panel as well as internal factors. The connections between the program and the robots are made through both the PLC and through Telnet communication using TCP/IP. The one-axis robot's controller does not support Ethernet connection, and as such, the program is initiated by connecting to the PLC and flipping the bit.

As mentioned earlier, there are two instances of this program that connect to the Laser Check system differently. One connects to the Laser Check system using the serial connection on the PC being used. The second involves hooking the serial connection to a serial server, which the program then accesses via an internet connection. The main difference between the two is that the first program requires a local connection, whereas the second can be performed remotely. The front panel's controls for both programs are shown in Figure 10.

There are minor differences between both front panels, which can be seen on the left side of each. These differences are due to differences in connecting and reading from the serial connection either locally or via Ethernet. The two buttons on the right are responsible for connecting to the robots and subsequently running their programs. The graph in the middle of the screen plots the 3D data that comes, and the table next to that displays all of the data as it comes in.

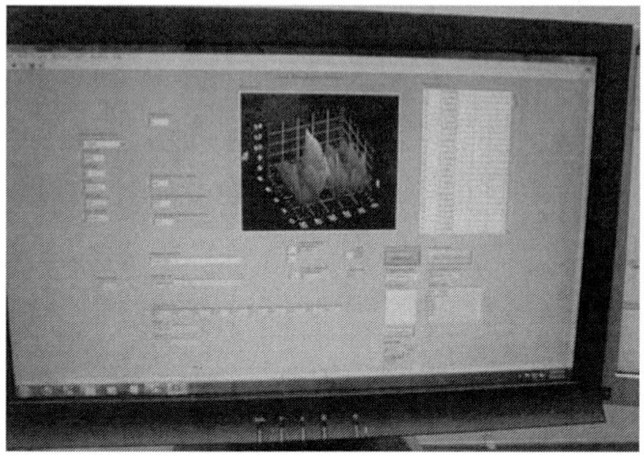

Figure 10: LabVIEW serial and Ethernet front panels respectively

Data Analysis/Discussion

To ensure that these values accurately represent the quantitative roughness of the pieces several calibration and comparison steps were taken. The Laser Check system was first calibrated using the calibration block specifically designed for this type of milled piece. Once completed, the Laser Check machine was again tested with calibrated pieces with a known roughness value to verify that the readings were within the tolerances. Then as an extra step, visual and physical inspection of the surfaces of each of the pieces was performed. The pieces were analyzed under a microscope in comparison with calibrated blocks. Also physical testing was done on the rougher piece and a calibrated piece using the fingernail test. Based on the sum of these steps, the system provides accurate surface roughness data.

The usefulness of the automated surface roughness measurement system described up to this point, dependents upon the quality of data that can be collected. Figure 11 shows example plots of the data, plotted with MATLAB, collected from the described system. The x and y axes correspond to the rows and columns of data collected, in this case 11 by 50. The z axis displays the measured R_a values. The choice in using solar cells is also a conscious one. The reason being, as measured data confirms, that using the laser system to measure the kind of solar cell yields certain characteristics unique to each type. The graph is taken from the profilometer data and are representative of the surface profile for the solar cell.

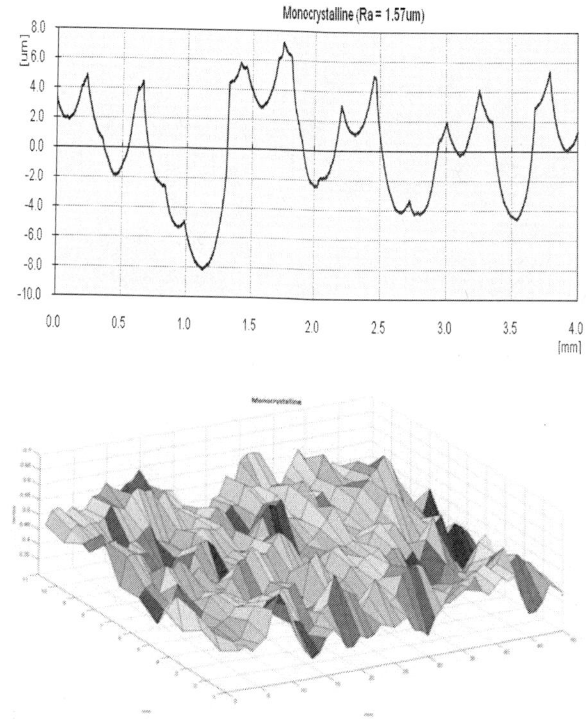

Solar cell (Ra = 1.57*um*)

Figure 11: surface roughness graph

CONCLUSIONS AND FUTURE WORK

For the past years, our focus has shifted towards incorporating green energy manufacturing topics in our laboratory development. We assign to our students the experimental topics related to renewable energy, power systems, or green energy manufacturing. These experiments provide multi-disciplinary collaboration and valuable hands-on experience to the students. In addition to useful lessons on teamwork and project management, the projects provide working demonstration of green energy manufacturing systems. To enhance the hands-on experience, the laboratory experiments have been restructured as a laboratory course. The goals of all the experiments are to explore and enhance student understanding of the fundamental concept of design-for-environment (dfE) and hands-on learning of green energy manufacturing.

ACKNOWLEDGEMENTS

This work has been supported by the US Department of Education under the joint DHSIP Program with the University of Texas at El Paso and Drexel University, PR/Award No.: P031S120131. The authors wish to express sincere gratitude for their financial support.

REFERENCES

1. Rosentrater, K. A. and Al-Kalaani, Y., "Renewable energy alternatives – a growing opportunity for engineering and technology education," *The Technology Interface*, 6(1), 2006.
2. P. Jennings, New directions in renewable energy education. International Journal of Renewable Energy, 34, 435-439, 2009.
3. Criteria for accrediting engineering programs: Effective for evaluations during the 2010–2011 accreditation cycle, ABET, Baltimore. *Accreditation Board for Engineering Technology* (ABET), Inc. (2010).
4. Allen, D. T., and Shonnard, D. R. (2003). Green engineering, *Prentice Hall*, Upper Saddle River, NJ, 397416, 2003.
5. ABET, Criteria for Accrediting Engineering Programs, 2013, http://www.abet.org/accreditation-criteria-policies-documents/
6. Felder, R. M., and R. Brent, (2004a), "The Intellectual Development of Science and Engineering Students. Part 1: Models and Challenges," *Journal of Engineering Education,* 93(4), pp. 269-277.
7. http://www.epa.gov/Sustainability/basicinfo.htm (accessed July 28, 2013).
8. Felder, R. M., and Brent, R., "The Intellectual Development of Science and Engineering Students. Part 2: Teaching to Promote Growth," *Journal of Engineering Education,* 93(4), pp. 269-277.
9. Gregory McNamee, "Careers in Renewable Energy," *PixyJack Press LLC*, 2008.
10. Walker, S, "Sustainable by Design: Explorations in Theory and Practice," *London: Earthscan Publications* Ltd, 2006.
11. Clift, R., "Sustainable development and its implications for chemical engineering," *Chem Eng Sci,* 61: 4179-4187, 2006.
12. Desha, C.J., Hargroves, K. and Smith, M.H., "Addressing the time lag dilemma in curriculum renewal towards engineering education for sustainable development," *Int J Sustain High Educ, vol.* 10(2): 184–199, 2009.
13. Desha, C.J. and Hargroves, K., "Engineering Education & Sustainable Development - A Guide for Rapid Curriculum Renewal," *London, England: Earthscan,* 2010.
14. Ehrenfeld, J., "Sustainability by Design," *New Haven, CT, USA: Yale University Press*, 2008.
15. DeLorenzo Energy: http://www.delorenzoenergy.com/index_en.html

This page left blank intentionally.

AUTHOR INDEX

Proceedings of the ASME 2014 International Mechanical Engineering Congress and Exposition Volume 5 – Education and Globalization